Stephan Russenschuck

Field Computation for Accelerator Magnets

Related Titles

Griffiths, D.

Introduction to Elementary Particles

470 pages with 106 figures and 12 tables, Softcover
2008
ISBN: 978-3-527-40601-2

Meunier, G.

The Finite Element Method for Electromagnetic Modeling

832 pages, Hardcover
2008
ISBN: 978-1-84821-030-1

Buckel, W., Kleiner, R.

Superconductivity
Fundamentals and Applications

475 pages with approx. 247 figures, Hardcover
2004
ISBN: 978-3-527-40349-3

Jackson, J. D.

Classical Electrodynamics

832 pages, Hardcover
2004
ISBN: 978-3-527-40349-3

Edwards, D. A., Syphers, M. J.

An Introduction to the Physics of High Energy Accelerators

304 pages with 103 figures, Hardcover
1993
ISBN: 978-0-471-55163-8

Stephan Russenschuck

Field Computation for Accelerator Magnets

Analytical and Numerical Methods for
Electromagnetic Design and Optimization

WILEY-VCH Verlag GmbH & Co. KGaA

The Author

Dr.-Ing. Stephan Russenschuck
CERN, Geneva, Switzerland
Stephan.Russenschuck@cern.ch

Cover

The reduced field from the iron magnetization in the MQXA magnets for the LHC inner triplet
(see also chapter 13)

Figures

The Field plots where created with the CERN-HIGZ graphics package and the sketches were drawn with Micrografx Designer. The photographs, if not taken by the author or as otherwise stated, were taken from the CERN document server with permission.

Library of Congress Card No.:
applied for

British Library Cataloguing-in-Publication Data
A catalogue record for this book is available from the British Library.

Bibliographic information published by the Deutsche Nationalbibliothek
The Deutsche Nationalbibliothek lists this publication in the Deutsche Nationalbibliografie; detailed bibliographic data are available on the Internet at <http://dnb.d-nb.de>.

Typesetting Uwe Krieg, Berlin
Printing and Binding Strauss GmbH, Moerlenbach

Printed in the Federal Republic of Germany
Printed on acid-free paper

ISBN: 978-3-527-40769-9

Contents

Field Computation for Accelerator Magnets. Stephan Russenschuck

ISBN: 978-3-527-40769-9

Preface

To every thing there is a season,
and a time to every purpose under the heaven:
A time to be silent, and a time to speak.

Ecclesiastes 3

This book aims at filling the need for a comprehensive and up-to-date reference work on the electromagnetic design of iron- and coil-dominated accelerator magnets. It provides the theoretical foundations required by the systems engineer, including back-of-the-envelope calculations for the conceptual design phase, and explains the use of numerical methods in optimization processes. The book's twenty chapters span the following key topics:

- vector algebra and analysis,
- analytical field computation,
- elementary beam optics,
- numerical field calculation,
- superconducting magnet design,
- mathematical optimization techniques.

The magnet design problems are arranged according to the mathematical methods required for their solution. This emphasizes the universal nature of the methods, which can be equally well applied to other challenging applications in electrical engineering. It also highlights the fact that a large number of subjects taught in graduate university courses on electromagnetic field theory have their role to play in real-world applications.

Finite-element and boundary-element techniques have become a standard tool in the design of electromagnetic equipment. The methods presented in this book have been implemented in the CERN field computation program ROXIE. The modeling capabilities of this program together with its mathematical optimization routines have inverted and largely superseded the classical

Field Computation for Accelerator Magnets. Stephan Russenschuck

ISBN: 978-3-527-40769-9

design process wherein numerical field calculations are performed for only a limited number of design variants. ROXIE is thus applied to the integrated design process of magnets, from the concept phase to field optimization, production follow-up, and hardware commissioning.

In the more than twenty years of R&D for the LHC magnet system, a large number of magnet variants were designed, constructed, and tested. An even larger number of designs were studied with numerical field calculation. Another aim of this book is therefore to share some of this experience whenever it serves to illustrate the described analytical and numerical concepts.

In the (likely) event that this first printing of the book contains errors there will be a list of errata kept at the website: http://www.wiley-vch.de/publish/dt/books/bySubjectPH00/ISBN3-527-40769-3. The website will also contain relevant links and extensions.

Acknowledgments

This book evolved from a set of notes prepared for lectures and training activities at the University of Vienna, the CERN Accelerator School (CAS), and the Joint Universities Accelerator School (JUAS). It contains results of nearly two decades of work for the LHC, during which I enjoyed a constructive and friendly atmosphere among students, colleagues, ROXIE program users, collaborators, supervisors, and friends. I wish to thank them all for their support, encouragement, and suggestions.

I have enormously benefited from the discussions with and feedback from many individuals from CERN, universities, and high-energy physics laboratories around the world. Some have influenced and aided me personally or professionally. Others have contributed to this book in particular, by providing graphs and measurements, by proofreading and editing, by contributing software, and by collaborating in code development. Without their assistance this book would not have been possible.

I owe special thanks to Karl Hubert Meß and Thomas Taylor for their expertise and encouragement, to Philippe Lebrun for his approval of this book project, and Professor Adalbert Prechtl, who invited me to Vienna as a Lecturer to present the thesis that later became the nucleus of the book.

I gratefully acknowledge the work and assistance by those students and fellows who have dedicated long hours to the ROXIE program development and LHC magnet design; Martin Aleksa, Riccardo de Maria, Christian Paul, Suitbert Ramberger, Nikolai Schwerg, and Christine Völlinger.

I much appreciated the help of Bernhard Holzer, Jean-Francois Ostiguy, Frank Schmidt, and Frank Zimmermann with the chapter on elementary beam physics. Mike Struik took the pleasing portrait shown on the back cover.

I am grateful to Professors Oszkár Biró and Kurt Preis, as well as Professor Stefan Kurz, for their collaboration and for making available their software packages for numerical field computation, and to Bernhard Auchmann, who not only kept the ROXIE program package alive during the busy times of the LHC installation, but who also made invaluable contributions to most aspects of this book, whether it be the content, structure, or details in the mathematical treatment.

I wish to express my appreciation to Piotr Komorowski for extensive proofreading and to Mike Strickler whose thoughtful editing of the English text has contributed immensely to the readability of this work. Christoph von Friedeburg, Uwe Krieg, Ulrike Werner, and the production team from Wiley-VCH were responsible for the quality reproduction of the manuscript.

Finally I wish to thank my wife Isabel, who deserves great credit for her patience with my neglect of her, the house, and the garden on those long evenings and weekends when I labored to complete this book.

Stephan Russenschuck

Notation

Mathematics is the art of giving the same name to different things,
poetry is the art of giving different names to the same thing.

Jules Henri Poincaré (1854–1912)

In this publication, we follow the recommendations of the International Electrotechnical Commission, IEC-CEI, 27-1: *Letter symbols to be used in electrical technology*. Consequently, we denote the vector fields, for example, **B** and **E**, by bold face capitals, even though in the mathematical literature capitals are used for the functional spaces and lower case for their elements. We follow this mathematical convention in Chapters 2 and 3, where we use precise mapping notations of the kind $\mathbf{b} : \Omega \to \mathbb{R}^3 : \mathbf{r} \mapsto \mathbf{b}(\mathbf{r})$. The path length of a curve, the area of a surface, and the volumetric extent[1] are defined in IEC 27 by s, A, and V, but we use the lower case a for the surface area to avoid confusion with the vector potential **A**. It might therefore follow that we should use lower case v for the volumetric extent, but this would clash with the usual notation for velocity. The point, straight line, curve, loop, surface, and volume are denoted $\mathscr{P}$, $\mathscr{L}$, $\mathscr{S}$, $\mathscr{C}$, $\mathscr{A}$, and $\mathscr{V}$, as special entities of the general space elements $\mathscr{M}$. For the electrical conductivity, we follow DIN 1324 and use the symbol $\varkappa$ instead of σ, which is also used for the electric surface charge density.

Owing to the habit of denoting the reference radius in accelerator magnets by r_0, we use r, φ, z for cylindrical coordinates and R, ϑ, φ for the spherical coordinates (where according to DIN 1324, ϑ is the polar angle and φ is the equatorial angle).

1 While there are two words for a line and its length (or the surface and its area) we invent the term volumetric extent in order to make the distinction.

Field Computation for Accelerator Magnets. Stephan Russenschuck

ISBN: 978-3-527-40769-9

We use ϕ instead of φ for the scalar potential and $V_{\rm m}$ (again according to DIN 1324) for the magnetomotive force (instead of the F recommended in IEC 27). In accordance with both norms, the magnetization M does not contain μ_0 and is therefore expressed in units of $\mathrm{A\,m^{-1}}$. As we use the symbol for the magnetic polarization only once in the entire text, we denote it, contrary to all norms by $\mathbf{P}_{\rm mag}$ instead of the $\mathbf{J}$, which is reserved for the current density.

We denote the position vectors of the source and field points by $\mathbf{r}'$ and $\mathbf{r}$. It is more difficult to maintain a consistent notation for their coordinates. We use $\mathbf{r}' = (r_{\rm c}, \varphi_{\rm c})$ with index c for "coil," and $\mathbf{r} = (r, \varphi)$ for the field point.

Complex numbers are written as $z = x + iy$ and the complex conjugate as $\overline{z} = x - iy$ in accordance with most mathematical texts. Note also that in some engineering texts, the complex number is written with an underline, i.e, $\underline{z}$, and j instead of i is used for $\sqrt{-1}$. The field point is denoted $z = x + iy$ and the source point $z_{\rm c} = x_{\rm c} + iy_{\rm c}$.

We write matrices in three different ways: as matrix $[A]$ to avoid confusion with the vector potential, as (a_{ij}) when we want to single out coefficients and as $\mathbf{A}$ when confusions with vector fields are excluded. In the same spirit we write column vectors as $\{x\} := (x_1, \ldots, x_N)^T$ but also as $\mathbf{x}$ in the mathematical chapters.

A final remark: We use italics in sub- and superscripts only for indices and coordinates. Descriptive add-ons are set in roman type characters. As an example, $\phi_{\mathrm{m},i}$ indicates the magnetic (roman type m) scalar potential in the ith material domain.

1
Magnets for Accelerators

If you want to build a ship, don't herd people together
to collect wood and don't assign them to tasks and work,
but rather teach them to long for the endless immensity of the sea.

Antoine de Saint-Exupéry (1900–1944).

A number of comprehensive books have been published on the design and construction of accelerator magnets, for example, by Wilson [68], Brechna [17], Meß, Schmüser, and Wolff [51], Iwasa [34], and Asner [2]. Other sources of information are the proceedings of the Magnet Technology (MT) conferences, which are usually published in the *IEEE Transactions on Applied Superconductivity*. The large amount of publications is not surprising inasmuch as the magnet systems and the cryogenic installations are the most expensive components of circular high-energy particle accelerators.

Like previous projects in high-energy physics, the *Large Hadron Collider* (LHC), built at CERN in Geneva, Switzerland, has greatly benefited from the developments in superconductivity and cryogenics. In turn, these fields have enormously gained through the R&D undertaken by CERN in collaboration with industry and national institutes, as well as by the production of components on an industrial scale.

This book concentrates on the mathematical foundations of field computation and its application to the electromagnetic design of accelerator magnets and solenoids. It is fitting, then, that we should briefly review a number of technological challenges that had to be met for the design, manufacture, construction, installation, and commissioning of the LHC main ring. A good overview is given in [27]. Challenges are found in all domains of physics and engineering. They comprise, among others:

- Material science aspects, such as the development of superconducting wires and cables, the specification of austenitic and magnetic steel, and the choice of radiation resistant insulation, among others.
- Mechanical engineering challenges, such as finding the appropriate force-restraining structure for the coils, the right level of prestress in

Field Computation for Accelerator Magnets. Stephan Russenschuck

ISBN: 978-3-527-40769-9

the coil/collar assembly, the design of manufacturing tooling, coldmass integration and welding techniques, cryostat integration, and magnet installation and interconnection, all made more difficult by the very tight tolerances required by the optics of the particle beam.

- The physics of superfluid helium and cryogenic engineering for helium distribution lines, refrigeration, and process control.
- Vacuum technology for insulation and the beam vacuum. The beam vacuum system must provide adequate beam lifetime in a cryogenic system, where heat flow to the 1.9 K helium circuit must be minimized.
- Metrology for magnet alignment in the tunnel.
- Electrical engineering challenges for power supplies (high current, low voltage), water-cooled cables, current leads using high T_c superconductors, superconducting busbars, diodes operating at cryogenic temperatures, magnet protection and energy extraction systems, and powering interlocks.
- Magnetic field quality measurements and powering tests.

We will review these engineering aspects after a brief introduction to the LHC project. Finally, we will turn to the challenges of electromagnetic design, which required the development of dedicated software for numerical field computation.

1.1 The Large Hadron Collider

With the Large Hadron Collider (LHC), the particle physics community aims at testing various grand unified theories by studying collisions of counter-rotating proton beams with center-of-mass energies of up to 14 teraelectronvolt (TeV). Physicists hope to prove the popular *Higgs mechanism*[1] for generating elementary particle masses of the quarks, leptons, and the W and Z bosons. Other research concerns supersymmetric (SUSY) partners of the particles, the apparent violations of the symmetry between matter and antimatter (CP-violation), extra dimensions indicated by the theoretical gravitons, and the nature of dark matter and dark energy. A general overview of this topic is given in [36].

The exploration of rare events in the LHC collisions requires both high beam energies and high beam intensities. The high beam intensities exclude the use of antiproton beams and thus imply two counter-rotating proton beams, requiring separate beam pipes and magnetic guiding fields of opposite polarity. Common sections are located only at the four *insertion regions* (IR), where

1 Peter Higgs, born in 1929.

the experimental detectors are located. The large number of bunches, 2808 for each proton beam, and a nominal bunch spacing of 25 ns creates 34 "parasitic" collision points in each experimental IR. Thus dedicated orbit bumps separate the two LHC beams left and right from the *interaction point* (IP) in order to avoid the parasitic collisions at these points. The number of events per second, generated in the LHC collisions, is given by $N = L\sigma$, where σ is the interaction cross section for the event and L the *machine luminosity*, $[L] = 1\ \mathrm{cm}^{-2}\,\mathrm{s}^{-1}$. In scattering theory and accelerator physics, luminosity is the number of events per unit area and unit time, multiplied by the opacity of the target. The machine luminosity of a collider depends only on the beam parameters and can be written for a Gaussian[2] beam distribution as

$$L = \frac{N_\mathrm{b}^2 n_\mathrm{b} f \gamma}{4\pi\epsilon_n \beta^*} F\,, \tag{1.1}$$

where N_b is the number of particles per bunch, n_b the number of bunches per beam, f the revolution frequency, γ the relativistic gamma factor, ϵ_n the normalized transverse beam emittance, and β^* the beta function at the collision point. The factor F accounts for the reduction of luminosity due to the crossing angle at the interaction point (IP):

$$F = \left(1 + \left(\frac{\theta\sigma_z}{2\sigma^*}\right)^2\right)^{-\frac{1}{2}}\,, \tag{1.2}$$

where θ is the full crossing angle at the interaction point, σ_z the RMS bunch length, and σ^* the transverse RMS beam size at the interaction point. All the above expressions assume equal beam parameters for the two circulating beams.

Two high-luminosity experiments, ATLAS [6] and CMS [20], aiming at a peak luminosity[3] of $10^{34}\ \mathrm{cm}^{-2}\,\mathrm{s}^{-1}$, will record the results of the particle collisions. In addition, the LHC has two low-luminosity experiments: LHCB [45] for B-physics aiming at a peak luminosity of $10^{32}\ \mathrm{cm}^{-2}\,\mathrm{s}^{-1}$ and TOTEM [63] for the detection of protons from elastic scattering at small angles aiming at a peak luminosity of $2 \times 10^{29}\ \mathrm{cm}^{-2}\,\mathrm{s}^{-1}$ for 156 bunches. LHCf is a special-purpose experiment for astroparticle physics. A seventh experiment, FP420, has been proposed that would add detectors to available spaces located 420 m on either side of the ATLAS and CMS detectors.

The LHC will also be able to collide heavy ions, such as lead ions, up to an energy level of about 1100 TeV. These collisions cause the phase transition of nuclear matter into quark-gluon plasma as it existed in the very early universe

2 Carl Friedrich Gauss (1777–1855).

3 The luminosity is not constant during a physics run, but decays due to the degradation of intensity and emittance of the beam.

around 10^{-6} s after the Big Bang. Heavy-ion physics is studied at the ALICE experiment. ALICE [7] aims at a peak luminosity of 10^{27} cm^{-2} s^{-1} for nominal Pb–Pb ion operation.

Figure 1.1 Layout of the LHC main ring with its physics experiments ATLAS, CMS, ALICE, LHC-B, and the radio frequency and beam dump insertions at IP 4 and 7.

The LHC reuses the civil engineering infrastructure of the Large Electron Positron Collider (LEP), which was operated between 1989 and 2000. The tunnel with a diameter of 3.8 m and a circumference of about 27 km straddles the Swiss/French boarder near Geneva, at a depth between 50 and 175 m underground. The layout of the LHC main ring with its physics experiments is shown in Figure 1.1. With a given circumference of the accelerator tunnel, the maximum achievable particle momentum is proportional to the operational field in the bending magnets. Superconducting dipole magnets cooled to 1.9 K, with a nominal field of 8.33 T, allow energies of up to 7 TeV per proton beam.[4]

A hypothetical 7 TeV collider using normal-conducting magnets, with a field limited to 1.8 T by the saturation of the iron yoke, would be 100 km in circumference. Moreover, it would require some 900 MW of electrical power, dissipated by ohmic heating in the magnet coils, instead of the 40 MW used by the cryogenic refrigeration system[5] of the superconducting LHC machine [42]. The operational magnetic field in the string of superconducting magnets de-

4 The conversion to degrees Celsius is $\{T\}_{°C} = \{T\}_K - 273.16$.

5 According to a National Institute of Standards and Technology (NIST) convention, cryogenics involves temperatures below -180 °C (93.15 K), i.e., below the boiling points of freon and other refrigerants.

pends, however, on the heat load and temperature margins of the cryomagnets and therefore on the beam losses in the machine during operation. Operating the superconducting magnets close to the *critical surface* of the superconductor therefore requires efficient operation with minimal beam losses.

The two counter-rotating beams require two separate magnetic channels with opposite magnetic fields.[6] The available space in the LHC tunnel does not allow for two separate rings of cryomagnets as was planned for the superconducting supercollider (SSC) [57]. The LHC main dipole and quadrupole magnets are twin-aperture designs with two sets of coils and beam channels within a common mechanical structure, iron yoke, and cryostat. The dipole cross section is shown in Figure 1.15. The distance between the beam channels is 194 mm at operational temperature.

The eight arcs of the LHC are composed of 23 regular arc-cells, each 106.9 m long, of the so-called *FODO structure* schematically shown in Figure 1.2. Each cell is made of two identical half-cells, each of which consists of a string of three 14.3-m-long main dipoles (MB) and one 3.10-m-long main quadrupole (MQ). Sextupole, decapole, and octupole correctors are located at the ends of the main dipoles. The quadrupoles are housed in the *short straight sections* (SSS), which also contain combined sextupole/dipole correctors, octupoles or trim quadrupoles, and beam position monitors. The role of the different correctors will be discussed in Chapter 11.

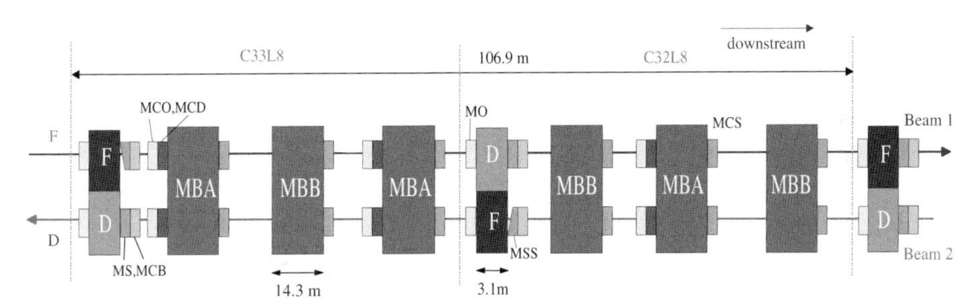

Figure 1.2 Layout of the FODO cells of the LHC main ring. Multipole corrector magnets are connected to the main dipoles (MBA and MBB), and lattice correctors are connected to the main quadrupoles in the short straight section (SSS). The lengths of the magnets are not to scale.

The two coil pairs in the dipole magnets are powered in series, and therefore all dipole magnets in one arc form a single electrical circuit. The quadrupoles in each arc form two electrical circuits: All focusing quadrupole magnets for Beam 1 and Beam 2 are powered in series (see Chapter 12), and all defocusing quadrupole magnets for both beams are powered in series. The optics

6 Beam 1 is defined as the clockwise circulating beam (seen from above), while Beam 2 circulates anticlockwise.

of Beam 1 and Beam 2 in the arc cells are therefore strictly coupled, via the powering of the main magnetic elements.

The eight *long straight sections*, each approximately 528 m long, are available for experimental insertions or utilities. IR 2 contains the injection systems for Beam 1, while the counter-rotating Beam 2 is fed into the LHC at IR 8. The insertion regions 3 and 7 each contain two *beam collimation systems* and use normal-conducting magnets that are more robust against the inevitable beam loss on the primary collimators. Insertion region 4 contains the radio frequency systems.

A total beam current of 0.584 A corresponds to a stored energy of approximately 362 MJ. This stored energy must be absorbed safely at the end of each physics run, in the event of a magnet quench or an emergency. At IR 6 the beams are extracted vertically from the machine using a combination of horizontally deflecting, fast-pulsed kicker magnets and vertically deflecting, normal-conducting septum magnets. In addition to the energy stored in the circulating beams, the LHC magnet system has a stored electromagnetic energy of approximately 600 MJ. As part of the magnet-protection system, an energy extraction system consisting of switches and *protection resistors* is installed outside the continuous cryostat.

Remanent magnetic fields in the bending magnets (from iron magnetization in normal-conducting magnets or superconducting filament magnetization in superconducting magnets) make it impossible to ramp accelerator magnets linearly from an arbitrarily small field level. The LHC uses an existing injector chain that includes many accelerators at CERN: The linear accelerator Linac 2 generates 50 MeV protons and feeds the Proton Synchrotron Booster (PSB). Protons are then injected at 1.4 GeV into the Proton Synchrotron (PS) where they are extracted at 26 GeV. The Super Proton Synchrotron (SPS) is used to increase the energy of protons to the LHC injection energy of 450 GeV.

Filling the LHC requires 12 cycles of the SPS and each SPS fill requires three to four cycles of the PS. The total LHC filling takes approximately 16 min. The minimum time required for ramping the beam energy in the LHC from 450 GeV to 7 TeV is approximately 20 min. After a beam abort at top energy it also takes 20 min to ramp the magnets down to the injection field level. Allowing for a 10 min check of all main systems, one obtains a theoretical turnaround time for the LHC of 70 min.[7]

7 The average time between the end of a luminosity run and a new beam at top energy in the HERA accelerator was about 6 h, compared to a theoretical minimum turnaround time of approximately 1 h.

A relativistic particle of charge e and mass m forced to move along a circular trajectory loses energy by emission of photons (synchrotron radiation) according to

$$\Delta E = \frac{1}{3\varepsilon_0} \frac{e^2 E^4}{(mc^2)^4 R}, \tag{1.3}$$

with every turn completed [66]. In Eq. (1.3), $\varepsilon_0 = 8.8542\ldots \times 10^{-12}\,\mathrm{F\,m^{-1}}$ is the permittivity of free space, R the radius of curvature of the particle trajectory, E the particle's energy, and c the speed of light in vacuum. A comparison between electron and proton beams of the same energy yields

$$\frac{\Delta E_\mathrm{p}}{\Delta E_\mathrm{e}} = \left(\frac{m_\mathrm{e} c^2}{m_\mathrm{p} c^2}\right)^4 = \left(\frac{0.511\,\mathrm{MeV}}{938.19\,\mathrm{MeV}}\right)^4 = 8.8 \times 10^{-14}. \tag{1.4}$$

Although the synchrotron radiation in hadron storage rings is small compared to that generated in electron rings, it still imposes practical limits on the maximum attainable beam intensities, as the radiation must be absorbed in a cryogenic system. This affects the installed power of the refrigeration system and is an important cost issue. Moreover, the synchrotron light impinges on the beam pipe walls in the form of a large number of hard UV photons. These in turn release absorbed gas molecules, which increase the residual gas pressure and liberate electrons; these are accelerated across the beam pipe by the positive electric field of the proton bunches.

The particle momentum in units of $\mathrm{GeV\,c^{-1}}$ is given by[8]

$$\{p_0\}_{\mathrm{GeV\,c^{-1}}} \approx 0.3\,\{R\}_\mathrm{m}\,\{B_0\}_\mathrm{T}. \tag{1.5}$$

The term $B_0 R$ is called the *magnetic rigidity* and a measure of the beam's stiffness in the bending field. For the LHC it is 1500 T m at injection and 23 356 T m at collision energy.

Table 1.1 shows a comparison of the maximum proton beam energy of different particle accelerators and the maximum flux density in the superconducting dipole magnets. Note that the effective radius is between 60% and 70% of the tunnel radius because of the dipole *filling factor* and the straight sections around the collision points.[9]

1.2 A Magnet Metamorphosis

The first superconducting magnets ever to be operated in an accelerator were the eight quadrupoles of the high-luminosity insertion at the CERN ISR [13].

8 For a proof see Section 11.2.

9 The filling factor is less than one because of dipole-free regions in the interconnections, space requests for focusing elements, etc.

Table 1.1 Comparison of the maximum proton beam energy in particle accelerators and the maximum flux density in their bending magnets.

Accelerator	Tevatron	HERA	UNK	SSC	RHIC	LHC
Laboratory	FNAL	DESY	IHEP	SSCL	BNL	CERN
Commissioning	1983	1990	canceled	canceled	2000	2008
Country	USA	Germany	Russia	USA	USA	Switzerland
Circumference (km)	6.3	6.3	21	87.0	3.8	27.0
Proton momentum (TeV/c)	0.9	0.92	3.0	20.0	0.1	7.0
Nominal dipole flux density (T)	4.4	5.8	5.11	6.79	3.45	8.33
Injection dipole flux density (T)	0.66	0.23	0.69	0.68	0.4	0.535
Nominal current (A)	4400	5640	5073	6553	5050	11850
Number of dipoles per ring	774	416	2168	3972	264	1232
Aperture (mm)	76.2	75	80	50	80	56
Magnetic length (m)	6.1	8.8	5.8	15.0	9.7	14.312
Dipole filling factor	0.75	0.58	0.59	0.68	0.67	0.65

They used epoxy-impregnated coils with conductors made of a niobium–titanium (Nb–Ti) wire. Housed in independent cryostats equipped with vapor-cooled current leads, they operated in a saturated bath of liquid helium at 4.3 K.

The first completely superconducting accelerator was the Tevatron at FNAL (USA). This proton synchrotron, with a circumference of 6.3 km, comprises 774 superconducting dipoles and 216 superconducting quadrupoles, wound with conductors made of a Nb–Ti composite wire, and operated in forced-flow supercritical helium at 4.4 K. The Tevatron was later converted into a proton–antiproton collider and is still in operation.

Another superconducting synchrotron of comparable size was the proton ring of the electron–proton collider HERA at DESY (Germany) [65]. The LHC also builds on experience from the magnet design work for the SSC project (USA) that was canceled in 1993 [57].

Figure 1.3 (left) shows the HERA accelerator tunnel at DESY, where a ring (bottom) of normal-conducting magnets steers the electron beam, and a ring of superconducting magnets (top) steers the counter-rotating 820 GeV proton beam. Its 416 superconducting dipoles (with a field of 5.2 T) and 224 quadrupoles were also wound with a Nb–Ti cable and operated in forced-flow supercritical helium at 4.4 K. Unlike those of the Tevatron, the magnets of HERA had their iron yoke positioned inside the helium vessel of the cryostat and therefore at cryogenic temperatures.

Now we will discuss the difference between normal– and superconducting magnets from the point of view of the electromagnetic design and optimization. To explain the design concepts and to account for some of the historical development of the technology, Figures 1.4–1.8 show the metamorphosis in the normal-conducting dipoles for LEP and two different designs of super-

Figure 1.3 Left: High Energy Ring Accelerator (HERA) at DESY [33], Hamburg, Germany, where in one ring the normal-conducting magnets steer the electron beam and in the other the superconducting magnets steer the counter-rotating proton beam. Right: LEP quadrupole and dipoles. The main dipoles have a bending field of 0.109 T at a beam energy of 100 GeV.

conducting twin-aperture magnets for counter-rotating high-energy proton beams.

All field calculations were performed using the CERN field computation program ROXIE, employing the numerical field methods described in Chapter 14. The color representation of the magnetic flux density in the iron yokes is identical in all cases, whereas the size of the field icons changes with the different field strengths. We always distinguish between the *source field* generated by the transport current in the coils alone, and the *total field* comprising both the coil field and the contribution from the iron yoke magnetization.[10] The total field, denoted B_t, is thus the sum of the source field B_s and the *reduced field* B_r. It is then easy to distinguish between iron-dominated and coil-dominated magnets. In the LHC dipole magnets the reduced field accounts only for around 20% of the total field; they are clearly coil-dominated.

As accelerator magnets are in general long with respect to the dimension of the aperture, we can limit ourselves here to 2D field computations. The effects of coil ends will be discussed in Chapters 15 and 19.

Figure 1.4 (left) shows the slightly simplified cross section of the C-shaped dipole of LEP. The advantage of C-shaped magnets is an easy access to the beam pipe. However, they have a higher fringe field and are mechanically less rigid than the H-type magnets as shown in Figure 1.4 (right). For the same air-gap flux, the iron yoke of the H-type magnet is smaller because the flux is guided through the two return paths. Additional pole shims can be mounted in order to improve the field quality in the aperture. The field of these magnets

10 Somewhat casually we often use the word field synonymously for magnetic flux density.

is dominated by the shape of the iron yoke; in the case of the H-magnet the source field is only 0.065 T, compared to a total field of 0.3 T. Figure 1.3 (right) shows the C-shaped dipole magnets and a quadrupole installed in the LEP tunnel.

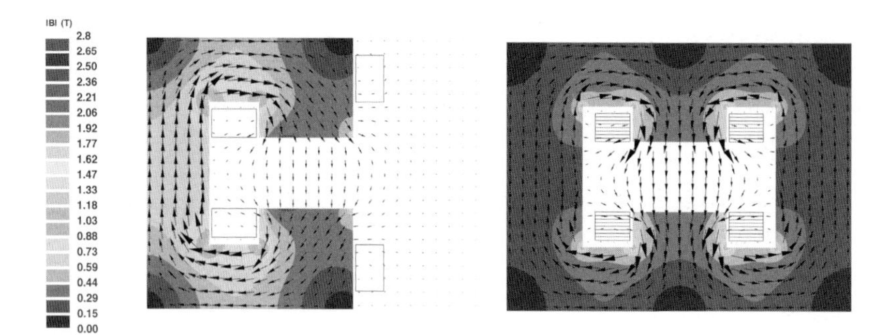

Figure 1.4 Left: C-shaped dipole of LEP ($N \cdot I = 2 \cdot 5250$ A, $B_t = 0.13$ T, $B_s = 0.042$ T). Filling factor of the yoke laminations of 0.27. Right: H-magnet as used in beam transfer lines. $N \cdot I = 24$ kA, $B_t = 0.3$ T, the source field from the coils is only $B_s = 0.065$ T, and the filling factor of the yoke laminations is 0.98.

The field quality in the aperture of the magnet is determined by the shape of the iron yoke and pole piece, which can be controlled by precision stamping of the laminations. Tolerances in the cable position have basically no effect on the magnetic field.

The LEP dipoles were ramped from 0.0218 T at injection energy (20 GeV) to 0.109 T at 100 GeV. In order to reduce the effect of remanent iron magnetization and achieve a more economical use of the steel, the yoke was laminated with a *stacking factor*[11] of only 0.27. The longitudinal spaces were filled with cement mortar, which ensured the mechanical rigidity of the yokes. This resulted, however, in unacceptable fluctuations in the bending field at low excitation, caused by the reduction of the maximum permeability due to *magnetostriction*. This phenomenon will be discussed in Section 4.6.4.

Superconducting technology allows the increase of the excitation current well above a density of 10 A mm^{-2}, which is the practical limit for normal-conducting (water-cooled) coils for DC magnets.[12] Disregarding for the moment the superconductor-specific phenomena such as magnetization, flux creep, and resistive transition, we can consider *engineering current densities* that are hundred times higher than those in water-cooled copper or aluminum

11 For a definition of the stacking factor see Section 4.6.3.
12 Pulsed septum magnets are operated with current densities of up to 300 A mm^{-2}.

coils.[13] Magnets in which the coils are superconducting but in which the magnetic field distribution is still dominated by the iron yoke are known as *superferric*. Figure 1.5 (left) shows a H-type magnet with increased excitation as can be achieved by superconducting coils. The poles are beginning to saturate, and the field quality in the aperture is degraded due to the increasing fringe field. An improved design, shown in Figure 1.5 (right), features tapered poles, a concave pole shape, and a small hole to equalize saturation effects during the field sweep from injection to nominal excitation.

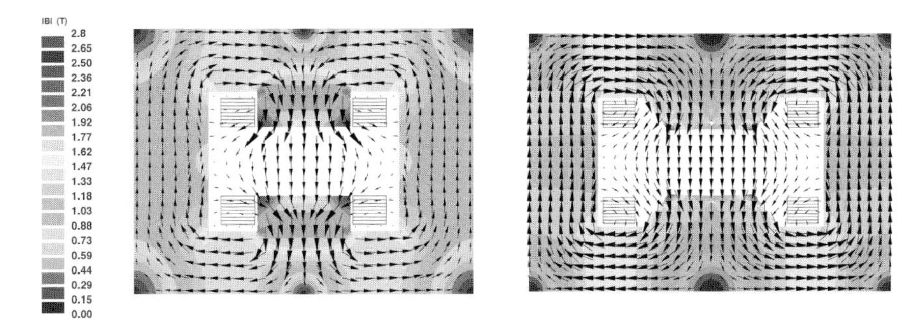

Figure 1.5 Left: H-magnet with increased excitation current ($N \cdot I = 96$ kA, $B_t = 1.17$ T, $B_s = 0.26$ T). Right: Improved design with tapered pole, concave pole shape, and a small hole to equalize iron saturation ($N \cdot I = 96$ kA, $B_t = 1.15$ T, $B_s = 0.208$ T).

Saturation of the pole can be avoided in window frame magnets as shown in Figure 1.6 (left). The disadvantage of window frame magnets is that synchrotron radiation is partly absorbed in the coils and access to the beam pipe is difficult. The advantages are that a better field quality is obtained and pole shims can be avoided. Superconducting window frame magnets have received considerable attention since the mid-1990s as a design alternative for high-field dipoles in the 14–16 T field range, taking advantage of easier coil winding and mechanical force retaining structures [31].

By superposition of two window frame magnets it is possible to increase the aperture field while reducing the magnetic flux density in the pole faces; see Figure 1.6 (right). A technical difficulty arises here in achieving the double current density in the square overlap area at the horizontal median plane [4].

At higher field levels, the field quality in the aperture of accelerator magnets is increasingly affected by the coil layout. It will be explained in Chapter 8 that a current distribution of $\cos n\varphi_c$ within a shell centered at the origin creates a perfectly homogeneous $2n$-polar field in the aperture. An advantage of this design is that for the dipole, $n = 1$, the field drops with r^{-2} outside the shell

13 The engineering current density is an overall current density considering cooper stabilization, filling factors, and insulation.

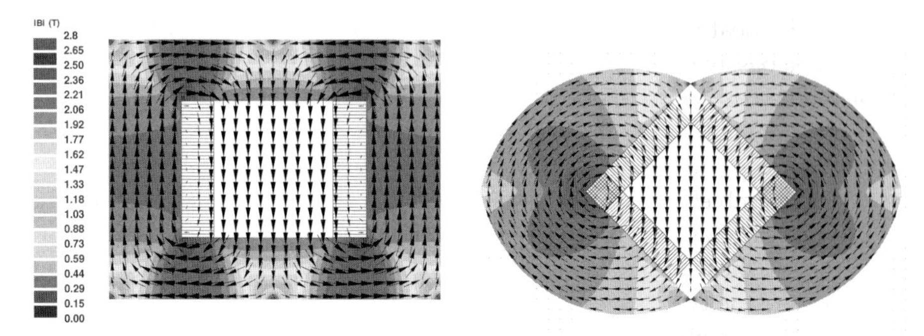

Figure 1.6 Left: Window frame geometry ($N{\cdot}I = 360$ kA, $B_t = 2.0$ T, $B_s = 1.04$ T). Notice the saturation of the poles in the H-magnet. Right: By superposition of two window frame magnets it is possible to increase the aperture field while reducing the magnetic flux density on the pole faces ($N{\cdot}I = 625$ kA, $B_t = 2.38$ T, $B_s = 1.36$ T).

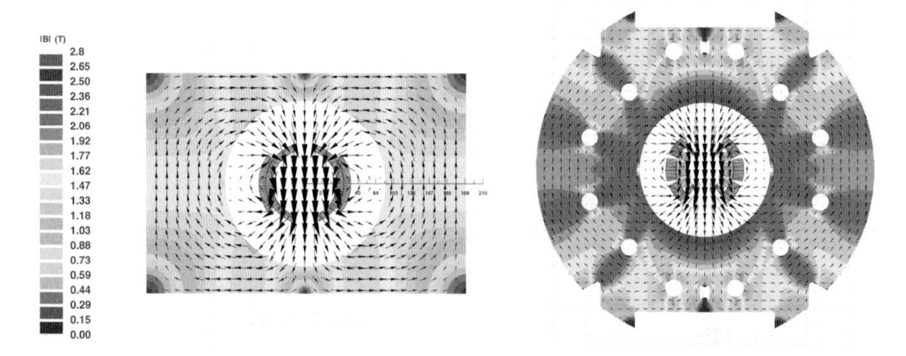

Figure 1.7 Left: Tevatron dipole with warm iron yoke and superconducting coil of the $\cos\varphi_c$ type ($N \cdot I = 471$ kA , $B_t = 4.16$ T, $B_s = 3.39$ T). Notice that even with increased aperture field the flux density in the yoke is reduced due to the $\cos\varphi_c$ coil. Right: LHC single-aperture coil-test facility ($N \cdot I = 960$ kA , $B_t = 8.33$ T, $B_s = 7.77$ T).

and saturation effects in the iron yoke are reduced. Figure 1.7 (left) shows the magnets for the Tevatron accelerator at FNAL [25] with a coil design approximating the ideal $\cos\varphi_c$ current distribution. The Tevatron dipole has an iron yoke at ambient temperature with the cryostat for the superconducting coil located inside the aperture of the yoke. The advantage of the solution is the low saturation in the yoke, up to the excitational limit set by the critical current of the superconductor. The disadvantage of the warm iron yoke is possible misalignment of the coil and its cryostat within the yoke, which is the source of unwanted multipole field errors. Misalignment also causes considerable net forces between the coil and the yoke, which requires many supports that increase the heat transfer from the warm iron yoke to the helium vessel. In addition, a passive quench protection system with parallel diodes is not easily implemented as it would require a parallel helium transfer line.

Figure 1.7 (right) shows the *coil-test facility* (CTF) used for validating the manufacturing process of the LHC magnets. The electromagnetic design of the CTF resembles the single-aperture dipole magnets proposed for the Superconducting Super Collider project, as well as the HERA and RHIC [60] dipoles. All these magnets feature iron yokes cooled to cryogenic temperatures, where the coil, collars, and yoke are enclosed in a helium-tight vessel, forming an assembly that is referred to as the *coldmass*. This principle was first developed for the main magnets of the ISABELLE project at Brookhaven and later adapted to the HERA accelerator magnets at DESY. This allows the positioning of the iron yoke closer to the superconducting coil and helps to increase the main field for a given amount of superconductor, while reducing the stored energy in the magnet. It also guarantees the centering of the coil and thus suppresses eccentricity forces on the coil. A disadvantage is the higher saturation-induced field distortion, which must be minimized using optimization methods coupled with FEM computations. Notice the large difference between the field in the aperture (8.3 T) and the field in the iron yoke (maximum 2.8 T) shown in Figure 1.7 (right). The iron yoke not only shields the stray fields, but also screens the beam from the influence of current busbars that are housed in groves on the outer rim of the iron yoke.

The source field generated by the superconducting coil alone is as high as 7.77 T and thus the magnetization of the iron yoke contributes only 10% to the total field. Higher contributions of the yoke magnetization to the central field can be obtained by replacing the collar with a fiberglass-phenolic spacer between the coil and yoke. This principle was used in the design of the RHIC dipole magnets in which the coil prestress was supplied by an outer cylindrical tube and the yoke laminations. This resulted in a 35% enhancement of the field due to the iron magnetization.

Figure 1.8 (left) shows the LHC main dipole in a twin-aperture design first proposed in [22]. It features two coils and two beam channels within a common mechanical structure and iron yoke. The mechanical structure and the cryostat are shown in Figure 1.15. The current-dependent multipole field errors can be controlled by optimal shape design of the yoke, including holes and notches.

On the right-hand side of Figure 1.8, an alternative twin-aperture magnet design is shown [31]. This construction allows for easier winding of the coil ends, as the coil blocks in the two apertures form a common coil with a minimum bending radius of half of the beam separation distance (97 mm). Disadvantages are the high iron saturation in the horizontal median plane and consequently a higher dependence of the field quality on the excitation level, as well as a strong crosstalk between the apertures. Summing up, we list the differences between normal-conducting and superconducting accelerator magnets.

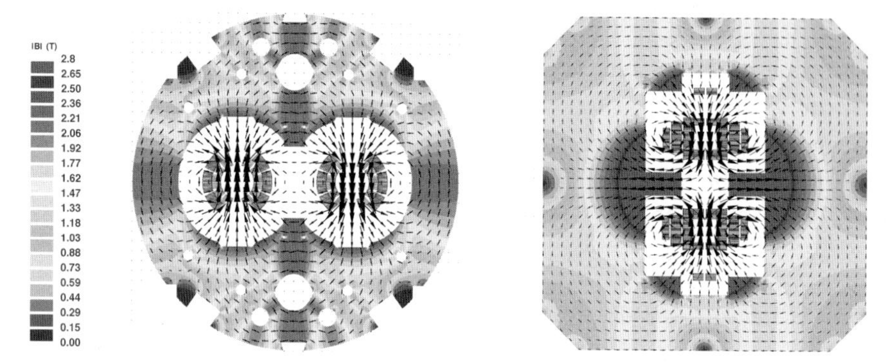

Figure 1.8 Left: Two-in-one LHC dipole with two coils in a common mechanical structure and iron yoke ($N \cdot I = 2 \cdot 944$ kA, $B_t = 8.32$ T, $B_s = 7.44$ T). Aperture diameter 56 mm. Right: Alternative design [31] with two apertures in a common coil design ($N \cdot I = 2 \cdot 1034$ kA, $B_t = 8.34$ T, $B_s = 7.35$ T). Aperture diameter 50 mm.

Normal-conducting magnets:

- The magnetic field is defined by the iron pole shape and limited to about 1.5 T. The conceptual design can be accomplished using one-dimensional field computation as described in Chapter 7.
- Normal-conducting magnets feature very high field quality because the yoke can be shaped with high precision. In addition, the field quality can be optimized by pole shims. Commercial finite-element software can be applied to the design as a "black box."
- Conductor placement is not critical, although the stray field can be reduced by bringing the coil close to the air gap.
- Ohmic losses in the coils (16 MW for all LEP dipoles) require water cooling, resulting in high operational costs.
- The voltage drop across the ohmic resistance must be considered, particularly in view of a series connection of a string of magnets.
- Electrical interconnections in strings of magnets are easy to make and to check.
- Hysteresis effects in the iron yoke must be modeled.

Superconducting magnets:

- The field is defined by the coil layout, which requires accurate coil modeling and adapted computational tools for optimization of the field quality.

- The shaping of the coils in the end region requires special attention to limit performance degradation. In addition, the effective magnetic length is shorter than the physical length.
- The high current density in the superconductor allows the building, on an industrial scale, of accelerator magnets with a maximum field of 9 T using Nb–Ti composite wire.
- The contribution of the magnetization in the iron yoke to the main field in the aperture is limited to 30%. Thus nonlinear variations in the field quality due to inhomogeneous saturation of the yoke can be limited for a wide range of excitation levels.
- The enormous electromagnetic forces (4 $\mathrm{MN\,m^{-1}}$ in the LHC main dipole at nominal excitation) require a careful mechanical design with adequate force-retaining structures.
- The voltage drop across the magnet terminals is limited to the inductive voltage during the ramping of the magnets.

A superconducting magnet system poses additional technological challenges in the domain of cooling and magnet protection:

- Operational stability must be guaranteed with heat transfer to the coolant, cryogenic installations (refrigerators), helium distribution lines, and insulation cryostats. Special designs for current feedthroughs from the room-temperature environment into the helium bath are required.
- Electrical interconnections of superconducting busbars are located inside the helium enclosure and cryostat and therefore impossible to verify once the accelerator is in operation.
- Protection against overheating during a resistive transition (quench) is required. This includes quench detection electronics, an energy extraction system with protection resistor, quench-back heaters, and cold bypass diodes, among other measures.
- Superconducting filament magnetization results in hysteresis effects and relatively large multipole field errors at injection field level. Magnetization-induced field errors are the principal reason for the installation of the spool-piece corrector magnets.

It is customary to refer to normal-conducting and superferric magnets as *iron-dominated* and superconducting magnets as *coil-dominated*. The latter can further be grouped into two classes [34]:

- Class 1 magnets for plasma confinement in fusion reactors, physics experiments, and magnetic energy storage (SMES), constructed for a field level of 4–5 T, and with large apertures in the range of meters. Owing to the magnet size and the large electromagnetic forces, the most challenging design aspect is the mechanical integrity.

- Class 2 magnets feature a high field and high current density but a relatively small aperture in the range of centimeters. Applications are nuclear magnetic resonance (NMR), magnetic resonance imaging (MRI), particle accelerators, superconducting motors and generators, magnetic separation, magnetic levitation, and high magnetic field research, among others. Critical issues are cable design, stability, magnet protection, and cooling.

1.3 Superconductor Technology

All superconducting synchrotron projects since the Tevatron at FNAL employ niobium–titanium (Nb–Ti) superconductors operated at cryogenic temperatures at or below 4.5 K. The advantage of the Nb–Ti alloy is the combination of good superconducting properties with favorable mechanical (ductility, tensile strength) and metallurgical properties that allow the coprocessing with different substrate materials such as copper and copper–nickel. Thus wires can be produced with the fine filaments necessary to control the field quality and to limit hysteresis losses in the magnets.

Even with Nb–Ti superconductors, the only way to obtain the required central field in the LHC main magnets is to apply cooling with superfluid helium II, a technology proven on a large scale with the Tore Supra Tokamak [8] built for fusion research at CEA (France).

After a worldwide industrial qualification program in the years 1994–1996, about 1300 tons of wire, extruded from more than 6000 billets, was produced and cabled to a total length of 7500 km.

1.3.1 Critical Current Density of Superconductors

The production of a Nb–Ti wire is nowadays achieved with a high homogeneity in the critical current density, above 1600 A mm^{-2} at 1.9 K in a 10 T applied field. Figure 1.9 shows the critical current density J_c of Nb–Ti as a function of the flux density B and temperature T. A magnet's working point on the load line is determined by the maximum flux and current densities in the coil for a given uniform temperature. The distance between the working point and the *critical surface* $J_c(B, T)$ along the load line determines the operational margin to *quench*.[14]

14 A quench is the transition between the superconducting and the resistive state of the material. The critical surface modeling is presented in Section 16.2. Quench simulation and magnet-protection schemes are discussed in Chapter 18.

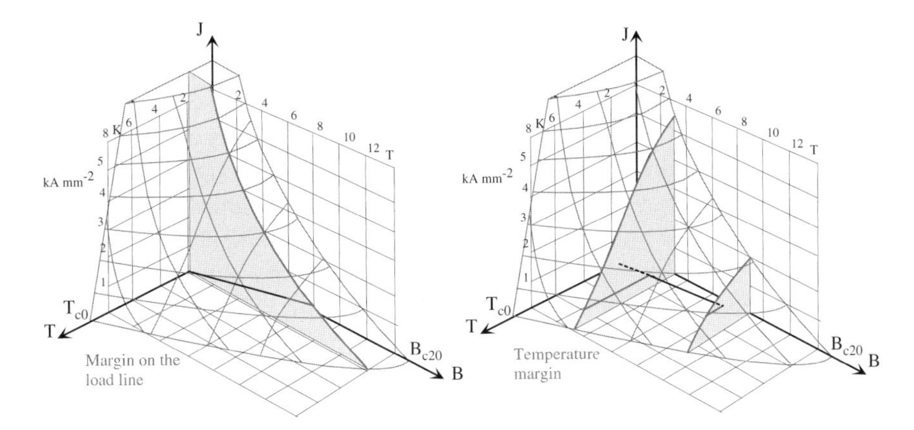

Figure 1.9 Critical surface of the Nb–Ti superconductor. Left: Operational margin of the magnet on the load line. Right: Temperature margin at a given field and current density.

Figure 1.9 (right) shows the temperature margin for a given flux density and current density. This temperature margin is important for the stable operation of the magnets at a given beam energy and intensity because of unavoidable beam losses and heat in-leaks through the magnet's cryostat.

Fields in excess of 14 T can, potentially, be achieved with accelerator-type magnets constructed with niobium–tin composite conductors. At the early stage of the LHC R&D phase, a mirror dipole magnet for operation at 4.5 K was constructed with Nb_3Sn conductors and reached a record value of 10 T [3]. Nevertheless, the Nb_3Sn magnet program was abandoned in 1991 due to the brittle nature of the material and because the stress-dependent critical current density results in performance degradation after coil winding. Furthermore, the minimum attainable filament sizes for this material are even nowadays too large to match the requirements of the LHC machine.[15] Conductors made of Nb_3Sn also require heat treatment, in which the niobium and tin react at a temperature of 700 °C to form the superconducting A15 phase of Nb_3Sn. The heat treatment can be performed before or after the coil winding; the two techniques are therefore known as react-and-wind and wind-and-react. While R&D is currently focused at the stress dependence of the material, short model magnets have achieved flux densities on the order of 14 T. Possible applications are magnets for the luminosity upgrade of the LHC and for next-generation hadron colliders. As a result of these efforts, the critical current density has doubled since the early 1990s, and record values for a single wire are as high as 3000 A mm^{-2} at 4.2 K and a 12 T applied field [24]. A good overview of magnet development using Nb_3Sn conductors is given in [32].

15 Smallest diameter of 25 μm for modern powder-in-tube wires [24].

High-temperature superconductor (multifilamentary BSCCO[16] 2223 tape stabilized with a silver gold alloy matrix) is used in the LHC project for the current leads that provide the transition from the room-temperature power cables to the superconducting busbars in the continuous cryostat. A total of 1030 leads, for current ratings between 600 A and 13 kA [10], are mounted in electrical feed boxes. Each lead is made of a section of high-temperature superconductor (HTS) material, operating between 4.5 and 50 K in the vapor generated by heat conduction, and a normal-conducting copper part operating between 50 K and ambient temperature. By using HTS current leads, the heat influx to the helium bath can be reduced by a factor of 10 with respect to normal-conducting copper leads. In total, 31 km of BSCCO 2223 tapes were vacuum soldered at CERN to form more than 10 000 stacks, containing four to nine tapes depending on the current rating [11].

For use in high-field magnets, however, the critical current density of $300\,\mathrm{A\,mm^{-2}}$ at 8 T and 4 K would be too low and the material too brittle and too expensive. More information on critical currents obtained in technical superconductors can be found in [48].

For the parametric representation of the critical surface of Nb–Ti we use the empirical relation [14]

$$J_c(B,T) = \frac{J_c^{\mathrm{ref}} C_0 B^{\alpha-1}}{(B_{c2})^{\alpha}} (1-b)^{\beta} \left(1 - t^{1.7}\right)^{\gamma}, \tag{1.6}$$

with the fit parameters α, β, γ, and the normalized temperature t and normalized field b defined by $t := T/T_{c0}$ and $b := B/B_{c2}(T)$. The critical field is scaled with $B_{c2} = B_{c20}\left(1 - t^{1.7}\right)$, where B_{c20} is the upper critical field at zero temperature. For the calculation of the margin on the load line, as well as the temperature margin with respect to heat deposits due to beam losses, we use the following parameters [14]: The reference value for the critical current density at 4.2 K and 5 T is $J_c^{\mathrm{ref}} = 3000\,\mathrm{A\,mm^{-2}}$, the upper critical field at zero temperature $B_{c20} = 14.5$ T, the critical temperature at zero flux density $T_{c0} = 9.2$ K, and the normalization constant $C_0 = 27.04$ T. The fit parameters are $\alpha = 0.57$, $\beta = 0.9$, and $\gamma = 2.32$. The maximum local error of the fit is in the range of 11% at low field and up to 5% at high field. It must be noted that critical current density values at a low field are determined from magnetization measurements and data at field levels above 3 T from I_c measurements of the strands. We will return to this point in Chapter 16.

Figure 1.10 (left) shows the critical current density of Nb–Ti at 1.9 K as a function of the applied field, together with the linear approximations

$$J_c(B) = J_{\mathrm{ref}} + c\,(B_{\mathrm{ref}} - B) \tag{1.7}$$

16 Bismuth strontium calcium copper oxide $Bi_2Sr_2Ca_2Cu_3O_{10}$ with $T_c =$ 107 K, discovered in 1988.

around the reference points for both cable types. The slopes are given by

$$c := -\left.\frac{\mathrm{d}J_\mathrm{c}}{\mathrm{d}B}\right|_{B_\mathrm{ref}} \tag{1.8}$$

and are in the range of 500 to 600 $\mathrm{A\,mm^{-2}\,T^{-1}}$. Figure 1.10 (left) also shows the load line of the LHC main dipoles. Nonlinearities due to iron saturation are disregarded. Note that the quench current of the magnet is limited by the critical current density in the coil, which is exposed to a 2–5% higher field than that of the magnet aperture. Important for the magnet performance is the engineering current density, J_E, which takes into account the copper matrix needed for stabilization, the filling factor of the cable, and the insulation thickness. Figure 1.10 (right) shows the temperature margin of the cables in the different coil blocks of the dipole magnet. The lines are the projections of the critical surface at a given flux density B onto the JT-plane; compare Figure 1.10 (right).

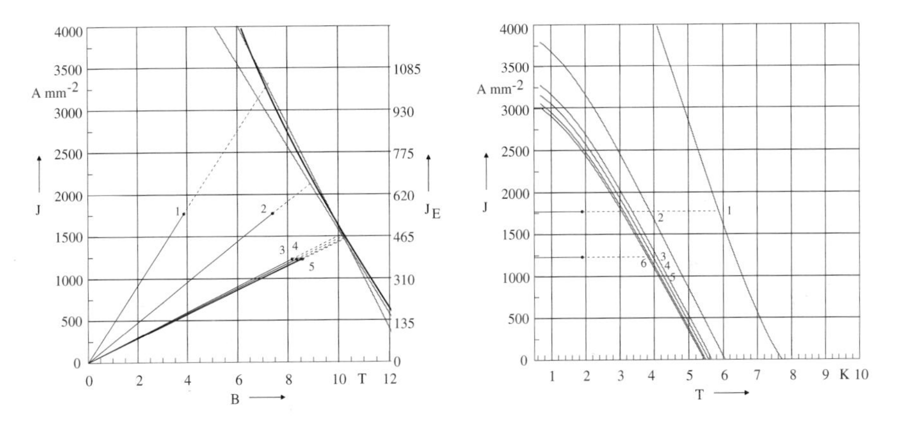

Figure 1.10 Left: Critical current density of Nb–Ti at 1.9 K as a function of the applied field (for linear approximations around reference points), together with the load line of the LHC main dipoles. Right: Temperature margin of the cables in the different coil blocks of the dipole magnet.

1.3.2 Strands

The *strands*[17] are made of thousands of Nb–Ti filaments embedded in a copper matrix that serves to stabilize the wire and to carry the current in the event of a quench, since superconductors have a high resistivity in the normal state.

17 This is common parlance for superconducting wires; the terms are used synonymously.

A microphotograph of a strand and its filaments is shown in Figure 1.11. The filaments are made as small as possible (constrained by the manufacturing costs and interfilament coupling effects) in order to reduce remanent magnetization effects and to increase the stability against flux jumps during excitation, that is, the release of fluxoids from their pinning centers.[18]

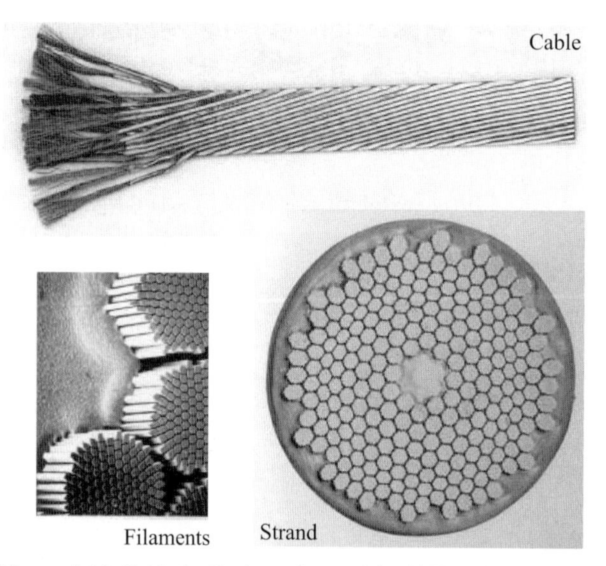

Figure 1.11 Cable for the inner layer of the LHC main dipole coils; microphotograph of the strand cross section and the superconducting filaments. Shown is a prototype wire produced with a double-stacking process.

The critical current density and the slope of the critical surface at the reference point (constant temperature of 1.9 K) are given for the strands used in LHC cables in Table 1.2. The critical current densities were measured for strands extracted from the cables, and therefore the degradation due to the cabling process is taken into account. Note that the values are scaled with the copper-to-superconductor area ratio, and are thus valid for the superconducting material only.

The critical current density of the superconductors (Figure 1.12) can be scaled with the strand filling factors

$$\lambda_{SC} = \frac{1}{1+\eta}, \qquad \lambda_{Cu} = \frac{\eta}{1+\eta}. \tag{1.9}$$

The quotient $\eta := a_{Cu}/a_{SC}$ is the area ratio of the copper cross section of the strand to the total superconducting filament cross section. The cross-sectional

18 The remanent magnetization effects are proportional to the critical current density and the filament size, as is explained in Chapter 16.

Table 1.2 Characteristic data for the strands used in the main dipole (MB) cables, the MQM and the MQY quadrupole cables, and in the auxiliary busbar cable (Line N)[a].

Magnet	Strand 1 MB inner	Strand 2 MB outer	Strand 5 MQM	Strand 6 MQY	Line N
Diameter of strands (mm)	1.065	0.825	0.48	0.735	1.6
Strand pitch length (mm)	25	25	15	15	25
Copper to SC area ratio	1.65	1.95	1.75	1.25	> 9
Filament diameter (μm)	7	6	6	6	58
Number of filaments/strand	8900	6500	2300	6580	17
T_{ref} (K)	1.9	1.9	1.9	4.5	4.2
B_{ref} (T)	10	9	8	5	1
$J_c(B_{\mathrm{ref}}, T_{\mathrm{ref}})$ $(\mathrm{A\,mm^{-2}})$	1433.3	1953.0	2872	2810	> 5000
$-\mathrm{d}J_c/\mathrm{d}B$ $(\mathrm{A\,mm^{-2}\,T})$	500.34	550.03	600	606	n.s.
$\rho(293\ \mathrm{K})/\rho(4.2\ \mathrm{K})$ of Cu	> 70	> 70	80	80	> 100

[a] The Line N cable is made from the same highly stabilized strands that are used in MRI magnets. The cable is used to power the lattice corrector magnets mounted in the short straight section. n.s. = not specified.

area of diffusion barriers and the strand coating is therefore disregarded. The copper-to-superconductor area ratio η is 1.65 for the inner-layer dipole cable, and 1.95 for the outer-layer dipole cable, which was also used for the production of the main quadrupoles.

In the case of Nb_3Sn composite conductors, with filaments including an inner core, the composition is more complex and one often refers to the ratio of copper to noncopper areas.[19] The brittleness of the material prevents it from being extruded from ingots. Instead, a diffusion technique is applied, where the elements of the compound are reacted at 700 °C for several hours. The reaction is performed after wire production or after the coil winding only. The ratio of copper to noncopper areas depends on the various production techniques, such as the most common bronze (Cu–Sn) diffusion technique, the external tin diffusion technique, and the powder-in-tube (PIT) process [26,28].

The manufacturing of multifilamentary Nb–Ti composite wires is accomplished by stacking hexagonal rods of Nb–Ti inside a sealed copper canister, an assembly usually referred to as a billet. Practical limitations are minimum rod diameters of approximately 1.5 mm and a maximum number of rods of 15 000. The billet is then hot-extruded, followed by conventional wire drawing in multiple stages to reach the final diameter. After the drawing process, the wire is annealed at 400 °C to form the dislocation cell structures needed for flux pinning and thus for enhancing the critical current. The filament size chosen for the LHC (7 μm for the strand in the inner-layer cable and 6 μm in the

19 Consequently, the noncopper critical current density J_c is the relevant quantity for magnet design.

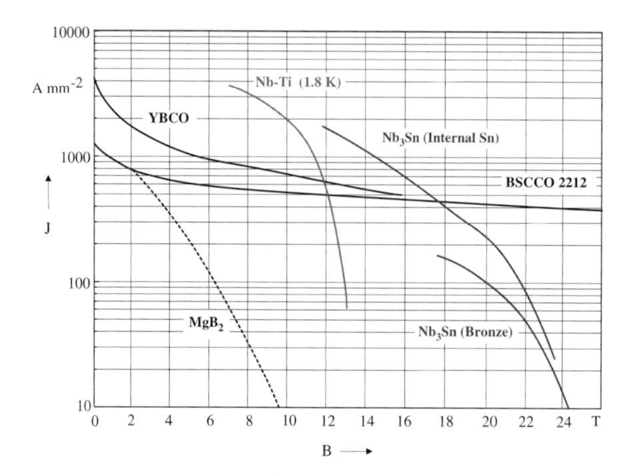

Figure 1.12 Critical current density of superconductors (at 4.2 K if not otherwise stated); source [48].

outer-layer cable) allows the fabrication of the wires in a single-stacking process. Even finer filaments can be obtained by a double-stacking process, where rods with a hexagonal section, resulting from a first extrusion, are stacked into a secondary billet, which is then extruded again and drawn to the final wire diameter. The filament pattern for a wire produced with the double-stacking process is shown in Figure 1.11.

Due to the extrusion and drawing process, the filament can deviate from the optimum round shape. The distortions may also vary along the length of the wire, a deformation usually referred to as *sausaging*. Distorted filaments show higher hysteresis losses and a wider resistive transition, resulting in an effective resistivity according to the empirical law

$$\rho = \frac{E_c}{J_c}\left(\frac{J}{J_c}\right)^{n-1} . \tag{1.10}$$

The factor n, called the *resistive transition index*, can be seen as a quality factor of the production. The field-dependent resistive transition index obtained for the LHC wire production is $n = 42$ at 10 T and $n = 48$ at 8 T [53].

1.3.3 Cables

According to Eq. (1.5) the trajectory radius of the particle increases with its momentum. As both the flux density and the aperture of the bending magnets are limited, the magnetic field must be ramped synchronously with the particle energy. The LHC main dipole magnets are ramped from the injection field level of 0.54 T to their nominal field of 8.33 T, according to the excitation

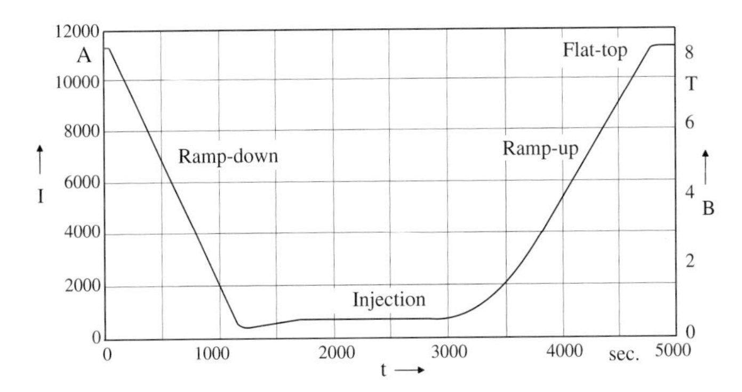

Figure 1.13 Excitation cycle [19] for the LHC main dipole circuits.

cycle shown in Figure 1.13. The mathematical description of the excitation cycle is given in Appendix C.

To ensure good tracking of the field with respect to the current and to reduce the number of power converters and current feedthroughs, the magnets in superconducting synchrotron accelerators are connected in series. Disregarding nonlinearities due to iron saturation in the magnet yokes, and the stored energy at an injection field level, the induced voltage during the ramping of an LHC magnet string is approximately

$$U \approx \frac{2E}{It}, \tag{1.11}$$

where E is the stored energy in the string of 154 dipoles at nominal current, t is the current rise time of 1200 s, and I is the nominal operating current of 11 800 A. The maximum voltage in the power supplies of the main dipole circuit can be calculated for a stored energy of 1.1×10^9 J (approximately 300 kW h) to 155 V. To avoid higher voltages, the coils of the LHC dipole and quadrupole magnets are wound from so-called *Rutherford cables* of trapezoidal shape. The cabling scheme is identical to the Roebel[20] bar known in the domain of electrical machines; see Figure 1.14.

It was shown in the Rutherford laboratory in the early 1970s that the cable could be produced, without wire or filament breakage at the cable edges, by rolling a hollow, twisted tube of wires. Two layers of fully transposed strands limit nonuniformities in the current distribution within the cable caused by the cable's self-field and flux linkage between the strands.[21] The Roebel scheme allows cable compaction of 88–94% without strand damage and good control of the dimensional accuracy on the order of 0.01 mm. The cable used

20 Ludwig Roebel (1878–1934), patent 1912.

21 Note that the strands in twisted litz wire are not fully transposed, as strands positioned in the center always remain there.

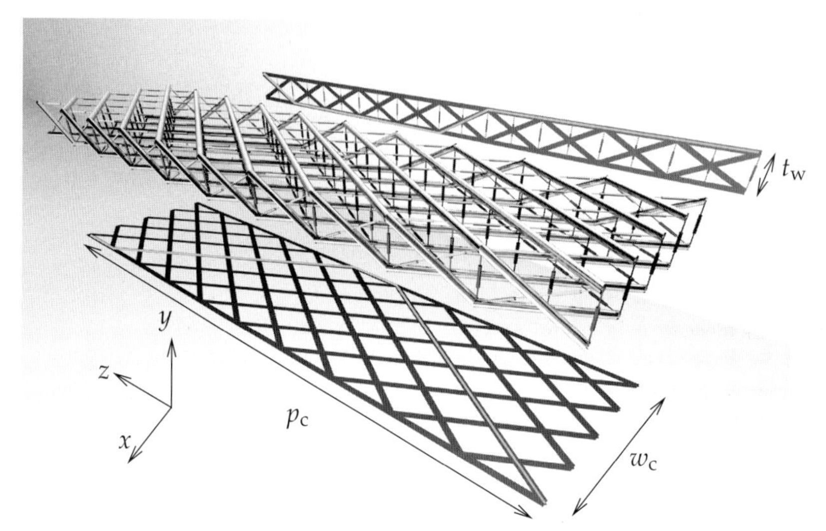

Figure 1.14 Winding scheme of a Rutherford cable (schematics with 10 strands). The blue path shows the full transposition of a superconducting strand. Depending on the transposition pitch and the cable compaction, the uninsulated strands are in contact, which gives rise to cross resistances (red) and adjacent resistances (yellow).

for the inner layer of the LHC main dipole coil contains 28 superconducting strands, while the cable used in the outer-layer dipole coil and in the main quadrupoles contains 36 strands.

For magnet design it is desirable that the cable be *keystoned*[22] at an angle that allows the winding of perfect arc segments. However, due to the critical current degradation during cabling, the keystone angle is limited; see Figure 8.1 for a cross-sectional view of a block of cables in the inner layer of the LHC main dipole coils. In [54] the packing factor at the cable's narrow edge is defined by

$$\lambda_{\mathrm{n}} := \frac{a_{\mathrm{s}}}{0.5\, t_{\mathrm{n}} d_{\mathrm{s}}} = \frac{\pi d_{\mathrm{s}}}{2 t_{\mathrm{n}}} \tag{1.12}$$

where a_s is the strand cross-sectional area, t_n is the cable thickness at its narrow edge, and d_s is the strand diameter. Measurements have revealed that the amount of degradation increases considerably for narrow-edge packing factors exceeding 0.98 because of local reduction in the strand cross section and breakage of filaments during cabling.[23] The narrow-edge packing factor for

22 A keystone is an architectural piece at the apex of an arch, of trapezoidal shape, locking the other stones into position.

23 In Nb_3Sn strands, breakage of antidiffusion barriers can lead to incomplete filament reaction [23].

Table 1.3 Cable characteristic data for inner-layer (IL) and outer-layer (OL) main dipole (MB) coils[a].

Magnet	Cable 1 MB (IL)	Cable 2 MB (OL)	Cable 4 MQM	Cable 5 MQY (OL)	Cable 6 MQY (IL)
Strand	1	2	5	5	6
Bare width (mm)	15.1	15.1	8.8	8.3	8.3
Bare thickness, thin edge (mm)	1.736	1.362	0.77	0.78	1.15
Bare thickness, thick edge (mm)	2.064	1.598	0.91	0.91	1.40
Mid thickness (mm)	1.9	1.48	0.84	0.845	1.275
Cable cross section (mm^2)	28.69	22.35	7.39	7.014	10.58
Keystone angle (degree)	1.25	0.9	0.91	0.89	1.72
Aspect ratio	7.95	10.2	10.47	9.82	6.51
Insulation narrow side (mm)	0.150	0.150	0.08	0.08	0.08
Insulation broad side (mm)	0.120	0.130	0.08	0.08	0.08
Insul. cable cross sec. (mm^2)	32.96	27.0	8.96	8.50	12.14
Cable transp. pitch length (mm)	115	100	66	66	66
Number of strands	28	36	36	34	22
Cross section of Cu (mm^2)	15.3	12.6	4.1	3.9	5.2
Cross section of SC (mm^2)	9.6	6.6	2.4	2.2	4.1
SC filling factor λ_{tot}	0.29	0.24	0.27	0.26	0.34

[a] Cable 3 for the main quadrupole (MQ) coil is identical with cable 2. Cables 5 and 6 are used for the insertion quadrupoles MQM and MQY.

cable 1 used in the LHC main dipoles is 96%. The cable thickness at its wide edge is chosen in such a way that the upper and the lower strands are in contact in order to maintain cable integrity during coil winding [54]. The cable packing factor is defined as the ratio of the bare strand volume to the cable volume,

$$\lambda_{\mathrm{c}} := \frac{n\pi d_{\mathrm{s}}^2}{2w_{\mathrm{c}}(t_{\mathrm{n}} + t_{\mathrm{w}})\cos\psi}, \tag{1.13}$$

where w_{c} is the cable width, t_{w} is the cable thickness at its wide edge, and ψ is the pitch angle. The pitch angle can be calculated from

$$\tan\psi = \frac{2w_{\mathrm{c}}}{p_{\mathrm{c}}}, \tag{1.14}$$

where p_{c} is the length of the transposition pitch.

For the 2D calculations we set $\cos\psi = 1$ in Eq. (1.13). Furthermore, we define λ_{i} as the quotient of the insulated cable cross section and the bare cable cross section. The engineering current density can therefore be calculated from

$$J_{\mathrm{E}} = \lambda_{\mathrm{f}}\,\lambda_{\mathrm{c}}\,\lambda_{\mathrm{i}}J_{\mathrm{c}} =: \lambda_{\mathrm{tot}}J_{\mathrm{c}}\,. \tag{1.15}$$

The total superconductor filling factor λ_{tot} is in the range of 30%. The values for the LHC cables are given in Table 1.3. Due to field variations in the coil ends and resistance differences in the solder joints between coils, the current may not be equally distributed within the Rutherford cable. For this reason the individual strands are not insulated. This gives rise to adjacent and transverse (cross) electrical contact resistances between the strands; see Figure 1.14. Across these resistances closed loops are formed in which coupling currents can be induced during the ramping of the magnets, as well as during the fast discharge in the event of a quench. While in the first case these coupling currents produce unwanted field distortions, in the latter case they help to distribute the stored magnetic energy evenly by inductive heating of the coil, a process known as *quench-back*.

The losses can be calculated by an electrical network model discussed in Chapter 17. They scale quadratically with the magnetic flux density, the transposition pitch, and the cable width and are inversely proportional to the electrical contact resistance[24] and the cable thickness [70]. In order to impose tight control over the contact resistance for the series production of the magnets, a silver tin coating on the strands guarantees a cross-contact resistance on the order of some tens of μΩ, a good compromise in terms of ramp-induced field errors, stability, and quench-back.

In the case of the main dipole, the cable is insulated with three layers of polyimide film. Two layers (in total 50.8 μm thick) are wrapped on the cable with a 50% overlap, and another, 68.8 μm thick, is wrapped around the cable with a spacing of 2 mm, and insulation scheme sometimes referred to as a barber-pole wrapping. An adhesive layer with a nominal thickness of 5 μm is applied to the outside of the barber-pole wrapping in order not to bond the insulation to the cable and thus avoid quenches due to energy release by bond failure. The insulation protects the cable from a turn-to-turn voltage of 50 V at quench, yet it has sufficient porosity and percolation for helium cooling. The main parameters of cable used in LHC magnets are given in Table 1.3.

Magnet cooling with superfluid helium at 1.8 K benefits from the very low viscosity and high thermal conductivity of the coolant. However, the heat capacity of the superconducting cables is reduced by nearly an order of magnitude compared to an operation at 4.5 K. This results in a higher temperature rise for a given deposit of energy. To avoid quenching below the so-called *short-sample limit*, all movement of the coil must be prevented by the use of an appropriate force-retaining structure. Because the forces and stored energy in the magnets increase with the square of the magnetic flux density, the mechanical design required an extensive R&D phase carried out at CERN during the years 1988 to 2001, in close collaboration with other HEP Institutes and with European industry.

24 The contact resistances are defined as the lumped element resistances in the network model; see Figure 1.14.

1.4
The LHC Dipole Coldmass

The coil-winding direction is defined by the first cable on the center-post when looking down on the winding mandrel (counter-clockwise for the inner-layer dipole coils, clockwise for the outer-layer coils). Experience has shown that best winding results are obtained if left-hand lay cables are wound in clockwise direction, while right-hand[25] lay cables are wound in counter-clockwise direction [52]. It is advantageous for making internal cable joints that the cables in the two coil layers have opposite pitch direction. Nevertheless, all Rutherford cables used in the LHC project are left-hand lay cables; see Figure 1.11.

As the field distribution is extremely sensitive to coil-positioning errors, each coil is polymerized after winding. The press and heating system of the mold allows the coil to be cured for 30 min at 190 °C under a maximum pressure of 80–90 MPa. This process activates the adhesive layer on the insulation to glue the turns together. In this way, the mechanical dimensions of the coils can be controlled within a tolerance of ±0.05 mm. The size and elastic modulus of each coil are then measured to determine the pole and coil-head shimming for the collaring procedure. The outer-layer coil is fitted onto the inner with a fiberglass-reinforced ULTEM® spacer between the two layers. The spacer gives a precise mechanical support for the outer-layer coil and it is slotted in order to provide channels for the superfluid helium. Because of its appearance, this spacer is also referred to as the *fishbone*.

The four coil packs for the twin-aperture magnet are assembled in a mechanical force-restraining structure, known as *collars*, made of preassembled packs of 3-mm-thick austenitic steel laminations (Nippon Steel YUS 130S) with a relative permeability of less than 1.003. A 3D rendering of the collar packs is shown in Figure 1.16. The required pole-shim thickness is calculated such that the compression under the collaring press is 120 MPa. After the locking rods are inserted into the collar stack and external pressure is released, the residual coil prestress is 50–60 MPa on both layers.

The collared coils are surrounded by an iron yoke, which not only enhances the magnetic field but also reduces the stored energy and shields the stray field. The stacking factor of the 5.8-mm-thick yoke laminations is 0.985 and thus provides for a helium buffer. The yoke also allows for sufficient helium flow in the magnet's axial direction and sufficient cross section for transverse heat conduction to the heat exchanger tube. The yoke laminations are made of low-carbon mild steel, hot-rolled and annealed, and precision punched with a tolerance of ±0.05 mm. The laminations are preassembled in 1.5-m-long

25 Left-hand lay cables (seen as the strand moves away from the viewpoint) are also said to have s-pitch; right-hand lay cables to have z-pitch.

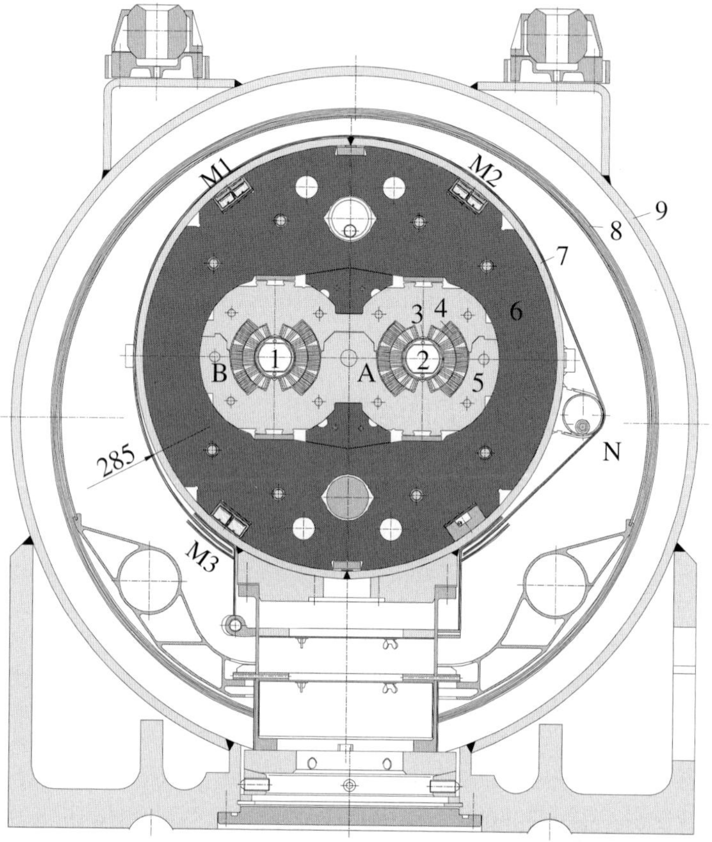

Figure 1.15 Cross section of an LHC cryodipole prototype. 1: Aperture 1 (outer ring). 2: Aperture 2 (inner ring). 3: Cold-bore and beam screen. 4: Superconducting coil. 5: Austenitic steel collar. 6: Iron yoke. 7: Shrinking cylinder. 8: Super-insulation, 9: Vacuum vessel. M1–M3: Busbars for the powering of the main dipole and quadrupole circuits. N: Auxiliary busbar for the powering of arc-corrector magnets.

packs, which are mounted into half-yokes. The collared coils, yoke, and busbars are enclosed in an austenitic steel pressure vessel, which acts as a helium enclosure. This construction forms the dipole coldmass shown in Figure 1.15, a containment filled with static, pressurized superfluid helium at 1.9 K. The principal components of the helium vessel are the main cylinder, composed of two half-shells, and the end covers. The half-shells are fused in a dedicated welding press designed to yield a circumferential stress of around 150 MPa at ambient temperature.

The aperture of the bending magnet must be large enough to contain the *sagitta* of the proton beam, or the magnet must be curved accordingly. To ob-

Figure 1.16 3D rendering of the collar packs for the LHC main dipole [56].

tain the nominal sagitta of 9.14 mm, the helium vessel is curved to 12 mm in the welding press in order to account for the spring-back after the coldmass is released from the press. After the assembly of the half-yokes and the welding of the helium vessel, the gap between the half-yokes is closed. The prestress has been chosen such that the gap between the half-yokes remains closed during cool-down and excitation of the magnet.

1.5 Superfluid Helium Physics and Cryogenic Engineering

In order to operate a superconducting magnet, the cooling agent must have a temperature well below the critical temperature of the superconductor. For Nb–Ti the critical temperature at zero flux density is 9.2 K and for Nb_3Sn it is 18.1 K.

Helium has the distinction of having no triple point; it solidifies only at pressures above 2.5 MPa, even at absolute zero. In addition, there are two liquid phases as indicated in Figure 1.17. Liquid ^{4}He below its lambda line, in a state called *helium II*, exhibits unusual characteristics. The transition between the two liquid phases is accompanied by a large peak in the heat capacity, but no latent heat. The lambda point refers to the anomaly in the heat capacity at 2.17 K, an effect also seen at the transition temperature of superconductors. When helium II flows at low velocity through capillaries even in the μm-diameter range, it exhibits no measurable viscosity. This frictionless flow, or *superfluidity*, was discovered independently by Allen and Misener [1], and by Kapitza [37] in 1938. The phenomenon was later explained theoretically by London,[26] who suggested a connection to Bose–Einstein[27] condensation, and by Tisza[28] who described a two-fluid model with condensed and non-condensed atoms being identified with the superfluid and the normal state, respectively. In 1941 Landau[29] suggested that the superfluid state can be understood in terms of phonons and rotons [29]. Introductions to superfluid helium physics can be found in [64] and [61].

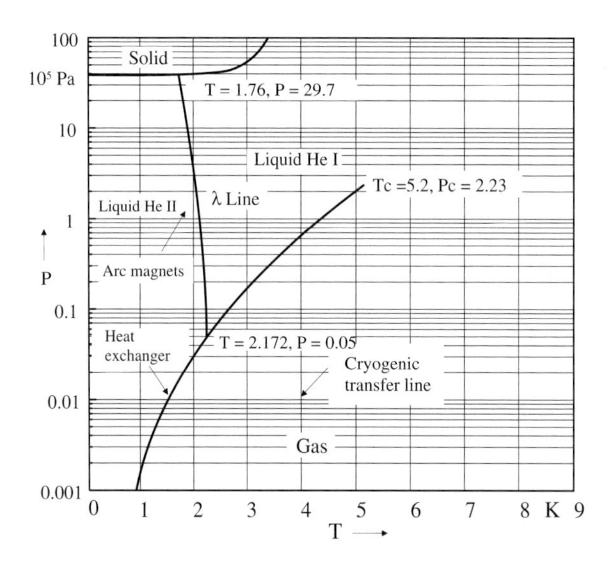

Figure 1.17 Phase diagram of ^{4}He and the thermodynamic states of the coolant in the LHC cryogenic system.

Superfluid helium has become a technical coolant subsequent to the work of Bon Mardion and others [49]. It is used in high-field magnets for condensed-matter research and nuclear magnetic resonance studies, and for magnetic confinement fusion in the Tore Supra Tokamak [8]. It also cools RF acceleration cavities in the CEBAF linear accelerator at the Jefferson Laboratory (USA) and the free electron laser, VUV-FEL at DESY.

The main reason for choosing superfluid helium as the coolant of the LHC magnets is the very low operating temperature, which increases the critical current density of the Nb–Ti superconductor. A disadvantage is the low specific heat capacity of the cable at superfluid helium temperature. This requires taking full advantage of the superfluid helium properties for thermal stabilization, heat extraction from the magnet windings, and heat transport to the

26 Fritz London (1900–1954).
27 Satyendra Nath Bose (1894–1974), Albert Einstein (1879–1955).
28 László Tisza (1907–2009).
29 Lew Landau (1908–1968).

cold source. With its low viscosity, superfluid helium can permeate the cables and buffer thermal transients, thanks to its high volumetric specific heat close to the lambda point, which is approximately 2000 times higher than that of the conductors. The excellent thermal conductivity of the fluid enables it to conduct heat without mass transport.[30]

In order to benefit from these unique properties, the electrical insulation of the cable must have sufficient porosity and percolation while preserving its mechanical resistance and dielectric strength [41]. These conflicting requirements are met by a multilayer wrapping of polyimide film with a partial overlap similar to the barber-pole wrapping described in Section 1.3.3.

As the thermal conductivity of superfluid helium remains finite, it is impossible to transport refrigeration power from one refrigerator across a 3.3-km-long LHC sector.[31] The LHC magnets operate in a static bath of pressurized superfluid helium close to atmospheric pressure at 0.13 MPa. Avoiding low-pressure operation in a large cryogenic system limits the risk of inward air leaks and helium contamination. In addition, saturated helium II exhibits low dielectric strength with the risk of electrical breakdowns at fairly low voltages.

The high-conductivity mono-phase liquid in the coldmasses is continuously cooled by heat exchange with saturated two-phase helium, flowing in a continuous heat exchanger tube. The deposited heat is absorbed quasi-isothermally by the latent heat of vaporization of the flowing helium. Advantages of this cooling scheme are the absence of convective flow in normal operation, the limited space it occupies in the magnet cross section, and the capacity to limit quench propagation between magnets in the string [42].

In view of the low saturation pressure of helium at 1.8 K, refrigeration by vapor compression requires a pressure ratio of 80 to bring the helium back to atmospheric pressure. To limit the volume-flow rate and hence the size of equipment, the large mass-flow rate in a high-power refrigerator must be processed at its highest density. This can only be done with contact-free, vane-free, nonlubricated (and therefore noncontaminating) cryogenic compressors. Hydrodynamic compressors of the axial-centrifugal type are used in multi-stage configurations. The LHC uses eight 1.8 K refrigeration units, each with a refrigeration power of 2.4 kW, based on multistage axial-centrifugal cryogenic compressors operating at high rotational speed on active magnetic bearings. This technology was developed in industry following CERN's specifications. The measured performance coefficient, that is, the ratio of electrical power to cooling power at 1.8 K is 950.

30 The thermal conductivity of He II is more than three orders of magnitude higher than the thermal conductivity of water while its viscosity is four orders of magnitude lower than that of water.

31 The thermal conductivity of superfluid helium is about 3000 times that of oxygen-free, high conductivity copper (OFHC) at room temperature.

The cryogenic system must cope with load variations during magnet ramping (ac losses in the cable) and stored-beam operation (synchrotron radiation, beam image currents, photoelectrons adsorbed by the beam screen, and random loss of particles, among others). The cryogenic system must also enable the cool-down of the magnet string in one sector in a maximum time of 15 days while avoiding thermal gradients exceeding 75 K per coldmass. In addition, the system must be able to handle the heat release during the resistive transition of magnets, by limiting the quench propagation to neighboring magnets and by containing the resulting pressure rise within the 2 MPa design pressure of the helium enclosure. Helium is discharged at high flow rate into a *header* with a large acceptance, which can thus act as a temporary storage, before it is discharged into 250 m^3 gas-storage vessels located at the surface areas of the LHC.

1.6 Cryostat Design and Cryogenic Temperature Levels

In view of the high thermodynamic cost of refrigeration at 1.8 K, the thermal design of the LHC cryogenic components aims at intercepting the largest fraction of applied heat loads at higher temperatures. The temperature levels are:

- 1.9 K for the quasi-isothermal superfluid helium for the magnet coldmass.
- 4 K at very low pressure in the cryogenic transfer line.
- 4.5 K normal saturated helium for cooling special superconducting magnets in the insertion regions and the superconducting accelerator cavities.
- 4.6–20 K for lower-temperature heat interception and for the cooling of the beam screens.
- 50–65 K for the thermal shields as first heat intercept level in the cryostat.
- 20–300 K for the cooling of the resistive section in the current leads.

The coldmass, weighing 28.5 tons, is assembled inside its cryostat, which comprises a support system, cryogenic piping, radiation insulation, and thermal shields, all contained within a vacuum vessel. The vacuum, at a pressure below 10^{-4} Pa, together with two thermal shields covered with super-insulation, minimizes inward heat conduction and radiation to the coldmass.

The cryostats combine several low-temperature insulation and heat intercept techniques, support posts made of fiberglass-epoxy composite for low thermal conductivity, low-impedance thermal contacts under vacuum for heat intercepts, and multilayer insulation blankets. The blankets, known as *super-insulation*, consist of alternating layers of highly reflective material and low-conductivity spacer material. The blankets are manufactured from sheets of

PET[32] film, coated with 40 μm aluminum on each side, and interleaved with polyester-net spacing sheets. The thermal shield, made of rolled high-purity aluminum, is equipped with two superimposed blankets of 15 layers each. The coldmasses and diode boxes are equipped with single blankets of 10 layers each.

1.7 Vacuum Technology

In general, vacuum conditions are obtained by removing gases from the contained volume by pumping (fast rotating turbo-molecular, and ion pumps) or by binding them via chemical or physical forces in the bulk of the pumping material (getters, sorption pumps) [39]. In the LHC straight sections that are at room temperature, the pressure requirements are fulfilled by ultra-high vacuum (UHV) technology with nonevaporable getter coatings (NEG), a technology developed at CERN [9].

The LHC has three distinct vacuum systems: The insulation vacuum systems for the continuous cryostat in the LHC arcs and the helium distribution line, and the beam vacuum system for the LHC arc and transfer lines for injection and beam dump. The insulation vacuum, covers a total volume of about 640 m^3; its room-temperature pressure need not be better than 10 Pa before cool-down. This is achieved with standard mechanical pumping groups. At cryogenic temperatures and in the absence of any significant air leak, the pressure will attain 10^{-4} Pa [44] by *cryosorption pumping* on the external surface of the magnet coldmasses. Vacuum separation in the nearly 3-km-long continuous arc-cryostat is achieved by vacuum barriers, such that the subsections can be individually commissioned, pumped, and leak tested.

The requirements for the beam vacuum are much more stringent. To ensure the required 100 h beam lifetime and a low background to the experimental areas, the equivalent hydrogen gas densities should remain below 10^{15} m^{-3} in the arc[33] and below 10^{13} m^{-3} in the interaction regions in order to minimize nuclear scattering of protons on the residual gas. Additional requirements result from magnet quench limits,[34] resistive power dissipation by beam image currents, beam-induced *multipacting*, beam loss by nuclear scattering, heat loads from beam gas scattering, and stimulated gas desorption from synchrotron radiation in the arcs.

32 Polyethylene terephthalate.

33 Corresponding to 7×10^{-8} Pa at 5 K.

34 The beam pipes in the center of the superconducting magnets are in direct contact with the 1.9 K helium bath and very close to the inner-layer dipole coil.

In the LHC arc beam pipe, the pumping of hydrogen and all other gas species (except helium) relies on cryosorption pumping on the 1.9 K cold bore. Gases will be adsorbed by the attractive van der Waals[35] forces exerted by a cold surface, that is, when the energy of evaporation is less than the adsorption energy on the surface. Cryosorption is limited by the saturated vapor pressure of the gas at a given temperature. Below 20 K, only neon, hydrogen, and helium have a significant saturated vapor pressure and require special precautions, for example, avoiding helium leaks from the cryogenic distribution lines or the coldmasses themselves.

It is essential to limit heat flow to the 1.9 K circuit due to synchrotron radiation (0.2 $\mathrm{W\,m^{-1}}$), nuclear scattering of the high energy protons on the residual gas (0.1 $\mathrm{W\,m^{-1}}$), beam image currents (0.1 $\mathrm{W\,m^{-1}}$), and electron clouds (0.2 $\mathrm{W\,m^{-1}}$) [44]. There is a limit to cryosorption set by beam-induced energetic particles generating desorption of bound gas species, which increases the outgassing rate and hence the pressure in the vacuum system. In the LHC, with its synchrotron radiation of 10^{17} photons per meter second, this recycling effect of adsorbed gas dominates the wall-pumping of hydrogen to such an extent, that any directly exposed surface loses all useful pumping efficiency [30].

Thus a racetrack-shaped *beam screen*, actively cooled to temperatures between 5 and 20 K for nominal cryogenic conditions, is inserted into the cold bore of all magnets and connection cryostats. The beam screen, shown in Figure 1.18 (right), intercepts the power deposited by the synchrotron radiation at a higher temperature level. It also intercepts the power deposited by electron clouds during the conditioning phase and limits the condensed gas coverage on the surface exposed to the impact of energetic particles. The manufacturing process of the beam screen starts by colaminating a specially developed low-permeability austenitic steel strip with a 75-μm-thick copper sheet, followed by the punching of pumping slots covering 4% of the surface area. This is a compromise aimed at keeping the radio frequency losses low while limiting the reduction of the net pumping speed for hydrogen. Without the pumping slots, the surface of the screen would not provide sufficient pumping capacity, as the equilibrium vapor density at 5 K for a monolayer of hydrogen exceeds the acceptable limits by several orders of magnitude. The coated steel strip is then rolled into the final shape of the beam screen. The structure is closed with a longitudinal laser weld on one side. Particular care was taken in the composition of the austenitic steel in order to avoid ferrite formations during welding [58]. As the power dissipated by the beam image currents depends on the resistivity of the beam screen, the heat load is reduced by the copper layer, profiting from the fact that the resistivity of high-purity copper at cryogenic temperatures is reduced by a factor of 100 with respect to room temperature.

35 Johannes Diderik van der Waals (1837–1923).

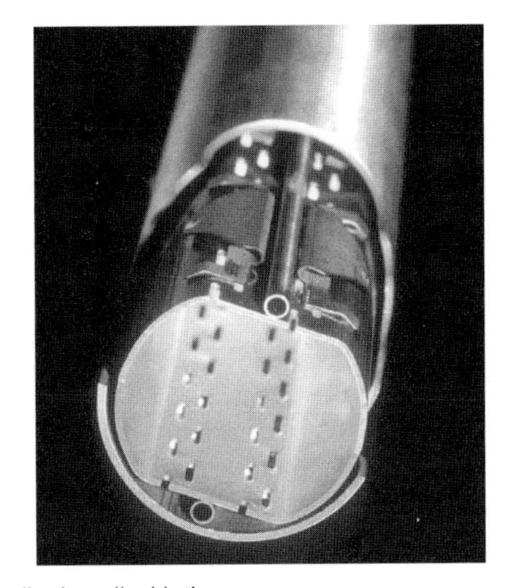

Figure 1.18 Left: Cross section of the LHC main dipole coil with the geometrical model of the cold bore and beam screen. Right: Photograph of the beam screen showing the pumping slots and the heat exchanger.

The beam screen is cooled by two austenitic steel heat-exchanger tubes, laser-welded onto the beam screen tube, allowing for an extraction of up to 1.13 W m^{-1} under nominal cryogenic conditions. The beam screen is fixed on the cold bore at one extremity of the cryomagnet. At the other extremity an expansion bellows is installed to allow for differential thermal expansion between the cold bore and the beam screen [44]. During ramping of the LHC machine, and in-particular when a quench occurs in the magnet, the beam screen is submitted to a horizontal expansion force due to the induced eddy currents.

We must also account for the effect of the beam screen on the field distribution in the magnet apertures; see Chapter 15. The model for the numerical field computation is shown in Figure 1.18 (left).

1.8 Powering and Electrical Quality Assurance

The magnet powering system for the LHC is complex. More than 10 000 magnet elements are connected in 1612 electrical circuit of 131 different types. The powering equipment includes:

- Current leads with HTS material for a 600–13 000 A current rating, as well as normal-conducting leads for a 60–120 A current rating.
- Electrical distribution feed-boxes to house the current leads.

- Superconducting power transmission lines in IPs 1, 3, and 5 of the LHC, where space constraints do not allow the installation of the electrical distribution feed-boxes close to the string of cryomagnets. The superconducting cables in these links are cooled by a flow of supercritical helium at 3.6×10^5 Pa and 4.5–6 K.
- Rigid busbars with thermal expansion loops for the 13 000 A main circuits, 120 km of multiwire cable for the corrector magnet circuits with a current rating of 600 A, and 600 A and 6 kA flexible busbar cables for the powering of the correctors in the short straight sections. The latter are routed in a separate cryogenic distribution line (line N) shown in Figure 1.15.
- Quench-protection equipment including quench protection electronics, quench-back heater power supplies, energy-extraction systems, high-current bypass diodes, and a supervision system.
- High-precision power converters for high currents (13 kA) at rather low voltages; 3–35 V in a steady state with peak voltages not exceeding 190 V.

Several tens of thousands of superconducting connections had to be made during the installation of the magnets in the LHC tunnel in the years 2005–2007. The power converters are connected to the current leads in the electrical distribution feed-boxes, the local current leads for orbit corrector magnets or directly to the magnet terminals of the normal-conducting magnets. Any incorrect magnet connection would seriously compromise LHC operation and is very difficult to correct, once the machine is in operation.

A rigorous Electrical Quality Assurance (ELQA) plan [16] was established for the LHC machine environment in order to ensure the safe and correct functioning of all superconducting circuits during hardware commissioning and machine operation. The steps in the electrical quality assurance are:

- Continuity, polarity, and electrical integrity verification during machine assembly.
- Measurement of electrical reference parameters at ambient temperatures for each individual electrical powering subsector.
- Online monitoring of the integrity of electrical insulation during the cool-down of the machine sector.
- Diagnostic measurements and verifications during sector commissioning and machine operation.
- Yearly verification (during shut-down periods) of cold electrical components such as the bypass diodes.
- Verification of in situ repairs of electrical circuit elements.

The photograph taken in the LHC tunnel during machine installation, Figure 1.19, shows the string of cryomagnets with an open interconnection be-

Figure 1.19 A view into the LHC tunnel during electrical tests of the interconnections between cryomagnets. (1) Interconnection board for the busbars powering the 600 A correctors in the short straight sections. (2) Beam tube. (3) Busbars for the spool piece corrector circuits. (4) Cryogenic service module. (5) Instrumentation feedthrough system.

tween them, at the time when tests were performed to verify the continuity of the electrical circuits.

Technical challenges for the power converter system result from the need to install the converters in underground areas close to the electrical feed-boxes. This is imposed by the high current rating and the inevitable ohmic heating in the water-cooled busbar systems. Only the normal-conducting magnets installed in IR 3 and 7 are powered from the surface, reusing the surface buildings and cabling from the LEP project.

Reduced-volume, high-efficiency power converters are required to fit the civil engineering infrastructure with the available radio-frequency galleries constructed for LEP. A severe design constraint was thus imposed on the Electro Magnetic Compatibility (EMC) of the equipment. The performance of the powering system is further dominated by the tolerance of 0.003 for the *Q-value*[36] of the machine. This issue arises from the segmentation of the ma-

36 See Section 11.8.

chine and the nonlinear *transfer function* of the magnets.[37] A resolution and short-term stability of the power converters on the order of a few units in 10^{-6} will be needed to allow precise cycling and fine adjustment of the magnetic fields [44].

1.9 Electromagnetic Design Challenges

A full treatment of the technical challenges sketched in the previous sections could easily fill a book on its own. The challenges posed by the stringent requirements for the field quality in the accelerator magnets gave rise to R&D in the domain of analytical and numerical field computation, electromagnetic design of magnets, and mathematical optimization techniques. Documenting the experience gained, and accounting for the methodological developments undertaken, is the main objective of this book.

The ideal current distribution in a coil-dominated magnet has a $\cos n\varphi$ dependence in order to produce a pure $2n$-polar magnetic flux density in the aperture. The ideal current distribution cannot, however, be technically realized with cables and a single power supply. Therefore, the magnetic design aims at an approximation of the ideal current distribution by using cables grouped in coil blocks as shown in Figure 1.20. In order to reduce degradation of the critical current density due to the cabling process, the keystoning is usually insufficient to build up arc-segments, and therefore copper wedges are inserted between blocks of cables. The optimal size and shape of these wedges yield the degree of freedom necessary for optimizing the field quality in the magnet. Spacers of variable thickness between the collars and the coil poles, called *pole shims*, can be used to compensate for coil-size variations in the production process and thus to ensure that the dipole magnets have practically identical magnetic characteristics.

The magnetic flux density varies considerably in the coil cross section; see Figure 1.20 (left). Designing coils with two layers of cables of the same width but of different thickness allows for approximately 40% higher current density in the outer-layer cable, which is exposed to a lower magnetic field. This principle is usually referred to as *current grading*. The LHC main dipole coils are wound from cables composed of 28 Nb–Ti multifilamentary strands of 1.065 mm diameter in the inner layer and 36 strands of 0.825 mm diameter in the outer layer.

37 The current to field correspondence influenced by superconductor magnetization and iron saturation.

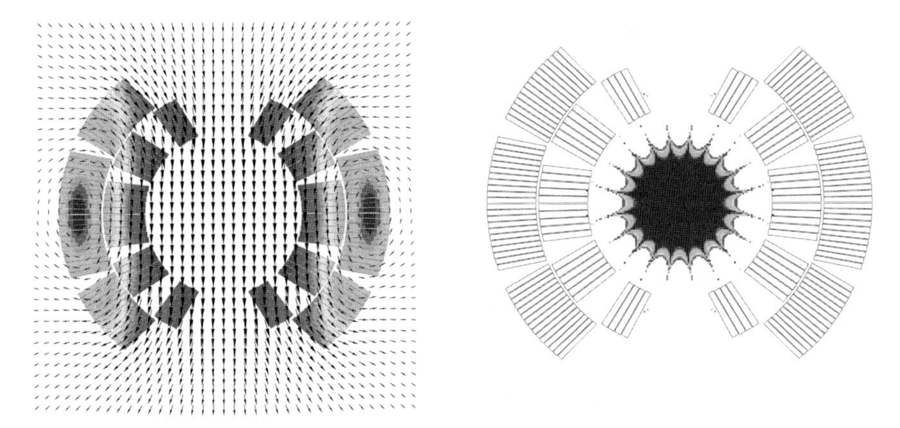

Figure 1.20 Coil cross section of an LHC model dipole [47]. Left: Field map. Right: Error on the B_y field component. $|1 - B_y / B_y^{\text{nom}}| < 0.1 \times 10^{-4}$ for the dark blue areas, 2×10^{-4} maximum.

The magnetic field in the aperture of the accelerator magnets can be described either by a field map or by relative deviation from the ideal field as visualized in Figure 1.20.

Neither method illustrated in Figure 1.20 is useful for field-quality optimization. Instead, the magnetic field errors in the apertures of accelerator magnets are usually expressed by the coefficients of the Fourier[38] series expansion of the radial field component at a given reference radius: Assuming that the radial component of the magnetic flux density B_r at a given reference radius $r = r_0$ inside the aperture of a magnet is measured or calculated as a function of the angular position φ, we obtain for the Fourier series expansion of the radial field component

$$B_r(r_0, \varphi) = \sum_{n=1}^{\infty} (B_n(r_0) \sin n\varphi + A_n(r_0) \cos n\varphi)$$
$$= B_N \sum_{n=1}^{\infty} (b_n(r_0) \sin n\varphi + a_n(r_0) \cos n\varphi), \quad (1.16)$$

where

$$A_n(r_0) = \frac{1}{\pi} \int_0^{2\pi} B_r(r_0, \varphi) \cos n\varphi \, d\varphi, \qquad n = 1, 2, 3, \ldots, \quad (1.17)$$

$$B_n(r_0) = \frac{1}{\pi} \int_0^{2\pi} B_r(r_0, \varphi) \sin n\varphi \, d\varphi, \qquad n = 1, 2, 3, \ldots. \quad (1.18)$$

38 Joseph Fourier (1768–1830).

As the magnetic flux density is divergence-free, $A_0 = 0$. The B_n are called the normal and the A_n the skew field components. The physical units are $[B_n] = [A_n] = 1\,\mathrm{T}$. The b_n are the normal relative, and a_n the skew relative multipole field coefficients. The latter are dimensionless and usually given in units of 10^{-4} at a 17 mm reference radius (about 2/3 of the LHC aperture). For a good field quality these multipole components must be smaller than a few units in 10^{-4}. In the three-dimensional case, the transverse field components are integrated over the entire length of the magnet. For beam tracking it is sufficient to consider the transverse field components, since the effect of the longitudinal component of the field (present only in the magnet ends) on the particle motion can be disregarded.

Two nonlinear effects influence the multipole field components: At low field the superconducting filament magnetization results in a screening of the coil field. At high excitation the saturation of the iron yoke influences mainly the lower-order multipole coefficients.

Figure 1.21 shows the transfer function and the most sensitive quadrupole field component (intrinsic to the two-in-one design) in the main bending magnets as a function of the excitation current. Numerical methods must be used for the calculation of the saturation effects in the iron yoke.

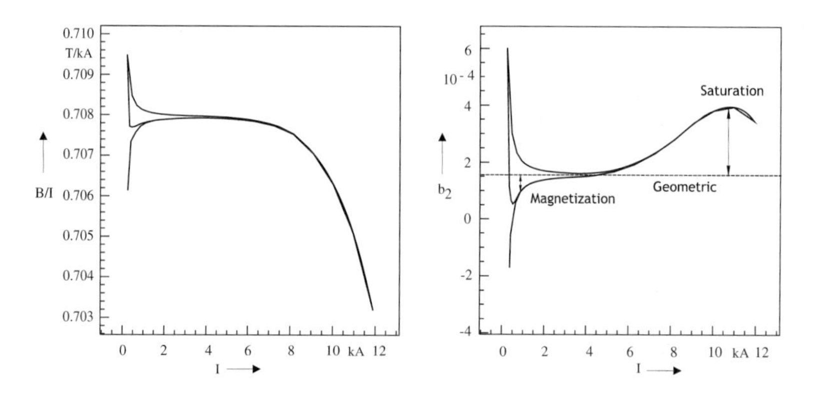

Figure 1.21 Variation in the transfer function B/I (left) and the relative quadrupole field component as a function of the excitation current (right). Notice the effect of the iron saturation at higher field levels and the superconductor magnetization at low excitation levels.

The three main sources of field errors (geometrical effects, superconductor magnetization, and time-transient effects) can be associated with three types of errors:

- Systematic errors inherent to the design geometry resulting in average errors over the whole LHC ring and in one single aperture. These er-

rors respect the coil and magnet symmetry; no skew field components appear in the aperture field.
- Uncertainty errors arising from variations in tooling used during magnet assembly, resulting in deviations of the systematic error per magnet production line. An appropriate sorting strategy limits the impact on accelerator performance.
- Random effects due to (uncorrelated) fabrication tolerance in the different components including the superconducting strands and cables.

Systematic errors can be classified as follows:

- Errors caused by the design of the coil windings that can only approximate the ideal $\cos n\varphi$ current distribution as shown in Section 8.1.1, called geometric errors.
- Remanent fields caused by persistent currents, induced in the superconducting filaments during the ramp of the magnets to their nominal field value. Time-transient effects such as *decay* and *snap-back* [15].
- Ramp-induced eddy currents in the multistrand cables, such as interstrand coupling currents and interfilament coupling currents.
- Errors from crosstalk in the two-in-one magnet with its common iron yoke, which is asymmetric with respect to the vertical plane of the proton beam.
- Cool-down of the structure and resulting deformations of the coil geometry.
- Eddy current and magnetization effects from the beam screen.
- Stray fields in the coil-end regions, including the effect of busbars and electrical interconnections.
- Coil deformations due to electromagnetic forces.

Interstrand coupling currents are inversely proportional to the interstrand contact resistance, which is high enough for the effects to be kept under control for slow LHC ramp rates. These will however present a challenge for fast-ramping accelerators. While the geometric errors are present at each stage of the LHC excitation cycle, the persistent current effects are largest at injection energy. The saturation effects mainly affect the first higher-order multipole (b_3 in the dipole and b_6 in the quadrupole) at nominal excitation.

Uncertainty errors include:

- Systematic perturbations arising from manufacturing tooling.
- Varying properties of the superconducting cable due to different manufacturing procedures.

- Varying properties of steel in yoke and collar laminations caused by batch-to-batch variations.
- Varying assembly procedures employed by the different coldmass manufacturers.
- Torsion and sagitta of the coldmass.

The random effects mainly arise from:

- Cable placement errors due to tolerances in coil parts, for example, insulation thickness, cable keystoning, and size of copper wedges.
- Tolerances in yoke parts, for example, collar outer shape and yoke laminations.
- Manufacturing tolerances and displacements of coil blocks due to varying elastic moduli of the coils, coil winding procedure, curing, collaring, and yoking, among others.
- Tolerances in the magnet alignment.

1.9.1 The CERN Field Computation Program ROXIE

In addition to the technical challenges imposed by a large accelerator project and its magnet system, there are computational challenges for the design of magnets. As previously mentioned, the electromagnetic design and optimization of accelerator magnets is dominated by the requirement of an extremely uniform field, which is mainly defined by the layout of the superconducting coil. For the field calculation it is necessary to account for even small geometrical effects, such as those produced by the keystoning of the cable, the insulation, and coil deformations due to collaring, cool down, and electromagnetic forces. If the coils had to be modeled in the finite-element mesh, as is the case in most commercial field computation software, it would be difficult to define the current density in the keystoned cable requiring further subdivision of the coil into a number of radial layers.

For the 3D case in particular, commercial software has proven hardly appropriate for the field optimization of the superconducting LHC magnets. The ROXIE (**R**outine for the **O**ptimization of magnet **X**-sections, **I**nverse field calculation and coil **E**nd design) program package was therefore developed at CERN with the following main objectives in mind [55]:

- To write an easy-to-use program for the design of superconducting coils in two and three dimensions, taking into account field quality, quench margin, and hysteresis effects from the persistent currents.
- To provide for accurate field calculation routines that are especially suited to the investigation of superconducting magnets: accurate cal-

culation of the field harmonics, field distribution within the superconducting coil, superconductor magnetization, and dynamic effects such as interstrand and interfilament coupling currents.
- To account for the mutual interdependence of physical effects such as cable eddy currents, magnetization currents, and iron saturation.
- To integrate the program into a mathematical optimization environment for field optimization and inverse problem solving.
- To integrate the program into the engineering design procedure through interfaces to Virtual Reality, to CAD/CAM systems (for the making of drawings and manufacturing of end spacers for the coil heads), and to commercial structural analysis programs.

The modeling capabilities of the ROXIE program, together with its interfaces to CAD/CAM and its mathematical optimization routines, have inverted the classical design process wherein numerical field calculation is performed for only a limited number of numerical models that only approximate the actual engineering design. ROXIE is now used as an approach toward an integrated design of superconducting magnets. The steps of the integrated design process, including both (semi-) analytical and numerical field computation methods, are as follows:

- Feature-based geometry modeling of the coil and yoke, both in two and three dimensions, requiring only a small number of meaningful input data to be supplied by the design engineer. This is a prerequisite for addressing these data as design variables of the optimization problem.
- Conceptual coil design using a genetic algorithm, which allows the treatment of combined discrete and continuous problems (for example, the change of the number of coil windings) and solving material distribution problems. The applied niching method provides the designer with a number of local optima, which can then be studied in detail.
- Minimization of iron-induced multipoles using a finite-element method with a reduced vector-potential formulation or the coupling method between boundary and finite elements (BEM–FEM).
- Subject to a varying magnetic field, so-called *persistent currents* are generated that screen the interior of the superconducting filaments. The relative field errors caused by these currents are highest at an injection field level and must be calculated to allow a subsequent partial compensation by geometrical field errors or magnetic shims. Different iteration schemes allow the consideration of nonlinear effects due to the saturation of these shims. Deterministic search algorithms are used for the final optimization of the shims and coil cross section.
- Use of electrical network models to calculate ramp-induced losses and field errors due to so-called *interfilament* and *interstrand coupling currents*.

- Simulation of time-transient effects in quenching superconducting magnets.
- Sensitivity analysis of the optimal design by means of Lagrange-multiplier estimation and the setup of payoff tables. This provides an evaluation of the hidden resources of the design.
- Tolerance analysis by calculating Jacobi matrices and estimation of the standard deviation of the multipole field errors.
- Generation of the coil-end geometry and shape of end spacers using methods of differential geometry. Field optimization including the modeling and optimization of the asymmetric connection side, ramp and splice regions as well as external connections.
- 3D field calculation of the saturated iron yoke using the BEM–FEM coupling method.
- Production of drawings by means of a DXF interface for both the cross sections and the 3D coil-end regions.
- End-spacer manufacture by means of interfaces to CAD/CAM, supporting rapid prototyping methods (laser sinter techniques), and computer controlled five-axis milling machines.
- Tracing of manufacturing errors from measured field imperfections by the minimization of a least-squares error function using the Levenberg–Marquard optimization algorithm.

1.9.2 Analytical and Numerical Field Computation

A feasible approach to structure this book would be to group the mathematical foundations and numerical methods described in this book according to the integrated design process described above. In order not to become too application-specific, we will take the opposite approach and group the applications according to the complexity of the mathematical methods required for their solution. In this way we emphasize the universal nature of methods that can be equally well applied to other large-scale applications in electrical engineering. It also highlights the fact that a large number of subjects taught in graduate university courses on electromagnetic field theory have their role to play in real-world applications:

- Harmonic fields, Fourier series, and Legendre[39] polynomials for the postprocessing of magnetic field measurement data and the definition of the field quality.

39 Adrien Marie Legendre (1752–1833).

- Green's[40] functions and the field of line-currents for the computation of coil fields in superconducting magnets.
- The image-current method for a first-order approximation of the effect of an iron yoke.
- Complex analysis methods for "back-of-the-envelope" calculations of 2D fields in the aperture of accelerator magnets.
- Numerical field computation for the calculation of iron magnetization in ferromagnetic yokes.
- Hysteresis modeling for superconducting filament magnetization.
- Coupled electromagnetic field, electric, and thermal network theory for quench simulation.
- Application of differential geometry to coil-end design.
- Mathematical optimization for shape design of coils and yokes, as well as inverse problem solving.

We will begin in the next two chapters with an introduction to linear algebra and vector analysis as foundations for the electromagnetic design of accelerator magnets.

40 George Green (1793–1841).

References

1 Allen, J.F., Misener, A.D.: Flow of liquid helium II, Nature 141, 1938

2 Asner, A.: High Field Superconducting Magnets, Clarendon Press, Oxford, 1999

3 Asner, A., Perin, R., Wenger, W., Zerobin, F.: First NB_3Sn superconducting dipole model magnets for the LHC break the 10 Tesla field threshold, Proceedings of the Magnet Technology Conference, MT-11, 1990

4 Asner A., Petrucci, G., Resegotti, L.: Some aspects of magnet design for high-energy physics, Proceedings of the International Conference on Magnet Technology, Published by the Rutherford Laboratory, 1967

5 Andreev, N. et al.: Status of the inner triplet quadrupole program at Fermilab, IEEE-Transactions on Applied Superconductivity, 2001

6 ATLAS Technical Proposal: CERN, LHCC 94-43, 1994

7 ALICE Technical Proposal: CERN, LHCC 95-71, 1995

8 Aymar, R., Claudet, G.: Tore Supra and He II cooling of large high-field magnets, Advances in Cryogenic Engineering, 35A, 1990

9 Benvenuti, C.: A new pumping approach for the large electron positron collider (LEP), Nuclear Instruments and Methods in Physics Research, 1983

10 Ballarino, A.: Large-capacity current leads, Physica C, 2008

11 Ballarino, A., Martini, L., Mathot, S., Taylor, T., Brambilla, R.: Large scale assembly and characterization of Bi-2223 HTS conductors, Transactions on Applied Superconductivity, 2007

12 Balewski, K., Degele, D., Horlitz, G., Kaiser, H., Lierl, H., Mess, K.-H., Wolff, S., Dustmann, C.-H., Schm§ser, P., Wiik, B. H.: Cold yoke dipole magnets for HERA, IEEE Transactions on Magnetics, 1987

13 Billan, J., Henrichsen, K.N., Laeger, H., Lebrun, Ph., Perin, R., Pichler, S., Pugin, P., Resegotti, L., Rohmig, P., Tortschanoff, T., Verdier, A., Walckiers, L., Wolf, R.: The eight superconducting quadrupoles for the ISR high-luminosity insertion, Proceedings of 11th International Conference on High-Energy Accelerators, Birkhäuser, 1980

14 Bottura, L.: A practical fit for the critical surface of NbTi, 16th International Conference on Magnet Technology (MT16), Florida, USA, 1999

15 Bottura, L., Pieloni, T., Sanfilippo, S., Ambrosio, G., Bauer, P., Haverkamp, M.: A scaling law for predicting snap-back in superconducting accelerator magnets, 9th European Particle Accelerator Conference, 2004

16 Bozzini, D. et al.: Electrical quality assurance of the superconducting circuits during LHC machine assembly, 11th European Particle Accelerator Conference, 2008

17 Brechna, H.: Superconducting Magnet Systems, Springer, Berlin, 1973

18 Bryant, P.J.: Basic theory of magnetic measurements, CAS, CERN Accelerator School on Magnetic Measurement and Alignment, CERN 92-05, Geneva, 1992

19 Burla, P., King, Q., Pett, J.G.: Optimization of the current ramp for the LHC, Proceedings of the PAC Conference, New York, 1999

20 CMS Technical proposal: CERN-LHCC-94-38, 1994

21 Chorowski, M., Lebrun, Ph., Serio, L., van Weelderen, R.: Thermohydraulics of quenches and helium recovery in the LHC prototype magnet strings, Cryogenics 38, 1998

22 Dahl, P. et. al: Performance of four 4.5-m two in one superconducting R&D dipoles for the SSC, IEEE Transactions on Nuclear Science, 32, 1985

23 Devred, A.: Practical low-temperature superconductors for electromagnets, CERN Yellow Report, 2004

24 Dieterich, D.R., Godeke A.: Nb_3Sn research and development in the USA – wires and cables, Cryogenics 48, 2008

25 Edwards, H.T.: The Tevatron energy doubler: a superconducting accelerator, Annual Review of Nuclear and Particle Science 35, 1985

26 Ekin, J.W.: Superconductors, in Reed R. P., Clark, A. F. (editors): Materials at low temperatures, American Society for Metals, OH, USA, 1983

27 Evans, L.: The Large Hadron Collider: a Marvel of Technology, EPFL Press, Lausanne, Switzerland, 2009

28 Flükiger, R., Uglietti, D., Senatore, C., Buta, F.: Microstructure, composition and critical current density of superconducting Nb_3Sn wires, Cryogenics, 2008

29 Feynman, R. P.: Superfluidity and superconductivity, Review of Modern Physics, 1957

30 Gröbner, O.: Vacuum and Cryopumping, CAS School on Superconductivity and Cryogenics for Accelerators and Detectors, CERN Yellow Report 2004-008

31 Gupta, R., Ramberger, S., Russenschuck, S.: Field quality optimization in a common coil magnet design, 16th International Conference on Magnet Technology, MT16, Fl, USA, 1999

32 Gurlay, S.A. et al.: Magnet R&D for the US LHC accelerator research program (LARP), IEEE Transactions on Applied Superconductivity, 2006

33 HERA tunnel, photograph reprinted with permission from DESY.

34 Iwasa, Y.: Case Studies in Superconducting Magnets, Design and Operational Issues, Plenum, New York, 1994

35 Jain, A. K.: Harmonic coils, CERN Accelerator School on Measurement and Alignment of Accelerator and Detector Magnets, CERN Yellow Report 98-05

36 Kane, G., Pierce, A. (editors): Perspectives on LHC Physics, World Scientific, Singapore, 2008

37 Kapitza, P.: Viscosity of liquid helium below the lambda point, Nature 141, 1938

38 Kuchnir, M., Walker, R. J., Fowler, W. B., Mantsch, P. M.: Spool piece testing facility, IEEE Transactions on Nuclear Science, 1981

39 Lafferty, J.M.: Foundations of Vacuum Science and Technology, John Wiley & Sons, New York, 1998

40 Lebrun, Ph.: Cryogenics for the large hadron collider, IEEE Transactions on Applied Superconductivity, 2000

41 Lebrun, Ph., Tavian, L.: The technology of superfluid helium, in Proceedings of the CERN-CAS School on Superconductivity and Cryogenics for Accelerators and Detectors, CERN-2004-008, 2004

42 Lebrun, Ph.: Advanced technology from and for basic science: superconductivity and superfluid helium at the large hadron collider, CERN/AT 2007-30, 2007

43 LEP Design Report: CERN-LEP 84-01, 1984 and CERN-AC 96-01 (LEP2), 1996

44 LHC Design Report, Vol. 1, The LHC main ring, CERN-2004-003, 2004

45 LHCB Technical Proposal: CERN, LHCC 98-4, 1998

46 The LHC Study Group, Large Hadron Collider, The accelerator project, CERN/AC/93-03

47 The LHC study group, The Large Hadron Collider, Conceptual Design, CERN/AC/95-05

48 Lee, P. J.: http://www.asc.wisc.edu/plot/plot.htm.

49 Bon Mardion, G., Claudet, G., Vallier, J.C.: Superfluid helium bath for superconducting magnets, Proceedings of the ICEC 6, IPC Science & Technology Press, 1976

50 Meinke, R.: Superconducting magnet system for HERA, IEEE Transaction on Magnetics, 1990

51 Meß, K.H., Schmüser, P., Wolff, S.: Superconducting Accelerator Magnets, World Scientific, Singapore, 1996

52 Morgan, G.H., Green, A., Jochen, G., Morgillo, A.: Winding Mandrel design for the wide cable SSC dipole, Applied Superconductivity Conference, 1990

53 Richter, D.: Private communications, 2008

54 Royet, J.M., Scanlan, R.M.: Development of scaling rules for Rutherford type superconducting cables, IEEE Transactions on Magnetics, 1991

55 Russenschuck, S.: A computer program for the design of superconducting accelerator magnets, 11th Annual Review of Progress In Applied Computational Electromagnetics, 1995, Monterey, CA, USA. LHC-Note 354, CERN, Geneva

56 Sahner, T.: Private communications, CERN, 2009

57 Sanford, J.R., Matthews, D.M. (editors): Site-Specific Conceptual Design of the

Superconducting Super Collider, Superconducting Super Collider Laboratory, 1990

58 Sgobba, S. et al.: Cryogenic properties of special welded austenitic steels for the beam screen of the large hadron collider, Proceedings of the 4th European Conference on Advanced Materials and Processes, 1995

59 Shintomi, T. et al.: Progress of the LHC low-b quadrupole magnets for the LHC insertions, IEEE Transactions of Applied Superconductivity, 2001

60 Thompson, P.A., Gupta, R.C., Kahn, S.A., Hahn, H., Morgan, G.H., Wanderer, P.J., Willen, E.: Revised cross section for RHIC dipole magnets, Conference Record of the 1991 Particle Accelerator Conference, 1991

61 Tilley D.R., Tilley, J.: Superfluidity and Superconductivity, Adam Hilger, UK, 1990

62 Todesco, E., Völlinger, C.: Corrective Actions on the LHC main dipole coil cross-section for steering systematic b3, b5, b7, private communication, 2003

63 TOTEM: Total Cross Section, Elastic Scattering and Diffractive Dissociation at the LHC, CERN, LHC 99-7, 1999

64 Van Sciver, S.: Helium Cryogenics, Plenum, New York, 1986

65 Wiik, B.H.: Progress with HERA, IEEE Transactions on Nuclear Science, 32, 1985

66 Wille, K.: The Physics of Particle Accelerators, Oxford University Press, Oxford, 2000

67 Willen, E. et al.: Superconducting dipole magnets for the LHC insertion regions, Proceedings of the EPAC Conference, Vienna, 2000

68 Wilson, M.N.: Superconducting Magnets, Oxford Science Publications, Oxford, 1983

69 Wilson, M.N.: Private communications, 2006

70 Wilson, M.N.: NbTi superconductors with low ac losses: a review, Cryogenics 48, 2008

2
Algebraic Structures and Vector Fields

The best way is to use the abstract field idea.
That it is abstract is unfortunate, but necessary.

R.P. Feynman (1918–1988), The Feynman Lectures on Physics.

The quote above nicely sets the stage for the content of this chapter: the introduction of the linear algebraic structures of the abstract field concept, principally the oriented Euclidean affine space and vector fields living therein. Together with the concepts of classical vector analysis we will lay the foundation for numerical field computation and simplified, back-of-the-envelope calculations for the design of iron-dominated magnets.

Readers who are "at home" in vector space may quickly browse the next two chapters as a refresher, possibly noticing some unfamiliar aspects here and there.

2.1
Mappings

To introduce the mathematical notation used in the next chapters, we shall first generalize the term function, which is often used synonymously with the word mapping. Let W and X be arbitrary nonempty sets. Suppose that to each element of X, called the *domain*, there is assigned a unique element of W (the codomain), and the collection of such assignments is called a *mapping* from X into W,

$$f : X \to W : x \mapsto f(x)\,, \tag{2.1}$$

which reads: $f(x)$ is the element of W that f assigns to $x \in X$. $f(x)$ is called the value of f at x or the *image* of x. In particular, the set of all images is called the *range* of f. The range of any function f is always contained in the codomain. The familiar squaring function in $\mathbb{R}$, $f(x) = x^2$ is written as

Field Computation for Accelerator Magnets. Stephan Russenschuck

ISBN: 978-3-527-40769-9

$f : \mathbb{R} \to \mathbb{R} : x \mapsto x^2$. This might not seem to be an improvement, but the following examples show the advantage of this notational rigor:

f	:	$[a, b] \to \mathbb{R}$	:	$x \mapsto f(x)$	Real function
$+$	:	$\mathbb{R} \times \mathbb{R} \to \mathbb{R}$	:	$(x, y) \mapsto x + y$	Addition
f	:	$\Omega \to \mathbb{R}$	:	$\mathscr{P} \mapsto f(\mathscr{P})$	Scalar field
$\mathbf{x} \cdot \mathbf{y}$	:	$\mathbb{R}^n \times \mathbb{R}^n \to \mathbb{R}$	:	$(\mathbf{x}, \mathbf{y}) \mapsto \mathbf{x} \cdot \mathbf{y}$	Scalar product
$\mathscr{S}$	:	$I \to E_3$	:	$t \mapsto \mathscr{S}(t)$	Space curve
$\mathbf{f}$	:	$\Omega \to \mathbb{R}^3$	:	$\mathscr{P} \mapsto \mathbf{f}(\mathscr{P})$	Vector field.

The symbol $\times$ in the mapping $f : \mathbb{R} \times \mathbb{R} \to \mathbb{R}$ denotes the Cartesian product[1] and is defined by the set of all ordered pairs (x, y) for $x, y \in \mathbb{R}$. Mappings f with $f : W \times W \to W$ are called *binary operations* in W, for example, multiplication $\cdot : W \times W \to W : (x, y) \mapsto xy$ and addition as in the table above. Let W, X, Y, Z all denote sets, and suppose we have mappings $f : X \to W$, $g : W \to Y$, $h : Y \to Z$. We call $h \circ g \circ f = h \circ (g \circ f) = (h \circ g) \circ f : X \to Z$ that carries elements of X into Z and obeys the associative law, the composition of h with g with f. Consider again the mapping (2.1):

1. If two distinct elements $x_1, x_2 \in X$ are always mapped into two distinct elements $w_1, w_2 \in W$, f is called *injective* (or one-to-one). If f is a one-to-one mapping, it follows from $f(x_1) = f(x_2)$ that $x_1 = x_2$.
2. If the range of the function f is exactly equal to the codomain W, f is called *surjective* (or onto). In this case the equation $f(x) = w$ has at least one solution for each $w \in W$.
3. If f is both injective and surjective, it is called *bijective* (or one-to-one onto). In this case the equation $f(x) = w$ has a unique solution for each $w \in W$ and there exists an inverse function $f^{-1} : W \to X$. This yields $f \circ f^{-1} = 1_W$ and $f^{-1} \circ f = 1_X$ with the identity mappings 1_W and 1_X. Precisely, $1_W : W \to W : a \mapsto f(a) = a$.

2.2 Groups, Rings, and Fields

A *groupoid* is a pair (G, μ) consisting of a nonempty set G and a binary operation μ in G, that is, $\mu : G \times G \to G$. A nonempty set G is called a *multiplicative group* if the groupoid $(G, \cdot)$ with $\cdot : G \times G \to G : (a, b) \mapsto ab$ contains an identity element I with $Ia = a$ for all $a \in G$, is associative $(ab)c = a(bc)$ for all $a, b, c \in G$, and every element $a \in G$ has an inverse $a^{-1} \in G$ such that $aa^{-1} = a^{-1}a = I$.

1 Rene Descartes (Renatus Cartesius) (1596–1650).

A nonempty set G is called an *additive group* if the groupoid $(G, +)$ with $+ : G \times G \to G : (a, b) \mapsto a + b$ contains a neutral element $a + 0 = a$ for all $a \in G$, is associative $(a + b) + c = a + (b + c)$ for all $a, b, c \in G$, and every element $a \in G$ has an element $-a \in G$ such that $a + (-a) = 0$.

Examples:

1. The additive group of integers $(\mathbb{Z}, +)$.
2. The rotation group $SO(3)$ of the 3×3 orthogonal matrices $[A]$ with $\det[A] = 1$ under matrix multiplication.
3. The multiplicative group of nonzero complex numbers $(\mathbb{C}, \cdot)$ with elements z, z_1, z_2 and the multiplication defined by $z_1 \cdot z_2 = (a + ib) \cdot (c + id) = (ac - bd) + i(ad + bc)$ and an inverse $z^{-1} = \frac{a}{a^2+b^2} - i\frac{b}{a^2+b^2}$. □

If all the binary operations in a group commute, the group is called an *Abelian group*[2] or commutative group. Elements of a group are called *generators* if any element in the group can be represented by finite sums or products of these generators. The minimal system of such generators in a group with a finite number of elements is called the *basis* of the group. Groups in which such a basis can be found are called *modules*.

A *ring* is a triple $(G, +, \cdot)$ consisting of a set and two operations so that $(G, +)$ is an Abelian group, the multiplication is associative and the two distributive laws $a(b + c) = ab + ac$ and $(a + b)c = ac + bc$ hold. If the ring contains the identity element with $Ia = aI = a$, it is called a *ring* with identity element. Examples are the ring of integers $(\mathbb{Z}, +, \cdot)$, the rational numbers $(\mathbb{Q}, +, \cdot)$, and the ring of square matrices.

A *field* is a triple $(\mathbb{F}, +, \cdot)$ with a set $\mathbb{F}$ and two binary operators $+, \cdot$ where both $(\mathbb{F}, +)$ and $(\mathbb{F} \setminus 0, \cdot)$ are Abelian groups and the distributive law $\lambda(a + b) = \lambda a + \lambda b$ holds for all $\lambda, a, b \in \mathbb{F}$. The most prominent fields are the real numbers $\mathbb{R}$, the rational numbers $\mathbb{Q}$, and the complex numbers $\mathbb{C}$.

2.3 Vector Space

We will now turn to the abstract definition of a vector space. Let $\mathbb{F}$ be a given field of scalars with elements $\lambda, \mu \in \mathbb{F}$ (for example, $\mathbb{F} = \mathbb{R}$ or $\mathbb{F} = \mathbb{C}$), and V be a nonempty set with rules of addition and scalar multiplication:

$$+ \; : \; V \times V \to V \; : \; (\mathbf{a}, \mathbf{b}) \mapsto \mathbf{a} + \mathbf{b} \,, \tag{2.2}$$

$$\cdot \; : \; \mathbb{F} \times V \to V \; : \; (\lambda, \mathbf{a}) \mapsto \lambda \mathbf{a} \,. \tag{2.3}$$

2 Niels Henrik Abel (1802–1829).

$(V, +, \cdot)$, shorthand V, is a *vector space* over $\mathbb{F}$ if the following axioms are fulfilled:

1. For any vectors $\mathbf{a}, \mathbf{b}, \mathbf{c} \in V : (\mathbf{a} + \mathbf{b}) + \mathbf{c} = \mathbf{a} + (\mathbf{b} + \mathbf{c})$.
2. There is a zero vector $\mathbf{0}$ for which $\mathbf{a} + \mathbf{0} = \mathbf{a}$ for any vector $\mathbf{a}$.
3. For each vector $\mathbf{a} \in V$ there is a vector $-\mathbf{a}$ in V for which $\mathbf{a} + (-\mathbf{a}) = \mathbf{0}$.
4. For any vectors $\mathbf{a}, \mathbf{b} \in V : \mathbf{a} + \mathbf{b} = \mathbf{b} + \mathbf{a}$.
5. For any scalar $\lambda \in \mathbb{F}$ and any vectors $\mathbf{a}, \mathbf{b} \in V$: $\lambda(\mathbf{a} + \mathbf{b}) = \lambda\mathbf{a} + \lambda\mathbf{b}$.
6. For any scalars $\lambda, \mu \in \mathbb{F}$ and any vector $\mathbf{a} \in V$: $(\lambda + \mu)\mathbf{a} = \lambda\mathbf{a} + \mu\mathbf{a}$.
7. For any scalars $\lambda, \mu \in \mathbb{F}$ and any vector $\mathbf{a} \in V$: $(\lambda\mu)\mathbf{a} = \lambda(\mu\mathbf{a})$.
8. For the unit scalar $1 \in \mathbb{F}$ and any vector $\mathbf{a} \in V$: $1\mathbf{a} = \mathbf{a}$.

Properties 1–4 (concerned only with the elements $\mathbf{a}, \mathbf{b}, \mathbf{c} \in V$) constitute the axioms for an Abelian group. Properties 5–8 for the scalar multiplication of vectors with elements in $\mathbb{F}$ constitute the second structural layer of the vector space. We shall denote vector spaces by capital letters, for example, V, W, W^* (omitting the notion of the binary operators) and the elements by lowercase bold face letters, for example, $\mathbf{a}$, $\mathbf{b}$, $\mathbf{x}_1$. One should be aware that vector is a generic name and may represent translations and positions; the name may also apply to other objects than the familiar two- and three-dimensional vectors of elementary geometry, provided these objects obey the vector space axioms. Therefore, vector spaces are frequently called *linear spaces*.

Examples:

1. Functional space $C^m(X)$ of all m-times continuously differentiable functions from any nonempty set $X \in \mathbb{R}^n$ into the field of real numbers $\mathbb{R}$ with rules of addition $(f + g)(x) = f(x) + g(x)$ and scalar multiplication $(\lambda f)(x) = \lambda f(x)$.
2. Matrix space M of all $m \times n$ matrices $[A] = (a_{ij}); i = 1, \ldots, m; j = 1, \ldots, n$, over a field $\mathbb{F}$ with addition $[A] + [B] = [C]$ defined by $c_{ij} = a_{ij} + b_{ij}$ and scalar multiplication $\lambda[B] = [C]$ defined by $c_{ij} = \lambda b_{ij}$.
3. Tuple space $\mathbb{R}^n = \{(a_1, a_2, \ldots, a_n) \,|\, a_1, \ldots, a_n \in \mathbb{R}\}$, where vector addition and scalar multiplication are defined by

$$(a_1, a_2, \ldots, a_n) + (b_1, b_2, \ldots, b_n) = (a_1 + b_1, a_2 + b_2, \ldots, a_n + b_n), \tag{2.4}$$

$$\lambda(a_1, a_2, \ldots, a_n) = (\lambda a_1, \lambda a_2, \ldots, \lambda a_n), \tag{2.5}$$

   and where the zero vector is defined by $\mathbf{0} := (0, 0, \ldots, 0)$.
4. Polynomial space $P(t)$ of all polynomials of grade m written in the form $p(t) = a_0 + a_1 t + a_2 t^2 + \cdots + a_m t^m$ and $q(t) = b_0 + b_1 t + b_2 t^2 + \cdots +$

$b_m t^m$ with coefficients $a_i, b_i \in \mathbb{F}$ with vector addition $p(t) + q(t) = (a_0 + b_0) + (a_1 + b_1)t + \cdots + (a_m + b_m)t^m$ and scalar multiplication $\lambda p(t) = \lambda a_0 + \lambda a_1 t + \cdots + \lambda a_m t^m$.

5. Space of position vectors in two and three dimensions with the geometrically defined rules of addition and scalar multiplication as shown in Figure 2.1.

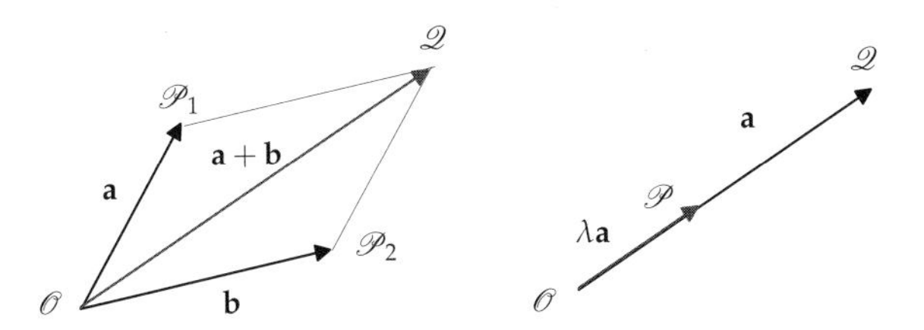

Figure 2.1 Geometrically defined rules of vector addition and scalar multiplication for position vectors represented by directed line segments.

□

2.3.1 Linear Independence and Basis

The vectors $\mathbf{x}_1, \mathbf{x}_2, \ldots, \mathbf{x}_n$ in V are said to be *linearly independent* if $\sum_{i=1}^{n} \lambda_i \mathbf{x}_i = 0$ holds only for the trivial solution where all the coefficients $\lambda_i = 0, i = 1, \ldots, n$. The $\sum_{i=1}^{n} \lambda_i \mathbf{x}_i$ are called linear combinations of the vectors $\mathbf{x}_1, \ldots, \mathbf{x}_n$. The vectors $\mathbf{x}_1, \ldots, \mathbf{x}_n$ in V are said to span V or to form a *spanning set* of V if every $\mathbf{v} \in V$ is a linear combination of these vectors. The vector space is n-dimensional if there exist n linearly independent vectors, and when $n+1$ vectors are always linearly dependent such that

$$\mu \mathbf{x} + \lambda_1 \mathbf{x}_1 + \lambda_2 \mathbf{x}_2 + \cdots + \lambda_n \mathbf{x}_n = 0, \qquad \text{with } \mu \neq 0. \tag{2.6}$$

Because of $\mu \neq 0$ we can express the $(n+1)$st vector through the n others in the form $\mathbf{x} = -\frac{1}{\mu} \sum_{i=1}^{n} \lambda_i \mathbf{x}_i$. The set $S = \{\mathbf{x}_1, \mathbf{x}_2, \ldots, \mathbf{x}_n\}$ of vectors is called a *basis* or *frame* of V_n and shall be denoted $\{\mathbf{g}_1, \mathbf{g}_2, \ldots, \mathbf{g}_n\}$. The basis is a minimal spanning set of V_n.

As a consequence of the linear independence of the basis vectors, there is exactly one way of expressing an element $\mathbf{x} \in V_n$ by

$$\mathbf{x} = \sum_{i=1}^{n} x^i \mathbf{g}_i \equiv x^i \mathbf{g}_i, \tag{2.7}$$

where the x^i are called the *coefficients* of $\mathbf{x}$ with respect to the basis. The terms $x^i \mathbf{g}_i$ are called the *components* of the vector.

Remark: The notation follows the convention of the *Ricci calculus*[3] with the coefficient's indices as superscript. The shorthand on the right-hand side of Eq. (2.7) is Einstein's summation convention. Other examples for the application of the summation convention in three dimensions are $z = x^i y_i \equiv \sum_{i=1}^{3} x^i y_i$ and $z^i = x^{ij} y_j \equiv \sum_{j=1}^{3} x^{ij} y_j$, the latter involving two sorts of indices; the index of summation j is called the dummy index and the index i, which may take any particular value between 1 and 3, is called the free index. □

By choosing a basis, we assign to every vector $\mathbf{x}$ the n-tuple of coefficients $(x^1, \ldots, x^n)$, a unique and reversible linear mapping, which is called a *basis-isomorphism* L_V between V_n and $\mathbb{R}^n$,

$$L_V : V_n \overset{\cong}{\rightarrow} \mathbb{R}^n \; : \; x^1 \mathbf{g}_1 + \cdots + x^n \mathbf{g}_n \cong (x^1, \ldots, x^n) \, . \tag{2.8}$$

The isomorphism allows us to identify all properties of the vector space with the properties of the tuple space $\mathbb{R}^n$. The choice of basis is the n-dimensional analogy of the choice of a unit for some physical quantity. Consequently, isomorphism does not mean an identification of V_n with $\mathbb{R}^n$ as there is no canonical (natural) way to associate V_n with $\mathbb{R}^n$, but we can state that V_n and $\mathbb{R}^n$ are structurally equivalent.

2.4 Linear Transformations

In Section 2.1, we discussed mappings between arbitrary nonempty sets. Now let U, V, and W be vector spaces over the same field $\mathbb{F}$. A mapping $T : V \rightarrow W$ is called a linear mapping, a linear transformation, or a *homomorphism*, if it satisfies the following two conditions: (1) For all vectors $\mathbf{x}, \mathbf{y} \in V$, $T(\mathbf{x} + \mathbf{y}) = T(\mathbf{x}) + T(\mathbf{y})$. (2) For each scalar $\lambda \in \mathbb{F}$ and vector $\mathbf{x} \in V$, $T(\lambda \mathbf{x}) = \lambda T(\mathbf{x})$. Thus a linear mapping is completely characterized by the condition

$$T(\lambda_1 \mathbf{x} + \lambda_2 \mathbf{y}) = \lambda_1 T(\mathbf{x}) + \lambda_2 T(\mathbf{y}) \tag{2.9}$$

for $\lambda_1, \lambda_2 \in \mathbb{F}$ and it preserves the two basic operations of a vector space; addition and scalar multiplication.

Examples:

- The projection mapping $T(x, y, z) = (x, y, 0)$ is linear because of

$$\begin{aligned} T(a(x_1, y_1, z_1) + b(x_2, y_2, z_2)) &= T(ax_1 + bx_2, ay_1 + by_2, az_1 + bz_2) \\ &= (ax_1 + bx_2, ay_1 + by_2, 0) = a(x_1, y_1, 0) + b(x_2, y_2, 0) \\ &= aT(x_1, y_1, z_1) + bT(x_2, y_2, z_2) \, . \end{aligned}$$

3 Gregorio Ricci-Curbastro (1853–1925).

- For linear mappings, it follows from condition 2 above that the zero vector is mapped into itself. Thus the translation mapping $F(x,y) = (x + x_0, y + y_0)$ is nonlinear because $F(0,0) = (x_0, y_0)$.
- The bijection between the vector space V_n and $\mathbb{R}^n$ preserves the vector space operations of vector addition and scalar multiplication and therefore the mapping $T : V_n \to \mathbb{R}^n$ is linear. Bijective linear mappings are *isomorphisms*.
- It is an easy task to show that the matrix mapping defined in example 2 of Section 2.3 preserves the addition and scalar multiplication in vector space and is therefore linear. If we consider the linear transformation $T : V_n \to W_m$, then there exists a matrix $[A] = (a_{ij}) \in \mathbb{R}^{m\times n}$ defined by $[A] = L_V^{-1} \circ T \circ L_W$ according to the commutative diagram

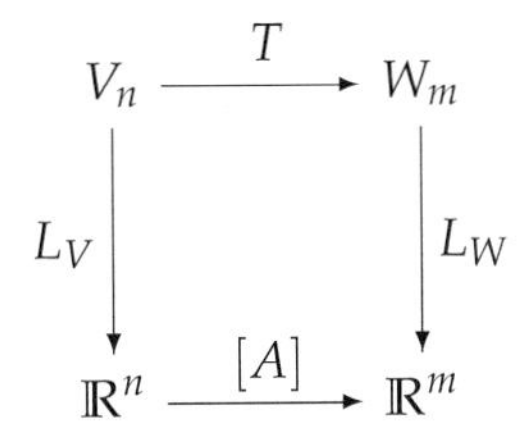

For $m = n$ and $\det[A] \neq 0$ there exists an inverse $[A]^{-1}$ and the group $GL_n : V_n \to V_n$ of linear transformations of a vector space into itself is isomorphic (via a selection of basis) to the group of $n \times n$ matrices. □

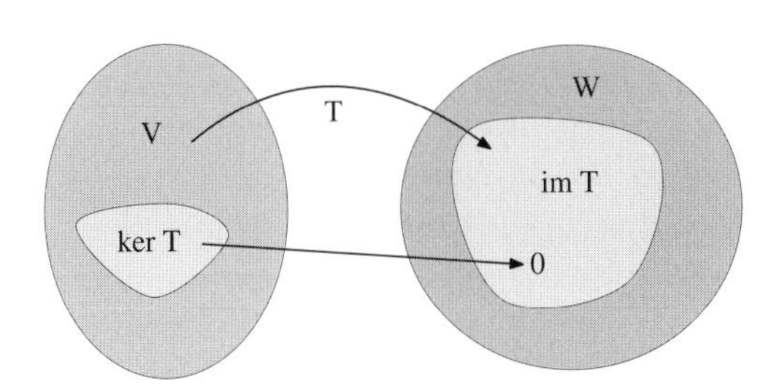

Figure 2.2 Domain V and codomain W, kernel (null-space) and image (range) of the linear transformation $T : V \to W$.

We will now define two important subspaces related to linear transformations: The collection of all vectors $T(\mathbf{x})$ in W is called the *image*, denoted im, or the range of T. The collection of all vectors $\mathbf{x}$ in V such that $T(\mathbf{x}) = \mathbf{0}$ is called the *kernel*, denoted ker T, or the *null space* of T. The kernel of any linear transformation is a subspace of the domain V, as illustrated in Figure 2.2. It holds that

$$\dim(\operatorname{im} T) = \dim V - \dim(\ker T). \tag{2.10}$$

The *rank* of a linear transformation is the dimension of its image. The nullity of T is the dimension of its kernel. A sequence

$$U \xrightarrow{T_1} V \xrightarrow{T_2} W \tag{2.11}$$

of homomorphisms is said to be *exact* if $\ker T_2 = \operatorname{im} T_1$.

Let $\{\mathbf{g}_1, \mathbf{g}_2, \ldots, \mathbf{g}_n\}$ be a basis for the domain V_n of T. Then the vectors $T(\mathbf{g}_1), \ldots, T(\mathbf{g}_n)$ in W span the range of T in W. The linear transformation is therefore entirely determined by its action on a given basis. However, the vectors $T(\mathbf{g}_1), \ldots, T(\mathbf{g}_n)$ are not always linearly independent and therefore may, or may not, be a basis for the range of T.

2.5 Affine Space

Consider a set A with members $\mathscr{P}$ called *points*, an associated vector space V of translations with scalars in some field $\mathbb{F}$, and an operation $+ : A \times V \to A$ satisfying the conditions

1. $\mathscr{P} + \mathbf{x} \in A$ if $\mathscr{P} \in A$ and $\mathbf{x} \in V$.
2. $(\mathscr{P} + \mathbf{x}) + \mathbf{y} = \mathscr{P} + (\mathbf{x} + \mathbf{y})$ for $\mathscr{P} \in A$ and $\mathbf{x}, \mathbf{y} \in V$.
3. There is a unique $\mathbf{x} \in V$ such that $\mathscr{P}_1 = \mathscr{P}_2 + \mathbf{x}$ for $\mathscr{P}_1, \mathscr{P}_2 \in A$.

These axioms define the affine point space associated with the vector space V denoted $(A, V, +)$, or for short the *affine space* A. Conditions 1 and 2 state that $(V, +)$, as an Abelian group, acts on A by $V \times A \to A$. Axiom 3 states that every pair of points defines a unique *free vector* $\mathbf{x} \in V$. Affine space inherits the dimension from the associated vector space. An affine space of dimension n is consequently denoted A_n. Depending on the dimension we may think of an open, connected subset of A_n as an unbounded line, surface or volume.

By selecting a specific origin $\mathscr{O} \in A$ an isomorphism between the (algebraic) vector space and the (physical) point space is established. Any point $\mathscr{P} \in A$ is then associated with a position vector $\mathbf{r}_{\mathscr{P}} = \mathbf{r}_{\mathscr{O}\mathscr{P}} \in V$ in the direction from $\mathscr{O}$ to $\mathscr{P}$. By means of the basis isomorphism (2.8) the position vector can in turn be represented by the tuple of real numbers:

$$\mathscr{P} \in A_n \xrightarrow{\text{Origin}} \mathbf{r} \in V_n \xrightarrow{\text{Basis}} (x^1, \ldots, x^n) \in \mathbb{R}^n . \tag{2.12}$$

It is therefore common practice not to distinguish between the elements of A and V or to mention the affine space but rather to start with the calculations in $\mathbb{R}^n$. In particular when dealing with vector fields, it will be instructive to consider the algebraic properties of affine space.

Remark: In magnetic field computation we refer to the field point as $\mathscr{P}$ and the source point as $\mathscr{Q}$ (the location of charges and currents) and write $\mathbf{r} := \mathbf{r}_{\mathscr{P}}$ and $\mathbf{r}' := \mathbf{r}_{\mathscr{Q}}$. The vector pointing from the source point to the field point is given by $\mathbf{r} - \mathbf{r}' = \mathbf{r}_{\mathscr{Q}\mathscr{P}}$. Vectors of that form are called vectors at $\mathbf{r}'$ or *bound vectors*. □

Affine combinations in A are the counterpart to linear combinations in V. Consider a family of points $\mathscr{P}_i \in A$ and a family of scalars $\lambda_i \in \mathbb{R}$. For any two points $\mathscr{Q}_1, \mathscr{Q}_2 \in A$ the following properties hold:

$$\mathscr{Q}_1 + \sum_i \lambda_i \mathbf{r}_{\mathscr{Q}_1 \mathscr{P}_i} = \mathscr{Q}_2 + \sum_i \lambda_i \mathbf{r}_{\mathscr{Q}_2 \mathscr{P}_i} \quad \text{if} \sum_i \lambda_i = 1 \,, \tag{2.13}$$

$$\sum_i \lambda_i \mathbf{r}_{\mathscr{Q}_1 \mathscr{P}_i} = \sum_i \lambda_i \mathbf{r}_{\mathscr{Q}_2 \mathscr{P}_i} \quad \text{if} \sum_i \lambda_i = 0 \,. \tag{2.14}$$

The proof can be found in [4]. Thus point $\mathscr{B}$, with

$$\mathscr{B} = \mathscr{Q} + \sum_i \lambda_i \mathbf{r}_{\mathscr{Q}\mathscr{P}_i} \quad \text{and} \quad \sum_i \lambda_i = 1, \tag{2.15}$$

is independent of $\mathscr{Q} \in A$. Point $\mathscr{B}$ is the *barycenter* of the points $\mathscr{P}_i$ and the λ_i are the weights associated with $\mathscr{P}_i$. The pair $(\mathscr{P}_i, \lambda_i)$ is called a *weighted point*. Physically, the barycenter is the center of mass of the family of weighted points, where the masses have been normalized such that $\sum_i \lambda_i = 1$. Using less rigorous notation we may write

$$\mathscr{B} = \sum_i \lambda_i \mathscr{P}_i \quad \text{with} \quad \sum_i \lambda_i = 1, \tag{2.16}$$

to be understood in the sense of Eq. (2.15) for arbitrary $\mathscr{Q}$. An affine subspace S of A is a subset of A for which every family of weighted points in S has a barycenter that belongs to S. Affine maps are transformations $T : A_n \to A_n$ that, by definition, preserve barycenters; formally

$$T\left(\sum_i \lambda_i \mathscr{P}_i\right) = \sum_i \lambda_i T(\mathscr{P}_i) \,. \tag{2.17}$$

Important special cases of affine maps are translations, which are not linear; an example is given in Section 2.4. Affine transformations that fix at least one point are called *orthogonal transforms*, or point groups, and include rotations and mirror reflections. However, affine transformations do not necessarily preserve distances and angles; a metric structure is needed for their definition. This will be the subject of Section 2.6.

2.5.1
Coordinates

A coordinate chart on A_n is a set ϕ of n functions that maps an open subset $\Omega \subset A_n$, called the domain of the coordinate chart, to an open subset $\phi(\Omega) \subset \mathbb{R}^n$ of the space of n-tuples:

$$\phi : \Omega \to \phi(\Omega) : \mathscr{P} \mapsto (\phi^1(\mathscr{P}), \phi^2(\mathscr{P}), \ldots, \phi^n(\mathscr{P})) . \tag{2.18}$$

The mapping (2.18) is a *homeomorphism* – it is bijective and both ϕ and ϕ^{-1} are continuous. A homeomorphism takes open (closed) sets into open (closed) sets; they are topologically equivalent. The notion of topology will allow us to discuss continuity and neighborhood in the absence of the distance concept; see Section 2.8 and Reference [9]. If $\Omega = A_n$ the coordinate chart is said to be globally defined; otherwise it is locally defined and (Ω, ϕ) is called a *coordinate patch* on A_n. When $\bigcup_i \Omega_i = A_n$, the family of patches is called an *atlas*. The inverse mapping $\phi^{-1} : \phi(\Omega) \to \Omega$ is called a parametric representation of Ω; see Figure 2.3 (left).

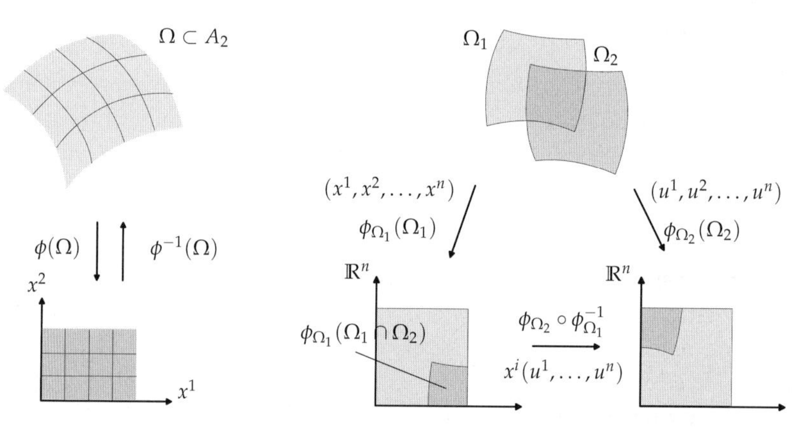

Figure 2.3 Left: Coordinate patch and its inverse, the parametric representation. Right: A point that lies in the overlap of two coordinate patches defined by $(\Omega_1, \phi_{\Omega_1})$ and $(\Omega_2, \phi_{\Omega_2})$ described by two sets of coordinates, denoted $(x^1, x^2, \ldots, x^n)$ and $(u^1, u^2, \ldots, u^n)$ and related by the coordinate transformation (2.19).

A bijective transformation $T : \mathbb{R}^n \to \mathbb{R}^n : \mathbf{u} \mapsto T(\mathbf{u})$ given in the component form as

$$x^i = x^i(u^1, u^2, \ldots, u^n), \qquad 1 \leq i \leq n \tag{2.19}$$

with functions x^i that map a given region $\Omega \in \mathbb{R}^n$ to reals is called a *coordinate transformation*. In the mathematical mapping notation the coordinate systems are denoted $\phi_{\Omega_1} : \Omega_1 \to \phi_{\Omega_1}(\Omega_1)$ and $\phi_{\Omega_2} : \Omega_2 \to \phi_{\Omega_2}(\Omega_2)$ and T is the bijection:

$$T := \phi_{\Omega_2} \circ \phi_{\Omega_1}^{-1}\Big|\phi_{\Omega_1}(\Omega_1 \cap \Omega_2) : \phi_{\Omega_1}(\Omega_1 \cap \Omega_2) \xrightarrow{\cong} \phi_{\Omega_2}(\Omega_1 \cap \Omega_2) \,, \tag{2.20}$$

cf. Figure 2.3 (right). In contrast, the notation of Eq. (2.19) is often referred to as the *anonymous notation*. If the x^i are the Cartesian coordinates,[4] usually denoted (x, y, z), the u^i, $i = 1, 2, 3$, are called *curvilinear coordinates*. If T is linear, the u^i are called *affine coordinates*.

Examples:

1. Polar coordinates. $T : \mathbb{R}^2 \to \mathbb{R}^2 : (r, \varphi) \mapsto (x, y)$ for $r > 0, 0 \le \varphi < 2\pi$ and $x = r\cos\varphi$, $y = r\sin\varphi$.
2. Spherical coordinates. $T : \mathbb{R}^3 \to \mathbb{R}^3 : (R, \vartheta, \varphi) \mapsto (x, y, z)$ for $R > 0$, $0 \le \vartheta < \pi$, $0 \le \varphi < 2\pi$ and $x = R\sin\vartheta\cos\varphi$, $y = R\sin\vartheta\sin\varphi$, $z = R\cos\vartheta$.
3. Affine coordinates, or barycentric coordinates, used in plane triangular finite elements with vertices $\mathscr{P}_1, \mathscr{P}_2, \mathscr{P}_3$. $T : \mathbb{R}^2 \to \mathbb{R}^2 : (\xi, \eta) \mapsto (x, y)$ for $0 \le \xi, \eta \le 1$ and

$$x = x(\mathscr{P}_3) + \big(x(\mathscr{P}_1) - x(\mathscr{P}_3)\big)\xi + \big(x(\mathscr{P}_2) - x(\mathscr{P}_3)\big)\eta , \tag{2.21}$$

$$y = y(\mathscr{P}_3) + \big(y(\mathscr{P}_1) - y(\mathscr{P}_3)\big)\xi + \big(y(\mathscr{P}_2) - y(\mathscr{P}_3)\big)\eta . \tag{2.22}$$

□

The examples show the necessity of defining the coordinate patch because the transformation is not bijective on the entire A_n. So far we have been vague regarding the objects requiring a parametric representation, which we will call *space elements* from now on. These k-dimensional surfaces in Euclidean n-space and their representations in local coordinates will be treated in Chapter 3.

Remark: In physics and engineering literature, little space is wasted either for the notion of the domain of the coordinate chart or for the coordinate map and the component functions. The coordinate functions ϕ^i are usually denoted x^i and the coordinates of a particular point $\mathscr{P}$ are then written as the n-tuple of real numbers $(x^1(\mathscr{P}), x^2(\mathscr{P}), \ldots, x^n(\mathscr{P})) \in \mathbb{R}^n$, which has the advantage that different notions for the point coordinates and the coordinate functions on

4 Coordinates in A_n are called rectangular or Cartesian if the distance between two arbitrary points $\mathscr{P}_1$ and $\mathscr{P}_2$ is given by $d(\mathscr{P}_1, \mathscr{P}_2) = \sqrt{\sum_{i=1}^{n} \left(x^i(\mathscr{P}_1) - x^i(\mathscr{P}_2)\right)^2}$. The distance concept will be discussed in the next section.

the domain Ω are not needed, and the disadvantage that they are not available. We will therefore call this the *anonymous notation*, in which a common practice is also to omit the notion of the point $\mathscr{P}$. In Chapter 3, which covers classical vector analysis, the space coordinates are usually denoted x, y, z and the variables of the parameter domain u and v. We will find notations such as $x = x(u, v)$ and $u = u(x, y, z)$, where u, v are the coordinate functions of $\mathscr{P}$. □

2.6
Inner Product Space

The abstract definition of vectors was given by way of the algebraic rules in Section 2.3, which do not involve the concepts of orientation and metric. The concepts of length and angle have not appeared in the definition of the vector spaces, and therefore the usual physical meaning of a vector "as a quantity having direction as well as magnitude" [12] is not yet completely defined.

We must now introduce the *inner product space* $(V, \langle\cdot,\cdot\rangle)$, shorthand V_n, a real vector space with an inner product $\langle \mathbf{a}, \mathbf{b}\rangle : V_n \times V_n \to \mathbb{R}$ that obeys bilinearity, symmetry, and positive definiteness:

1. $\langle \mathbf{a}+\mathbf{b}, \mathbf{c}\rangle = \langle \mathbf{a}, \mathbf{c}\rangle + \langle \mathbf{b}, \mathbf{c}\rangle$ and $\langle \mathbf{a}, \lambda\mathbf{b} + \mu\mathbf{c}\rangle = \lambda\langle \mathbf{a}, \mathbf{b}\rangle + \mu\langle \mathbf{a}, \mathbf{c}\rangle$.
2. $\langle \mathbf{a}, \mathbf{b}\rangle = \langle \mathbf{b}, \mathbf{a}\rangle$.
3. $\langle \mathbf{a}, \mathbf{a}\rangle > 0$ and $\langle \mathbf{a}, \mathbf{a}\rangle = 0$ if and only if $\mathbf{a} = \mathbf{0}$.

A finite-dimensional real vector space V_n with an inner product declared on it is called inner product space or a pre-Hilbert space.[5] As $\langle \mathbf{a}, \mathbf{a}\rangle$ is always positive, a real square root exists, which is called the *Euclidean norm* of $\mathbf{a}$:

$$\| \mathbf{a} \| = \sqrt{\langle \mathbf{a}, \mathbf{a}\rangle}, \tag{2.23}$$

for which the following triangular inequality holds:

$$\| \mathbf{a}+\mathbf{b} \| \leq \| \mathbf{a} \| + \| \mathbf{b} \| . \tag{2.24}$$

Note, however, that there are norms that are not induced by an inner product. The angle between two vectors $\mathbf{a} \neq \mathbf{0}$ and $\mathbf{b} \neq \mathbf{0}$ is defined by

$$\cos\alpha(\mathbf{a}, \mathbf{b}) := \frac{\langle \mathbf{a}, \mathbf{b}\rangle}{\| \mathbf{a} \| \| \mathbf{b} \|}, \qquad 0 \leq \alpha \leq \pi. \tag{2.25}$$

5 David Hilbert (1862–1943).

That the angle is well defined follows from the *Cauchy–Schwarz inequality*[6]

$$| \langle \mathbf{a}, \mathbf{b} \rangle | \leq \| \mathbf{a} \| \, \| \mathbf{b} \|, \tag{2.26}$$

which may be written in the form

$$-1 \leq \frac{\langle \mathbf{a}, \mathbf{b} \rangle}{\| \mathbf{a} \| \, \| \mathbf{b} \|} \leq 1 \tag{2.27}$$

and carries the angle concept into arbitrary real inner product spaces.

Examples:

1. The tuple space $\mathbb{R}^3$ equipped with an origin and with the ordinary *scalar product*, or dot product, defined by

$$\mathbf{a} \cdot \mathbf{b} := a^1 b^1 + a^2 b^2 + a^3 b^3 \tag{2.28}$$

   is called the *Euclidean space*. The norm induced by this scalar product is usually written as $| \cdot |$ instead of $\| \cdot \|$.
2. Instead of the positive definiteness one can impose the nondegeneration: If for a fixed $\mathbf{a}$, $\langle \mathbf{a}, \mathbf{b} \rangle = 0$ for all $\mathbf{b}$, then $\mathbf{a} = \mathbf{0}$. This space is called pseudo-Euclidean. In the four-dimensional Minkowski[7] space M^4 of special relativity, the Lorentz inner product is given by

$$\langle \mathbf{a}, \mathbf{b} \rangle := a^1 b^1 + a^2 b^2 + a^3 b^3 - (c)^2 a^4 b^4, \tag{2.29}$$

   where c is the speed of light.
3. In the (infinite-dimensional) vector space of integrable functions $f, g : [-1, 1] \mapsto \mathbb{R}$, an inner product is given by

$$\langle f, g \rangle := \int_{-1}^{1} f(x) g(x) \mathrm{d}x. \tag{2.30}$$

□

Any finite-dimensional vector space is turned into an inner product space by selection of a frame and then defining the inner product $\langle \mathbf{g}_i, \mathbf{g}_j \rangle = g_{ij}$ for $g_{ij} \in \mathbb{R}$. The inner product of the two vectors $\mathbf{a} = a^i \mathbf{g}_i$ and $\mathbf{b} = b^j \mathbf{g}_j$ can be constructed by a bilinear extension:

$$\begin{aligned}
\langle \mathbf{a}, \mathbf{b} \rangle = {} & a^1 b^1 \langle \mathbf{g}_1, \mathbf{g}_1 \rangle + a^1 b^2 \langle \mathbf{g}_1, \mathbf{g}_2 \rangle + a^1 b^3 \langle \mathbf{g}_1, \mathbf{g}_3 \rangle \\
& + a^2 b^1 \langle \mathbf{g}_2, \mathbf{g}_1 \rangle + a^2 b^2 \langle \mathbf{g}_2, \mathbf{g}_2 \rangle + a^2 b^3 \langle \mathbf{g}_2, \mathbf{g}_3 \rangle \\
& + a^3 b^1 \langle \mathbf{g}_3, \mathbf{g}_1 \rangle + a^3 b^2 \langle \mathbf{g}_3, \mathbf{g}_2 \rangle + a^3 b^3 \langle \mathbf{g}_3, \mathbf{g}_3 \rangle,
\end{aligned} \tag{2.31}$$

6 Augustin Cauchy (1789–1857), Herman Schwarz (1843–1921).
7 Hermann Minkowski (1864–1909).

which can be written with the summation convention as

$$\langle \mathbf{a}, \mathbf{b} \rangle = \sum_{i=1}^{3} \sum_{j=1}^{3} a^i b^j \langle \mathbf{g}_i, \mathbf{g}_j \rangle \equiv a^i b^j \langle \mathbf{g}_i, \mathbf{g}_j \rangle =: a^i b^j g_{ij}, \tag{2.32}$$

where $[G] := (g_{ij})$ is called the *metric*. The norm of a vector **a** in the general system is $\| \mathbf{a} \| = \sqrt{\langle \mathbf{a}, \mathbf{a} \rangle} = \sqrt{a^i a^j g_{ij}}$, and the angle between two nonnull vectors **a** and **b** can be calculated from

$$\cos \alpha = \frac{\langle \mathbf{a}, \mathbf{b} \rangle}{\| \mathbf{a} \| \, \| \mathbf{b} \|} = \frac{a^i b^j g_{ij}}{\sqrt{a^p a^q g_{pq}} \sqrt{b^r b^s g_{rs}}} \,. \tag{2.33}$$

2.6.1
Metric Space

An affine space is turned into a metric space (A, d) by means of the distance $d : A \times A \to \mathbb{R}$ between two points $\mathscr{P}_1$ and $\mathscr{P}_2$:

$$d(\mathscr{P}_1, \mathscr{P}_2) := \| \mathbf{r}_{\mathscr{P}_1} - \mathbf{r}_{\mathscr{P}_2} \| \tag{2.34}$$

satisfying

1. $d(\mathscr{P}_1, \mathscr{P}_2) = d(\mathscr{P}_2, \mathscr{P}_1) \geq 0 \; \forall \mathscr{P}_1, \mathscr{P}_2$,
2. $d(\mathscr{P}_1, \mathscr{P}_2) > 0$ if $\mathscr{P}_1 \neq \mathscr{P}_2$,
3. $d(\mathscr{P}_1, \mathscr{P}_3) + d(\mathscr{P}_3, \mathscr{P}_2) \geq d(\mathscr{P}_1, \mathscr{P}_2) \; \forall \mathscr{P}_1, \mathscr{P}_2, \mathscr{P}_3$.

The distance concept is compatible with the affine structure because it is translation invariant, that is, $d(\mathscr{P}_1 + \mathbf{v}, \mathscr{P}_2 + \mathbf{v}) = d(\mathscr{P}_1, \mathscr{P}_2)$, for any displacement vector **v**. Note that not all distance concepts are induced by a norm, for example, the trivial metric for two distinct points $d(\mathscr{P}_1, \mathscr{P}_1) = 0$, $d(\mathscr{P}_1, \mathscr{P}_2) = 1$ or the French railroad metric, where $\mathscr{Q}$ is a fixed point (Paris): $d(\mathscr{P}_1, \mathscr{P}_2) = 0$ for $\mathscr{P}_1 = \mathscr{P}_2$, $d(\mathscr{P}_1, \mathscr{P}_2) = \| \mathbf{r}_{\mathscr{P}_1} - \mathbf{r}_{\mathscr{P}_2} \|$ if $\mathscr{Q} = \mathscr{P}_1 + \lambda(\mathbf{r}_{\mathscr{P}_2} - \mathbf{r}_{\mathscr{P}_1})$ and $d(\mathscr{P}_1, \mathscr{P}_2) = \| \mathbf{r}_{\mathscr{P}_1} - \mathbf{r}_{\mathscr{Q}} \| + \| \mathbf{r}_{\mathscr{P}_2} - \mathbf{r}_{\mathscr{Q}} \|$ otherwise.

Consider two metric spaces (X, d_x) and (Y, d_y) and the mapping $T : X \to Y$, with the property that for all $\mathscr{P}_1, \mathscr{P}_2 \in X$

$$d_x(\mathscr{P}_1, \mathscr{P}_2) = d_y(T(\mathscr{P}_1), T(\mathscr{P}_2)) \,. \tag{2.35}$$

This mapping is called an *isometry* and the two metric spaces are called isometric if a bijective isometry exists between them.

Employing the metric (g_{ij}), we can express the distance between two arbitrary points $\mathscr{P}_1$ and $\mathscr{P}_2$ in the affine space as

$$d(\mathscr{P}_1, \mathscr{P}_2) = \sqrt{(x^i(\mathscr{P}_1) - x^i(\mathscr{P}_2))(x^j(\mathscr{P}_1) - x^j(\mathscr{P}_2)) g_{ij}} \,. \tag{2.36}$$

Affine geometry is thus a generalization of Euclidean geometry with slant and scale distortions.

Example: The following example can be found in [11]: Carpenters taking distance measurements in a room notice that there are no right angles at the corner they use as reference. The measures of the angles are given in Figure 2.4. Using the unit vectors **a**, **b**, **c** along the oblique axes, the angle α can be calculated from Eq. (2.33) by

$$\cos\alpha = \frac{\delta_1^i \delta_2^j g_{ij}}{\sqrt{\delta_1^p \delta_1^q g_{pq}}\sqrt{\delta_2^r \delta_2^s g_{rs}}} = \frac{g_{12}}{\sqrt{g_{11}}\sqrt{g_{22}}} = g_{12}, \tag{2.37}$$

since $g_{11} = g_{22} = g_{33} = 1$. The symbol δ is called the *Kronecker delta*,[8] which has the effect of annihilating the index terms in the double summation. In the same way we obtain $\cos\beta = g_{13}$ and $\cos\gamma = g_{23}$. The metric is thus

$$[G] = \begin{pmatrix} 1 & \cos\alpha & \cos\beta \\ \cos\alpha & 1 & \cos\gamma \\ \cos\beta & \cos\gamma & 1 \end{pmatrix} \tag{2.38}$$

and the carpenters must use the distance formula (2.36) with the g_{ij} from Eq. (2.38). □

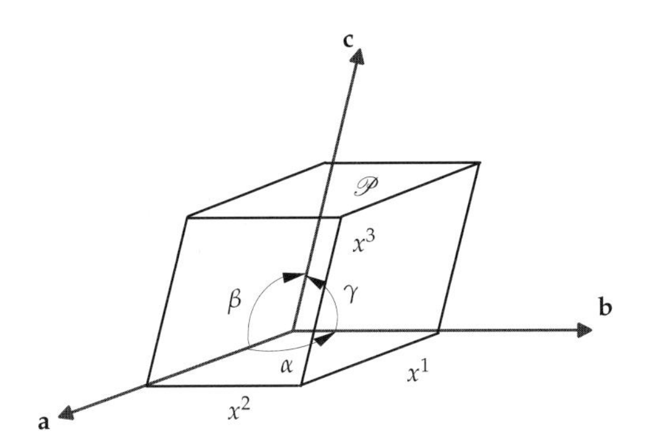

Figure 2.4 Distance measurements in affine coordinates.

2.6.2 Orthonormal Bases

A set of basis vectors $S = \{\mathbf{g}_1, \mathbf{g}_2, \ldots, \mathbf{g}_n\}$ is said to be *orthogonal* if the inner product of each pair i, j; $i \neq j$ of vectors is zero. The set is *orthonormal* if S is orthogonal and each vector has unit length:

$$\langle \mathbf{g}_i, \mathbf{g}_j \rangle = g_{ij} = \delta_{ij} = \begin{cases} 0 \text{ for } i \neq j \\ 1 \text{ for } i = j \end{cases} . \tag{2.39}$$

8 Leopold Kronecker (1823–1891).

It is common practice to denote orthonormal basis vectors instead of $\mathbf{g}_i$ with the symbol $\mathbf{e}_i$. The matrix $(g_{ij}) = \mathrm{diag}(1, \ldots, 1)$ is called the Euclidean metric, under which the *Pythagorean relation*[9] holds for right triangles,

$$d(\mathscr{P}_1, \mathscr{P}_2) = \sqrt{\sum_{i=1}^{n} (x^i(\mathscr{P}_1) - x^i(\mathscr{P}_2))^2}\,. \tag{2.40}$$

The inner product then takes the usual Euclidean form

$$\langle \mathbf{a}, \mathbf{b} \rangle = \mathbf{a} \cdot \mathbf{b} = \mathbf{b} \cdot \mathbf{a} = a^1 b^1 + a^2 b^2 + a^3 b^3 + \cdots + a^n b^n. \tag{2.41}$$

If we restricted ourselves to the use of orthonormal bases, there would be no need to introduce the matrix (g_{ij}), as is the case in elementary linear algebra.

Let V be an inner product space. The vectors $\mathbf{a}, \mathbf{b} \in V$ are said to be orthogonal if $\langle \mathbf{a}, \mathbf{b} \rangle = 0$.

Examples:

1. The trigonometric functions $\sin\varphi$ and $\cos\varphi$ in the vector space of smooth functions on the closed interval $[-\pi, \pi]$ denoted $C^\infty[-\pi, \pi]$, with $\langle \sin\varphi, \cos\varphi \rangle = \int_{-\pi}^{\pi} \sin\varphi \cos\varphi \mathrm{d}\varphi = \frac{1}{2}\sin^2\varphi|_{-\pi}^{\pi} = 0$, are orthogonal functions in the vector space $C^\infty[-\pi, \pi]$.
2. The Legendre polynomials $P_n(x)$ of degree $n \geq 0$, with $P_0(x) = 1$, $P_1(x) = x$, and

$$P_n(x) := \frac{1}{2^n n!} \frac{\mathrm{d}^n}{\mathrm{d}x^n} (x^2 - 1)^n \quad \forall n > 1, \tag{2.42}$$

are orthogonal with respect to the integral inner product $\langle P_n(x), P_k(x) \rangle = \int_{-1}^{1} P_n(x) P_k(x) \mathrm{d}x$ for all $n \neq k$. □

2.6.3 The Erhard Schmidt Orthogonalization

Consider two nonzero vectors $\mathbf{x}, \mathbf{y} \in V_n$. It follows that the *projection* of $\mathbf{x}$ along $\mathbf{y}$ is by definition the scaled vector $\lambda\mathbf{y}$ of $\mathbf{y}$ such that the vector $\mathbf{x} - \lambda\mathbf{y}$ is orthogonal to $\mathbf{y}$; see Figure 2.5 (left). Consequently

$$\langle \mathbf{x} - \lambda\mathbf{y}, \mathbf{y} \rangle = \langle \mathbf{x}, \mathbf{y} \rangle - \lambda \langle \mathbf{y}, \mathbf{y} \rangle = 0, \tag{2.43}$$

and therefore $\lambda = \frac{\langle \mathbf{x}, \mathbf{y} \rangle}{\langle \mathbf{y}, \mathbf{y} \rangle}$. The projection is defined as the mapping

$$\mathrm{proj} : V_n \times V_n \to V : (\mathbf{x}, \mathbf{y}) \mapsto \lambda\mathbf{y} = \frac{\langle \mathbf{x}, \mathbf{y} \rangle}{\langle \mathbf{y}, \mathbf{y} \rangle}\mathbf{y}\,. \tag{2.44}$$

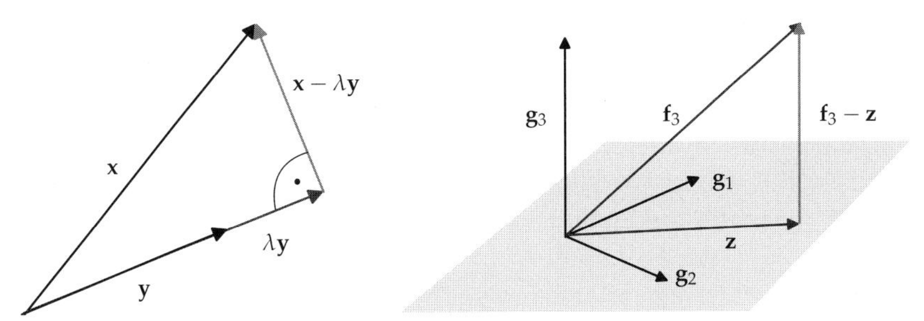

Figure 2.5 Left: Projection of the vector x along y. Right: Erhard Schmidt orthogonalization step for the special case of three dimensions, when two orthogonal vectors $\mathbf{g}_1$ and $\mathbf{g}_2$ have already been found. $\mathbf{z} = \mathrm{proj}(\mathbf{f}_3, \mathbf{g}_1) + \mathrm{proj}(\mathbf{f}_3, \mathbf{g}_2)$.

The scalar λ is unique and is called the *generalized Fourier coefficient* of $\mathbf{x}$ with respect to $\mathbf{y}$.

Suppose now that $\{\mathbf{f}_1, \mathbf{f}_2, \ldots, \mathbf{f}_n\}$ is a basis of an n-dimensional inner product space. It is an important property of inner product spaces that we can construct an orthogonal basis $\{\mathbf{g}_1, \mathbf{g}_2, \ldots, \mathbf{g}_n\}$ of V_n by means of the *Erhard Schmidt orthogonalization*[10] process, illustrated in Figure 2.5 (right):

$$
\begin{aligned}
\mathbf{g}_1 &= \mathbf{f}_1 , \\
\mathbf{g}_2 &= \mathbf{f}_2 - \frac{\langle \mathbf{f}_2, \mathbf{g}_1 \rangle}{\langle \mathbf{g}_1, \mathbf{g}_1 \rangle} \mathbf{g}_1 , \\
\mathbf{g}_3 &= \mathbf{f}_3 - z = \mathbf{f}_3 - \frac{\langle \mathbf{f}_3, \mathbf{g}_1 \rangle}{\langle \mathbf{g}_1, \mathbf{g}_1 \rangle} \mathbf{g}_1 - \frac{\langle \mathbf{f}_3, \mathbf{g}_2 \rangle}{\langle \mathbf{g}_2, \mathbf{g}_2 \rangle} \mathbf{g}_2
\end{aligned}
\tag{2.45}
$$

for $k = 2, 3, \ldots, n$, that is,

$$
\mathbf{g}_k = \mathbf{f}_k - \sum_{i=1}^{k-1} \mathrm{proj}(\mathbf{f}_k, \mathbf{g}_i) . \tag{2.46}
$$

An orthonormal basis $\{\mathbf{e}_1, \mathbf{e}_2, \ldots, \mathbf{e}_n\}$ can be found by dividing the orthogonal projection at each step by its norm to obtain the unit vector $\mathbf{e}_n$ orthogonal to $\mathbf{f}_{n-1}$:

$$
\mathbf{e}_k = \frac{\mathbf{f}_k - \sum_{i=1}^{k-1} \mathrm{proj}(\mathbf{f}_k, \mathbf{g}_i)}{\| \mathbf{f}_k - \sum_{i=1}^{k-1} \mathrm{proj}(\mathbf{f}_k, \mathbf{g}_i) \|} . \tag{2.47}
$$

We emphasize that in an inner product space with a given basis, it is always possible to generate an orthonormal basis.

9 Pythagoras of Samos (569–475 BC).
10 Erhard Schmidt (1876–1959).

2.7 Orientation

Global variables of physics are associated with space elements $\mathscr{M}$ such as points $\mathscr{P}$, curves $\mathscr{S}$, surfaces $\mathscr{A}$, and volumes $\mathscr{V}$. The physical variables are classified as configuration variables (displacement, position, strain, velocity, electric field, magnetic flux), source variables (force, mass, pressure, electric flux, heat flux), and energy variables obtained by the product of a configuration variable and a source variable (power, work, heat) [13]. It turns out that configuration variables require inner orientation of the space elements while source variables require outer orientation. Constitutive equations link configuration with source variables.

Consider two frames $\{\mathbf{g}_1, \ldots, \mathbf{g}_n\}$ and $\{\mathbf{f}_1, \ldots, \mathbf{f}_n\}$ in V_n. One may express the basis vectors $\mathbf{g}_i$ as linear combinations of the $\mathbf{f}_j$ by means of a *transition matrix* $[A]$ such that $\mathbf{g}_i = a_i^j \mathbf{f}_j$. Two frames are said to have the same orientation if the determinant of the transition matrix is positive.

The matching of the orientation defines on the set of all frames in V_n an equivalence relation that does not depend on the basis and is therefore intrinsic to the structure of V_n. The equivalence relation separates the frames in V_n into two *equivalence classes*.[11] Each of these classes, generally denoted Or and $-Or$, defines an orientation on V_n.

Orienting a vector space consists of declaring one of the orientation classes as positively oriented. For each vector space there are two oriented vector spaces: (V_n, Or), which is called the positively-oriented space (abbreviated ${}^+V_n$) with frames therein called *direct* and $(V_n, -Or)$, also denoted ${}^-V_n$, with frames therein called *skew*. In three dimensions the direct frame can geometrically be described by the right-handed screw. James Clerk Maxwell (1831–1879) wrote: "Prof. W.H. Miller has suggested to me that as the tendrils of the vine are right-handed screws and those of the hop left-handed, the two systems of relations in space might be called those of the vine and the hop respectively." Although this notation is used, for example, in [1], there are vines with left-handed tendrils on the southern hemisphere and therefore we are not adopting this notation.

Affine space is oriented by orienting its associated vector space, in other words, by choosing direct or skew frames at all points $\mathscr{P}$ in a consistent[12] way. Connected subspaces of A_3, subsequently referred to as space elements, such

11 The concepts of equivalence classes and quotient spaces are best explained with the example of the quotient group of integers mod 2, $\mathbb{Z}_2 = \mathbb{Z}/2\mathbb{Z}$ that consists of the equivalence class $\tilde{0}$ of even integers $(0, \pm2, \pm4, \ldots)$ and the equivalence class $\tilde{1}$ of odd integers $(\pm1, \pm3, \ldots)$.

12 The consistency requirement rules out nonorientable objects such as the famous Möbius strip. August Möbius (1790–1868).

as line segments or polygonal faces, are oriented by orienting their *tangent planes*;[13] see Figure 2.6 (left).

- A point is oriented by attributing a plus or minus sign in accordance with the incoming lines considered positive or negative. The point acts as a sink or a source.
- A line is oriented by selecting a vector parallel to it and in this way defining a direction along the line.
- A surface is oriented by selecting a sense of rotation.
- A volume is oriented by deciding on the consistent orientation of its faces.

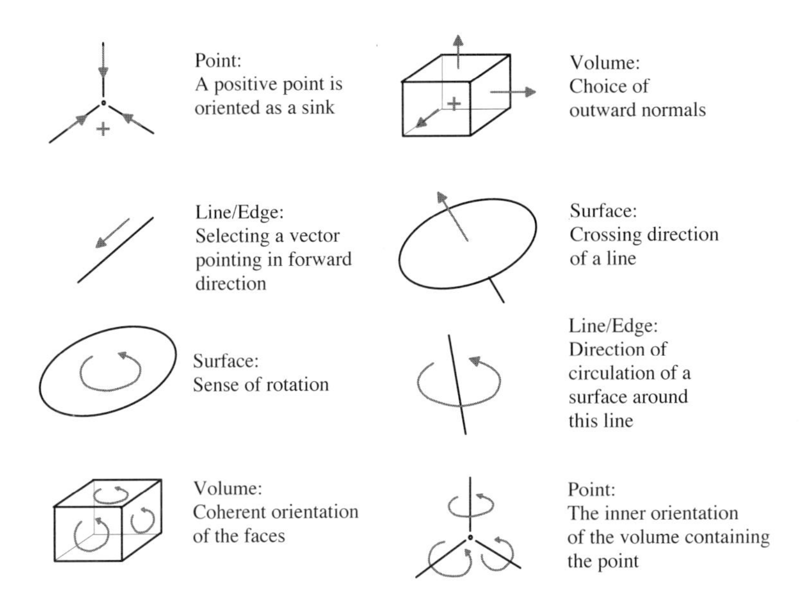

Figure 2.6 Inner (left) and outer (right) orientation of space elements [15]. As an example, inner orienting a line means selecting a vector parallel to it, and outer orienting a line means to inner orient a transverse surface, that is, to make a choice between the two ways to turn around the line.

The orientations as described above are *inner orientations*, wherein the object need not be embedded in a higher-dimensional space. In contrast, defining a crossing direction through a surface, for example, is called *outer orientation*. Generally, when a space element of dimension p is embedded in a higher-dimensional space A_n, an outer orientation of the tangent plane at a point defines an inner orientation of its $n - p$ dimensional complement.

13 The term has to make intuitive sense; the discussion is deferred to Chapter 3.

- A point is outer oriented by attributing an inner orientation to the volume that encloses the point.
- A line is outer oriented in coherence with the inner orientation of a transverse surface. A choice is made between the two ways to turn around the line. In [15] we read: "A one-lane street is a line with an inner orientation while the axis of rotation of Earth is a line with outer orientation because of its direction of rotation around the axis."
- A surface is outer oriented by selection of a crossing direction through it. This is equivalent to selecting one face of the surface.
- A volume is outer oriented by selection of a direction from inside out.

Inner and outer orientation of space elements are visualized in Figure 2.6. To ensure consistent orientation of a p-dimensional space element and its $(n-p)$-dimensional complement in an oriented, encompassing space A_n, take a frame in the vector space W_p consisting of p vectors, and add (append) the $n-p$ vectors of a positively oriented frame in the associated vector space U_{n-p} to the complement. Then check if the resulting frame in V_n has the same orientation as the frame in W_p; see Figure 2.9 (left). In this case we do not distinguish between inner and outer orientation. In [17] we read: "... in some ways it hides the essential feature; it gives rise to the well-known 'swimming-rules' in electrodynamics, which in no wise signify that there is a unique direction of twist in the space in which electro-dynamic events occur ..."

2.8 A Glimpse on Topological Concepts

Although topological concepts can be introduced without reference to a metric employing set theory [9], each metric space is an example of a topological space. Let (A, d) denote a metric affine space: An open ball centered at $\mathscr{P}$ is defined by $B_{\mathscr{P},\varepsilon} := \{\mathscr{Q} \in (A, d) \mid d(\mathscr{P}, \mathscr{Q}) < \varepsilon\}$. A subset $\Omega \subset (A, d)$ is called *open* if

$$B_{\mathscr{P},\varepsilon} \in \Omega, \forall \mathscr{P} \in \Omega. \tag{2.48}$$

The set of all open subsets of (A, d) is called the *topology* $\mathcal{T}$ induced by the metric d. In other words, a metric space (A, d) is a topological space $(A, \mathcal{T})$ with the topology induced by a metric. We will henceforth omit the notation of d and $\mathcal{T}$.

A subset of a topological space $\Omega \subset A$ is said to be disconnected if there exist open subsets X and Y such that $X \cap \Omega$ and $Y \cap \Omega$ are disjoint nonempty sets whose union is Ω. A set is connected if it is not disconnected. A connected

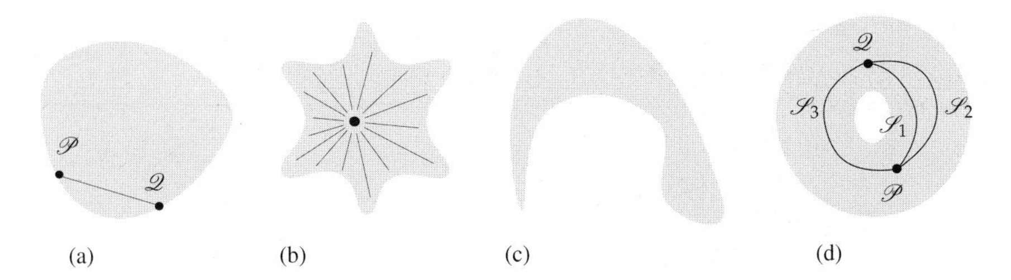

Figure 2.7 Topology of two-dimensional domains. (a) Convex domain. (b) Star-shaped domain. The bullet marks an eligible center. (c) Domain that is not star-shaped. No center can be found such that lines from the center to any other point $\mathscr{P} \in \Omega$ are entirely inside the domain. (d) Multiply-connected domain; not contractible. Paths $\mathscr{S}_1$ and $\mathscr{S}_3$ constitute a loop. Note that in three dimensions the domain must be simply-connected with connected boundary to be contractible (absence of bubbles).

affine subspace is called a *domain*. A domain is *star-shaped* if there exists a point $\mathscr{P}$ such that for all $\mathscr{Q} \in \Omega$, the line

$$\mathscr{P} + \lambda(\mathscr{Q} - \mathscr{P}) = (1-t)\mathscr{Q} + t\mathscr{P}, \qquad \lambda, t \in [0,1],\ t = 1 - \lambda \tag{2.49}$$

is entirely in Ω. A domain is said to be *convex* if Eq. (2.49) holds for any two points $\mathscr{P}$ and $\mathscr{Q}$ in Ω; an example is shown in Figure 2.7.

2.8.1 Homotopy

Let $\mathscr{S}_1 : [0,1] \to A$ and $\mathscr{S}_2 : [0,1] \to A$ be two paths with the same initial and terminal points, denoted $\mathscr{P}$ and $\mathscr{Q}$. $\mathscr{S}_1$ is *homotopic* to $\mathscr{S}_2$, written as $\mathscr{S}_1 \simeq \mathscr{S}_2$, if there exists a continuous function (a homotopy from $\mathscr{S}_1$ to $\mathscr{S}_2$) $H : [0,1] \times [0,1] \to A : (s,t) \mapsto H(s,t)$ such that

$$H(s,0) = \mathscr{S}_1(s), \qquad H(s,1) = \mathscr{S}_2(s), \tag{2.50}$$

and $H(0,t) = \mathscr{P}$, $H(1,t) = \mathscr{Q}$. Figure 2.8 (left) shows homotopic paths with fixed initial and terminal points generated by continuous deformations. In particular it yields $\mathscr{S}_1 \simeq \mathscr{S}_1$, and from $\mathscr{S}_1 \simeq \mathscr{S}_2$ follows $\mathscr{S}_2 \simeq \mathscr{S}_1$.

A path with identical initial and terminal points, $\mathscr{S}(0) = \mathscr{S}(1) = \mathscr{P}$, is said to be closed at $\mathscr{P}$. It is also referred to as a *loop*. In particular, the constant path $\mathscr{S}_C : [0,1] \to A : \mathscr{S}_C(s) \mapsto \mathscr{P}$ is a loop. A loop is said to be *contractible* to a point, or zero-homotopic, if it is homotopic to the constant path as shown in Figure 2.8 (right).

A domain is *simply-connected* if all loops in the domain are zero-homotopic. In particular, a star-shaped domain, shown in Figure 2.7, is simply-connected, because each loop $\mathscr{S}$ at $\mathscr{P}$ is zero-homotopic by

$$H(s,t) = (1-t)\mathscr{S}(s) + t\mathscr{P}, \qquad s,t \in [0,1]. \tag{2.51}$$

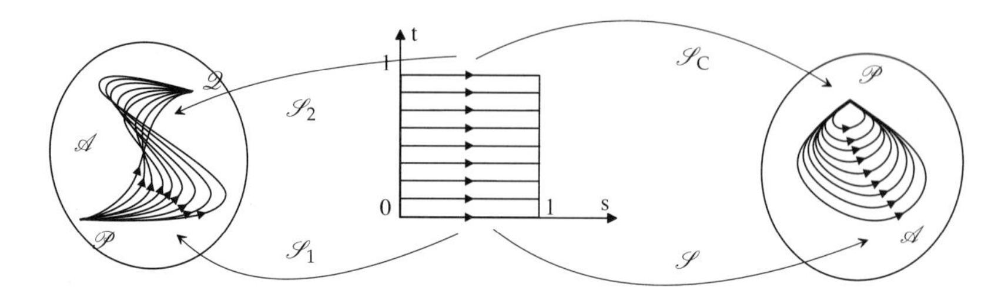

Figure 2.8 Left: Homotopy from $\mathcal{S}_1$ to $\mathcal{S}_2$. Right: Homotopy from a loop to the constant path (zero-homotopy).

There are, however, simply-connected domains which are not star-shaped; see Figure 2.7. A domain is said to be contractible if it is simply-connected with a connected boundary, that is, there exists a contraction mapping f onto the center $\mathcal{P}$ such that for each point $\mathcal{Q}$ on the boundary we have

$$d(f(\mathcal{P}) - f(\mathcal{Q})) < \lambda d(\mathcal{P}, \mathcal{Q}), \qquad 0 \leq \lambda < 1. \tag{2.52}$$

The distance between the images of the points is less than the distance between the points.

A simply-connected, two-dimensional domain is contractible. The inside of a ring is connected but not simply-connected. In three dimensions, the inside of two nested spheres is simply-connected while the boundary is not connected and therefore the domain is not contractible.

2.8.2
The Boundary Operator

The distance $d(\mathcal{P}, \Omega)$ of a point $\mathcal{P}$ to a subset $\Omega \in A$ is defined by $\inf\{d(\mathcal{P}, \mathcal{Q})\}$ for all $\mathcal{Q} \in \Omega$. The interior of Ω is the set of all points $\mathcal{P} \in A$ such that $d(\mathcal{P}, A \setminus \Omega) > 0$. The *boundary* $\partial\Omega$ of Ω is the set of all points for which the distances $d(\mathcal{P}, \Omega) = 0$ and $d(\mathcal{P}, A \setminus \Omega) = 0$.

With these preliminaries out of the way we can turn to the special cases of lines, surfaces, and volumes. The boundary $\partial\mathcal{S}$ of a connected, inner-oriented curve $\mathcal{S}$ consists of the initial point $\mathcal{P}_1$ and the terminal point $\mathcal{P}_2$, i.e., $\partial\mathcal{S} = \{\mathcal{P}_1, \mathcal{P}_2\}$. As the space difference of a function between two points is given by $(+1)f(\mathcal{P}_2) + (-1)f(\mathcal{P}_1)$, the plus and minus signs can be conceived as the *incidence numbers* between the curve and the boundary points oriented as sinks. For an incidence number of +1 the orientations of the curve and the point are said to be *consistent* or *compatible*. The curve can be composed of a number of nonconnected parts. In this case a multiple of points define the boundary. Closed curves (loops), denoted $\mathcal{C}$, have no boundary, $\partial\mathcal{C} = \emptyset$.

An orientable surface $\mathscr{A}$ has a boundary $\partial\mathscr{A}$, which is called a *contour*. The inner orientation of the contour is said to be consistent if the movement along the contour fits the sense of rotation of the surface; see Figure 2.6 (left). A closed surface $\mathscr{A}_c$ has no contour, $\partial\mathscr{A}_c = \emptyset$.

The boundary $\partial\mathscr{V}$ of a volume $\mathscr{V}$ is a closed surface, which is consistently oriented if the sense of rotation of $\partial\mathscr{V}$ matches the sense of the screw orientation of $\mathscr{V}$ when the surface is approached from the inside of the domain.

More generally, for a consistent orientation of the boundary $\partial\mathscr{M}$ of an inner-oriented space element $\mathscr{M}$ of dimension $p \geq 2$, take $p-1$ tangent vectors $\mathbf{v}_1, \ldots, \mathbf{v}_{p-1}$ at a point of $\partial\mathscr{M}$ and append them to the outward normal vector $\mathbf{v}$. Then verify that the frame $\{\mathbf{v}, \mathbf{v}_1, \ldots, \mathbf{v}_{p-1}\}$ is in the same orientation class as $\mathscr{M}$; see Figure 2.9 (right).

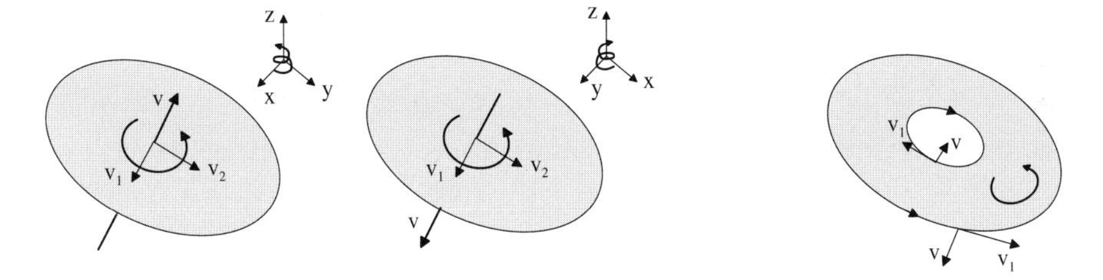

Figure 2.9 Left: Consistent inner and outer orientation of a surface in an oriented encompassing space: $\{\mathbf{v}_1, \mathbf{v}_2, \mathbf{v}\}$ has the same orientation as $\{\mathbf{e}_x, \mathbf{e}_y, \mathbf{e}_z\}$. Right: Consistent orientation of the boundary of an inner-oriented surface: $\{\mathbf{v}, \mathbf{v}_1\}$ has the same orientation as the surface.

For consistent outer orientation, first establish the consistent inner orientation by means of the orientation of ambient space and then check the inner orientation of the space element and its boundary. Since the boundary of a surface is a closed curve and the boundary of a volume is a closed surface, we have the two important topological properties,

$$\partial(\partial\mathscr{V}) = \emptyset, \qquad \partial(\partial\mathscr{A}) = \emptyset, \tag{2.53}$$

that is, closed curves and closed surfaces have no boundary. In general, $\partial(\partial\mathscr{M}) = \emptyset$.

2.9 Exterior Products

The alternating *exterior product* is defined as a bilinear mapping that assigns to each ordered pair of vectors $\mathbf{a}$ and $\mathbf{b}$ a so-called *2-vector*; a mapping $V_n \times V_n \rightarrow \wedge^2 V_n : (\mathbf{a}, \mathbf{b}) \mapsto \mathbf{a} \wedge \mathbf{b}$ which obeys

$$\mathbf{a} \wedge (\lambda \mathbf{b} + \mu \mathbf{c}) = \lambda(\mathbf{a} \wedge \mathbf{b}) + \mu(\mathbf{a} \wedge \mathbf{c}), \tag{2.54}$$

$$(\lambda \mathbf{a} + \mu \mathbf{b}) \wedge \mathbf{c} = \lambda(\mathbf{a} \wedge \mathbf{c}) + \mu(\mathbf{b} \wedge \mathbf{c}), \tag{2.55}$$

and which is alternating,

$$\mathbf{a} \wedge \mathbf{b} = -(\mathbf{b} \wedge \mathbf{a}) . \tag{2.56}$$

This yields

$$\mathbf{a} \wedge \mathbf{a} = 0. \tag{2.57}$$

The usefulness of such alternations as an appropriate algebraic structure for integrands will be shown in Section 3.13. Sometimes the exterior product is referred to as the polar product [1] or the wedge product. If we select a frame, n products out of the n^2 possible products of the basis vectors are zero because of the property (2.57). Only half the remaining $n(n-1)$ products are linearly independent due to the property (2.56); therefore,

$$\dim(\wedge^2 V_n) = n(n-1)/2 . \tag{2.58}$$

In particular for $n = 3$, the dimension of $\wedge^2 V_3$ is 3. Because of linearity we obtain

$$\begin{aligned} \mathbf{a} \wedge \mathbf{b} &= (a^1 \mathbf{g}_1 + a^2 \mathbf{g}_2 + a^3 \mathbf{g}_3) \wedge (b^1 \mathbf{g}_1 + b^2 \mathbf{g}_2 + b^3 \mathbf{g}_3) \\ &= (a^1 b^2 - a^2 b^1) \mathbf{g}_1 \wedge \mathbf{g}_2 + (a^2 b^3 - a^3 b^2) \mathbf{g}_2 \wedge \mathbf{g}_3 + (a^3 b^1 - a^1 b^3) \mathbf{g}_3 \wedge \mathbf{g}_1 \\ &=: (a^1 b^2 - a^2 b^1) \mathbf{g}_{12} + (a^2 b^3 - a^3 b^2) \mathbf{g}_{23} + (a^3 b^1 - a^1 b^3) \mathbf{g}_{31}, \end{aligned} \tag{2.59}$$

which can also be expressed by means of a determinant

$$\mathbf{a} \wedge \mathbf{b} = \begin{vmatrix} \mathbf{g}_{23} & \mathbf{g}_{31} & \mathbf{g}_{12} \\ a^1 & a^2 & a^3 \\ b^1 & b^2 & b^3 \end{vmatrix} . \tag{2.60}$$

In contrast to the inner product, the exterior product can act on three vectors by assigning to each triple of vectors in V_n a 3-vector: $V_n \times V_n \times V_n \rightarrow \wedge^3 V_n : (\mathbf{a}, \mathbf{b}, \mathbf{c}) \mapsto \mathbf{a} \wedge \mathbf{b} \wedge \mathbf{c}$. This product is alternating in the sense that it becomes zero if two of the three factors are identical, that is, $\mathbf{a} \wedge \mathbf{b} \wedge \mathbf{b} = \mathbf{a} \wedge \mathbf{b} \wedge \mathbf{a} =$

$\mathbf{a} \wedge \mathbf{a} \wedge \mathbf{b} = 0$ and $\mathbf{a} \wedge \mathbf{b} \wedge \mathbf{c} = \mathbf{b} \wedge \mathbf{c} \wedge \mathbf{a} = \mathbf{c} \wedge \mathbf{a} \wedge \mathbf{b} = -\mathbf{c} \wedge \mathbf{b} \wedge \mathbf{a} = -\mathbf{a} \wedge \mathbf{c} \wedge \mathbf{b} = -\mathbf{b} \wedge \mathbf{a} \wedge \mathbf{c}$, or generally

$$\mathbf{a}_i \wedge \mathbf{a}_j \wedge \mathbf{a}_k = \epsilon_{ijk}\, \mathbf{a}_1 \wedge \mathbf{a}_2 \wedge \mathbf{a}_3\,, \tag{2.61}$$

where ϵ_{ijk} is the *Levi-Civita symbol*[14] defined by

$$\epsilon_{ijk} = \begin{cases} 1 & \text{if (i,j,k) = (1,2,3) or (3,1,2) or (2,3,1),} \\ -1 & \text{if (i,j,k) = (3,2,1) or (1,3,2) or (2,1,3),} \\ 0 & \text{for two or more equal indices.} \end{cases} \tag{2.62}$$

Selecting a frame $\{\mathbf{g}_1, \mathbf{g}_2, \mathbf{g}_3\}$ only the

$$\frac{n(n-1)(n-2)}{3!} \tag{2.63}$$

basis vectors are linearly independent. For $n = 3$ the dimension of $\wedge^3 V_3$ is 1 and the space $\wedge^3 V_3$ is isomorphic to V. We obtain by linear extension

$$\begin{aligned} \mathbf{a} \wedge \mathbf{b} \wedge \mathbf{c} &= a^i b^j c^k (\mathbf{g}_i \wedge \mathbf{g}_j \wedge \mathbf{g}_k) = a^i b^j c^k \epsilon_{ijk} (\mathbf{g}_1 \wedge \mathbf{g}_2 \wedge \mathbf{g}_3) \\ &= \det(\mathbf{a}, \mathbf{b}, \mathbf{c}) \mathbf{g}_1 \wedge \mathbf{g}_2 \wedge \mathbf{g}_3 =: \det(\mathbf{a}, \mathbf{b}, \mathbf{c}) \mathbf{g}_{123} \end{aligned} \tag{2.64}$$

with the determinant

$$\det(\mathbf{a}, \mathbf{b}, \mathbf{c}) := \begin{vmatrix} a^1 & a^2 & a^3 \\ b^1 & b^2 & b^3 \\ c^1 & c^2 & c^3 \end{vmatrix}\,. \tag{2.65}$$

As in three dimensions $\dim(\wedge^2 V_3)$ is 3, the linear space $\wedge^2 V_3$ is isomorphic to V_3 and there is a one-to-one mapping of 2-vectors to the ordinary 1-vectors of V_3; it is the one that represents the surface of the parallelogram spanned by the 2-vector. Obviously, both metric and orientation are required.

Given two basis vectors $\mathbf{g}_1$ and $\mathbf{g}_2$, the *cross product* $\mathbf{f} = \mathbf{g}_1 \times \mathbf{g}_2$ is defined as a vector orthogonal to both of them, i.e., $\langle \mathbf{f}, \mathbf{g}_1 \rangle = \langle \mathbf{f}, \mathbf{g}_2 \rangle = 0$ and of a length representing the surface of the parallelogram spanned by the two vectors

$$\begin{aligned} \| \mathbf{f} \| &= \| \mathbf{g}_1 \times \mathbf{g}_2 \| = \sqrt{\| \mathbf{g}_1 \|^2 \, \| \mathbf{g}_2 \|^2 - \langle \mathbf{g}_1, \mathbf{g}_2 \rangle^2} \\ &= \| \mathbf{g}_1 \| \, \| \mathbf{g}_2 \| \sqrt{1 - \cos^2 \alpha} = \| \mathbf{g}_1 \| \, \| \mathbf{g}_2 \| \sin \alpha\,, \end{aligned} \tag{2.66}$$

14 Tullio Levi-Civita (1873–1941).

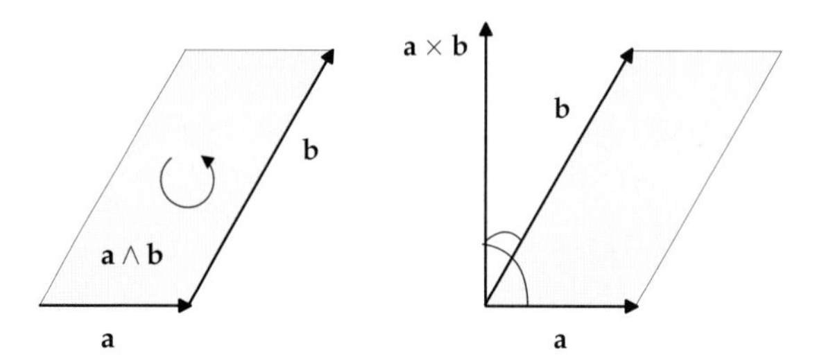

Figure 2.10 Visualization of the wedge and cross products in three dimensions. Whereas $\mathbf{a} \wedge \mathbf{b}$ "is" the oriented parallelogram spanned by the two vectors, the cross product yields a vector orthogonal to both $\mathbf{a}$ and $\mathbf{b}$ with its length "representing" the surface of the parallelogram.

such that the frame $\{\mathbf{g}_1, \mathbf{g}_2, \mathbf{f}\}$ is direct; see Figure 2.10. For a direct orthonormal basis[15] $\{\mathbf{e}_1, \mathbf{e}_2, \mathbf{e}_3\}$ it yields:

$$
\begin{aligned}
\mathbf{e}_1 \times \mathbf{e}_2 &= -\mathbf{e}_2 \times \mathbf{e}_1 = \mathbf{e}_3 \,, \\
\mathbf{e}_2 \times \mathbf{e}_3 &= -\mathbf{e}_3 \times \mathbf{e}_2 = \mathbf{e}_1 \,, \\
\mathbf{e}_3 \times \mathbf{e}_1 &= -\mathbf{e}_3 \times \mathbf{e}_1 = \mathbf{e}_2 \,.
\end{aligned}
\tag{2.67}
$$

The cross product between two vectors $\mathbf{a}$ and $\mathbf{b}$ can then be written as

$$
\mathbf{a} \times \mathbf{b} = (a^2b^3 - a^3b^2)\mathbf{e}_1 + (a^3b^1 - a^1b^3)\mathbf{e}_2 + (a^1b^2 - a^2b^1)\mathbf{e}_3. \tag{2.68}
$$

Cross and triple product $\mathbf{abc} := \mathbf{a} \cdot (\mathbf{b} \times \mathbf{c})$ can be written in the form of a determinant or with the Ricci notation as

$$
\mathbf{a} \times \mathbf{b} = \begin{vmatrix} \mathbf{e}_1 & \mathbf{e}_2 & \mathbf{e}_3 \\ a^1 & a^2 & a^3 \\ b^1 & b^2 & b^3 \end{vmatrix} = \epsilon^i_{\ jk} a^j b^k \mathbf{e}_i \,, \tag{2.69}
$$

$$
\mathbf{abc} = \begin{vmatrix} a^1 & a^2 & a^3 \\ b^1 & b^2 & b^3 \\ c^1 & c^2 & c^3 \end{vmatrix} = \epsilon_{ijk} a^i b^j c^k \,. \tag{2.70}
$$

If the three vectors are linearly dependent, the triple product is zero. The triple product is positive if $\{\mathbf{a}, \mathbf{b}, \mathbf{c}\}$ is positively oriented. It is in absolute value equal to the volume of a parallelepiped with sides $\mathbf{a}$, $\mathbf{b}$, and $\mathbf{c}$.

15 As explained, an orthonormal basis can always be obtained in inner-product spaces by means of the Erhard Schmidt orthogonalization procedure.

2.10 Identities of Vector Algebra

We shall summarize without proof some identities of vector algebra in three-dimensional, oriented Euclidean space:

$$(\lambda \mathbf{a}) \cdot \mathbf{b} = \mathbf{a} \cdot (\lambda \mathbf{b}) = \lambda \mathbf{a} \cdot \mathbf{b}, \tag{2.71}$$

$$(\lambda \mathbf{a}) \times \mathbf{b} = \mathbf{a} \times (\lambda \mathbf{b}) = \lambda \mathbf{a} \times \mathbf{b}, \tag{2.72}$$

$$\mathbf{a} \cdot (\mathbf{b} + \mathbf{c}) = \mathbf{a} \cdot \mathbf{b} + \mathbf{a} \cdot \mathbf{c}, \tag{2.73}$$

$$\mathbf{a} \times (\mathbf{b} + \mathbf{c}) = \mathbf{a} \times \mathbf{b} + \mathbf{a} \times \mathbf{c}, \tag{2.74}$$

$$(\mathbf{a} + \mathbf{b}) \times (\mathbf{c} + \mathbf{d}) = (\mathbf{a} \times \mathbf{c}) + (\mathbf{b} \times \mathbf{c}) + (\mathbf{a} \times \mathbf{d}) + (\mathbf{b} \times \mathbf{d}), \tag{2.75}$$

$$\mathbf{a} \cdot (\mathbf{b} \times \mathbf{c}) = \mathbf{b} \cdot (\mathbf{c} \times \mathbf{a}) = \mathbf{c} \cdot (\mathbf{a} \times \mathbf{b}), \tag{2.76}$$

$$(\mathbf{a} \times \mathbf{b}) \cdot (\mathbf{c} \times \mathbf{d}) = (\mathbf{a} \cdot \mathbf{c})(\mathbf{b} \cdot \mathbf{d}) - (\mathbf{a} \cdot \mathbf{d})(\mathbf{b} \cdot \mathbf{c}), \tag{2.77}$$

$$\mathbf{a} \times (\mathbf{b} \times \mathbf{c}) = \mathbf{b}(\mathbf{a} \cdot \mathbf{c}) - \mathbf{c}(\mathbf{a} \cdot \mathbf{b}), \tag{2.78}$$

$$\mathbf{x}(\mathbf{abc}) = \mathbf{a}(\mathbf{xbc}) + \mathbf{b}(\mathbf{axc}) + \mathbf{c}(\mathbf{abx}), \tag{2.79}$$

$$\mathbf{x}(\mathbf{abc}) = (\mathbf{a} \cdot \mathbf{x})(\mathbf{b} \times \mathbf{c}) + (\mathbf{b} \cdot \mathbf{x})(\mathbf{c} \times \mathbf{a}) + (\mathbf{c} \cdot \mathbf{x})(\mathbf{a} \times \mathbf{b}), \tag{2.80}$$

$$(\mathbf{abc})^2 = (\mathbf{a} \times \mathbf{b})(\mathbf{b} \times \mathbf{c})(\mathbf{c} \times \mathbf{a}). \tag{2.81}$$

Equation (2.78) serves for the calculation of the tangential and normal components of a vector to the boundary of a domain. With $\mathbf{n}$ denoting the surface normal vector, it follows that $\mathbf{n} \times (\mathbf{n} \times \mathbf{a}) = \mathbf{n}(\mathbf{n} \cdot \mathbf{a}) - \mathbf{a}(\mathbf{n} \cdot \mathbf{n})$ and therefore

$$\mathbf{a} = \mathbf{n}(\mathbf{n} \cdot \mathbf{a}) - \mathbf{n} \times (\mathbf{n} \times \mathbf{a}), \tag{2.82}$$

where the first term on the right-hand side gives the normal component and the second term the tangential component of the vector $\mathbf{a}$.

2.11 Vector Fields

The pair $\{\mathscr{P}, \mathbf{x}\}$ which consists of a point $\mathscr{P} \in \Omega \subset A$ and a vector from the associated vector space $\mathbf{x} \in V$ is called a *bound vector*.[16] Note that it makes no sense to add $\{\mathscr{P}_1, \mathbf{x}\}$ and $\{\mathscr{P}_2, \mathbf{y}\}$ unless $\mathscr{P}_1 = \mathscr{P}_2$, in which case

$$\lambda\{\mathscr{P}, \mathbf{x}\} = \{\mathscr{P}, \lambda \mathbf{x}\} \qquad \mathscr{P} \in \Omega, \lambda \in \mathbb{R}, \tag{2.83}$$

$$\{\mathscr{P}, \mathbf{x}\} + \{\mathscr{P}, \mathbf{y}\} = \{\mathscr{P}, \mathbf{x} + \mathbf{y}\} \qquad \mathscr{P} \in \Omega, \mathbf{x}, \mathbf{y} \in V. \tag{2.84}$$

16 Note that Burke [3] uses free for what we call bound in accordance with [2,8].

The bound vectors form a vector space at point $\mathscr{P}$, which is called the tangent space to Ω in $\mathscr{P}$, denoted $T_{\mathscr{P}}\Omega$. In other words, the tangent space is the set of all possible vectors located at $\mathscr{P}$.

In electromagnetics we study not only one single bound vector but *vector fields*, which assign to each point $\mathscr{P} \in \Omega \subset A$ an element of its tangent space. Different representations of such vector fields are shown in Figure 4.15. Of course, one cannot plot an icon at all points of the problem domain, but it should be understood this way. Indeed, all plots in Figure 4.15 represent the same physical field. The z-component of the magnetic vector potential for the 2D field problem can be displayed by means of a color representation as well by the classical field lines; both shown in Section 4.8.

A vector field is defined by the mapping

$$\mathbf{x} : \Omega \to \bigcup_{\mathscr{P} \in \Omega} T_{\mathscr{P}}\Omega : \mathscr{P} \mapsto \mathbf{x}(\mathscr{P}), \tag{2.85}$$

which assigns to each point of $\Omega \subset A$ one and only one bound vector $\mathbf{x}(\mathscr{P}) := \{\mathscr{P}, \mathbf{x}\}$ from its tangent space $T_{\mathscr{P}}\Omega$. This union of the tangent spaces is called the *tangent bundle*.

Electric fields represented by the gradient of a scalar field require A_3 with a scalar product of the associated vector space. Magnetic fields also require orientation. Thus the framework for classical electromagnetism is the Euclidean affine space, denoted E_3:

- E_3 is endowed with the structure of the affine point space A_3,
- carries the vector space structure of its associated vector space, and
- is equipped with a metric that gives rise to concepts such as distance and angles between points.

We define the space E_3 to be endowed with an orientation (necessary for all the operations involving cross products) in the sense of the right-handed screw. The constitutive equations also require a metric. As there is no need to address relativity, we will consider time as a universal parameter and thus limit ourselves to three dimensions. We will further consider only space elements embedded in the encompassing space E_3.

By selection of an origin $\mathscr{O} \in E_3$ each point $\mathscr{P} \in E_3$ is associated with a position vector $\mathbf{r}_{\mathscr{P}} = \mathbf{r}_{\mathscr{O}\mathscr{P}} \in V$ in direction from $\mathscr{O}$ to $\mathscr{P}$. The projection of the position vector onto the canonical basis[17] $\{\mathbf{e}_1, \mathbf{e}_2, \mathbf{e}_3\}$ yields the coordinate expression $\mathbf{r}_{\mathscr{P}} = (x_1, x_2, x_3)$. The notation of $\mathscr{P}$ in the subscript of $\mathbf{r}$ is usually dropped, a practice we will follow from here on.

17 Also denoted $\{\mathbf{e}_x, \mathbf{e}_y, \mathbf{e}_z\}$ or $\{\mathbf{i}, \mathbf{j}, \mathbf{k}\}$.

The canonical basis can be made to a globally defined *basis field* by means of a translation into the point $\mathscr{P}$. A vector field on $\Omega \subset E_3$ can then be written as

$$\mathbf{a} : \Omega \to \mathbb{R}^3 : \mathbf{r} \mapsto \mathbf{a}(\mathbf{r}) : \mathbf{a}(\mathbf{r}) = \left(a^1(\mathbf{r}), a^2(\mathbf{r}), a^3(\mathbf{r})\right), \tag{2.86}$$

where the $a^i(\mathbf{r})$ are the Cartesian coordinates of the vector field depending not only on the choice of basis, but also on the point $\mathscr{P}$. The coordinates must thus be interpreted as smooth component functions, $a^i \in C^\infty$. All smooth vector fields form an infinite-dimensional vector space, a *functional space*, discussed in Chapter 3.16. We denote it by

$$\mathcal{V}(\Omega) := C^\infty(\Omega, \mathbb{R}^3) \tag{2.87}$$

in accordance with [8]. But in addition to the operations declared on vector spaces, vector fields can also be multiplied by functions. Thus vector fields constitute a *module* over the ring of smooth functions.[18] A *scalar field* is a smooth mapping

$$\phi : \Omega \to \mathbb{R} : \mathbf{r} \mapsto \phi(\mathbf{r}), \tag{2.88}$$

and the vector space of the C^∞ functions ϕ is denoted

$$\mathcal{S}(\Omega) := C^\infty(\Omega, \mathbb{R}). \tag{2.89}$$

Different means of displaying a scalar field in two dimensions are shown in Figure 4.15. Note that we have not defined the smoothness of ϕ and the component functions of the vector field. Thus we must turn to (classical) vector analysis and some aspects of infinite-dimensional spaces in the next chapter.

2.12 Phase Portraits

There are vector fields on domains other than subspaces of Euclidean affine space. As an example we present the phase portraits of ordinary differential equations. Consider the nonlinear equation of motion of a mathematical pendulum,[19]

$$\varphi'' + \alpha\varphi' + \frac{g}{l}\sin\varphi = 0, \tag{2.90}$$

where φ is the angle from the perpendicular, α the damping term, g the gravitational constant, l the length of the pendulum, and where the prime indicates

18 A module is a generalization of the concept of vector space where the scalars may lie in an arbitrary ring instead of a field.

19 A mathematical pendulum is a point with mass m, connected with a massless bar of length l to a frictionless pivot point.

the time derivative. The position angle and angular velocity are combined in a column vector

$$\mathbf{x}(t) := (x_1(t), x_2(t))^T = (\varphi(t), \varphi'(t))^T. \tag{2.91}$$

The space of all these vectors is called the two-dimensional *phase space* P_2 with coordinates x_1 and x_2. Hence

$$\begin{aligned}\mathbf{x}'(t) = \mathbf{v}(t) &= \left(x_1'(t),\, x_2'(t)\right)^T = \left(\varphi'(t),\, \varphi''(t)\right)^T \\ &= \left(x_2(t),\, -\alpha x_2(t) - \frac{g}{l}\sin x_1(t)\right)^T.\end{aligned} \tag{2.92}$$

Thus $\mathbf{v}$ seen as a mapping

$$\mathbf{v} : P_2 \to \bigcup_{\mathbf{x}\in P_2} T_{\mathscr{P}} P_2 : \mathbf{x}(t) \mapsto \mathbf{v}(\mathbf{x}(t)), \tag{2.93}$$

with

$$\mathbf{v}(\mathbf{x}(t)) = (x_1'(t),\, x_2'(t))^T = (x_2(t),\, -\alpha x_2(t) - \frac{g}{l}\sin x_1(t))^T, \tag{2.94}$$

which assigns to every point $\mathbf{x}$ in P_2 the velocity vector $\mathbf{v}(\mathbf{x}(t)) \in T_{\mathscr{P}}P_2$, is a vector field on the phase space. The vector field is displayed for the damped and the undamped mathematical pendulum in Figure 2.11.

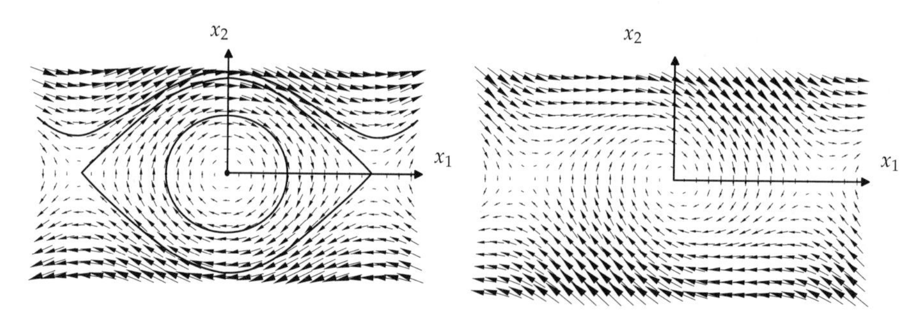

Figure 2.11 Left: Vector field and integrator lines on the phase space of the differential equation for the undamped mathematical pendulum. Right: Vector field for the damped mathematical pendulum.

Given a vector field $\mathbf{v}$ on P_2, the dynamical system $\mathbf{x}'(t) = \mathbf{v}(\mathbf{x}(t))$, written in the component form as

$$\begin{aligned}x_1'(t) &= \mathbf{v}_1(x_1(t),\, x_2(t)) \\ x_2'(t) &= \mathbf{v}_2(x_1(t),\, x_2(t)),\end{aligned} \tag{2.95}$$

is called an autonomous differential equation system[20] of first order for the unknown functions $x_1(t)$, $x_2(t)$.

We are concerned with the dynamics of a system in the vector field of phase space velocities. To this end we study *flow maps*

$$\phi(t,\mathbf{x}) \,:\, P_2 \to P_2 \,:\, \mathbf{x}(0) \mapsto \mathbf{x}(t)\,, \tag{2.96}$$

with the group properties

$$\phi(0,\mathbf{x}) = \mathbf{x}, \tag{2.97}$$

$$\phi(t+s,\mathbf{x}) = \phi(s,\phi(t,\mathbf{x})), \tag{2.98}$$

for $t,s \in \mathbb{R}$, $\mathbf{x} \in P_2$. The point $\mathbf{x}$ is called a *periodic* point of ϕ if $\mathbf{x}(T) = \mathbf{x}(0)$ with period $T > 0$. The *integrator line* or *flux line* of the vector field is defined by the space curve $\gamma_{\mathbf{x}} \,:\, \mathbb{R} \to P_2 \,:\, t \mapsto \phi(t,\mathbf{x})$ that satisfies $\gamma'_{\mathbf{x}}(t) = \mathbf{v}(\mathbf{x}(t))$. The graph of this space curve is called the *orbit* of $\mathbf{x}$. We distinguish fixed, periodic (closed), and nonclosed orbits; examples are shown Figure 2.11 (left). The flux line is the solution of the dynamical system for the initial value $\mathbf{x}$. The flux axioms (2.97)–(2.98) state that the orbit launched from a particular point in phase space is unique. The flow map of the vector field $\mathbf{v}$ on P_2, and sometimes the vector field itself, is called the *phase portrait* of the differential equation $x'' = f(x,x')$; see Figure 2.11.

Given a vector field $\mathbf{v}$ on P_2, the dynamical system $\mathbf{x}'(t) = \mathbf{v}(\mathbf{x}(t))$ written in component form as

$$\begin{aligned} x_1'(t) &= a_{11}x_1(t) + a_{12}x_2(t) \\ x_2'(t) &= a_{21}x_1(t) + a_{22}x_2(t)\,, \end{aligned} \tag{2.99}$$

is called a system of two linear homogeneous differential equations with constant coefficients. The linear approximation of Eq. (2.90), with $\sin\varphi \approx \varphi$ for small angles, yields

$$\begin{aligned} x_1'(t) &= x_2(t) \\ x_2'(t) &= -\frac{g}{l}x_1(t) - \alpha x_2(t)\,, \end{aligned} \tag{2.100}$$

which can be written in the matrix form as

$$\begin{pmatrix} x_1'(t) \\ x_2'(t) \end{pmatrix} = \begin{pmatrix} 0 & 1 \\ -\frac{g}{l} & -\alpha \end{pmatrix} \begin{pmatrix} x_1(t) \\ x_2(t) \end{pmatrix}. \tag{2.101}$$

20 The system is autonomous if it does not depend on the independent variable t, i.e., $\mathbf{v} = \mathbf{v}(\mathbf{x}(t))$ and not $\mathbf{v} = \mathbf{v}(\mathbf{x}(t),t)$.

The trace of the matrix, which is the sum of its diagonal elements, is $-\alpha$. If there is no damping, $\alpha = 0$, the area in phase space encompassed by the orbit is preserved. With damping ($\alpha > 0$) the phase space area will be reduced.

Note that the concepts presented in this section are not limited to two dimensions.

2.13
The Physical Dimension System

Before we leave this chapter, we must address the third aspect of a physical vector defined as a quantity having direction as well as magnitude [12], that is, the physical dimension of a vector field.

Mathematical (algebraic) objects that represent physical quantities must be equipped with a structure that represents their physical dimension, such as length or mass. The physical dimensions can be derived from the basis dimensions mass, length, and time, and form the basis $\{\mathsf{M}, \mathsf{L}, \mathsf{T}\}$ of the dimension[21] system $\mathcal{D}$.

Clearly, the physical dimensions are independent of the units used to represent them, which are defined by the SI standard [6]. The physical quantity of velocity may be expressed in meters per second or kilometers per hour, but its dimension is always length divided by time, LT^{-1}. Some quantities are said to be dimensionless, for example, the scalar arguments to trigonometric functions, exponential functions, and logarithms are of the identity dimension 1_{D}. This follows from the fact that the Taylor series expansion of these functions must be dimensionally homogeneous, that is, the square of the argument must be of the same dimension as the argument itself. Dimensionless quantities may have units, however. The angles, for example, may be measured in units of radians or degrees.

Remark: We comply with the ISO [15] standards, expressing a physical quantity as $G = \{G\}[G]$ where $\{G\}$ denotes the numerical value of the quantity G expressed in the physical unit $[G]$ that is appropriate for the measurement of that quantity. As an example we refer to Eq. (1.5). We write $B = 5\ T$ and $[B] = 1$ T, and **not** $B\,[\mathrm{T}] = \ldots$. It is important not to use italics for the unit symbols in order to avoid confusion with the physical quantities (T could denote torque or the thermodynamic temperature, for example). The factor 1 indicates that a basis unit within the coherent unit system is used. The physical quantity must obviously be invariant to the change of unit $[\mathrm{G}]_{\mathrm{old}} = a[G]_{\mathrm{new}}$. Hence $G = \{G\}_{\mathrm{old}}[G]_{\mathrm{old}} = a\{G\}_{\mathrm{old}}[G]_{\mathrm{new}} = \{G\}_{\mathrm{new}}[G]_{\mathrm{new}}$ from which follows $\{G\}_{\mathrm{old}} = 1/a\,\{G\}_{\mathrm{new}}$. □

21 This of course has nothing to do with the algebraic dimension of the vector space as introduced in Section 2.3.1.

Quantities of different dimensions do not form an algebraic field, as they are not closed under addition; adding 1 m and 1 kg does not make sense.[22] However, algebraic vectors as introduced in this chapter are isomorphic to tuples of real numbers, which are indeed closed under addition.

A *dimension system* $\mathcal{D}$ has the structure of a free, multiplicative module over a ring R with the neutral dimension denoted 1_D. A multiplicative module is called *free* if from $\mathsf{D}_1^{q_1}\mathsf{D}_2^{q_2}\cdots\mathsf{D}_r^{q_r} = 1_\mathsf{D}$ it always follows that $q_1 = q_2 = \cdots = q_r = 0$ for all $q_i \in R$. For our purpose R will be $\mathbb{Z}$ or $\mathbb{Q}$, in which case the module is an Abelian group or a finite-dimensional vector space, respectively. We require that

$$\{\mathsf{D}_1^{q_1}\mathsf{D}_2^{q_2}\cdots\mathsf{D}_r^{q_r} \mid q_i \in R\} = \mathcal{D}, \tag{2.102}$$

where r is the rank of the module. A free module has therefore at least one basis $\mathcal{B}$. The set $\{\mathsf{D}_1, \mathsf{D}_2, \ldots, \mathsf{D}_r\} \subset \mathcal{D}$ is the basis of the dimension system $\mathcal{D}$ and every dimension can be expressed as a unique product of basis dimensions raised to powers of q_i, that is,

$$\mathsf{D} = \mathsf{D}_1^{q_1}\mathsf{D}_2^{q_2}\cdots\mathsf{D}_r^{q_r}. \tag{2.103}$$

The axioms of the module read

$$(\mathsf{D}^{q_1})^{q_2} = \mathsf{D}^{q_1 q_2}, \tag{2.104}$$

$$(\mathsf{D}_1\mathsf{D}_2)^{q} = \mathsf{D}_1^{q}\mathsf{D}_2^{q}, \tag{2.105}$$

$$\mathsf{D}^{q_1+q_2} = \mathsf{D}^{q_1}\mathsf{D}^{q_2}, \tag{2.106}$$

where $\mathsf{D}, \mathsf{D}_1, \mathsf{D}_2 \in \mathcal{D}$ and $q, q_1, q_2 \in R$.

Remark: The need for fractional powers of the basis dimensions might not be obvious at first glance. Consider Coulomb's law and Ampère's force law for two parallel wires [7]

$$F_1 = k_1 \frac{Q_1 Q_2}{r^2} \quad \text{and} \quad F_2 = 2k_2 I_1 I_2 \frac{l}{d}, \tag{2.107}$$

where r is the distance between the charges Q_1 and Q_2, d the distance between the wires and l the length of the wires. The choice of the proportionality constants k_1 and k_2 is arbitrary, but k_1/k_2 is the square of the speed of light in vacuum.

In SI units the electric current is chosen as a fourth basis dimension. The charge has dimension IT and k_2 has dimension $\mathsf{MLI}^{-2}\mathsf{T}^{-2}$. If k_2 is taken dimensionless, as in the *so-called electromagnetic (emu) system*, then it results in the current having dimension $\mathsf{M}^{\frac{1}{2}}\mathsf{L}^{\frac{1}{2}}\mathsf{T}^{-1}$. □

22 In the physical science we may not add "apples and oranges", but we may multiply them.

Dimension mappings A dimension mapping assigns a physical dimension to a set $\mathcal{S}$ of mathematical objects:

$$pd : \mathcal{S} \to \mathcal{D} : x \mapsto pd(x). \tag{2.108}$$

As examples we have for the magnetic flux density $pd(\mathbf{B}) = \mathsf{UTL}^{-2}$, for the electric flux density $pd(\mathbf{D}) = \mathsf{ITL}^{-2}$, for the magnetic field $pd(\mathbf{H}) = \mathsf{IL}^{-1}$, and for the permeability $pd(\mu_0) = \mathsf{UTI}^{-1}\mathsf{L}^{-1}$. Note that the dimension mapping is required to be neither one-to-one nor onto. A set $\mathcal{S}$ with a constant dimension mapping is called *dimensionally uniform*; each element from this set having the same physical dimension.

An *additive dimension group*, denoted $(\mathcal{S}, +, \mathcal{D}, pd)$, consists of an Abelian group $(\mathcal{S}, +)$, a mapping

$$+ : \mathcal{S} \times \mathcal{S} \to \mathcal{S} : \quad (x, y) \mapsto x + y, \tag{2.109}$$

a dimension system $\mathcal{D}$, and a dimension mapping pd that is dimensionally uniform. This would be an appropriate structure for dealing with additions in the same (tangent) vector spaces, for example, for adding the reduced field $\mathbf{B}_\mathrm{r}$ and the source field $\mathbf{B}_\mathrm{s}$ in a field point $\mathscr{P}$ (with position vector $\mathbf{r}$): $\mathbf{B}(\mathbf{r}) = \mathbf{B}_\mathrm{r}(\mathbf{r}) + \mathbf{B}_\mathrm{s}(\mathbf{r})$.

A *multiplicative dimension group*, denoted $(\mathcal{S}, *, \mathcal{D}, pd)$, consists of a group $(\mathcal{S}, *)$, a mapping

$$* : \mathcal{S} \times \mathcal{S} \to \mathcal{S} : (x, y) \mapsto x * y, \tag{2.110}$$

a dimension system $\mathcal{D}$, and a dimension mapping pd with the property

$$pd(x * y) = pd(x) * pd(y) \tag{2.111}$$

for all $x, y \in \mathcal{S}$. This is an appropriate structure for, e.g., the integral of a vector field along a path. A *dimension ring*, denoted $(\mathcal{S}, +, *, \mathcal{D}, pd)$, consists of the set $\mathcal{S}$, with the two mappings $+$ and $*$, a dimension system $\mathcal{D}$, and the dimension mapping that is necessarily dimensionally neutral:

$$pd(x) = 1_\mathsf{D}, \tag{2.112}$$

for all $x \in \mathcal{S}$, because of properties (2.109) and (2.111). In other words, every ring element necessarily has the physical dimension 1_D. A *linear dimension space* consists of the additive dimension group and the operation

$$\mathbb{F} \times \mathcal{S} \to \mathcal{S} : (\lambda, x) \mapsto \lambda x \tag{2.113}$$

($\mathbb{F} = \mathbb{R}$ or $\mathbb{C}$) for which the axioms 5–8 of linear spaces are fulfilled; see Section 2.3. Due to the properties of the additive dimension group, the linear dimension space is dimensionally uniform. In other words, the elements of the linear spaces that we study in this chapter are necessarily of the same physical dimension. The linear structure allows us to attribute the physical dimensions to the basis vectors. If all computations were limited to addition and scalar multiplication of dimensionally uniform, physical vectors, then the linear space approach would be adequate.

However, there are vectors that are not dimensionally uniform, for example, voltage/current tuples in network analysis or vectors in the phase space of mechanical systems. The magnitude and the dot-product operations are not defined for these physical vectors (the scalar product in linear algebra maps to reals, while the scalar product of physical vectors maps to dimensioned scalars). In these cases we may drop the physical dimension and place only their related values as elements in vectors and matrices. Less trivial approaches, which exceed the scope of this book, can be found in [5] and [16].

References

1 Blaschke, W.: Einführung in die Differentialgeometrie, Springer, Berlin 1960

2 Bossavit, A.: On the geometry of electromagnetism, Journal of the Japan Society of Applied Electromagnetism & Mechanics, 6, 1998

3 Burke, W. L.: Applied Differential Geometry, Cambridge University Press, Cambridge, 1985

4 Gallier, J.: Geometric Methods and Applications for Computer Science and Engineering, Springer, New York, 2001

5 Hart, G.W.: Multidimensional Analysis: Algebras and Systems for Science and Engineering, Springer, Berlin, 1998

6 ISO Standards Handbook 2, Units of Measurement, 1982

7 Jackson, J.D.: Classical Electrodynamic, John Wiley & Sons, New York, 1997

8 Jänich, K. Vectoranalysis, Springer, Berlin, 2001

9 Jänich, K. Topologie, Springer, Berlin, 2001

10 Jänich, K. Lineare Algebra , Springer, Berlin, 2001

11 Kay, D.C.: Tensor Calculus, Schaum's Outlines Series, McGraw-Hill, New York, 1988

12 Lipschutz, S., Lipson, M.: Linear Algebra, Schaum's Outline Series, McGraw-Hill, New York, 2001

13 Penfield, P., Haus, H.: Electrodynamics of Moving Media, MIT Press, Cambridge, MA, 1967

14 Prechtl, A.: Physicalische Dimensionen mathematischer Objekte, Colloquium, TU–Vienna, 2001

15 Tonti, E.: On the Geometrical Structure of the Electromagnetism, Pitagora Editrice, Bologna, 1995

16 Whitney, H.: The Mathematics of Physical Quantities, Part 1 and 2, American Mathematical Monthly, 1968

17 Weyl, H.: Space, Time, Matter, Dover, New York, 1922

3
Classical Vector Analysis

Walk one mile east, then north, then west, then south.
Have you really returned?

T. Frankel, The Geometry of Physics.

The foundations of the analytical and numerical field computation methods presented in this book, are Maxwell's equations in their local form, based on classical vector analysis with its differential operators grad, div, and curl. These operators act on smooth scalar and vector fields "living" on connected subsets of the Euclidean affine space discussed in Chapter 2.

In this chapter we begin with the theory of space curves and their local coordinate frames. Our specific applications are magnets built of tilted helices, coil ends of superconducting magnets, and the transverse motion of charged particles in circular accelerators.

After studying the directional derivative, the differential as the best linear approximation for the mapping between space elements, and the classical integral theorems, we will discuss under which conditions a curl-free or a divergence-free vector field can be expressed as the gradient of a scalar potential and the curl of a vector potential, respectively. This leads to the lemmata of Poincaré[1] and some aspects of de Rham cohomology.[2]

In classical vector analysis the coordinates on the space elements (points, curves, surfaces, and volumes denoted $\mathscr{P}$, $\mathscr{S}$, $\mathscr{A}$, $\mathscr{V}$, respectively) are given by the mapping ϕ^{-1}, as explained in Section 2.5.1. In the case of a surface the mapping is defined by the three coordinate functions $x(u,v)$, $y(u,v)$, $z(u,v)$, where u, v are the local coordinates. The space metric is assumed to be the Euclidean metric, and the coordinates are denoted (x,y,z) or (u_1,u_2,u_3) instead of (x^1,x^2,x^3). Only in Section 3.11 we will return to the Ricci notation when we present the operators of classical vector analysis in orthogonal curvilinear coordinates.

1 Jules Henri Poincaré (1854–1912).
2 Georges de Rham (1903–1990).

Field Computation for Accelerator Magnets. Stephan Russenschuck

ISBN: 978-3-527-40769-9

Finally, we will treat the convergence behavior of Fourier series in the framework of the Hilbert spaces l^2 and L^2.

3.1 Space Curves

If the position vector[3] $\mathbf{r}$ is given as a smooth function of the real parameter t,

$$\mathbf{r}(t) = x(t)\mathbf{e}_x + y(t)\mathbf{e}_y + z(t)\mathbf{e}_z\,, \tag{3.1}$$

and if t changes in some interval $I = [a, b] \subset \mathbb{R}$, the locus of $\mathbf{r}$ traces a *space curve* written in the mapping notation as

$$\mathscr{S} : I \to E_3 : t \mapsto \mathbf{r}(t)\,. \tag{3.2}$$

The curve is inner oriented by the direction it traces from a to b; see Figure 3.1. A curve is said to be *closed* if $\mathbf{r}(a) = \mathbf{r}(b)$. A closed curve is called a *loop*. A loop is *simple* if it has no points of self-intersection,

$$\left(x(t_1) - x(t_2)\right)^2 + \left(y(t_1) - y(t_2)\right)^2 + \left(z(t_1) - z(t_2)\right)^2 \neq 0 \tag{3.3}$$

for $t_1 - t_2 \neq 0$ and $t_1, t_2 \in I$. Simple loops are referred to as *Jordan curves*.[4] The derivative $\mathrm{d}\mathbf{r}(t)/\mathrm{d}t$ at $\mathbf{r}(t)$,

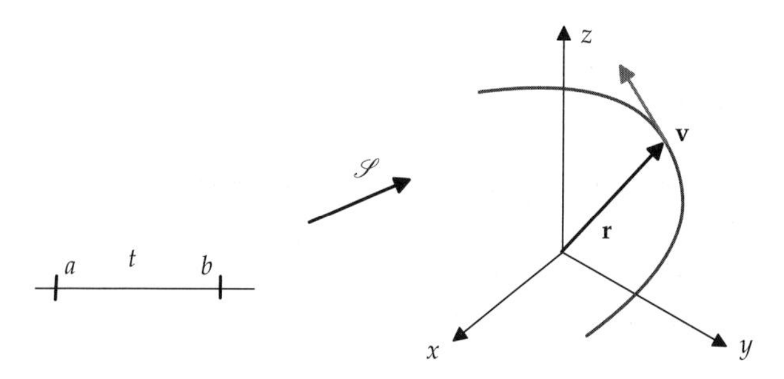

Figure 3.1 Space curve as a mapping $\mathscr{S} : I \to E_3 : t \mapsto \mathbf{r}(t)$.

$$\frac{\mathrm{d}\mathbf{r}(t)}{\mathrm{d}t} = \lim_{\Delta t \to 0} \frac{\Delta \mathbf{r}}{\Delta t} = \lim_{\Delta t \to 0} \frac{\mathbf{r}(t + \Delta t) - \mathbf{r}(t)}{\Delta t} \tag{3.4}$$

is used to define the tangent to the space curve at $\mathbf{r}(t)$. In component form it yields

$$\frac{\mathrm{d}\mathbf{r}(t)}{\mathrm{d}t} = \frac{\mathrm{d}x}{\mathrm{d}t}\mathbf{e}_x + \frac{\mathrm{d}y}{\mathrm{d}t}\mathbf{e}_y + \frac{\mathrm{d}z}{\mathrm{d}t}\mathbf{e}_z\,. \tag{3.5}$$

3 This is the vector $\mathbf{r}_{\mathscr{O}\mathscr{P}}$ but we will drop the subscript in the notation from now on.

4 Camille Jordan (1838–1922).

If we regard the parameter t as the time, the tangent vector has the physical meaning of the velocity $\mathbf{v}(t) = \mathrm{d}\mathbf{r}(t)/\mathrm{d}t$. The acceleration is

$$\frac{\mathrm{d}\mathbf{v}(t)}{\mathrm{d}t} = \mathbf{a}(t) = \frac{\mathrm{d}^2 x}{\mathrm{d}t^2}\mathbf{e}_x + \frac{\mathrm{d}^2 y}{\mathrm{d}t^2}\mathbf{e}_y + \frac{\mathrm{d}^2 z}{\mathrm{d}t^2}\mathbf{e}_z \,. \tag{3.6}$$

The simple expression (3.5) holds only for Cartesian coordinates and consequently for basis vectors, which are independent of the position. Take the example of a space curve expressed in cylindrical coordinates (r, φ, z):

$$\frac{\mathrm{d}\mathbf{r}(t)}{\mathrm{d}t} = \mathbf{v}(t) = \frac{\mathrm{d}}{\mathrm{d}t}(r\mathbf{e}_r + z\mathbf{e}_z) = \frac{\mathrm{d}r}{\mathrm{d}t}\mathbf{e}_r + r\frac{\mathrm{d}\mathbf{e}_r}{\mathrm{d}t} + \frac{\mathrm{d}z}{\mathrm{d}t}\mathbf{e}_z + z\frac{\mathrm{d}\mathbf{e}_z}{\mathrm{d}t}\,, \tag{3.7}$$

where $\mathrm{d}\mathbf{e}_r/\mathrm{d}t = (\mathrm{d}\varphi/\mathrm{d}t)\mathbf{e}_\varphi$. A proof can be found in Section 3.13.2. The derivative of the unit vector is orthogonal to it. Because of $\mathrm{d}\mathbf{e}_z/\mathrm{d}t = 0$ we obtain

$$\mathbf{v}(t) = \frac{\mathrm{d}r}{\mathrm{d}t}\mathbf{e}_r + r\frac{\mathrm{d}\varphi}{\mathrm{d}t}\mathbf{e}_\varphi + \frac{\mathrm{d}z}{\mathrm{d}t}\mathbf{e}_z \,. \tag{3.8}$$

The *tangent vector* $\mathbf{T}(t)$ is defined as the unit vector in the direction of $\mathbf{v}(t)$. With the convention[5] $v(t) := |\,\mathbf{v}(t)\,|$ it yields

$$\mathbf{T}(t) := \frac{\mathbf{v}(t)}{v(t)} = T_x(t)\mathbf{e}_x + T_y(t)\mathbf{e}_y + T_z(t)\mathbf{e}_z \,. \tag{3.9}$$

A curve is said to be *regular* if it is differentiable and if its velocity is nonzero at all points of the interval I. Clearly, if all functions x, y, z are constants, the curve would degenerate into a single point. If $v(t) = 1$ the curve is said to be of velocity 1.

A curve is oriented by means of the parametric representation of the curve, with t increasing from a to b in the interval I. A regular oriented curve is said to be an equivalence class of parametric representations where any two parameters, for example, t and s, are related by an allowable[6] change of parameter $s = s(t)$ such that $\mathrm{d}s/\mathrm{d}t > 0$.

The direction of the tangent vector is independent of the parametric representation. An oriented space curve traced from t_0 to t can be expressed as a function of its arc length:

$$s(t) : [t_0, t] \to \mathbb{R} : t \mapsto \int_{t_0}^{t} |\mathbf{v}(t)|\,\mathrm{d}t \,. \tag{3.10}$$

By the fundamental theorem of calculus we obtain

$$\frac{\mathrm{d}s(t)}{\mathrm{d}t} = \frac{\mathrm{d}}{\mathrm{d}t}\int_{t_0}^{t} |\mathbf{v}(t)|\,\mathrm{d}t = \left|\frac{\mathrm{d}\mathbf{r}(t)}{\mathrm{d}t}\right|\,, \tag{3.11}$$

5 We use henceforth the symbol $|\cdot|$ for the norm induced by the Euclidean scalar product.

6 $s = s(t)$ must be at least 1-smooth with $\mathrm{d}s/\mathrm{d}t \neq 0$.

and by the differentiation rule of composite functions,

$$\frac{df(s(t))}{dt} = \frac{df}{ds}\frac{ds}{dt} = v\frac{df}{ds} . \tag{3.12}$$

For $v \neq 0$ this yields $\frac{d}{ds} = \frac{1}{v}\frac{d}{dt}$. For the x-component of $\mathbf{T}$ we obtain $T_x(s) = \frac{1}{v}\frac{dx}{dt} = \frac{dx}{ds}$. With similar expressions for the y- and z-components it yields in vector notation:

$$\mathbf{T}(s) = \frac{1}{v}\frac{d\mathbf{r}(t)}{dt} = \frac{d\mathbf{r}(s)}{ds} . \tag{3.13}$$

Thus a curve with velocity 1 is said to be represented in terms of arc length.

If $\mathbf{a}$ and $\mathbf{b}$ are smooth vector-valued functions of a scalar t ($\mathbf{a}, \mathbf{b} : \mathbb{R} \rightarrow V$), and if ϕ is a smooth scalar function of t ($\phi : \mathbb{R} \rightarrow \mathbb{R}$), then

$$\frac{d}{dt}(\mathbf{a}+\mathbf{b}) = \frac{d\mathbf{a}}{dt} + \frac{d\mathbf{b}}{dt} , \tag{3.14}$$

$$\frac{d}{dt}(\mathbf{a}\cdot\mathbf{b}) = \mathbf{a}\cdot\frac{d\mathbf{b}}{dt} + \frac{d\mathbf{a}}{dt}\cdot\mathbf{b} , \tag{3.15}$$

$$\frac{d}{dt}(\mathbf{a}\times\mathbf{b}) = \mathbf{a}\times\frac{d\mathbf{b}}{dt} + \frac{d\mathbf{a}}{dt}\times\mathbf{b} , \tag{3.16}$$

$$\frac{d}{dt}(\phi\mathbf{a}) = \phi\frac{d\mathbf{a}}{dt} + \frac{d\phi}{dt}\mathbf{a} . \tag{3.17}$$

3.1.1 The Frenet Frame of Space Curves

Consider a smooth curve of velocity 1, represented in terms of arc length s, as a mapping $\mathscr{S} : I \rightarrow E_3 : s \mapsto \mathbf{r}(s)$ on the open interval $I = (a, b) \subset \mathbb{R}$. The s-dependence of the curvature parameters and the vectors will be omitted in the notation. Moreover, we will use the prime notation for the s-derivative, $\mathbf{r}' \equiv d\mathbf{r}/ds$. The normalized s-derivative of the tangent vector to the space curve is called the *principle normal vector* $\mathbf{N}$:

$$\mathbf{T}' = \mathbf{r}'' = |\mathbf{T}'|\ \mathbf{N} =: \kappa\mathbf{N} \tag{3.18}$$

with

$$\mathbf{N} = \frac{\mathbf{r}''}{\kappa} = \frac{1}{\kappa}\left(x''\mathbf{e}_x + y''\mathbf{e}_y + z''\mathbf{e}_z\right) , \tag{3.19}$$

where κ is called the curvature and $\rho = 1/\kappa$ is the radius of curvature. The curvature is a measure for the turning of the tangent vector $\mathbf{T}$ around the normal vector as it moves along the curve by ds. The direction of $\mathbf{T}'$ at each point on the curve is normal to the curve at that point. This can be proved as follows: Since $\mathbf{T}$ has a constant magnitude, $\mathbf{T}\cdot\mathbf{T}$ = constant. Hence

$$\frac{\mathrm{d}}{\mathrm{d}s}(\mathbf{T}\cdot\mathbf{T}) = \mathbf{T}\cdot\mathbf{T}' + \mathbf{T}'\cdot\mathbf{T} = 2\,\mathbf{T}\cdot\mathbf{T}' = 0. \tag{3.20}$$

The plane spanned by the tangent vector and the principle normal vector ($\mathbf{T}$ and $\mathbf{N}$) to the curve at point $\mathscr{P}$ is called the *osculating plane*. The normal vector $\mathbf{N}$ may be assigned to either of the two opposite directions along the principle normal. In general, the curvature is regarded as positive if $\mathbf{N}$ is directed from the point $\mathscr{P}$ on the curve into the concave side of this curve. The vector

$$\mathbf{B} = \mathbf{T}\times\mathbf{N} \tag{3.21}$$

is called the *binormal vector*; it is orthogonal to both the tangent and the principle normal vectors. The plane perpendicular to the tangent plane (spanned by $\mathbf{N}$ and $\mathbf{B}$) is called the *normal plane*. The *rectifying plane* is the plane through $\mathscr{P}$ spanned by $\mathbf{B}$ and $\mathbf{T}$. The vectors form a right-handed, orthonormal frame $\{\mathbf{T},\mathbf{N},\mathbf{B}\}$, called the *triad* or *Frenet frame*.[7] The triad moves with the point on the space curve; see Figure 3.2. To this end, consider the relations

$$\mathbf{B}' = (\mathbf{T}\times\mathbf{N})' = \mathbf{T}'\times\mathbf{N} + \mathbf{T}\times\mathbf{N}' = \kappa\mathbf{N}\times\mathbf{N} + \mathbf{T}\times\mathbf{N}' = \mathbf{T}\times\mathbf{N}'. \tag{3.22}$$

As $\mathbf{B}$ is a unit vector, $\mathbf{B}'$ is orthogonal to $\mathbf{B}$ and lies in the surface spanned by $\mathbf{T}$ and $\mathbf{N}$. The s-derivative of the binormal $\mathbf{B}'$ is also orthogonal to $\mathbf{T}$, see Eq. (3.22), and therefore it holds:

$$\mathbf{B}' = -\tau\mathbf{N}. \tag{3.23}$$

7 Jean Frenet (1816–1900).

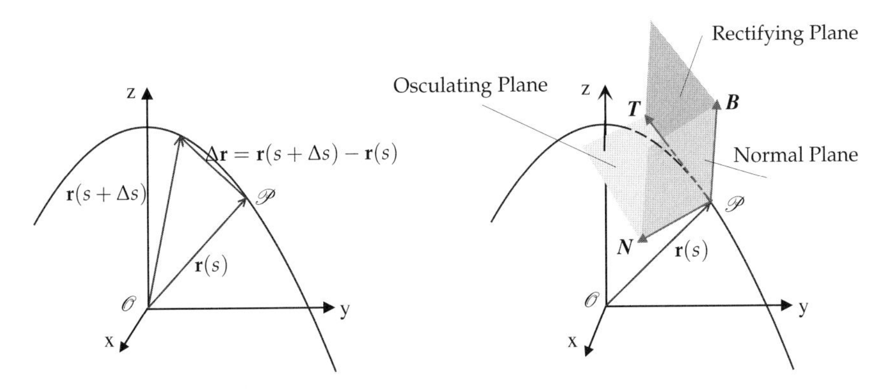

Figure 3.2 Space curve in Cartesian coordinates and the Frenet frame $\mathbf{T},\mathbf{N},\mathbf{B}$.

The measure of the binormal vector's turning arc rate τ is called the *torsion*. Furthermore,

$$\mathbf{N}' = (\mathbf{B} \times \mathbf{T})' = \mathbf{B}' \times \mathbf{T} + \mathbf{B} \times \mathbf{T}' = -\tau\,\mathbf{N} \times \mathbf{T} + \kappa\,\mathbf{B} \times \mathbf{N} = \tau\,\mathbf{B} - \kappa\,\mathbf{T}\,. \quad (3.24)$$

The relations between the Frenet frame at $s + \mathrm{d}s$ and s are thus given by the *Frenet–Serret equations*:[8]

$$\mathbf{T}' = \kappa\,\mathbf{N}\,, \qquad \mathbf{N}' = \tau\,\mathbf{B} - \kappa\,\mathbf{T}\,, \qquad \mathbf{B}' = -\tau\,\mathbf{N}\,, \quad (3.25)$$

where the coefficients form a skew-symmetric rotational matrix. This can be seen in the matrix notation

$$\begin{pmatrix} \mathbf{T}' \\ \mathbf{N}' \\ \mathbf{B}' \end{pmatrix} = \begin{pmatrix} 0 & \kappa & 0 \\ -\kappa & 0 & \tau \\ 0 & -\tau & 0 \end{pmatrix} \begin{pmatrix} \mathbf{T} \\ \mathbf{N} \\ \mathbf{B} \end{pmatrix}\,. \quad (3.26)$$

The *Darboux vector*[9] $\mathbf{D}$ at some point of the curve $\mathbf{r}$ with velocity 1 is given by

$$\mathbf{D} = \tau\,\mathbf{T} + \kappa\,\mathbf{B}, \quad (3.27)$$

which is the axis of rotation of the triad as described by Eq. (3.26). For a given Darboux vector, the Frenet–Serret equations can be derived from

$$\mathbf{T}' = \mathbf{D} \times \mathbf{T}\,, \qquad \mathbf{N}' = \mathbf{D} \times \mathbf{N}\,, \qquad \mathbf{B}' = \mathbf{D} \times \mathbf{B}\,. \quad (3.28)$$

The curvature, torsion, and Frenet frame for a space curve with the general parameter t are given with the notation $\dot{\mathbf{a}}(t) := \mathrm{d}\mathbf{a}(t)/\mathrm{d}t$ by

$$\mathbf{T} = \frac{\mathbf{v}(t)}{|\mathbf{v}(t)|}\,, \qquad \mathbf{N} = \mathbf{B} \times \mathbf{T}\,, \qquad \mathbf{B} = \frac{\mathbf{v}(t) \times \mathbf{a}(t)}{|\mathbf{v}(t) \times \mathbf{a}(t)|}\,, \quad (3.29)$$

and

$$\kappa = \frac{|\mathbf{v}(t) \times \mathbf{a}(t)|}{|\mathbf{v}(t)|^3}\,, \qquad \tau = \frac{(\mathbf{v}(t) \times \mathbf{a}(t)) \cdot \dot{\mathbf{a}}(t)}{|\mathbf{v}(t) \times \mathbf{a}(t)|^2}\,. \quad (3.30)$$

Proof. We show only the proof for the curvature κ. From Eq. (3.10) it follows $\dot{s} = |\dot{\mathbf{r}}|$. Hence

$$\ddot{s} = \frac{\mathrm{d}}{\mathrm{d}t}|\dot{\mathbf{r}}| = \frac{\mathrm{d}}{\mathrm{d}t}\sqrt{\dot{\mathbf{r}} \cdot \dot{\mathbf{r}}} = \frac{2\dot{\mathbf{r}} \cdot \ddot{\mathbf{r}}}{2\sqrt{\dot{\mathbf{r}} \cdot \dot{\mathbf{r}}}} = \frac{\dot{\mathbf{r}} \cdot \ddot{\mathbf{r}}}{|\dot{\mathbf{r}}|}\,, \quad (3.31)$$

and therefore

$$\mathbf{T}' = \frac{\mathrm{d}\mathbf{T}}{\mathrm{d}t}\frac{\mathrm{d}t}{\mathrm{d}s} = \dot{\mathbf{T}}\frac{1}{|\dot{\mathbf{r}}|} = \frac{\ddot{\mathbf{r}}|\dot{\mathbf{r}}| - \frac{\dot{\mathbf{r}} \cdot \ddot{\mathbf{r}}}{|\dot{\mathbf{r}}|}\dot{\mathbf{r}}}{|\dot{\mathbf{r}}|^2}\frac{1}{|\dot{\mathbf{r}}|} = \frac{|\dot{\mathbf{r}}|^2\ddot{\mathbf{r}} - (\dot{\mathbf{r}} \cdot \ddot{\mathbf{r}})\dot{\mathbf{r}}}{|\dot{\mathbf{r}}|^4}\,. \quad (3.32)$$

8 Joseph Serret (1819–1885).
9 Gaston Darboux (1842–1917).

The curvature can now be calculated simply with $\kappa = |\mathbf{T}'|$. The easier expression (3.30) is obtained from $|\mathbf{T} \times \mathbf{T}'| = |\mathbf{T}||\mathbf{T}'| = \kappa$ with

$$\kappa = |\mathbf{T} \times \mathbf{T}'| = \left| \frac{\dot{\mathbf{r}}}{|\dot{\mathbf{r}}|} \times \frac{|\dot{\mathbf{r}}|^2 \ddot{\mathbf{r}} - (\dot{\mathbf{r}} \cdot \ddot{\mathbf{r}}) \dot{\mathbf{r}}}{|\dot{\mathbf{r}}|^4} \right| = \frac{|\dot{\mathbf{r}} \times \ddot{\mathbf{r}}|}{|\dot{\mathbf{r}}|^3} . \tag{3.33}$$

□

Example: Consider the tilted helix with a parametric representation in $t \in (0, 2N\pi)$ as shown in Figure 3.3 (right):

$$\mathbf{r}(t) = R_1 \cos t \, \mathbf{e}_x + R_2 \sin t \, \mathbf{e}_y + \left(R_2 \sin t \, \tan \alpha + \frac{p}{2\pi} t \right) \mathbf{e}_z , \tag{3.34}$$

with the ellipse semiaxes R_1 and R_2, tilt angle α, and pitch length p. For $q = p/2\pi$ we obtain

$$\mathbf{v} = \frac{\mathrm{d}\mathbf{r}}{\mathrm{d}t} = -R_1 \sin t \, \mathbf{e}_x + R_2 \cos t \, \mathbf{e}_y + (R_2 \tan \alpha \cos t + q) \, \mathbf{e}_z , \tag{3.35}$$

$$\mathbf{a} = \frac{\mathrm{d}\mathbf{v}}{\mathrm{d}t} = -R_1 \cos t \, \mathbf{e}_x - R_2 \sin t \, \mathbf{e}_y - (R_2 \tan \alpha \sin t) \, \mathbf{e}_z , \tag{3.36}$$

$$\frac{\mathrm{d}\mathbf{a}}{\mathrm{d}t} = R_1 \sin t \, \mathbf{e}_x - R_2 \cos t \, \mathbf{e}_y - (R_2 \tan \alpha \cos t) \, \mathbf{e}_z , \tag{3.37}$$

$$\mathbf{v} \times \mathbf{a} = R_2 q \sin t \, \mathbf{e}_x - (R_1 R_2 \tan \alpha + R_1 q \cos t) \, \mathbf{e}_y + R_1 R_2 \, \mathbf{e}_z . \tag{3.38}$$

Therefore it yields:

$$\mathbf{T} = \frac{-R_1 \sin t \, \mathbf{e}_x + R_2 \cos t \, \mathbf{e}_y + (R_2 \tan \alpha \cos t + q) \, \mathbf{e}_z}{\sqrt{R_1^2 \sin^2 t + R_2^2 \cos^2 t + (R_2 \tan \alpha \cos t + q)^2}} , \tag{3.39}$$

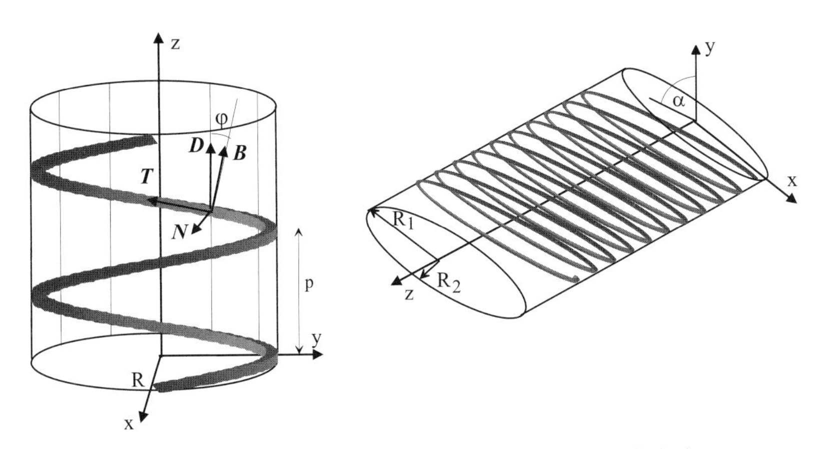

Figure 3.3 Left: Helix and Darboux frame. Right: Generalized elliptical helix with tilted turns.

$$\mathbf{B} = \frac{R_2 q \sin t \, \mathbf{e}_x - (R_1 R_2 \tan\alpha + R_1 q \cos t) \, \mathbf{e}_y + R_1 R_2 \, \mathbf{e}_z}{\sqrt{(R_2 q)^2 \sin^2 t + (R_1 R_2 \tan\alpha + R_1 q \cos t)^2 + (R_1 R_2)^2}} . \tag{3.40}$$

The normal vector can be calculated from $\mathbf{N} = \mathbf{B} \times \mathbf{T}$:

$$\mathbf{N} = \frac{1}{|\mathbf{v}| \, |\mathbf{v} \times \mathbf{a}|} \begin{vmatrix} \mathbf{e}_x & \mathbf{e}_y & \mathbf{e}_z \\ R_2 q \sin t & -R_1 R_2 \tan\alpha - R_1 q \cos t & R_1 R_2 \\ -R_1 \sin t & R_2 \cos t & R_2 \tan\alpha \cos t + q \end{vmatrix} . \tag{3.41}$$

The special case of $\alpha = 0$, $R_1 = R_2 = R$, and

$$\mathbf{r}(t) = R \cos t \, \mathbf{e}_x + R \sin t \, \mathbf{e}_y + \frac{p}{2\pi} t \, \mathbf{e}_z \tag{3.42}$$

is the circular helix of radius R and pitch length p as shown in Figure 3.3 (left). It follows that

$$\kappa = \frac{R}{R^2 + q^2} , \qquad \tau = \frac{2q}{R^2 + q^2} , \tag{3.43}$$

and $\mathbf{N} = \mathbf{B} \times \mathbf{T} = -\cos t \, \mathbf{e}_x - \sin t \, \mathbf{e}_y = -\mathbf{e}_r$. The Darboux vector takes the form

$$\mathbf{D} = \tau \, \mathbf{T} + \kappa \, \mathbf{B} = \frac{1}{\sqrt{R^2 + q^2}} \mathbf{e}_z . \tag{3.44}$$

□

In the case of the circular helix the curvature and torsion are both constant, and consequently their ratio is constant. Therefore, the tangent and binormal vectors are inclined at constant angles φ to the generators of the cylindrical surface onto which the helix is drawn; see Figure 3.3 (left). It is an important property of all helices that the curvature and the torsion are everywhere at a constant ratio.

Proof. Differentiating the relation $\mathbf{N} \cdot \mathbf{U}$ with respect to t (where $\mathbf{U}$ is a constant vector parallel to the generators of the cylinder) yields with Eq. (3.25)

$$(\tau \, \mathbf{B} - \kappa \, \mathbf{T}) \cdot \mathbf{U} = 0 . \tag{3.45}$$

Thus $\mathbf{U}$ is perpendicular to the vector $\tau \, \mathbf{B} - \kappa \, \mathbf{T}$. However, $\mathbf{U}$ is parallel to the rectifying plane spanned by $\mathbf{T}$ and $\mathbf{B}$ and must therefore also be parallel to the vector $\tau \, \mathbf{T} - \kappa \, \mathbf{B}$. This is the Darboux vector, inclined at a constant angle to $\mathbf{T}$.

□

3.2 The Directional Derivative

Suppose we are given a curve according to Eq. (3.1), with a parametric representation such that it passes through the point $\mathscr{P}$ at $t = 0$.

If we consider a smooth scalar field $\phi \in \mathcal{S}(\Omega)$ on an open domain $\Omega \subset E_3$ including the curve, we can express $\phi(\mathbf{r}(t))$ as an implicit function of t. The *directional derivative*, denoted $\partial_{\mathbf{v}}$, is the derivative of ϕ in the direction of the unit vector $\mathbf{v}$ at point $\mathscr{P}$, defined by

$$\partial_{\mathbf{v}}\phi := \frac{\mathrm{d}}{\mathrm{d}t}\phi(\mathbf{r}+t\mathbf{v})\bigg|_{t=0} = \lim_{t\to 0}\frac{\phi(\mathbf{r}+t\mathbf{v})-\phi(\mathbf{r})}{t}, \tag{3.46}$$

which is an intrinsic definition and does not depend on the choice of coordinates. If the directional derivative at $\mathscr{P}$ exists in every direction, ϕ is called *Gâteaux differentiable.*[10] Now suppose the canonical basis is given. The derivative in the direction of the basis vector $\mathbf{e}_x$ is known as the partial derivative of the scalar field with respect to the x-coordinate at $\mathscr{P}$:

$$\partial_{\mathbf{e}_x}\phi := \frac{\partial\phi}{\partial x} = \lim_{\Delta x\to 0}\frac{\phi(x+\Delta x,y,z)-\phi(x,y,z)}{\Delta x}. \tag{3.47}$$

The scalar field ϕ is said to be *Fréchet differentiable*[11] (see Section 3.6) at $\mathscr{P}$ if it is Gâteaux differentiable and linear in $\mathbf{v}$, that is, $\partial_{\mathbf{v}}\phi = \alpha_1\partial_{\mathbf{v}_1}\phi + \alpha_2\partial_{\mathbf{v}_2}\phi$ is fulfilled for all $\mathbf{v} = \alpha_1\mathbf{v}_1 + \alpha_2\mathbf{v}_2$, $\alpha_1, \alpha_2 \in \mathbb{R}$. If ϕ is Fréchet differentiable at $\mathscr{P}$, then ϕ is continuous there. Using the differentiation rule of composite functions we can express the rate at which the function $\phi(t)$ varies with changing parameter t by

$$\partial_{\mathbf{v}}\phi = \frac{\mathrm{d}}{\mathrm{d}t}\phi(\mathbf{r}(t)) = \frac{\partial\phi}{\partial x}\frac{\mathrm{d}x}{\mathrm{d}t} + \frac{\partial\phi}{\partial y}\frac{\mathrm{d}y}{\mathrm{d}t} + \frac{\partial\phi}{\partial z}\frac{\mathrm{d}z}{\mathrm{d}t}. \tag{3.48}$$

The directional derivative is the projection of the *gradient*

$$\operatorname{grad}\phi := \frac{\partial\phi}{\partial x}\mathbf{e}_x + \frac{\partial\phi}{\partial y}\mathbf{e}_y + \frac{\partial\phi}{\partial z}\mathbf{e}_z \tag{3.49}$$

onto the tangent vector $\mathbf{v}$ to the curve at $\mathscr{P}$,

$$\partial_{\mathbf{v}}\phi = \operatorname{grad}\phi\cdot\mathbf{v}. \tag{3.50}$$

Furthermore, we find for two scalar fields $\phi, \psi \in \mathcal{S}(\Omega)$ and a scalar $\lambda \in \mathbb{R}$:

$$\partial_{\mathbf{v}}(\phi+\psi) = \partial_{\mathbf{v}}\phi + \partial_{\mathbf{v}}\psi, \tag{3.51}$$

$$\partial_{\mathbf{v}}(\lambda\phi) = \lambda\partial_{\mathbf{v}}\phi, \tag{3.52}$$

$$\partial_{\mathbf{v}}(\phi\psi) = \psi\partial_{\mathbf{v}}\phi + \phi\partial_{\mathbf{v}}\psi. \tag{3.53}$$

10 René Gâteaux (1889–1914).
11 Maurice Fréchet (1878–1973).

Remark: From the Cauchy–Schwarz relation, Eq. (2.26), we obtain $|\partial_{\mathbf{v}}\phi| \leq |\operatorname{grad}\phi|\,|\mathbf{v}|$, which implies that the directional derivative is maximal when $\mathbf{v}$ points in the direction of $\operatorname{grad}\phi$. The gradient points in the direction of the steepest ascent of ϕ. In particular, $\operatorname{grad}\phi$ is normal to the surface of equipotential. This is an important result that we will need in Section 7.6 for the calculation of ideal pole shapes of conventional accelerator magnets. □

3.3 Gradient, Divergence, and Curl

For the vector differential operator[12] *nabla* expressed in Cartesian coordinates as

$$\nabla := \frac{\partial}{\partial x}\mathbf{e}_x + \frac{\partial}{\partial y}\mathbf{e}_y + \frac{\partial}{\partial z}\mathbf{e}_z\,, \tag{3.54}$$

it follows that $\operatorname{grad}\phi := \nabla\phi = \frac{\partial\phi}{\partial x}\mathbf{e}_x + \frac{\partial\phi}{\partial y}\mathbf{e}_y + \frac{\partial\phi}{\partial z}\mathbf{e}_z$ for $\phi \in \mathcal{S}(\Omega)$. The *gradient* is a mapping from a scalar to a vector field, $\operatorname{grad} : \mathcal{S}(\Omega) \to \mathcal{V}(\Omega)$. The nabla operator can also act on smooth vector fields by $\nabla\cdot : \mathcal{V}(\Omega) \to \mathcal{S}(\Omega) : \mathbf{a} \mapsto \nabla\cdot\mathbf{a}$ according to

$$\operatorname{div}\mathbf{a} := \nabla\cdot\mathbf{a} = \frac{\partial a_x}{\partial x} + \frac{\partial a_y}{\partial y} + \frac{\partial a_z}{\partial z}\,, \tag{3.55}$$

and $\nabla\times : \mathcal{V}(\Omega) \to \mathcal{V}(\Omega) : \mathbf{a} \mapsto \nabla\times\mathbf{a}$ according to

$$\operatorname{curl}\mathbf{a} := \nabla\times\mathbf{a} = \left(\frac{\partial a_z}{\partial y} - \frac{\partial a_y}{\partial z}\right)\mathbf{e}_x + \left(\frac{\partial a_x}{\partial z} - \frac{\partial a_z}{\partial x}\right)\mathbf{e}_y + \left(\frac{\partial a_y}{\partial x} - \frac{\partial a_x}{\partial y}\right)\mathbf{e}_z\,. \tag{3.56}$$

The nabla operator can furthermore act as $(\mathbf{a}\cdot\nabla) : \mathcal{V}(\Omega)\times\mathcal{V}(\Omega) \to \mathcal{V}(\Omega) : \mathbf{a},\mathbf{b} \mapsto (\mathbf{a}\cdot\nabla)\mathbf{b}$ with

$$\begin{aligned}(\mathbf{a}\cdot\operatorname{grad})\mathbf{b} := (\mathbf{a}\cdot\nabla)\mathbf{b} &= \left(a_x\frac{\partial}{\partial x} + a_y\frac{\partial}{\partial y} + a_z\frac{\partial}{\partial z}\right)\mathbf{b}\\ &= (a_x\frac{\partial b_x}{\partial x} + a_y\frac{\partial b_x}{\partial y} + a_z\frac{\partial b_x}{\partial z})\mathbf{e}_x + (\cdots)\mathbf{e}_y + (\cdots)\mathbf{e}_z\\ &= (\mathbf{a}\cdot\operatorname{grad} b_x)\mathbf{e}_x + (\mathbf{a}\cdot\operatorname{grad} b_y)\mathbf{e}_y + (\mathbf{a}\cdot\operatorname{grad} b_z)\mathbf{e}_z\,.\end{aligned} \tag{3.57}$$

12 An operator is a common designation for a linear transformation that links objects of different types.

We shall also recall the *Laplace operator*[13] defined by

$$\nabla^2 := \nabla \cdot \nabla = \frac{\partial^2}{\partial x^2} + \frac{\partial^2}{\partial y^2} + \frac{\partial^2}{\partial z^2}. \tag{3.58}$$

The Laplace operator maps both the scalar and vector fields to themselves.

Remark: One should not conclude that calculations with the nabla operator can be performed in the strict algebraic sense that the notation suggests. For example, we have $\mathbf{x} \cdot \mathbf{y} = \mathbf{y} \cdot \mathbf{x}$ but not $\nabla \cdot \mathbf{y} = \mathbf{y} \cdot \nabla$. Moreover, the vector field $\mathbf{b} = \nabla \times \mathbf{a}$ is not everywhere perpendicular to the vector field $\mathbf{a}$. In fact, $\mathbf{b}$ can span any angle with $\mathbf{a}$. For this reason we use grad, div, and curl as notation for the differential operators. If we had wanted to emphasize the role of ∇ as a differential operator, the goal could have been met more directly by using the exterior differential of Cartan's[14] calculus [5]. Experts may prefer the ∇ notation because of the *nabla calculus* [4], so we briefly summarize its rules: Let f_i denote scalar fields or vector fields on a domain Ω and let $\lambda_i \in \mathbb{R}$. For the nabla operator acting on a linear combination of the f_i we obtain

$$\nabla \sum_i \lambda_i f_i = \sum_i \lambda_i \nabla f_i \tag{3.59}$$

and for the operator acting on a product of fields

$$\nabla(f_1 f_2 f_3) = \nabla(\overset{\downarrow}{f_1} f_2 f_3) + \nabla(f_1 \overset{\downarrow}{f_2} f_3) + \nabla(f_1 f_2 \overset{\downarrow}{f_3}). \tag{3.60}$$

The products in the parentheses have to be rewritten according to the rules of vector algebra such that the marked field follows directly the ∇ operator. As an example, consider

$$\nabla \times (\phi \mathbf{a}) = \nabla \times (\overset{\downarrow}{\phi} \mathbf{a}) + \nabla \times (\phi \overset{\downarrow}{\mathbf{a}}) = -\mathbf{a} \times \nabla \overset{\downarrow}{\phi} + (\nabla \times \overset{\downarrow}{\mathbf{a}})\phi, \tag{3.61}$$

where we have used Eq. (2.72). Equation (3.61) proves the relation

$$\operatorname{curl}(\phi \mathbf{a}) = \phi \operatorname{curl} \mathbf{a} - \mathbf{a} \times \operatorname{grad} \phi \tag{3.62}$$

for $\phi \in \mathcal{S}(\Omega)$ and $\mathbf{a} \in \mathcal{V}(\Omega)$. □

13 Pierre-Simon Laplace (1749–1827).
14 Elie Cartan (1895–1951).

3.4 Identities of Vector Analysis

Useful identities of classical vector analysis for $\mathbf{a}, \mathbf{b} \in \mathcal{V}(\Omega)$, $\phi, \psi \in \mathcal{S}(\Omega)$, $\lambda \in \mathbb{R}$ are summarized hereafter.

$$\operatorname{grad}(\phi + \psi) = \operatorname{grad}\phi + \operatorname{grad}\psi\,, \tag{3.63}$$

$$\operatorname{grad}(\phi\psi) = \psi \operatorname{grad}\phi + \phi \operatorname{grad}\psi\,, \tag{3.64}$$

$$\operatorname{grad}(\mathbf{a}\cdot\mathbf{b}) = (\mathbf{a}\cdot \operatorname{grad})\mathbf{b} + (\mathbf{b}\cdot\operatorname{grad})\mathbf{a} + \mathbf{a}\times\operatorname{curl}\mathbf{b} + \mathbf{b}\times\operatorname{curl}\mathbf{a}\,, \tag{3.65}$$

$$\operatorname{div}(\mathbf{a}+\mathbf{b}) = \operatorname{div}\mathbf{a} + \operatorname{div}\mathbf{b}\,, \tag{3.66}$$

$$\operatorname{div}\lambda\mathbf{a} = \lambda\operatorname{div}\mathbf{a} + \mathbf{a}\cdot\operatorname{grad}\lambda\,, \tag{3.67}$$

$$\operatorname{div}(\mathbf{a}\times\mathbf{b}) = \mathbf{b}\cdot\operatorname{curl}\mathbf{a} - \mathbf{a}\cdot\operatorname{curl}\mathbf{b}\,, \tag{3.68}$$

$$\operatorname{curl}\lambda\mathbf{a} = \lambda\operatorname{curl}\mathbf{a} - \mathbf{a}\times\operatorname{grad}\lambda\,, \tag{3.69}$$

$$\operatorname{curl}(\mathbf{a}+\mathbf{b}) = \operatorname{curl}\mathbf{a} + \operatorname{curl}\mathbf{b}\,, \tag{3.70}$$

$$\operatorname{curl}(\mathbf{a}\times\mathbf{b}) = \mathbf{a}\operatorname{div}\mathbf{b} - \mathbf{b}\operatorname{div}\mathbf{a} + (\mathbf{b}\cdot\operatorname{grad})\mathbf{a} - (\mathbf{a}\cdot\operatorname{grad})\mathbf{b}\,, \tag{3.71}$$

$$\operatorname{curl}\operatorname{curl}\mathbf{a} = \operatorname{grad}\operatorname{div}\mathbf{a} - \nabla^2\mathbf{a}\,, \tag{3.72}$$

$$\operatorname{div}\operatorname{grad}\phi = \nabla^2\phi \tag{3.73}$$

$$\operatorname{div}\operatorname{curl}\mathbf{a} = 0\,, \tag{3.74}$$

$$\operatorname{curl}\operatorname{grad}\phi = 0\,. \tag{3.75}$$

Equations (3.74) and (3.75) state that the curl of an arbitrary vector field is source-free, and that an arbitrary gradient field is curl-free (irrotational). These statements and their reversals are the *lemmata of Poincaré*.

3.5 Surfaces in E_3

We discussed a space curve as the locus of points whose coordinates depend on a single parameter t (or s). In the *Gauss form*,[15] a surface $\mathscr{A}$ is defined as the locus of a point whose coordinates are functions of two independent parameters u, v in some domain $U = [u_a, u_b] \times [v_a, v_b] \subset \mathbb{R}^2$,

$$\mathscr{A} : U \to E_3 : u, v \mapsto \mathbf{r}(u, v) \tag{3.76}$$

15 This is the inverse mapping to the coordinate map introduced in Section 2.5.1 from $\phi(U)$ into Ω on the surface.

The position vector to any point on that surface is given by

$$\mathbf{r}(u,v) = x(u,v)\,\mathbf{e}_x + y(u,v)\,\mathbf{e}_y + z(u,v)\,\mathbf{e}_z. \tag{3.77}$$

If u and v are eliminated from the component functions of $\mathbf{r}$, this yields

$$f(x,y,z) = 0, \tag{3.78}$$

which is known as *Monge's*[16] *form* of the surface equation. The vector $\partial\mathbf{r}/\partial u$ at a given point $\mathscr{P}$ with parameters (u_0, v_0) is obtained by differentiating $\mathbf{r}$ with respect to u while keeping v constant. From the theory of space curves we know that $\partial\mathbf{r}/\partial u$ represents a tangent vector to the curve $v = v_0$. Similarly $\partial\mathbf{r}/\partial v$ represents a tangent vector to the curve $u = u_0$. Consequently,

$$\mathbf{n} = \frac{\partial\mathbf{r}}{\partial u} \times \frac{\partial\mathbf{r}}{\partial v} \tag{3.79}$$

yields the normal vector to the surface. The equation of the *tangent plane* is given by $(\mathbf{r}_0 - \mathbf{r}) \cdot \mathbf{n} = 0$, where $\mathbf{r}$ is the position of an arbitrary point on that tangent plane. If we assign this point to $\mathbf{r} = \mathbf{r}_0 + \lambda\mathbf{T}$, $\lambda \in \mathbb{R}$, where $\mathbf{T}$ denotes the tangent vector to a curve traced out on the surface, we see that $(\mathbf{r}_0 - \mathbf{r})$ is orthogonal to $\mathbf{n}$, as expected.

Example: As an example we will derive the equation of the tangent plane to the paraboloid $x = u, y = v, z = u^2 + v^2$ in Cartesian coordinates at point $(1,1)$. The position vector to any point on the paraboloid is

$$\mathbf{r}(u,v) = u\mathbf{e}_x + v\mathbf{e}_y + (u^2 + v^2)\mathbf{e}_z. \tag{3.80}$$

Therefore,

$$\left.\frac{\partial\mathbf{r}}{\partial u}\right|_{(1,1)} = (\mathbf{e}_x + 2u\mathbf{e}_z)|_{(1,1)} = \mathbf{e}_x + 2\mathbf{e}_z\,, \tag{3.81}$$

$$\left.\frac{\partial\mathbf{r}}{\partial v}\right|_{(1,1)} = (\mathbf{e}_y + 2v\mathbf{e}_z)|_{(1,1)} = \mathbf{e}_y + 2\mathbf{e}_z\,. \tag{3.82}$$

The surface normal $\mathbf{n}$ at this point is

$$\mathbf{n}|_{(1,1)} = \left.\left(\frac{\partial\mathbf{r}}{\partial u} \times \frac{\partial\mathbf{r}}{\partial v}\right)\right|_{(1,1)} = -2\mathbf{e}_x - 2\mathbf{e}_y + \mathbf{e}_z\,. \tag{3.83}$$

Eliminating the parameters u, v from the parametric form of the surface yields $x^2 + y^2 - z = 0$. The equation of the tangent plane can then be derived:

$$\begin{aligned}(\mathbf{r}_0 - \mathbf{r}) \cdot \mathbf{n} &= ((1-x)\mathbf{e}_x + (1-y)\mathbf{e}_y + (2-z)\mathbf{e}_z) \cdot (-2\mathbf{e}_x - 2\mathbf{e}_y + \mathbf{e}_z) \\ &= 2x + 2y - z - 2 = 0.\end{aligned} \tag{3.84}$$

□

16 Gaspard Monge (1746–1818).

3.6 The Differential

If a function of a single real variable is smooth at a point, it can be linearly approximated in a neighborhood of that point. We will extend this concept to define the differentiability of a mapping $\mathbf{f} : \Omega_1 \to \Omega_2$ for open $\Omega_1, \Omega_2 \subset E_3$. The mapping is said to be differentiable at $\mathscr{P}$ (with position vector $\mathbf{r}$) if there exists a linear function $T(t\mathbf{v})$ such that

$$\mathbf{f}(\mathbf{r} + t\mathbf{v}) = \mathbf{f}(\mathbf{r}) + T(t\mathbf{v}) + R(\mathbf{r}, t\mathbf{v}), \tag{3.85}$$

with an error term $R(\mathbf{r}, t\mathbf{v})$ vanishing in the sense of $\lim_{t\to 0} \frac{R(\mathbf{r},t\mathbf{v})}{t} = 0$. Since T is linear, we can write

$$\frac{\mathbf{f}(\mathbf{r} + t\mathbf{v}) - \mathbf{f}(\mathbf{r})}{t} = T(\mathbf{v}) + \frac{R(\mathbf{r}, t\mathbf{v})}{t}. \tag{3.86}$$

It follows that the derivative of $\mathbf{f}$ at $\mathscr{P}$ exists in every direction $\mathbf{v}$ and is given by

$$\lim_{t\to 0} \frac{\mathbf{f}(\mathbf{r} + t\mathbf{v}) - \mathbf{f}(\mathbf{r})}{t} = T(\mathbf{v}) + \lim_{t\to 0} \frac{R(\mathbf{r}, t\mathbf{v})}{t}. \tag{3.87}$$

The linear part T is called the *total differential* of $\mathbf{f}$ at $\mathbf{r}$ and is written as $\mathbf{df}|_{\mathbf{r}}(\mathbf{v})$. We obtain

$$\mathbf{df}|_{\mathbf{r}}(\mathbf{v}) := T(\mathbf{v}) = \frac{\mathrm{d}}{\mathrm{d}t}\mathbf{f}(\mathbf{r} + t\mathbf{v})\bigg|_{t=0} = |\mathbf{v}|\, \partial_{\frac{\mathbf{v}}{|\mathbf{v}|}} \mathbf{f}(\mathbf{r}), \tag{3.88}$$

which is for $|\mathbf{v}| = 1$ identical to the directional derivative of the mapping $\mathbf{f}$ in the direction of $\mathbf{v}$ at $\mathbf{r}$. We will omit the notation of the point $\mathscr{P}$ and its position vector $\mathbf{r}$. Assume that a Cartesian basis is fixed at $\mathscr{P}$ and $\mathbf{f}(\mathscr{P})$. The directional derivative in the direction of one of the basis vectors, for example, $\mathbf{e}_x$ is

$$T(\mathbf{e}_x) = \partial_{\mathbf{e}_x} \mathbf{f} = \frac{\partial \mathbf{f}}{\partial x}, \tag{3.89}$$

where similar equations hold for $T(\mathbf{e}_y)$ and $T(\mathbf{e}_z)$. Because of the linearity of $\mathbf{df}$ in $\mathbf{v}$, we can write in the component form

$$\begin{aligned}\mathbf{df}(\mathbf{v}) &= \mathbf{df}(v_x\mathbf{e}_x + v_y\mathbf{e}_y + v_z\mathbf{e}_z) = v_x\mathbf{df}(\mathbf{e}_x) + v_y\mathbf{df}(\mathbf{e}_y) + v_z\mathbf{df}(\mathbf{e}_z) \\ &= v_x\frac{\partial \mathbf{f}}{\partial x} + v_y\frac{\partial \mathbf{f}}{\partial y} + v_z\frac{\partial \mathbf{f}}{\partial z}.\end{aligned} \tag{3.90}$$

Writing $\mathbf{f} = f_x\mathbf{e}_x + f_y\mathbf{e}_y + f_z\mathbf{e}_z$ we obtain

$$\frac{\partial \mathbf{f}}{\partial x} = \frac{\partial f_x}{\partial x}\mathbf{e}_x + \frac{\partial f_y}{\partial x}\mathbf{e}_y + \frac{\partial f_z}{\partial x}\mathbf{e}_z, \tag{3.91}$$

and equivalent equations for y and z. Thus Eq. (3.90) yields

$$\begin{aligned}\mathbf{df}(\mathbf{v}) &= v_x\left(\frac{\partial f_x}{\partial x}\mathbf{e}_x + \frac{\partial f_y}{\partial x}\mathbf{e}_y + \frac{\partial f_z}{\partial x}\mathbf{e}_z\right) + v_y\left(\frac{\partial f_x}{\partial y}\mathbf{e}_x + \frac{\partial f_y}{\partial y}\mathbf{e}_y + \frac{\partial f_z}{\partial y}\mathbf{e}_z\right)\\ &\quad + v_z\left(\frac{\partial f_x}{\partial z}\mathbf{e}_x + \frac{\partial f_y}{\partial z}\mathbf{e}_y + \frac{\partial f_z}{\partial z}\mathbf{e}_z\right)\\ &= \left(\frac{\partial f_x}{\partial x}v_x + \frac{\partial f_x}{\partial y}v_y + \frac{\partial f_x}{\partial z}v_z\right)\mathbf{e}_x + \left(\frac{\partial f_y}{\partial x}v_x + \frac{\partial f_y}{\partial y}v_y + \frac{\partial f_y}{\partial z}v_z\right)\mathbf{e}_y\\ &\quad + \left(\frac{\partial f_z}{\partial x}v_x + \frac{\partial f_z}{\partial y}v_y + \frac{\partial f_z}{\partial z}v_z\right)\mathbf{e}_z\,. \end{aligned} \tag{3.92}$$

Hence

$$\mathrm{d}f_x(\mathbf{v}) = \frac{\partial f_x}{\partial x}v_x + \frac{\partial f_x}{\partial y}v_y + \frac{\partial f_x}{\partial z}v_z\,, \tag{3.93}$$

$$\mathrm{d}f_y(\mathbf{v}) = \frac{\partial f_y}{\partial x}v_x + \frac{\partial f_y}{\partial y}v_y + \frac{\partial f_y}{\partial z}v_z\,, \tag{3.94}$$

$$\mathrm{d}f_z(\mathbf{v}) = \frac{\partial f_z}{\partial x}v_x + \frac{\partial f_z}{\partial y}v_y + \frac{\partial f_z}{\partial z}v_z\,. \tag{3.95}$$

The matrix representation of $\mathbf{df}$ is therefore given by the *Jacobi matrix*[17]

$$[J] = \begin{pmatrix} \frac{\partial f_x}{\partial x} & \frac{\partial f_x}{\partial y} & \frac{\partial f_x}{\partial z}\\ \frac{\partial f_y}{\partial x} & \frac{\partial f_y}{\partial y} & \frac{\partial f_y}{\partial z}\\ \frac{\partial f_z}{\partial x} & \frac{\partial f_z}{\partial y} & \frac{\partial f_z}{\partial z} \end{pmatrix}. \tag{3.96}$$

The coordinate functions $x = x(\mathbf{r})$, $y = y(\mathbf{r})$, $z = z(\mathbf{r})$ are differentiable functions, and their differentials $\mathrm{d}x$, $\mathrm{d}y$, $\mathrm{d}z$, satisfy $\mathrm{d}x(\mathbf{v}) = v_x$, etc. for any vector $\mathbf{v}$. Hence

$$\mathbf{df}(\mathbf{v}) = \mathrm{d}x(\mathbf{v})\frac{\partial \mathbf{f}}{\partial x} + \mathrm{d}y(\mathbf{v})\frac{\partial \mathbf{f}}{\partial y} + \mathrm{d}z(\mathbf{v})\frac{\partial \mathbf{f}}{\partial z} \tag{3.97}$$

which yields the formula

$$\mathbf{df} = \frac{\partial \mathbf{f}}{\partial x}\mathrm{d}x + \frac{\partial \mathbf{f}}{\partial y}\mathrm{d}y + \frac{\partial \mathbf{f}}{\partial z}\mathrm{d}z\,. \tag{3.98}$$

In particular, for scalar functions we obtain

$$\mathrm{d}f = \frac{\partial f}{\partial x}\mathrm{d}x + \frac{\partial f}{\partial y}\mathrm{d}y + \frac{\partial f}{\partial z}\mathrm{d}z = \operatorname{grad} f \cdot \mathrm{d}\mathbf{r}\,. \tag{3.99}$$

It is an easy task to extend the above concept to mappings between $\mathbb{R}^n$ and $\mathbb{R}^m$.

17 Carl Jacobi (1804–1851).

Examples:

1. Consider a column vector $\{b\} \in \mathbb{R}^m$ and the linear mapping represented by the transition matrix $[A] \in \mathbb{R}^{m\times n}$. The affine mapping $\mathbf{f} : \mathbb{R}^n \to \mathbb{R}^m : \{r\} \mapsto [A]\{r\} + \{b\}$ is differentiable, and at all points we have $\mathrm{d}\mathbf{f}|_{\mathbf{r}} = [A]$.
2. The Jacobi matrix at $t \in I$ of a curve $\mathscr{S} : I \to E_3 : t \mapsto \mathbf{r}(t)$ is the special case for $n = 1$; $\mathrm{d}\mathbf{r}|_t$ is the linear approximation for the mapping $\Delta t \mapsto \mathbf{r}(t + \Delta t) - \mathbf{r}(t)$. Hence $\mathrm{d}\mathbf{r}(\Delta t) = [J]\Delta t$ and consequently

$$\mathrm{d}\mathbf{r}(1) = [J] = \begin{pmatrix} \frac{\partial x}{\partial t} \\ \frac{\partial y}{\partial t} \\ \frac{\partial z}{\partial t} \end{pmatrix} \tag{3.100}$$

is the tangent vector to the curve of velocity 1. The differential $\mathrm{d}\mathbf{r}$ is called the *vectorial line element*, for which holds

$$\mathrm{d}\mathbf{r} = \mathbf{v}\mathrm{d}t = \frac{\mathbf{v}}{v} v\mathrm{d}t = \mathbf{T}\,\mathrm{d}s\,, \tag{3.101}$$

and where s is the arc length of the curve and $\mathbf{T}$ the tangent vector field.
3. For $m = 1$ we have the special case of the scalar field $f : E_3 \to \mathbb{R}$. It follows that $\mathrm{d}f$ is the linear approximation for the mapping $\mathbf{v} \mapsto f(\mathbf{r} + \mathbf{v}) - f(\mathbf{r})$ given by $\mathrm{d}f(\mathbf{v}) = \sum_{i=1}^{3} \frac{\partial f}{\partial x_i} v_i$ and thus by the Jacobi matrix $[J] = \left(\frac{\partial f}{\partial x_1}, \frac{\partial f}{\partial x_2}, \frac{\partial f}{\partial x_3}\right)$, which is the gradient of f.
4. Consider the parametric representation of a surface, $\mathscr{A} : U \to E_3 : u, v \mapsto \mathbf{r}(u, v)$ with $U = [u_1, u_2] \times [v_1, v_2] \subset \mathbb{R}^2$ as shown in Figure 3.4. The basis vectors $\partial\mathbf{r}/\partial u$ and $\partial\mathbf{r}/\partial v$ spanning the tangent plane are the columns of the Jacobi matrix of the parametric representation. Using $\mathrm{d}\mathbf{r} = \frac{\partial \mathbf{r}}{\partial u}\mathrm{d}u + \frac{\partial \mathbf{r}}{\partial v}\mathrm{d}v$ we obtain for the arc length on the surface

$$\begin{aligned} \mathrm{d}s^2 = \mathrm{d}\mathbf{r}\cdot\mathrm{d}\mathbf{r} &= \frac{\partial \mathbf{r}}{\partial u}\cdot\frac{\partial \mathbf{r}}{\partial u}\mathrm{d}u^2 + 2\frac{\partial \mathbf{r}}{\partial u}\cdot\frac{\partial \mathbf{r}}{\partial v}\mathrm{d}u\mathrm{d}v + \frac{\partial \mathbf{r}}{\partial v}\cdot\frac{\partial \mathbf{r}}{\partial v}\mathrm{d}v^2 \\ &=: E\mathrm{d}u^2 + 2F\mathrm{d}u\mathrm{d}v + G\mathrm{d}v^2\,. \end{aligned} \tag{3.102}$$

This is the first quadratic differential form of the surface. The quantities $E, F,$ and G are called the *fundamental magnitudes* of first order. The arc length along the curve v = const. ($\mathrm{d}v$=0) is equal to $\mathrm{d}s_v = \sqrt{E}\mathrm{d}u$. Similarly $\mathrm{d}s_u = \sqrt{G}\mathrm{d}v$. The tangent vectors to the curves v = const. and u = const. are therefore

$$\mathbf{T}_v = \frac{1}{\sqrt{E}}\frac{\partial \mathbf{r}}{\partial u}\,, \qquad \mathbf{T}_u = \frac{1}{\sqrt{G}}\frac{\partial \mathbf{r}}{\partial v}\,. \tag{3.103}$$

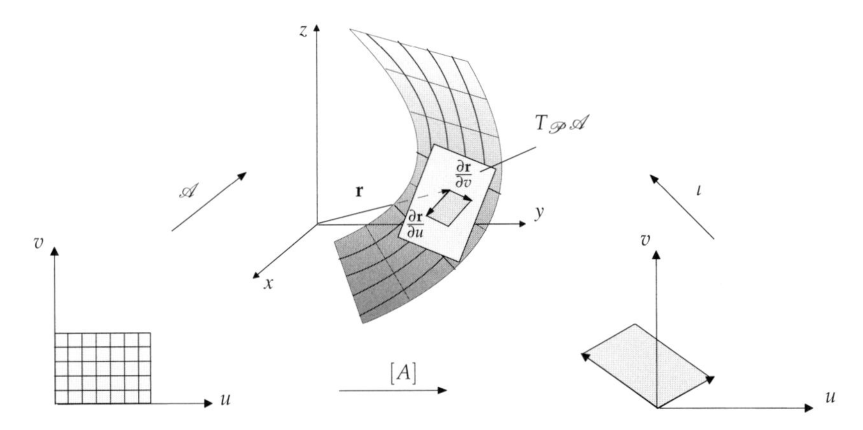

Figure 3.4 The differential as the linear approximation for the parametric representation of the surface. The surface element is spanned by the basis vectors $\partial\mathbf{r}/\partial u$ and $\partial\mathbf{r}/\partial v$ in the tangent space $T_{\mathscr{P}}\mathscr{A}$ at point $\mathscr{P}$.

The surface area can be calculated by summing up the areas of the parallelepipeds spanned by the vectors $\mathrm{d}\mathbf{u} = \frac{\partial \mathbf{r}}{\partial u}\mathrm{d}u$ and $\mathrm{d}\mathbf{v} = \frac{\partial \mathbf{r}}{\partial v}\mathrm{d}v$:

$$\mathrm{d}a = \left|\frac{\partial \mathbf{r}}{\partial u} \times \frac{\partial \mathbf{r}}{\partial v}\right| \mathrm{d}u\mathrm{d}v = \sqrt{EG - F^2}\,\mathrm{d}u\mathrm{d}v\,. \tag{3.104}$$

The vectorial surface element is given by

$$\mathrm{d}\mathbf{a} = \mathbf{n}\,\mathrm{d}a = \left(\frac{\partial \mathbf{r}}{\partial u} \times \frac{\partial \mathbf{r}}{\partial v}\right) \mathrm{d}u\mathrm{d}v\,, \tag{3.105}$$

where the modulus of the cross product is the surface of the parallelepiped spanned by the basis vectors in the tangent space to $\mathscr{A}$.

5. The linear approximation for the parametric representation of a volume $\mathscr{V} : U \to E_3 : u_1, u_2, u_3 \mapsto \mathbf{r}(u_1, u_2, u_3)$ for $U = [u_{1,a}, u_{1,b}] \times [u_{2,a}, u_{2,b}] \times [u_{3,a}, u_{3,b}] \subset \mathbb{R}^3$ is represented by the Jacobi matrix $[J] \in \mathbb{R}^{3\times 3}$, and the volume element is given by

$$\mathrm{d}V = |\det[J]|\,\mathrm{d}u_1\mathrm{d}u_2\mathrm{d}u_3\,. \tag{3.106}$$

□

Remark: In modern mathematics $\mathrm{d}\mathbf{f}$ is not interpreted as an infinitesimal increment of the vector function, but rather as a linear operator. In particular, if $\mathbf{f}$ is a mapping between two space elements $\mathscr{M}$ and $\mathscr{N}$, the differential is the mapping $\mathrm{d}\mathbf{f}|_{\mathscr{P}} : T_{\mathscr{P}}\mathscr{M} \to T_{\mathbf{f}(\mathscr{P})}\mathscr{N}$ between their tangent spaces (their linear approximations in $\mathscr{P}$) and is given in coordinates by the Jacobi matrix. □

3.7
Differential Operators on Scalar and Vector Fields in $\mathbf{r}$ and $\mathbf{r}'$

Consider a scalar field $\phi \in \mathcal{S}(\Omega)$ that is given as a function of the position vector $\phi : \Omega \to \mathbb{R} : \mathbf{r} \mapsto \phi(r)$, where $r := |\mathbf{r}|$. The differential of the norm of the position vector is

$$\mathrm{d}r = \operatorname{grad} r \cdot \mathrm{d}\mathbf{r} = \mathrm{d}\sqrt{\mathbf{r}\cdot\mathbf{r}} = \frac{1}{2}\frac{1}{\sqrt{\mathbf{r}\cdot\mathbf{r}}}(\mathbf{r}\cdot\mathrm{d}\mathbf{r} + \mathrm{d}\mathbf{r}\cdot\mathbf{r}) = \frac{\mathbf{r}}{r}\cdot\mathrm{d}\mathbf{r}, \tag{3.107}$$

from which follows

$$\operatorname{grad} r = \frac{\mathbf{r}}{r} = \mathbf{e}_r\,. \tag{3.108}$$

In a similar way we find

$$\mathrm{d}\frac{1}{r} = \operatorname{grad}\frac{1}{r}\cdot\mathrm{d}\mathbf{r} = \mathrm{d}\frac{1}{\sqrt{\mathbf{r}\cdot\mathbf{r}}} = -\frac{1}{2}\frac{1}{\sqrt{\mathbf{r}\cdot\mathbf{r}}^{3}}(\mathbf{r}\cdot\mathrm{d}\mathbf{r} + \mathrm{d}\mathbf{r}\cdot\mathbf{r}) = -\frac{\mathbf{r}}{r^3}\cdot\mathrm{d}\mathbf{r}, \tag{3.109}$$

which yields

$$\operatorname{grad}\frac{1}{r} = -\frac{\mathbf{r}}{r^3} = -\frac{1}{r^2}\mathbf{e}_r\,. \tag{3.110}$$

We state without proof that

$$\operatorname{div}\mathbf{r} = 3 \quad \text{and} \quad \operatorname{curl}\mathbf{r} = 0\,. \tag{3.111}$$

With these results it is easy to show that

$$\nabla^2\frac{\mathbf{r}}{r^3} = \operatorname{div}\operatorname{grad}\frac{1}{r} = \operatorname{div}\frac{\mathbf{r}}{r^3} = \frac{1}{r^3}\operatorname{div}\mathbf{r} + \mathbf{r}\operatorname{grad}\frac{1}{r^3} = \frac{3}{r^3} - 3\frac{\mathbf{r}\cdot\mathbf{e}_r}{r^4} = 0\,. \tag{3.112}$$

In Chapter 5 we will employ differential operators that act on scalar and vector fields expressed as functions of the distance between the field point $\mathbf{r}$ and the source point $\mathbf{r}'$. Thus we need to be more specific about whether the differential operators act on the coordinates of the field point or the source point. More precisely, do the differential operators act on scalar and vector fields of the form $\mathbf{r} \mapsto \phi(|\mathbf{r}-\mathbf{r}'|)$ and $\mathbf{r} \mapsto \mathbf{a}(|\mathbf{r}-\mathbf{r}'|)$, or do they operate on fields of the form $\mathbf{r}' \mapsto \phi(|\mathbf{r}-\mathbf{r}'|)$ and $\mathbf{r}' \mapsto \mathbf{a}(|\mathbf{r}-\mathbf{r}'|)$? In the latter case the differential operators are indexed with the symbol $\mathbf{r}'$ while in the first case the index $\mathbf{r}$ is omitted. We find

$$\operatorname{grad}\phi(|\mathbf{r}-\mathbf{r}'|) = -\operatorname{grad}_{\mathbf{r}'}\phi(|\mathbf{r}-\mathbf{r}'|)\,, \tag{3.113}$$

$$\operatorname{div}\mathbf{a}(|\mathbf{r}-\mathbf{r}'|) = -\operatorname{div}_{\mathbf{r}'}\mathbf{a}(|\mathbf{r}-\mathbf{r}'|)\,, \tag{3.114}$$

$$\operatorname{curl}\mathbf{a}(|\mathbf{r}-\mathbf{r}'|) = -\operatorname{curl}_{\mathbf{r}'}\mathbf{a}(|\mathbf{r}-\mathbf{r}'|)\,, \tag{3.115}$$

$$\nabla^2\phi(|\mathbf{r}-\mathbf{r}'|) = \nabla^2_{\mathbf{r}'}\phi(|\mathbf{r}-\mathbf{r}'|)\,. \tag{3.116}$$

More specifically,

$$\operatorname{grad}\frac{1}{|\mathbf{r}-\mathbf{r}'|} = -\frac{\mathbf{r}-\mathbf{r}'}{|\mathbf{r}-\mathbf{r}'|^3} = -\operatorname{grad}_{\mathbf{r}'}\frac{1}{|\mathbf{r}-\mathbf{r}'|}, \tag{3.117}$$

$$\operatorname{grad}|\mathbf{r}-\mathbf{r}'| = \frac{\mathbf{r}-\mathbf{r}'}{|\mathbf{r}-\mathbf{r}'|} = -\operatorname{grad}_{\mathbf{r}'}|\mathbf{r}-\mathbf{r}'|, \tag{3.118}$$

$$\operatorname{div}\frac{\mathbf{r}-\mathbf{r}'}{|\mathbf{r}-\mathbf{r}'|^3} = 4\pi\delta(|\mathbf{r}-\mathbf{r}'|) = -\operatorname{div}_{\mathbf{r}'}\frac{\mathbf{r}-\mathbf{r}'}{|\mathbf{r}-\mathbf{r}'|^3}, \tag{3.119}$$

$$\nabla^2\frac{1}{|\mathbf{r}-\mathbf{r}'|} = -4\pi\delta(|\mathbf{r}-\mathbf{r}'|) = \nabla^2_{\mathbf{r}'}\frac{1}{|\mathbf{r}-\mathbf{r}'|}, \tag{3.120}$$

$$\nabla^2|\mathbf{r}-\mathbf{r}'| = \frac{2}{|\mathbf{r}-\mathbf{r}'|} = \nabla^2_{\mathbf{r}'}|\mathbf{r}-\mathbf{r}'|. \tag{3.121}$$

3.8 The Path Integral of a Vector Field

Let $\mathbf{r}(t) = x(t)\mathbf{e}_x + y(t)\mathbf{e}_y + z(t)\mathbf{e}_z$ be the parametric representation of a curve with parameter t on a closed interval $[t_1, t_2]$, and let $x(t)$, $y(t)$, $z(t)$ be continuous on that interval. Then

$$\int_{t_1}^{t_2}\mathbf{r}(t)\mathrm{d}t = \int_{t_1}^{t_2}x(t)\mathrm{d}t\,\mathbf{e}_x + \int_{t_1}^{t_2}y(t)\mathrm{d}t\,\mathbf{e}_y + \int_{t_1}^{t_2}z(t)\mathrm{d}t\,\mathbf{e}_z \tag{3.122}$$

is called the *definite vector integral*. Let further $\mathbf{a} \in \mathcal{V}(\Omega)$ be a smooth vector field in an open domain including the curve

$$\mathbf{a} = a_x(x,y,z)\mathbf{e}_x + a_y(x,y,z)\mathbf{e}_y + a_z(x,y,z)\mathbf{e}_z\,. \tag{3.123}$$

The integral of the tangential component of $\mathbf{a}$ along the curve from $\mathscr{P}_1(t = t_1)$ to $\mathscr{P}_2(t = t_2)$, written as

$$\begin{aligned}\int_{\mathscr{P}_1}^{\mathscr{P}_2}\mathbf{a}\cdot\mathrm{d}\mathbf{r} &= \int_{\mathscr{P}_1}^{\mathscr{P}_2}a_x\mathrm{d}x + a_y\mathrm{d}y + a_z\mathrm{d}z\\ &= \int_{t_1}^{t_2}\left(a_x(\mathbf{r}(t))\frac{\partial x}{\partial t} + a_y(\mathbf{r}(t))\frac{\partial y}{\partial t} + a_z(\mathbf{r}(t))\frac{\partial z}{\partial t}\right)\mathrm{d}t\\ &= \int_{t_1}^{t_2}\mathbf{a}(\mathbf{r}(t))\cdot\frac{\partial\mathbf{r}(t)}{\partial t}\mathrm{d}t = \int_{t_1}^{t_2}\mathbf{a}(\mathbf{r}(t))\cdot\mathbf{v}(t)\mathrm{d}t\,,\end{aligned} \tag{3.124}$$

is called the *path integral* or the *line integral* of *Pfaff's*[18] *form*

$$\omega := a_x\mathrm{d}x + a_y\mathrm{d}y + a_z\mathrm{d}z\,, \tag{3.125}$$

18 Johann-Friedrich Pfaff (1765–1826).

with

$$\int_{\mathscr{S}} \omega = \int_{t_1}^{t_2} \mathbf{a}(\mathbf{r}(t)) \cdot \frac{\partial \mathbf{r}(t)}{\partial t} \mathrm{d}t \tag{3.126}$$

for $\mathscr{S} : [t_1, t_2] \to E_3 : t \mapsto \mathbf{r}(t)$. The coefficients in Eq. (3.125) are position-dependent. For the reversal of the orientation of $\mathscr{S}$ we obtain

$$\int_{-\mathscr{S}} \omega = -\int_{\mathscr{S}} \omega . \tag{3.127}$$

The line integral is an invariant with respect to the parametric representation of the curve, that is, $\int_{\mathscr{S}} \omega = \int_{\mathscr{S} \circ \varphi} \omega$, where $\varphi : [u_1, u_2] \to [t_1, t_2]$ is a 1-smooth change of parametric representation. If the curve is only piecewise smooth, we shall define

$$\int_{\mathscr{S}} \omega := \sum_{k=1}^{K} \int_{\mathscr{S}_k} \omega . \tag{3.128}$$

The Pfaff forms themselves constitute, with rules of addition and scalar multiplication, a vector space, where $\mathrm{d}x, \mathrm{d}y, \mathrm{d}z$ take the role of basis vectors.

3.9
Coordinate-Free Definitions of the Differential Operators

Gradient Consider a vector field $\mathbf{a} = \operatorname{grad} \phi$ for $\phi(x, y, z) \in \mathcal{S}(\Omega)$ and the open domain Ω including the curve $\mathscr{S} : [t_1, t_2] \to E_3$ with $\mathscr{P}_1 = \mathbf{r}(t_1)$ and $\mathscr{P}_2 = \mathbf{r}(t_2)$. It follows that

$$\int_{\mathscr{P}_1}^{\mathscr{P}_2} \mathbf{a} \cdot \mathbf{dr} = \int_{\mathscr{P}_1}^{\mathscr{P}_2} \operatorname{grad} \phi \cdot \mathbf{dr} = \int_{\mathscr{P}_1}^{\mathscr{P}_2} \mathrm{d}\phi = \phi(\mathscr{P}_2) - \phi(\mathscr{P}_1), \tag{3.129}$$

which is the higher-dimensional analog of the fundamental theorem of calculus. Equation (3.129) states that the integral is independent of the curve. In particular

$$\int_{\mathscr{C}} \operatorname{grad} \phi \cdot \mathbf{dr} = 0 \tag{3.130}$$

on a loop $\mathscr{C}$. Vector fields with the property (3.130) are called *conservative*. Equation (3.129) yields a coordinate-free definition of the gradient, $\operatorname{grad} : \mathcal{S}(\Omega) \to \mathcal{V}(\Omega)$, for an open domain $\Omega \subset E_3$:

$$\partial_{\mathbf{v}} \phi = \mathbf{v} \cdot \operatorname{grad} \phi = \lim_{s \to 0} \frac{\phi(\mathscr{P}_2) - \phi(\mathscr{P}_1)}{s} = \lim_{\Delta t \to 0} \frac{\phi(\mathbf{r}(t + \Delta t)) - \phi(\mathbf{r}(t))}{\Delta t} , \tag{3.131}$$

In accordance with the fundamental theorem of calculus, Eq. (3.131), can be interpreted as the integral over the boundary $\partial\mathscr{S}$ related to the length s of the path $\mathscr{S}$.

Curl A coordinate-free definition of the curl operator, $\operatorname{curl} : \mathcal{V}(\Omega) \to \mathcal{V}(\Omega)$, for an open domain $\Omega \subset E_3$, is given by

$$\mathbf{n} \cdot \operatorname{curl} \mathbf{g} = \lim_{a \to 0} \frac{\int_{\partial\mathscr{A}} \mathbf{g} \cdot d\mathbf{r}}{a}, \tag{3.132}$$

which relates the path integral of the vector field along the boundary $\partial\mathscr{A}$ to the area a of the surface $\mathscr{A} \subset \Omega$. The surface and its boundary need to be consistently oriented. For Cartesian coordinates the method of calculating the

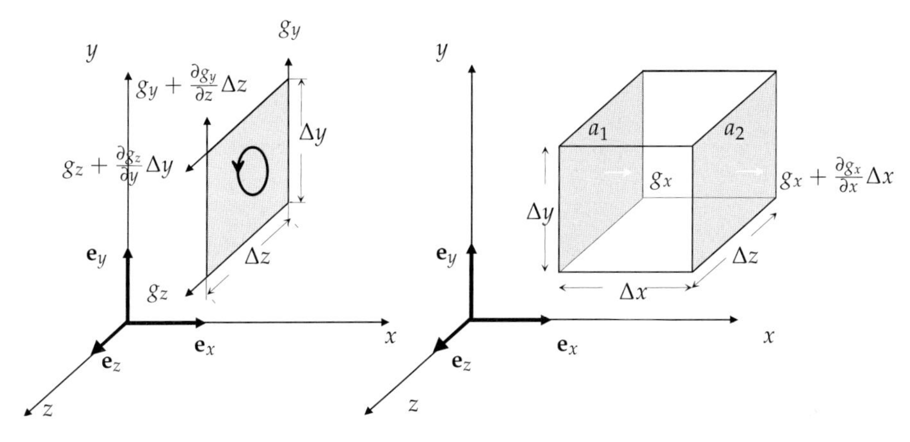

Figure 3.5 On the calculation of $\operatorname{curl} \mathbf{g}$ (left) and $\operatorname{div} \mathbf{g}$ (right) in Cartesian coordinates.

x-component of $\operatorname{curl} \mathbf{g}$ is explained by means of Figure 3.5 (left). Integrating along the rim of the rectangle yields

$$\begin{aligned}
\mathbf{e}_x \cdot \operatorname{curl} \mathbf{g} &= \lim_{a \to 0} \frac{\int_{\partial\mathscr{A}} \mathbf{g} \cdot d\mathbf{r}}{a} \\
&= \lim_{\Delta y, \Delta z \to 0} \frac{g_y \Delta y + \left(g_z + \frac{\partial g_z}{\partial y}\Delta y\right)\Delta z - \left(g_y + \frac{\partial g_y}{\partial z}\Delta z\right)\Delta y - g_z \Delta z}{\Delta y \, \Delta z} \\
&= \frac{\partial g_z}{\partial y} - \frac{\partial g_y}{\partial z}.
\end{aligned} \tag{3.133}$$

Repeating this step for the other two components results in

$$\operatorname{curl} \mathbf{g} = \left(\frac{\partial g_z}{\partial y} - \frac{\partial g_y}{\partial z}\right)\mathbf{e}_x + \left(\frac{\partial g_x}{\partial z} - \frac{\partial g_z}{\partial x}\right)\mathbf{e}_y + \left(\frac{\partial g_y}{\partial x} - \frac{\partial g_x}{\partial y}\right)\mathbf{e}_z. \tag{3.134}$$

Divergence A coordinate-free definition of the divergence operator, div : $\mathcal{V}(\Omega) \rightarrow \mathcal{S}(\Omega)$, for an open domain $\Omega \subset E_3$, is given by

$$\operatorname{div} \mathbf{g} = \lim_{V \to 0} \frac{\int_{\partial \mathscr{V}} \mathbf{g} \cdot \mathbf{da}}{V}, \tag{3.135}$$

which relates the integral over the boundary surface $\partial \mathscr{V}$ of the volume $\mathscr{V}$ to the volumetric extent V of this volume. For the gray pair of surfaces shown in Figure 3.5 (right) we can write

$$\lim_{V \to 0} \frac{\int_{\mathscr{A}_1, \mathscr{A}_2} g_x \mathrm{d}a}{V} = \lim_{\Delta x, \Delta y, \Delta z \to 0} \frac{-g_x \Delta y \Delta z + \left(g_x + \frac{\partial g_x}{\partial x} \Delta x\right) \Delta y \Delta z}{\Delta x \Delta y \Delta z} = \frac{\partial g_x}{\partial x}. \tag{3.136}$$

Repetition for the other two pairs of surfaces and summing up yields

$$\operatorname{div} \mathbf{g} = \frac{\partial g_x}{\partial x} + \frac{\partial g_y}{\partial y} + \frac{\partial g_z}{\partial z}. \tag{3.137}$$

3.10 Integral Theorems

3.10.1 The Kelvin–Stokes Theorem

Consider an oriented, simply-connected and piecewise-smooth surface $\mathscr{A}$ with a simply-connected, closed, and piecewise-smooth boundary $\mathscr{S} = \partial \mathscr{A}$, traced in a positive sense with respect to the surface; see Figure 2.6. For a smooth vector field $\mathbf{g} \in \mathcal{V}(\Omega)$, the *Kelvin–Stokes*[19] *theorem* is given by

$$\int_{\mathscr{A}} \operatorname{curl} \mathbf{g} \cdot \mathbf{da} = \int_{\partial \mathscr{A}} \mathbf{g} \cdot \mathbf{dr}, \tag{3.138}$$

with the vectorial surface and line elements defined by $\mathbf{da} := \mathbf{n}\, \mathrm{d}a$ and $\mathbf{dr} := \mathbf{T}\, \mathrm{d}s$.

Proof. According to Figure 3.6 (left) we can split the surface $\mathscr{A}$ into two surfaces $\mathscr{A}_1$ and $\mathscr{A}_2$ with boundaries $\mathscr{S}_1 = \mathscr{S}_{11} + \mathscr{S}_{12}$ and $\mathscr{S}_2 = \mathscr{S}_{21} + \mathscr{S}_{22}$, respectively. It follows that

$$\int_{\partial \mathscr{A}} \mathbf{g} \cdot \mathbf{dr} = \int_{\mathscr{S}_1} \mathbf{g} \cdot \mathbf{dr} + \int_{\mathscr{S}_2} \mathbf{g} \cdot \mathbf{dr} = \int_{\mathscr{S}_{11}} \mathbf{g} \cdot \mathbf{dr} + \int_{\mathscr{S}_{22}} \mathbf{g} \cdot \mathbf{dr}, \tag{3.139}$$

19 William Thomson (Lord Kelvin) (1824–1907), Sir Georg Stokes (1819–1903).

because the line integrals over the paths $\mathscr{S}_{12}$ and $\mathscr{S}_{21}$ cancel out. If the surface is now repeatedly segmented, we obtain

$$\begin{aligned}\int_{\partial\mathscr{A}} \mathbf{g}\cdot \mathrm{d}\mathbf{r} &= \lim_{I\to\infty}\sum_{i=1}^{I}\int_{\partial\mathscr{A}_i}\mathbf{g}\cdot\mathrm{d}\mathbf{r} = \lim_{I\to\infty}\sum_{i=1}^{I}\Delta a_i \frac{1}{\Delta a_i}\int_{\partial\mathscr{A}_i}\mathbf{g}\cdot\mathrm{d}\mathbf{r} \\ &= \lim_{I\to\infty}\sum_{i=1}^{I}(\operatorname{curl}\mathbf{g})_i\cdot\mathbf{n}\,\Delta a_i = \int_{\mathscr{A}}\operatorname{curl}\mathbf{g}\cdot\mathrm{d}\mathbf{a}\,. \end{aligned} \tag{3.140}$$

□

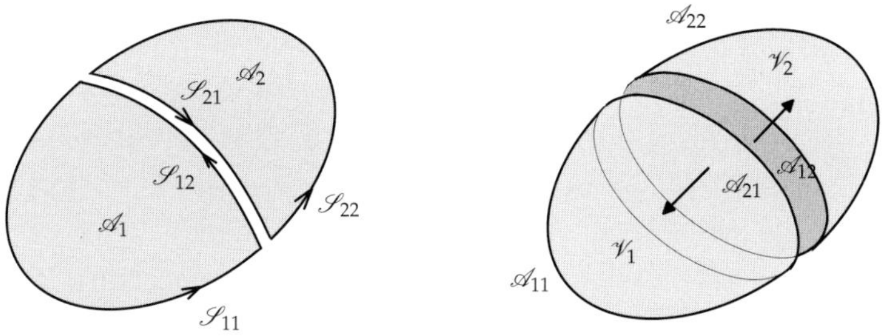

Figure 3.6 Left: Segmentation of a surface for the proof of the Kelvin–Stokes theorem. Right: Segmentation of a volume for the proof of the Gauss–Ostrogradski theorem.

3.10.2
Green's Theorem in the Plane

Consider the special case of an area $\mathscr{A}$ in the xy-plane, with its consistently oriented unit normal vector $\mathbf{n} = \mathbf{e}_z$, and with a closed, piecewise-smooth boundary $\partial\mathscr{A}$. Any two-dimensional, smooth vector field $\mathbf{g}$ in this plane can be expressed as $\mathbf{g} = g_x\mathbf{e}_x + g_y\mathbf{e}_y$ with g_x and g_y being smooth functions in x and y. With the position vector on the boundary given as $\mathbf{r} = x\mathbf{e}_x + y\mathbf{e}_y$ it follows that $\mathbf{g}\cdot\mathrm{d}\mathbf{r} = g_x\mathrm{d}x + g_y\mathrm{d}y$. From

$$\operatorname{curl}\mathbf{g}\cdot\mathbf{n} = \left(-\frac{\partial g_y}{\partial z}\mathbf{e}_x + \frac{\partial g_x}{\partial z}\mathbf{e}_y + \left(\frac{\partial g_y}{\partial x} - \frac{\partial g_x}{\partial y}\right)\mathbf{e}_z\right)\cdot\mathbf{e}_z = \frac{\partial g_y}{\partial x} - \frac{\partial g_x}{\partial y} \tag{3.141}$$

and the Kelvin–Stokes theorem, $\int_{\mathscr{A}}\operatorname{curl}\mathbf{g}\cdot\mathbf{n}\mathrm{d}a = \int_{\partial\mathscr{A}}\mathbf{g}\cdot\mathrm{d}\mathbf{r}$, we derive

$$\int_{\mathscr{A}}\left(\frac{\partial g_y}{\partial x} - \frac{\partial g_x}{\partial y}\right)\mathrm{d}x\mathrm{d}y = \int_{\partial\mathscr{A}} g_x\mathrm{d}x + g_y\mathrm{d}y, \tag{3.142}$$

which is called *Green's theorem in the plane* or the *Green–Gauss theorem.*

3.10.3
The Gauss–Ostrogradski Divergence Theorem

Let $\mathscr{V}$ be a volume that is bounded by the closed (piecewise-smooth and consistently oriented) surface $\mathscr{A} = \partial\mathscr{V}$ and consider a smooth vector field $\mathbf{g} \in \mathcal{V}(\Omega)$. The *Gauss–Ostrogradski divergence theorem*,[20] or for short the Gauss theorem, is given by

$$\int_{\mathscr{V}} \operatorname{div} \mathbf{g} \, \mathrm{d}V = \int_{\partial\mathscr{V}} \mathbf{g} \cdot \mathrm{d}\mathbf{a} \,, \tag{3.143}$$

where the surface normal vector of the surface $\mathscr{A} = \partial\mathscr{V}$ points to the outward direction.

Proof. According to Figure 3.6 (right) we may split the volume $\mathscr{V}$ into two volumes $\mathscr{V}_1$ and $\mathscr{V}_2$ with surfaces $\mathscr{A}_1 = \mathscr{A}_{11} + \mathscr{A}_{12}$ and $\mathscr{A}_2 = \mathscr{A}_{21} + \mathscr{A}_{22}$, respectively. It follows that

$$\int_{\partial\mathscr{V}} \mathbf{g} \cdot \mathrm{d}\mathbf{a} = \int_{\mathscr{A}_1} \mathbf{g} \cdot \mathrm{d}\mathbf{a} + \int_{\mathscr{A}_2} \mathbf{g} \cdot \mathrm{d}\mathbf{a} = \int_{\mathscr{A}_{11}} \mathbf{g} \cdot \mathrm{d}\mathbf{a} + \int_{\mathscr{A}_{22}} \mathbf{g} \cdot \mathrm{d}\mathbf{a} \tag{3.144}$$

because the surface integrals over the surfaces $\mathscr{A}_{12}$ and $\mathscr{A}_{21}$ cancel out. If the volume is now repeatedly segmented, we obtain

$$\begin{aligned} \int_{\partial\mathscr{V}} \mathbf{g} \cdot \mathrm{d}\mathbf{a} &= \lim_{I\to\infty} \sum_{i=1}^{I} \int_{\partial\mathscr{V}_i} \mathbf{g} \cdot \mathrm{d}\mathbf{a} = \lim_{I\to\infty} \sum_{i=1}^{I} \Delta V_i \frac{1}{\Delta V_i} \int_{\partial\mathscr{V}_i} \mathbf{g} \cdot \mathrm{d}\mathbf{a} \\ &= \lim_{I\to\infty} \sum_{i=1}^{I} (\operatorname{div} \mathbf{g})_i \, \Delta V_i = \int_{\mathscr{V}} \operatorname{div} \mathbf{g} \, \mathrm{d}V \,. \end{aligned} \tag{3.145}$$

□

3.10.4
A Variant of the Gauss Theorem

We will now derive a variant of the Gauss theorem, for a position-dependent vector field **g** and a uniform vector field **d**:

$$\begin{aligned} \int_{\mathscr{V}} \operatorname{div}(\mathbf{g} \times \mathbf{d}) \, \mathrm{d}V &= \int_{\partial\mathscr{V}} (\mathbf{g} \times \mathbf{d}) \cdot \mathrm{d}\mathbf{a} \,, \\ \int_{\mathscr{V}} \mathbf{d} \cdot \operatorname{curl} \mathbf{g} \, \mathrm{d}V - \int_{\mathscr{V}} \mathbf{g} \cdot \operatorname{curl} \mathbf{d} \, \mathrm{d}V &= -\int_{\partial\mathscr{V}} \mathbf{d} \cdot \mathbf{g} \times \mathrm{d}\mathbf{a} \,, \\ \mathbf{d} \cdot \int_{\mathscr{V}} \operatorname{curl} \mathbf{g} \, \mathrm{d}V &= -\mathbf{d} \cdot \int_{\partial\mathscr{V}} \mathbf{g} \times \mathrm{d}\mathbf{a} \,, \\ \int_{\mathscr{V}} \operatorname{curl} \mathbf{g} \, \mathrm{d}V &= -\int_{\partial\mathscr{V}} \mathbf{g} \times \mathrm{d}\mathbf{a} \,. \end{aligned} \tag{3.146}$$

20 Michel Vassilievitch Ostrogradski (1801–1862).

3.10.5 Green's First Identity

Green's first theorem can be derived from the Gauss theorem (3.143) by setting $\mathbf{g} = \psi \operatorname{grad} \phi$ for scalar fields $\psi, \phi \in \mathcal{S}(\Omega)$:

$$\int_{\mathcal{V}} \operatorname{div} \mathbf{g} \, \mathrm{d}V = \int_{\partial\mathcal{V}} \psi \operatorname{grad} \phi \cdot \mathrm{d}\mathbf{a} \,,$$

$$\int_{\mathcal{V}} \operatorname{div} (\psi \operatorname{grad} \phi) \, \mathrm{d}V = \int_{\mathcal{V}} (\psi \operatorname{div} \operatorname{grad} \phi + \operatorname{grad} \psi \cdot \operatorname{grad} \phi) \, \mathrm{d}V \,. \tag{3.147}$$

Interchanging ψ and ϕ yields

$$\int_{\mathcal{V}} (\phi \operatorname{div} \operatorname{grad} \psi + \operatorname{grad} \phi \cdot \operatorname{grad} \psi) \, \mathrm{d}V = \int_{\partial\mathcal{V}} \phi \operatorname{grad} \psi \cdot \mathrm{d}\mathbf{a} \,. \tag{3.148}$$

Using $\operatorname{div} \operatorname{grad} \psi = \nabla^2 \psi$ and the notation

$$\operatorname{grad} \psi \cdot \mathbf{a} = \partial_{\mathbf{n}} \psi \, a \tag{3.149}$$

for the directional derivative, Eq. (3.148) can be rewritten as

$$\int_{\mathcal{V}} \left(\operatorname{grad} \phi \cdot \operatorname{grad} \psi + \phi \nabla^2 \psi \right) \, \mathrm{d}V = \int_{\partial\mathcal{V}} \phi \, \partial_{\mathbf{n}} \psi \, \mathrm{d}a \,, \tag{3.150}$$

which is Green's first identity. Notice the removal of the second derivative of ψ. For $\phi = \psi$ it follows that

$$\int_{\mathcal{V}} \left((\operatorname{grad} \phi)^2 + \phi \nabla^2 \phi \right) \, \mathrm{d}V = \int_{\partial\mathcal{V}} \phi \partial_{\mathbf{n}} \phi \, \mathrm{d}a \,. \tag{3.151}$$

Equation (3.151) further reduces in two dimensions to

$$\int_{\mathcal{A}} \left((\operatorname{grad} \phi)^2 + \phi \nabla^2 \phi \right) \, \mathrm{d}a = \int_{\partial\mathcal{A}} \phi \, \partial_{\mathbf{n}} \phi \, \mathrm{d}r \tag{3.152}$$

and in 1D to

$$\int_{x_1}^{x_2} \left(\phi \phi'' + \phi'^2 \right) \, \mathrm{d}x = \left[\phi \phi' \right]_{x_1}^{x_2} \,, \tag{3.153}$$

where the prime denotes the x-derivative. Equation (3.153) is nothing but the *integration-by-parts* rule, better known in the form

$$\int_a^b f(x) g'(x) \, \mathrm{d}x = \left[g(x) f(x) \right]_a^b - \int_a^b g(x) f'(x) \, \mathrm{d}x \tag{3.154}$$

for the C^1 functions $f, g : I \to \mathbb{R}$.

3.10.6
Green's Second Identity (Green's Theorem)

Subtracting Eq. (3.147) from Eq. (3.148), and substituting div grad $\phi = \nabla^2\phi$, yields *Green's theorem*

$$\int_{\mathscr{V}} \left(\phi\nabla^2\psi - \psi\nabla^2\phi\right) \, \mathrm{d}V = \int_{\partial\mathscr{V}} (\phi \operatorname{grad} \psi - \psi \operatorname{grad} \phi) \cdot \mathrm{d}\mathbf{a}. \tag{3.155}$$

The right-hand side of Eq. (3.155) can again be written with the notation (3.149) for the directional derivative as

$$\int_{\partial\mathscr{V}} (\phi \operatorname{grad} \psi - \psi \operatorname{grad} \phi) \cdot \mathrm{d}\mathbf{a} = \int_{\partial\mathscr{V}} (\phi\, \partial_{\mathbf{n}}\psi - \psi\, \partial_{\mathbf{n}}\phi) \, \mathrm{d}a\,. \tag{3.156}$$

Green's theorem reduces in two dimensions to

$$\int_{\mathscr{A}} \left(\phi\nabla^2\psi - \psi\nabla^2\phi\right) \, \mathrm{d}a = \int_{\partial\mathscr{A}} (\phi\, \partial_{\mathbf{n}}\psi - \psi\, \partial_{\mathbf{n}}\phi) \, \mathrm{d}r \tag{3.157}$$

and reads in one dimension:

$$\int_{x_1}^{x_2} (\phi\, \psi'' - \psi\phi'') \, \mathrm{d}x = \left[\phi\psi' - \psi\phi'\right]_{x_1}^{x_2}. \tag{3.158}$$

3.10.7
Vector Form of Green's Theorem

From div $(\mathbf{a} \times \mathbf{b}) = \mathbf{b} \cdot \operatorname{curl} \mathbf{a} - \mathbf{a} \cdot \operatorname{curl} \mathbf{b}$ we obtain

$$\int_{\mathscr{V}} \mathbf{a} \cdot \operatorname{curl} \mathbf{b} \, \mathrm{d}V = \int_{\mathscr{V}} \mathbf{b} \cdot \operatorname{curl} \mathbf{a} \, \mathrm{d}V - \int_{\mathscr{V}} \operatorname{div} (\mathbf{a} \times \mathbf{b}) \, \mathrm{d}V. \tag{3.159}$$

Applying the Gauss theorem (3.143) to the right-hand side of Eq. (3.159), writing $\mathrm{d}\mathbf{a}$ as $\mathbf{n}\, \mathrm{d}a$, and remembering that $\mathbf{n} \cdot (\mathbf{a} \times \mathbf{b}) = \mathbf{a} \cdot (\mathbf{b} \times \mathbf{n})$ yields

$$\int_{\mathscr{V}} \mathbf{a} \cdot \operatorname{curl} \mathbf{b} \, \mathrm{d}V = \int_{\mathscr{V}} \mathbf{b} \cdot \operatorname{curl} \mathbf{a} \, \mathrm{d}V - \int_{\partial\mathscr{V}} \mathbf{a} \cdot (\mathbf{b} \times \mathbf{n}) \, \mathrm{d}a. \tag{3.160}$$

3.10.8
Generalization of the Integration-by-Parts Rule

The generalization of the integration-by-parts rule can be obtained from the divergence product rule:

$$\operatorname{div} (\phi \cdot \mathbf{a}) = \phi \operatorname{div} \mathbf{a} + \mathbf{a} \cdot \operatorname{grad} \phi \tag{3.161}$$

and the Gauss theorem for $\phi \cdot \mathbf{a}$,

$$\int_{\mathscr{V}} \operatorname{div} (\phi \cdot \mathbf{a}) \mathrm{d}V = \int_{\partial\mathscr{V}} \phi(\mathbf{a} \cdot \mathbf{n}) \mathrm{d}a. \tag{3.162}$$

Therefore

$$-\int_{\mathcal{V}} \mathbf{a} \cdot \operatorname{grad} \phi \, \mathrm{d}V = \int_{\mathcal{V}} \phi \operatorname{div} \mathbf{a} \, \mathrm{d}V - \int_{\partial \mathcal{V}} \phi (\mathbf{a} \cdot \mathbf{n}) \, \mathrm{d}a. \tag{3.163}$$

3.10.9
The Stratton Theorems

From the Gauss theorem we derive

$$\int_{\mathcal{V}} \operatorname{div}(\mathbf{a} \times \operatorname{curl} \mathbf{b}) \mathrm{d}V = \int_{\partial \mathcal{V}} (\mathbf{a} \times \operatorname{curl} \mathbf{b}) \cdot \mathbf{n} \, \mathrm{d}a \,. \tag{3.164}$$

Using Eq. (3.68) yields directly the *first theorem by Stratton* [8],

$$\int_{\mathcal{V}} (\operatorname{curl} \mathbf{a} \operatorname{curl} \mathbf{b} - \mathbf{a} \operatorname{curl} \operatorname{curl} \mathbf{b}) \, \mathrm{d}V = \int_{\partial \mathcal{V}} (\mathbf{a} \times \operatorname{curl} \mathbf{b}) \cdot \mathbf{n} \, \mathrm{d}a \,, \tag{3.165}$$

which is a fundamental theorem for the proof of uniqueness of the vector-potential formulation for divergence-free fields. Interchanging **a** and **b** in the first theorem and subtracting from Eq. (3.165) yields the second theorem

$$\int_{\mathcal{V}} (\mathbf{a} \operatorname{curl} \operatorname{curl} \mathbf{b} - \mathbf{b} \operatorname{curl} \operatorname{curl} \mathbf{a}) \, \mathrm{d}V = \int_{\partial \mathcal{V}} (\mathbf{b} \times \operatorname{curl} \mathbf{a} - \mathbf{a} \times \operatorname{curl} \mathbf{b}) \cdot \mathbf{n} \, \mathrm{d}a \,. \tag{3.166}$$

3.11
Curvilinear Coordinates

Let a coordinate transformation from (x^1, x^2, x^3) to (u^1, u^2, u^3) be given in the component form as

$$x^i = x^i(u^1, u^2, u^3), \qquad i = 1, 2, 3. \tag{3.167}$$

In this section we will again use the Ricci notation and Einstein's summation convention. We recall from Section 2.5.1 that the anonymous notation (3.167) does not distinguish between the coordinates x^i and the smooth coordinate functions $x^i(u^1, u^2, u^3)$ and little space is waisted in defining the domain $U = (u_a^1, u_b^1) \times (u_a^2, u_b^2) \times (u_a^3, u_b^3)$.

As all functions in Eqs. (3.167) are assumed to be single-valued and smooth, the correspondence between the x^i and u^j is unique with a nonzero Jacobian:

$$\det[J] = \begin{vmatrix} \frac{\partial x^1}{\partial u^1} & \frac{\partial x^1}{\partial u^2} & \frac{\partial x^1}{\partial u^3} \\ \frac{\partial x^2}{\partial u^1} & \frac{\partial x^2}{\partial u^2} & \frac{\partial x^2}{\partial u^3} \\ \frac{\partial x^3}{\partial u^1} & \frac{\partial x^3}{\partial u^2} & \frac{\partial x^3}{\partial u^3} \end{vmatrix} \neq 0 \,. \tag{3.168}$$

Thus Eqs. (3.167) can be solved for u^1, u^2, u^3:

$$u^1 = u^1(x^1, x^2, x^3), \qquad u^2 = u^2(x^1, x^2, x^3), \qquad u^3 = u^3(x^1, x^2, x^3). \tag{3.169}$$

If the $x^i, i = 1, 2, 3$, are the Cartesian coordinates x, y, z, the u^j are called curvilinear coordinates. Let $\mathbf{r}(x^1, x^2, x^3)$ be the position vector of a point $\mathscr{P}$ that can be expressed as a function of the curvilinear coordinates u^j as

$$\begin{aligned}\mathbf{r}(x^1, x^2, x^3) &= \tilde{\mathbf{r}}(u^1, u^2, u^3) \\ &= x^1(u^1, u^2, u^3)\,\mathbf{e}_1 + x^2(u^1, u^2, u^3)\,\mathbf{e}_2 + x^3(u^1, u^2, u^3)\,\mathbf{e}_3 \,. \end{aligned} \tag{3.170}$$

Then the tangent vector to the coordinate curve

$$u^1 \mapsto \tilde{\mathbf{r}}(u^1, \beta, \gamma)\,, \tag{3.171}$$

where β and γ are constants, is given by

$$\mathbf{T}_1 = \frac{\partial \tilde{\mathbf{r}}}{\partial u^1} = \frac{\partial x^1(u^1, \beta, \gamma)}{\partial u^1}\,\mathbf{e}_1 + \frac{\partial x^2(u^1, \beta, \gamma)}{\partial u^1}\,\mathbf{e}_2 + \frac{\partial x^3(u^1, \beta, \gamma)}{\partial u^1}\,\mathbf{e}_3 \,. \tag{3.172}$$

The three tangent vectors are linearly independent and thus form a *covariant basis* $\{\mathbf{T}_1, \mathbf{T}_2, \mathbf{T}_3\}$ of the tangent space $T_{\mathscr{P}}E_3$. We write

$$\mathbf{g}_i := \mathbf{T}_i = \frac{\partial \tilde{\mathbf{r}}}{\partial u^i} \tag{3.173}$$

and call $h_i := |\mathbf{g}_i|$ the *scale factors*. The unit basis vector in the curvilinear coordinate system is denoted

$$\mathbf{e}_{u_i} := \frac{\mathbf{g}_i}{h_i} \,. \tag{3.174}$$

The coordinate surface

$$u^1, u^2 \mapsto \tilde{\mathbf{r}}(u^1, u^2, \gamma), \tag{3.175}$$

can be expressed in Monge's form as $u^3(x^1, x^2, x^3) = \gamma$, where γ is a constant. The normal vector to the coordinate surface at $\mathscr{P}$ is given by

$$\mathbf{N}_3 = \operatorname{grad} u^3 = \frac{\partial u^3(x^1, x^2, x^3)}{\partial x^1}\,\mathbf{e}_1 + \frac{\partial u^3(x^1, x^2, x^3)}{\partial x^2}\,\mathbf{e}_2 + \frac{\partial u^3(x^1, x^2, x^3)}{\partial x^3}\,\mathbf{e}_3 \,. \tag{3.176}$$

The basis $\{\mathbf{N}_1, \mathbf{N}_2, \mathbf{N}_3\}$ is called the *contravariant basis*. We can write

$$\mathbf{g}^i := \mathbf{N}_i = \operatorname{grad} u^i, \tag{3.177}$$

and define the scale factors by $h^i := |\mathbf{g}^i| = |\operatorname{grad} u^i|$.

3.11.1
Components of a Vector Field

At each point of the curvilinear coordinate system, there exist two sets of local basis vectors: the covariant $\{\mathbf{g}_1, \mathbf{g}_2, \mathbf{g}_3\}$ tangent to the coordinate lines and the contravariant $\{\mathbf{g}^1, \mathbf{g}^2, \mathbf{g}^3\}$ normal to the coordinate surfaces; Figure 3.7 illustrates these frames. They form dual sets of basis vectors, $\mathbf{g}^i \cdot \mathbf{g}_j = \delta^i_j$, which can be proved as follows:

Proof. The differential of $\tilde{\mathbf{r}}(u^1, u^2, u^3)$ is

$$\mathrm{d}\tilde{\mathbf{r}} = \frac{\partial \tilde{\mathbf{r}}}{\partial u^1}\mathrm{d}u^1 + \frac{\partial \tilde{\mathbf{r}}}{\partial u^2}\mathrm{d}u^2 + \frac{\partial \tilde{\mathbf{r}}}{\partial u^3}\mathrm{d}u^3 . \tag{3.178}$$

Scalar multiplication by $\operatorname{grad} u^1$ gives

$$\begin{aligned}\operatorname{grad} u^1 \cdot \mathrm{d}\tilde{\mathbf{r}} = \mathrm{d}u^1 &= (\operatorname{grad} u^1 \cdot \frac{\partial \tilde{\mathbf{r}}}{\partial u^1})\mathrm{d}u^1 + (\operatorname{grad} u^1 \cdot \frac{\partial \tilde{\mathbf{r}}}{\partial u^2})\mathrm{d}u^2 \\ &+ (\operatorname{grad} u^1 \cdot \frac{\partial \tilde{\mathbf{r}}}{\partial u^3})\mathrm{d}u^3\end{aligned} \tag{3.179}$$

and therefore

$$\operatorname{grad} u^1 \cdot \frac{\partial \tilde{\mathbf{r}}}{\partial u^1} = 1, \qquad \operatorname{grad} u^1 \cdot \frac{\partial \tilde{\mathbf{r}}}{\partial u^2} = 0, \qquad \operatorname{grad} u^1 \cdot \frac{\partial \tilde{\mathbf{r}}}{\partial u^3} = 0. \tag{3.180}$$

Repeating these steps for $\operatorname{grad} u^2$ and $\operatorname{grad} u^3$ proves that $\mathbf{g}^i \cdot \mathbf{g}_j = \delta^i_j$. □

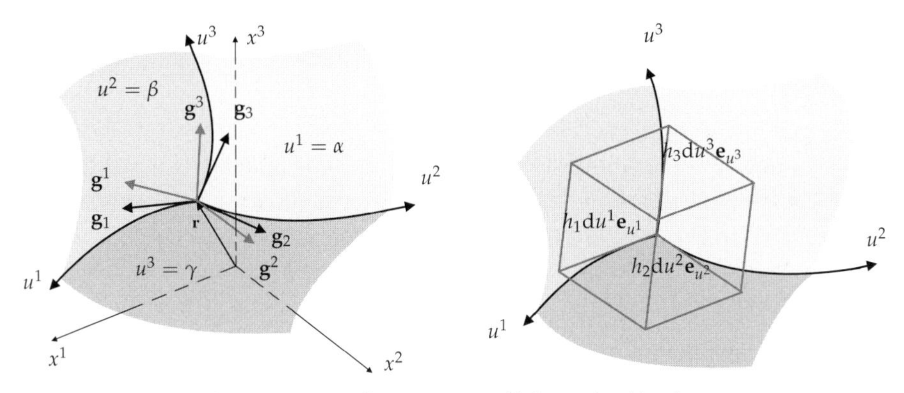

Figure 3.7 Left: Curvilinear coordinate system with two sets of basis vectors. Right: Volume element for an orthogonal curvilinear coordinate system.

The bases become identical if and only if the curvilinear coordinate system is orthogonal. A vector field $\mathbf{a}$ can be represented in terms of the co- or contravariant basis as

$$\mathbf{a} = c^i \mathbf{g}_i = c^i \frac{\partial \tilde{\mathbf{r}}}{\partial u^i}, \tag{3.181}$$

$$\mathbf{a} = c_i \mathbf{g}^i = c_i \operatorname{grad} u^i, \tag{3.182}$$

where the c_i are called the covariant and c^i the contravariant coefficients of $\mathbf{a}$. The notion derives from the type of transformation of coefficients and bases under changes of coordinates.

When the vector field $\mathbf{a}$ is differentiated, we must bear in mind that the coordinate lines are not straight and therefore the basis vectors change their direction. Consequently, the basis vectors must be differentiated as well:

$$\frac{\partial \mathbf{a}}{\partial u^j} = \frac{\partial c^i}{\partial u^j} \mathbf{g}_i + c^i \frac{\partial \mathbf{g}_i}{\partial u^j} . \tag{3.183}$$

3.11.2
Contravariant Coefficients

Let another set of curvilinear coordinates be given by (v^1, v^2, v^3) that is smoothly related to the u^i by $u^i = u^i(v^1, v^2, v^3)$ for $i = 1, 2, 3$, with nonvanishing Jacobian. Then

$$\mathrm{d}\tilde{\mathbf{r}} = \frac{\partial \tilde{\mathbf{r}}}{\partial u^i} \mathrm{d}u^i = \mathrm{d}u^i \mathbf{g}_i, \tag{3.184}$$

and

$$\mathrm{d}\bar{\mathbf{r}} = \frac{\partial \bar{\mathbf{r}}}{\partial v^i} \mathrm{d}v^i = \mathrm{d}v^i \mathbf{h}_i. \tag{3.185}$$

Substituting

$$\mathrm{d}u^i = \frac{\partial u^i}{\partial v^j} \mathrm{d}v^j, \tag{3.186}$$

into Eq. (3.184), and comparing the coefficients of $\mathrm{d}v^i$, we obtain

$$\mathbf{h}_i = \frac{\partial u^j}{\partial v^i} \mathbf{g}_j =: T^j_i \mathbf{g}_j . \tag{3.187}$$

The change of coordinates induces a change of basis in the tangent space. The vector can now be expressed in terms of the two bases by

$$\mathbf{a} = d^i \mathbf{h}_i = c^j \mathbf{g}_j . \tag{3.188}$$

Combining Eq. (3.188) with (3.187) yields

$$d^i = \left(T^{-1}\right)^i_j c^j ; \tag{3.189}$$

the coefficients transform the other way (contravariant) with respect to the basis vectors.

3.11.3 Covariant Coefficients

A similar calculation shows that the covariant coefficients of **a** transform the same way as the basis vectors. We write the vector in the covariant form as

$$\mathbf{a} = c_i \operatorname{grad} u^i = c_i \mathbf{g}^i = d_i \operatorname{grad} v^i = d_i \mathbf{h}^i . \tag{3.190}$$

With Eq. (3.176) we obtain, after a comparison of the coefficients,

$$c_j \frac{\partial u^j}{\partial x^i} = d_j \frac{\partial v^j}{\partial x^i} . \tag{3.191}$$

Application of the chain rule yields

$$\frac{\partial u^j}{\partial x^i} = \frac{\partial u^j}{\partial v^1}\frac{\partial v^1}{\partial x^i} + \frac{\partial u^j}{\partial v^2}\frac{\partial v^2}{\partial x^i} + \frac{\partial u^j}{\partial v^3}\frac{\partial v^3}{\partial x^i} . \tag{3.192}$$

Substituting the terms (3.192) into Eq. (3.191) and comparing the coefficients of $\partial v^j / \partial x^i$ results in

$$d_i = c_j \frac{\partial u^j}{\partial v^i} =: T^j_i c_j , \tag{3.193}$$

in other words, the covariant coefficients transform the same way as the basis vectors. As an important theorem, it follows that the generalized inner product is an invariant with respect to coordinate transformations,

$$E = d^k d_k = \left(\left(T^{-1}\right)^k_i c^i\right)\left(T^l_k c_l\right) = \left(T^{-1}\right)^k_i T^l_k c^i c_l = \delta^l_i c^i c_l = c^i c_i . \tag{3.194}$$

3.12 Integration on Space Elements

With the preliminaries of the previous section behind us we can turn to p-dimensional space elements $\mathscr{M} : U \to E_n : u^1, \ldots, u^p \mapsto \mathbf{r}(u^1, \ldots, u^p)$ on which lives a potential $\phi : \mathscr{M} \to \mathbb{R}$. We will study integrals of the form

$$\int_{\mathscr{M}} \phi \, \mathrm{d}S_p \in \mathbb{R} , \tag{3.195}$$

where dS_p denotes the p-dimensional surface element on $\mathscr{M}$. This yields for $n = 3$ the ordinary line, surface, and volume elements $dS_1 = ds$, $dS_2 = da$, and $dS_3 = dV$, respectively. Consider the Jacobi matrix $[J] := (\frac{\partial x^i}{\partial u^j})$ of the coordinate transformation (3.167) from a general coordinate system (u^i) to the Cartesian system (x^i). The metric tensor $[G] := (g_{ij})$ of the general system (u^i) with its coefficients

$$g_{ij} := \mathbf{g}_i \cdot \mathbf{g}_j = \frac{\partial \tilde{\mathbf{r}}}{\partial u^i} \cdot \frac{\partial \tilde{\mathbf{r}}}{\partial u^j} = \frac{\partial x^1}{\partial u^i}\frac{\partial x^1}{\partial u^j} + \frac{\partial x^2}{\partial u^i}\frac{\partial x^2}{\partial u^j} + \frac{\partial x^3}{\partial u^i}\frac{\partial x^3}{\partial u^j} \tag{3.196}$$

is given by

$$[G] = [J]^T [J] , \tag{3.197}$$

and the determinant of this metric tensor,

$$g := \det(g_{ij}) , \tag{3.198}$$

is the *Gram determinant*.[21]

Example: Cylindrical coordinates with $T : \mathbb{R}^3 \to \mathbb{R}^3 : (r, \varphi, z) \mapsto (x, y, z)$ for $r > 0, 0 \le \varphi < 2\pi$ and defined by $x = r\cos\varphi$, $y = r\sin\varphi$, $z = z$. Then

$$[G] = [J]^T[J] = \begin{pmatrix} \cos\varphi & \sin\varphi & 0 \\ -r\sin\varphi & r\cos\varphi & 0 \\ 0 & 0 & 1 \end{pmatrix} \begin{pmatrix} \cos\varphi & -r\sin\varphi & 0 \\ \sin\varphi & r\cos\varphi & 0 \\ 0 & 0 & 1 \end{pmatrix}$$

$$= \begin{pmatrix} 1 & 0 & 0 \\ 0 & r^2 & 0 \\ 0 & 0 & 1 \end{pmatrix} , \tag{3.199}$$

and the Gram determinant is $\det[G] = r^2$. □

Consider the situation shown in Figure 3.4. The parallelepiped spanned by the basis vectors in the tangent space $T_{\mathscr{P}}\mathscr{A}$ is the image of the unit cube under the linear approximation for the parameter mapping $\mathscr{A}$. In the case of $p = n = 3$ we obtain a quadratic Jacobi matrix of this mapping, while for $p < n$ the matrix contains only p columns and is not quadratic. Thus we consider an orthonormal basis $(\mathbf{v}_1, \ldots, \mathbf{v}_p)$ in $T_{\mathscr{P}}\mathscr{A}$ and employ the basis isomorphism $\iota : \mathbb{R}^p \cong T_{\mathscr{P}}\mathscr{A}$. This brings the parallelepiped "down" into the parameter space $\mathbb{R}^p$. The parallelepiped is then the image of the unit cube under the linear mapping $[A] := \iota^{-1} \circ d\mathbf{r} : \mathbb{R}^p \to \mathbb{R}^p$ and its volume

21 Jörgen Pederson Gram (1876–1916).

is $|\det[A]|$. Because $\mathrm{d}\mathbf{r}(\mathbf{v}_i) = \iota([A]\{\mathbf{v}_i\})$ it follows [7] that the metric tensor (g_{ij}) is equal to the matrix product $[A]^t[A]$. Therefore $g = \det[A]^t \det[A]$ and, because $\det[A]^t = \det[A]$, it yields $|\det[A]| = \sqrt{g}$. The integral of a potential on the p-dimensional space elements is thus

$$\int_{\mathscr{A}} \phi \mathrm{d}S_p = \int_{\mathscr{A}} \phi(\tilde{\mathbf{r}}(u^1, \ldots, u^p)) \sqrt{g(u^1, \ldots, u^p)} \mathrm{d}u^1 \ldots \mathrm{d}u^p \,, \tag{3.200}$$

where S_p is the p-dimensional surface element. For the special cases of $n = 3$ we obtain for $p = 1$ (line)

$$\sqrt{g} = \left| \frac{\mathrm{d}\mathbf{r}(t)}{\mathrm{d}t} \right| . \tag{3.201}$$

For $p = 2$ (surface) we derive

$$\sqrt{g} = \sqrt{EG - F^2} \,, \tag{3.202}$$

where the parameters of the first quadratic differential form are given by

$$\begin{pmatrix} E & F \\ F & G \end{pmatrix} = \begin{pmatrix} \frac{\partial \mathbf{r}}{\partial u} \cdot \frac{\partial \mathbf{r}}{\partial u} & \frac{\partial \mathbf{r}}{\partial u} \cdot \frac{\partial \mathbf{r}}{\partial v} \\ \frac{\partial \mathbf{r}}{\partial v} \cdot \frac{\partial \mathbf{r}}{\partial u} & \frac{\partial \mathbf{r}}{\partial v} \cdot \frac{\partial \mathbf{r}}{\partial v} \end{pmatrix} . \tag{3.203}$$

For $p = 3$ (volume) we get

$$\sqrt{g} = |\det J| \,. \tag{3.204}$$

3.13 Orthogonal Coordinate Systems

For orthogonal coordinate systems in three dimensions we find

$$\frac{\partial \tilde{\mathbf{r}}}{\partial u^1} \cdot \frac{\partial \tilde{\mathbf{r}}}{\partial u^2} = \frac{\partial \tilde{\mathbf{r}}}{\partial u^2} \cdot \frac{\partial \tilde{\mathbf{r}}}{\partial u^3} = \frac{\partial \tilde{\mathbf{r}}}{\partial u^1} \cdot \frac{\partial \tilde{\mathbf{r}}}{\partial u^3} = 0 \,. \tag{3.205}$$

The metric becomes a diagonal matrix of the form

$$[G] = (g_{ij}) = \left(\frac{\partial \tilde{\mathbf{r}}}{\partial u^i} \cdot \frac{\partial \tilde{\mathbf{r}}}{\partial u^j} \right) = \mathrm{diag}(h_1^2, h_2^2, h_3^2) \,. \tag{3.206}$$

Writing $\tilde{\mathbf{r}}(u^1, u^2, u^3)$ as $x^1(u^1, u^2, u^3)\, \mathbf{e}_1 + x^2(u^1, u^2, u^3)\, \mathbf{e}_2 + x^3(u^1, u^2, u^3)\, \mathbf{e}_3$ it yields

$$\begin{aligned} \mathrm{d}\tilde{\mathbf{r}} = \mathbf{g}_i \mathrm{d}u^i &= \frac{\partial \tilde{\mathbf{r}}}{\partial u^1} \mathrm{d}u^1 + \frac{\partial \tilde{\mathbf{r}}}{\partial u^2} \mathrm{d}u^2 + \frac{\partial \tilde{\mathbf{r}}}{\partial u^3} \mathrm{d}u^3 \\ &= h_1 \mathrm{d}u^1\, \mathbf{e}_{u^1} + h_2 \mathrm{d}u^2\, \mathbf{e}_{u^2} + h_3 \mathrm{d}u^3\, \mathbf{e}_{u^3} \,, \end{aligned} \tag{3.207}$$

and the differential arc length $\mathrm{d}s$ is determined by

$$\mathrm{d}s^2 = \mathrm{d}\tilde{\mathbf{r}} \cdot \mathrm{d}\tilde{\mathbf{r}} = g_{ij}\mathrm{d}u^i\mathrm{d}u^j \,. \tag{3.208}$$

For the orthogonal coordinate system this yields

$$h_i^2 = g_{ii}^2 = \left(\frac{\partial x^1}{\partial u^i}\right)^2 + \left(\frac{\partial x^2}{\partial u^i}\right)^2 + \left(\frac{\partial x^3}{\partial u^i}\right)^2 , \tag{3.209}$$

and therefore Eq. (3.208) reduces to

$$\mathrm{d}s^2 = h_1^2\,\mathrm{d}(u^1)^2 + h_2^2\,\mathrm{d}(u^2)^2 + h_3^2\,\mathrm{d}(u^3)^2 , \tag{3.210}$$

which is known as the *Pythagorean theorem in the small.* The position-dependent, unit basis vectors are thus given by

$$\mathbf{e}_{u^i}(\tilde{\mathbf{r}}) = \frac{1}{h_i}\frac{\partial \tilde{\mathbf{r}}}{\partial u^i} \,. \tag{3.211}$$

With $\mathrm{d}s_i = h_i\mathrm{d}u^i$ the surface element of a coordinate surface $\tilde{\mathbf{r}}(u^1, u^2, \gamma)$ is

$$\mathrm{d}a_3 = \left|\frac{\partial \tilde{\mathbf{r}}}{\partial u^1} \times \frac{\partial \tilde{\mathbf{r}}}{\partial u^2}\right| \mathrm{d}u^1\mathrm{d}u^2 = h_1h_2\mathrm{d}u^1\mathrm{d}u^2. \tag{3.212}$$

The differential volume element in orthogonal curvilinear coordinates is given by

$$\mathrm{d}V = \left|(h_1\mathrm{d}u^1\,\mathbf{e}_{u^1}) \cdot [(h_2\mathrm{d}u^2\,\mathbf{e}_{u^2}) \times (h_3\mathrm{d}u^3\,\mathbf{e}_{u^3})]\right| = h_1h_2h_3\mathrm{d}u^1\mathrm{d}u^2\mathrm{d}u^3 \,. \tag{3.213}$$

Using the Jacobian of the transformation $T : \mathbb{R}^3 \to \mathbb{R}^3 : (u^1, u^2, u^3) \mapsto (x, y, z)$ we obtain

$$\mathrm{d}x\,\mathrm{d}y\,\mathrm{d}z = \begin{vmatrix} \frac{\partial x}{\partial u^1} & \frac{\partial x}{\partial u^2} & \frac{\partial x}{\partial u^3} \\ \frac{\partial y}{\partial u^1} & \frac{\partial y}{\partial u^2} & \frac{\partial y}{\partial u^3} \\ \frac{\partial z}{\partial u^1} & \frac{\partial z}{\partial u^2} & \frac{\partial z}{\partial u^3} \end{vmatrix} \mathrm{d}u^1\,\mathrm{d}u^2\,\mathrm{d}u^3 \,. \tag{3.214}$$

If we set $x = y$ or $x = z$ or $y = z$, the Jacobian has equal rows and thus vanishes. Odd permutations of x, y, z change the sign of the determinant while even permutations do not. Hence we obtain

$$\mathrm{d}x\mathrm{d}y\mathrm{d}z = \mathrm{d}y\mathrm{d}z\mathrm{d}x = \mathrm{d}z\mathrm{d}x\mathrm{d}y = -\mathrm{d}y\mathrm{d}x\mathrm{d}z = -\mathrm{d}x\mathrm{d}z\mathrm{d}y = -\mathrm{d}z\mathrm{d}y\mathrm{d}x \,. \tag{3.215}$$

3.13.1 Differential Operators

Gradient The coordinate-free definition (3.131) of the gradient makes it possible to calculate a component of $\operatorname{grad}\phi$ in an arbitrary direction, from the difference of the scalar function in two neighboring points in this direction, divided by the distance between them. Consequently, for the component in the direction of the coordinate axis u^1 with $\Delta s_1 = h_1 \Delta u^1$ we obtain

$$\mathbf{e}_{u^1} \cdot \operatorname{grad}\phi = \lim_{\Delta u^1 \to 0} \frac{\phi(u^1 + \Delta u^1, u^2, u^3) - \phi(u^1, u^2, u^3)}{h_1 \Delta u^1} = \frac{1}{h_1} \frac{\partial \phi}{\partial u^1} . \tag{3.216}$$

Thus the gradient of a scalar field $\phi \in \mathcal{S}(\Omega)$ can be expressed as

$$\operatorname{grad}\phi = \frac{1}{h_1} \frac{\partial \phi}{\partial u^1} \mathbf{e}_{u^1} + \frac{1}{h_2} \frac{\partial \phi}{\partial u^2} \mathbf{e}_{u^2} + \frac{1}{h_3} \frac{\partial \phi}{\partial u^3} \mathbf{e}_{u^3} . \tag{3.217}$$

The direction of the basis vectors is a function of the point coordinates, and the components of the gradient are the projections of the vector $\operatorname{grad}\phi$ onto these basis vectors.

Curl Using the coordinate-free definition of the curl operator, we can calculate the line integral of a vector field $\mathbf{g} \in \mathcal{V}(\Omega)$ along the two edges $\mathscr{S}_{12}, \mathscr{S}_{34} \subset \Omega$ shown in Figure 3.8 (right).

$$\begin{aligned}
\lim_{a \to 0} \frac{\int_{\mathscr{S}_{12}, \mathscr{S}_{34}} \mathbf{g} \cdot d\tilde{\mathbf{r}}}{a} &= \lim_{\Delta u^2, \Delta u^3 \to 0} \frac{g_2 h_2 \Delta u^2 - \frac{\Delta u^3}{2} \frac{\partial}{\partial u^3}(g_2 h_2) \Delta u^2}{h_1 h_2 \Delta u^2 \Delta u^3} \\
&\quad - \lim_{\Delta u^2, \Delta u^3 \to 0} \frac{g_2 h_2 \Delta u^2 + \frac{\Delta u^3}{2} \frac{\partial}{\partial u^3}(g_2 h_2) \Delta u^2}{h_1 h_2 \Delta u^2 \Delta u^3} \\
&= -\frac{1}{h_1 h_2} \frac{\partial}{\partial u^3}(h_2 g_2) .
\end{aligned} \tag{3.218}$$

Employing the contribution of the line integral along the other two edges yields

$$\mathbf{e}_{u^1} \cdot \operatorname{curl}\mathbf{g} = \frac{1}{h_2 h_3} \left(\frac{\partial}{\partial u^2}(h_3 g_3) - \frac{\partial}{\partial u^3}(h_2 g_2) \right) . \tag{3.219}$$

With equivalent relations for the other two components it is convenient to write the curl of a vector field in form of a determinant:

$$\operatorname{curl}\mathbf{g} = \frac{1}{h_1 h_2 h_3} \begin{vmatrix} h_1 \mathbf{e}_{u^1} & h_2 \mathbf{e}_{u^2} & h_3 \mathbf{e}_{u^3} \\ \frac{\partial}{\partial u^1} & \frac{\partial}{\partial u^2} & \frac{\partial}{\partial u^3} \\ h_1 g_1 & h_2 g_2 & h_3 g_3 \end{vmatrix} . \tag{3.220}$$

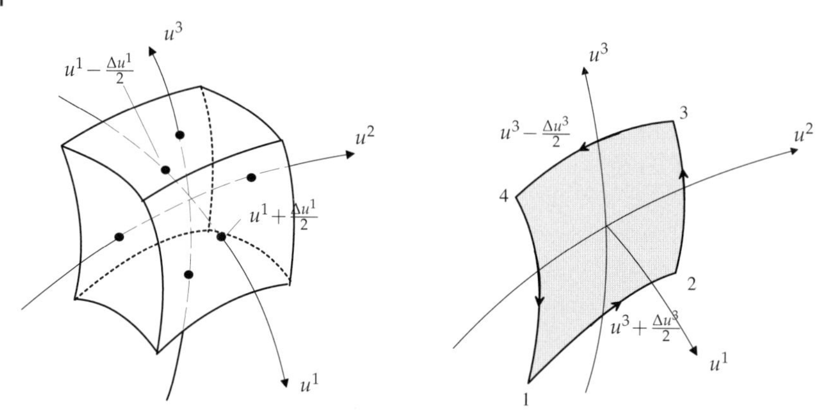

Figure 3.8 Left: On the calculation of the divergence in orthogonal curvilinear coordinates. Right: On the calculation of curl in curvilinear coordinates.

Divergence At the central point $\mathscr{P}$ of the volume element shown in Figure 3.8 (left), the line elements in the direction of the coordinate lines are $\mathrm{d}s_2 = h_2 \mathrm{d}u^2$ and $\mathrm{d}s_3 = h_3 \mathrm{d}u^3$. Consider the surface $\mathscr{A} \subset \Omega$ through $\mathscr{P}$ that is normal to u^1. Its surface area a is given by $\mathrm{d}s_2 \mathrm{d}s_3 = h_2 h_3\, \mathrm{d}u^2 \mathrm{d}u^3$. The surface integral of a vector field $\mathbf{g} \in \mathcal{V}(\Omega)$ on $\mathscr{A}$ is therefore

$$\int_{\mathscr{A}} \mathbf{g} \cdot \mathbf{da} = \int_{u_a^3}^{u_b^3} \int_{u_a^2}^{u_b^2} g_1 h_2 h_3 \mathrm{d}u^2 \mathrm{d}u^3. \tag{3.221}$$

The other components of the vector field $\mathbf{g}$ do not contribute to the integral as they are orthogonal to $\mathscr{A}$. On the front and back sides, the values of the surface integrals change not only because of the vector coefficients but also because the scale factors h_2 and h_3 depend on u^1. Therefore the front and back sides contribute

$$\begin{aligned}
\lim_{V\to 0} \frac{\int_{\mathscr{A}_1,\mathscr{A}_2} \mathbf{g} \cdot \mathbf{da}}{V} &= \lim_{\Delta u^1,\Delta u^2,\Delta u^3 \to 0} \frac{-\left(g_1 h_2 h_3 - \frac{\partial}{\partial u^1}(g_1 h_2 h_3)\frac{\Delta u^1}{2}\right)\Delta u^2 \Delta u^3}{h_1 h_2 h_3 \Delta u^1 \Delta u^2 \Delta u^3} \\
&\quad + \lim_{\Delta u^1,\Delta u^2,\Delta u^3 \to 0} \frac{\left(g_1 h_2 h_3 + \frac{\partial}{\partial u^1}(g_1 h_2 h_3)\frac{\Delta u^1}{2}\right)\Delta u^2 \Delta u^3}{h_1 h_2 h_3 \Delta u^1 \Delta u^2 \Delta u^3} \\
&= \frac{1}{h_1 h_2 h_3} \frac{\partial}{\partial u^1}(g_1 h_2 h_3).
\end{aligned} \tag{3.222}$$

Employing the four other sides yields

$$\operatorname{div} \mathbf{g} = \frac{1}{h_1 h_2 h_3}\left[\frac{\partial}{\partial u^1}(h_2 h_3 g_1) + \frac{\partial}{\partial u^2}(h_3 h_1 g_2) + \frac{\partial}{\partial u^3}(h_1 h_2 g_3)\right]. \tag{3.223}$$

The Laplace operator From $\nabla^2\phi = \operatorname{div}\operatorname{grad}\phi$ we obtain for a scalar field $\phi \in \mathcal{S}(\Omega)$:

$$\nabla^2\phi = \frac{1}{h_1h_2h_3}\left[\frac{\partial}{\partial u^1}\left(\frac{h_2h_3}{h_1}\frac{\partial\phi}{\partial u^1}\right) + \frac{\partial}{\partial u^2}\left(\frac{h_3h_1}{h_2}\frac{\partial\phi}{\partial u^2}\right) + \frac{\partial}{\partial u^3}\left(\frac{h_1h_2}{h_3}\frac{\partial\phi}{\partial u^3}\right)\right]. \tag{3.224}$$

From $\nabla^2\mathbf{g} = \operatorname{grad}\operatorname{div}\mathbf{g} - \operatorname{curl}\operatorname{curl}\mathbf{g}$ we derive for a vector field $\mathbf{g} \in \mathcal{V}(\Omega)$:

$$\begin{aligned}\nabla^2\mathbf{g} = &\left[\frac{1}{h_1}\frac{\partial\Gamma}{\partial u^1} + \frac{h_1}{h_1h_2h_3}\left(\frac{\partial\Gamma_2}{\partial u^3} - \frac{\partial\Gamma_3}{\partial u^2}\right)\right]\mathbf{e}_{u^1} \\ &+ \left[\frac{1}{h_2}\frac{\partial\Gamma}{\partial u^2} + \frac{h_2}{h_1h_2h_3}\left(\frac{\partial\Gamma_3}{\partial u^1} - \frac{\partial\Gamma_1}{\partial u^3}\right)\right]\mathbf{e}_{u^2} \\ &+ \left[\frac{1}{h_3}\frac{\partial\Gamma}{\partial u^3} + \frac{h_3}{h_1h_2h_3}\left(\frac{\partial\Gamma_2}{\partial u^1} - \frac{\partial\Gamma_1}{\partial u^2}\right)\right]\mathbf{e}_{u^3},\end{aligned} \tag{3.225}$$

where

$$\Gamma = \frac{1}{h_1h_2h_3}\left[\frac{\partial}{\partial u^1}(h_2h_3g_1) + \frac{\partial}{\partial u^2}(h_1h_3g_2) + \frac{\partial}{\partial u^3}(h_1h_2g_3)\right] \tag{3.226}$$

$$\Gamma_1 = \frac{h_1^2}{h_1h_2h_3}\left(\frac{\partial(h_3g_3)}{\partial u^2} - \frac{\partial(h_2g_2)}{\partial u^3}\right), \tag{3.227}$$

$$\Gamma_2 = \frac{h_2^2}{h_1h_2h_3}\left(\frac{\partial(h_1g_1)}{\partial u^3} - \frac{\partial(h_3g_3)}{\partial u^1}\right), \tag{3.228}$$

$$\Gamma_3 = \frac{h_3^2}{h_1h_2h_3}\left(\frac{\partial(h_2g_2)}{\partial u^1} - \frac{\partial(h_1g_1)}{\partial u^2}\right). \tag{3.229}$$

3.13.2 Cylindrical Coordinates

For cylindrical coordinates (see Figure 3.9) we have $(u^1, u^2, u^3) = (r, \varphi, z)$, and the transformation is given by $T : \mathbb{R}^3 \to \mathbb{R}^3 : (r, \varphi, z) \mapsto (x, y, z)$ for $r \geq 0$, $0 \leq \varphi < 2\pi$, $-\infty < z < \infty$ and $x = r\cos\varphi$, $y = r\sin\varphi$, $z = z$. The position vector becomes $\tilde{\mathbf{r}} = r\cos\varphi\,\mathbf{e}_x + r\sin\varphi\,\mathbf{e}_y + z\,\mathbf{e}_z$. The unit tangent vectors to the r, φ, z lines are given according to Eq. (3.211) as

$$\mathbf{e}_r = \cos\varphi\,\mathbf{e}_x + \sin\varphi\,\mathbf{e}_y\,, \qquad \mathbf{e}_\varphi = -\sin\varphi\,\mathbf{e}_x + \cos\varphi\,\mathbf{e}_y\,, \tag{3.230}$$

and $\mathbf{e}_z$. A space curve has the form[22]

$$\tilde{\mathbf{r}}(t) = r(t)\,\mathbf{e}_r + z(t)\,\mathbf{e}_z\,. \tag{3.231}$$

22 In cylindrical coordinates $|\mathbf{r}|$ is not identical to the r-coordinate, which is therefore often denoted ρ.

Table 3.1 Relations between the Cartesian, cylindrical, and spherical coordinates[a].

Cartesian	Cylindrical	Spherical
x	$r\cos\varphi$	$R\sin\vartheta\cos\varphi$
y	$r\sin\varphi$	$R\sin\vartheta\sin\varphi$
z	z	$R\cos\vartheta$
$\sqrt{x^2+y^2}$	r	$R\sin\vartheta$
$\arctan\frac{y}{x}+\alpha$	φ	φ
z	z	$R\cos\vartheta$
$\sqrt{x^2+y^2+z^2}$	$\sqrt{r^2+z^2}$	R
$\arctan\frac{\sqrt{x^2+y^2}}{z}+\alpha$	$\arctan\frac{r}{z}+\alpha$	ϑ
$\arctan\frac{y}{x}+\alpha$	φ	φ

[a]For $\arctan(a/b)$ we have $\alpha=0$ for $b\geq 0, a\geq 0$, $\alpha=\pi$ for $b<0$, and $\alpha=2\pi$ for $b\geq 0, a<0$.

With

$$\frac{\mathrm{d}\mathbf{e}_r}{\mathrm{d}t} = -(\sin\varphi)\frac{\mathrm{d}\varphi}{\mathrm{d}t}\,\mathbf{e}_x + (\cos\varphi)\frac{\mathrm{d}\varphi}{\mathrm{d}t}\,\mathbf{e}_y = \frac{\mathrm{d}\varphi}{\mathrm{d}t}\mathbf{e}_\varphi\,, \tag{3.232}$$

$$\frac{\mathrm{d}\mathbf{e}_\varphi}{\mathrm{d}t} = -(\cos\varphi)\frac{\mathrm{d}\varphi}{\mathrm{d}t}\,\mathbf{e}_x - (\sin\varphi)\frac{\mathrm{d}\varphi}{\mathrm{d}t}\,\mathbf{e}_y = -\frac{\mathrm{d}\varphi}{\mathrm{d}t}\,\mathbf{e}_r\,, \tag{3.233}$$

the velocity vector is given by

$$\mathbf{v} = \frac{\mathrm{d}\tilde{\mathbf{r}}}{\mathrm{d}t} = \frac{\mathrm{d}r}{\mathrm{d}t}\,\mathbf{e}_r + r\frac{\mathrm{d}\,\mathbf{e}_r}{\mathrm{d}t} + \frac{\mathrm{d}z}{\mathrm{d}t}\,\mathbf{e}_z = \frac{\mathrm{d}r}{\mathrm{d}t}\mathbf{e}_r + r\frac{\mathrm{d}\varphi}{\mathrm{d}t}\mathbf{e}_\varphi + \frac{\mathrm{d}z}{\mathrm{d}t}\mathbf{e}_z\,. \tag{3.234}$$

The scale factors h_i are

$$h_1 = \sqrt{\cos^2\varphi + \sin^2\varphi} = 1\,, \tag{3.235}$$

$$h_2 = \sqrt{(r\sin\varphi)^2 + (r\cos\varphi)^2} = r\,, \tag{3.236}$$

$$h_3 = 1\,. \tag{3.237}$$

The differential operators in cylinder coordinates are listed in the center column of Table 3.3.

3.13.3 Spherical Coordinates

For the spherical coordinates $(u^1,u^2,u^3)=(R,\vartheta,\varphi)$ (see Figure 3.9) the transformation is given by $T:\mathbb{R}^3\to\mathbb{R}^3:(x,y,z)\mapsto(R,\vartheta,\varphi)$ for $R\geq 0, 0\leq\vartheta<\pi$, $0\leq\varphi<2\pi$, and $x=R\sin\vartheta\cos\varphi$, $y=R\sin\vartheta\sin\varphi$, $z=R\cos\vartheta$. The position

Table 3.2 Relations between the unit vectors in Cartesian, cylindrical, and spherical coordinates.

Cartesian	Cylindrical	Spherical
$\mathbf{e}_x$	$\cos\varphi\mathbf{e}_r - \sin\varphi\mathbf{e}_\varphi$	$\sin\vartheta\cos\varphi\mathbf{e}_R + \cos\vartheta\cos\varphi\mathbf{e}_\vartheta - \sin\varphi\mathbf{e}_\varphi$
$\mathbf{e}_y$	$\sin\varphi\mathbf{e}_r + \cos\varphi\mathbf{e}_\varphi$	$\sin\vartheta\sin\varphi\mathbf{e}_R + \cos\vartheta\sin\varphi\mathbf{e}_\vartheta + \cos\varphi\mathbf{e}_\varphi$
$\mathbf{e}_z$	$\mathbf{e}_z$	$\cos\vartheta\mathbf{e}_R - \sin\vartheta\mathbf{e}_\vartheta$
$\cos\varphi\mathbf{e}_x + \sin\varphi\mathbf{e}_y$	$\mathbf{e}_r$	$\sin\vartheta\mathbf{e}_r + \cos\vartheta\mathbf{e}_\vartheta$
$-\sin\varphi\mathbf{e}_x + \cos\varphi\mathbf{e}_y$	$\mathbf{e}_\varphi$	$\mathbf{e}_\varphi$
$\mathbf{e}_z$	$\mathbf{e}_z$	$\cos\vartheta\mathbf{e}_r - \sin\vartheta\mathbf{e}_\vartheta$
$\sin\vartheta\cos\varphi\mathbf{e}_x + \sin\vartheta\sin\varphi\mathbf{e}_y + \cos\vartheta\mathbf{e}_z$	$\sin\vartheta\mathbf{e}_r + \cos\vartheta\mathbf{e}_z$	$\mathbf{e}_R$
$\cos\vartheta\cos\varphi\mathbf{e}_x + \cos\vartheta\sin\varphi\mathbf{e}_y - \sin\vartheta\mathbf{e}_z$	$\cos\vartheta\mathbf{e}_r - \sin\vartheta\mathbf{e}_z$	$\mathbf{e}_\vartheta$
$-\sin\varphi\mathbf{e}_x + \cos\varphi\mathbf{e}_y$	$\mathbf{e}_\varphi$	$\mathbf{e}_\varphi$

vector becomes $\tilde{\mathbf{r}} = R\sin\vartheta\cos\varphi\,\mathbf{e}_x + R\sin\vartheta\sin\varphi R\,\mathbf{e}_y + R\cos\vartheta\,\mathbf{e}_z$. The unit tangent vectors to the R, ϑ, φ lines yield the basis vectors

$$\mathbf{e}_R = \sin\vartheta\cos\varphi\,\mathbf{e}_x + \sin\vartheta\sin\varphi\,\mathbf{e}_y + \cos\vartheta\,\mathbf{e}_z\,, \tag{3.238}$$

$$\mathbf{e}_\vartheta = \cos\vartheta\cos\varphi\,\mathbf{e}_x + \cos\vartheta\sin\varphi\,\mathbf{e}_y - \sin\vartheta\,\mathbf{e}_z\,, \tag{3.239}$$

$$\mathbf{e}_\varphi = -\sin\varphi\,\mathbf{e}_x + \cos\varphi\,\mathbf{e}_y\,. \tag{3.240}$$

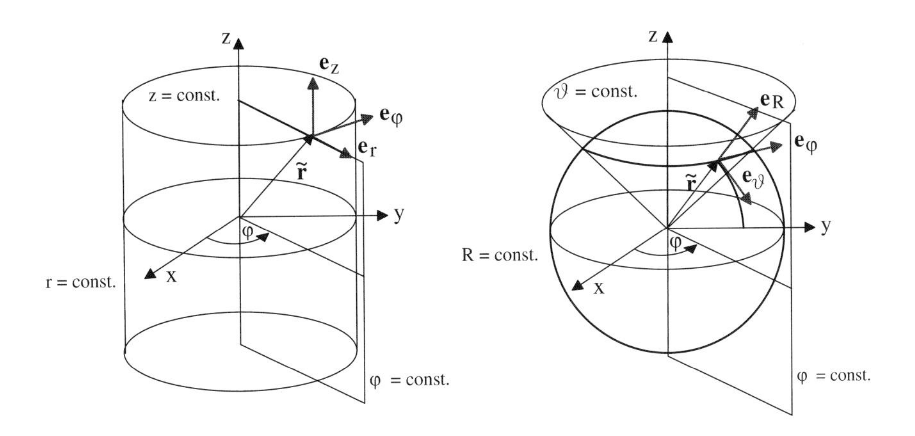

Figure 3.9 Left: Cylindrical coordinate system. Right: Spherical coordinate system.

Table 3.3 Differential operators in orthogonal coordinate systems.

	Cartesian	Cylindrical	Spherical
u^1, u^2, u^3	x, y, z	r, φ, z	R, ϑ, φ
Scale factors	$h_1 = 1$ $h_2 = 1$ $h_3 = 1$	$h_1 = 1$ $h_2 = r$ $h_3 = 1$	$h_1 = 1$ $h_2 = R$ $h_3 = R \sin\vartheta$
$\mathrm{d}s^2$	$\mathrm{d}x^2 + \mathrm{d}y^2 + \mathrm{d}z^2$	$\mathrm{d}r^2 + r^2\mathrm{d}\varphi^2 + \mathrm{d}z^2$	$\mathrm{d}R^2 + R^2\mathrm{d}\vartheta^2 + R^2\sin^2\vartheta\mathrm{d}\varphi^2$
$\mathrm{d}V$	$\mathrm{d}x\mathrm{d}y\mathrm{d}z$	$r\mathrm{d}r\mathrm{d}\varphi\mathrm{d}z$	$R^2\sin\vartheta\mathrm{d}R\mathrm{d}\vartheta\mathrm{d}\varphi$
$\operatorname{grad}\phi$	$\frac{\partial\phi}{\partial x}\mathbf{e}_x + \frac{\partial\phi}{\partial y}\mathbf{e}_y + \frac{\partial\phi}{\partial z}\mathbf{e}_z$	$\frac{\partial\phi}{\partial r}\mathbf{e}_r + \frac{1}{r}\frac{\partial\phi}{\partial\varphi}\mathbf{e}_\varphi + \frac{\partial\phi}{\partial z}\mathbf{e}_z$	$\frac{\partial\phi}{\partial R}\mathbf{e}_R + \frac{1}{R}\frac{\partial\phi}{\partial\vartheta}\mathbf{e}_\vartheta + \frac{1}{R\sin\vartheta}\frac{\partial\phi}{\partial\varphi}\mathbf{e}_\varphi$
$\operatorname{div}\mathbf{g}$	$\frac{\partial g_x}{\partial x} + \frac{\partial g_y}{\partial y} + \frac{\partial g_z}{\partial z}$	$\frac{1}{r}\frac{\partial}{\partial r}(r g_r) + \frac{1}{r}\frac{\partial g_\varphi}{\partial\varphi} + \frac{\partial g_z}{\partial z}$	$\frac{1}{R^2}\frac{\partial}{\partial R}(R^2 g_R)$ $+\frac{1}{R\sin\vartheta}\frac{\partial}{\partial\vartheta}(\sin\vartheta g_\vartheta)$ $+\frac{1}{R\sin\vartheta}\frac{\partial g_\varphi}{\partial\varphi}$
$\operatorname{curl}\mathbf{g}$	$(\frac{\partial g_z}{\partial y} - \frac{\partial g_y}{\partial z})\mathbf{e}_x$ $+(\frac{\partial g_x}{\partial z} - \frac{\partial g_z}{\partial x})\mathbf{e}_y$ $+(\frac{\partial g_y}{\partial x} - \frac{\partial g_x}{\partial y})\mathbf{e}_z$	$(\frac{1}{r}\frac{\partial g_z}{\partial\varphi} - \frac{\partial g_\varphi}{\partial z})\mathbf{e}_r$ $+(\frac{\partial g_r}{\partial z} - \frac{\partial g_z}{\partial r})\mathbf{e}_\varphi$ $+(\frac{1}{r}\frac{\partial}{\partial r}(r g_\varphi) - \frac{1}{r}\frac{\partial g_r}{\partial\varphi})\mathbf{e}_z$	$\frac{1}{R\sin\vartheta}[\frac{\partial}{\partial\vartheta}(\sin\vartheta g_\varphi) - \frac{\partial g_\vartheta}{\partial\varphi}]\mathbf{e}_R$ $+[\frac{1}{R\sin\vartheta}\frac{\partial g_R}{\partial\varphi} - \frac{1}{R}\frac{\partial}{\partial R}(R g_\varphi)]\mathbf{e}_\vartheta$ $+\frac{1}{R}[\frac{\partial}{\partial R}(R g_\vartheta) - \frac{\partial g_R}{\partial\vartheta}]\mathbf{e}_\varphi$
$\nabla^2\phi$	$\frac{\partial^2\phi}{\partial x^2} + \frac{\partial^2\phi}{\partial y^2} + \frac{\partial^2\phi}{\partial z^2}$	$\frac{1}{r}\frac{\partial}{\partial r}(r\frac{\partial\phi}{\partial r}) + \frac{1}{r^2}\frac{\partial^2\phi}{\partial\varphi^2} + \frac{\partial^2\phi}{\partial z^2}$	$\frac{1}{R^2}\frac{\partial}{\partial R}\left(R^2\frac{\partial\phi}{\partial R}\right)$ $+\frac{1}{R^2\sin\vartheta}\frac{\partial}{\partial\vartheta}(\sin\vartheta\frac{\partial\phi}{\partial\vartheta})$ $+\frac{1}{R^2\sin^2\vartheta}\frac{\partial^2\phi}{\partial\varphi^2}$
$\nabla^2\mathbf{g}$	$(\nabla^2 g_x)\mathbf{e}_x$ $+(\nabla^2 g_y)\mathbf{e}_y$ $+(\nabla^2 g_z)\mathbf{e}_z$	$(\nabla^2 g_r - \frac{1}{r^2}g_r - \frac{2}{r^2}\frac{\partial g_\varphi}{\partial\varphi})\mathbf{e}_r$ $+(\nabla^2 g_\varphi - \frac{1}{r^2}g_\varphi + \frac{2}{r^2}\frac{\partial g_r}{\partial\varphi})\mathbf{e}_\varphi$ $+\nabla^2 g_z\mathbf{e}_z$	See Eq. (3.225) with Eqs. (3.226) and (3.227)–(3.229)

Differentiation in t of the basis vectors $\mathbf{e}_R$, $\mathbf{e}_\vartheta$, and $\mathbf{e}_\varphi$ along a space curve with the general parameter t yields

$$\frac{\mathrm{d}\mathbf{e}_R}{\mathrm{d}t} = \frac{\mathrm{d}\vartheta}{\mathrm{d}t}\mathbf{e}_\vartheta + \sin\vartheta\frac{\mathrm{d}\varphi}{\mathrm{d}t}\mathbf{e}_\varphi, \tag{3.241}$$

$$\frac{\mathrm{d}\mathbf{e}_\vartheta}{\mathrm{d}t} = -\frac{\mathrm{d}\vartheta}{\mathrm{d}t}\mathbf{e}_R + \cos\vartheta\frac{\mathrm{d}\varphi}{\mathrm{d}t}\mathbf{e}_\varphi, \tag{3.242}$$

$$\frac{\mathrm{d}\mathbf{e}_\varphi}{\mathrm{d}t} = -\sin\vartheta\frac{\mathrm{d}\varphi}{\mathrm{d}t}\mathbf{e}_R - \cos\vartheta\frac{\mathrm{d}\varphi}{\mathrm{d}t}\mathbf{e}_\vartheta. \tag{3.243}$$

The scale factors h_i are

$$h_1 = \sqrt{\sin^2\vartheta\cos^2\varphi + \sin^2\vartheta\sin^2\varphi + \cos^2\vartheta} = 1, \tag{3.244}$$

$$h_2 = \sqrt{R^2\cos^2\vartheta\cos^2\varphi + R^2\cos^2\vartheta\sin^2\varphi + R^2\sin^2\vartheta} = R, \tag{3.245}$$

$$h_3 = \sqrt{R^2\sin^2\vartheta\sin^2\varphi + R^2\sin^2\vartheta\cos^2\varphi} = R\sin\vartheta\,. \tag{3.246}$$

The differential operators in spherical coordinates are listed in the right column of Table 3.3.

3.14 The Lemmata of Poincaré

We will now study under which conditions it is possible to find scalar and vector potentials over an open domain $\Omega \in E_3$. Can a given curl-free vector field $\mathbf{g} \in \mathcal{V}(\Omega)$ be expressed as the gradient of a scalar field? Can a divergence-free vector field be expressed as the curl of another vector field (vector potential)? In case the scalar or vector potential exists, is it uniquely defined? The answers are found in the *lemmata of Poincaré* and the *De Rham cohomology*. The first Poincaré lemma states that an arbitrary gradient field $\operatorname{grad}\phi$ for $\phi \in \mathcal{S}(\Omega)$ is curl-free,

$$\operatorname{curl}\operatorname{grad}\phi = 0, \tag{3.247}$$

and that the curl of an arbitrary vector field $\mathbf{g} \in \mathcal{V}(\Omega)$ is source free,

$$\operatorname{div}\operatorname{curl}\mathbf{g} = 0\,. \tag{3.248}$$

Proof. We will prove Eq. (3.247) in orthogonal curvilinear coordinates.

$$\begin{aligned}\operatorname{curl}\operatorname{grad}\phi &= \operatorname{curl}\left[\frac{1}{h_1}\frac{\partial\phi}{\partial u^1}\mathbf{e}_{u^1} + \frac{1}{h_2}\frac{\partial\phi}{\partial u^2}\mathbf{e}_{u^2} + \frac{1}{h_3}\frac{\partial\phi}{\partial u^3}\mathbf{e}_{u^3}\right]\\ &= \frac{1}{h_2h_3}\left(\frac{\partial^2\phi}{\partial u^2\partial u^3} - \frac{\partial^2\phi}{\partial u^3\partial u^2}\right)\mathbf{e}_{u^1}\\ &\quad + \frac{1}{h_3h_1}\left(\frac{\partial^2\phi}{\partial u^3\partial u^1} - \frac{\partial^2\phi}{\partial u^1\partial u^3}\right)\mathbf{e}_{u^2}\\ &\quad + \frac{1}{h_1h_2}\left(\frac{\partial^2\phi}{\partial u^1\partial u^2} - \frac{\partial^2\phi}{\partial u^2\partial u^1}\right)\mathbf{e}_{u^3} = 0\,,\end{aligned} \tag{3.249}$$

which shows how cumbersome the calculation with coordinates can be. □

Note that we restricted ourselves to orthogonal coordinate systems. We will now turn to a more general and elegant proof employing the boundary operator. Consider that the boundary surface of a closed volume has no contour, and that the contour of a surface is always closed,

$$\partial(\partial\mathscr{V}) = \emptyset, \qquad \partial(\partial\mathscr{A}) = \emptyset, \tag{3.250}$$

where $\emptyset$ denotes the empty set. We obtain

$$\int_{\mathscr{A}} \operatorname{curl} \operatorname{grad} \phi \cdot \mathbf{da} = \int_{\partial\mathscr{A}} \operatorname{grad} \phi \cdot \mathbf{dr} = \phi|_{\partial(\partial\mathscr{A})} = 0, \tag{3.251}$$

which holds for arbitrary surfaces only if $\operatorname{curl} \operatorname{grad} \phi = 0$. On the other hand,

$$\int_{\mathscr{V}} \operatorname{div} \operatorname{curl} \mathbf{g} dV = \int_{\partial\mathscr{V}} \operatorname{curl} \mathbf{g} \cdot \mathbf{da} = \int_{\partial(\partial\mathscr{V})} \mathbf{g} \cdot \mathbf{dr} = 0, \tag{3.252}$$

which holds for arbitrary volumes only if $\operatorname{div} \operatorname{curl} \mathbf{g} = 0$. Reversal of the first lemma yields the second lemma of Poincaré: A source-free, smooth vector field $\mathbf{b} \in \mathcal{V}(\Omega)$ over an open contractible[23] domain $\Omega \subset E_3$ can always be expressed as a smooth vector potential $\mathbf{a} \in \mathcal{V}(\Omega)$,

$$\operatorname{div} \mathbf{b} = 0 \quad \rightarrow \quad \mathbf{b} = \operatorname{curl} \mathbf{a}. \tag{3.253}$$

A curl-free, smooth vector field $\mathbf{h} \in \mathcal{V}(\Omega)$ over an open, contractible domain $\Omega \subset E_3$ can always be expressed as a smooth scalar potential $\phi \in \mathcal{S}(\Omega)$.

$$\operatorname{curl} \mathbf{h} = 0 \quad \rightarrow \quad \mathbf{h} = \operatorname{grad} \phi. \tag{3.254}$$

The vector field $\mathbf{h}$ is then called a gradient field or *conservative field*.

3.15
De Rham Cohomology

Figure 3.10 shows the *de Rham complex* for contractible domains. The image of $\operatorname{grad} \phi$ is identical to the kernel of the curl operator. The image of $\operatorname{curl} \mathbf{a}$ is identical to the kernel of the div operator. This is called an *exact sequence*; the second lemma of Poincaré always holds.

For contractible domains we can express any curl-free vector field $\mathbf{h}$ as a scalar potential ϕ. However, as can be seen from Figure 3.10, ϕ can be determined only up to an additive constant from a given vector field $\mathbf{h}$ because the constant lies in the kernel of the grad operator. In the case where the vector field $\mathbf{d}$ is sought, such that a given scalar field ρ can be derived from $\operatorname{div} \mathbf{d} = \rho$, both $\mathbf{d}$ and $\mathbf{d} + \operatorname{curl} \mathbf{v}$ are possible solutions. In the case where the

23 Thus in particular over an open star-shaped domain.

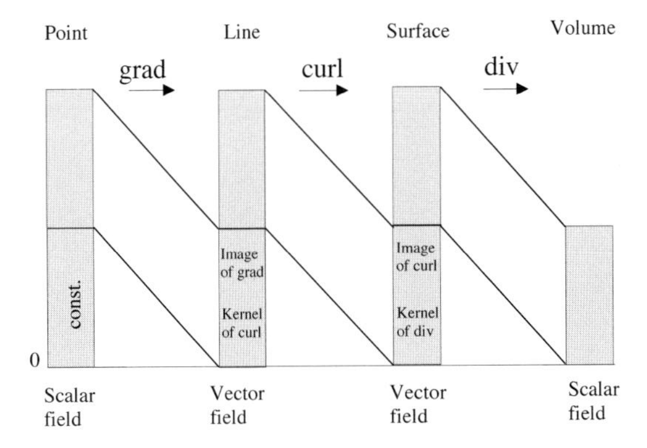

Figure 3.10 De Rham complex for contractible domains. The diagram is parsed as follows: The image of $\operatorname{grad}\phi$ is identical to the kernel of the curl operator. The image of $\operatorname{curl}\mathbf{a}$ is identical to the kernel of the div operator. The corresponding spaces of scalar and vector potentials constitute together with the grad, curl, and div operators an exact sequence: $\operatorname{curl}\operatorname{grad}\phi = 0$ and $\operatorname{div}\operatorname{curl}\mathbf{a} = 0$. In this case, the second lemma of Poincaré always holds.

vector field $\mathbf{a}$ is sought, such that a given vector field $\mathbf{b}$ can be derived from $\operatorname{curl}\mathbf{a} = \mathbf{b}$, then both $\mathbf{a}$ and $\mathbf{a} + \operatorname{grad}\phi$ are possible solutions. It is clear that the knowledge of the sources or the curls alone is not sufficient to determine the vector field. However, any vector field on a closed contractible domain Ω can be uniquely determined through all its sources and curls if the normal component of the field is prescribed on the domain boundary.[24]

In noncontractible domains there may exist curl-free vector fields, which cannot be expressed as the gradient of a scalar function; see Figure 3.11. There may be also divergence-free vector fields, which cannot be expressed as the curl of a vector field.

Because $\operatorname{curl}\operatorname{grad}\phi = 0$ and $\operatorname{div}\operatorname{curl}\mathbf{g} = 0$ we have $\operatorname{im}(\operatorname{grad}) \subset \ker(\operatorname{curl})$ and $\operatorname{im}(\operatorname{curl}) \subset \ker(\operatorname{div})$. By how much is the image of the gradient larger than the kernel of curl? And how does the image of curl compare to the kernel of div? This is measured by the first and second de Rham cohomology spaces of the open domain $\Omega \subset E_3$ defined as the quotient spaces

$$\mathcal{H}^1(\Omega) := \frac{\ker(\operatorname{curl})}{\operatorname{im}(\operatorname{grad})}, \qquad \mathcal{H}^2(\Omega) := \frac{\ker(\operatorname{div})}{\operatorname{im}(\operatorname{curl})}. \tag{3.255}$$

24 This is the theorem of Hermann von Helmholtz (1812–1894).

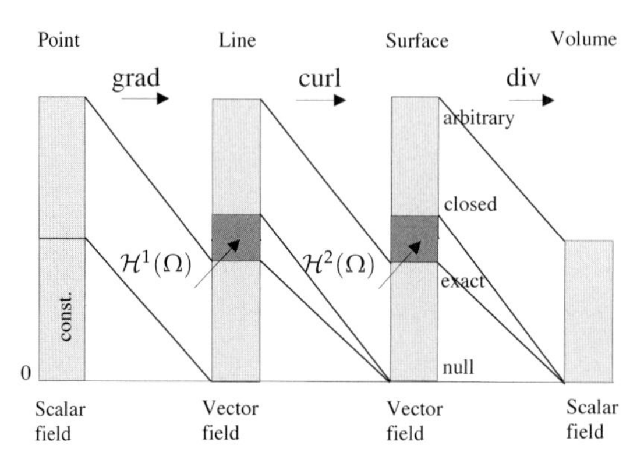

Figure 3.11 In topologically noncontractible domains there may exist curl-free vector fields that cannot be expressed as the gradient of a scalar function. There may also be source-free vector fields that cannot be expressed as the curl of a vector field. These two vector fields constitute the first and second *de Rham cohomology space*, denoted $\mathcal{H}^1(\Omega)$ and $\mathcal{H}^2(\Omega)$.

Examples:

1. Consider a two-dimensional disk with a hole at the origin of the polar coordinate system (r, φ). The magnetic field strength of a line-current at the origin is given by $H_\varphi = I/2\pi r$, which is curl-free at $r > 0$:

$$\operatorname{curl} \mathbf{H} = \frac{1}{r}\frac{\partial}{\partial r}(rH_\varphi) = 0. \tag{3.256}$$

   Nevertheless, this field cannot be expressed as the gradient of a scalar potential in the form $\mathbf{H} = \operatorname{grad}\phi$. This can be shown by comparing $\int_{\mathscr{A}_c} \mathbf{H} \cdot \mathbf{dr} = I$, where $\mathscr{A}_c$ is a loop around the origin, with $\int_{\mathscr{A}_c} \operatorname{grad}\phi \cdot \mathbf{dr} = 0$, which are contradictory. The vector field $\mathbf{H}$ is said to lie in the first de Rham cohomology space $\mathcal{H}^1(\Omega)$. It is said to be *closed* but not exact [5].

2. Consider the domain between two nested spheres centered at the origin. The domain is simply-connected but has nonconnected boundaries and is thus not contractible. The electric flux density of a charge Q located at the origin given by $D_R = Q/(4\pi R^2)$ is divergence-free at $R > 0$:

$$\operatorname{div} \mathbf{D} = \frac{1}{r^2}\frac{\partial}{\partial R}(R^2 D_R) = 0\,. \tag{3.257}$$

   Nevertheless, this field cannot be expressed as the curl of a vector potential. This can be shown by comparing $\int_{\mathscr{A}} \mathbf{D} \cdot \mathbf{da} = Q$ with[25]

25 $\mathscr{A}$ is the surface of a sphere centered at the origin.

$\int_{\mathscr{A}} \operatorname{curl} \mathbf{v} \cdot d\mathbf{a} = \int_{\partial\mathscr{A}} \mathbf{v} \cdot d\mathbf{r} = 0$, which are contradictory. The vector field $\mathbf{D}$ is said to lie in the second de Rham cohomology space $\mathcal{H}^2(\Omega)$. □

The second Poincaré lemma can now be rewritten as follows:

- A curl-free, smooth vector field $\mathbf{h}$ can always be expressed as a smooth scalar potential ϕ if $\mathcal{H}^1(\Omega) = \emptyset$. Consequently, there exists no solution that is curl-free and yet cannot be expressed as the gradient of a scalar field.
- A divergence-free, smooth vector field $\mathbf{b}$ can always be expressed through a smooth vector potential $\mathbf{a}$ if $\mathcal{H}^2(\Omega) = \emptyset$. Consequently there exists no solution that is divergence-free and yet cannot be expressed as the curl of a vector field.

Following [6] we state alternatively:

- If $\{\mathbf{h}_1, \ldots, \mathbf{h}_{b_1}\}$ is a basis of $\mathcal{H}^1(\Omega)$, then any curl-free, smooth vector field $\mathbf{h}$ can be expressed in the form $\mathbf{h} = \sum_{i=1}^{b_1} \lambda_i \mathbf{h}_i + \operatorname{grad} \phi$.
- If $\{\mathbf{b}_1, \ldots, \mathbf{b}_{b_2}\}$ is a basis of $\mathcal{H}^2(\Omega)$, then any divergence-free, smooth vector field $\mathbf{b}$ can be expressed in the form $\mathbf{b} = \sum_{i=1}^{b_2} \lambda_i \mathbf{b}_i + \operatorname{curl} \mathbf{a}$,

where b_1 and b_2 are the dimensions of the first and second cohomology spaces and are called first and second *Betti numbers*:[26]

$$b_p(\Omega) := \dim \mathcal{H}^p(\Omega)\,. \tag{3.258}$$

The Betti numbers depend only on the topology of the domains, that is, on the so-called *homology class*. For trivial domains we obtain $b_1 = b_2 = 0$.

In Example 1 above, the two-dimensional domain Ω is not simply-connected, and therefore it is not possible to find a surface bounded by the loop $\mathscr{C}$ which is entirely in Ω. In this case $b_1 = 1$. In other words, b_1 is the number of holes in the domain. In Example 2, it is not possible to find a volume bounded by the surface $\mathscr{A}$, which is entirely in $\mathscr{V}$. The volume is simply-connected but with a nonconnected boundary. Thus $b_2 = 1$. The second Betti number represents the number of cavities in the domain.

3.16 Fourier Series

Before we will end this chapter we will make a brief study of the classical Fourier series and the orthogonal projections as their generalizations. Let

26 Enrico Betti (1823–1892).

$f : \mathbb{R} \to \mathbb{R}$ be a 2π-periodic function. The classical Fourier series is given by

$$f(t) = \frac{a_0}{2} + \sum_{n=1}^{\infty} (b_n \sin nt + a_n \cos nt) \tag{3.259}$$

with the Fourier coefficients

$$a_n = \frac{1}{\pi} \int_{-\pi}^{\pi} f(t) \cos nt \, \mathrm{d}t, \qquad n = 1, 2, 3, \ldots, \tag{3.260}$$

$$b_n = \frac{1}{\pi} \int_{-\pi}^{\pi} f(t) \sin nt \, \mathrm{d}t, \qquad n = 1, 2, 3, \ldots. \tag{3.261}$$

In particular, if f is a 2π-periodic, odd function, then $a_n = 0$ for all n. If the function is even, all the b_n vanish.

There are Fourier series of continuous functions that do not converge to $f(t)$ at all points $t \in \mathbb{R}$. It turns out, however, that any function in the *Hilbert space* $L^2(\Omega)$, $\Omega = [-\pi, \pi]$ can be developed into a Fourier series that converges in the mean-square sense

$$\lim_{m \to \infty} \int_{-\pi}^{\pi} \left(f(t) - s_m(t)\right)^2 \mathrm{d}t = 0, \tag{3.262}$$

where s_m is the so-called *Fourier polynomial*

$$s_m(t) := \frac{a_0}{2} + \sum_{n=1}^{m} (b_n \sin nt + a_n \cos nt), \tag{3.263}$$

and where the trigonometric functions, together with a_0, form a complete orthogonal function set. The convergence concept (3.262) requires some basics of functional analysis, which we will provide in this section.

Consider a metric space (M, d) with a sequence $(x_n) := (x_1, x_2, \ldots, x_n, \ldots)$. This is called a *Cauchy sequence*[27] if for each $\epsilon > 0$, there exists a $N(\epsilon)$ such that for all $n, m \geq N(\epsilon)$ the distance between x_n and x_m is less than ϵ:

$$d(x_n - x_m) < \epsilon. \tag{3.264}$$

A metric space in which every Cauchy sequence has a limit in (M, d) is considered *complete*. As an example, the set of real numbers is complete as it can be constructed by Cauchy sequences of rational numbers. However, the space of real numbers in the open interval $0 < x < 1$ is not complete, since the limit of the Cauchy sequence $x_n = 1/(n+1) \to 0$ is not contained in the space itself. A series $\sum_{n=1}^{\infty} x_n$ is convergent if and only if the sequence of partial

27 Augustin Cauchy (1789–1857).

sums $s_m = \sum_{n=1}^{m} x_n$ converges. It is easy to determine whether the sequence of partial sums is a Cauchy sequence because for $p > q$ we obtain

$$s_p - s_q = \sum_{n=q+1}^{p} x_n \,. \tag{3.265}$$

A mapping $\|\cdot\|: V \to \mathbb{R}$ is called a *norm* on the vector space V if it satisfies the following axioms for $\mathbf{x}, \mathbf{y} \in V$:

1. $\| \mathbf{x} \| \geq 0$.
2. $\| \mathbf{x} \| = 0$ if and only if $\mathbf{x} = \mathbf{0}$.
3. $\| \lambda \mathbf{x} \| = |\lambda| \, \| \mathbf{x} \|$.
4. $\| \mathbf{x} + \mathbf{y} \| \leq \| \mathbf{x} \| + \| \mathbf{y} \|$.

Property 4 is called the triangular inequality. A vector space that has a norm defined on it is called a *normed space* and denoted $(V, \|\cdot\|)$.

Examples:

1. The Euclidean space $\mathbb{R}^n$ with its elements $\mathbf{x} = (x_1, x_2, \ldots, x_n)$ and the Euclidean norm $\| \mathbf{x} \| = \sqrt{\sum_{i=1}^{n} |x_i|^2}$.
2. The Euclidean space $\mathbb{R}^n$ with the *Manhattan norm*[28] $\| \mathbf{x} \|_1 = \sum_{i=1}^{n} |x_i|$.
3. The vector space $C^0[a,b]$ of all continuous functions $f : [a,b] \to \mathbb{R}$ with the supremum norm $\| f \| = \sup\{|f(t)|\}$ for $t \in [a,b]$. The space is complete because the limit of an absolutely convergent sequence of continuous functions is again continuous. □

A normed space is complete if the limit of every Cauchy sequence with respect to the metric $d(\mathbf{x}, \mathbf{y}) = \| \mathbf{x} - \mathbf{y} \|$ is contained in the space itself. A complete, normed vector space is called a *Banach space*.[29] A vector space equipped with an inner product is a *pre-Hilbert space*, $(V, \langle\cdot,\cdot\rangle)$. A *Hilbert space* is a complete vector space with a norm induced by the scalar product $\|\cdot\| = \sqrt{\langle\cdot,\cdot\rangle}$. Every finite-dimensional, inner-product space is complete, but there are infinite-dimensional pre-Hilbert spaces that are not:

Example: The convergent sequence of truncated Fourier series $s_m(t)$ of smooth periodic time functions with period T,

$$s_m(t) = \sum_{k=1}^{m} \frac{4}{(k-1)(2k+1)\pi} \cos \frac{(2k+1)\pi t}{T} \tag{3.266}$$

28 The name stems from the distance that a Manhattan taxi must travel from the origin to a point $\mathbf{x}$ in the rectangular street grid of Manhattan.

29 Stefan Banach (1892–1945).

approaches the unit-amplitude square wave as m increases. Although every term s_m is smooth, the sequence does not converge to a differentiable function. Thus the differentiable periodic functions constitute an inner product space but not a Hilbert space. □

Every Hilbert space is also a Banach space, but the converse is not true, unless the parallelogram identity holds for the norm

$$\| \mathbf{x}+\mathbf{y} \|^2 + \| \mathbf{x}-\mathbf{y} \|^2 = 2\left(\| \mathbf{x} \|^2 + \| \mathbf{y} \|^2\right), \tag{3.267}$$

and in which case the inner product is uniquely determined by the norm

$$\langle \mathbf{x}, \mathbf{y} \rangle = \frac{\| \mathbf{x}+\mathbf{y} \|^2 - \| \mathbf{x}-\mathbf{y} \|^2}{4}. \tag{3.268}$$

Given a set W, for example, $\mathbb{N}$, the *sequence space* l^2 over W is defined by

$$l^2(W) := \left\{ x : W \to \mathbb{R} \text{ and } \sum_{t \in W} |x(t)|^2 < \infty \right\}. \tag{3.269}$$

This space is a Hilbert space with the inner product $\langle x, y \rangle = \sum_{t \in W} x(t)y(t)$ for all $x, y \in l^2(W)$. If $W = \mathbb{N}$ the notation of the set W is usually omitted.

In the Hilbert space $L^2(\Omega)$ of square *Lebesgue integrable*[30] functions $f, g \in \Omega$, the scalar product is constructed with

$$\langle f, g \rangle = \int_\Omega f(t)g(t)\mathrm{d}t, \tag{3.270}$$

which gives rise to the norm

$$\| f \| = \sqrt{\langle f, f \rangle} = \sqrt{\int_\Omega |f(t)|^2 \mathrm{d}t}, \tag{3.271}$$

and the distance

$$d(f, g) = \| f - g \| = \sqrt{\int_\Omega |f(t) - g(t)|^2 \mathrm{d}t}. \tag{3.272}$$

In [3] we read: "Hilbert spaces are those spaces in which notions and concepts of ordinary Euclidean geometry hold, without any restriction on the dimension." In spaces that are complete, the convergence of sequences can be studied without *a priori* knowledge of the limit. The corresponding algebraic structures between the finite-dimensional vector spaces and the Hilbert space are summarized in Table 3.4.

30 Henri Lebesgue (1875–1941). Roughly speaking, the Lebesgue integral allows the computation of Fourier coefficients of functions such as the Dirichlet function that is 0 if x is rational and 1 if x is irrational. For the Dirichlet function, all coefficients are zero, as the set of rational numbers has measure zero; see [2] for an introduction. Formally, the Lebesgue integral is needed to ensure completeness.

Table 3.4 Comparison of algebraic structures in E_3 and Hilbert spaces.

Structure	E_3	Hilbert
Vector	$\mathbf{x}$	$g(t)$
Basis	$\{\mathbf{e}_1, \mathbf{e}_2, \mathbf{e}_3\}$	$\{g_n(t)\}$
Orthonormality	$\mathbf{e}_n \cdot \mathbf{e}_k = \delta_{nk}$	$\langle g_n, g_k \rangle = \delta_{nk}$
Expansion	$\mathbf{x} = \sum_{n=1}^{3} x_n \mathbf{e}_n$	$f(t) = \sum_{n=1}^{\infty} x_n g_n(t)$
Coefficients	$x_n = \mathbf{x} \cdot \mathbf{e}_n$	$x_n = \langle g_n, f \rangle$

Remark: It is of particular importance in numerical field calculation that functions with jump discontinuities be nonetheless square integrable; they must be L^2 functions even though they are not C^0. Therefore to check for L^2 (and consequently finite energy) is a more general way of measuring smoothness of functions than is establishing the C^m class. For example, the electrical field $\mathbf{E}$ should belong to $L^2(\Omega)$ so that the integral $\int_\Omega |\mathbf{E}|^2 \mathrm{d}\Omega$ is finite. For similar reasons one requires that $\operatorname{curl}\mathbf{E}$ is also in $L^2(\Omega)$. Hence the notation $H^1(\operatorname{curl}, \Omega)$ is used for all square-integrable vector fields whose curl is in $L^2(\Omega)$. Fields of this kind obey the tangential continuity at material interfaces, as is required of $\mathbf{E}$. □

Every element of a vector space can be expressed as a sum of basis vectors of the vector space. Extension of this concept to infinite dimensions is found within the Hilbert spaces l^2 and L^2. The Fourier series expansion can then be interpreted as a linear approximation for a periodic function[31] $f(t)$ on the interval $[-\pi, \pi]$, by a projection onto the trigonometric functions

$$g_n(t) := \cos nt \quad \text{and} \quad g_{-n}(t) := \sin nt \tag{3.273}$$

for $n \geq 1$ that form a Hilbert basis together with $g_0 = \sqrt{1/2}$. We can expand

$$f(t) = \sum_{n=1}^{\infty} c_n g_n \tag{3.274}$$

and obtain, because of the orthogonality of the trigonometric functions $\langle g_n(t), g_m(t) \rangle = \delta_{nm}$,

$$\langle f(t), g_m(t) \rangle = \sum_{n=1}^{\infty} c_n \langle g_n(t), g_m(t) \rangle = \sum_{n=1}^{\infty} c_n \delta_{nm} = c_m\,; \tag{3.275}$$

31 As an element of L^2 with the scalar product $\langle f, g \rangle$ and the norm $\| f \| = \sqrt{\langle f, f \rangle}$.

thus $c_n = \langle f(t), g_n(t) \rangle$. The $c_n \in \mathbb{R}$ are called the generalized Fourier coefficients. The countable orthonormal system $\{g_0, g_1, g_2, \dots\}$ is complete in L^2 if the Fourier polynomial

$$s_m : \sum_{n=0}^{m} c_n g_n = \sum_{n=0}^{m} \langle f, g_n \rangle g_n \tag{3.276}$$

converges for all $f \in L^2$, that is, $f = \lim_{m \to \infty} s_m$.

Theorem 3.1 *The Fourier polynomial constitutes among all finite linear combinations $\sum_{k=0}^{m} \gamma_k g_k$ the best approximation for f:*

$$\| f - s_m \|^2 = \| f \| - \sum_{k=0}^{m} |c_k|^2 \leq \| f - \sum_{k=0}^{m} \gamma_k g_k \|^2 \tag{3.277}$$

for all $m = 0, 1, 2, \dots$.

Proof.

$$\begin{aligned} \| f - \sum_{k=0}^{m} \gamma_k g_k \|^2 &= \left\langle f - \sum_{k=0}^{m} \gamma_k g_k , f - \sum_{k=0}^{m} \gamma_k g_k \right\rangle \\ &= \langle f, f \rangle - 2 \sum_{k=0}^{m} \gamma_k \langle f, g_k \rangle + \sum_{k=0}^{m} \gamma_k^2 \\ &= \| f \|^2 - \sum_{k=0}^{m} |\langle f, g_k \rangle|^2 + \sum_{k=0}^{m} |\langle f, g_k \rangle - \gamma_k|^2 , \end{aligned} \tag{3.278}$$

which takes its minimum when $\gamma_k = c_k = \langle f, g_k \rangle$ for $k = 0, 1, 2, \dots$. □

We finally arrive at the convergence criterion for the orthonormal system $\{g_n\}$ in the Hilbert space L^2:

Theorem 3.2 *The series $\sum_{n=0}^{\infty} c_n g_n$ with $c_n \in \mathbb{R}$ converges if $\sum_{n=0}^{\infty} |c_n|^2$ converges.*

Proof. Due to the orthogonality of the g_n it yields:

$$\begin{aligned} \| s_{m+k} - s_m \|^2 &= \left\langle \sum_{n=m+1}^{m+k} c_n g_n , \sum_{i=m+1}^{m+k} c_i g_i \right\rangle \\ &= \sum_{n=m+1}^{m+k} \| c_n g_n \|^2 = \sum_{n=m+1}^{m+k} |c_n|^2 \end{aligned} \tag{3.279}$$

for all $m, k = 1, 2, 3, \dots$. If the series $\sum_{n=0}^{\infty} |c_n|^2$ converges, (s_m) is a Cauchy sequence and therefore $\sum_{n=0}^{\infty} c_n g_n$ converges. If on the other hand we assume that $\sum_{n=0}^{\infty} c_n g_n$ converges, (s_m) is a Cauchy sequence and $\sum_{n=0}^{\infty} |c_n|^2$ converges. □

From the *Bessel inequality*[32]

$$\sum_{n=0}^{m} |c_n|^2 \leq \| f \|^2 \tag{3.280}$$

it follows that the series $\sum_{n=0}^{\infty} \langle f, g_n \rangle g_n$ converges and $(\langle f, g_n \rangle) \in l^2$. The Fourier series expansion $\mathcal{F}$ is thus given by

$$\mathcal{F} : L^2 \rightarrow l^2 : f \mapsto \mathcal{F}\{f\} = (\langle f, g_n \rangle)_{n=0}^{\infty}, \tag{3.281}$$

which together with the inverse mapping

$$\mathcal{F}^{-1} : l^2 \rightarrow L^2 : (c_n)_{n=0}^{\infty} \mapsto \mathcal{F}^{-1}\{(c_n)_{n=0}^{\infty}\} = \sum_{n=0}^{\infty} c_n g_n \tag{3.282}$$

is a linear, bijective, and isometric mapping from L^2 to l^2. Due to the linearity of the mapping we obtain

$$\mathcal{F}\{\lambda(f_1 + f_2)\} = \lambda(\mathcal{F}\{f_1\} + \mathcal{F}\{f_2\}). \tag{3.283}$$

In addition, $\langle f, g \rangle = \langle \mathcal{F}\{f\}, \mathcal{F}\{g\} \rangle$ and $\| f \| = \| \mathcal{F}\{f\} \|$.

Remark: In field computation, it makes therefore no difference whether we first sum up the field contributions of the current sources and the iron magnetization before performing the Fourier series expansion or we proceed the other way around. We can in fact reach the same result by analyzing each field's contribution separately and adding the Fourier coefficients. □

32 Friedrich Wilhelm Bessel (1784–1846).

References

1 Bamberg, P., Sternberg, S.: A Course in Mathematics for Students of Physics, Academic Press, San Diego, 1983

2 Bear, H. S.: A Primer of Lebesgue Integration, Academic Press, San Diego, 1995

3 Bossavit, A.: Computational Electromagnetism, Academic Press, San Diego, 1998

4 Bronstein, I.N., Semendjajew, K.A., Musiol, G., Muehlig, H.: Handbook of Mathematics, Springer, 2007

5 Frankel, T.: The Geometry of Physics, an Introduction, Cambride University Press, Cambride, 1997

6 Jänich, K.: Vectoranalysis, Springer, Berlin, 2001

7 Jänich, K.: Mathematik 2, Springer, Berlin, 2002

8 Stratton, J. A.: Electromagnetic Theory, A Classic Reissue, IEEE-Press, Wiley-Interscience, New York, 2007

4
Maxwell's Equations and Boundary Value Problems in Magnetostatics

War es ein Gott der diese Zeichen schrieb,
die mit geheimnisvoll verborg'nem Trieb
die Kräfte der Natur um mich enthüllen
und mir das Herz mit stiller Freude füllen.

Ludwig Boltzmann (1844–1906).[1]

In this chapter, we will present the foundations of electromagnetism, starting with Maxwell's[2] equations in the global, integral, and local forms of classical vector analysis. Before we can solve physically meaningful problems we must first study the constitutive equations, as well as discuss the boundary and interface conditions. The concepts described in Chapters 2 and 3 are now required, in particular vector fields in E_3, and the differential operators grad, curl, and div.

The nonlinear magnetic properties of iron laminations will be described by a phenomenological model, and a scaling law for the permeability of stacked iron laminations will be derived.

After a classification of electromagnetic field problems we will introduce the magnetic vector and scalar potentials and then arrange the field equations in a classification diagram known as "Maxwell's house." We will close by reviewing the boundary value problems in magnetostatics resulting from different formulations, such as the reduced and total scalar potentials, as well as the reduced and total vector potentials.

1 *Was it a god whose inspirations led him to write these fine equations, nature's field to me he shows and so my heart with pleasure glows.* Translation by J.P. Blewett.

2 James Clerk Maxwell (1831–1879).

Field Computation for Accelerator Magnets. Stephan Russenschuck

ISBN: 978-3-527-40769-9

4.1
Maxwell's Equations

Different embodiments of Maxwell's equations, depending on the descriptions of the field intensities, are known from the literature. For our purposes it will suffice to present Maxwell's equations in the global, integral, and local form of classical vector analysis.[3] The original, component form is shown in Section 4.1.4.

4.1.1
The Global Form

We first summarize the governing laws of electromagnetism in their global form for all geometrical objects at rest:

$$V_{\mathrm{m}}(\partial\mathscr{A}) = I(\mathscr{A}) + \frac{\mathrm{d}}{\mathrm{d}t}\Psi(\mathscr{A})\,, \tag{4.1}$$

$$U(\partial\mathscr{A}) = -\frac{\mathrm{d}}{\mathrm{d}t}\Phi(\mathscr{A})\,, \tag{4.2}$$

$$\Phi(\partial\mathscr{V}) = 0\,, \tag{4.3}$$

$$\Psi(\partial\mathscr{V}) = Q(\mathscr{V})\,. \tag{4.4}$$

Equation (4.1) is Ampère's magnetomotive force law with the displacement-current term added by Maxwell, and Eq. (4.2) is Faraday's[4] law of electromagnetic induction. Equation (4.3) is the magnetic flux conservation law, and Eq. (4.4) is Gauss' fundamental theorem of electrostatics. These are topological laws that depend neither on the nature of the medium nor on its dimension. This property is reflected by the absence of length in the dimension system. In SI units:

- Q is the *electric charge* in an outer-oriented volume V, $[Q] = 1\,\mathrm{C} = 1\,\mathrm{A\,s}$.
- $I(\mathscr{A}) = \lim_{\Delta t \to 0} \frac{\Delta Q}{\Delta t} = \frac{\mathrm{d}Q}{\mathrm{d}t}$ is the *electric current* across the surface $\mathscr{A}$ which is thus outer oriented. The physical unit is $[I] = 1\,\mathrm{A}$. Note that although we speak of a direction of the current (positive when positive charges are moving through the surface), the electric current is not a geometrically directed quantity. The key point is that every global physical variable referring to a space element reverses its sign when the orientation of the space element is reversed. This is called the *oddness principle* [32].
- $U(\mathscr{S})$ denotes the *electric voltage* between the start and end points of an inner-oriented line $\mathscr{S}$, defined by the mechanical work done by a

3 There are additional versions employing quaternions, differential forms, or 4-vectors.
4 Michael Faraday (1791–1867).

displacement of a charge along this line, $U(\mathscr{S}) = W(\mathscr{S})/Q$, reflected by the physical unit $[U] = 1\,\mathrm{V} = 1\,\mathrm{J\,C^{-1}}$. If $\mathscr{A}$ is an inner-oriented surface and $\partial\mathscr{A}$ its boundary, then $U(\partial\mathscr{A})$ is called the *circulation voltage*, which is zero if the surface is not penetrated by a time-transient magnetic flux.

- V_m is the *magnetomotive force* along the boundary $\partial\mathscr{A}$ of an outer-oriented surface $\mathscr{A}$. The physical unit is $[V_\mathrm{m}] = 1\,\mathrm{A}$. Outer orientation of the surface is required because the sign of V_m depends on the sign of the current crossing the surface.
- The *electric flux* through an outer-oriented surface $\mathscr{A}$ is denoted Ψ with the physical unit coulomb, $[\Psi] = 1\,\mathrm{C} = 1\,\mathrm{A\ s}$. It is a measure of the electric charge that can be influenced on a surface, and consequently requires the choice of one face of the surface, that is, its outer orientation. Although the quantity is called flux, it is important to note that nothing actually flows in a physical sense; the name is inspired rather by the mathematical structure of the quantity.
- The *magnetic flux* through a surface $\mathscr{A}$ is denoted Φ with $[\Phi] = 1\,\mathrm{Wb} = 1\,\mathrm{V\,s}$ and is a measure of the electric voltage that can be induced along $\partial\mathscr{A}$. Thus, the term magnetic flux refers to a surface endowed with inner orientation.

Modern approaches to field computation use these global variables directly while approximating the space domain with so-called *cell complexes*. Links to recent works in this area can be found in [33].

4.1.2 The Integral Form

The mathematical treatment of field problems has for long been based on differential models employing vector fields. These yield the integral forms of Maxwell's equations, which read in SI units:

$$\int_{\partial\mathscr{A}} \mathbf{H}\cdot\mathrm{d}\mathbf{r} = \int_{\mathscr{A}} \mathbf{J}\cdot\mathrm{d}\mathbf{a} + \frac{\mathrm{d}}{\mathrm{d}t}\int_{\mathscr{A}} \mathbf{D}\cdot\mathrm{d}\mathbf{a}, \tag{4.5}$$

$$\int_{\partial\mathscr{A}} \mathbf{E}\cdot\mathrm{d}\mathbf{r} = -\frac{\mathrm{d}}{\mathrm{d}t}\int_{\mathscr{A}} \mathbf{B}\cdot\mathrm{d}\mathbf{a}, \tag{4.6}$$

$$\int_{\partial\mathscr{V}} \mathbf{B}\cdot\mathrm{d}\mathbf{a} = 0, \tag{4.7}$$

$$\int_{\partial\mathscr{V}} \mathbf{D}\cdot\mathrm{d}\mathbf{a} = \int_{\mathscr{V}} \rho\,\mathrm{d}V. \tag{4.8}$$

The vector fields **E** and **H** are the electric and magnetic field intensities, and **D** and **B** are the *electric* and *magnetic flux densities*, respectively.[5] The sources are the *electric charge density* ρ and *electric current density* **J**. In classical electrodynamics, the vector fields and sources are assumed to be finite in the entire domain and to be continuous functions of position and time. Discontinuities in the field distribution may occur, however, on surfaces between domains with different material properties. Such discontinuities must therefore be excluded until we have treated the interface conditions in Section 4.5. Localized distributions of sources will be approximated by point- and surface-charges, and by line- and surface-currents. These concepts will be discussed in Section 4.5.

The relations between the electromagnetic fields and their corresponding global physical values are summarized in Table 4.1.

Table 4.1 Relation between the electromagnetic fields and their integral values (MMF = magnetomotive force).

Global quantity	SI unit	Relation	SI unit	Field
MMF	$1\,\mathrm{A}$	$V_{\mathrm{m}}(\mathscr{S}) = \int_{\mathscr{S}} \mathbf{H} \cdot \mathbf{dr}$	$1\,\mathrm{A\,m^{-1}}$	Magnetic field
Electric voltage	$1\,\mathrm{V}$	$U(\mathscr{S}) = \int_{\mathscr{S}} \mathbf{E} \cdot \mathbf{dr}$	$1\,\mathrm{V\,m^{-1}}$	Electric field
Magnetic flux	$1\,\mathrm{V\,s}$	$\Phi(\mathscr{A}) = \int_{\mathscr{A}} \mathbf{B} \cdot \mathbf{da}$	$1\,\mathrm{V\,s\,m^{-2}}$	Magnetic flux density
Electric flux	$1\,\mathrm{A\,s}$	$\Psi(\mathscr{A}) = \int_{\mathscr{A}} \mathbf{D} \cdot \mathbf{da}$	$1\,\mathrm{A\,s\,m^{-2}}$	Electric flux density
Electric current	$1\,\mathrm{A}$	$I(\mathscr{A}) = \int_{\mathscr{A}} \mathbf{J} \cdot \mathbf{da}$	$1\,\mathrm{A\,m^{-2}}$	Electric current density
Electric charge	$1\,\mathrm{A\,s}$	$Q(\mathscr{V}) = \int_{\mathscr{V}} \rho \cdot dV$	$1\,\mathrm{A\,s\,m^{-3}}$	Electric charge density

The global variables can be reconstructed by an integration process. The field intensities **E** and **H** are integrated along a line, evident from the physical units $[\mathbf{E}] = 1\,\mathrm{V\,m^{-1}}$, $[\mathbf{H}] = 1\,\mathrm{A\,m^{-1}}$, whereas the flux and current densities **D**, **B**, and **J** are integrated over a surface, $[\mathbf{D}] = 1\,\mathrm{A\,s\,m^{-2}}$, $[\mathbf{B}] = 1\,\mathrm{V\,s\,m^{-2}}$, $[\mathbf{J}] = 1\,\mathrm{A\,m^{-2}}$. The vectorial surface element appearing in this integration process is given by $\mathbf{da} = \mathbf{n}\,da$, where **n** is the surface normal vector directed such that it matches, together with the inner orientation of the surface, the right-handed screw orientation of the ambient space; see Section 2.7.

The electric charge density ρ is integrated on a volume, as evident from the physical unit $[\rho] = 1\,\mathrm{A\,s\,m^{-3}}$. As noted by Maxwell in his treatise, the field intensities and the fluxes have different natures. They are physical quantities associated with lines or cross-sectional areas. Their product yields an energy density associated with a volume.

5 From now on we will use the capital bold face characters according to the IEC norm [18] for the vector fields in $\mathcal{V}(\Omega)$ and assume that Ω is the entire E_3 if nothing else is stated.

4.1.3 The Local Form

As long as the necessary conditions[6] hold for the application of the Kelvin–Stokes and Gauss theorems, the field equations can be rewritten for the stationary case as

$$\int_{\mathscr{A}} \operatorname{curl} \mathbf{H} \cdot \mathrm{d}\mathbf{a} = \int_{\mathscr{A}} \left(\mathbf{J} + \frac{\partial}{\partial t} \mathbf{D} \right) \cdot \mathrm{d}\mathbf{a}, \tag{4.9}$$

$$\int_{\mathscr{A}} \operatorname{curl} \mathbf{E} \cdot \mathrm{d}\mathbf{a} = - \int_{\mathscr{A}} \frac{\partial}{\partial t} \mathbf{B} \cdot \mathrm{d}\mathbf{a}, \tag{4.10}$$

$$\int_{\mathscr{V}} \operatorname{div} \mathbf{B} \, \mathrm{d}V = 0, \tag{4.11}$$

$$\int_{\mathscr{V}} \operatorname{div} \mathbf{D} \, \mathrm{d}V = \int_{\mathscr{V}} \rho \, \mathrm{d}V. \tag{4.12}$$

These equations hold for arbitrary volumes and surfaces only if the following equations hold for their integrands:

$$\operatorname{curl} \mathbf{H} = \mathbf{J} + \frac{\partial}{\partial t} \mathbf{D}, \tag{4.13}$$

$$\operatorname{curl} \mathbf{E} = - \frac{\partial}{\partial t} \mathbf{B}, \tag{4.14}$$

$$\operatorname{div} \mathbf{B} = 0, \tag{4.15}$$

$$\operatorname{div} \mathbf{D} = \rho. \tag{4.16}$$

This is the classical, local form of Maxwell's equations. The notation in Eqs. (4.13)–(4.16) is usually attributed to O. Heaviside and J.W. Gibbs[7], who eliminated the vector and the scalar potentials in Maxwell's original set of equations, presented in the next section. Note that

- Eqs. (4.13)–(4.16) have too many unknowns: Additional material relations must be taken into account.
- The vector differential operators require the vector fields to be at least 1-smooth. However, on material boundaries of permeable media, both $\mathbf{B}$ and $\mathbf{H}$ are discontinuous and therefore $\operatorname{curl} \mathbf{H}$ and $\operatorname{div} \mathbf{B}$ cease to have any meaning there.
- Consider the static case ($\partial/\partial t = 0$) for $\mathbf{J} = 0$. Both $\mathbf{H}$ and $\mathbf{B}$ are zero. But this is not implied by Eqs. (4.13)–(4.16) as for $\mathbf{H} = \operatorname{grad} \phi$ the equations

6 Smooth vector fields, smooth surfaces with simply connected, closed, piecewise-smooth and consistently oriented boundaries, and volumes with piecewise-smooth, closed and consistently oriented surfaces.

7 Oliver Heaviside (1850–1925), Josiah Willard Gibbs (1839–1903).

are fulfilled even though there is no source to create the nonzero static field.

4.1.4 Maxwell's Original Set of Equations

We list the equations from Maxwell's books [27]. A glossary for the notation is given in Table 4.2. In component form, Maxwell's equations are written as[8]

$$a = \frac{\partial H}{\partial y} - \frac{\partial G}{\partial z}, \qquad b = \frac{\partial F}{\partial z} - \frac{\partial H}{\partial x}, \qquad c = \frac{\partial G}{\partial x} - \frac{\partial F}{\partial y}, \tag{4.17}$$

which reads $\mathbf{B} = \operatorname{curl} \mathbf{A}$ in vector notation. The equations

$$P = -\frac{\partial F}{\partial t} - \frac{\partial \varphi}{\partial x}, \qquad Q = -\frac{\partial G}{\partial t} - \frac{\partial \varphi}{\partial y}, \qquad R = -\frac{\partial H}{\partial t} - \frac{\partial \varphi}{\partial z} \tag{4.18}$$

yield $\mathbf{E} = -\frac{\partial}{\partial t}\mathbf{A} - \operatorname{grad} \varphi$ in local form, and

$$p + \frac{\partial f}{\partial t} = \frac{\partial \gamma}{\partial y} - \frac{\partial \beta}{\partial z}, \qquad q + \frac{\partial g}{\partial t} = \frac{\partial \alpha}{\partial z} - \frac{\partial \gamma}{\partial x}, \qquad r + \frac{\partial h}{\partial t} = \frac{\partial \beta}{\partial x} - \frac{\partial \alpha}{\partial y} \tag{4.19}$$

can be expressed as $\mathbf{J} + \frac{\partial}{\partial t}\mathbf{D} = \operatorname{curl} \mathbf{H}$. Finally,

$$\rho = \frac{\partial f}{\partial x} + \frac{\partial g}{\partial y} + \frac{\partial h}{\partial z} \tag{4.20}$$

corresponds to $\rho = \operatorname{div} \mathbf{D}$.

Table 4.2 Symbols used in Maxwell's original set of equations (Version of 1891).

Original notation		Constituents	Vector notation	
Radius vector of a point	ρ	x, y, z	Position vector	$\mathbf{r}$
Electromagnetic momentum	$\mathcal{U}$	F, G, H	Magnetic vector potential	$\mathbf{A}$
Magnetic induction	$\mathcal{B}$	a, b, c	Magnetic flux density	$\mathbf{B}$
Intensity of magnetization	$\mathcal{J}$	A, B, C	Magnetization	$\mathbf{M}$
Total electric current	$\mathcal{C}$	u, v, w	Current density	$\mathbf{J}$
Magnetic force	$\mathcal{H}$	α, β, γ	Magnetic field strength	$\mathbf{H}$
Electric displacement	$\mathcal{D}$	f, g, h	Electric field strength	$\mathbf{D}$
Electromotive intensity	$\mathcal{E}$	P, Q, R	Electric flux density	$\mathbf{E}$
Current of conduction	$\mathcal{R}$	p, q, r	Current density	$\mathbf{J}_0$

8 See Reference [27], page 255 ff.

4.2 Kirchhoff's Laws

From the first Poincaré lemma, div curl $\mathbf{g} = 0$, it follows directly that

$$\operatorname{div}\left(\mathbf{J} + \frac{\partial}{\partial t}\mathbf{D}\right) = \operatorname{div}\mathbf{J} + \frac{\partial}{\partial t}\rho = 0\,, \tag{4.21}$$

which is the *conservation of charge law*. The commutation of the div and $\partial/\partial t$ operators is admissible if the fields and charge distributions are smooth. The law can be written in integral form as

$$\int_{\partial\mathscr{V}} \mathbf{J}\cdot \mathrm{d}\mathbf{a} + \frac{\mathrm{d}}{\mathrm{d}t}\int_{\mathscr{V}} \rho\, \mathrm{d}V = 0 \tag{4.22}$$

or in the global form as

$$I(\partial\mathscr{V}) + \frac{\mathrm{d}}{\mathrm{d}t}Q(\mathscr{V}) = 0\,. \tag{4.23}$$

If at every point within a volume $\mathscr{V}$ the charge density is constant in time, we obtain

$$\int_{\mathscr{V}} \operatorname{div}\mathbf{J}\,\mathrm{d}V = 0\,, \qquad \int_{\partial\mathscr{V}} \mathbf{J}\cdot\mathrm{d}\mathbf{a} = 0\,, \qquad I(\partial\mathscr{V}) = 0\,, \tag{4.24}$$

the latter known as *Kirchhoff's current law*.[9]

Remark: We can derive Kirchhoff's current law directly by applying Ampère's magnetomotive force law (4.1) to a closed surface $\mathscr{A} = \partial\mathscr{V}$, which together with Eq. (4.4) yields:

$$V_{\mathrm{m}}(\partial(\partial\mathscr{V})) = 0 = I(\partial\mathscr{V}) + \frac{\mathrm{d}}{\mathrm{d}t}Q(\mathscr{V})\,. \tag{4.25}$$

Because the ideal node in network theory cannot carry charges we have $I(\partial\mathscr{V}) = 0$. □

For the sake of completeness we note that for surfaces not penetrated by time-varying magnetic flux we obtain

$$U(\partial\mathscr{A}) = 0, \tag{4.26}$$

which is *Kirchhoff's voltage law*.

4.3 Conversion of Energy in Electromagnetic Fields

From Eqs. (4.13) and (4.14) we obtain

$$\mathbf{H}\cdot\operatorname{curl}\mathbf{E} - \mathbf{E}\cdot\operatorname{curl}\mathbf{H} = -\mathbf{H}\cdot\frac{\partial}{\partial t}\mathbf{B} - \mathbf{E}\cdot\frac{\partial}{\partial t}\mathbf{D} - \mathbf{E}\cdot\mathbf{J}\,. \tag{4.27}$$

9 Gustav Kirchhoff (1824–1878).

The left-hand side can be expressed as

$$\mathbf{H} \cdot \operatorname{curl} \mathbf{E} - \mathbf{E} \cdot \operatorname{curl} \mathbf{H} = \operatorname{div}(\mathbf{E} \times \mathbf{H}) =: \operatorname{div} \mathbf{S}, \tag{4.28}$$

which defines the *Poynting vector*[10] $\mathbf{S}$ where $[\mathbf{S}] = 1\,\mathrm{J\,m^{-2}\,s}$. Eq. (4.27) can be rewritten as

$$\operatorname{div} \mathbf{S} + \mathbf{E} \cdot \frac{\partial}{\partial t}\mathbf{D} + \mathbf{H} \cdot \frac{\partial}{\partial t}\mathbf{B} = -\mathbf{E} \cdot \mathbf{J}, \tag{4.29}$$

known as *Poynting's theorem*, which describes the energy flow in an electromagnetic field. At the same time it postulates the conservation of electromagnetic energy when the right-hand side is zero. The integral form of Poynting's theorem for linear lossless media is given by

$$\int_{\partial \mathcal{V}} \mathbf{S} \cdot \mathrm{d}\mathbf{a} + \frac{\mathrm{d}}{\mathrm{d}t} \int_{\mathcal{V}} \frac{1}{2} (\mathbf{E} \cdot \mathbf{D} + \mathbf{H} \cdot \mathbf{B})\, \mathrm{d}V = -\int_{\mathcal{V}} \mathbf{E} \cdot \mathbf{J}\, \mathrm{d}V. \tag{4.30}$$

The Poynting vector is a measure of the density of the local energy flow. The term $\int_{\partial \mathcal{V}} \mathbf{S} \cdot \mathrm{d}\mathbf{a}$ is the power of the energy flow through the closed boundary surface of the volume. $\mathbf{E} \cdot \mathbf{J}$ is the electromagnetic power per volume element that is converted into mechanical energy or heat loss. The Poynting theorem states that the time rate of change of electromagnetic energy within a certain volume, plus the energy per unit time flowing through the boundary surface of that volume, equals the total work done by the fields on the sources in the volume.

4.4 Constitutive Equations

Maxwell's equations consist of two vector equations (two times three equations) and two scalar equations; all together eight equations for the sixteen unknown components of $\mathbf{E}$, $\mathbf{D}$, $\mathbf{H}$, $\mathbf{B}$, $\mathbf{J}$, and ρ. If we also consider that Eq. (4.15) follows from Eq. (4.14), we can see that Maxwell's equations can only be solved with nine additional material relations, which are known as the *constitutive equations*. These result from the aim to treat the electromagnetic subsystem by itself without resorting to coupled systems of partial differential equations, for example, Maxwell-Vlasov[11] for coupled electromagnetic/charge-motion subsystems [10]. The complex interactions of fields and matter are thus modeled within the electromagnetic subsystem by the macroscopic description of the constitutive equations. The most basic form of these constitutive equations holds only for linear (field-independent), homogeneous

10 John Henry Poynting (1852–1914).
11 Anatoly Vlasov (1908–1975).

(position-independent), isotropic (direction-independent), lossless, and stationary media:

$$\mathbf{B} = \mu \mathbf{H}, \qquad \mathbf{D} = \varepsilon \mathbf{E}, \qquad \mathbf{J} = \varkappa \mathbf{E}, \tag{4.31}$$

where $\mu, \varepsilon, \varkappa$ are the *permeability*, *permittivity*, and *conductivity*.[12] Their physical units are:

$$[\mu] = 1\,\mathrm{V\,s\,A^{-1}\,m^{-1}} = 1\,\mathrm{H\,m^{-1}}, \tag{4.32}$$

$$[\varepsilon] = 1\,\mathrm{A\,s\,V^{-1}\,m^{-1}}, \tag{4.33}$$

$$[\varkappa] = 1\,\mathrm{A\,V^{-1}m^{-1}} = 1\,\Omega^{-1}\,\mathrm{m^{-1}}. \tag{4.34}$$

The unit of conductance Ω^{-1} is called siemens[13] or mho. For the resistivity we have $\rho = 1/\varkappa$ where $[\varkappa] = 1\,\mathrm{V\,m\,A^{-1}} = 1\,\Omega\,\mathrm{m}$.

In isotropic media it is customary to express the permeability and permittivity as a function of the free-space (vacuum) field constants:

$$\mu = \mu_r \mu_0, \qquad \varepsilon = \varepsilon_r \varepsilon_0, \tag{4.35}$$

with

$$\mu_0 = 4\pi \times 10^{-7}\,\mathrm{H\,m^{-1}}, \qquad \varepsilon_0 = 8.8542\ldots \times 10^{-12}\,\mathrm{F\,m^{-1}}, \tag{4.36}$$

which are related by the velocity of light in vacuum:

$$c_0 = \frac{1}{\sqrt{\varepsilon_0 \mu_0}} = 299\,792\,458\,\mathrm{m\,s^{-1}}. \tag{4.37}$$

In the more general case, for a permanent magnetic or electric polarization, which are volume densities of magnetic and electric dipole momenta, respectively, it will prove convenient to introduce new vectors: the *electric polarization* $\mathbf{P}_{\mathrm{el}}$, the *magnetic polarization* $\mathbf{P}_{\mathrm{m}}$, and the impressed current density $\mathbf{J}_{\mathrm{S}}$. The magnetic polarization is often replaced by the *magnetization* $\mathbf{M}$ in units of $\mathrm{A\,m^{-1}}$. The material relations can then be expressed as

$$\mathbf{B} = \mu_0 \mathbf{H} + \mathbf{P}_{\mathrm{m}}(\mathbf{H}) = \mu_0 (\mathbf{H} + \mathbf{M}(\mathbf{H})), \tag{4.38}$$

$$\mathbf{D} = \varepsilon_0 \mathbf{E} + \mathbf{P}_{\mathrm{el}}(\mathbf{E}), \tag{4.39}$$

$$\mathbf{J} = \varkappa \mathbf{E} + \mathbf{J}_{\mathrm{S}}. \tag{4.40}$$

12 The international IEC standard recommends to use the symbol σ for the conductivity which is, however, also used for the surface-charge. We therefore use the symbol $\varkappa$ as proposed in DIN 1324.

13 Werner von Siemens (1816–1892).

Equation (4.38) also holds for permanent magnets where remanent magnetization is present without an excitation field. The definition of the magnetization is not unique in the literature, where **M** sometimes contains μ_0 [26]. This has also consequences for the definition of the *magnetic dipole moment* **m** in Eq. (5.76). The polarization vectors are associated with the electro-mechanical properties of matter and thus vanish in free space. For linear isotropic material, the polarization vectors are parallel to the field vectors and are found to be proportional to the field: $\mathbf{P}_{\mathrm{el}} = \chi_e \epsilon_0 \mathbf{E}$ and $\mathbf{M} = \chi_{\mathrm{m}} \mathbf{H}$, so that for magnetic materials

$$\mathbf{B} = \mu_0 \, \mathbf{H} + \mu_0 \chi_{\mathrm{m}} \, \mathbf{H} = \mu_0 (1 + \chi_{\mathrm{m}}) \mathbf{H} = \mu_0 \mu_{\mathrm{r}} \, \mathbf{H} = \mu \, \mathbf{H}, \tag{4.41}$$

where $\mu_{\mathrm{r}} = 1 + \chi_{\mathrm{m}}$ is the *relative permeability*, and χ_{m} the *magnetic susceptibility*. Both are dimensionless quantities: $[\mu_{\mathrm{r}}] = 1_{\mathrm{U}}$, $[\chi_{\mathrm{m}}] = 1_{\mathrm{U}}$. In the general case, for example, inside permanent magnets, **B** and **H** have different directions, as illustrated in Figure 4.12. In the free space outside the material we have simply $\mathbf{B} = \mu_0 \mathbf{H}$.

4.5 Boundary and Interface Conditions

Consider the models of two typical problems in magnet design, as shown in Figure 4.1. These consist of (generally two- or three-dimensional) iron regions Ω_{i} and air regions Ω_{a}. Further, let us define the entire problem domain as $\Omega = \Omega_{\mathrm{a}} \cup \Omega_{\mathrm{i}}$. The regions are connected to each other at the interface Γ_{ai}. The nonconductive air regions may also contain a number of conductors that do not intersect the iron boundaries. The field quantities **B** and **H** satisfy boundary conditions on the at least piecewise-smooth boundary $\Gamma = \partial\Omega$ of the domain Ω. Two types of boundary conditions, prescribed on the two disjoint boundaries, denoted Γ_{H} and Γ_{B} with $\Gamma = \Gamma_{\mathrm{H}} \cup \Gamma_{\mathrm{B}}$, cover all practical cases:

- On the part Γ_{B} of the boundary, the normal component of the magnetic flux density is imposed. On symmetry planes parallel to the field, on far boundaries, or on outer boundaries of iron yokes surrounded by air,[14] the normal component B_{n} of the flux density is zero. In other cases, B_{n} may be imposed along a physical surface, for example, a sinusoidal flux distribution in the air gap of an electrical machine. All these boundary conditions can be written in the form

$$\mathbf{B} \cdot \mathbf{n} = \sigma_{\mathrm{m}} \qquad \text{on } \Gamma_{\mathrm{B}}, \tag{4.42}$$

14 If the iron yoke is unsaturated it can often be assumed that no flux escapes the outer boundary.

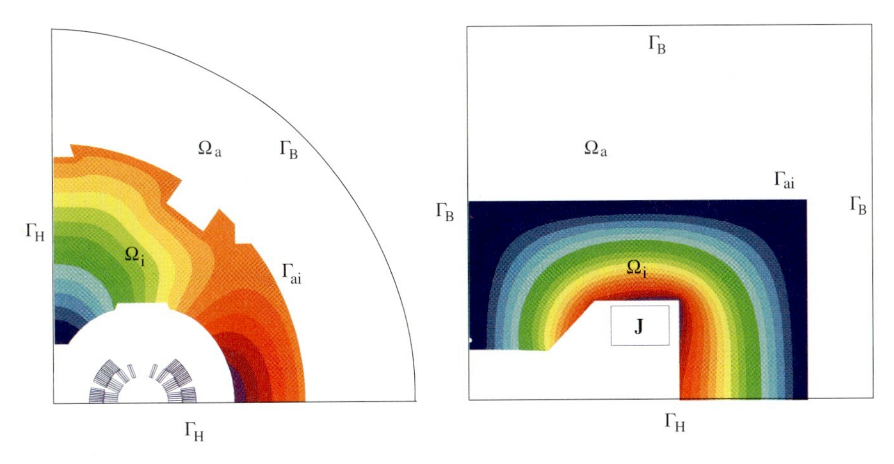

Figure 4.1 Two-dimensional, elementary model problems in magnet design. The nonconductive air region Ω_a contains areas with an impressed current density $\mathbf{J}$ that do not intersect the boundary Γ_{ai} between iron and air.

where σ_m is the fictitious *magnetic surface-charge* and $\mathbf{n}$ is the outward unit normal vector.

- On the part Γ_H of the boundary, the tangential components of the magnetic field are imposed. In many cases, as on symmetry planes perpendicular to the field and on infinitely permeable iron poles (where the field enters at a right angle), the tangential component H_t of the field is zero. The tangential component H_t can also be determined by a real or fictitious *surface-current* $\boldsymbol{\alpha}$. All these boundary conditions can be written in the form

$$\mathbf{H} \times \mathbf{n} = \boldsymbol{\alpha} \qquad \text{on } \Gamma_H, \tag{4.43}$$

The condition that the tangential component be zero on Γ_H implies according to Eq. (2.82)

$$\mathbf{n} \times (\mathbf{H} \times \mathbf{n}) = \mathbf{0} \qquad \text{on } \Gamma_H. \tag{4.44}$$

The interface conditions require a definition of the fictitious magnetic surface-charge and the surface-current:

- Consider a thin layer of thickness h carrying a fictitious magnetic charge of density ρ_m; see Figure 4.2 (left). In an area $\Delta x \Delta y$ of this layer there is a charge $\Delta Q_m = \Delta x \Delta y \, h \rho_m$, which corresponds to $h\rho_m$ per unit area. If $h \to 0$ and $\rho_m \to \infty$ so that $h\rho_m$ remains finite, we obtain the magnetic surface-charge $\sigma_m = h\rho_m$. The physical unit of σ_m is $1\,\mathrm{V\,s\,m^{-2}}$.

- Consider a thin layer of thickness h in which a current of density J flows as shown in Figure 4.2 (right). In a length Δl of this layer flows the current $\Delta I = Jh\Delta l$, which corresponds to Jh per unit length. If $h \to 0$ and $J \to \infty$ so that Jh remains finite, we obtain the surface-current $\boldsymbol{\alpha} = \mathbf{J}\,h$. The physical unit is $[\boldsymbol{\alpha}] = 1\,\mathrm{A\,m^{-1}}$.

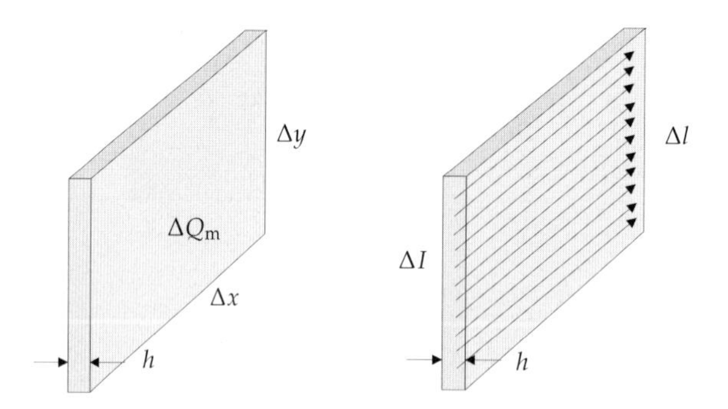

Figure 4.2 Left: Fictitious magnetic surface-charge. Right: Surface-current.

We will now establish the interface conditions on a smooth surface Γ_{12} between two domains Ω_1 and Ω_2 of different magnetic properties (permeabilities μ_1 and μ_2). The surface is oriented by the crossing direction $\mathbf{n}$ from Ω_2 to Ω_1, as shown in Figure 4.3.

Consider a surface element in the interface, where the vector $\mathrm{d}\mathbf{a}$ lies in the interface plane, as shown in Figure 4.3 (left). Applying Ampère's law $\int_{\partial\mathscr{A}} \mathbf{H} \cdot \mathrm{d}\mathbf{r} = \int_{\mathscr{A}} \mathbf{J} \cdot \mathrm{d}\mathbf{a}$ to the rectangular loop, yields for $\delta \to 0$

$$\int_{\mathscr{S}_2} \mathbf{H}_2 \cdot \mathrm{d}\mathbf{r} + \int_{\mathscr{S}_1} \mathbf{H}_1 \cdot \mathrm{d}\mathbf{r} = \int_{\mathscr{S}} (\mathbf{H}_1 - \mathbf{H}_2) \cdot \mathrm{d}\mathbf{r} = -\int_{\mathscr{S}} (\mathbf{n} \times \boldsymbol{\alpha}) \cdot \mathrm{d}\mathbf{r}, \qquad (4.45)$$

where the surface normal vector $\mathbf{n}$ points from Ω_2 to Ω_1 as shown in Figure 4.3. Equation (4.45) holds for any curve $\mathscr{S}$ if the integrands are equal or differ only by a normal component $\lambda\mathbf{n}$ [25]:

$$(\mathbf{H}_1 - \mathbf{H}_2) + \lambda\mathbf{n} = -\mathbf{n} \times \boldsymbol{\alpha}\,. \qquad (4.46)$$

Using $\mathbf{n} \times (\mathbf{n} \times \boldsymbol{\alpha}) = (\mathbf{n} \cdot \boldsymbol{\alpha})\mathbf{n} - (\mathbf{n} \cdot \mathbf{n})\boldsymbol{\alpha} = -\boldsymbol{\alpha}$ we obtain

$$\boldsymbol{\alpha} = \mathbf{n} \times (\mathbf{H}_1 - \mathbf{H}_2) = \mathbf{n} \times [\![\mathbf{H}]\!]_{12}\,. \qquad (4.47)$$

The jump $[\![.]\!]_{12}$ introduced in Eq. (4.47) is a mapping on $\Omega_1 \cup \Omega_2 \cup \Gamma_{12}$ defined by

$$[\![\mathbf{g}]\!]_{12}(\mathbf{x}) := \lim_{\substack{\mathbf{x}_1 \to \mathbf{x} \\ \mathbf{x}_1 \in \Omega_1}} \mathbf{g}(\mathbf{x}_1) - \lim_{\substack{\mathbf{x}_2 \to \mathbf{x} \\ \mathbf{x}_2 \in \Omega_2}} \mathbf{g}(\mathbf{x}_2), \qquad \mathbf{x} \in \Gamma_{12}\,, \qquad (4.48)$$

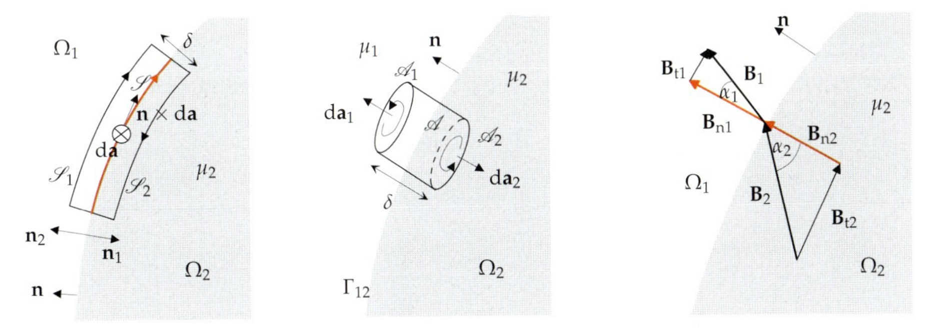

Figure 4.3 Conventions for deriving the interface conditions for permeable media.

and is thus defined as the strength of the vector field $\mathbf{g}$ just before the jump, minus the value just after it. If $\mathbf{g}$ is continuous at the interface, we obtain $[\![\mathbf{g}]\!] = [\![\mathbf{g}]\!]_{12} = [\![\mathbf{g}]\!]_{21} = 0$.

In the absence of real or fictitious surface-currents, the tangential component of the magnetic field strength is continuous at the interface

$$H_{t1} = H_{t2} \quad \equiv \quad \mathbf{n} \times (\mathbf{H}_1 - \mathbf{H}_2) = \mathbf{0} \quad \equiv \quad \mathbf{n} \times [\![\mathbf{H}]\!]_{12} = \mathbf{0}. \tag{4.49}$$

Now consider the volume of the "pill box" as shown in Figure 4.3 (middle). Employing the *flux conservation law* $\int_{\partial \mathscr{V}} \mathbf{B} \cdot \mathbf{da} = \int_{\mathscr{V}} \rho_m \, dV$, which holds for any closed simply connected surface, we have for $\delta \to 0$,

$$\int_{\mathscr{A}} \sigma_m \, da = \int_{\mathscr{A}_1} \mathbf{B}_1 \cdot \mathbf{da} + \int_{\mathscr{A}_2} \mathbf{B}_2 \cdot \mathbf{da} = \int_{\mathscr{A}} (\mathbf{B}_1 - \mathbf{B}_2) \cdot \mathbf{da} \,. \tag{4.50}$$

Equation (4.50) holds for any surface $\mathscr{A}$ if the integrands obey

$$\sigma_m = \mathbf{n} \cdot (\mathbf{B}_1 - \mathbf{B}_2) = \mathbf{n} \cdot [\![\mathbf{B}]\!]_{12} \,. \tag{4.51}$$

In the absence of a magnetic surface-charge, the normal component of the magnetic flux density is continuous at the interface:

$$B_{n1} = B_{n2} \quad \equiv \quad \mathbf{n} \cdot (\mathbf{B}_1 - \mathbf{B}_2) = 0 \quad \equiv \quad \mathbf{n} \cdot [\![\mathbf{B}]\!]_{12} = 0. \tag{4.52}$$

For a boundary of isotropic materials, free of surface-currents and magnetic surface-charges, it follows that

$$\frac{\tan \alpha_1}{\tan \alpha_2} = \frac{\frac{B_{t1}}{B_{n1}}}{\frac{B_{t2}}{B_{n2}}} = \frac{\mu_1 \frac{H_{t1}}{B_{n1}}}{\mu_2 \frac{H_{t2}}{B_{n2}}} = \frac{\mu_1 H_{t1}}{\mu_2 H_{t2}} = \frac{\mu_1}{\mu_2} \tag{4.53}$$

for all points $\mathscr{P} \in \Gamma_{12}$. For $\mu_2 \gg \mu_1$ we distinguish two cases: For $\alpha_1 \approx 0$, the field exits vertically from a highly permeable medium, as can be seen in

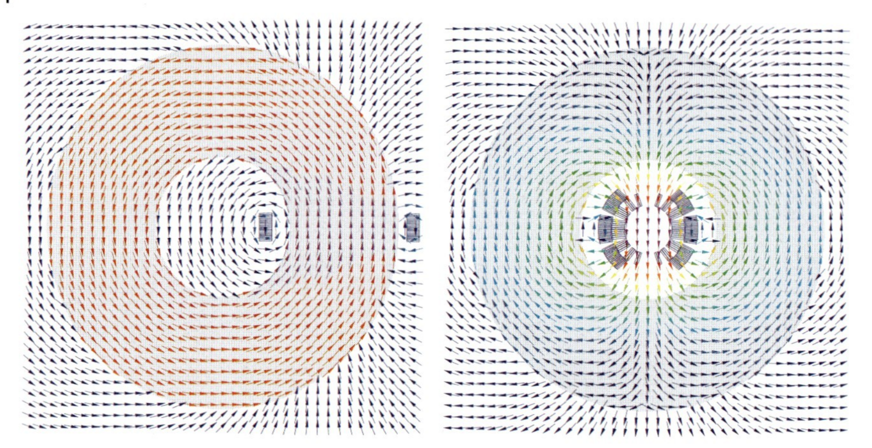

Figure 4.4 Either the magnetic flux density exits vertically from a highly permeable medium or the flux density is tangential to the interface inside the highly permeable medium. The field strength is represented by the color scheme of the icons. Dark blue icons indicate near zero field strengths.

Figure 4.4 (right). For $\alpha_2 \approx \pi/2$, the field inside the domain Ω_2 is tangential to the interface; see Figure 4.4 (left). We will come back to the first case when we discuss ideal pole shapes of normal-conducting magnets in Section 7.6.

We state without proof that the electric flux density and field obey

$$\mathbf{n} \cdot [\![\mathbf{D}]\!]_{12} = \sigma \, , \qquad \mathbf{n} \times [\![\mathbf{E}]\!]_{12} = \boldsymbol{\alpha} \, , \tag{4.54}$$

and therefore it yields at interfaces, free of electric charges and surface-currents,

$$\frac{\tan \alpha_1}{\tan \alpha_2} = \frac{\epsilon_1}{\epsilon_2} \, . \tag{4.55}$$

Remark: Both $\mathbf{B}$ and $\mathbf{H}$ are discontinuous at Γ_{12} and therefore $\operatorname{curl} \mathbf{H}$ and $\operatorname{div} \mathbf{B}$ cease to have any meaning there, although some authors [23, 30] propose preserving the concept using the definition of a surface divergence and a surface rotation,

$$\operatorname{Div} \mathbf{B} := \lim_{a \to 0} \frac{\int_{\partial \mathscr{V}} \mathbf{B} \cdot \mathrm{d}\mathbf{a}}{a} = \mathbf{n} \cdot [\![\mathbf{B}]\!]_{12} = \sigma_\mathrm{m} \, , \tag{4.56}$$

$$\operatorname{Curl} \mathbf{H} := \lim_{s \to 0} \frac{\int_{\partial \mathscr{A}} \mathbf{H} \cdot \mathrm{d}\mathbf{r}}{s} = \mathbf{n} \times [\![\mathbf{H}]\!]_{12} = \boldsymbol{\alpha} \, . \tag{4.57}$$

□

4.6 Magnetic Material

In *diamagnetic* substances, such as copper, zinc, silver, and gold, the orbit and spin magnetic momenta cancel in the absence of external magnetic fields. In the presence of an applied field the spin magnetic momenta slightly exceed the angular momenta, resulting in a small net magnetic moment which opposes the applied field. In this case, the permeability is less than μ_0. In the case of water the magnetic susceptibility χ_m is -8.8×10^{-6}. Superconductors in the Meißner[15] phase represent the limiting case of $\mu = 0$, the ideal diamagnet with a complete shielding of the external field. Diamagnetic samples brought to either pole of a magnet will be repelled.

The diamagnetic effect in materials is so low that it is easily overwhelmed in materials where the spin and angular momenta are unequal. In ideal *paramagnets* the individual magnetic momenta do not interact with each other and take random orientation in space due to thermal agitation. When an external field is applied, the magnetic momenta align themselves[16] in the field direction resulting in positive susceptibilities χ_m on the order of 10^{-5} to 10^{-3} nearly independent of field strength and without hysteresis. For small applied fields the magnetization of paramagnets follows the *Curie law*[17]

$$\mathbf{M} = \chi_m \mathbf{H} = \frac{C}{T}\mathbf{H}, \tag{4.58}$$

where T is the absolute temperature measured in kelvin and C is the material-dependent Curie constant. Paramagnetic substances include the *rare-earth* elements[18] platinum, sodium, and oxygen.

In *antiferromagnetic* materials, for instance chromium and iron manganese, the magnetic momenta align in regular patterns, but with neighboring momenta in opposite direction. Above the *Néel temperature*,[19] the materials usually turn paramagnetic. With no external field the net magnetization is zero. In an applied field, a ferrimagnetic interaction leads to spin magnetic momenta that are unequal in the two directions. The resulting net magnetization has a maximum at the Néel temperature T_N, which itself depends on the chemical

15 Fritz Walter Meißner (1882–1974).
16 The origin of this alignment can only be described by the quantum mechanical properties of spin and angular momentum.
17 Pierre Curie (1859–1906).
18 Rare-earth elements are scandium and yttrium together with the fifteen lanthanoids that include the better known neodymium, samarium, and terbium.
19 Louis Néel (1904–2000).

composition of the alloy. For iron manganese T_N can be estimated with the empirical formula [22] to an RMS deviation of 0.99 K:

$$\{T_N\}_K = 99.81 - 137\,\lambda_{Cr} + 314\,\lambda_{Ni} + 883\,\lambda_{Mn} - 1268\,\lambda_{Si} + 448\,\lambda_{Mo} - 3245\,\lambda_C - 3386\,\lambda_N\,, \quad (4.59)$$

where the λ are the mass fractions of the alloying elements. The Néel temperature, defined as the temperature at the peak of the $\chi_m(T)$ curve, should be high enough ($T_N > 100$ K) to yield low and stable magnetization levels near the operation temperature of the superconducting magnet.

During the annealing a carbon steel transforms from austenite to a mixture of ferrite and iron carbide. If the annealing is very rapid, or the material is cold worked, it may undergo martensitic transformation into a very hard form of crystalline structure that is ferromagnetic. A high chrome, nickel, and manganese content in austenitic steel suppresses this transformation. The collars of the LHC main dipoles are made of YUS 130S austenitic steel, produced by the Nippon Steel Corporation, with the chemical composition[20] 18Cr-7Ni-11Mn-0.3N-0.09C. Close monitoring of the production yielded an average μ_r of 1.0024 above 0.1 T at 4.2 K and an integrated thermal contraction of 0.0025 from room temperature to 15 K [4]. Another material with excellent properties in terms of permeability and thermal contraction is the KHMN30L grade (Kawasaki Steel Company) with the chemical composition 7Cr-1Ni-28Mn-0.1N-0.1C. The material has a permeability of $\mu_r < 1.002$ throughout the range between 1.9 K and room temperature, and an integrated thermal contraction as small as 0.0018 [29].

4.6.1 Ferromagnetism

Although all materials are either ferro-, dia-, or paramagnetic, it is customary to speak of magnetic material only in the case of ferromagnetic behavior with either a wide *hysteresis*[21] curve (hard ferromagnetic material and permanent magnets) or soft ferromagnetic material with narrow hysteresis as used for yoke laminations.

Ferromagnetic substances (which include iron, nickel, and cobalt as well as alloys of these elements) cannot be characterized by simple, single-valued constitutive laws because different $B(H)$ relations may exist, depending on the history of the excitation.

The properties of the magnetization curves $M(H)$ are governed by two mechanisms known as *exchange coupling* and *anisotropy*. Exchange coupling between electron orbitals in the crystal lattice favors long-range spin ordering

20 To be parsed as 18% (weight) chrome, 7% (weight) nickel, etc.
21 A precise definition will follow.

over macroscopic distances and is isotropic in space. At temperatures above the Curie temperature (for iron 770 °C), exchange coupling disappears. Magnetocrystalline anisotropy from spin/orbit interactions favors spin orientation along certain symmetry axes of the lattice and thus results in a preferential direction (the so-called *easy axis*) for its magnetic moment. The study of the quantum origin of these mechanisms is not needed in our phenomenological treatment of the material properties in field computation.

The principal characteristics of the magnetization curve, seen in the three sections shown in Figure 4.5, can be described by means of the domain theory by Weiß;[22] see Reference [11]. Within the domains of a size of about 10^{-5} to 10^{-3} m, the magnetic momenta are directed in parallel. The boundaries between the *Weiß domains* are known as *Bloch walls*.[23] In an unmagnetized (and unstrained) piece of iron the directions in which the domains are magnetized are either distributed at random (in parallel to one of the six crystal axes) or in such a way that the resultant magnetization of the specimen is zero. Application of a magnetic field changes only the direction of the magnetization in a given volume and not the magnitude. This change of magnitude is attained by a reversible, and later irreversible, boundary displacement of the domains. Saturation at high fields is attained by a reversible process of rotation (alignment with the excitation field) within the domains.

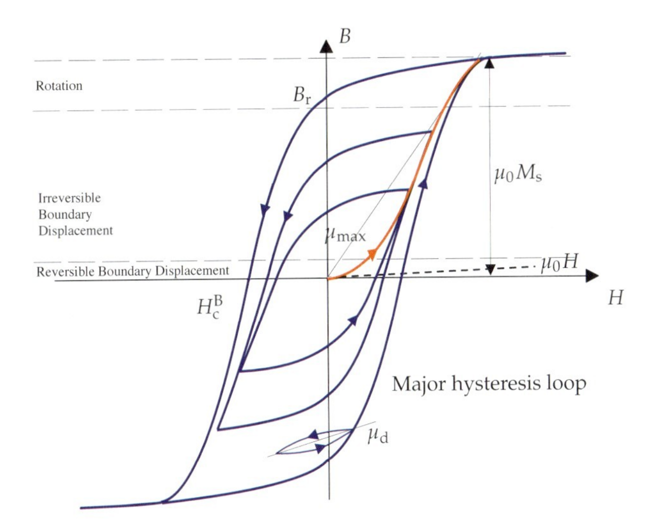

Figure 4.5 Hysteresis curve for a ferromagnetic material. Coercive field H_c^B, remanence B_r. Saturation magnetization M_s.

22 Pierre Ernest Weiß (1865–1940).
23 Felix Bloch (1905–1983).

If the field applied to a specimen is increased to saturation and then decreased, the decrease in flux density is not as rapid as the increase along the *normal* or *initial magnetization curve*. The initial magnetization curve is often called the *virgin curve*, although some authors reserve the term for the case where the initial state is obtained after thermal demagnetization. When H reaches zero there remains a residual flux density or remanence B_r. In order to reduce B to zero, a negative field $-H_c^B$, called the *coercive field*, must be applied.

Hysteresis is the phenomenon that causes B to lag behind H, so that the magnetization curves for increasing and decreasing fields are different. Hysteresis nonlinearities are defined as deterministic, rate-independent operators according to Ewing[24] [5]: "When there are two quantities M and N, such that cyclic variation in N causes cyclic variation in M, then if the changes of M lag behind those of N, we may say that there is hysteresis in the relation of M to N." Hysteresis behavior is not limited to ferromagnetism and can be found in mechanics (plastic hysteresis), phase transitions, hydrology (soil-moisture hysteresis), and control theory, among others.

A hysteresis loop can be represented in terms $B(H)$ or $\mu_0 M(H)$. In a soft ferromagnet, the fields involved in the hysteresis loop are much smaller than the corresponding magnetization, and plotting $B(H)$ instead of $M(H)$ makes only a small difference. However, in permanent magnet material H and M have the same order of magnitude, and the $B(H)$ loop differs considerably from the $M(H)$ curve, as illustrated in Figure 4.10 (left). We therefore need to distinguish between the two coercive fields H_c^B and H_c^M; in all cases $H_c^M > H_c^B$.

As there is no mention of time in the hysteresis curves, the relationship is rate-independent, that is, invariant with respect to time scaling. This implies that at any instant t, the magnetic flux density $B(t)$ depends only on $H([0,t])$ and on the history that the values attained before t. Because of its rate-independence, the output of a hysteresis operator is completely determined by the relative minima and maxima of the input variable in the time interval $[0,t]$ and thus by the sequence of $\{H(t_i)\}$ for $t_i < t$ at which H inverted its monotonicity.[25] This sequence $\{H(t_i)\}$ is named a *memory sequence* [5].

It must be emphasized that the description of the magnetic state of the specimen is based on its average magnetization over volumes larger than the sizes of the *Weiss domains* and therefore the material can be regarded as homogeneous. The so-called *saturation state* is obtained through the application of a field amplitude large enough to wipe out the domain structure that depends on the magnetization history prior to saturation. After saturation is reached,

24 Sir James Ewing (1855–1935).
25 $B(H)$ is monotonic if $\int_0^T [B(H_1) - B(H_2)](H_2 - H_1)\mathrm{d}t \geq 0$ for all H_1 and H_2 in the domain of B.

a further increase in H causes a linear $B(H)$-dependence with a differential permeability $\mu_\mathrm{d} := \mathrm{d}B/\mathrm{d}H$ that approaches μ_0:

$$\lim_{H\to\infty} \mu_\mathrm{d} = \lim_{H\to\infty} \frac{\mathrm{d}B}{\mathrm{d}H} = \mu_0 \,. \tag{4.60}$$

The coercive field H_c^B is on the order of 50–100 $\mathrm{A\,m^{-1}}$ in nonoriented Si–Fe alloys and low-carbon steels used in electrical motors. The low-carbon steel used for the LHC yoke laminations is specified to have a coercivity of less than 60 $\mathrm{A\,m^{-1}}$ and to have $B > 1.5$ T at $H = 1200\,\mathrm{A\,m^{-1}}$. The chemical composition of the material is 0.02Ni-0.02S-0.02Sn-0.01P. The thickness of the laminations is 5 mm with a tolerance of ± 0.2 mm. Low-carbon steels are a good choice for yoke laminations because they are easy to handle (draw, bend, and punch) and are fairly inexpensive. Their relative permeability is around 1000 at 1.6 T. The 3% Ni–Fe material used in transformer cores has a μ_r of 1000 at 1.4 T, but with the coercive field decreased to 20–60 $\mathrm{A\,m^{-1}}$.

Extremely soft magnetic materials with very high permeability below 0.5 T can be obtained from nickel alloys (with a maximum at around 80% Ni). The most common examples are mumetal and 78 permalloy (78% Ni). These substances are, however, very sensitive to heat treatment and the degree of cold working. A classification [7] of magnetic materials is given in Figure 4.6.

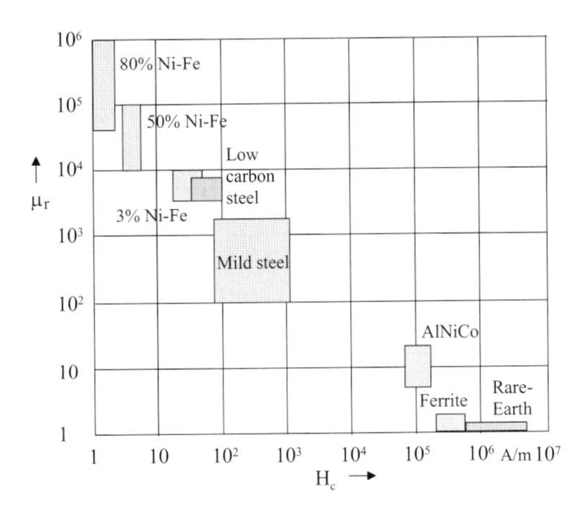

Figure 4.6 Classification of magnetic materials showing the correlation between the initial relative permeability and the coercive field [7].

4.6.2 Measurement of Hysteresis Curves

A specimen that is to be measured can either have a closed shape, for example, stacked rings, or be mounted in a flux return yoke of a highly permeable material. In both cases it is assumed that the $B(H)$ relationship is scalar, that is, $\mathbf{B}$ and $\mathbf{H}$ have the same direction. The material must be either isotropic or magnetized along a high-symmetry direction. For strongly oriented, polycrystalline materials, the tensor relationship between the field and flux density vectors may depend on the shape of the specimen. In this case it is customary to apply an *Epstein test frame* [19], wherein four identical strips of material are arranged in a frame-like assembly carrying the primary and secondary windings.

The magnetization curves of yoke steel can be measured by means of *permeameters* containing a homogeneous, weld-free annulus of steel with an inner radius r_1 and outer radius r_2. To maintain uniform magnetization within the specimen the ratio between the outer and the inner radii must stay below 1.1. A more practical approach is to calculate the mean value of the excitation field $H = NI/2\pi r$ within the specimen, which is

$$\overline{H} = \frac{NI}{2\pi(r_2 - r_1)} \ln\left(\frac{r_2}{r_1}\right) . \tag{4.61}$$

The induced voltage in the pickup coil (which is wound directly onto the specimen) is proportional to the rate of change of the flux

$$U = \frac{\mathrm{d}}{\mathrm{d}t}\Phi = \frac{\mathrm{d}}{\mathrm{d}t}\overline{B}a, \tag{4.62}$$

where a is the cross sectional area of the ring and $\overline{B}$ is the average magnetic flux density in the specimen. Time integration ($\int U \mathrm{d}t = \overline{B}\,a$) using a fluxmeter yields, after calibration [20], the hysteresis curve for $\overline{H}$ and $\overline{B}$. As permeability is strongly nonlinear, the flux density throughout the sample cannot easily be estimated, and therefore the normal permeability is calculated from $\mu = \overline{B}/\overline{H}$ where $\overline{B} = \overline{B}_{\mathrm{meas}} + B_{\mathrm{corr}}$. The correction term B_{corr} is iteratively determined from repeated measurements of the hysteresis loop for cycles between positive and negative excitation fields such that the modulus of the excitation currents at $\overline{B} = 0$ are identical for the up- and down-ramp branches. Each cycle is completed in a time between 30 and 60 s to avoid eddy currents and a drift in the measuring system.[26] The normal magnetization curve, shown in Figure 4.5, is measured by demagnetizing the material with cycles of decreasing amplitudes [7]. Under these conditions the material responds equally when the field is applied in either of the two opposite directions.

For easy exchange of specimens the permeameters are made of split coils for $B(H)$ measurements; the device is shown in Figure 4.7 (left). However, the total resistance of the excitation coil is inherently high due to the large number of contacts (two per turn), resulting in high power dissipations at large currents. In addition, the field strength generated by the excitation coil is limited due to the fixed number of turns, 90 in the case of the permeameter shown in Figure 4.7 (left). Thus for high-field measurements at cryogenic temperatures,

26 The flux penetration depth is $\delta = (\pi \varkappa \mu_r \mu_0 f)^{-0.5}$ with $f = 1/T$.

Figure 4.7 Left: Split coil permeameter for the warm measurements of yoke iron samples (180 turns for the excitation coil, 90 turns for the pickup coil). Right: Superconducting excitation coil (3000 turns) and pickup coil wound directly onto a glass epoxy box containing the specimen for $B(H)$ measurements at cryogenic measurements.

3000 turns of superconducting wire are wound directly onto the glass epoxy casing of a toroidal specimen [24]. For the machining of the specimen, a technique that reduces cold working of the material is preferred, for example, wire electrical discharge machining (EDM) or laser cutting. A sample prepared for measurements is shown in Figure 4.7 (right).

Figure 4.8 (right) shows the normal magnetization curve of the yoke laminations used for the normal-conducting, combined-function magnets of the CERN PS accelerator. The figure also shows the magnetization curve of the yoke laminations of the LHC main dipole and quadrupole magnets. The measurements were performed at 4.2 K for a ring specimen, a toroidal superconducting excitation coil, and a copper search coil in magnetic flux densities of up to 7.4 T. The third curve is the $B(H)$ curve of the steel used for an LHC model dipole built in Japan [24]. The steel was cooled to 4.5 K and prestressed with an aluminum ring to 20 MPa. The corresponding $\mu_r(H)$ curves are given in Figure 4.8 (left). The figure also shows the effect of the stacking factor[27] on the average permeability of the PS magnet laminations as well as the comparison of the permeability of the LHC yoke steel at room temperature and at cryogenic temperatures [2].

Table 4.3 shows the measured temperature and stress-dependence on the coercive field, remanence, and maximum permeability of yoke laminations used in an LHC model magnet [24]. The maximum permeability drops by 10% at 4.2 K but increases slightly at saturation field levels.

27 For a definition of the stacking factor see Section 4.6.3.

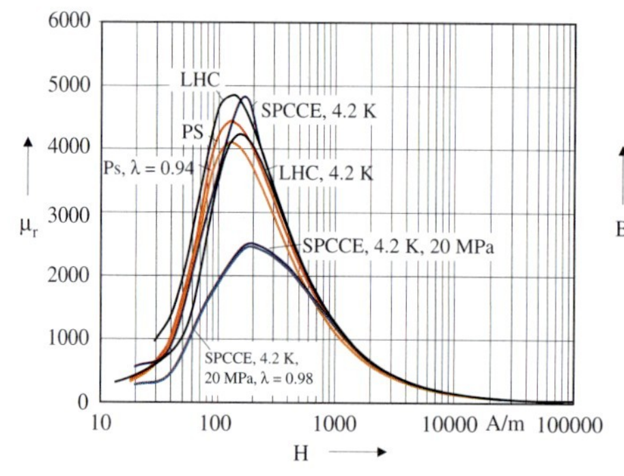

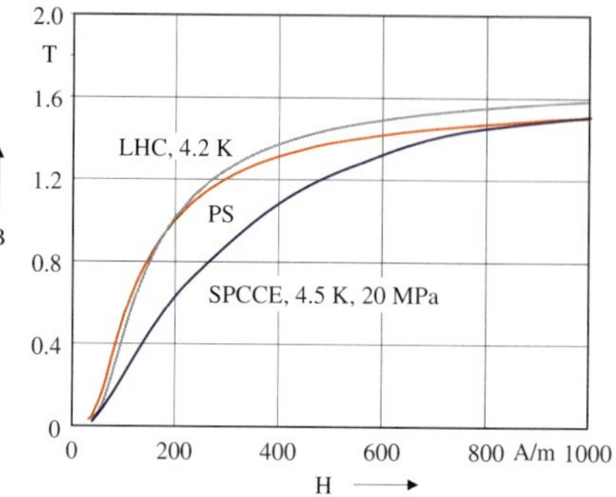

Figure 4.8 Left: $\mu_r(H)$ measured for the steel lamination of the normal-conducting PS magnets (with and without taking the yoke stacking factor of 0.94 into account), for the steel for the series production of the LHC main dipole and quadrupole yokes (both at room temperature and at 4.2 K) and of the Kawasaki steel (SPCCE) used for a model dipole magnet and measured with and without prestress at cryogenic temperature. Right: $B(H)$ curves for the PS magnet steel, the LHC magnet steel at 4.2 K and the Kawasaki steel under 20 MPa of prestress.

Table 4.3 Measured temperature and stress-dependence on the coercive field H_c^B, remanence B_r, and maximum relative permeability μ_r of the LHC yoke laminations [24].

Temperature T	Stress	Coercive field H_c^B	Remanence B_r	max μ_r
K	MPa	$\mathrm{A\,m^{-1}}$	T	
300	0	68.4	1.07	5900
77	0	79.6	1.12	5600
4.2	0	85.1	1.06	4800
4.2	20	110	0.67	2460

Following Wlodarski [35], an *anhysteretic* magnetization curve of iron can be approximated by the empirical relation

$$M(H) = M_\mathrm{a} L\left(\frac{H}{a}\right) + M_\mathrm{b} \tanh\left(\frac{|H|}{b}\right) L\left(\frac{H}{b}\right), \tag{4.63}$$

where

$$L\left(\frac{H}{a}\right) := \coth\left(\frac{H}{a}\right) - \left(\frac{a}{H}\right). \tag{4.64}$$

The fit is consistent with the phenomenological description of magnetization, which distinguishes between the reversible and the irreversible boundary displacements of the Weiß domains. The parameters M_a and M_b denote the reversible and irreversible components of the saturation magnetization, and a, b determine the rate of their approach to saturation. The sum of M_a and M_b equals the saturation magnetization M_s, and therefore the fit has only three free parameters. Figure 4.9 (left) shows the best fit for the magnetization curve of the low-carbon steel [3] used for the LHC magnet yokes, measured at 4 K without prestress. The parameters of the fit are $\mu_0 M_a = 0.46628\,\text{T}$, $a = 9337.7\,\text{A}\,\text{m}^{-1}$, $\mu_0 M_b = 1.7218\,\text{T}$, $b = 89.434\,\text{A}\,\text{m}^{-1}$, and $\mu_0 M_s = \mu_0(M_a + M_b) = 2.188\,\text{T}$. Note that for very small values of H the numerical implementation of Eq. (4.58) becomes unstable.

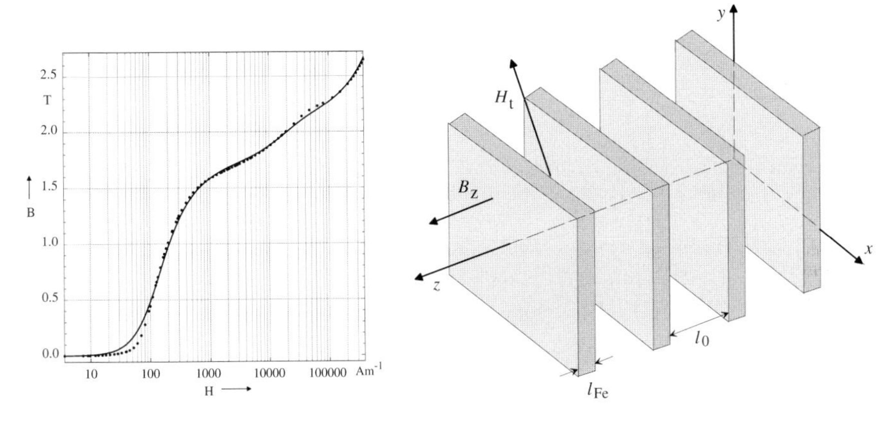

Figure 4.9 Left: Measurement of the normal magnetization curve (dots) for the yoke laminations of the LHC main magnets and its approximation by Eq. (4.63). Right: Conventions for the calculation of the μ tensor for laminated materials. The transversal dimensions are large with respect to l_0 and l_{Fe}.

4.6.3 Magnetic Anisotropy in Laminated Iron Yokes

In many materials, such as rolled metal sheets, the fabrication process produces some regularity in the crystal structure and consequently a dependence of the magnetic properties on the direction of rolling. The best-known (and strongest) *anisotropy* in magnetic materials can be achieved by laminating the iron yokes. Between each of the ferromagnetic laminations of thickness l_{Fe} (magnetically isotropic) there is a nonmagnetic ($\mu = \mu_0$) layer of thickness l_0, as shown schematically in Figure 4.9 (right). The permeability has the form of a diagonal rank-2 tensor, so that $\mathbf{B} = [\mu]\,\mathbf{H}$, where

$$[\mu] = \begin{pmatrix} \mu_x & 0 & 0 \\ 0 & \mu_y & 0 \\ 0 & 0 & \mu_z \end{pmatrix} . \qquad (4.65)$$

Consider a lamination in z-direction and the field components H_t in the xy-plane. Because of the continuity condition $H_\mathrm{t}^0 = H_\mathrm{t}^\mathrm{Fe} = \overline{H}_\mathrm{t}$ for the effective macroscopic tangential flux density we obtain

$$\overline{B}_\mathrm{t} = \frac{1}{l_\mathrm{Fe} + l_0} \left(l_\mathrm{Fe} \mu \overline{H}_\mathrm{t} + l_0 \mu_0 \overline{H}_\mathrm{t} \right) . \tag{4.66}$$

As the normal component of the magnetic flux density is continuous, $B_z^0 = B_z^\mathrm{Fe} = \overline{B}_z$, the average magnetic field intensity can be calculated from

$$\overline{H}_z = \frac{1}{l_\mathrm{Fe} + l_0} \left(l_\mathrm{Fe} \frac{\overline{B}_z}{\mu} + l_0 \frac{\overline{B}_z}{\mu_0} \right) . \tag{4.67}$$

We now define the *stacking factor* by

$$\lambda := \frac{l_\mathrm{Fe}}{l_\mathrm{Fe} + l_0} , \tag{4.68}$$

which is 0.985 for the LHC dipole yokes. The result for the average permeability in the plane of the laminations is then

$$\overline{\mu}_\mathrm{t} = \lambda \mu + (1 - \lambda) \mu_0 . \tag{4.69}$$

Normal to the plane of the laminations it yields

$$\overline{\mu}_z = \left(\frac{\lambda}{\mu} + \frac{1 - \lambda}{\mu_0} \right)^{-1} . \tag{4.70}$$

We have obtained a simple equation for the stacking-factor scaling of the material characteristic. For laminations in the x- and y-directions, with the plane of the laminations normal to the 2D cross section, the laminations have a strong directional effect and the stacking-factor scaling is no longer useful. A macroscopic model for this case and a modified iteration scheme for its numerical solution is developed in [8].

4.6.4 Magnetostriction

The phenomenon in which a ferromagnetic specimen changes minutely its dimensions when it is magnetized is called *magnetostriction* (positive for expansion and negative for contraction). The effect is due to magnetocrystalline anisotropy, which gives rise to energy variations when the relative positions of magnetic ions in the lattice are modified. The saturation magnetostriction is defined by

$$\lambda_s := \frac{\Delta l}{l_0} , \tag{4.71}$$

where the original length of the specimen is denoted l_0. λ_s is in the range of 1–4 $\times 10^{-5}$ for alloys of iron and nickel. With some rare-earth materials, for example, $TbFe_2$ an alloy from terbium and iron known as terfenol, values of up to 2×10^{-3} can be obtained. Terfenol is therefore used to construct micro actuators.

Magnetostriction has a negligible effect on the field quality in the LHC main dipoles, where the field is dominated by the coil layout and where the yoke is mostly saturated at nominal field level.

A small stacking factor of 0.27 was chosen to reduce hysteresis effects in the LEP dipoles, which operated at a low nominal[28] field of 0.135 T. The small stacking factor was realized by regularly spacing magnetic steel laminations and filling the spaces with cement mortar. This solution provided for mechanical rigidity at low cost. Mortar shrinkage accompanying *hydration*[29] had an effect on the longitudinal magnet geometry, which was well controlled by means of four tie rods. In the transverse plane, however, the steel laminations opposed the shrinkage of the mortar layers, resulting in a buildup of tensile stresses in the mortar (about 10 MPa, near the upper limit of the mortar yield strength) and compressive stresses in the iron laminations (at 30 MPa due to a different elastic modulus and thickness of the layers). This in turn resulted in unacceptable fluctuations in the bending field at low excitation, caused by the reduction of the maximum permeability due to magnetostriction. The chosen solution was a hydraulic system, by which transverse forces were exerted along the length of the poles to relieve the mortar-induced compressive stresses in the yoke laminations [6].

4.6.5 Permanent Magnets

Designing accelerator magnets that are excited by permanent magnets, we must focus on the section of the hysteresis loop in the second quadrant of the $B(H)$ and $M(H)$ diagrams. If the loop is the major hysteresis loop, it is called the *demagnetization curve*; see Figure 4.10 (right).

It is desirable that the material have a high remanence for the maximum possible flux density in a circuit. Another goal is a high coercive field H_c^M, so that the permanent magnet will not be easily demagnetized. The maximum product $(BH)_{max}$ is therefore a good figure of merit. The shape of the demagnetization curve can be characterized by the *fullness factor* [11]

$$\gamma := \frac{(BH)_{max}}{B_r H_c^B}, \tag{4.72}$$

28 For an operation energy of 60 GeV.

29 Hydration is the technical term for the chemical process of solidification of cement.

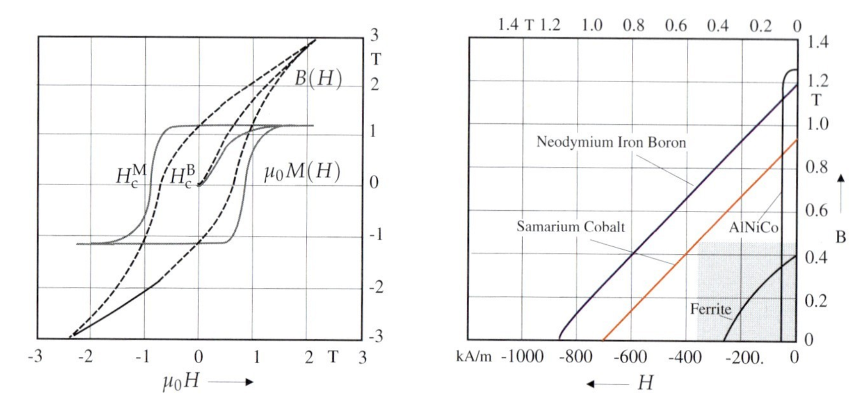

Figure 4.10 Left: Typical $B(H)$ and $M(H)$ curves for hard ferromagnets. Right: Demagnetization curves for different permanent magnet materials at room temperature. The gray shaded area represents the ideal $(BH)_{\text{max}}$.

which varies between 0.25 for the straight line and 1 for the rectangular loop. The calculation of the operation point of permanent magnets in magnetic iron circuits is presented in Section 7.9.

Rare-earth materials such as samarium cobalt ($SmCo_5$) and neodymium iron boron (NdFeB) are sintered from a powder with grain sizes in the range of 5 μm. These grains are magnetically highly anisotropic along one crystalline direction. The powder is pressed and exposed to a strong magnetic field so that the grains rotate until their magnetically preferred axis is aligned to the magnetic field. The material is then subjected to a high pressure and sintered. Finally, the sintered and machined material is exposed to a very strong magnetization field in parallel to the previously established direction. A typical $B(H)$ relationship for B and H parallel to the magnetically preferred axis of the grains is shown in Figure 4.10 (left). The demagnetization curve for rare-earth materials is basically a straight line with a differential permeability of $\mathrm{d}B/\mathrm{d}H$ between 1.04 μ_0 and 1.08 μ_0, so that the coercive field $\mu_0 H_{\mathrm{c}}^{\mathrm{B}}$ is 4–8% less than B_{r}. In the perpendicular direction, typical values for the relative differential permeability are in the range of 1.02 to 1.08. Most of the grain-oriented materials have hysteresis loops that are close to the ideal, with a differential permeability close to μ_0. These materials may thus be treated as vacuum with an impressed (field-dependent) magnetization. This gives rise to the definition of the ideal $(BH)_{\text{max}}^{\text{id}}$,

$$(BH)_{\text{max}}^{\text{id}} := \frac{B_{\mathrm{r}}^2}{4\mu_0}, \tag{4.73}$$

as the demagnetization curve is assumed to be a straight line defined by $B = B_{\mathrm{r}} + \mu_0 H$. For samarium cobalt magnets, the $(BH)_{\text{max}}^{\text{id}}$ is represented by the gray shaded area in Figure 4.10 (right). Typical parameters for permanent-magnet material are given in Table 4.4.

Table 4.4 Typical magnetic parameters for permanent-magnet material; source [34].

	B_r	$\mu_0 H_c^M$	H_c^B	$(BH)_{max}$	$(BH)_{max}^{id}$	T_c
	T	T	T	$kJ\,m^{-3}$	$kJ\,m^{-3}$	°C
AlNiCo	1.3	0.06	0.06	50	336	857
Ferrite	0.4	0.4	0.37	30	32	447
$SmCo_5$	0.9	2.5	0.87	160	161	727
Sm_2Co_{17}	1.1	1.3	0.97	220	241	827
NdFeB	1.3	1.5	1.25	320	336	313

Aluminum nickel cobalt (AlNiCo) magnets feature low coercive fields and are easily demagnetized. However, due to their thermal stability they are often used in measurement devices. Ferrites show a good long-term stability and corrosion resistance and are fairly inexpensive. Samarium cobalt magnets are the most expensive and have a high remanence and nearly ideal demagnetization curve. This is also true for the less expensive NdFeB magnets, which feature a low Curie temperature T_c, however.

4.6.6 Magnetization Currents and Fictitious Magnetic Charges

Homogeneous magnetization in ferromagnets can be represented by a magnetic surface-charge or a surface-current. This is a simplified model because the magnetization depends on the demagnetization field. The demagnetization field can be calculated analytically only for ellipsoids in 3D, and for cylinders in 2D as will be shown in Section 7.9. Defining $\mathbf{J}_m := \operatorname{curl}\mathbf{M}$ we obtain

$$\frac{1}{\mu_0}\operatorname{curl}\mathbf{B} = \mathbf{J}_s + \mathbf{J}_m = \mathbf{J}_s + \operatorname{curl}\mathbf{M} \tag{4.74}$$

and therefore

$$\operatorname{curl}\mathbf{H} = \operatorname{curl}\left(\frac{\mathbf{B}}{\mu_0} - \mathbf{M}\right) = \mathbf{J}_s. \tag{4.75}$$

Hence $\mathbf{H}$ is curl-free if there are no impressed (source) currents. However, $\mathbf{B}$ is not curl-free. The magnetic flux density $\mathbf{B}$ is always source-free, but this does not necessarily hold for the magnetic field $\mathbf{H}$ as

$$\operatorname{div}\mu_0\mathbf{H} = \operatorname{div}(\mathbf{B} - \mu_0\mathbf{M}) = -\operatorname{div}\mu_0\mathbf{M}, \tag{4.76}$$

which gives rise to the definition of a fictitious magnetic charge density

$$\rho_m = -\operatorname{div}\mu_0\mathbf{M}, \tag{4.77}$$

where $[\rho_\mathrm{m}] = 1\,\mathrm{V\,s\,m^{-3}}$. The magnetic charges allow the application of techniques known from electrostatics, by using the magnetic scalar potential with the magnetic charges as its sources. With $\mathbf{H} = -\operatorname{grad}\phi_\mathrm{m}$ we can derive the magnetic Poisson equation[30]

$$\operatorname{div}(-\operatorname{grad}\phi_\mathrm{m}) = -\nabla^2\phi_\mathrm{m} = \frac{\rho_\mathrm{m}}{\mu_0}\,. \tag{4.78}$$

For a homogeneously magnetized specimen Ω_2, the magnetic field in free space can be calculated by employing a magnetic surface-charge (Figure 4.11), which is

$$\sigma_\mathrm{m} = -\mu_0\mathbf{n}\cdot[\![\mathbf{M}]\!]_{12} = -\operatorname{Div}\mu_0\mathbf{M}\,, \tag{4.79}$$

where $[\sigma_\mathrm{m}] = 1\,\mathrm{V\,s\,m^{-2}}$. The surface normal vector $\mathbf{n}$ points from Ω_2 to Ω_1. We can employ also the electric surface-current shown in Figure 4.12. This yields

$$\alpha = -\mathbf{n}\times[\![\mathbf{M}]\!]_{12} = -\operatorname{Curl}\mathbf{M}\,. \tag{4.80}$$

30 Simeon Poisson (1781–1840).

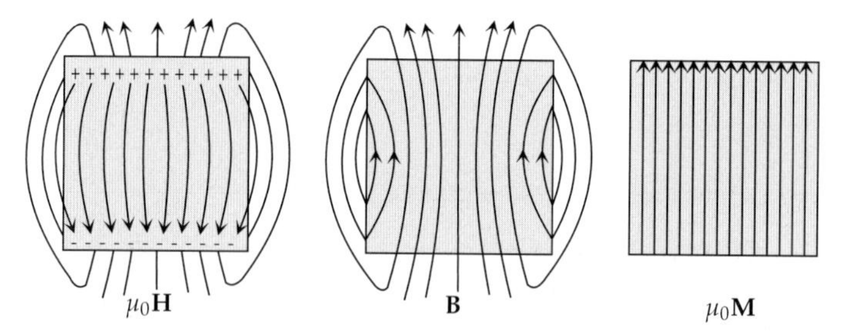

Figure 4.11 Field distribution, magnetic flux density, and magnetization in a permanent magnet (schematic). Inside the permanent magnet $\mathbf{B}$ and $\mathbf{H}$ have different directions. Outside the magnet, the simple expression $\mathbf{B} = \mu_0\mathbf{H}$ applies.

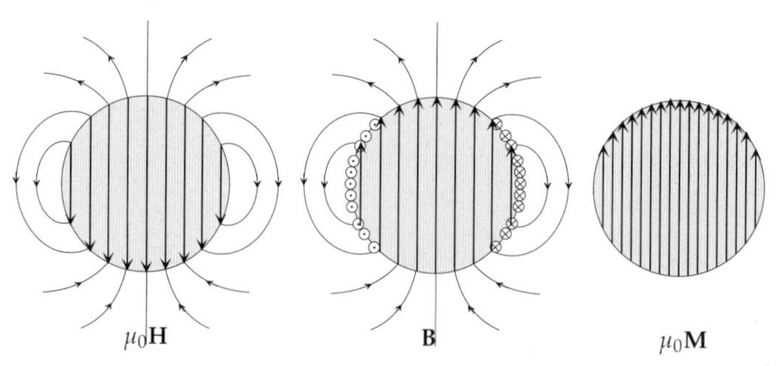

Figure 4.12 Field distribution, magnetic flux density, and magnetization in an infinitely long permanent magnet of cylindrical shape (schematic). Inside the permanent magnet $\mathbf{B}$ and $\mathbf{H}$ have exactly opposite directions.

4.7 Classification Diagrams for Electromagnetism

Employing the second lemma of Poincaré in contractible domains Ω, we can express the magnetic flux density by means of a *magnetic vector potential* $\mathbf{A} \in \mathcal{V}(\Omega)$ with

$$\mathbf{B} = \operatorname{curl} \mathbf{A}, \tag{4.81}$$

as the magnetic flux density is source-free. The physical unit of the magnetic vector potential is $[\mathbf{A}] = 1\,\mathrm{V\,s\,m^{-1}}$. If $\mathbf{B}$ in the Faraday law (4.14) is expressed by the magnetic vector potential, we obtain

$$\operatorname{curl}\left(\mathbf{E} + \frac{\partial}{\partial t}\mathbf{A}\right) = \mathbf{0}, \tag{4.82}$$

and therefore the electric field is

$$\mathbf{E} = -\operatorname{grad}\phi - \frac{\partial}{\partial t}\mathbf{A}. \tag{4.83}$$

In the electrostatic case, $\partial/\partial t = 0$, the electric field is curl-free (irrotational), $\operatorname{curl}\mathbf{E} = \mathbf{0}$, and therefore

$$\mathbf{E} = -\operatorname{grad}\phi\,. \tag{4.84}$$

Since the curl of the magnetic field is in general nonzero, it cannot always be written as the gradient of a scalar potential. If, however, a vector field $\mathbf{T}$ is found such that $\operatorname{curl}\mathbf{T} = \mathbf{J}$, then the vector field $\mathbf{H} - \mathbf{T}$ is curl free, $\operatorname{curl}(\mathbf{H} - \mathbf{T}) = \mathbf{0}$, and the magnetic field can be expressed as

$$\mathbf{H} = -\operatorname{grad}\phi_{\mathrm{m}}^{\mathrm{red}} + \mathbf{T}\,. \tag{4.85}$$

$\mathbf{T}$ is known as the *electric vector potential* in [12] in the context of the so-called $\mathbf{T}-\Omega$ method for steady-state field problems, and $\phi_{\mathrm{m}}^{\mathrm{red}}$ is the *reduced magnetic scalar potential*. Several options for the choice of $\mathbf{T}$ exist [9]. The straightforward one is to use the Biot–Savart[31] field $\mathbf{H}_s$ computed from the impressed current distribution.

The structure of the Maxwell equations and the electromagnetic potentials is revealed in a classification diagram (Figure 4.13), which has been called *Maxwell's house* in [10]. The relations between $\mathbf{B}$ and $\mathbf{E}$ on the left wall of the house are known as the *Faraday complex*. The relations between $\mathbf{D}$, $\mathbf{J}$, and $\mathbf{H}$ on the right wall constitute the *Ampère–Maxwell complex*. The facade contains electric fields and charges, the backside magnetic flux density, field and currents. The vertical links do not depend on the nature of the medium, do not contain physical constants, and are therefore called topological laws. The horizontal links are the constitutive (or material) equations.

31 Jean-Baptiste Biot (1774–1862), Felix Savart (1791–1841).

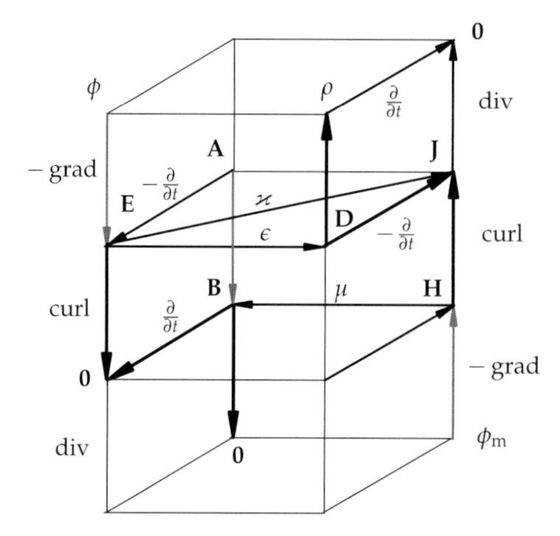

Figure 4.13 Classification diagram, which has been called Maxwell's house in [10], showing the relations between the fields and potentials, as well as the material relations in classical vector analytical form.

An alternative view of this classification diagram [32] is shown in Figure 4.14.

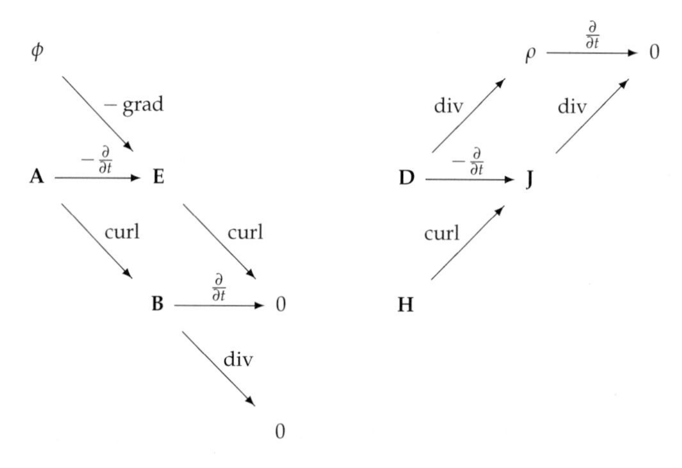

Figure 4.14 Classification diagram for the field variables of electromagnetism [32].

4.8 Field Lines

A graphical representation of magnetic fields starts with the observation that $\mathbf{dr}$ is the tangent vector field to a space curve $\mathscr{S} : I \to E_3 : t \mapsto \mathbf{r}(t)$ and thus

$$\mathbf{dr} \times \mathbf{B}(\mathbf{r}) = 0 \tag{4.86}$$

yields a differential equation for the field lines parallel to the magnetic flux density. Written in coordinates, Eq. (4.86) results in

$$\frac{\mathrm{d}x}{B_x} = \frac{\mathrm{d}y}{B_y} = \frac{\mathrm{d}z}{B_z} \,. \tag{4.87}$$

In particular for plane fields, independent of z,

$$\frac{\mathrm{d}y}{\mathrm{d}x} = \frac{B_y}{B_x} \,, \tag{4.88}$$

the solution of which can be written as $f(x,y) = p$. This yields the one parameter family of field lines, which are equipotentials of the vector potential A_z [28]:

$$\begin{aligned} \mathbf{dr} \times \mathbf{B} &= \mathbf{dr} \times \operatorname{curl}(A_z \mathbf{e}_z) = -\mathbf{dr} \times (\mathbf{e}_z \times \operatorname{grad} A_z) \\ &= -\mathbf{e}_z(\mathbf{dr} \cdot \operatorname{grad} A_z) + (\mathbf{e}_z \cdot \mathbf{dr}) \operatorname{grad} A_z = -\mathbf{e}_z \mathrm{d}A_z = 0 \,. \end{aligned} \tag{4.89}$$

Figure 4.15 shows the z-component of the magnetic vector potential for the 2D field problem by means of a color representation (bottom, left) as well as the classical field line representation (bottom, right). The top figures represent the vector field by icons with size scaling (right) and color scaling (left). To avoid the impression of asymmetries, the icons are centered at the field point.

4.8.1 Classification of Electromagnetic Field Problems

Electrostatics For electrostatic problems ($\partial/\partial t = 0$), the electric and magnetic phenomena are decoupled, and only the facade of Maxwell's house remains to be studied. Maxwell's equations reduce to

$$V_\mathrm{m}(\partial\mathscr{A}) = 0 \,, \qquad \Psi(\partial\mathscr{V}) = Q(\mathscr{V}) \tag{4.90}$$

in the global form and

$$\operatorname{curl} \mathbf{E} = \mathbf{0} \,, \qquad \operatorname{div} \mathbf{D} = \rho \,, \tag{4.91}$$

in the local form. The remaining constitutive equation is

$$\mathbf{D} = \varepsilon \mathbf{E} \,. \tag{4.92}$$

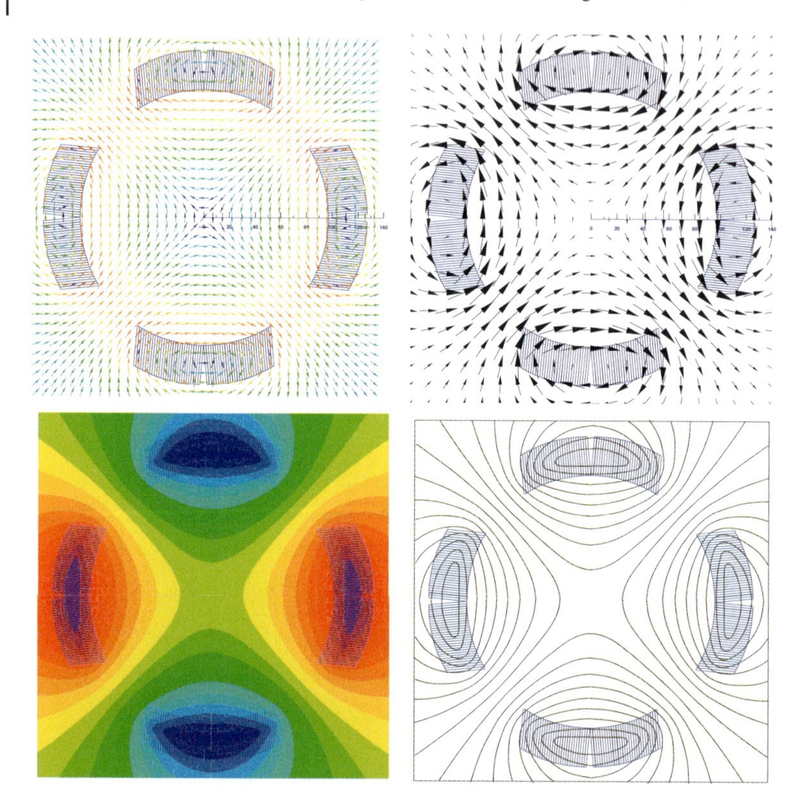

Figure 4.15 Top: Representation of vector fields (magnetic flux density in a quadrupole magnet) by means of icons. Top left: Representation of the field strength by means of a color scheme. Top right: Representation of the field strength by a size scaling of the icons. Bottom: Representation of scalar fields (here the z-component of the magnetic vector potential for the 2D field problem) by means of a color representation (left) and the classical field lines (right).

Magnetostatics For magnetostatic problems with vanishing time derivative ($\partial/\partial t = 0$), only the backside of Maxwell's house remains to be studied. Maxwell's equations reduce to

$$V_{\mathrm{m}}(\partial\mathscr{A}) = I(\mathscr{A}), \qquad \Phi(\partial\mathscr{V}) = 0 \tag{4.93}$$

in the global form and

$$\operatorname{curl}\mathbf{H} = \mathbf{J}, \qquad \operatorname{div}\mathbf{B} = 0, \tag{4.94}$$

in the local form. The remaining constitutive equation reads

$$\mathbf{B} = \mu\mathbf{H} \quad \text{or} \quad \mathbf{B} = \mu_0(\mathbf{H} + \mathbf{M}). \tag{4.95}$$

Current flow If the current density is not prescribed but is coupled to the electric field through the constitutive equation $\mathbf{J} = \varkappa\mathbf{E}$ in electrically conducting media, electric and magnetic phenomena are coupled through the set of equations in global form

$$U(\partial\mathscr{A}) = 0\,, \qquad I(\partial\mathscr{V}) = 0\,, \tag{4.96}$$

and in local form

$$\operatorname{curl}\mathbf{E} = \mathbf{0}\,, \qquad \operatorname{div}\mathbf{J} = 0\,. \tag{4.97}$$

The constitutive equation reads

$$\mathbf{J} = \varkappa\mathbf{E}\,. \tag{4.98}$$

The current flow problem is both divergence and curl-free.

Magnetic diffusion We recall from Eq. (4.2) the Faraday law of electromagnetic induction, which links the induced voltage on the rim of a surface to the time derivative of the magnetic flux through that surface:

$$U(\partial\mathscr{A}) = -\frac{\mathrm{d}}{\mathrm{d}t}\Phi(\mathscr{A})\,. \tag{4.99}$$

The minus sign stems from the *Lenz law*,[32] which states that the induced current and its accompanying magnetic flux oppose the change of flux through the surface. Let us consider a thin conductor along the closed rim of that surface and refer to it as a circuit. The time-transient magnetic flux may be due to the varying flux density in the stationary circuit and/or some translation of the circuit within a constant field. Employing the *convective derivative*, which is defined by

$$\frac{\mathrm{d}}{\mathrm{d}t}(\phi(\mathbf{r},t)) = \frac{\partial}{\partial t}\phi + \partial_{\mathbf{v}}\phi \tag{4.100}$$

for scalar fields and by

$$\frac{\mathrm{d}}{\mathrm{d}t}(\mathbf{x}(\mathbf{r},t)) = \frac{\partial}{\partial t}\mathbf{x} + (\mathbf{v}\cdot\operatorname{grad})\mathbf{x} \tag{4.101}$$

for vector fields, the total time derivative of flux through the moving circuit is given by

$$\frac{\mathrm{d}}{\mathrm{d}t}\Phi(\mathscr{A}(t)) = \int_{\mathscr{A}(t)}\frac{\partial}{\partial t}\mathbf{B}\cdot\mathrm{d}\mathbf{a} + \int_{\mathscr{A}(t)}(\mathbf{v}\cdot\operatorname{grad})\mathbf{B}\cdot\mathrm{d}\mathbf{a}\,. \tag{4.102}$$

32 Heinrich Lenz (1804–1865).

Using the relation

$$(\mathbf{v}\cdot\ \mathrm{grad}\,)\mathbf{B} = \mathbf{v}\,\mathrm{div}\,\mathbf{B} - \ \mathrm{curl}\,(\mathbf{v}\times\mathbf{B})\,, \tag{4.103}$$

which is derived from Eq. (3.71) with **v** treated as a fixed vector in the differentiation, and considering that div $\mathbf{B} = 0$, the induced voltage can be calculated:

$$\begin{aligned} U &= -\int_{\mathscr{A}(t)} \frac{\partial}{\partial t}\mathbf{B}\cdot \mathrm{d}\mathbf{a} + \int_{\mathscr{A}(t)} \mathrm{curl}\,(\mathbf{v}\times\mathbf{B})\cdot \mathrm{d}\mathbf{a} \\ &= \int_{\partial\mathscr{A}(t)} \mathbf{E}\cdot\mathrm{d}\mathbf{r} + \int_{\partial\mathscr{A}(t)} (\mathbf{v}\times\mathbf{B})\cdot\mathrm{d}\mathbf{r}\,. \end{aligned} \tag{4.104}$$

The induced voltage can be expressed also as the path integral of an electric field $\mathbf{E}'$ in the moving frame (in which $\mathrm{d}\mathbf{r}$ is at rest) by

$$U = \int_{\partial\mathscr{A}(t)} \mathbf{E}'\cdot\mathrm{d}\mathbf{r}\,. \tag{4.105}$$

The assumption of *Galilean invariance*[33] implies that the voltages in the above equations are equal, and thus that the electric field in the moving frame is

$$\mathbf{E}' = \mathbf{E} + \mathbf{v}\times\mathbf{B}\,. \tag{4.106}$$

From Stoke's theorem we obtain

$$\begin{aligned} U = \int_{\partial\mathscr{A}(t)} \mathbf{E}'\cdot\mathrm{d}\mathbf{r} &= \int_{\mathscr{A}(t)} \mathrm{curl}\,\mathbf{E}'\cdot\mathrm{d}\mathbf{a} \\ &= \int_{\mathscr{A}(t)} \mathrm{curl}\,\mathbf{E}\cdot\mathrm{d}\mathbf{a} + \int_{\mathscr{A}(t)} \mathrm{curl}\,(\mathbf{v}\times\mathbf{B})\cdot\mathrm{d}\mathbf{a}\,. \end{aligned} \tag{4.107}$$

The terms in $\mathbf{v}\times\mathbf{B}$ cancel out in Eqs. (4.104) and (4.107), which are equal for any surface $\mathscr{A}$ if

$$\mathrm{curl}\,\mathbf{E} = -\frac{\partial}{\partial t}\mathbf{B}\,. \tag{4.108}$$

The local form of Maxwell's equations for magnetic diffusion problems reduces to

$$\mathrm{curl}\,\mathbf{H} = \mathbf{J}\,, \tag{4.109}$$

$$\mathrm{curl}\,\mathbf{E} = -\frac{\partial}{\partial t}\mathbf{B}\,, \tag{4.110}$$

$$\mathrm{div}\ \mathbf{B} = 0\,, \tag{4.111}$$

$$\mathrm{div}\ \mathbf{D} = \rho\,. \tag{4.112}$$

33 Galileo Galilei (1564–1642).

The constitutive equations take the form:

$$\mathbf{B} = \mu_0(\mathbf{H} + \mathbf{M}), \tag{4.113}$$

and $\mathbf{J} = \varkappa \mathbf{E}'$, that is,

$$\mathbf{J} = \varkappa(\mathbf{E} + \mathbf{v} \times \mathbf{B}), \tag{4.114}$$

which is known as Ohm's law for moving conductors. Hence, the $\mathbf{v} \times \mathbf{B}$ term becomes a part of the constitutive equations and not of the Maxwell equations.

4.9 Boundary Value Problems 1: Magnetostatic

Consider again the elementary model problem shown in Figure 4.1, consisting of the iron domain Ω_i of permeability $\mu(H)$ and the air region Ω_a of permeability μ_0. Let us further define Ω as $\Omega_a \cup \Omega_i$. The regions are connected at the interface Γ_{ai}. Furthermore, each domain is bounded by a surface $\Gamma = \partial\Omega$, itself consisting of two different parts Γ_H and Γ_B. In the following sections, we exclude magnetic surface-charges and surface-currents.

4.9.1 Scalar-Potential Formulations

The reduced scalar potential For magnetostatic problems the vector field $\mathbf{H} \in \mathcal{V}(\Omega)$ can be split into the field $\mathbf{H}_s \in \mathcal{V}(\Omega)$ generated by the source currents ($\operatorname{curl} \mathbf{H}_s = \mathbf{J}$) and the field arising from the magnetization in ferromagnetic materials, denoted $\mathbf{H}_m \in \mathcal{V}(\Omega)$,

$$\mathbf{H} = \mathbf{H}_s + \mathbf{H}_m. \tag{4.115}$$

Many definitions for $\mathbf{H}_s$ exist, as the gradient fields lie in the kernel of the curl operator; see Section 3.15:

$$\operatorname{curl} \mathbf{H}_s = \operatorname{curl}(\mathbf{H}_s + \operatorname{grad} \psi). \tag{4.116}$$

Since $\operatorname{curl} \mathbf{H}_m = \mathbf{0}$ by definition, it follows that $\mathbf{H} - \mathbf{H}_s$ is curl-free and

$$\mathbf{H} - \mathbf{H}_s = -\operatorname{grad} \phi_m^{red}. \tag{4.117}$$

The potential $\phi_m^{red} \in \mathcal{S}(\Omega)$ is called the reduced magnetic scalar potential, $[\phi_m^{red}] = 1$ A. Its definition originates with Maxwell and Noether[34] with applications to computational electromagnetism by Carpenter [12]. Eq. (4.117)

34 Emmy Noether (1882–1935).

automatically satisfies Ampère's law, so the flux conservation law must be imposed:

$$\operatorname{div} \mathbf{B} = 0\,,$$

$$\operatorname{div} \mu \left(-\operatorname{grad} \phi_{\mathrm{m}}^{\mathrm{red}} + \mathbf{H}_{\mathrm{s}} \right) = 0\,,$$

$$\operatorname{div} \left(\mu \operatorname{grad} \phi_{\mathrm{m}}^{\mathrm{red}} \right) = \operatorname{div}\, \mu \mathbf{H}_{\mathrm{s}}\,. \tag{4.118}$$

The reduced field $\mathbf{H}_{\mathrm{s}}$ can be determined by means of the Biot–Savart law, which will be derived in Chapter 5. The boundary conditions $H_{\mathrm{t}} = 0$ and $B_{\mathrm{n}} = 0$ become

$$\operatorname{grad} \phi_{\mathrm{m}}^{\mathrm{red}} \times \mathbf{n} = \mathbf{H}_{\mathrm{s}} \times \mathbf{n} \quad \text{on } \Gamma_H\,, \tag{4.119}$$

$$\mu \partial_{\mathbf{n}} \phi_{\mathrm{m}}^{\mathrm{red}} = \mu \mathbf{H}_{\mathrm{s}} \cdot \mathbf{n} \quad \text{on } \Gamma_B\,. \tag{4.120}$$

At the interface Γ_{ai} the continuity conditions of B_{n} and H_{t} must be satisfied:

$$\mathbf{n} \cdot [\![\mathbf{B}]\!]_{\mathrm{ai}} = 0, \tag{4.121}$$

$$\mathbf{n} \times [\![\mathbf{H}]\!]_{\mathrm{ai}} = \mathbf{0}, \tag{4.122}$$

where $\mathbf{n}$ points from the iron to the air domain. Equations (4.121) and (4.122) may be expressed in terms of the reduced magnetic scalar potential:

$$[\![-\mu \partial_{\mathbf{n}} \phi_{\mathrm{m}}^{\mathrm{red}}]\!]_{\mathrm{ai}} + \mathbf{n} \cdot [\![\mu \mathbf{H}_{\mathrm{s}}]\!]_{\mathrm{ai}} = 0, \tag{4.123}$$

$$-\mathbf{n} \times [\![\operatorname{grad} \phi_{\mathrm{m}}^{\mathrm{red}}]\!]_{\mathrm{ai}} + \mathbf{n} \times [\![\mathbf{H}_{\mathrm{s}}]\!]_{\mathrm{ai}} = \mathbf{0}. \tag{4.124}$$

We gauge the reduced magnetic scalar potential by imposing $\phi_{\mathrm{m}}^{\mathrm{red}}(\mathbf{r}_0) = \phi_0$ at an arbitrary point $\mathbf{r}_0 \in \Omega$.

Remark: Inside highly permeable domains $\mathbf{H}_{\mathrm{m}}$ and $\mathbf{H}_{\mathrm{s}}$ are of similar magnitude but opposite direction. Cancellation errors occur if the magnetic scalar potential is calculated numerically, while the source term $\mathbf{H}_{\mathrm{s}}$ is computed from the Biot–Savart law. A numerical example is presented in [8]. An alternative is to interpolate $\mathbf{H}_{\mathrm{s}}$ by means of the same functions that are used for the approximation of $\operatorname{grad} \phi_{\mathrm{m}}^{\mathrm{red}}$ [9]. However, this technique has proven insufficiently accurate for field optimization of superconducting magnets. □

The total magnetic scalar potential A method for avoiding cancellation errors is the use of the *total magnetic scalar potential* in the domain Ω, now assumed to be free of induced or impressed currents. Hence $\operatorname{curl} \mathbf{H} = 0$, and thus the field can be represented as

$$\mathbf{H} = -\operatorname{grad} \phi_{\mathrm{m}}\,, \tag{4.125}$$

where $\phi_{\mathrm{m}} \in \mathcal{S}(\Omega)$ is the *total magnetic scalar potential*, $[\phi_{\mathrm{m}}] = 1$ A. This choice automatically fulfills Ampère's law, so the flux conservation law must be imposed:

$$\operatorname{div}(\mu \operatorname{grad} \phi_{\mathrm{m}}) = 0 \qquad \text{in } \Omega. \tag{4.126}$$

The boundary conditions become

$$\mathbf{n} \times (\operatorname{grad} \phi_{\mathrm{m}} \times \mathbf{n}) = \mathbf{0} \qquad \text{on } \Gamma_H, \tag{4.127}$$

$$\mu \partial_{\mathbf{n}} \phi_{\mathrm{m}} = 0 \qquad \text{on } \Gamma_B. \tag{4.128}$$

The interface conditions at the boundary between iron and air can be expressed as

$$[\![\mu \partial_{\mathbf{n}} \phi_{\mathrm{m}}]\!]_{\mathrm{ai}} = 0, \tag{4.129}$$

$$\mathbf{n} \times [\![\operatorname{grad} \phi_{\mathrm{m}}]\!]_{\mathrm{ai}} = \mathbf{0}. \tag{4.130}$$

Further, we gauge the magnetic scalar potential by imposing $\phi_{\mathrm{m}}(\mathbf{r}_0) = \phi_0$ at an arbitrary point $\mathbf{r}_0$ in Ω. Equation (4.126) reduces to $\mu_0 \operatorname{div} \operatorname{grad} \phi_{\mathrm{m}} = 0$ in free space Ω_{a} and for $\mathbf{J} = 0$, and consequently

$$\nabla^2 \phi_{\mathrm{m}} = 0, \tag{4.131}$$

which is Laplace's equation for the magnetic scalar potential.

Two scalar potentials The *two scalar potentials* method [8] couples the reduced magnetic scalar potential $\phi_{\mathrm{m}}^{\mathrm{red}} \in \mathcal{S}(\Omega_{\mathrm{a}})$ in the current-carrying region Ω_{a} with the total scalar potential $\phi_{\mathrm{m}} \in \mathcal{S}(\Omega_{\mathrm{i}})$ in the current-free region Ω_{i} by the interface conditions

$$-\mu \partial_{\mathbf{n}} \phi_{\mathrm{m}} = \mu_0 \left(\mathbf{n} \cdot \mathbf{H}_{\mathrm{s}} - \partial_{\mathbf{n}} \phi_{\mathrm{m}}^{\mathrm{red}} \right), \tag{4.132}$$

$$\mathbf{n} \times \operatorname{grad} \phi_{\mathrm{m}} = \mathbf{n} \times \mathbf{H}_{\mathrm{s}} - \mathbf{n} \times \operatorname{grad} \phi_{\mathrm{m}}^{\mathrm{red}}, \tag{4.133}$$

on Γ_{ai} and $\phi_{\mathrm{m}}(\mathbf{r}_0) = \phi_0$ at an arbitrary point $\mathbf{r}_0$ in Ω_{i}.

4.9.2 Vector-Potential Formulations

Different *total vector-potential formulations* have been proposed. One is based on the solution of the so-called *curl–curl equation* [16]. Uniqueness requires the gauging of the vector potential and the introduction of additional boundary conditions. The curl–curl formulation results in a vector Poisson equation [13] after the incorporation of the Coulomb gauge. Current sources appear explicitly on the right-hand side of the equations. When numerical methods are employed for the solution of the boundary value problems, it is necessary that the current-carrying conductors be modeled in the finite-element mesh. This is particularly difficult in three dimensions. The reduced vector-potential formulation decomposes the magnetic vector potential into one part due to the iron magnetization and another attributable to the source currents in free space. Using a Biot-Savart-type integral, the source vector potential can be calculated with the precision required for the design of superconducting magnets.

The curl–curl equation Because the divergence of the magnetic flux density $\mathbf{B} \in \mathcal{V}(\Omega)$ is zero, it can be written as the curl of a magnetic vector potential $\mathbf{A} \in \mathcal{V}(\Omega)$:

$$\mathbf{B} = \operatorname{curl} \mathbf{A}\,, \tag{4.134}$$

with the physical unit $[\mathbf{A}] = 1\,\mathrm{T\,m}$. This choice automatically fulfills the flux conservation law, and so Ampère's law remains to be solved:

$$\operatorname{curl} \frac{1}{\mu} \operatorname{curl} \mathbf{A} = \mathbf{J}\,. \tag{4.135}$$

In terms of the magnetic vector potential, the boundary conditions can be written

$$\frac{1}{\mu} \operatorname{curl} \mathbf{A} \times \mathbf{n} = \mathbf{0} \qquad \text{on } \Gamma_H, \tag{4.136}$$

$$\operatorname{curl} \mathbf{A} \cdot \mathbf{n} = 0 \qquad \text{on } \Gamma_B. \tag{4.137}$$

Simpler expressions result for 2D field problems using Eq. (2.78) and $\operatorname{curl}(g_z \mathbf{e}_z) = \operatorname{grad} g_z \times \mathbf{e_z}$, which can be derived from Eq. (3.69):

$$\partial_{\mathbf{n}} A_z = 0 \qquad \text{on } \Gamma_H, \tag{4.138}$$

$$A_z = 0 \qquad \text{on } \Gamma_B, \tag{4.139}$$

where $\partial_{\mathbf{n}}$ denotes the *normal derivative*, that is, the directional derivative in the direction of the normal vector to the boundary. Equation (4.138) is known as the homogeneous *Neumann boundary condition*,[35] and Eq. (4.139) as the *Dirichlet boundary condition*.[36]

The far-field boundary is also part of Γ_B. Because $B_n = 0$ we obtain

$$\Phi = 0 = \int_{\Gamma_B} \mathbf{B} \cdot \mathbf{da} = \int_{\Gamma_B} \operatorname{curl} \mathbf{A} \cdot \mathbf{da} = \int_{\partial \Gamma_B} \mathbf{A} \cdot \mathbf{ds} = \int_{\partial \Gamma_B} \mathbf{A} \cdot \mathbf{T}\, \mathrm{d}s\,, \tag{4.140}$$

35 Carl Neumann (1832–1925).
36 Gustav Lejeune Dirichlet (1805–1859).

and consequently $A_t = 0$ on Γ_B. Condition (4.137) is equivalent to

$$A_\mathrm{t} = 0 \quad \rightarrow \quad \mathbf{n} \times (\mathbf{A} \times \mathbf{n}) = \mathbf{0} \quad \text{on } \Gamma_B ; \tag{4.141}$$

see Eq. (2.82). The interface conditions in terms of the vector potential are written as

$$\mathbf{n} \times \left[\!\left[\frac{1}{\mu} \,(\operatorname{curl} \mathbf{A}) \right]\!\right]_{\mathrm{ai}} = \mathbf{0}\,, \tag{4.142}$$

$$[\![\, \mathbf{A} \,]\!]_{\mathrm{ai}} = \mathbf{0}. \tag{4.143}$$

The solution of the boundary value problem is not unique, however. The gradient of any smooth scalar field ψ can be added without changing the curl of $\mathbf{A}$,

$$\mathbf{A} \rightarrow \mathbf{A}' : \mathbf{A}' = \mathbf{A} + \operatorname{grad} \psi. \tag{4.144}$$

Equation (4.144) is called a *gauge transformation* between $\mathbf{A}'$ and $\mathbf{A}$. $\mathbf{B}$ is gauge invariant, because the transformation from $\mathbf{A}$ to $\mathbf{A}'$ does not change $\mathbf{B}$. The freedom given by the gauge transformation (only the curl of the field is defined) can be used to set the divergence of $\mathbf{A}'$ to zero

$$\operatorname{div} \mathbf{A}' = 0. \tag{4.145}$$

Proof. We must show that if $\mathbf{A}$ is a solution of the curl–curl equation, then an $\mathbf{A}'$ can be found that is also a solution of the curl–curl equation and obeys $\operatorname{div} \mathbf{A}' = q$, for which $q = 0$ is the special case of the *Coulomb gauge*.[37] The sources of $\mathbf{A}'$ can be defined at will. From the gauge transformation (4.144) we obtain $q = \operatorname{div} \mathbf{A} + \nabla^2 \psi$ with the solution[38]

$$\psi = \int_\Omega \frac{\operatorname{div} \mathbf{A} - q}{4\pi |\mathbf{r} - \mathbf{r}'|} \,\mathrm{d}V\,, \tag{4.146}$$

which proves the assertion that we can find a scalar field ψ such that the divergence of $\mathbf{A}'$ is zero [25]. □

The Coulomb gauge, Eq. (4.145), leads to a Poisson-type equation for the magnetic vector potential as shown in Section 4.9.2. In Maxwell's treatise we find the gauge transformation equation

$$\mathbf{A} = \mathbf{A}' - \operatorname{grad} \varphi \tag{4.147}$$

written in component form (Eq. (7), page 256 of [27]). Maxwell writes: "The quantity φ disappears from equations[39] … and is not related to any physical

37 Charles de Coulomb (1736–1806).
38 See Chapter 5.
39 We list them in Eq. (4.17).

phenomenon." Therefore, it would be historically more correct to speak of the Maxwell gauge rather than the Coulomb gauge. Enforcing the Coulomb gauge on the vector potential,

$$\frac{1}{\mu} \operatorname{div} \mathbf{A} = 0 , \tag{4.148}$$

and considering the additional boundary condition

$$\mathbf{A} \cdot \mathbf{n} = 0 \quad \text{on } \Gamma_H , \tag{4.149}$$

it can be proved that the resulting boundary value problem has a unique solution:

Proof. Let $\mathbf{A}_1$ and $\mathbf{A}_2$ be two vector potentials, with the difference $\mathbf{A}_1 - \mathbf{A}_2 =: \mathbf{A}_0$. Both vector potentials fulfill the conditions $\operatorname{curl} \mathbf{A}_{1,2} = \mathbf{B}$ and $\operatorname{div} \mathbf{A}_{1,2} = 0$ in the domain. Consequently, $\operatorname{curl} \mathbf{A}_0 = 0$ and $\mathbf{A}_0$ can be expressed in terms of a scalar field u_0 as $\mathbf{A}_0 = \operatorname{grad} u_0$. It follows that $\operatorname{div} \operatorname{grad} u_0 = \nabla^2 u_0 = 0$ in Ω, and the scalar field is said to be *harmonic*. For the boundary conditions this yields $\operatorname{grad} u_0 \cdot \mathbf{n} = 0$ on Γ_H and $\mathbf{n} \times (\operatorname{grad} u_0 \times \mathbf{n}) = 0$ on Γ_B. The latter implies that u_0 is constant on Γ_B. Using Green's first theorem we obtain

$$\int_{\Omega} (\operatorname{grad} u_0)^2 \mathrm{d}V = -\int_{\Omega} (u_0 \nabla^2 u_0) \mathrm{d}V + \int_{\partial\Omega} u_0 \operatorname{grad} u_0 \cdot \mathbf{n} \, \mathrm{d}a . \tag{4.150}$$

Because of the boundary conditions on $\partial\Omega$, and because u_0 is harmonic, the right-hand side of the equation vanishes. The remaining integral $\int_{\Omega} (\operatorname{grad} u_0)^2 \mathrm{d}V$ is identical to zero if and only if $\operatorname{grad} u_0 = 0$. Hence $\mathbf{A}_0 = \mathbf{A}_1 - \mathbf{A}_2 = \operatorname{grad} u_0 = 0$; the vector potential is unique. □

Chari [13] adopts the Coulomb gauge and solves Eqs. (4.135) and (4.148) together. Biro [31] introduces a penalty term to subtract from Eq. (4.135), which yields

$$\operatorname{curl} \frac{1}{\mu} \operatorname{curl} \mathbf{A} - \operatorname{grad} \frac{1}{\mu} \operatorname{div} \mathbf{A} = \mathbf{J} . \tag{4.151}$$

The penalty term is chosen such that for constant permeability the left-hand side of Eq. (4.151) results in

$$\operatorname{curl} \operatorname{curl} \mathbf{A} = -\nabla^2 \mathbf{A} . \tag{4.152}$$

This yields a unique solution of the boundary value problem if, in addition to the boundary conditions (4.136), (4.141), and (4.149), the conditions

$$\operatorname{div} \mathbf{A} = 0 \text{ on } \Gamma_B \quad \text{and} \quad \left[\!\left[\frac{1}{\mu} \operatorname{div} \mathbf{A} \right]\!\right]_{\mathrm{ai}} = 0 \tag{4.153}$$

are fulfilled.

The vector Poisson equation An equivalent formulation for the total vector potential can be obtained from

$$\operatorname{curl}\mathbf{A} = \mu_0(\mathbf{H}+\mathbf{M}),$$

$$\frac{1}{\mu_0}\operatorname{curl}\operatorname{curl}\mathbf{A} = \mathbf{J} + \operatorname{curl}\mathbf{M},$$

$$\frac{1}{\mu_0}(-\nabla^2\mathbf{A} + \operatorname{grad}\operatorname{div}\mathbf{A}) = \mathbf{J} + \operatorname{curl}\mathbf{M}, \tag{4.154}$$

where the iron magnetization is accounted for by the introduction of the magnetization vector $\mathbf{M}$. After incorporating the Coulomb gauge it follows from Eq. (4.154) that

$$\nabla^2\mathbf{A} = -\mu_0(\mathbf{J} + \operatorname{curl}\mathbf{M}). \tag{4.155}$$

The boundary conditions become

$$\mathbf{n}\times\left(\frac{1}{\mu_0}\operatorname{curl}\mathbf{A}\times\mathbf{n}\right) = \mathbf{0} \quad \text{on } \Gamma_H, \tag{4.156}$$

$$\operatorname{curl}\mathbf{A}\cdot\mathbf{n} = 0 \quad \text{on } \Gamma_B. \tag{4.157}$$

The interface conditions can be expressed as

$$\mathbf{n}\times\left[\!\left[\frac{1}{\mu}\,(\operatorname{curl}\mathbf{A})\right]\!\right]_{\mathrm{ai}} + \mathbf{n}\times[\![\mathbf{M}]\!]_{\mathrm{ai}} = \mathbf{0}, \tag{4.158}$$

$$[\![\mathbf{A}]\!]_{\mathrm{ai}} = \mathbf{0}. \tag{4.159}$$

The additional boundary and interface conditions

$$\mathbf{A}\cdot\mathbf{n} = 0 \quad \text{on } \Gamma_H, \tag{4.160}$$

$$\operatorname{div}\mathbf{A} = 0 \quad \text{on } \Gamma_B, \tag{4.161}$$

$$[\![\operatorname{div}\mathbf{A}]\!]_{\mathrm{ai}} = 0, \tag{4.162}$$

must be imposed in order to guarantee the uniqueness of the vector potential. From vector analysis we know that the Laplace operator acting on a vector field expressed in Cartesian coordinates yields a vector field that can be obtained through the application of the operator to each of the components:

$$\begin{aligned}\nabla^2\mathbf{A} &= \left(\frac{\partial^2 A_x}{\partial x^2} + \frac{\partial^2 A_x}{\partial y^2} + \frac{\partial^2 A_x}{\partial z^2}\right)\mathbf{e}_x + \left(\frac{\partial^2 A_y}{\partial x^2} + \frac{\partial^2 A_y}{\partial y^2} + \frac{\partial^2 A_y}{\partial z^2}\right)\mathbf{e}_y \\ &\quad + \left(\frac{\partial^2 A_z}{\partial x^2} + \frac{\partial^2 A_z}{\partial y^2} + \frac{\partial^2 A_z}{\partial z^2}\right)\mathbf{e}_z \\ &= (\nabla^2 A_x)\mathbf{e}_x + (\nabla^2 A_y)\mathbf{e}_y + (\nabla^2 A_z)\mathbf{e}_z\,.\end{aligned} \tag{4.163}$$

The vector Poisson equation thus reduces to three scalar equations,

$$\nabla^2 A_i = -\mu_0 (J_i + (\operatorname{curl} \mathbf{M})_i) , \tag{4.164}$$

where $i = 1, 2, 3$ stand for the x-, y-, and z-components of the vector fields.

Remark: This simple expression does not hold for local coordinates because the basis vectors are themselves a function of the point in space. The vector Laplace operator is given for cylindrical coordinates in Table 3.3. But even in Cartesian coordinates the components of the vector field might be coupled through curvilinear boundaries [13]. □

The reduced vector potential The vector potential $\mathbf{A}$ is now split into two parts:

$$\mathbf{A} = \mathbf{A}_s + \mathbf{A}_r, \tag{4.165}$$

where $\mathbf{A}_r$ is the *reduced vector potential* due to the iron magnetization and $\mathbf{A}_s$ is the impressed vector potential due to the source currents in free space. Both parts are defined on the entire problem domain, that is, $\mathbf{A}_s, \mathbf{A}_r \in \mathcal{V}(\Omega)$. Hence,

$$\mathbf{B} = \mu_0 \mathbf{H}_s + \operatorname{curl} \mathbf{A}_r . \tag{4.166}$$

The source vector potential can be calculated with *Biot–Savart-type integrals* from the coil current distribution, as we will discuss in the next chapter. The boundary value problem can then be derived from Eq. (4.151), replacing $\mathbf{A}$ with $\mathbf{A}_s + \mathbf{A}_r$,

$$\operatorname{curl} \frac{1}{\mu} \operatorname{curl} (\mathbf{A}_r + \mathbf{A}_s) - \operatorname{grad} \frac{1}{\mu} \operatorname{div} (\mathbf{A}_r + \mathbf{A}_s) = \mathbf{J}, \tag{4.167}$$

from which follows

$$\operatorname{curl} \frac{1}{\mu} \operatorname{curl} \mathbf{A}_r - \operatorname{grad} \frac{1}{\mu} \operatorname{div} \mathbf{A}_r = \mathbf{J} - \operatorname{curl} \frac{1}{\mu} \operatorname{curl} \mathbf{A}_s,$$

$$\operatorname{curl} \frac{1}{\mu} \operatorname{curl} \mathbf{A}_r - \operatorname{grad} \frac{1}{\mu} \operatorname{div} \mathbf{A}_r = \operatorname{curl} \mathbf{H}_s - \operatorname{curl} \frac{\mu_0}{\mu} \mathbf{H}_s \tag{4.168}$$

in Ω. For $\mu = \mu_0$ inside the aperture of a magnet the right-hand side of Eq. (4.168) vanishes. Now we assume that the current-carrying conductors do not intersect the iron/air interface. Thus, when the problem domain is discretized with a finite-element mesh, the coil embedded in Ω_a does not need to be represented by the mesh, as the right-hand side of the equations system is zero. In the iron domain the required source field $\mathbf{H}_s$ can be calculated with high precision using the Biot–Savart law. The boundary conditions become

$$\frac{1}{\mu_0} \operatorname{curl} \mathbf{A}_r \times \mathbf{n} = -\mathbf{H}_s \times \mathbf{n} \qquad \text{on } \Gamma_H, \tag{4.169}$$

$$\operatorname{curl} \mathbf{A}_r \cdot \mathbf{n} = -\mu_0 \mathbf{H}_s \cdot \mathbf{n} \qquad \text{on } \Gamma_B. \tag{4.170}$$

The interface conditions expressed in terms of the reduced vector potential can be rewritten as

$$\mathbf{n} \times \left[\!\left[\frac{1}{\mu} \left(\operatorname{curl} \mathbf{A}_r \right) \right]\!\right]_{\text{ai}} = \mathbf{0}, \tag{4.171}$$

$$[\![\mathbf{A}_r]\!]_{\text{ai}} = \mathbf{0}. \tag{4.172}$$

The different formulations and their boundary conditions on Γ_H and Γ_B are summarized in Table 4.5.

Table 4.5 Different formulations of the magnetostatic boundary value problem.

Formulation		$\Gamma_H (H_t = 0)$	$\Gamma_B (B_n = 0)$
Reduced magnetic scalar potential	$\operatorname{div}(\mu \operatorname{grad} \varphi_m^{\text{red}}) = \operatorname{div}(\mu \mathbf{H}_s)$	$\operatorname{grad} \varphi_m^{\text{red}} \times \mathbf{n} = \mathbf{H}_s \times \mathbf{n}$	$\mu \partial_{\mathbf{n}} \varphi_m^{\text{red}} = \mu \mathbf{H}_s \cdot \mathbf{n}$
Total magnetic scalar potential	$\operatorname{div}(\mu \operatorname{grad} \varphi_m) = 0$	$\mathbf{n} \times (\operatorname{grad} \varphi_m \times \mathbf{n}) = \mathbf{0}$	$\mu \partial_{\mathbf{n}} \varphi_m = 0$
Vector potential (curl–curl)	$\operatorname{curl} \frac{1}{\mu} \operatorname{curl} \mathbf{A} - \operatorname{grad} \frac{1}{\mu} \operatorname{div} \mathbf{A} = \mathbf{J}$	$\frac{1}{\mu} \operatorname{curl} \mathbf{A} \times \mathbf{n} = \mathbf{0}$	$\operatorname{curl} \mathbf{A} \cdot \mathbf{n} = 0$
Vector potential (Vector Poisson)	$\nabla^2 \mathbf{A} = \mu_0 (\mathbf{J} + \operatorname{curl} \mathbf{M})$	$\mathbf{n} \times (\frac{1}{\mu_0} \operatorname{curl} \mathbf{A} \times \mathbf{n}) = \mathbf{0}$	$\operatorname{curl} \mathbf{A} \cdot \mathbf{n} = 0$
Reduced vector potential	$\operatorname{curl} \frac{1}{\mu} \operatorname{curl} (\mathbf{A}_r + \mathbf{A}_s) - \operatorname{grad} \frac{1}{\mu} \operatorname{div}(\mathbf{A}_r + \mathbf{A}_s) = \mathbf{J}$	$\frac{1}{\mu_0} \operatorname{curl} \mathbf{A}_r \times \mathbf{n} = -\mathbf{H}_s \times \mathbf{n}$	$\operatorname{curl} \mathbf{A}_r \cdot \mathbf{n} = -\mu_0 \mathbf{H}_s \cdot \mathbf{n}$

4.9.3 The Scalar Laplace Equation in 2D

In the two-dimensional case, with $\partial/\partial z = 0$ and all impressed currents flowing in the z-direction ($\mathbf{J} = J_z \mathbf{e}_z$), the vector potential has only a z-component, which depends on the transversal coordinates $\mathbf{A} = A_z(x, y)\mathbf{e}_z$. Hence the Coulomb gauge $\operatorname{div} \mathbf{A} = 0$ is automatically fulfilled. We obtain the scalar Poisson equation in the air domain,

$$\nabla^2 A_z = -\mu_0 J_z, \tag{4.173}$$

$A_z, J_z \in \mathcal{S}(\Omega_a)$. For current-free regions, Eq. (4.173) reduces to the Laplace equation,

$$\nabla^2 A_z = 0, \tag{4.174}$$

which can be written in Cartesian coordinates as

$$\frac{\partial^2 A_z}{\partial x^2} + \frac{\partial^2 A_z}{\partial y^2} = 0 \tag{4.175}$$

and in circular coordinates as

$$r^2 \frac{\partial^2 A_z}{\partial r^2} + r \frac{\partial A_z}{\partial r} + \frac{\partial^2 A_z}{\partial \varphi^2} = 0. \tag{4.176}$$

4.10 Boundary Value Problems 2: Magnetic Diffusion Problems

The complete set of Maxwell's equations contains the displacement–current term $\partial \mathbf{D}/\partial t$ and the induction term $\partial \mathbf{B}/\partial t$. Field diffusion inside conducting media is described by the *quasistationary approximation*, which omits the displacement current term. This approximation is valid inside conducting media if the frequency ω obeys the condition

$$\omega \ll \frac{\varkappa}{\epsilon}. \tag{4.177}$$

This can be better understood if we consider all fields varying sinusoidally in time, that is, $\mathbf{H}$, $\mathbf{E}$, $\mathbf{B}$, $\mathbf{D} \propto \mathrm{e}^{i\omega t}$. Hence $\partial/\partial t = i\omega$, which yields

$$\operatorname{curl} \mathbf{H} = i\omega\epsilon \mathbf{E}, \tag{4.178}$$

$$\operatorname{curl} \mathbf{E} = -i\omega\mu \mathbf{H}, \tag{4.179}$$

from which follows

$$\nabla^2 \mathbf{E} + \mu\omega^2 \left(\epsilon - i\frac{\varkappa}{\omega}\right) \mathbf{E} = \mathbf{0}. \tag{4.180}$$

Thus for condition (4.177) the first term in the parentheses in the wave equation (4.180), corresponding to the displacement current, can be neglected.

Now with the $\partial \mathbf{D}/\partial t$ term omitted, we return to the local form of Maxwell's equations for time-transient fields, $\operatorname{curl} \mathbf{H} = \mathbf{J}$, $\operatorname{curl} \mathbf{E} = -\partial \mathbf{B}/\partial t$, div $\mathbf{B} = 0$, and div $\mathbf{D} = \rho$, together with the constitutive equations for stationary media and an impressed current density $\mathbf{J}_\mathrm{s}$:

$$\mathbf{B} = \mu_0(\mathbf{H} + \mathbf{M}), \qquad \mathbf{J} = \varkappa \mathbf{E} + \mathbf{J}_\mathrm{s}. \tag{4.181}$$

We consider once more the elementary model problem shown in Figure 4.1, consisting of the iron domain Ω_i with a magnetization $\mathbf{M}$, but now also of a finite conductivity $\varkappa$. As before, we denote the air region as Ω_a and define $\Omega = \Omega_\mathrm{a} \cup \Omega_\mathrm{i}$. The regions are connected at the interface Γ_{ai}. Moreover, each domain is bounded by a surface $\Gamma = \partial\Omega$ itself consisting of two different parts Γ_H and Γ_B. Since div curl $\mathbf{g} = 0$ for all vector fields $\mathbf{g} \in \mathcal{V}(\Omega)$, the magnetic flux density $\mathbf{B}$ can be written in terms of the vector potential as $\mathbf{B} = \operatorname{curl} \mathbf{A}$. We obtain from Faraday's law

$$\operatorname{curl} \mathbf{E} = -\frac{\partial}{\partial t} \operatorname{curl} \mathbf{A},$$

$$\operatorname{curl} \left(\mathbf{E} + \frac{\partial}{\partial t}\mathbf{A} \right) = \mathbf{0}. \tag{4.182}$$

As $\mathbf{E} + \partial\mathbf{A}/\partial t$ is curl-free, it can be expressed in a contractible domain as the gradient of a scalar field $\phi \in \mathcal{S}(\Omega)$, and thus as

$$\mathbf{E} = -\frac{\partial}{\partial t}\mathbf{A} - \operatorname{grad} \phi. \tag{4.183}$$

The induced eddy current is

$$\mathbf{J}_\mathrm{e} = \varkappa\mathbf{E} = -\varkappa \left(\frac{\partial}{\partial t}\mathbf{A} + \operatorname{grad} \phi \right). \tag{4.184}$$

From the constitutive equation (4.181) we derive

$$\operatorname{curl} \mathbf{A} = \mu_0 \left(\mathbf{H} + \mathbf{M} \right),$$

$$\frac{1}{\mu_0} \operatorname{curl} \operatorname{curl} \mathbf{A} = \mathbf{J} + \operatorname{curl} \mathbf{M},$$

$$\frac{1}{\mu_0} \operatorname{curl} \operatorname{curl} \mathbf{A} = \varkappa\mathbf{E} + \mathbf{J}_\mathrm{s} + \operatorname{curl} \mathbf{M},$$

$$-\frac{1}{\mu_0}\nabla^2\mathbf{A} - \frac{1}{\mu_0} \operatorname{grad} \operatorname{div} \mathbf{A} = \varkappa \left(-\frac{\partial}{\partial t}\mathbf{A} - \operatorname{grad} \phi \right) + \mathbf{J}_\mathrm{s} + \operatorname{curl} \mathbf{M}, \tag{4.185}$$

and after incorporation of the Coulomb gauge $\operatorname{div} \mathbf{A} = 0$,

$$-\frac{1}{\mu_0}\nabla^2\mathbf{A} + \varkappa\frac{\partial}{\partial t}\mathbf{A} + \varkappa \operatorname{grad} \phi = \operatorname{curl} \mathbf{M} + \mathbf{J}_\mathrm{s}. \tag{4.186}$$

Imposing $\operatorname{div} \mathbf{J} = 0$ yields

$$\operatorname{div} \left(\varkappa\mathbf{E} + \mathbf{J}_\mathrm{s} \right) = \operatorname{div} \left(-\varkappa\frac{\partial}{\partial t}\mathbf{A} - \varkappa \operatorname{grad} \phi + \mathbf{J}_\mathrm{s} \right) = 0. \tag{4.187}$$

To establish the boundary value problem we must consider

$$\frac{1}{\mu_0}(\operatorname{curl}\mathbf{A}) \times \mathbf{n} = \mathbf{0} \quad \text{on } \Gamma_{\mathrm{H}}, \tag{4.188}$$

$$\mathbf{A} \times \mathbf{n} = \mathbf{0} \quad \text{on } \Gamma_{\mathrm{B}}, \tag{4.189}$$

$$\mathbf{n} \times \left[\!\left[\frac{1}{\mu_0}(\operatorname{curl}\mathbf{A}) \right]\!\right]_{\mathrm{ai}} + \mathbf{n} \times [\![\mathbf{M}]\!]_{\mathrm{ai}} = \mathbf{0}, \tag{4.190}$$

$$[\![\mathbf{A}]\!]_{\mathrm{ai}} = \mathbf{0}, \tag{4.191}$$

$$\mathbf{n} \cdot [\![\varkappa \operatorname{grad} \phi + \varkappa \frac{\partial}{\partial t}\mathbf{A}]\!]_{\mathrm{ai}} = 0 \tag{4.192}$$

and $\phi(\mathbf{r}_0) = \phi_0$ at an arbitrary point $\mathbf{r}_0 \in \Omega_{\mathrm{i}}$. Equation (4.186) can be simplified if we account only for conducting, amagnetic domains (such as aluminum and copper), with the assumption of a negligible free charge and of the time-varying magnetic flux density as the sole source of the electric field. This yields the *diffusion equation* for the magnetic vector potential

$$\nabla^2 \mathbf{A} = \mu_0 \varkappa \frac{\partial}{\partial t}\mathbf{A}. \tag{4.193}$$

If the conductivity is constant in the domain, it follows that the current and magnetic flux densities satisfy similar diffusion equations [21],

$$\nabla^2 \mathbf{J} = \mu_0 \varkappa \frac{\partial}{\partial t}\mathbf{J} \qquad \text{and} \qquad \nabla^2 \mathbf{B} = \mu_0 \varkappa \frac{\partial}{\partial t}\mathbf{B}. \tag{4.194}$$

The latter can be derived also in the following way:

$$\operatorname{curl}\operatorname{curl}\mathbf{H} = \operatorname{curl}\mathbf{J},$$

$$\frac{1}{\mu_0}\operatorname{curl}\operatorname{curl}\mathbf{B} = \varkappa \operatorname{curl}\mathbf{E},$$

$$-\frac{1}{\mu_0}\operatorname{div}\operatorname{grad}\mathbf{B} = -\varkappa \frac{\partial}{\partial t}\mathbf{B},$$

$$\nabla^2 \mathbf{B} = \mu_0 \varkappa \frac{\partial}{\partial t}\mathbf{B}. \tag{4.195}$$

In the two-dimensional case, the current density has only one component (e.g., the z-component):

$$\mathbf{J} = \mathbf{J}_{\mathrm{e}} + \mathbf{J}_{\mathrm{s}} = J_z \mathbf{e}_z \tag{4.196}$$

with

$$J_z = -\varkappa \left(\frac{\partial A_z}{\partial t} + \frac{\partial \phi}{\partial z} \right) + J_{z,\mathrm{s}}. \tag{4.197}$$

Integrating over the cross section of the conducting domain $\mathscr{A}$ yields the total measurable current

$$I = \int_{\mathscr{A}} J_z \, da = \int_{\mathscr{A}} \varkappa \left(\frac{\partial A_z}{\partial t} + \frac{\partial \phi}{\partial z} \right) da + \int_{\mathscr{A}} J_{z,s} \, da \,. \tag{4.198}$$

Because $I = \int_{\mathscr{A}} \mathbf{J}_s \cdot \mathbf{da}$, it follows that

$$\int_{\mathscr{A}} \varkappa \left(\frac{\partial A_z}{\partial t} + \frac{\partial \phi}{\partial z} \right) da = 0 \,. \tag{4.199}$$

As in two dimensions $E_z = \partial\phi/\partial z = (\phi(z) - \phi(0))/z$ is constant, the $\operatorname{grad}\phi$ term is the mean value of $\partial A_z/\partial t$ in the conducting domain and therefore proportional to the sum current in this domain. The $\operatorname{grad}\phi$ term can therefore be used to modify the eddy current distribution in the conductor. If the line $A_z = 0$ coincides with the central line of the eddy current loop, a zero-sum current is imposed and the $\operatorname{grad}\phi$ term can be omitted.

References

1 Ansorge, W., Billan, J., Gourber, J.P., Haren, P. Henrichsen, K.N., Verdier, A.: Proceedings of the 6th International Conference on Magnet Technology (MT6), Bratislava, Czechoslovakia, 1977

2 Babic, S., Comel, S., Beckers, F., Brixhe, F., Peiro, G., Verbeek, T.: Toward the production of 50 000 tonnes of low-carbon steel sheet for the LHC superconducting dipole and quadrupole magnets, IEEE-TAS, 2002

3 Bertinelli, F., Comel, S., Harlet, P., Peiro, G., Russo, A., Taquet, A.: Production of low-carbon magnetic steel for the LHC superconducting dipole and quadrupole magnets, IEEE Transactions on Applied Superconductivity, 2006

4 Bertinelli, F., Fudanoki, F., Komori, T., Peiro, G., Rossi, L.: Production of austenitic steel for the LHC superconducting dipole magnets, IEEE Transactions on Applied Superconductivity, 2006

5 Bertotti, G., Mayergoyz I.: The Science of Hysteresis I–III, Elsevier, Academic Press, The Netherlands, San Diago, 2006

6 Billan, J.: Influence of mortar induced stresses on magnetic characteristics, Conference on Magnet Technology, Boston, 1987

7 Billan, J.: Materials, CAS School on Magnetic Measurement and Alignment, CERN 92-05, 1992

8 Binns, K.J., Lawrenson, P.J., Trowbridge, C.W.: The Analytical and Numerical Solution of Electric and Magnetic Fields, John Wiley & Sons, New York, 1992

9 Biro, O., Preis, K., Vrisk, G., Richter, K.R., Ticar, I.: Computation of 3D magnetostatic fields using reduced scalar potential, IEEE Transactions on Magnetics, 1993

10 Bossavit, A.: Computational Electromagnetism, Academic Press, New York, 1998

11 Bozorth, R.M.: Ferromagnetism, IEEE Press, Piscataway, NJ, 1978

12 Carpenter, C. J.: Comparison of alternative formulations of 3-dimensional magnetic-field and eddy current problems at power frequencies, Proceedings IEE, 1977

13 Chari, M.V.K., Silvester, P., Konrad, A., Csendes, Z.J., Palmo, M. A.: Three-dimensional magnetostatic field analysis of electrical machinery by the finite element method, IEEE Transactions, PAS-100, 1981

14 Csendes, Z. J., Weiss, J., Hoole, S.R.H.: Alternative vector potential formulations of 3-D magnetostatic field problems, IEEE Transactions on Magnetics, Vol. 18, 1982

15 Cullity, B.D., Graham, C.D.: Introduction to Magnetic Materials, Wiley, New York, 2009

16 Demerdash, N.A., Nehl, T.W., Fouad, F.A., Mohammed, O.A.: Three-dimensional finite element vector potential formulation of magnetic fields in electrical apparatus, IEEE Transactions, PAS-100, 1981

17 Fetzer, J.: Die Lösung statischer und quasistationärer elektromagnetischer Feldprobleme mit Hilfe der Kopplung der Methode der finiten Elemente und der Randelementmethode, Fortschritt Berichte VDE, Reihe 21, VDE Verlag, 1992

18 IEC-CEI International standard, Letter symbols to be used in electrical technology, 27-1

19 IEC-CEI International standard, Magnetic materials, Part 2, Methods of measurement of the magnetic properties of electrical steel sheet and strip by means of an Epstein frame, 1996

20 IEC-CEI International standard, Magnetic materials, Part 4, Methods of measurement of d.c. magnetic properties of iron and steel, 404-4, 1995

21 Jackson, J.D.: Classical Electrodynamic, John Wiley & Sons, New York, 1997

22 Jones, E.R., Datta, T., Almasan, C, Edwards, D., Ledbetter, H. M.: Low temperature magnetic properties of F.C.C Fe–Cr–Ni alloys: effects of manganese and interstitial carbon and nitrogen, material science and engineering, 1987

23 Joos, G.: Lehrbuch der Theoretischen Physik, Akademische Verlagsgesellschaft, Leipzig, 1942

24 Kawabata, S.: Magnetic permeability of the iron yoke in high field superconducting magnets. Nuclear Instruments and Methods in Physics Research A329, 1993

25 Kurz, S.: Vorlesung ueber Theoretische Elektrotechnik, Hamburg 2004, private communication

26 Lehner, G.: Elektromagnetische Feldtheorie für Ingenieure und Physiker, Springer, Berlin, 1990

27 Maxwell, J.C.: A treatise on electricity & magnetism, Dover, New York, 1954

28 Nethe, A., Stahlmann, H.-D. (editors): Einführung in die Feldtheorie, Verlag Dr. Köster, Berlin, 2003

29 Ozaki, Y., Furukimi, O., Kakihara, S., Shiraishi, M., Morito, N., Nohara, K.: Development of non-magnetic high manganese cryogenic steel for the construction of the LHC project's superconducting magnets, IEEE Transactions on Applied Superconductivity, 2002

30 Piefke, G.: Feldtheorie I, BI Hochschultaschenbuecher, Band 771

31 Preis, K., Biro, O., Magele, C.A., Renhart, W., Richer, K.R., Vrisk, G.: Numerical analysis of 3D magnetostatic fields, IEEE Transactions on Magnetics, 1991

32 Tonti, E: On the Geometrical Structure of the Electromagnetism, Pitagora Editrice Bologna, 1995

33 Tonti, E: http://discretephysics.dic.univ.trieste.it/.

34 du Tremolet de la Cheisserie, E.: Magnetism. Fundamentals, Materials and Applications, Kluwer, Dordrecht, 2002

35 Wlodarski, Z.: Analytical description of magnetization curves, Physica B, Elsevier, The Netherlands, 2005

5
Fields and Potentials of Line-Currents

In Section 1.2, we showed that the field in superconducting magnets is dominated by the coil, with the iron-yoke magnetization contributing only about 20%. The electromagnetic design of superconducting magnets can thus be split into two tasks: First the layout and optimization of the coil, and secondly, the numerical field calculation of the magnetization in the iron yoke, which mainly affects the lower-order multipole field errors.

In this chapter, we start with the fundamental solution of the Laplace equation given by the Green's functions in two and three dimensions. We will then present the proofs of some important theorems on harmonic functions. The *Liouville theorem*,[1] derived from the Gauss mean-value theorem, states that any bounded harmonic function defined on the entire Euclidean space is in fact a constant function. Therefore, we must also study isolated singularities of harmonic functions.

The coil field can be calculated with a high degree of accuracy by applying the *Biot–Savart law* to the current loops formed by the strands. In 2D the situation is even easier, as the optimization can be performed semianalytically using Ampère's law and the *image-current method*. We will limit ourselves to current sources flowing in very thin wires represented by closed space curves and denoted as *line-currents*. After deriving the Biot–Savart law we will calculate the vector potential and magnetic flux density of ring currents and straight line-current segments, which can be used to calculate the field of polygonal conductors.

Section 5.10 treats the global quantities of the field solutions: stored energy, self and mutual inductances, and electromagnetic forces in linear and nonlinear circuits. In particular, mutual inductances and forces in combined-function magnets will be reviewed. We will finally derive an equation for the differential inductance needed for comparing calculations and measurements of inductive voltages in nonlinear magnetic circuits.

1 Josef Liouville (1809–1882).

Field Computation for Accelerator Magnets. Stephan Russenschuck

ISBN: 978-3-527-40769-9

5.1
Green Functions

The *Green function* of a linear differential operator[2] $\mathcal{L}_{\mathbf{r}'}$ that acts on distributions over a domain $\mathscr{V} \subseteq \Omega$ is any solution of the differential equation

$$\mathcal{L}_{\mathbf{r}'} G(\mathbf{r}, \mathbf{r}') = -\delta(\mathbf{r} - \mathbf{r}') , \tag{5.1}$$

where $\delta(\mathbf{r} - \mathbf{r}')$ is the *Dirac delta distribution*[3] centered at $\mathbf{r}'$. From $\delta(\mathbf{r} - \mathbf{r}') = 0$ for $\mathbf{r} \neq \mathbf{r}'$ and $\int_{-\infty}^{\infty} \delta(\mathbf{r} - \mathbf{r}') \, \mathrm{d}V = 1$ it follows that

$$f(\mathbf{r}') = \int_{\mathscr{V}} f(\mathbf{r}) \delta(\mathbf{r} - \mathbf{r}') \, \mathrm{d}V . \tag{5.2}$$

This property can be used to solve the operator equation

$$\mathcal{L}_{\mathbf{r}'} \phi(\mathbf{r}') = -f(\mathbf{r}') \tag{5.3}$$

for the potential $\phi(\mathbf{r}) \in \mathcal{S}(\Omega)$ because we obtain

$$\int_{\mathscr{V}} \mathcal{L}_{\mathbf{r}'} G(\mathbf{r}, \mathbf{r}') \, f(\mathbf{r}) \, \mathrm{d}V = -\int_{\mathscr{V}} \delta(\mathbf{r} - \mathbf{r}') f(\mathbf{r}) \, \mathrm{d}V = -f(\mathbf{r}') . \tag{5.4}$$

Thus,

$$\mathcal{L}_{\mathbf{r}'} \phi(\mathbf{r}') = \int_{\mathscr{V}} \mathcal{L}_{\mathbf{r}'} G(\mathbf{r}, \mathbf{r}') f(\mathbf{r}) \, \mathrm{d}V = \mathcal{L}_{\mathbf{r}'} \int_{\mathscr{V}} G(\mathbf{r}, \mathbf{r}') f(\mathbf{r}) \, \mathrm{d}V , \tag{5.5}$$

resulting from the linearity of the operator $\mathcal{L}_{\mathbf{r}'}$ acting on $\mathbf{r}'$. From Eq. (5.5) it yields

$$\phi(\mathbf{r}') = \int_{\mathscr{V}} G(\mathbf{r}, \mathbf{r}') f(\mathbf{r}) \, \mathrm{d}V . \tag{5.6}$$

Equation (5.3) is known as the *Poisson problem*. If the source is confined to a point $\mathscr{P}$, line $\mathscr{S}$, or surface $\mathscr{A}$ in a homogeneous medium, the resulting fields can be obtained from the Green functions, employing the integral of Eq. (5.6).

The problem is now reduced to finding the Green function of the operator $\mathcal{L}_{\mathbf{r}'}$. For an unbounded three-dimensional domain,

$$G_3(\mathbf{r}, \mathbf{r}') = \frac{1}{4\pi |\mathbf{r} - \mathbf{r}'|} \tag{5.7}$$

is the *fundamental solution* of the Laplace operator or the free-space Green function of the Laplace operator. In an unbounded two-dimensional domain it is given by

$$G_2(\mathbf{r}, \mathbf{r}') = \frac{1}{2\pi} \ln \left(\frac{|\mathbf{r} - \mathbf{r}'|}{r_{\mathrm{ref}}} \right) , \tag{5.8}$$

2 For our purpose $\mathcal{L}_{\mathbf{r}'}$ stands for the Laplace operator.
3 Paul Dirac (1902–1984).

where r_{ref} is an arbitrary reference radius. Heuristically, the Green function of free space can be seen as the purely geometric proportionality factor of Coulomb's law. The symmetry of the Green function, $G(\mathbf{r}, \mathbf{r}') = G(\mathbf{r}', \mathbf{r})$, allows us to exchange $\mathbf{r}$ and $\mathbf{r}'$ and obtain, from Eq. (5.6),

$$\phi(\mathbf{r}) = \int_{\mathscr{V}} G(\mathbf{r}, \mathbf{r}') f(\mathbf{r}') \, \mathrm{d}V' . \tag{5.9}$$

By convention, $\mathbf{r}'$ is the position vector of the source point $\mathscr{Q}$, and $\mathbf{r}$ is the position vector of the field point $\mathscr{P}$. Consequently, the vector $\mathbf{r} - \mathbf{r}'$ points from the source point to the field point as shown in Figure 5.2. The term $|\mathbf{r} - \mathbf{r}'|^{-1}$ is known as the *reciprocal distance* between these points. Obviously, the function

$$G^*(\mathbf{r}, \mathbf{r}') = G(\mathbf{r}, \mathbf{r}') + g(\mathbf{r}, \mathbf{r}') \tag{5.10}$$

is also a fundamental solution of the Laplace operator in case $g(\mathbf{r}, \mathbf{r}')$ is harmonic[4] in $\mathbf{r}$.

5.2 Potentials on Bounded Domains

Often the source distribution is known, but additional boundary conditions, for example, on interfaces of infinitely permeable domains, must be taken into account. The expression (5.9) for the free-space Green function is generally not a solution of the boundary value problem. Because of the freedom given by Eq. (5.10) we can add a solution of the Laplace equation such that the boundary conditions are fulfilled. This approach is equivalent to replacing the boundary with image sources, a technique that will be explained later.

Recall Green's second identity (3.156) for scalar fields $\phi, \psi \in \mathcal{S}(\Omega)$,

$$\int_{\mathscr{V}} (\phi \nabla^2 \psi - \psi \nabla^2 \phi) \, \mathrm{d}V = \int_{\partial \mathscr{V}} (\phi \, \partial_{\mathbf{n}} \psi - \psi \, \partial_{\mathbf{n}} \phi) \, \mathrm{d}a, \tag{5.11}$$

where $\mathbf{n}$ is the normal direction to the domain boundary $\partial\mathscr{V}$ of $\mathscr{V} \subset \Omega$, and where $\partial_{\mathbf{n}}$ denotes the normal derivative. For $\psi = G(\mathbf{r}, \mathbf{r}')$ the identity (5.11) yields

$$\int_{\mathscr{V}} \Big(-\phi(\mathbf{r}) \delta(\mathbf{r} - \mathbf{r}') + G(\mathbf{r}, \mathbf{r}') f(\mathbf{r}) \Big) \, \mathrm{d}V$$
$$= \int_{\partial\mathscr{V}} \Big(\phi(\mathbf{r}) \, \partial_{\mathbf{n}} G(\mathbf{r}, \mathbf{r}') - G(\mathbf{r}, \mathbf{r}') \, \partial_{\mathbf{n}} \phi(\mathbf{r}) \Big) \, \mathrm{d}a . \tag{5.12}$$

4 That is, $\nabla^2 g(\mathbf{r}, \mathbf{r}') = 0$.

Using Eq. (5.2), we obtain

$$\phi(\mathbf{r}') = \int_{\mathcal{V}} G(\mathbf{r},\mathbf{r}') f(\mathbf{r})\, \mathrm{d}V \\ + \int_{\partial\mathcal{V}} \Big(-\phi(\mathbf{r})\, \partial_{\mathbf{n}} G(\mathbf{r},\mathbf{r}') + G(\mathbf{r},\mathbf{r}')\, \partial_{\mathbf{n}}\phi(\mathbf{r}) \Big)\, \mathrm{d}a \,. \quad (5.13)$$

The symmetry of the Green function allows us again to exchange the vectors $\mathbf{r}$ and $\mathbf{r}'$ and we obtain

$$\phi(\mathbf{r}) = \int_{\mathcal{V}} G(\mathbf{r},\mathbf{r}') f(\mathbf{r}')\, \mathrm{d}V' \\ + \int_{\partial\mathcal{V}} \Big(-\phi(\mathbf{r}')\, \partial_{\mathbf{n}'} G(\mathbf{r},\mathbf{r}') + G(\mathbf{r},\mathbf{r}')\, \partial_{\mathbf{n}'}\phi(\mathbf{r}') \Big)\, \mathrm{d}a' \,. \quad (5.14)$$

This is *Kirchhoff's theorem*, or Green's third identity, wherein boundary values of ϕ and $\partial_{\mathbf{n}}\phi$ are involved that represent sources outside the volume $\mathcal{V}$. Equation (5.14) constitutes the basis for the boundary-element method treated in Chapter 15. If the boundary surface goes to infinity, and if the potential decays faster than $|\mathbf{r} - \mathbf{r}'|^{-1}$, the surface integral vanishes and the expression (5.14) reduces to the result (5.9).

Note that Eq. (5.14) holds for observation points inside the volume. For exterior problems ($\mathbf{r} \in \mathcal{V}_{\mathrm{e}}$), consider that the domain $\mathcal{V}_{\mathrm{e}}$ is bounded by $\partial\mathcal{V}$ (the boundary to $\mathcal{V}$) and the far field boundary Γ_∞. The surface integral term over Γ_∞ vanishes and the surface integral over $\partial\mathcal{V}$ changes its sign in Eq. (5.14).

We will later see that the additional terms in the brackets of Eq. (5.14) represent the potentials due to single and double-layer sources on the domain boundary. For Neumann problems, where $\partial_{\mathbf{n}'}\phi(\mathbf{r}') = 0$ on the domain boundary $\mathbf{r}' \in \partial\mathcal{V}$, Eq. (5.14) reduces to

$$\phi(\mathbf{r}) = \int_{\mathcal{V}} G(\mathbf{r},\mathbf{r}') f(\mathbf{r}')\, \mathrm{d}V' - \int_{\partial\mathcal{V}} \phi(\mathbf{r}')\, \partial_{\mathbf{n}'} G(\mathbf{r},\mathbf{r}')\, \mathrm{d}a' \,, \quad (5.15)$$

which can be solved for boundary values of $\phi(\mathbf{r}')$. For Dirichlet problems, where $\phi(\mathbf{r}') = 0$ on the domain boundary, Eq. (5.14) reduces to

$$\phi(\mathbf{r}) = \int_{\mathcal{V}} G(\mathbf{r},\mathbf{r}') f(\mathbf{r}')\, \mathrm{d}V' + \int_{\partial\mathcal{V}} G(\mathbf{r},\mathbf{r}')\, \partial_{\mathbf{n}'}\phi(\mathbf{r}')\, \mathrm{d}a' \,, \quad (5.16)$$

which can be solved for normal derivatives of the potential $\partial_{\mathbf{n}'}\phi(\mathbf{r}')$ on the domain boundary. Consequently, we may not prescribe both ϕ and $\partial_{\mathbf{n}'}\phi$ in Kirchhoff's theorem because the problem would be overdetermined.[5]

5 This would be equivalent to prescribing both the current and the voltage to the terminals of an electrical circuit.

For 3D magnetostatic problems we use the special version of Eq. (5.14) for the magnetic scalar potential,

$$\phi_{\mathrm{m}}(\mathbf{r}) = \frac{1}{4\pi\mu_0}\int_{\mathscr{V}} \frac{\rho_{\mathrm{m}}}{|\mathbf{r}-\mathbf{r}'|}\,\mathrm{d}V' + \frac{1}{4\pi}\int_{\partial\mathscr{V}}\left(\phi_{\mathrm{m}}\frac{(\mathbf{r}-\mathbf{r}')\cdot\mathbf{n}'}{|\mathbf{r}-\mathbf{r}'|^3} + \partial_{\mathbf{n}'}\phi_{\mathrm{m}}(\mathbf{r}')\frac{1}{|\mathbf{r}-\mathbf{r}'|}\right)\mathrm{d}a', \quad (5.17)$$

where Eq. (3.110) is used, and for a Cartesian component A_i of the magnetic vector potential

$$A_i(\mathbf{r}) = \frac{\mu_0}{4\pi}\int_{\mathscr{V}} \frac{J_i}{|\mathbf{r}-\mathbf{r}'|}\,\mathrm{d}V' + \frac{1}{4\pi}\int_{\partial\mathscr{V}}\left(A_i\frac{(\mathbf{r}-\mathbf{r}')\cdot\mathbf{n}'}{|\mathbf{r}-\mathbf{r}'|^3} + \partial_{\mathbf{n}'}A_i(\mathbf{r}')\frac{1}{|\mathbf{r}-\mathbf{r}'|}\right)\mathrm{d}a'. \quad (5.18)$$

For 2D problems the corresponding equations are

$$\phi_{\mathrm{m}}(\mathbf{r}) = \frac{1}{2\pi\mu_0}\int_{\mathscr{A}} \rho_{\mathrm{m}}\ln\left(\frac{|\mathbf{r}-\mathbf{r}'|}{r_{\mathrm{ref}}}\right)\mathrm{d}a' + \frac{1}{2\pi}\int_{\partial\mathscr{A}}\left(-\phi_{\mathrm{m}}\frac{(\mathbf{r}-\mathbf{r}')\cdot\mathbf{n}'}{|\mathbf{r}-\mathbf{r}'|^2} + \partial_{\mathbf{n}'}\phi_{\mathrm{m}}(\mathbf{r}')\ln\left(\frac{|\mathbf{r}-\mathbf{r}'|}{r_{\mathrm{ref}}}\right)\right)\mathrm{d}\mathbf{r}', \quad (5.19)$$

$$A_z(\mathbf{r}) = \frac{\mu_0}{2\pi}\int_{\mathscr{A}} J_z\ln\left(\frac{|\mathbf{r}-\mathbf{r}'|}{r_{\mathrm{ref}}}\right)\mathrm{d}a' + \frac{1}{2\pi}\int_{\partial\mathscr{A}}\left(-A_z\frac{(\mathbf{r}-\mathbf{r}')\cdot\mathbf{n}'}{|\mathbf{r}-\mathbf{r}'|^2} + \partial_{\mathbf{n}'}A_z(\mathbf{r}')\ln\left(\frac{|\mathbf{r}-\mathbf{r}'|}{r_{\mathrm{ref}}}\right)\right)\mathrm{d}\mathbf{r}'. \quad (5.20)$$

5.3 Properties of Harmonic Fields

A scalar field $\phi \in \mathcal{S}(\Omega)$, which obeys the Laplace equation $\nabla^2\phi = 0$ and converges uniformly to ϕ_∞ at infinity,[6] is said to be harmonic in Ω.

A vector field $\mathbf{g} \in \mathcal{V}(\Omega)$, which obeys the equations $\operatorname{div}\mathbf{g} = 0$ and $\operatorname{curl}\mathbf{g} = 0$ and converges uniformly to $\mathbf{g}_\infty$ at infinity, is said to be harmonic in Ω. Harmonic vector fields are the 3D generalizations of the holomorphic functions on the complex plane.[7] The set of harmonic fields is the kernel of the Laplace operator and therefore a vector space. Thus sums, differences, and scalar multiples of harmonic functions are harmonic.

6 Often $\phi_\infty = 0$.

7 For more discussion on holomorphic functions and applications to magnet design, see Chapter 9.

This is then the right moment to present the proofs of some important theorems on harmonic scalar fields.

Theorem 5.1 *If ϕ is harmonic in the closed contractible volume $\mathcal{V} \subset \Omega$ bounded by the surface $\partial\mathcal{V}$, the surface integral of the normal derivative of ϕ vanishes.*

Proof. From Green's first identity (see Section 3.10.5) we obtain

$$\int_{\partial\mathcal{V}} \partial_{\mathbf{n}}\phi \,\mathrm{d}a = \int_{\mathcal{V}} \nabla^2\phi \,\mathrm{d}V = 0\,. \tag{5.21}$$

□

Theorem 5.2 *If ϕ is harmonic in the closed, contractible volume $\mathcal{V} \subset \Omega$, bounded by the surface $\partial\mathcal{V}$, with the same magnitude at all points on that surface, then ϕ is constant throughout $\mathcal{V}$ and equal to its value ϕ_0 on the boundary.*

Proof. Again Green's first identity yields

$$\phi_0 \int_{\partial\mathcal{V}} \partial_{\mathbf{n}}\phi \,\mathrm{d}a = \int_{\mathcal{V}} |\operatorname{grad}\phi|^2 \,\mathrm{d}V + \int_{\mathcal{V}} \phi\nabla^2\phi \,\mathrm{d}V\,. \tag{5.22}$$

Hence, using Theorem 5.1 yields $\int_{\mathcal{V}} |\operatorname{grad}\phi|^2 \,\mathrm{d}V = 0$. Since $|\operatorname{grad}\phi|$ is positive or zero in the entire domain, it must be zero if the volume integral is zero. Thus $\operatorname{grad}\phi$ is zero in the domain and consequently $\phi = \phi_0$ in $\mathcal{V}$. □

It follows from the same arguments that if the normal derivative $\partial_{\mathbf{n}}\phi$ is zero on the closed boundary, the potential is constant within the volume.

Theorem 5.3 *If ϕ is harmonic in the closed contractible volume $\mathcal{V} \subset \Omega$ bounded by $\partial\mathcal{V}$ and its value is specified at each point of that boundary, then ϕ is uniquely determined at all points inside the volume.*

Proof. Let $\psi \in \mathcal{S}(\Omega)$ be another scalar field that is harmonic in $\mathcal{V} \subset \Omega$, with the same magnitude as ϕ on the boundary $\partial\mathcal{V}$. In this case $\psi - \phi = 0$ at each point of the boundary, and $\nabla^2(\psi - \phi) = \nabla^2\psi - \nabla^2\phi = 0$ in $\mathcal{V}$. Consequently,

$$\int_{\mathcal{V}} |\operatorname{grad}(\psi - \phi)| \,\mathrm{d}V = \int_{\mathcal{V}} |\operatorname{grad}\psi - \operatorname{grad}\phi| \,\mathrm{d}V = 0\,. \tag{5.23}$$

Thus $\operatorname{grad}\psi = \operatorname{grad}\phi$ at all interior points of $\mathcal{V}$, and ϕ and ψ can differ only by a constant. But $\phi = \psi$ on the boundary, and therefore ϕ and ψ must be equal in the domain. Consequently, ϕ is uniquely determined. □

Theorem 5.4 (Maximum principle) *If ϕ is harmonic in a closed contractible volume $\mathcal{V} \subset \Omega$, ϕ cannot take a maximum or a minimum at any interior point of $\mathcal{V}$.*

Proof. Suppose ϕ has a maximum at an interior point $\mathscr{P}$ and $\partial\mathscr{V}_\mathrm{s}$ is the boundary surface of a sphere $\mathscr{V}_\mathrm{s}$ of radius R centered at $\mathscr{P}$. For a sufficiently small R the normal derivative $\partial_\mathbf{n}\phi$ must be negative for $\mathbf{n}$ pointing from inside out. But now

$$\int_{\mathscr{V}_\mathrm{s}} \partial_\mathbf{n}\phi \,\mathrm{d}V = \int_{\partial\mathscr{V}_\mathrm{s}} \operatorname{grad}\phi \cdot \mathrm{d}\mathbf{a} = \int_{\mathscr{V}_\mathrm{s}} \nabla^2\phi \,\mathrm{d}V = 0\,. \tag{5.24}$$

Hence the assumption that ϕ is maximum at $\mathscr{P}$ involves a contradiction. The reasoning holds for ϕ taking its minimum value at $\mathscr{P}$. □

Theorem 5.5 (Liouville) *If ϕ is a harmonic scalar field in E_n with an upper (or lower) bound, ϕ is constant.*

Liouville's theorem is a direct consequence of the mean value theorem and states that harmonic fields in an unbounded domain are trivial. We must therefore study singularities of harmonic fields. For magnetostatic problems, these are given by line-currents and magnetic double layers. Less rigorously, we can say that harmonic functions are determined by their singularities.

From mechanics we know that a massive object cannot achieve stable equilibrium unless its potential energy is a minimum. This theorem can be adapted to the equilibrium position of a positive point charge with energy proportional to the electric potential from other charges. But because this potential cannot have a minimum in $\mathscr{V}$, we obtain Theorem 5.6:

Theorem 5.6 (Earnshaw)[8] *A point charge can never be maintained in a stable stationary equilibrium by the electrostatic interaction of charges alone.*

5.4 The Biot–Savart Law

In three-dimensional Cartesian coordinates the vector *Poisson equation*,[9]

$$\nabla^2\mathbf{A} = -\mu_0\mathbf{J}\,, \tag{5.25}$$

can be separated into three scalar equations with the free-space solutions

$$A_i(\mathbf{r}) = \frac{\mu_0}{4\pi}\int_{\mathscr{V}} \frac{J_i(\mathbf{r}')}{|\mathbf{r}-\mathbf{r}'|}\,\mathrm{d}V'\,, \tag{5.26}$$

of the Cartesian components of the vector potential and the current density, A_i and J_i, respectively.

8 Samuel Earnshaw (1805–1888).
9 See Section 4.9.2.

Proof. Taking

$$A_i(\mathbf{r}) = \frac{\mu_0}{4\pi} \int_{\mathcal{V}} \frac{J_i(\mathbf{r}')}{|\mathbf{r}-\mathbf{r}'|} \, \mathrm{d}V' \tag{5.27}$$

and applying the Laplace operator yields[10]

$$\begin{aligned} \nabla^2 A_i(\mathbf{r}) &= \frac{\mu_0}{4\pi} \int_{\mathcal{V}} \nabla^2 \frac{J_i(\mathbf{r}')}{|\mathbf{r}-\mathbf{r}'|} \mathrm{d}V' = \frac{\mu_0}{4\pi} \int_{\mathcal{V}} J_i(\mathbf{r}') \nabla^2_{\mathbf{r}} \frac{1}{|\mathbf{r}-\mathbf{r}'|} \, \mathrm{d}V' \\ &= \frac{\mu_0}{4\pi} \int_{\mathcal{V}} J_i(\mathbf{r}')(-4\pi)\delta(\mathbf{r}-\mathbf{r}') \, \mathrm{d}V' = -\mu_0 J_i(\mathbf{r}) \,. \end{aligned} \tag{5.28}$$

□

Assembling the components of Eq. (5.26) gives

$$\mathbf{A}(\mathbf{r}) = A_x \mathbf{e_x} + A_y \mathbf{e_y} + A_z \mathbf{e_z} = \frac{\mu_0}{4\pi} \int_{\mathcal{V}} \frac{\mathbf{J}(\mathbf{r}')}{|\mathbf{r}-\mathbf{r}'|} \, \mathrm{d}V'. \tag{5.29}$$

From Eqs. (5.29) and (3.69) we can derive,

$$\begin{aligned} \mathbf{B}(\mathbf{r}) &= \operatorname{curl} \mathbf{A}(\mathbf{r}) = \frac{\mu_0}{4\pi} \int_{\mathcal{V}} \operatorname{curl} \left(\frac{\mathbf{J}(\mathbf{r}')}{|\mathbf{r}-\mathbf{r}'|} \right) \mathrm{d}V' \\ &= \frac{\mu_0}{4\pi} \int_{\mathcal{V}} \left(\frac{1}{|\mathbf{r}-\mathbf{r}'|} \operatorname{curl} \mathbf{J}(\mathbf{r}') - \mathbf{J}(\mathbf{r}') \times \operatorname{grad} \left(\frac{1}{|\mathbf{r}-\mathbf{r}'|} \right) \right) \mathrm{d}V' \\ &= \frac{\mu_0}{4\pi} \int_{\mathcal{V}} \frac{\mathbf{J}(\mathbf{r}') \times (\mathbf{r}-\mathbf{r}')}{|\mathbf{r}-\mathbf{r}'|^3} \, \mathrm{d}V' \,. \end{aligned} \tag{5.30}$$

Remark: In Cartesian coordinates, the vector Laplace equation decomposes into three scalar equations. Thus Eq. (5.30) can be used directly to calculate the three Cartesian components of the magnetic flux density. However, in using curvilinear coordinates one must account for the position-dependence of the basis vectors. This can be accomplished by writing

$$\mathbf{A}(\mathbf{r}) = \sum_{i=1}^{3} A_i(\mathbf{r}) \mathbf{e}_i(\mathbf{r}) \,, \qquad \mathbf{J}(\mathbf{r}') = \sum_{k=1}^{3} J_k(\mathbf{r}') \mathbf{e}_k(\mathbf{r}') \,. \tag{5.31}$$

Substituting Eq. (5.31) into Eq. (5.29) yields

$$\mathbf{A}(\mathbf{r}) = \sum_{i=1}^{3} A_i(\mathbf{r}) \mathbf{e}_i(\mathbf{r}) = \frac{\mu_0}{4\pi} \int_{\mathcal{V}} \frac{1}{|\mathbf{r}-\mathbf{r}'|} \sum_{k=1}^{3} J_k(\mathbf{r}') \mathbf{e}_k(\mathbf{r}') \, \mathrm{d}V' \,. \tag{5.32}$$

10 See Section 3.7 for differential operators acting on functions of $\mathbf{r}$ and $\mathbf{r}'$.

For $A_i(\mathbf{r}) = \mathbf{e}_i(\mathbf{r}) \cdot \mathbf{A}(\mathbf{r})$, we finally obtain the general formula for the ith component function of the vector potential in a curvilinear coordinate system:

$$A_i(\mathbf{r}) = \frac{\mu_0}{4\pi} \int_{\mathscr{V}} \frac{1}{|\mathbf{r} - \mathbf{r}'|} \sum_{k=1}^{3} J_k(\mathbf{r}')(\mathbf{e}_i(\mathbf{r}) \cdot \mathbf{e}_k(\mathbf{r}'))\, \mathrm{d}V' . \tag{5.33}$$

For Cartesian coordinates this yields Eq. (5.26), as expected. □

Poisson's equation (5.25) was derived for free space incorporating Coulomb's gauge ($\operatorname{div} \mathbf{A} = 0$). It must therefore be proved that the vector potential of Eq. (5.29) is indeed source free: Using the identity (3.67) we obtain

$$\begin{aligned}
\operatorname{div} \mathbf{A}(\mathbf{r}) &= \frac{\mu_0}{4\pi} \int_{\mathscr{V}} \operatorname{div} \left(\frac{\mathbf{J}(\mathbf{r}')}{|\mathbf{r} - \mathbf{r}'|} \right) \mathrm{d}V' \\
&= \frac{\mu_0}{4\pi} \int_{\mathscr{V}} \left(\mathbf{J}(\mathbf{r}') \cdot \operatorname{grad} \left(\frac{1}{|\mathbf{r} - \mathbf{r}'|} \right) + \frac{1}{|\mathbf{r} - \mathbf{r}'|} \operatorname{div} \mathbf{J}(\mathbf{r}') \right) \mathrm{d}V' \\
&= \frac{\mu_0}{4\pi} \int_{\mathscr{V}} \mathbf{J}(\mathbf{r}') \cdot \operatorname{grad} \left(\frac{1}{|\mathbf{r} - \mathbf{r}'|} \right) \mathrm{d}V' \\
&= -\frac{\mu_0}{4\pi} \int_{\mathscr{V}} \mathbf{J}(\mathbf{r}') \cdot \operatorname{grad}_{\mathbf{r}'} \left(\frac{1}{|\mathbf{r} - \mathbf{r}'|} \right) \mathrm{d}V' \\
&= -\frac{\mu_0}{4\pi} \int_{\mathscr{V}} \left(\operatorname{div}_{\mathbf{r}'} \left(\frac{\mathbf{J}(\mathbf{r}')}{|\mathbf{r} - \mathbf{r}'|} \right) - \frac{1}{|\mathbf{r} - \mathbf{r}'|} \operatorname{div}_{\mathbf{r}'} \mathbf{J}(\mathbf{r}') \right) \mathrm{d}V' \\
&= -\frac{\mu_0}{4\pi} \int_{\mathscr{V}} \operatorname{div}_{\mathbf{r}'} \left(\frac{\mathbf{J}(\mathbf{r}')}{|\mathbf{r} - \mathbf{r}'|} \right) \mathrm{d}V' = -\frac{\mu_0}{4\pi} \int_{\partial\mathscr{V}} \frac{\mathbf{J}(\mathbf{r}')}{|\mathbf{r} - \mathbf{r}'|} \cdot \mathrm{d}\mathbf{a}' .
\end{aligned} \tag{5.34}$$

We therefore must ensure that no current is escaping the volume $\mathscr{V}$.

If the current is confined to a thin wire with the loop $\mathscr{C}$ as its center-line, $\mathbf{J}(\mathbf{r}')$ is parallel to $\mathrm{d}\mathbf{r}'$ and thus[11]

$$\mathrm{d}\mathbf{a}' \times (\mathbf{J}(\mathbf{r}') \times \mathrm{d}\mathbf{r}') = (\mathrm{d}\mathbf{a}' \cdot \mathrm{d}\mathbf{r}')\, \mathbf{J}(\mathbf{r}') - (\mathbf{J}(\mathbf{r}') \cdot \mathrm{d}\mathbf{a}')\, \mathrm{d}\mathbf{r}' = 0 , \tag{5.35}$$

from which follows

$$\int_{\mathscr{V}} \mathbf{J}(\mathbf{r}')\, \mathrm{d}V' = \int_{\mathscr{C}} \left(\int_{\mathscr{A}} \mathbf{J}(\mathbf{r}') \cdot \mathrm{d}\mathbf{a}' \right) \mathrm{d}\mathbf{r}' = \int_{\mathscr{C}} I\, \mathrm{d}\mathbf{r}', \tag{5.36}$$

where $\mathscr{A}$ is a disk orthogonal to the loop $\mathscr{C}$. Equations (5.29) and (5.30) take the form

$$\mathbf{A}(\mathbf{r}) = \frac{\mu_0 I}{4\pi} \int_{\mathscr{C}} \frac{\mathrm{d}\mathbf{r}'}{|\mathbf{r} - \mathbf{r}'|} \tag{5.37}$$

11 Employing the relation (2.78).

and

$$\mathbf{B}(\mathbf{r}) = \frac{\mu_0 I}{4\pi} \int_{\mathscr{C}} \frac{\mathrm{d}\mathbf{r}' \times (\mathbf{r} - \mathbf{r}')}{|\mathbf{r} - \mathbf{r}'|^3}, \tag{5.38}$$

which is the *Biot–Savart law*.

In numerical field computation, current loops are often approximated by straight line-current segments. It is important to note that these segments must always form a closed polygonal approximation for the loop $\mathscr{C}$.

We are now in a position to formalize the calculation of the magnetomotive force (ampere-turns). Consider Ampère's law $\int_{\partial\mathscr{A}} \mathbf{H} \cdot \mathrm{d}\mathbf{r} = \int_{\mathscr{A}} \mathbf{J} \cdot \mathbf{da} = N I$ where $\partial\mathscr{A}$ is the boundary of an outer-oriented surface $\mathscr{A}$. Further let K currents of the same magnitude be confined to the loops $\mathscr{C}_k$, $k = 1, \ldots, K$, which intersect the surface $\mathscr{A}$ but do not intersect its boundary $\partial\mathscr{A}$; see Figure 5.1. In that case $N I$ is determined by the oriented *linking numbers* of $\partial\mathscr{A}$ and $\mathscr{C}_k$, which can in turn be expressed as the sum of oriented intersections of the loops $\mathscr{C}_k$ with the surface $\mathscr{A}$:

$$N I = I \sum_{k=1}^{K} \mathrm{link}(\partial\mathscr{A}, \mathscr{C}_k) = I \sum_{k=1}^{K} \mathrm{int}(\mathscr{A}, \mathscr{C}_k), \tag{5.39}$$

where $\mathrm{int}(\mathscr{A}, \mathscr{C}_k) = \sum_{\mathscr{A} \cap \mathscr{C}_k} \pm 1$ is the intersection number. Inserting the Biot–Savart law into Ampère's law yields

$$\mathrm{link}(\partial\mathscr{A}, \mathscr{C}_k) = \frac{1}{4\pi} \int_{\partial\mathscr{A}} \int_{\mathscr{C}_k} \frac{\mathrm{d}\mathbf{r}' \times (\mathbf{r} - \mathbf{r}')}{|\mathbf{r} - \mathbf{r}'|^3} \cdot \mathrm{d}\mathbf{r}, \tag{5.40}$$

an expression easy to program but costly to compute for real-world applications.

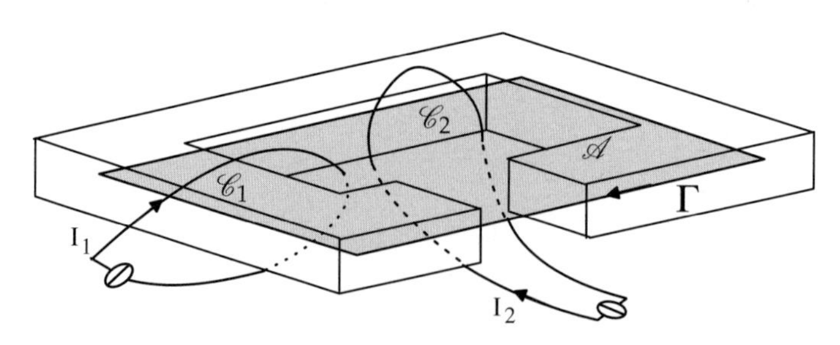

Figure 5.1 On the oriented linking number of curves $\partial\mathscr{A}$ and $\mathscr{C}_k$.

5.5 Field of a Straight Line-Current Segment

Consider a line-current segment along the z-axis from $z_c = a$ to $z_c = b$ pointing in the positive z-direction as shown in Figure 5.2 (left), so that $A_x = A_y = 0$. From Eq. (5.37) we obtain

$$
\begin{aligned}
A_z(x, y, z) &= \frac{\mu_0 I}{4\pi} \int_a^b \frac{\mathrm{d}z_c}{|\mathbf{r} - \mathbf{r}'|} = \frac{\mu_0 I}{4\pi} \int_a^b \frac{\mathrm{d}z_c}{\sqrt{x^2 + y^2 + (z - z_c)^2}} \\
&= \frac{-\mu_0 I}{4\pi} \ln \left((z - z_c) + \sqrt{x^2 + y^2 + (z - z_c)^2} \right) \Bigg|_a^b \\
&= \frac{\mu_0 I}{4\pi} \ln \frac{z - a + \sqrt{x^2 + y^2 + (z - a)^2}}{z - b + \sqrt{x^2 + y^2 + (z - b)^2}} .
\end{aligned}
\tag{5.41}
$$

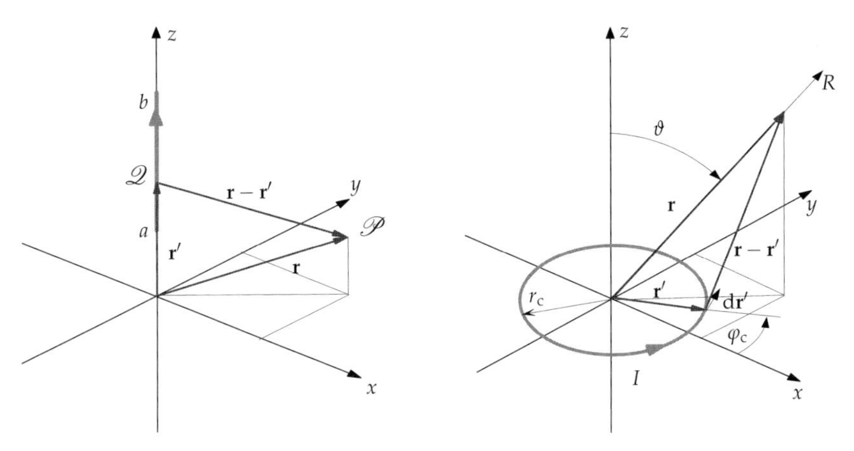

Figure 5.2 Geometrical relations for a straight line-current segment (left) and a ring current (right) centered at the origin.

For $a \to -\infty, b \to +\infty$ one obtains the limiting value[12]

$$
\begin{aligned}
\lim_{a,b \to \pm\infty} \ln \frac{z - a + \sqrt{x^2 + y^2 + (z - a)^2}}{z - b + \sqrt{x^2 + y^2 + (z - b)^2}} &= \lim_{a,b \to \pm\infty} \ln \frac{-a + |a|\sqrt{1 + \frac{x^2+y^2}{a^2}}}{-b + |b|\sqrt{1 + \frac{x^2+y^2}{b^2}}} \\
= \lim_{a,b \to \pm\infty} \ln \frac{-a - a(1 + \frac{x^2+y^2}{2a^2} + \cdots)}{-b + b(1 + \frac{x^2+y^2}{2b^2} + \cdots)} &= \lim_{a,b \to \pm\infty} \ln \frac{-2a}{-b + b + \frac{x^2+y^2}{2b}} \\
= \lim_{a,b \to \pm\infty} \ln \frac{-4ab}{x^2 + y^2} .
\end{aligned}
\tag{5.42}
$$

12 Using the binomial series $\sqrt{1+x} = 1 + \frac{1}{2}x - \frac{1}{8}x^2 + \cdots$ for $|x| \leq 1$ and writing $a, b \to \pm\infty$ instead of $a \to -\infty, b \to +\infty$.

An infinitely long line-current cannot have a finite vector potential at a finite point $\mathscr{P}$. Just like the homogeneous field on the entire E_3, this is an example of a nonregular field.[13]

Introducing an arbitrarily large reference radius r_{ref} with $r_{\text{ref}}^2 = x_0^2 + y_0^2$, we can extract an infinite constant and write

$$A_z(x,y) = \lim_{a,b \to \pm\infty} \frac{\mu_0 I}{4\pi} \ln \left(\frac{-4ab}{x_0^2 + y_0^2} \right) - \frac{\mu_0 I}{4\pi} \ln \left(\frac{x^2 + y^2}{x_0^2 + y_0^2} \right) . \tag{5.43}$$

The constant term can be gauged out and the vector potential for each line current can be written as

$$\mathbf{A}(x,y) = -\frac{\mu_0 I}{4\pi} \ln \left(\frac{x^2 + y^2}{x_0^2 + y_0^2} \right) \mathbf{e}_z = -\frac{\mu_0 I}{2\pi} \ln \left(\frac{r}{r_{\text{ref}}} \right) \mathbf{e}_z \,, \tag{5.44}$$

where r is the distance between the field point $\mathscr{P}$ with coordinates (x,y) and the line-current at the z-axis. The reference radius is exactly the radius at which the potential has dropped to zero. From an engineering point of view, we need not be troubled with the size of the reference radius, as long as the sum of all currents vanishes.

From the vector potential (5.43) the magnetic field components can be calculated:

$$B_x(x,y) = \frac{\partial A_z}{\partial y} - \frac{\partial A_y}{\partial z} = -\frac{\mu_0 I}{2\pi} \frac{y}{x^2 + y^2} \,, \tag{5.45}$$

$$B_y(x,y) = \frac{\partial A_x}{\partial z} - \frac{\partial A_z}{\partial x} = \frac{\mu_0 I}{2\pi} \frac{x}{x^2 + y^2} \,, \tag{5.46}$$

and $B_z(x,y) = \frac{\partial A_y}{\partial x} - \frac{\partial A_x}{\partial y} = 0$. Using the transformations $B_r = B_x \cos\varphi + B_y \sin\varphi$ and $B_\varphi = -B_x \sin\varphi + B_y \cos\varphi$, and the relations

$$\cos\varphi = \frac{x}{r} = \frac{x}{\sqrt{x^2 + y^2}} \,, \qquad \sin\varphi = \frac{y}{r} = \frac{y}{\sqrt{x^2 + y^2}} \,, \tag{5.47}$$

the field can be expressed in cylindrical coordinates as

$$B_r(r,\varphi) = 0, \qquad B_\varphi(r,\varphi) = \frac{\mu_0 I}{2\pi r}, \qquad B_z(r,\varphi) = 0. \tag{5.48}$$

We can also calculate the magnetic vector potential perpendicular to the surface spanned by the line-current segment $\mathscr{L}$ and the field point $\mathscr{P}$ according to the geometrical relations shown in Figure 5.3. We find $r_{\mathscr{P}\mathscr{Q}} := |\mathbf{r} - \mathbf{r}'| = R/\cos\alpha$ and thus

$$\mathrm{d}r' = \frac{R}{\cos^2\alpha} \mathrm{d}\alpha \,, \qquad \mathrm{d}\mathbf{r}' \times (\mathbf{r} - \mathbf{r}') = r_{\mathscr{Q}\mathscr{P}} \cos\alpha \, \mathbf{n} \, \mathrm{d}r' \,. \tag{5.49}$$

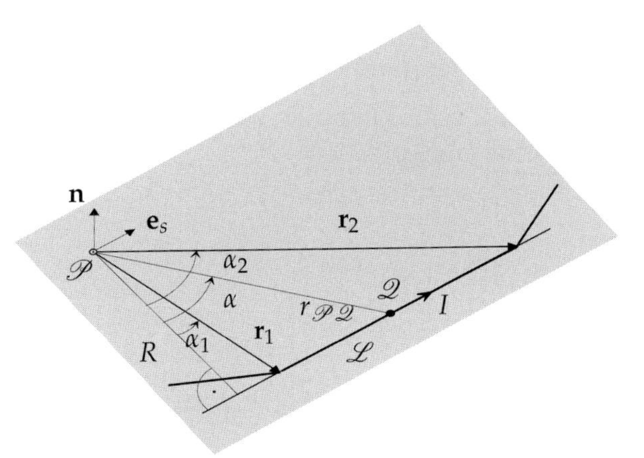

Figure 5.3 Geometrical relations for the calculation of the field in the plane spanned by the line-current segment and the field point $\mathscr{P}$.

The vector potential for the line-current segment $\mathscr{L}$ is

$$\begin{aligned} \mathbf{A}(\mathscr{P}) &= \frac{\mu_0 I}{4\pi} \int_{\mathscr{L}} \frac{1}{r_{\mathscr{P}\mathscr{Q}}} \, \mathrm{d}\mathbf{r}' = \frac{\mu_0 I}{4\pi} \mathbf{e}_s \int_{\alpha_1}^{\alpha_2} \frac{1}{\cos\alpha} \, \mathrm{d}\alpha \\ &= \frac{\mu_0 I}{4\pi} \ln \tan\left(\frac{\alpha}{2} + \frac{\pi}{4}\right)\Bigg|_{\alpha_1}^{\alpha_2} \mathbf{e}_s = \frac{\mu_0 I}{4\pi} \ln \sqrt{\frac{1-\sin\alpha_1}{1+\sin\alpha_1} \frac{1+\sin\alpha_2}{1-\sin\alpha_2}} \mathbf{e}_s \,. \end{aligned} \tag{5.50}$$

For the magnetic flux density, we derive

$$\begin{aligned} \mathbf{B}(\mathscr{P}) &= \frac{\mu_0 I}{4\pi} \int_{\mathscr{L}} \frac{\cos\alpha}{r^2_{\mathscr{P}\mathscr{Q}}} \, \mathrm{d}\mathbf{r}' = \frac{\mu_0 I}{4\pi R} \mathbf{n} \int_{\alpha_1}^{\alpha_2} \cos\alpha \, \mathrm{d}\alpha = \frac{\mu_0 I}{4\pi R} (\sin\alpha_2 - \sin\alpha_1) \, \mathbf{n} \\ &= \frac{\mu_0 I}{4\pi} \frac{\cos\alpha_2 + \cos\alpha_1}{R} \frac{\sin\alpha_2 - \sin\alpha_1}{\cos\alpha_2 + \cos\alpha_1} \mathbf{n} \\ &= \frac{\mu_0 I}{4\pi} \left(\frac{1}{|\mathbf{r}_1|} + \frac{1}{|\mathbf{r}_2|}\right) \frac{\sin(\alpha_2 - \alpha_1)}{1 + \cos(\alpha_2 - \alpha_1)} \mathbf{n} \\ &= \frac{\mu_0 I}{4\pi} \left(\frac{1}{|\mathbf{r}_1|} + \frac{1}{|\mathbf{r}_2|}\right) \frac{\sin(\alpha_2 - \alpha_1)}{1 + \frac{\mathbf{r}_1 \cdot \mathbf{r}_2}{|\mathbf{r}_1| \, |\mathbf{r}_2|}} \frac{\mathbf{r}_1 \times \mathbf{r}_2}{|\mathbf{r}_1| \, |\mathbf{r}_2| \sin(\alpha_2 - \alpha_1)} \\ &= \frac{\mu_0 I}{4\pi} \frac{|\mathbf{r}_1| + |\mathbf{r}_2|}{|\mathbf{r}_1| \, |\mathbf{r}_2| + \mathbf{r}_1 \cdot \mathbf{r}_2} \frac{\mathbf{r}_1 \times \mathbf{r}_2}{|\mathbf{r}_1| \, |\mathbf{r}_2|} \,, \end{aligned} \tag{5.51}$$

an easy equation to program, employing only the vectors from the field point to the boundary points of the line-current segment.

13 A regular vector field has a potential for which $\lim_{r\to\infty} |\mathbf{r}| \mathbf{A}(\mathbf{r})$ remains finite.

5.6
Field of a Ring Current

Consider a circular loop current $\mathscr{C}$ of radius r_c in the xy-plane according to Figure 5.2 (right), with its parametric representation

$$\mathbf{r}' = \cos\varphi_c r_c\,\mathbf{e}_x + \sin\varphi_c r_c\,\mathbf{e}_y \tag{5.52}$$

for $\varphi_c \in [0, 2\pi]$. The vectorial line element is

$$\mathrm{d}\mathbf{r}' = -\sin\varphi_c r_c\,\mathrm{d}\varphi_c \mathbf{e}_x + \cos\varphi_c r_c\,\mathrm{d}\varphi_c \mathbf{e}_y\,. \tag{5.53}$$

Because of the symmetry, we can set the field point at any angular position, and thus without loss of generality set $\varphi = 0$. Assuming $z_c = 0$ this yields[14]

$$\begin{aligned}
|\mathbf{r} - \mathbf{r}'| &= \sqrt{(x - x_c)^2 + (y - y_c)^2 + z^2} \\
&= \sqrt{(r\cos\varphi - r_c\cos\varphi_c)^2 + (r\sin\varphi - r_c\sin\varphi_c)^2 + z^2} \\
&= \sqrt{r^2 + r_c^2 + z^2 - 2rr_c\cos\varphi_c}\,,
\end{aligned} \tag{5.54}$$

and the components of the vector potential can be calculated from

$$A_y(r,z) = \frac{\mu_0 I r_c}{2\pi}\int_0^{\pi} \frac{\cos\varphi_c\,\mathrm{d}\varphi_c}{\sqrt{r^2 + r_c^2 + z^2 - 2rr_c\cos\varphi_c}}\,, \tag{5.55}$$

$$A_x(r,z) = \frac{\mu_0 I r_c}{2\pi}\int_0^{\pi} \frac{-\sin\varphi_c\,\mathrm{d}\varphi_c}{\sqrt{r^2 + r_c^2 + z^2 - 2rr_c\cos\varphi_c}} = 0\,. \tag{5.56}$$

Now we can return to the components in cylindrical coordinates, and following Lehner [12] substitute $\psi := (\pi + \varphi_c)/2$, so that $\cos\varphi_c = 2\sin^2\psi - 1$. Employing the definition

$$k^2 := \frac{4rr_c}{(r + r_c)^2 + z^2} \tag{5.57}$$

we obtain from Eq. (5.55)

$$A_\varphi(r,z) = \frac{\mu_0 I r_c}{\pi\sqrt{(r + r_c)^2 + z^2}}\int_0^{\pi/2} \frac{2\sin^2\psi - 1}{\sqrt{1 - k^2\sin^2\psi}}\,\mathrm{d}\psi\,. \tag{5.58}$$

Remark 1: We have referred to the canonical basis isomorphism, Eq. (5.26), in order to avoid calculation errors due to the position-dependence of the curvilinear basis; we refer to the remark in Section 5.4 . An alternative method is

14 Here r is not $|\mathbf{r}|$, and therefore ρ is often used for this coordinate. See Figure 5.2 (right) for the geometrical relations.

to apply Eq. (5.33) while considering that $\mathbf{e}_\varphi(\mathbf{r}) \cdot \mathbf{e}_\varphi(\mathbf{r}') = \cos(\varphi(\mathbf{r}) - \varphi(\mathbf{r}')) = \cos\varphi_c$. □

The integral on the right-hand side of Eq. (5.58) can be written as

$$\begin{aligned}
&\int_0^{\pi/2} \frac{2\sin^2\psi - 1 + \left(\frac{2}{k^2} - \frac{2}{k^2}\right)}{\sqrt{1-k^2\sin^2\psi}}\,\mathrm{d}\psi = \\
&= -\frac{2}{k^2}\int_0^{\pi/2} \frac{1-k^2\sin^2\psi}{\sqrt{1-k^2\sin^2\psi}}\,\mathrm{d}\psi + \left(\frac{2}{k^2}-1\right)\int_0^{\pi/2}\frac{\mathrm{d}\psi}{\sqrt{1-k^2\sin^2\psi}} \\
&= \left(\frac{2}{k^2}-1\right)\int_0^{\pi/2}\frac{\mathrm{d}\psi}{\sqrt{1-k^2\sin^2\psi}} - \frac{2}{k^2}\int_0^{\pi/2}\sqrt{1-k^2\sin^2\psi}\,\mathrm{d}\psi\,. \qquad (5.59)
\end{aligned}$$

Using the definitions

$$K\left(\frac{\pi}{2},k\right) := \int_0^{\pi/2}\frac{\mathrm{d}\psi}{\sqrt{1-k^2\sin^2\psi}}\,, \qquad (5.60)$$

$$E\left(\frac{\pi}{2},k\right) := \int_0^{\pi/2}\sqrt{1-k^2\sin^2\psi}\,\mathrm{d}\psi\,, \qquad (5.61)$$

this yields

$$A_\varphi(r,z) = \frac{\mu_0 I}{2\pi r}\sqrt{(r+r_c)^2+z^2}\left[\left(1-\frac{k^2}{2}\right)K\left(\frac{\pi}{2},k\right) - E\left(\frac{\pi}{2},k\right)\right]. \qquad (5.62)$$

Equations (5.60) and (5.61) are the *complete elliptic integrals* of the first and second kind [2] in the *Legendre form*;[15] see Figure 5.4. The parameter $k \in [0,1]$ in these expressions is called the *modulus*. No closed-form solution of these integrals exists. We will therefore continue to study special cases.

Remark 2: Why are the expressions (5.60) known as elliptic integrals? This is because they appear when calculating the arc length of an ellipse traced out by $\mathbf{r}(t) = a\cos t\,\mathbf{e}_x + b\sin t\,\mathbf{e}_y$. With the norm of the velocity vector to this space curve given by $|\mathbf{v}| = \sqrt{a^2\sin^2 t + b^2\cos^2 t}$, we obtain the integral for the arc length

$$\begin{aligned}
s(t) &= \int_0^t |\mathbf{v}(\tau)|\mathrm{d}\tau = \int_0^t b\sqrt{1+\frac{a^2-b^2}{b^2}\sin^2\tau}\,\mathrm{d}\tau \\
&= \int_0^t b\sqrt{1-e^2\sin^2\tau}\,\mathrm{d}\tau\,, \qquad (5.63)
\end{aligned}$$

15 Adrien–Marie Legendre (1752–1833).

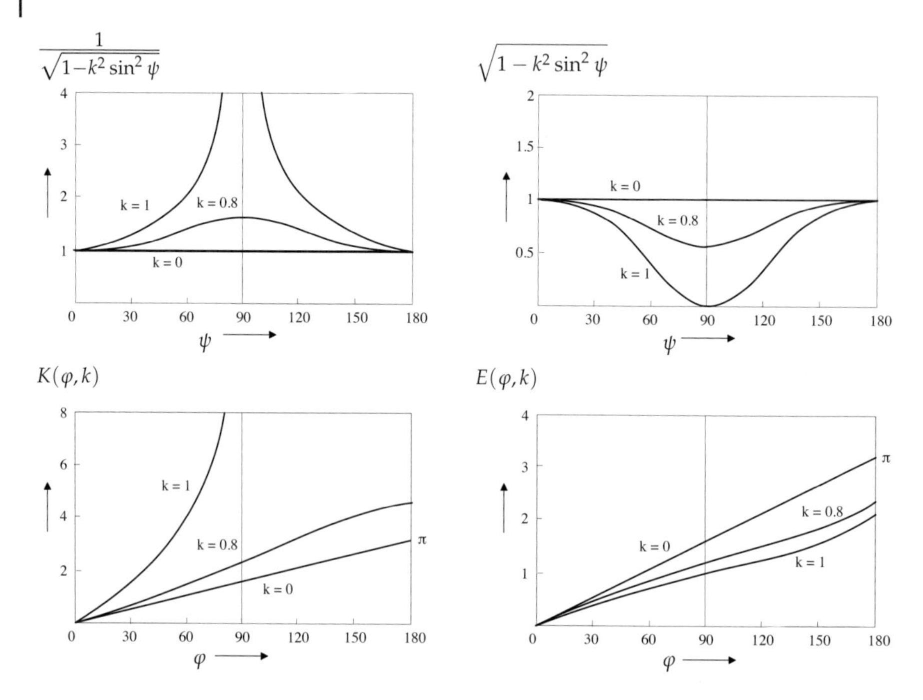

Figure 5.4 Top: Integrands of the elliptic integrals of the first (left) and of the second kind (right). Bottom: Elliptic integrals of the first (left) and the second kind (right) for the limits 0 to φ.

where

$$e := \sqrt{1 - \frac{a^2}{b^2}} \tag{5.64}$$

is called the *elliptic modulus* or the *eccentricity* with $0 \leq e < 1$ for $b \geq a$. □

The general case By a series expansion of the integrand and integration of the elements we obtain

$$K\left(\frac{\pi}{2},k\right) = \frac{\pi}{2}\left[1 + \left(\frac{1}{2}\right)^2 k^2 + \left(\frac{1\cdot 3}{2\cdot 4}\right)^2 k^4 + \cdots + \left(\frac{(2n)!}{2^{2n}(n!)^2}\right)^2 k^{2n} + \cdots\right], \tag{5.65}$$

$$E\left(\frac{\pi}{2},k\right) = \frac{\pi}{2}\left[1 - \left(\frac{1}{2}\right)^2 k^2 - \left(\frac{1\cdot 3}{2\cdot 4}\right)^2 \frac{k^4}{3} - \cdots - \left(\frac{(2n)!}{2^{2n}(n!)^2}\right)^2 \frac{k^{2n}}{2n-1} - \cdots\right], \tag{5.66}$$

for $k^2 < 1$ and $k^2 \leq 1$, respectively. This yields approximations for the complete elliptic integrals, which can be used for the calculation of the excitation field of thick solenoidal magnets; see Section 8.7.

Differentiation of the elliptic integrals with respect to the modulus yields

$$\frac{\mathrm{d}K\left(\frac{\pi}{2},k\right)}{\mathrm{d}k}=\frac{E\left(\frac{\pi}{2},k\right)}{k(1-k^2)}-\frac{K\left(\frac{\pi}{2},k\right)}{k}, \tag{5.67}$$

$$\frac{\mathrm{d}E\left(\frac{\pi}{2},k\right)}{\mathrm{d}k}=\frac{E\left(\frac{\pi}{2},k\right)-K\left(\frac{\pi}{2},k\right)}{k}. \tag{5.68}$$

The components of the magnetic flux density can be calculated from $B_r(r,z)=-\frac{\partial A_\varphi}{\partial z}$ and $B_z(r,z)=\frac{1}{r}\frac{\partial A}{\partial r}(rA_\varphi)$, which yields

$$B_r(r,z)=\frac{\mu_0 I}{2\pi r}\frac{z}{\sqrt{(r+r_\mathrm{c})^2+z^2}}\left[-K\left(\frac{\pi}{2},k\right)+\frac{r_\mathrm{c}^2+r^2+z^2}{(r_\mathrm{c}-r)^2+z^2}E\left(\frac{\pi}{2},k\right)\right], \tag{5.69}$$

$$B_z(r,z)=\frac{\mu_0 I}{2\pi}\frac{1}{\sqrt{(r+r_\mathrm{c})^2+z^2}}\left[K\left(\frac{\pi}{2},k\right)+\frac{r_\mathrm{c}^2-r^2-z^2}{(r_\mathrm{c}-r)^2+z^2}E\left(\frac{\pi}{2},k\right)\right]. \tag{5.70}$$

On-axis field For $r \ll r_\mathrm{c}$ it follows that $k \ll 1$. Because the vector potential does not depend on φ or ψ we obtain

$$A_\varphi(r,z)=\frac{\mu_0 I r_\mathrm{c}^2}{4}\frac{r}{\left(r_\mathrm{c}^2+z^2\right)^{\frac{3}{2}}}, \qquad B_z(z)=\frac{\mu_0 I}{2}\frac{r_\mathrm{c}^2}{\left(r_\mathrm{c}^2+z^2\right)^{\frac{3}{2}}}. \tag{5.71}$$

In the center of the circle plane at $z=0$ this yields the simple expression

$$B_z(z=0)=\frac{\mu_0 I}{2r_\mathrm{c}}. \tag{5.72}$$

5.7 The Magnetic Dipole Moment

We will now calculate the magnetic flux density for a field point far from the ring current centered at the origin; $\sqrt{r^2+z^2} \gg r_\mathrm{c}$, $k \ll 1$. Hence we use only the first elements of the series expansions (5.65) and (5.66),

$$A_\varphi(r,z)\approx\frac{\mu_0 I}{2\pi r}\sqrt{(r+r_\mathrm{c})^2+z^2}\cdot\frac{\pi}{2}\left\{\left(1-\frac{k^2}{2}\right)\left[1+\frac{1}{4}k^2+\frac{9}{64}k^4\right]-\left[1-\frac{1}{4}k^2-\frac{3}{64}k^4\right]\right\}, \tag{5.73}$$

which can be reduced to

$$A_\varphi(r,z)\approx\frac{\mu_0 I}{4r}\sqrt{(r+r_\mathrm{c})^2+z^2}\,\frac{k^4}{16}=\frac{\mu_0 I}{4r}\frac{r^2 r_\mathrm{c}^2}{(r^2+z^2)^{3/2}}. \tag{5.74}$$

Expressing this far-field approximation in spherical coordinates R, ϑ, φ, with $R = \sqrt{r^2 + z^2}$ and $\sin\vartheta = r/R$, yields

$$A_\varphi(R,\vartheta) \approx \frac{\mu_0 I r_c^2 \pi}{4\pi} \frac{\sin\vartheta}{R^2} = \frac{\mu_0 m}{4\pi} \frac{\sin\vartheta}{R^2}, \tag{5.75}$$

where the *magnetic dipole moment*

$$m := I r_c^2 \pi \tag{5.76}$$

has been introduced; $[m] = 1\,\mathrm{A\,m^2}$. The field components can be calculated with $\mathbf{B} = \operatorname{curl}\mathbf{A}$ as

$$B_R(R,\vartheta) = \frac{1}{R\sin\vartheta}\frac{\partial}{\partial\vartheta}(\sin\vartheta A_\varphi) = \frac{\mu_0 m}{2\pi}\frac{\cos\vartheta}{R^3}, \tag{5.77}$$

$$B_\vartheta(R,\vartheta) = -\frac{1}{R}\frac{\partial}{\partial R}(R A_\varphi) = \frac{\mu_0 m}{4\pi}\frac{\sin\vartheta}{R^3}, \tag{5.78}$$

and $B_\varphi = 0$. The components of m can be arranged in a vector to give

$$\mathbf{m} = I\mathbf{a}, \tag{5.79}$$

where $\mathbf{a}$ is the surface vector of the ring current. It happens that Eq. (5.79) holds for any plane loop regardless of its shape. The surface is inner oriented by the sense of rotation of the current in the loop, and the surface normal vector is directed according to the right-handed screw rule. The more general definition of the magnetic moment is

$$\mathbf{m} := \frac{1}{2}\int_{\mathcal{V}} \mathbf{r} \times \mathbf{J}(\mathbf{r})\,\mathrm{d}V, \tag{5.80}$$

which for line-currents yields,

$$\mathbf{m} = \frac{I}{2}\int_{\mathcal{C}} \mathbf{r} \times \mathrm{d}\mathbf{r}, \tag{5.81}$$

where $\mathcal{C}$ denotes the current loop. The magnetization can be introduced as the volume density of a magnetic dipole moment [7],

$$\mathbf{M}(\mathbf{r}) := \frac{\mathrm{d}\mathbf{m}}{\mathrm{d}V} = \frac{1}{2}\mathbf{r} \times \mathbf{J}(\mathbf{r}), \tag{5.82}$$

where $[\mathbf{M}] = 1\,\mathrm{A\,m^{-1}}$. At a sufficiently large distance, the field of any source (currents or magnetic material) can be reduced to that of a magnetic dipole moment. This accords with Ampère's interpretation of magnetism as infinitesimal circulating currents.

5.8 The Magnetic Double Layer

Consider a current confined to a thin wire along a loop Γ and connected across a battery as shown in Figure 5.5. The domain of the wire is denoted Ω. In the domain $E_3 \setminus \Omega$ (outside the current-carrying wire) the field is irrotational ($\operatorname{curl} \mathbf{H} = 0$) and may therefore be expressed as the gradient of a magnetic scalar potential $\mathbf{H} = -\operatorname{grad} \phi_\mathrm{m}$. Right? First we note that both the exterior domain $E_3 \setminus \Omega$ as well as the wire domain Ω are not simply-connected. From

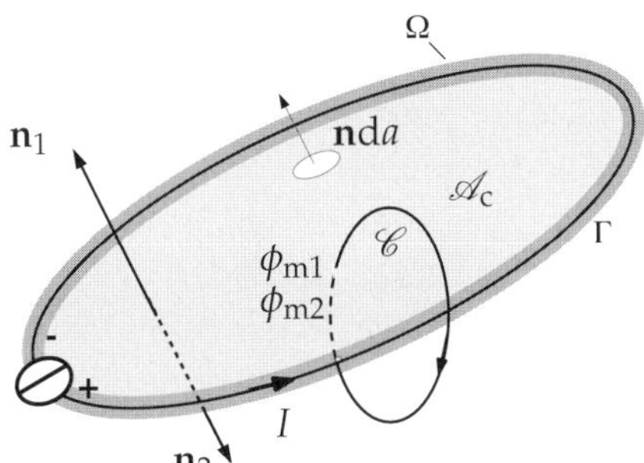

Figure 5.5 Magnetic double layer.

Ampère's law we obtain for any closed curve $\mathscr{C}$:

$$\int_{\mathscr{C}} \mathbf{H} \cdot \mathrm{d}\mathbf{r} = I \operatorname{link}(\Gamma, \mathscr{C}), \tag{5.83}$$

but

$$\int_{\mathscr{C}} \mathbf{H} \cdot \mathrm{d}\mathbf{r} = -\int_{\mathscr{C}} \operatorname{grad} \phi_\mathrm{m} \cdot \mathrm{d}\mathbf{r} = \int_{\partial \mathscr{C}} \phi_\mathrm{m} = 0, \tag{5.84}$$

which is a contradiction, and thus no single-valued scalar potential exists. It is usually said that the condition for representing a curl–free vector field with a single-valued scalar potential is to make the problem domain simply-connected by introducing a cut-surface. But it is not necessary that the cuts make the domain simply-connected. As a counter-example the cut-surface of a trefoil knot is discussed in [4]. It is sufficient to make a cut to prevent any path $\mathscr{C}$ from linking a current. If we introduce a cut-surface $\mathscr{A}_\mathrm{c}$ with $\partial \mathscr{A}_\mathrm{c} = \Gamma$, the homology space $\mathcal{H}_1(E_3 \setminus \mathscr{A}_\mathrm{c} \setminus \Omega)$ is zero. Thus the magnetic field can be represented by the gradient of a scalar field, which features a jump whenever $\mathscr{A}_\mathrm{c}$ is traversed in the direction of its normal vector, such that

$$[\![\phi_\mathrm{m}]\!]_{21} = \phi_{\mathrm{m}2} - \phi_{\mathrm{m}1} = I\,. \tag{5.85}$$

The jump discontinuity of the magnetic scalar potential across the surface is equivalent to the current in the surrounding loop.

An arbitrary current loop can be regarded as a homogeneous distribution of magnetic dipoles in such a way that all the currents on the inner surfaces cancel and only the loop current I remains on the outer boundary. The homogeneous surface density of the magnetic dipole momenta is given by

$$\frac{\mathrm{d}m}{\mathrm{d}a} = \frac{\mathrm{d}(Ia)}{\mathrm{d}a} = I\,. \tag{5.86}$$

In $E_3 \setminus \mathscr{A}_\mathrm{c} \setminus \Omega$ the magnetic field $\mathbf{H}$ can indeed be represented by a single-valued scalar magnetic potential ϕ_m, and therefore

$$\mathbf{B}(\mathbf{r}) = \frac{\mu_0 I}{4\pi} \int_{\partial \mathscr{A}_\mathrm{c}} \frac{\mathrm{d}\mathbf{r}' \times (\mathbf{r} - \mathbf{r}')}{|\mathbf{r} - \mathbf{r}'|^3} = \mu_0 \mathbf{H} = -\mu_0 \operatorname{grad} \phi_\mathrm{m}. \tag{5.87}$$

As we will show in the next section, this result is closely related to the *solid angle* that the current loop subtends at the field point. The solid angle is thus a means for calculating the magnetic scalar potential.

In an analogy to surface densities of electrical dipoles, which can be represented by two oppositely charged surfaces forming a *double layer*, the distribution of magnetic dipoles is referred to as a *magnetic double layer*. However, while electrical double layers are physically possible, magnetic double layers are a mathematical abstraction only. This is reflected by the fact that the cut-surfaces are somewhat arbitrary because only their boundary is prescribed.

5.8.1 The Solid Angle

Whereas a plane angle is defined as the ratio of the length l of an arc segment to its radius r

$$\beta := \frac{l}{r}, \tag{5.88}$$

the solid angle Θ is defined as the ratio of the surface of the sphere segment a to the square of the radius R

$$\Theta := \frac{a}{R^2}. \tag{5.89}$$

The solid angle that an object subtends at the field point is equal to the projective surface area a of that object on a sphere of unit radius. Hence

$$\Theta = \int_{\mathscr{A}} \frac{\cos\gamma}{R^2}\,\mathrm{d}a = \int_{\mathscr{A}} \frac{(\mathbf{r} - \mathbf{r}') \cdot \mathbf{n}}{|\mathbf{r} - \mathbf{r}'|^3}\,\mathrm{d}a\,, \tag{5.90}$$

where $\mathscr{A} = \partial\mathscr{V}$ is the boundary of the object. In Eq. (5.90), $\cos\gamma$ is the angle between $\mathbf{r} - \mathbf{r}'$ and the normal vector to the object's surface as illustrated in Figure 5.6 (left).

Following Maxwell [13] we can determine the solid angle by integrating $\mathrm{d}\Theta$ for the field point displaced by $\mathrm{d}\mathbf{l}$. This is equivalent to considering the

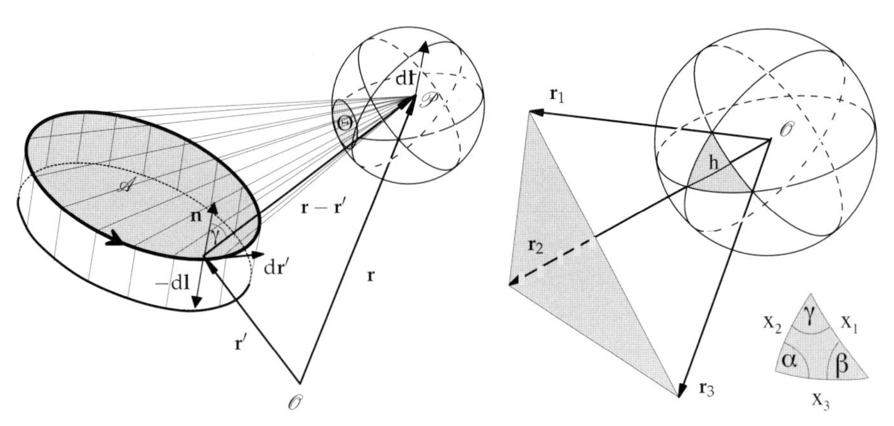

Figure 5.6 Left: Geometrical relations to calculate the difference in solid angle that a current loop subtends at the field point displaced by $\mathrm{d}\mathbf{l}$. Right: The solid angle of a plane triangle.

current loop shown in Figure 5.6 (left), displaced by $-\mathrm{d}\mathbf{l}$. Hence $\mathrm{d}\Theta$ is the sum of all changes of the surfaces of the parallelepipeds, $-\mathrm{d}\mathbf{l} \times \mathrm{d}\mathbf{r}'$, projected onto the direction of $\mathbf{r} - \mathbf{r}'$ and divided by the square of the distance $|\mathbf{r} - \mathbf{r}'|$. This yields

$$\mathrm{d}\Theta = -\int_{\partial\mathscr{A}} \frac{1}{|\mathbf{r} - \mathbf{r}'|^2} (\mathrm{d}\mathbf{l} \times \mathrm{d}\mathbf{r}') \cdot \mathbf{e}_R = -\int_{\partial\mathscr{A}} \frac{(\mathbf{r} - \mathbf{r}')}{|\mathbf{r} - \mathbf{r}'|^3} \cdot (\mathrm{d}\mathbf{l} \times \mathrm{d}\mathbf{r}')$$
$$= -\mathrm{d}\mathbf{l} \int_{\partial\mathscr{A}} \frac{\mathrm{d}\mathbf{r}' \times (\mathbf{r} - \mathbf{r}')}{|\mathbf{r} - \mathbf{r}'|^3} . \tag{5.91}$$

If the observation point is moved through the cut-surface of a loop, the total change of Θ is merely the linking number according to Eq. (5.40), multiplied by 4π. In other words, Θ increases by 4π each time the paths are linked [4]. Expressing $\mathrm{d}\Theta$ as $\operatorname{grad}\Theta \cdot \mathrm{d}\mathbf{l}$ yields

$$\operatorname{grad}\Theta = -\int_{\partial\mathscr{A}} \frac{\mathrm{d}\mathbf{r}' \times (\mathbf{r} - \mathbf{r}')}{|\mathbf{r} - \mathbf{r}'|^3} . \tag{5.92}$$

Comparing this with the Biot–Savart law we notice that

$$\mathbf{B} = \frac{\mu_0 I}{4\pi} \int_{\partial\mathscr{A}} \frac{\mathrm{d}\mathbf{r}' \times (\mathbf{r} - \mathbf{r}')}{|\mathbf{r} - \mathbf{r}'|^3} = -\frac{\mu_0 I}{4\pi} \operatorname{grad}\Theta . \tag{5.93}$$

But $\mathbf{B} = \mu_0 \mathbf{H} = -\mu_0 \operatorname{grad}\phi_{\mathrm{m}}$, and therefore

$$\phi_{\mathrm{m}}(\mathbf{r}) = \frac{I}{4\pi}\Theta . \tag{5.94}$$

The magnetic scalar potential, at any observation point $\mathscr{P}$ (position vector $\mathbf{r}$), due to a current loop is proportional to the solid angle that the current loop subtends at $\mathscr{P}$.

5.8.2
Approximating the Solid Angle of a Current Loop

The solid angle of a current loop can be approximated by the projection of plane triangles onto the unit sphere. The plane triangles have one vertex at the current-loop center and the other two on the space curve defining the loop. To calculate the solid angle that the plane triangle subtends at the field point, we place the field point, without a loss of generality, into the origin of the coordinate system as shown in Figure 5.6 (right). It follows that the three vertices of the plane triangle have position vectors $\mathbf{r}_1$, $\mathbf{r}_2$, $\mathbf{r}_3$ of lengths r_1, r_2, r_3. The *spherical triangle* (or *Euler triangle*) has an angle sum larger than π. The difference

$$\epsilon := \alpha + \beta + \gamma - \pi \tag{5.95}$$

is called the *spherical excess*. The surface area of the spherical triangle is determined by

$$a = R^2 \epsilon , \tag{5.96}$$

where R is the radius of the sphere. Comparing this result with Eq. (5.89) reveals that $\epsilon = \Theta$. Consider the triangle with two sides from the pole to the equator, and the third side along $\pi/2$ on the equator. This triangle has three right angles, and therefore its spherical excess is $\pi/2$.

Adding the spherical excesses of all the projected plane triangles thus results in the solid angle that the current loop subtends at the origin. Van Oosterom and Strackee [14] derive a computing-time efficient expression from the equation

$$\cos\left(\frac{\Theta}{2}\right) = \frac{1 + \sum_{i=1}^{3} \cos x_i}{4 \prod_{i=1}^{3} \cos\left(\frac{x_i}{2}\right)} , \tag{5.97}$$

where x_1, x_2, x_3 are the arc-segments as shown in Figure 5.6 (right). Using the trigonometric relations

$$\cos^2\left(\frac{x}{2}\right) = \frac{1}{2}(\cos x + 1) , \qquad \tan^2\left(\frac{\Theta}{2}\right) = \frac{1}{\cos^2(\frac{x}{2})} - 1 , \tag{5.98}$$

we obtain

$$\tan\left(\frac{\Theta}{2}\right) = \frac{\sqrt{1 + 2\prod_{i=1}^{3} \cos x_i - \sum_{i=1}^{3} \cos^2 x_i}}{1 + \sum_{i=1}^{3} \cos x_i} = \frac{\sin x_1 \sin h}{1 + \sum_{i=1}^{3} \cos x_i} , \tag{5.99}$$

where h is the arc length shown in Figure 5.6 (right). The triple product $\mathbf{r}_1 \cdot (\mathbf{r}_2 \times \mathbf{r}_3)$ represents the volume of the parallelepiped spanned by the three vectors and thus equals $V = r_1 r_2 r_3 \sin x_1 \sin h$. Employing the relation $\mathbf{r}_i \cdot \mathbf{r}_j = r_i r_j \cos(\alpha(\mathbf{r}_i, \mathbf{r}_j))$ we finally obtain

$$\tan\left(\frac{\Theta}{2}\right) = \frac{\mathbf{r}_1 \cdot (\mathbf{r}_2 \times \mathbf{r}_3)}{r_1 r_2 r_3 + (\mathbf{r}_1 \cdot \mathbf{r}_2) r_3 + (\mathbf{r}_1 \cdot \mathbf{r}_3) r_2 + (\mathbf{r}_2 \cdot \mathbf{r}_3) r_1} . \tag{5.100}$$

5.9 The Image-Current Method

The image-current method, traceable to Kelvin and Maxwell, can be used to obtain solutions for field problems in bounded domains. By modeling the effect of the boundary by means of *image currents*, it avoids the need for a formal solution of the Poisson equation. The field in the problem domain is calculated as the sum of the applied field and the field generated by the image currents.

Consider a current loop in a domain Ω_1 and a highly permeable domain Ω_2 with $\mu_2 \rightarrow \infty$. The field in the domain Ω_1, which is generated by the excitation-current loop, is calculated by means of an image current of suitable strength, polarity, and position within the domain Ω_2, such that the interface conditions for **H** and **B** at the material boundary Γ_{12} are fulfilled.

Following Kurz [10] the image-current method can be formalized employing an image mapping (T, λ) in order to establish the link to the Green functions in bounded domains. The aim is to find a $G^*(\mathbf{r}, \mathbf{r}') = G(\mathbf{r}, \mathbf{r}') + \lambda(\mathbf{r}') G(\mathbf{r}, T\mathbf{r}')$ such that the boundary conditions at Γ_{12} are fulfilled, and where $G(\mathbf{r}, T\mathbf{r}')$ is a harmonic function with a singularity in the domain Ω_2. The crucial point here is that the potential of the image current at the field point $\mathscr{P}$ is identical to the potential of the current loop in the imaged field point $\mathscr{P}^*$. This allows us to check the image mapping without explicit verification of the interface conditions for **H** and **B**.

To find the image current, we require a smooth mapping T that assigns an image point to each source point $\mathscr{Q}$ (position vector $\mathbf{r}'$) by

$$T : E_3 \rightarrow E_3 : \mathbf{r}' \mapsto T\mathbf{r}' . \tag{5.101}$$

It is essential that all points in Ω_1 be mapped to image points in Ω_2 as no additional singularities are allowed in the field domain:

$$T\Omega_1 \subset \Omega_2 . \tag{5.102}$$

The mapping T is its own inverse; in other words, each point is the image of its image:

$$(T \circ T)\mathbf{r}' = \mathbf{r}' . \tag{5.103}$$

Moreover, we require that images of boundary points lie on the boundary; they are fixed points with respect to the mapping T: $T\mathbf{r}' = \mathbf{r}'$ for all source points $\mathscr{Q} \in \partial\Omega_2$. The strength of image currents must in general be scaled. This is expressed by the smooth mapping

$$\lambda : \Omega_1 \cup \partial\Omega_2 \to \mathbb{R} : \mathbf{r}' \mapsto \lambda(\mathbf{r}') \tag{5.104}$$

with the condition that

$$\lambda(\mathbf{r}) = 1, \tag{5.105}$$

for $\mathscr{P} \in \partial\Omega_2$. The symmetry of the free-space Green functions allows us to state

$$\lambda(\mathbf{r}')G(\mathbf{r}, T\mathbf{r}') = \lambda(\mathbf{r})G(T\mathbf{r}, \mathbf{r}') \tag{5.106}$$

for $\mathscr{P}, \mathscr{Q} \in \Omega_1 \cup \partial\Omega_2$. Equation (5.106) implies that the magnetic scalar potential is zero at the material interface. The modified Green function

$$G^*(\mathbf{r}, \mathbf{r}') = G(\mathbf{r}, \mathbf{r}') + \lambda(\mathbf{r}')G(\mathbf{r}, T\mathbf{r}') = G(\mathbf{r}, \mathbf{r}') + \lambda(\mathbf{r})G(T\mathbf{r}, \mathbf{r}') \tag{5.107}$$

is a problem-specific version, which can be used together with the expressions (5.7) or (5.8).

The condition (5.106) implies that the potential of the image-current loop in the field point is identical to the potential of the excitation current loop at the imaged field point; see Figure 5.7.

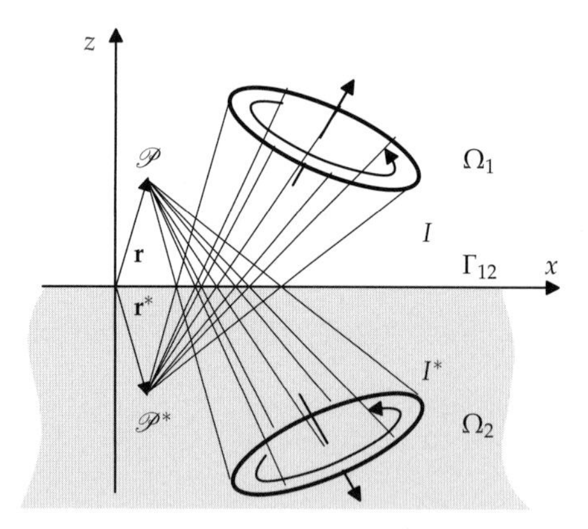

Figure 5.7 Imaging of a current loop at a highly permeable half space. Notice the skew (left-handed) orientation of the image-current loop I^*. The solid angle that the image-current loop subtends at the field point $\mathscr{P}$ equals the solid angle that the excitation-current loop subtends at the imaged field point $\mathscr{P}^*$.

5.9.1 Plane Boundaries

It is an easy task to show that the imaging at a plane, defined by[16]

$$T : E_3 \to E_3 : \mathbf{r}' \mapsto T\mathbf{r}' = \mathbf{r}' - 2\mathbf{n}\,(\mathbf{n} \cdot \mathbf{r}'), \tag{5.108}$$

fulfills the above conditions for $\lambda(\mathbf{r}) = 1$. The surface normal vector $\mathbf{n}$ points from Ω_2 to Ω_1. The condition (5.106) is fulfilled because we have $(T \circ T)\mathbf{r} = \mathbf{r}$, $|T\mathbf{r}| = |\mathbf{r}|$ and, because T is linear,

$$|\mathbf{r} - T\mathbf{r}'| = |(T \circ T)\mathbf{r} - T\mathbf{r}'| = |T(T\mathbf{r} - \mathbf{r}')| = |T\mathbf{r} - \mathbf{r}'|. \tag{5.109}$$

We now generalize the method for a current loop inside any permeable domain Ω_1 at $z > 0$ (permeability μ_1) above a plane boundary to a domain Ω_2 at $z < 0$ (permeability μ_2). The magnetic scalar potential due to the current loop in Ω_1 (see Figure 5.7) is

$$\phi_{\mathrm{m}}(\mathbf{r}) = \frac{I}{4\pi}\Theta = \frac{I}{4\pi}\int_{\mathscr{A}} \frac{(\mathbf{r} - \mathbf{r}') \cdot \mathbf{n}}{|\mathbf{r} - \mathbf{r}'|^3}\, da\,. \tag{5.110}$$

With the observations of the previous sections in mind we can avoid the explicit calculation of the fields and potentials on the material interface. To this end, we write the magnetic scalar potential $\phi_{\mathrm{m1}}(\mathbf{r})$ for field points $\mathscr{P} \in \Omega_1$ as a weighted sum of a primary potential from the loop current and a secondary potential at the imaged field point $\mathscr{P}^*$. For field points $\mathscr{P} \in \Omega_2$ we consider a potential due to the loop in Ω_1 with the current scaled by a factor β. The primary and secondary potentials are denoted $\phi_{\mathrm{m}}(\mathbf{r})$ and $\phi_{\mathrm{m}}(\mathbf{r}^*)$, respectively.

$$\phi_{\mathrm{m1}}(\mathbf{r}) = \phi_{\mathrm{m}}(\mathbf{r}) + \alpha\,\phi_{\mathrm{m}}(\mathbf{r}^*) \qquad z > 0, \tag{5.111}$$

$$\phi_{\mathrm{m2}}(\mathbf{r}) = \beta\,\phi_{\mathrm{m}}(\mathbf{r}) \qquad z < 0. \tag{5.112}$$

At the interface between the two permeable domains Γ_{12} (at $z = 0$), the continuity conditions

$$B_{\mathrm{n1}} = B_{\mathrm{n2}}, \qquad H_{\mathrm{t1}} = H_{\mathrm{t2}}, \tag{5.113}$$

must be fulfilled. In terms of the magnetic scalar potential, they can be expressed[17] as

$$\partial_{\mathbf{n}}(\mu_1\phi_{\mathrm{m1}} - \mu_2\phi_{\mathrm{m2}}) = 0\,, \qquad \mathbf{n} \times \operatorname{grad}(\phi_{\mathrm{m1}} - \phi_{\mathrm{m2}}) = \mathbf{0}\,. \tag{5.114}$$

16 The expression $\mathbf{n}(\mathbf{n} \cdot \mathbf{r}')$ yields the normal component of the position vector, as explained in Section 2.10.

17 See Eqs. (4.129) and (4.130).

This yields

$$\frac{\partial \phi_{\text{m1}}}{\partial x}(\mathbf{r}) = \frac{\partial \phi_{\text{m2}}}{\partial x}(\mathbf{r}), \tag{5.115}$$

$$\frac{\partial \phi_{\text{m1}}}{\partial y}(\mathbf{r}) = \frac{\partial \phi_{\text{m2}}}{\partial y}(\mathbf{r}), \tag{5.116}$$

$$\frac{\partial \phi_{\text{m1}}}{\partial z}(\mathbf{r}) = \frac{\mu_2}{\mu_1}\frac{\partial \phi_{\text{m2}}}{\partial z}(\mathbf{r}). \tag{5.117}$$

Thus using Eqs. (5.111) and (5.112),

$$\frac{\partial \phi_{\text{m}}}{\partial z}(\mathbf{r}) + \alpha\frac{\partial \phi_{\text{m}}}{\partial z}(\mathbf{r}^*) = \beta\frac{\mu_2}{\mu_1}\frac{\partial \phi_{\text{m}}}{\partial z}(\mathbf{r}), \tag{5.118}$$

$$\frac{\partial \phi_{\text{m}}}{\partial x}(\mathbf{r}) + \alpha\frac{\partial \phi_{\text{m}}}{\partial x}(\mathbf{r}^*) = \beta\frac{\partial \phi_{\text{m}}}{\partial x}(\mathbf{r}), \tag{5.119}$$

$$\frac{\partial \phi_{\text{m}}}{\partial y}(\mathbf{r}) + \alpha\frac{\partial \phi_{\text{m}}}{\partial y}(\mathbf{r}^*) = \beta\frac{\partial \phi_{\text{m}}}{\partial y}(\mathbf{r}). \tag{5.120}$$

Now we make use of the fact that the solid angle that the image-current loop subtends at the field point $\mathscr{P}$ equals the solid angle that the excitation current loop subtends at the imaged field point $\mathscr{P}^*$. This fact is illustrated in Figure 5.7. At the material interface we thus obtain

$$\frac{\partial \phi_{\text{m}}}{\partial x}(\mathbf{r}) = \frac{\partial \phi_{\text{m}}}{\partial x}(\mathbf{r}^*), \tag{5.121}$$

$$\frac{\partial \phi_{\text{m}}}{\partial y}(\mathbf{r}) = \frac{\partial \phi_{\text{m}}}{\partial y}(\mathbf{r}^*), \tag{5.122}$$

$$\frac{\partial \phi_{\text{m}}}{\partial z}(\mathbf{r}) = -\frac{\partial \phi_{\text{m}}}{\partial z}(\mathbf{r}^*). \tag{5.123}$$

It follows that $1+\alpha = \beta$ and $\mu_1(1-\alpha) = \mu_2\beta$, or

$$\alpha = \frac{\mu_1 - \mu_2}{\mu_1 + \mu_2}, \qquad \beta = \frac{2\mu_1}{\mu_1 + \mu_2}. \tag{5.124}$$

For free space ($\mu_1 = \mu_2 = \mu_0$) this yields $\alpha = 0$ and $\beta = 1$, as expected.

Figure 5.8 (left) shows the field distribution in a quadrupole magnet. The symmetry suggests that the lower coil can be replaced by a "mirror" plate of high permeability as is shown on the right-hand side. Inversely, the field in the upper part of the half-quadrupole in Figure 5.8 (right) can be calculated by means of the image-current method.

Half-quadrupoles, albeit normal-conducting ones, were installed in the interaction regions of the HERA electron proton collider [18] and the KEK B-factory [16], an electron/positron collider. The magnets are installed in the

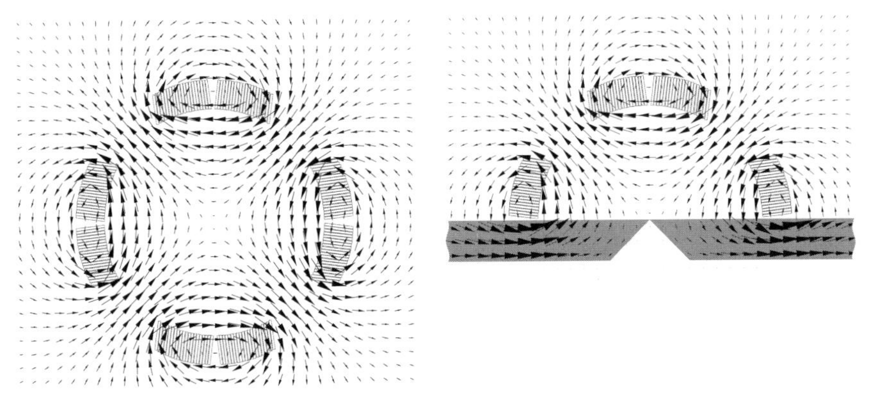

Figure 5.8 Left: Field distribution in a superconducting quadrupole magnet. Right: Half-quadrupole with flux plate. This configuration can be calculated for low excitation by means of the image-current method at plane boundaries.

region where the beam separation is small. The magnets provide focusing or defocusing fields for one beam and a nearly field-free area for the second beam on the opposite side of the septum, that is, on the lower side of the mirror plate shown in Figure 5.8 (right).

5.9.2 Circular Boundaries

Consider the two-dimensional field problem of a single line-current in the circular aperture of an iron yoke with constant relative permeability μ_{r}. The inner yoke radius is denoted r_{y}. The magnetization in the iron yoke can be taken into account by means of the imaging transformation

$$T : E_2 \to E_2 : \mathbf{r}' \mapsto T\mathbf{r}' = \frac{r_{\mathrm{y}}^2}{|\mathbf{r}'|^2}\mathbf{r}' , \tag{5.125}$$

where $\mathbf{r}' = (r_{\mathrm{c}}, \varphi_{\mathrm{c}})$ according to Figure 5.9. The image current must be scaled according to

$$I^* = \lambda_\mu I := \frac{\mu_{\mathrm{r}} - 1}{\mu_{\mathrm{r}} + 1} I . \tag{5.126}$$

With the original and the image current in place, the iron boundary can be disregarded and the field problem reduced to the field calculation of line-currents in free space. However, the solution is valid only inside the aperture domain. The image mappings, Eqs. (5.125) and (5.126), require a proof that we extend to a circular domain Ω_1 (permeability μ_1) inside Ω_2 (permeability μ_2) as shown in Figure 5.9.

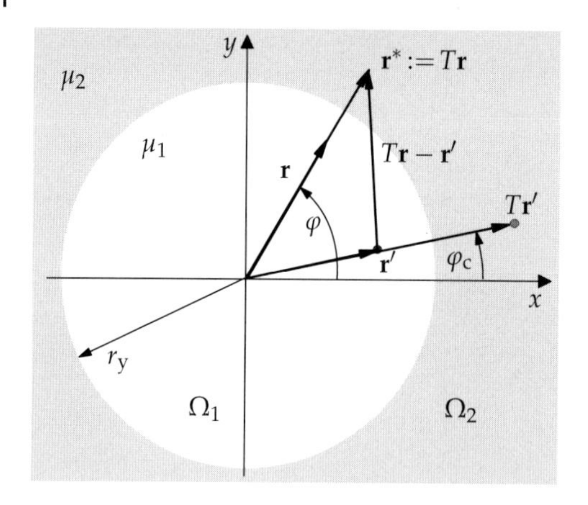

Figure 5.9 Configuration of a line-current within a circular domain Ω_1, of permeability μ_1, inside another permeable domain Ω_2 of permeability μ_2.

Again, we make use of the fact that the magnetic vector potential of the imaged line-current is identical to the vector potential of the excitation line-current at the imaged field point. Thus the explicit calculation of the potential can be avoided. It must be verified that the mapping (5.125) obeys the condition

$$G(\mathbf{r}, T\mathbf{r}') = G(T\mathbf{r}, \mathbf{r}') . \tag{5.127}$$

From

$$\begin{aligned} |T\mathbf{r} - \mathbf{r}'|^2 &= \left| \frac{r_y^2}{|\mathbf{r}|^2}\mathbf{r} - \mathbf{r}' \right|^2 = \frac{r_y^4}{|\mathbf{r}|^2} - \frac{2r_y^2}{|\mathbf{r}|^2}\mathbf{r}\cdot\mathbf{r}' + |\mathbf{r}'|^2 \\ &= \frac{|\mathbf{r}'|^2}{|\mathbf{r}|^2}\left(|\mathbf{r}|^2 - \frac{2r_y^2}{|\mathbf{r}'|^2}\mathbf{r}\cdot\mathbf{r}' + \frac{r_y^4}{|\mathbf{r}'|^2} \right) = \frac{|\mathbf{r}'|^2}{|\mathbf{r}|^2}\,|\mathbf{r} - T\mathbf{r}'|^2 , \end{aligned} \tag{5.128}$$

it follows that

$$\begin{aligned} G(\mathbf{r}, T\mathbf{r}') &= -\frac{1}{2\pi}\ln\left(\frac{|\mathbf{r} - T\mathbf{r}'|}{r_{\mathrm{ref}}}\right) = -\frac{1}{2\pi}\ln\left(\frac{|T\mathbf{r} - \mathbf{r}'|\,|\mathbf{r}|}{r_{\mathrm{ref}}\,|\mathbf{r}'|}\right) \\ &= -\frac{1}{2\pi}\ln\left(\frac{|T\mathbf{r} - \mathbf{r}'|}{r_{\mathrm{ref}}}\right) - \frac{1}{2\pi}\ln\left(\frac{|\mathbf{r}|}{r_{\mathrm{ref}}}\right) + \frac{1}{2\pi}\ln\left(\frac{|\mathbf{r}'|}{r_{\mathrm{ref}}}\right) \\ &= G(T\mathbf{r}, \mathbf{r}') + G(\mathbf{r}, 0) - G(0, \mathbf{r}') . \end{aligned} \tag{5.129}$$

We need not worry about the term $G(0, \mathbf{r}')$ as this yields a constant potential on the circular interface between the two permeable domains. However, the term $G(\mathbf{r}, 0)$ gives rise to the potential of a line-current positioned at the central axis. Condition (5.127) is fulfilled only for the vanishing sum current within the domain Ω_1. It is the situation shown in Figure 4.4 (right), where the magnetic flux density enters vertically into the highly permeable medium, for the limiting case of $\mu_1 = \mu_0$ and $\mu_2 \to \infty$. Configurations shown in Figure 4.4 (left) are excluded.[18]

The remaining part of the proof is straightforward. As in Section 5.9.1 we write the vector potential $A_{z1}(\mathbf{r})$ in Ω_1 as a weighted sum of a primary potential from the excitation line-current at the field point $\mathbf{r} = (r, \varphi)$ and a secondary potential at the imaged field point $\mathbf{r}^* = (r^*, \varphi)$. In Ω_2 we consider a potential from the scaled line-current. The primary and secondary potentials are denoted $A_z(\mathbf{r})$ and $A_z(\mathbf{r}^*)$, respectively.

$$A_{z1}(\mathbf{r}) = A_z(\mathbf{r}) + \alpha\, A_z(\mathbf{r}^*) \qquad r > r_y, \tag{5.130}$$

$$A_{z2}(\mathbf{r}) = \beta\, A_z(\mathbf{r}) \qquad r < r_y. \tag{5.131}$$

The continuity conditions at the interface Γ_{12} of two permeable domains (at $r = r_y$) are given by

$$B_{r1} = B_{r2}, \qquad H_{\varphi 1} = H_{\varphi 2}, \tag{5.132}$$

and are expressed in terms of the vector potential as

$$\frac{\partial A_{z1}}{\partial \varphi}(\mathbf{r}) = \frac{\partial A_{z2}}{\partial \varphi}(\mathbf{r}), \qquad \frac{1}{\mu_1}\frac{\partial A_{z1}}{\partial r}(\mathbf{r}) = \frac{1}{\mu_2}\frac{\partial A_{z2}}{\partial r}(\mathbf{r}). \tag{5.133}$$

Thus,

$$\frac{\partial A_z}{\partial \varphi}(\mathbf{r}) + \alpha\frac{\partial A_z}{\partial \varphi}(\mathbf{r}^*) = \beta\frac{\partial A_z}{\partial \varphi}(\mathbf{r}), \tag{5.134}$$

$$\frac{\partial A_z}{\partial r}(\mathbf{r}) + \alpha\frac{\partial A_z}{\partial r}(\mathbf{r}^*) = \beta\frac{\partial A_z}{\partial r}(\mathbf{r}). \tag{5.135}$$

At the material interface we find

$$\frac{\partial A_z}{\partial \varphi}(\mathbf{r}^*) = \frac{\partial A_z}{\partial \varphi}(\mathbf{r}), \qquad \frac{\partial A_z}{\partial r}(\mathbf{r}^*) = -\frac{\partial A_z}{\partial r}(\mathbf{r}), \tag{5.136}$$

from which follows $1 + \alpha = \beta$ and $\mu_2(1 - \alpha) = \mu_1\beta$, or

$$\alpha = \frac{\mu_2 - \mu_1}{\mu_1 + \mu_2}, \qquad \beta = \frac{2\mu_2}{\mu_1 + \mu_2}. \tag{5.137}$$

Notice the sign changes in the results for α and β with respect to Section 5.9.1. For the special case of free space in Ω_1 ($\mu_1 = \mu_0$) and high permeability in the outer domain $\mu_2 = \mu_r\mu_0$, we obtain the weighting factor used in Eq. (5.126) for the imaged line-current:

$$\lambda_\mu := \frac{\mu_r - 1}{\mu_r + 1}. \tag{5.138}$$

18 These can be also solved by the image-current method, but the proof is less elegant.

Figure 5.10 (left) shows the field distribution in a superconducting magnet, when the iron yoke is represented by the image currents. A comparison to Figure 5.10 (right) shows that the image currents excite a completely different field in the iron domain.

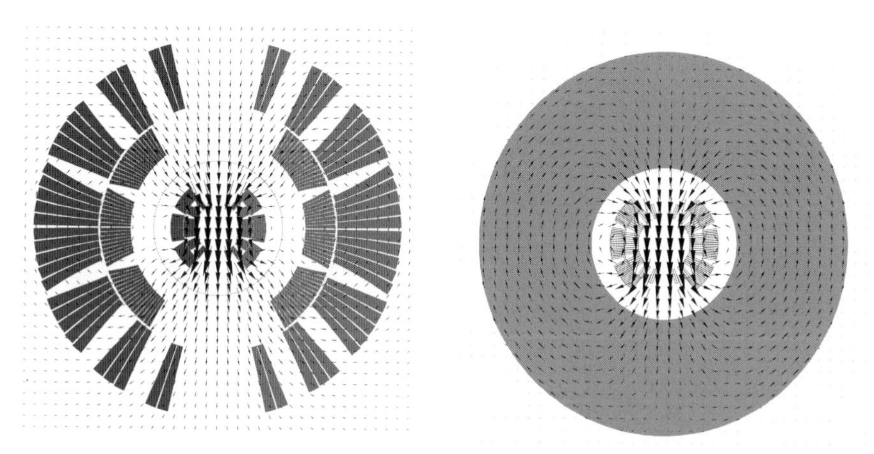

Figure 5.10 Left: Field distribution in a superconducting magnet calculated by means of the image-current method. Field domain Ω_1 inside and Ω_2 outside the circle. Right: Original configuration consisting of a ring-shaped yoke of constant relative permeability $\mu_r = 2000$.

5.10
Stored Energy in a Magnetostatic Field

In the linear, static case the stored magnetic energy in a contractible volume $\mathscr{V}$ is given by the integral

$$W = \int_{\mathscr{V}} w_\mathrm{m} \,\mathrm{d}V = \frac{1}{2}\int_{\mathscr{V}} \mathbf{H}\cdot\mathbf{B}\,\mathrm{d}V = \frac{1}{2}\int_{\mathscr{V}} \mathbf{H}\cdot \operatorname{curl}\mathbf{A}\,\mathrm{d}V\,, \tag{5.139}$$

where $[W] = 1\,\mathrm{V\,A\,s} = 1\,\mathrm{J}$ and $[w_\mathrm{m}] = 1\,\mathrm{J\,m^{-3}}$. Equation (5.139) follows from Poynting's theorem (4.30). Because of $\operatorname{div}(\mathbf{A}\times\mathbf{H}) = \mathbf{H}\cdot\operatorname{curl}\mathbf{A} - \mathbf{A}\cdot\operatorname{curl}\mathbf{H}$, Eq. (5.139) can be rewritten as [12]:

$$\begin{aligned} W &= \frac{1}{2}\int_{\mathscr{V}} \operatorname{div}(\mathbf{A}\times\mathbf{H})\,\mathrm{d}V + \frac{1}{2}\int_{\mathscr{V}} \mathbf{A}\cdot\operatorname{curl}\mathbf{H}\,\mathrm{d}V \\ &= \frac{1}{2}\int_{\partial\mathscr{V}} (\mathbf{A}\times\mathbf{H})\cdot\mathrm{d}\mathbf{a} + \frac{1}{2}\int_{\mathscr{V}} \mathbf{A}\cdot\operatorname{curl}\mathbf{H}\,\mathrm{d}V\,. \end{aligned} \tag{5.140}$$

The term $\frac{1}{2}\int_{\partial\mathscr{V}}(\mathbf{A}\times\mathbf{H})\cdot\mathrm{d}\mathbf{a}$ vanishes on the far-field boundary because $A \propto 1/r, H \propto 1/r^2, \mathrm{d}a \propto r^2$, as can be seen in the Biot–Savart–type integrals.

Inner surfaces, which must be considered from both sides to keep the volume contractible, do not contribute to the total energy if they carry no surface-currents ($\boldsymbol{\alpha} = 0$). This is true in the case of finite conductivity and excitation without jump discontinuity:

$$W = \frac{1}{2}\int_{\partial\mathscr{V}} (\mathbf{A} \times \mathbf{H}) \cdot \mathbf{da} = \frac{1}{2}\sum_i \int_{\mathscr{A}_i} [\mathbf{A} \times (\mathbf{H}_2 - \mathbf{H}_1)] \cdot \mathbf{n}_2 \, \mathrm{d}a$$
$$= \frac{1}{2}\sum_i \int_{\mathscr{A}_i} \mathbf{A} \cdot [(\mathbf{H}_2 - \mathbf{H}_1) \times \mathbf{n}_2] \, \mathrm{d}a = \frac{1}{2}\sum_i \int_{\mathscr{A}_i} \mathbf{A} \cdot \boldsymbol{\alpha} \, \mathrm{d}a \,, \tag{5.141}$$

which is zero for $\boldsymbol{\alpha} = 0$. The magnetic energy can now be calculated as follows:

$$W = \frac{1}{2}\int_{\mathscr{V}} \mathbf{A} \cdot \operatorname{curl} \mathbf{H} \, \mathrm{d}V = \frac{1}{2}\int_{\mathscr{V}} \mathbf{A} \cdot \mathbf{J} \, \mathrm{d}V$$
$$= \frac{1}{2}\int_{\mathscr{V}_{\text{coil}}} \mathbf{A}(\mathbf{r}) \cdot \mathbf{J}(\mathbf{r}) \, \mathrm{d}V \,. \tag{5.142}$$

Remark: We have to think about the gauge invariance of this result: What will be the calculated magnetic energy for $A' = A + \operatorname{grad} \psi$?

$$W = \frac{1}{2}\int_{\mathscr{V}} \mathbf{A}' \cdot \mathbf{J} \, \mathrm{d}V = \frac{1}{2}\int_{\mathscr{V}} \mathbf{A} \cdot \mathbf{J} \, \mathrm{d}V + \frac{1}{2}\int_{\mathscr{V}} \operatorname{grad} \psi \cdot \mathbf{J} \, \mathrm{d}V$$
$$= \frac{1}{2}\int_{\mathscr{V}} \mathbf{A} \cdot \mathbf{J} \, \mathrm{d}V + \frac{1}{2}\int_{\partial\mathscr{V}} \psi (\mathbf{J} \cdot \mathbf{n}) \, \mathrm{d}a - \frac{1}{2}\int_{\mathscr{V}} \psi \operatorname{div} \mathbf{J} \, \mathrm{d}V \,. \tag{5.143}$$

We must therefore impose the condition that no current leave the boundary of the integration domain and that all current-carrying conductors be closed within the domain. □

Combining Eqs. (5.29) and (5.142) yields

$$W = \frac{\mu_0}{8\pi}\int_{\mathscr{V}}\int_{\mathscr{V}'} \frac{\mathbf{J}(\mathbf{r}) \cdot \mathbf{J}(\mathbf{r}')}{|\mathbf{r} - \mathbf{r}'|} \, \mathrm{d}V' \, \mathrm{d}V. \tag{5.144}$$

For line-currents, the energy is infinite as $|\mathbf{r} - \mathbf{r}'|$ tends to zero. This result reflects the fact that a line-current is not technically feasible. The problem can be circumvented by employing the equation[19]

$$W = \sum_{k=1}^{K} \frac{1}{2}\int_{\mathscr{C}_k} I_k \mathbf{A}_k(\mathbf{r}) \, \mathrm{d}\mathbf{r} \,, \tag{5.145}$$

for coil cross sections made of K strands, modeled as line-currents defined by the closed loops $\mathscr{C}_k$. The $\mathbf{A}_k$ terms refer to the vector potentials due to the currents (other than the I_k in the strand) that produce the field $\mathbf{B}_k$ in that strand.

19 See also Eq. (5.36).

For 3D calculations this approach results in a small error whose magnitude depends on the number of line-current segments.

For 2D the energy per unit length in z can be calculated as

$$\frac{W}{l} = \frac{1}{2} \sum_{k=1}^{K} A_{z,k} I_k + \frac{W_{\text{strand}}}{l}, \tag{5.146}$$

where we add the energy per unit length of the strand:

$$\frac{W_{\text{strand}}}{l} \approx \frac{\mu_0 \, I_k^2}{8\pi} \left(\ln \frac{1}{r_0} + 1 \right), \tag{5.147}$$

where r_0 is the radius of the strand. The result is derived using the concept of geometric mean distance [5].

5.10.1
Self and Mutual Inductance

For a set of n closed current loops with current densities $\mathbf{J}_i(\mathbf{r}), i = 1, 2, \ldots, n$, we obtain from Eq. (5.144),

$$W = \sum_{i=1}^{n} \sum_{j=1}^{n} W_{ij} = \frac{\mu_0}{8\pi} \sum_{i=1}^{n} \sum_{j=1}^{n} \int_{\mathcal{V}} \int_{\mathcal{V}'} \frac{\mathbf{J}_i(\mathbf{r}) \cdot \mathbf{J}_j(\mathbf{r}')}{|\mathbf{r} - \mathbf{r}'|} \, \mathrm{d}V' \mathrm{d}V$$

$$= \frac{\mu_0}{8\pi} \sum_{i=1}^{n} \sum_{j=1}^{n} I_i I_j \int_{\mathcal{V}} \int_{\mathcal{V}'} \frac{\mathbf{J}_i(\mathbf{r}) \cdot \mathbf{J}_j(\mathbf{r}')}{I_i I_j |\mathbf{r} - \mathbf{r}'|} \, \mathrm{d}V' \mathrm{d}V. \tag{5.148}$$

The *mutual inductances* are defined by

$$L_{ij} := \frac{\mu_0}{4\pi I_i I_j} \int_{\mathcal{V}} \int_{\mathcal{V}'} \frac{\mathbf{J}_i(\mathbf{r}) \cdot \mathbf{J}_j(\mathbf{r}')}{|\mathbf{r} - \mathbf{r}'|} \, \mathrm{d}V' \, \mathrm{d}V \tag{5.149}$$

and expressed in units of henry,[20] with $[L] = 1\,\mathrm{H} = 1\,\mathrm{V\,s\,A^{-1}}$. Eq. (5.148) can then be rewritten as

$$W = \frac{1}{2} \sum_{i=1}^{n} \sum_{j=1}^{n} L_{ij} I_i I_j. \tag{5.150}$$

Note that this result crucially depends on the principle of linear superposition and the inductance so defined is referred to as *apparent inductance* [3]. We have just seen from Eq. (5.149) that the inductance depends only on the coil geometry and has the property

$$L_{ij} = L_{ji}. \tag{5.151}$$

For $i = j$ the coefficient (5.149) is called the *self inductance*. The inductances can be calculated directly using Eq. (5.149) or by calculating the stored energy and comparing it with Eq. (5.150).

20 Joseph Henry (1797–1878)

Example: Consider the nested skew-dipole/sextupole corrector as it is shown in Figure 5.11. The two dipole coils are connected in series; we

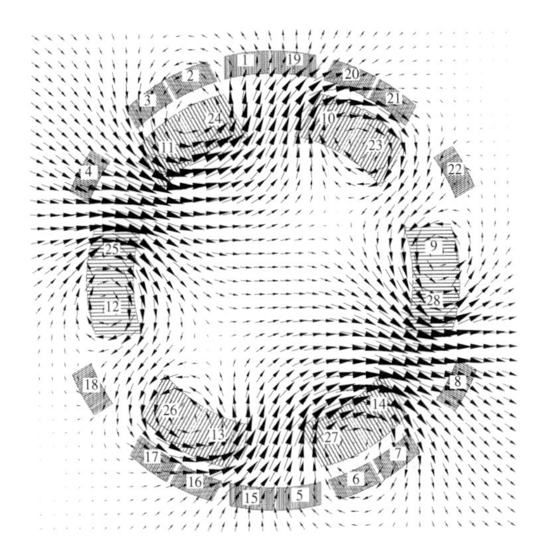

Figure 5.11 Cross section of a dipole/sextupole corrector magnet. Coil 1 = blocks 1–4 and 15–18; coil 2 = blocks 5–8 and 19–22; coil 3 = blocks 9, 23; coil 4 = blocks 10, 24; coil 5 = blocks 11, 25, etc.

shall call them coil 1 and coil 2. The six coils of the sextupole magnet are also connected in series and are referred to as coils 3–8. The self and mutual inductances can be derived by powering a single coil at a time, for instance, coil i, containing k individual wires and total current I, and calculating the stored magnetic energy W_{ii} according to Eq. (5.146). This gives

$$L_{ii} = \frac{2W_{ii}}{I^2} \,. \tag{5.152}$$

Subsequently, powering any two coils i and j with the same current I yields because of the symmetry of the mutual inductances

$$L_{ij} = \frac{1}{2}\left(\frac{2W_{ij}}{I^2} - L_{ii} - L_{jj}\right) \,. \tag{5.153}$$

The self and mutual inductances for the nested corrector are given in Table 5.1. □

Table 5.1 Self and mutual inductances per unit length in $\mathrm{mH\,m^{-1}}$ for the combined dipole and sextupole corrector magnet.

Coil	1	2	3	4	5	6	7	8
1	12.601	6.517	−0.245	0.252	0.478	−0.478	−0.252	0.245
2	6.517	12.601	−0.478	−0.252	0.245	−0.245	0.252	0.478
3	−0.245	−0.478	0.136	0.027	−0.010	0.009	−0.010	0.027
4	0.252	−0.252	0.027	0.136	0.027	−0.010	0.009	−0.010
5	0.478	0.245	−0.010	0.027	0.136	0.027	−0.010	0.009
6	−0.478	−0.245	0.009	−0.010	0.027	0.136	0.027	−0.010
7	−0.252	0.252	−0.010	0.009	−0.010	0.027	0.136	0.027
8	0.245	0.478	0.027	−0.010	0.009	−0.010	0.027	0.136

5.10.2
The Geometric Mean Distance

Equation (5.149) for the calculation of self and mutual inductances can be rewritten as

$$L_{ij} = \frac{\mu_0}{4\pi}\frac{1}{a_i a_j}\int_{\mathscr{V}_i}\int_{\mathscr{V}_j}\frac{w_i \cdot w_j}{r_{ij}}\,\mathrm{d}V_i \mathrm{d}V_j$$
$$= \frac{\mu_0}{4\pi}\frac{1}{a_i a_j}\int_{\mathscr{A}_i}\int_{\mathscr{A}_j}\int_{\mathscr{L}_i}\int_{\mathscr{L}_j}\frac{w_i \cdot w_j}{r_{ij}}\,\mathrm{d}s_i\,\mathrm{d}s_j\,\mathrm{d}a_i\,\mathrm{d}a_j, \tag{5.154}$$

where $r_{ij} := |\mathbf{r} - \mathbf{r}'|$ and w_i are weighting factors in $\{0, 1\}$. For line-currents, Eq. (5.154) reduces to

$$L_{ij} = \frac{\mu_0}{4\pi}\int_{\mathscr{L}_i}\int_{\mathscr{L}_j}\frac{w_i \cdot w_j}{r_{ij}}\,\mathrm{d}s_i\,\mathrm{d}s_j. \tag{5.155}$$

The self inductance of a solid wire can be calculated by modeling it as a set of parallel line-currents, and thus by solving the integral in Eq. (5.155):

$$\mathrm{L}\left(\frac{l}{d}\right) = \frac{\mu_0}{4\pi}\int_0^l\int_0^l \frac{1}{\sqrt{d^2 + (s_i - s_j)^2}}\,\mathrm{d}s_i\,\mathrm{d}s_j$$
$$= \frac{\mu_0}{4\pi}2l\left[\ln\left(\frac{l}{d} + \sqrt{1 + \frac{l^2}{d^2}}\right) - \sqrt{1 + \frac{d^2}{l^2}} + \frac{d}{l},\right], \tag{5.156}$$

where d is the distance between the line-currents and l is the wire length. The mean value of the distances between points in the wire cross section is obtained by calculating the integral term

$$\frac{1}{a^2}\int_{\mathscr{A}}\int_{\mathscr{A}}\mathrm{L}\left(\frac{l}{d}\right)\,\mathrm{d}a\,\mathrm{d}a. \tag{5.157}$$

This integral cannot be computed easily because $\mathrm{L}\,(l/d)$ diverges for short distances. Following Grover [5], Eq. (5.156) can be approximated for wires with $l/d \gg 1$, by

$$\mathrm{L_a}\left(\frac{l}{d}\right) := \frac{\mu_0}{4\pi}2l\left(\ln\left(\frac{2l}{d}\right) - 1\right). \tag{5.158}$$

The integral (5.157) can be calculated numerically with

$$\frac{1}{a^2}\int_{\mathscr{A}}\int_{\mathscr{A}}\ln\left(\frac{2l}{d}\right)\,\mathrm{d}a\,\mathrm{d}a \approx \frac{1}{N}\sum_{n=1}^{N}\ln\frac{2l}{d_n} = \ln\sqrt[N]{\prod_{n=1}^{N}\frac{2l}{d_i}}, \tag{5.159}$$

where d_i is the distance between any pair of points in the surface $\mathscr{A}$. Using the expression for the *geometric mean distance* (GMD),

$$d_g := \lim_{N\to\infty} \sqrt[N]{\prod_{n=1}^{N} d_i}, \tag{5.160}$$

we can write for $l/d \gg 1$:

$$L \approx \frac{1}{a^2}\int_{\mathscr{A}}\int_{\mathscr{A}} L_a\left(\frac{l}{d}\right)\, da\, da = L_a\left(\frac{l}{d_g}\right). \tag{5.161}$$

For a round wire of radius r, the GMD is

$$d_g = K_g r = e^{-1/4} r \approx 0.7788\, r, \tag{5.162}$$

where K_g is called the *GMD factor*. This approximation cannot be used for the calculation of inductances in Rutherford-type cables because the strand thickness and the branch length are nearly equal. Nevertheless, we can define a shape-dependent GMD, denoted d_e, by

$$\frac{1}{a^2}\int_{\mathscr{A}}\int_{\mathscr{A}} F\left(\frac{l}{d}\right)\, da\, da =: F\left(\frac{l}{d_e}\right), \tag{5.163}$$

where $F(l/d)$ is the dimensionless quantity

$$F\left(\frac{l}{d}\right) = \ln\left(\frac{l}{d} + \sqrt{1+\frac{l^2}{d^2}}\right) - \sqrt{1+\frac{d^2}{l^2}} + \frac{d}{l}. \tag{5.164}$$

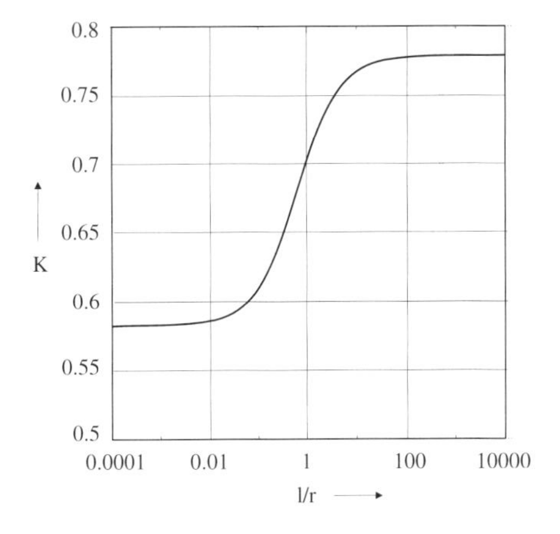

Figure 5.12 Equivalent mean distance factor for a round wire.

Solving the integral on the left-hand side of Eq. (5.163) and inverting $F(l/d_e)$, the *equivalent mean distance* d_e can be found for any shape and length. For a cylinder of radius r, the quantity $F(l/d_e)$ depends only on l/r. Hence d_e is

$$d_e = K\left(\frac{l}{r}\right) r, \tag{5.165}$$

where K is the equivalent mean distance factor and

$$F\left(\frac{l}{d_e}\right) = F\left(\frac{1}{K\left(\frac{l}{r}\right)}\frac{l}{r}\right). \tag{5.166}$$

The integral (5.157) can be solved numerically by choosing Gauss points close to, but not on the singular surface. The result $K(l/r)$ is plotted in Figure 5.12 for a round wire. For $l \gg r$, K approaches K_g, the GMD factor for the thin-wire approximation.

5.10.3 Magnetic Flux

Outside current-carrying conductors, the field can be represented by a magnetic scalar potential $\mathbf{H} = -\operatorname{grad}\phi_m$ if a cut-surface $\mathscr{A}_c$ prevents all paths from linking a current. If we now assume a single closed loop of a wire with negligible cross section ($\mathscr{V}$ does not contain the conductor nor therefore its inner energy), the stored magnetic energy can be calculated by integrating:

$$W = \frac{\mu_0}{2}\int_{\mathscr{V}} \mathbf{H}\cdot\mathbf{H}\,\mathrm{d}V = \frac{\mu_0}{2}\int_{\mathscr{V}} (\operatorname{grad}\phi_m)^2 \mathrm{d}V, \tag{5.167}$$

where $\mathscr{V}$ is the current-free external volume. Using the identity

$$\operatorname{div}(\phi_m \operatorname{grad}\phi_m) = \phi_m \nabla^2 \phi_m + (\operatorname{grad}\phi_m)^2, \tag{5.168}$$

taking into account that $\nabla^2\phi_m = 0$, and applying the Gauss theorem yields

$$W = \frac{\mu_0}{2}\int_{\mathscr{V}} \operatorname{div}(\phi_m \operatorname{grad}\phi_m)\,\mathrm{d}V = \frac{\mu_0}{2}\int_{\partial\mathscr{V}} \phi_m \operatorname{grad}\phi_m \cdot \mathrm{d}\mathbf{a}. \tag{5.169}$$

Now the surface $\partial\mathscr{V}$ is split up into two surfaces: one at infinite distance and the other at the current loop $\mathscr{A}_c$, which consists of an upper surface (with outward normal $\mathbf{n}_1$) and a lower surface (with outward normal $\mathbf{n}_2$); see Figure 5.5. Hence from Eq. (5.169),

$$\begin{aligned} W &= \frac{\mu_0}{2}\int_{\mathscr{A}_\infty} \phi_\mathrm{m}\,\mathrm{grad}\,\phi_\mathrm{m}\cdot\mathrm{d}\mathbf{a} + \frac{\mu_0}{2}\int_{\mathscr{A}_\mathrm{c}} \phi_\mathrm{m}\,\mathrm{grad}\,\phi_\mathrm{m}\cdot\mathrm{d}\mathbf{a} \\ &= 0 + \frac{\mu_0}{2}\int_{\mathscr{A}_\mathrm{c}} \left[\phi_\mathrm{m1}\,\mathrm{grad}\,\phi_\mathrm{m}\cdot\mathbf{n}_1 + \phi_\mathrm{m2}\,\mathrm{grad}\,\phi_\mathrm{m}\cdot\mathbf{n}_2\right]\mathrm{d}a \\ &= \frac{\mu_0}{2}\int_{\mathscr{A}_\mathrm{c}} (\phi_\mathrm{m1}-\phi_\mathrm{m2})\,\mathrm{grad}\,\phi_\mathrm{m}\cdot\mathbf{n}\,\mathrm{d}a \\ &= \frac{1}{2}\int_{\mathscr{A}_\mathrm{c}} (\phi_\mathrm{m2}-\phi_\mathrm{m1})B_\mathrm{n}\,\mathrm{d}a\,. \end{aligned} \tag{5.170}$$

The surface of the current loop $\mathscr{A}_\mathrm{c}$ can be regarded as a double layer of fictitious magnetic charges [12], on which the difference $\phi_\mathrm{m2} - \phi_\mathrm{m1}$ is constant; $\phi_\mathrm{m2} - \phi_\mathrm{m1} = I$, and therefore

$$W = \frac{1}{2}\int_{\mathscr{A}_\mathrm{c}} I B_\mathrm{n}\,\mathrm{d}a = \frac{1}{2}I\Phi\,. \tag{5.171}$$

The concept of the magnetic double layer is explained in Section 5.8. Comparing the result (5.171) with

$$W = \frac{1}{2}L_{11}I^2 = \frac{1}{2}(L_{11}I)I \tag{5.172}$$

yields $\Phi = L_{11}I$. For multiple conductors we obtain from Eq. (5.171) the expression $W = \frac{1}{2}\sum_{i=1}^{n} I_i\Phi_i$. Substituting Eq. (5.150) in slightly edited form,

$$W = \frac{1}{2}\sum_{i=1}^{n} I_i \sum_{j=1}^{n} L_{ij}I_j\,, \tag{5.173}$$

yields

$$\Phi_i = \sum_{j=1}^{n} L_{ij}I_j. \tag{5.174}$$

The apparent inductance relates the flux-linkage to the current in a coil winding. From Eqs. (5.142) and (5.171) we find for a single coil $\Phi = \frac{\int_{\mathscr{V}}\mathbf{A}\cdot\mathbf{J}\,\mathrm{d}V}{I}$. If the currents are time-dependent, the voltage will be

$$U_i = \frac{\mathrm{d}\Phi_i}{\mathrm{d}t} = \sum_{j=1}^{n} L_{ij}\frac{\mathrm{d}I_j}{\mathrm{d}t}\,, \tag{5.175}$$

and for a single coil

$$U = L\frac{\mathrm{d}I}{\mathrm{d}t}\,. \tag{5.176}$$

Remark: In the nested corrector magnet shown in Figure 5.11, a time transient field in the skew-dipole coil does not induce a voltage across the six series-connected coils of the sextupole coil (and vice-versa), as can be verified by adding the mutual inductances from Table 5.1:

$$U_{\text{Dipole}} = U_1 + U_2 = \sum_{j=3}^{8} L_{1j} \frac{\mathrm{d}I_j}{\mathrm{d}t} + \sum_{j=3}^{8} L_{2j} \frac{\mathrm{d}I_j}{\mathrm{d}t} = 0\,. \tag{5.177}$$

In general, for $n \neq m$, a coil generating a $\cos m\varphi$ field distribution cannot induce a voltage in a nested coil that generates a perfect $\cos n\varphi$-dependent field. □

5.11 Magnetic Energy in Nonlinear Circuits

In nonlinear circuits, the stored magnetic energy can be calculated by integrating the magnetic energy density,

$$w_{\mathrm{m}} = \int_0^B \mathbf{H}(\mathbf{B}) \cdot \mathrm{d}\mathbf{B} = \int_0^1 \mathbf{H}(\lambda \mathbf{B}) \cdot \mathbf{B}\, \mathrm{d}\lambda\,, \tag{5.178}$$

over the volume $\mathscr{V}$. This result is obtained from the differential of the magnetic energy density, Eq. (5.139), under the integrability condition,

$$\frac{\partial H_i}{\partial B_j} = \frac{\partial H_j}{\partial B_i}\,, \qquad i, j = x, y, z, \tag{5.179}$$

and the assumption of lossless media, which is a necessary condition to make the integral in Eq. (5.178) path-independent. If we express Eq. (5.178) again in terms of the magnetic vector potential and the current density we obtain

$$W = \int_{\mathscr{V}_{\text{coil}}} \left(\int_0^A \mathbf{J}(\mathbf{r}) \cdot \mathrm{d}\mathbf{A}(\mathbf{r}) \right) \mathrm{d}V. \tag{5.180}$$

5.11.1 Differential Inductance

The usual definition of the apparent inductance, relating winding flux-linkages to winding currents (see Eq. (5.174)), breaks down in the nonlinear case for iron saturation. A more general definition links the current rate of change to the induced voltage:

$$U = L^{\mathrm{d}} \frac{\mathrm{d}I}{\mathrm{d}t}\,. \tag{5.181}$$

The inductance so defined is called *incremental* or *differential inductance* [3] and determined by the voltage measured during the ramping of a magnet. Subject to the conditions that the system be causal and nonhysteretic, and does not contain secondary loops, we find

$$U(t) = \frac{\mathrm{d}\Phi}{\mathrm{d}t} = \frac{\mathrm{d}(L\,I)}{\mathrm{d}t} = L\frac{\mathrm{d}I}{\mathrm{d}t} + I\frac{\mathrm{d}L}{\mathrm{d}t}\,. \tag{5.182}$$

The total differential of L is

$$\mathrm{d}L = \frac{\partial L}{\partial I}\,\mathrm{d}I + \frac{\partial L}{\partial t}\,\mathrm{d}t\,. \tag{5.183}$$

Therefore

$$U(t) = \left(\frac{\partial L}{\partial I}\,I + L\right)\frac{\mathrm{d}I}{\mathrm{d}t} + I\frac{\partial L}{\partial t}\,. \tag{5.184}$$

The term $I\frac{\partial L}{\partial t}$ accounts for time-varying inductances, for example, those due to rotor movements in electrical machines or armature movements in actuators. For stationary magnet systems the term is zero and therefore

$$L^{\mathrm{d}} = L + I\frac{\partial L}{\partial I} = \frac{\mathrm{d}\Phi}{\mathrm{d}I}\,. \tag{5.185}$$

The apparent and differential inductances are equal in the absence of magnetic saturation, as the apparent inductance becomes independent of the winding currents.

Remark: The sign of the induced voltage depends on whether the induction loop is seen as a source or a load. We must ensure that the *Lenz law*[21] is obeyed, in other words, that the induced current creates a field that weakens the coupled magnetic flux. For this purpose, consider the electromagnetic circuit element [11] as shown in Figure 5.13 (right), consisting of a superconducting loop $\mathscr{S}_1$ intercepting a time-transient magnetic flux Φ. The loop is closed via two

21 Heinrich Lenz (1804–1865).

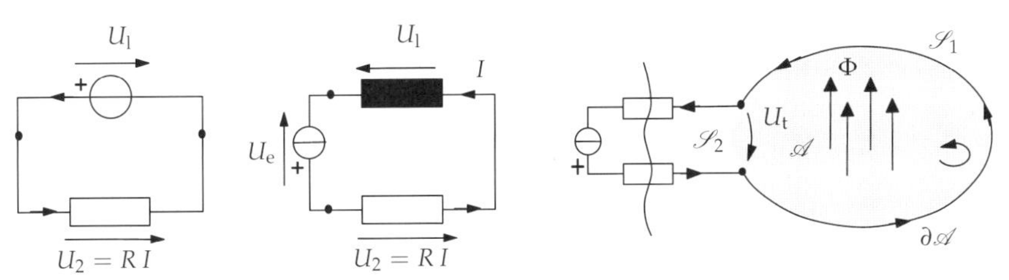

Figure 5.13 Lenz's law and the polarity of $\mathrm{d}\Phi/\mathrm{d}t$. Left: Equivalent circuit diagram in the generator convention. Middle: Equivalent circuit diagram in the load convention. Right: Electromagnetic circuit element.

resistive current leads and connected to a current source. Faraday's law of induction in global form is written as $U(\partial\mathscr{A}) = -\frac{\mathrm{d}}{\mathrm{d}t}\Phi(\mathscr{A})$, where $\mathscr{A}$ is the gray shaded surface with boundary $\partial\mathscr{A} = \mathscr{S}_1 + \mathscr{S}_2$. Because the loop is superconducting, the voltage drop on $\mathscr{S}_1$ is zero and consequently the terminal voltage is $U_\mathrm{t} = -\frac{\mathrm{d}}{\mathrm{d}t}\Phi(\mathscr{A})$. Now consider the generator case,[22] shown in Figure 5.13 (left). We obtain

$$U_1 = U_2 = U_\mathrm{t} = -\frac{\mathrm{d}}{\mathrm{d}t}\Phi(\mathscr{A}) . \tag{5.186}$$

For the load case according to Figure 5.13 (middle), we obtain

$$U_1 = U_\mathrm{e} - U_2 = -U_\mathrm{t} = \frac{\mathrm{d}}{\mathrm{d}t}\Phi(\mathscr{A}) . \tag{5.187}$$

In Eq. (5.187) U_e is the voltage across the external current source. □

Example: Figure 5.14 shows the z-component of the vector potential in the iron yoke of the LHC insertion quadrupole MQXA and the apparent and differential inductances as a function of the excitation current. □

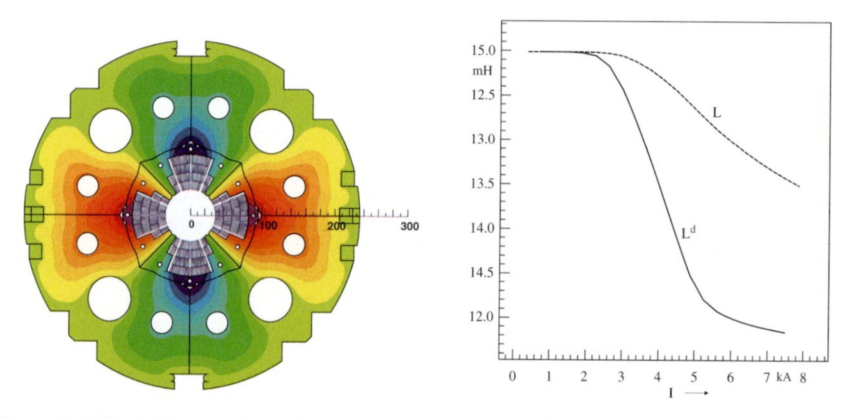

Figure 5.14 Left: Vector potential in the iron yoke of the LHC insertion quadrupole MQXA. Right: Apparent and differential inductances per meter length of the insertion quadrupole.

In nonlinear circuits with multiple coils, the inductances cannot be calculated using Eq. (5.153). Instead, the concept of the differential inductance applied to these circuits yields the *differential mutual inductances*, which can be calculated with

$$L^\mathrm{d}_{jk}(I_1,\dots,I_j,\dots,I_k,\dots,I_n) = \frac{\partial\Phi_k(I_1,\dots,I_j,\dots,I_k,\dots,I_n)}{\partial I_j} . \tag{5.188}$$

22 For simplicity, the lead resistances are combined in a single resistance R in the circuit diagram.

This is the rate-of-change of the flux in coil k, due to a change of excitation in coil j. In a nonlinear circuit, this rate-of-change depends, however, on the excitation of all the n coils in the circuit.

When determining winding flux linkages it can be quite difficult to deal with surface integrals of the magnetic flux density or line integrals of the magnetic vector potential. There is thus a benefit in relating the currents to the stored energy by yet another definition of inductance [11]:

$$W = \frac{1}{2} L^{\mathrm{W}} I^2 . \tag{5.189}$$

Subject to the conditions that the system be causal and nonhysteretic, and does not contain secondary loops, we find

$$\begin{aligned} W &= \int_0^t U\, I \,\mathrm{d}\tau = \int_0^t \frac{\mathrm{d}\Phi(I(\tau))}{\mathrm{d}\tau} I(\tau)\,\mathrm{d}\tau = \int_0^t \frac{\mathrm{d}\Phi(I(\tau))}{\mathrm{d}I}\frac{\mathrm{d}I(\tau)}{\mathrm{d}\tau} I(\tau)\,\mathrm{d}\tau \\ &= \int_0^{I(t)} \frac{\mathrm{d}\Phi}{\mathrm{d}I} I \,\mathrm{d}I = \int_0^{I(t)} L^{\mathrm{d}}\, I \,\mathrm{d}I , \end{aligned} \tag{5.190}$$

from which directly follows,

$$L^{\mathrm{W}} = \frac{2}{I(t)^2}\int_0^{I(t)} L^{\mathrm{d}}\, I\,\mathrm{d}I = \int_0^1 2\lambda L^{\mathrm{d}}(\lambda I)\,\mathrm{d}\lambda . \tag{5.191}$$

5.12 Magnetic Forces and the Maxwell Stress Tensor

The electromagnetic force on a current-carrying conductor is given by

$$F_{\mathrm{m}} = \int_{\mathcal{V}} \mathbf{J} \times \mathbf{B}\,\mathrm{d}V . \tag{5.192}$$

In the case of line-currents coaxial to the z-axis this is modified to

$$\frac{F_{\mathrm{m}}}{\ell} = \sum_{i=1}^{n} I_i\, \mathbf{e}_z \times \mathbf{B} , \tag{5.193}$$

where n is the total number of line-currents. For a domain that is free of all magnetic materials, Eq. (5.192) yields

$$F_{\mathrm{m}} = \int_{\mathcal{V}} \frac{1}{\mu_0}(\operatorname{curl}\mathbf{B}) \times \mathbf{B}\,\mathrm{d}V . \tag{5.194}$$

Because

$$(\operatorname{curl}\mathbf{B}) \times \mathbf{B} = \begin{vmatrix} \mathbf{e}_x & \mathbf{e}_y & \mathbf{e}_z \\ \frac{\partial B_z}{\partial y} - \frac{\partial B_y}{\partial z} & \frac{\partial B_x}{\partial z} - \frac{\partial B_z}{\partial x} & \frac{\partial B_y}{\partial x} - \frac{\partial B_x}{\partial y} \\ B_x & B_y & B_z \end{vmatrix} , \tag{5.195}$$

the x-component of this vector field is given by

$$[(\operatorname{curl} \mathbf{B}) \times \mathbf{B}] \cdot \mathbf{e}_x = B_z \frac{\partial B_x}{\partial z} - B_z \frac{\partial B_z}{\partial x} - B_y \frac{\partial B_y}{\partial x} + B_y \frac{\partial B_x}{\partial y}$$

$$= \frac{\partial}{\partial x} \left(B_x^2 - \frac{1}{2} |\mathbf{B}|^2 \right) + \frac{\partial}{\partial y} (B_y B_x) + \frac{\partial}{\partial z} (B_z B_x) - B_x \operatorname{div} \mathbf{B}. \quad (5.196)$$

By repeating this step for the remaining components, Eq. (5.195) can be rewritten as

$$(\operatorname{curl} \mathbf{B}) \times \mathbf{B} = \operatorname{div} \mathbf{S}_{\mathrm{m}} - \mathbf{B} \operatorname{div} \mathbf{B}, \quad (5.197)$$

where the magnetic *Maxwell stress tensor* is defined by

$$\mathbf{S}_{\mathrm{m}} := (\sigma_{ij}) = \begin{pmatrix} B_x^2 - \frac{1}{2}|\mathbf{B}|^2 & B_x B_y & B_x B_z \\ B_y B_x & B_y^2 - \frac{1}{2}|\mathbf{B}|^2 & B_y B_z \\ B_z B_x & B_z B_y & B_z^2 - \frac{1}{2}|\mathbf{B}|^2 \end{pmatrix}. \quad (5.198)$$

The divergence of a tensor is defined by the expression

$$(\operatorname{div} \mathbf{S}_{\mathrm{m}})_i := \sum_{j=1}^{3} \frac{\partial \sigma_{ij}}{\partial x_j}. \quad (5.199)$$

Therefore, using Eqs. (5.194) and (5.197), $\operatorname{div} \mathbf{B} = 0$, and the generalized divergence theorem, we obtain the expression for the electromagnetic force:

$$\mathbf{F}_{\mathrm{m}} = \int_{\mathcal{V}} \frac{1}{\mu_0} \operatorname{div} \mathbf{S}_{\mathrm{m}} \, \mathrm{d}V = \int_{\partial \mathcal{V}} \frac{1}{\mu_0} \mathbf{S}_{\mathrm{m}} \cdot \mathbf{n} \, \mathrm{d}a \,. \quad (5.200)$$

Introducing the tensor components (5.198) yields

$$\mathbf{S}_{\mathrm{m}} \cdot \mathbf{n} = (\mathbf{B} \cdot \mathbf{n})\mathbf{B} - \frac{1}{2} |\mathbf{B}|^2 \mathbf{n} \quad (5.201)$$

and

$$\mathbf{F}_{\mathrm{m}} = \int_{\partial \mathcal{V}} \left(\frac{1}{\mu_0} (\mathbf{B} \cdot \mathbf{n})\mathbf{B} - \frac{1}{2\mu_0} |\mathbf{B}|^2 \mathbf{n} \right) \mathrm{d}a \,. \quad (5.202)$$

From Eq. (5.200), the force components acting on a number of objects inside a closed surface, can be calculated with

$$F_i = \int_{\partial \mathcal{V}} \sum_j \sigma_{ij} n_j \mathrm{d}a \,, \quad (5.203)$$

where F_i denotes the force components in an orthogonal coordinate system (x_1, x_2, x_3), $\mathrm{d}a$ is the surface element, n_j are the components of the surface normal vector $\mathbf{n}$, and σ_{ij} the components of the Maxwell stress tensor.

Integration along a straight line In the 2D case the surface integral in Eq. (5.203) reduces to an integral on a closed line. To this end, consider a straight line given by the parametric representation

$$\mathbf{r}(s) = (x_0 + s\cos\alpha)\,\mathbf{e}_x + (y_0 + s\sin\alpha)\,\mathbf{e}_y\,, \tag{5.204}$$

where α is the angle with respect to the x-axis and $s \in [s_1, s_2]$. The normal vector to $\mathbf{r}$ is $\mathbf{n} = \pm(\sin\alpha\,\mathbf{e}_x - \cos\alpha\,\mathbf{e}_y)$, with the sign chosen such that $\mathbf{n}$ points to the outward direction of the enclosed area. The contribution to the force components per unit length l in the z-direction is

$$\begin{aligned}
\frac{\mathbf{F}_{\text{line}}}{\ell}\cdot\mathbf{e}_x &= \pm\frac{1}{\mu_0}\int_{s_1}^{s_2}\left(-B_x(\mathbf{r})B_y(\mathbf{r})\cos\alpha + \frac{1}{2}\left(B_x(\mathbf{r})^2 - B_y(\mathbf{r})^2\right)\sin\alpha\right)\mathrm{d}s\,,\\
\frac{\mathbf{F}_{\text{line}}}{\ell}\cdot\mathbf{e}_y &= \pm\frac{1}{\mu_0}\int_{s_1}^{s_2}\left(B_x(\mathbf{r})B_y(\mathbf{r})\sin\alpha - \frac{1}{2}\left(B_y(\mathbf{r})^2 - B_x(\mathbf{r})^2\right)\cos\alpha\right)\mathrm{d}s\,.
\end{aligned} \tag{5.205}$$

If the line is parallel to the x-axis ($\alpha = 0$), we obtain the well-known equations

$$\begin{aligned}
\frac{\mathbf{F}_{\text{line}}}{\ell}\cdot\mathbf{e}_x &= \pm\frac{1}{2\mu_0}\int_{s_1}^{s_2}\left(2B_x(\mathbf{r})B_y(\mathbf{r})\right)\mathrm{d}s\,,\\
\frac{\mathbf{F}_{\text{line}}}{\ell}\cdot\mathbf{e}_y &= \pm\frac{1}{2\mu_0}\int_{s_1}^{s_2}\left(B_y(\mathbf{r})^2 - B_x(\mathbf{r})^2\right)\mathrm{d}s\,.
\end{aligned} \tag{5.206}$$

In order to obtain a closed integration path several line contributions must be added. Another possibility is to close the integration path far away from the sources, where the components of the Maxwell stress tensor vanish.

Example: Consider a force computation for a high gradient ($450\,\mathrm{T\,m^{-1}}$) quadrupole with permanent magnets in a so-called *zero-clearance* design; cf. Figure 5.15. The principle is known also as a circular *Halbach array*.[23] Integrating the Maxwell stress tensor for different displacements of one of the sectors gives an estimate of the forces acting during the assembly of the device.

The symmetry causes the normal force components to cancel for the two integrals over the lines shown in Figure 5.15, whereas the tangential components add up. The end points of the lines were chosen sufficiently far outside the magnet for field components to vanish. Figure 5.15 (right) shows the tangential force as a function of the sector displacement. □

23 Klaus Halbach (1925–2000).

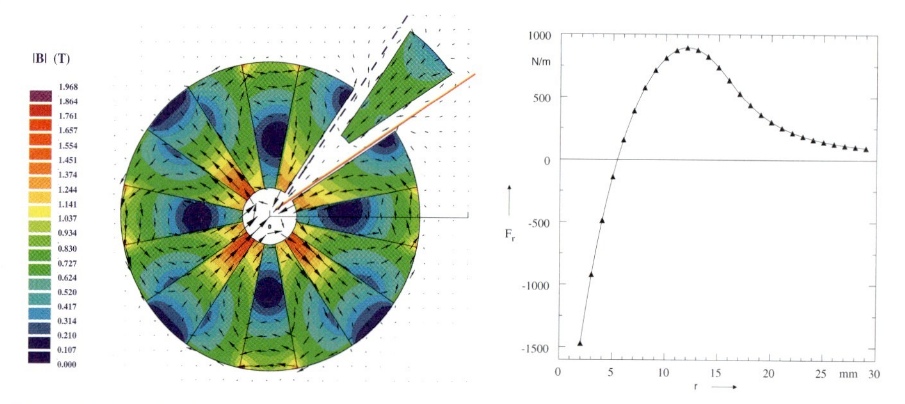

Figure 5.15 Left: Cross section of a permanent magnet quadrupole in zero-clearance design. One sector is radially displaced by 10 mm as it may happen during the assembly process. Right: Radial force component (F_r) as a function of the radial displacement of one permanent magnet sector.

Integration on a circle Consider a circle in the plane $z = 0$, centered at (x_0, y_0), of radius R, and parametric representation

$$\mathbf{r}(\varphi) = (x_0 + R\cos\varphi)\,\mathbf{e}_x + (y_0 + R\sin\varphi)\,\mathbf{e}_y \tag{5.207}$$

for $\varphi \in [\pi, \pi]$. The normal vector $\mathbf{n}$ to this loop is $\cos\varphi\,\mathbf{e}_x + \sin\varphi\,\mathbf{e}_y$. For the force components per unit length ℓ in the z-direction Eq. (5.203) yields

$$\frac{\mathbf{F}_{\text{circle}}}{\ell} \cdot \mathbf{e}_x = \frac{R}{\mu_0} \int_{-\pi}^{\pi} \left(B_x(\mathbf{r}) B_y(\mathbf{r}) \sin\varphi + \frac{1}{2} \left(B_x(\mathbf{r})^2 - B_y(\mathbf{r})^2 \right) \cos\varphi \right) \mathrm{d}\varphi\,, \tag{5.208}$$

$$\frac{\mathbf{F}_{\text{circle}}}{\ell} \cdot \mathbf{e}_y = \frac{R}{\mu_0} \int_{-\pi}^{\pi} \left(B_x(\mathbf{r}) B_y(\mathbf{r}) \cos\varphi + \frac{1}{2} \left(B_y(\mathbf{r})^2 - B_x(\mathbf{r})^2 \right) \sin\varphi \right) \mathrm{d}\varphi\,. \tag{5.209}$$

5.13
Fields and Potentials of Magnetization Currents

Consider a magnetostatic problem for a given magnetization $\mathbf{M}$ and for $\mathbf{J} = 0$. The flux-conservation law can be written as $\operatorname{div}\mathbf{B} = \mu_0 \operatorname{div}(\mathbf{H} + \mathbf{M}) = 0$. With the field expressed by the magnetic scalar potential, $\mathbf{H} = -\operatorname{grad}\phi_{\mathrm{m}}$, it follows that

$$\operatorname{div}(\operatorname{grad}\phi_{\mathrm{m}}) = \nabla^2 \phi_{\mathrm{m}} = -\frac{\rho_{\mathrm{m}}}{\mu_0}, \tag{5.210}$$

where

$$\rho_{\mathrm{m}} = -\operatorname{div} \mu_0 \mathbf{M} . \tag{5.211}$$

If there are no boundary surfaces, the solution is given in Eq. (5.17):

$$\phi_{\mathrm{m}}(\mathbf{r}) = \frac{1}{4\pi\mu_0} \int_{\mathcal{V}} \frac{\rho_{\mathrm{m}}}{|\mathbf{r}-\mathbf{r}'|} \, \mathrm{d}V' . \tag{5.212}$$

M is usually not given explicitly. In hard ferromagnets we may specify the magnetization inside a volume and assume that it jumps to zero at the volume's boundary. In this case we must also consider an effective magnetic surface charge density

$$\sigma_{\mathrm{m}} = \mu_0 \, \mathbf{n} \cdot \mathbf{M} ; \tag{5.213}$$

see Section 4.6.6, and the solution (5.212) must be augmented to

$$\phi_{\mathrm{m}}(\mathbf{r}) = \frac{1}{4\pi\mu_0} \int_{\mathcal{V}} \frac{\rho_{\mathrm{m}}}{|\mathbf{r}-\mathbf{r}'|} \, \mathrm{d}V' + \frac{1}{4\pi\mu_0} \int_{\partial\mathcal{V}} \frac{\sigma_{\mathrm{m}}}{|\mathbf{r}-\mathbf{r}'|} \, \mathrm{d}a' . \tag{5.214}$$

In the special case of uniform magnetization within the volume, the first term vanishes and only the surface integral over σ_{m} remains. In the absence of boundary surfaces, and thus for smooth distributions of **M**, the second term vanishes. Using the relation

$$\frac{\operatorname{div} \mathbf{M}}{|\mathbf{r}-\mathbf{r}'|} = \operatorname{div} \left(\frac{\mathbf{M}}{|\mathbf{r}-\mathbf{r}'|} \right) - \mathbf{M} \cdot \operatorname{grad} \left(\frac{1}{|\mathbf{r}-\mathbf{r}'|} \right) , \tag{5.215}$$

which can be derived from Eq. (3.67), substituting Eqs. (5.211) and (5.213) into the solution (5.214), and applying the Gauss theorem yields

$$\phi_{\mathrm{m}}(\mathbf{r}) = \frac{1}{4\pi} \int_{\mathcal{V}} \mathbf{M} \cdot \operatorname{grad} \left(\frac{1}{|\mathbf{r}-\mathbf{r}'|} \right) \mathrm{d}V' = \frac{1}{4\pi} \int_{\mathcal{V}} \frac{\mathbf{M} \cdot (\mathbf{r}-\mathbf{r}')^3}{|\mathbf{r}-\mathbf{r}'|} \, \mathrm{d}V' . \tag{5.216}$$

For an observation point far from the nonvanishing magnetization the following relation holds:

$$\phi_{\mathrm{m}}(\mathbf{r}) = \frac{\mathbf{m}}{4\pi} \cdot \frac{\mathbf{r}-\mathbf{r}'}{|\mathbf{r}-\mathbf{r}'|^3} = \frac{m \cos \vartheta}{|\mathbf{r}-\mathbf{r}'|^2} , \tag{5.217}$$

see Figure 5.2 for the definition of the angle ϑ. For arbitrary, localized current-distributions, Eq. (5.217) has only approximate validity, because of the use of only the first term in the series expansion of $|\mathbf{r}-\mathbf{r}'|^{-1}$. In Eq. (5.217), the elementary magnetic moment **m** acts as a point-like source of the potential [7].

We can now derive the magnetic vector potential due to magnetization currents. Again consider a magnetostatic problem with a given magnetization **M** and $\mathbf{J} = 0$. It follows that Ampère's law can be written as

$$\operatorname{curl} \mathbf{H} = \operatorname{curl} \left(\frac{\mathbf{B}}{\mu_0} - \mathbf{M} \right) = \mathbf{0} . \tag{5.218}$$

Setting $\mathbf{B} = \operatorname{curl}\mathbf{A}$ it follows that

$$\nabla^2\mathbf{A} = -\mu_0 \operatorname{curl}\mathbf{M} = -\mu_0\mathbf{J}_\mathrm{m}\,. \tag{5.219}$$

The solution is known from Eq. (5.29),

$$\mathbf{A}(\mathbf{r}) = \frac{\mu_0}{4\pi}\int_{\mathcal{V}} \frac{\mathbf{J}_\mathrm{m}(\mathbf{r}')}{|\mathbf{r}-\mathbf{r}'|}\,\mathrm{d}V', \tag{5.220}$$

if there are no boundary surfaces. Otherwise, we must account for additional surface-currents $\boldsymbol{\alpha} = -\mathbf{n}\times\mathbf{M}$; see Section 4.6.6. In this case, the solution (5.220) must be augmented and yields

$$\mathbf{A}(\mathbf{r}) = \frac{\mu_0}{4\pi}\int_{\mathcal{V}} \frac{\mathbf{J}_\mathrm{m}(\mathbf{r}')}{|\mathbf{r}-\mathbf{r}'|}\,\mathrm{d}V' + \frac{\mu_0}{4\pi}\int_{\partial\mathcal{V}} \frac{\boldsymbol{\alpha}}{|\mathbf{r}-\mathbf{r}'|}\,\mathrm{d}a'\,. \tag{5.221}$$

In the special case of uniform magnetization within the volume, the first term vanishes and only the surface integral over $\boldsymbol{\alpha}$ contributes to $\mathbf{A}(\mathbf{r})$. In the absence of boundary surfaces the second term vanishes. Using the relation

$$\frac{\operatorname{curl}\mathbf{M}}{|\mathbf{r}-\mathbf{r}'|} = \operatorname{curl}\left(\frac{\mathbf{M}}{|\mathbf{r}-\mathbf{r}'|}\right) + \mathbf{M}\times\operatorname{grad}\left(\frac{1}{|\mathbf{r}-\mathbf{r}'|}\right), \tag{5.222}$$

derived from Eq. (3.69), and xpressing the solution (5.221) in terms of $\mathbf{M}$ yields

$$\mathbf{A}(\mathbf{r}) = -\frac{\mu_0}{4\pi}\int_{\mathcal{V}} \mathbf{M}\times\operatorname{grad}\left(\frac{1}{|\mathbf{r}-\mathbf{r}'|}\right)\mathrm{d}V' = \frac{\mu_0}{4\pi}\int_{\mathcal{V}} \frac{\mathbf{M}\times(\mathbf{r}-\mathbf{r}')}{|\mathbf{r}-\mathbf{r}'|^3}\,\mathrm{d}V'\,. \tag{5.223}$$

For an observation point far from the nonvanishing magnetization we derive, approximately,

$$\mathbf{A}(\mathbf{r}) = \frac{\mu_0\mathbf{m}\times(\mathbf{r}-\mathbf{r}')}{4\pi|\mathbf{r}-\mathbf{r}'|^3} = -\frac{\mu_0\mathbf{m}}{4\pi}\times\operatorname{grad}\left(\frac{1}{|\mathbf{r}-\mathbf{r}'|}\right) = \frac{\mu_0}{4\pi}\operatorname{curl}\left(\frac{\mathbf{m}}{|\mathbf{r}-\mathbf{r}'|}\right). \tag{5.224}$$

At any point far from the nonvanishing magnetization the magnetic flux density can be calculated as the curl of Eq. (5.224), which results in the dipolar field,

$$\mathbf{B} = \frac{\mu_0}{4\pi}\left(\frac{3\mathbf{r}(\mathbf{r}\cdot\mathbf{m})}{|\mathbf{r}-\mathbf{r}'|^5} - \frac{\mathbf{m}}{|\mathbf{r}-\mathbf{r}'|^3}\right). \tag{5.225}$$

5.14 Magnetic Levitation

The force on a magnetized specimen of volume $\mathcal{V}$ is given by

$$\mathbf{F} = \int_{\mathcal{V}} \mathbf{J}_\mathrm{m}\times\mathbf{B}\,\mathrm{d}V\,, \tag{5.226}$$

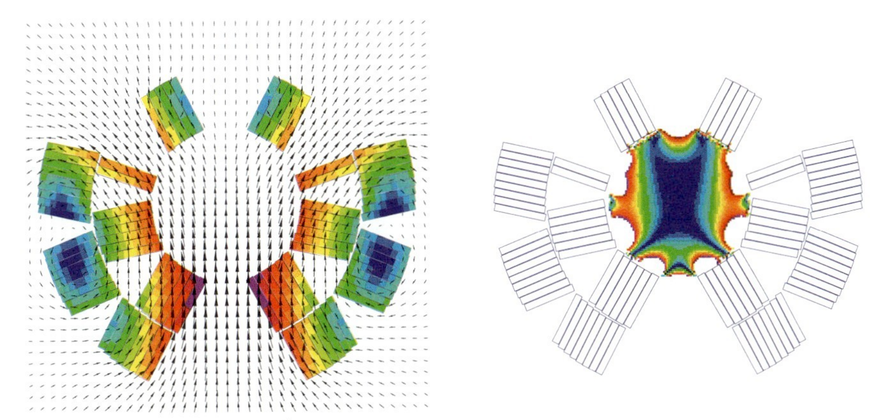

Figure 5.16 Left: Field distribution in a combined-function magnet producing a constant gradient field in a domain used for microgravity experiments. Right: Domain in which 10 $\mathrm{T}^2\,\mathrm{m}^{-1}$ is achieved required for the levitation of water [15].

where $\mathbf{B}$ is the applied field. The total torque about the barycenter of $\mathscr{V}$ is given by [7],

$$\mathbf{T} = \mathbf{m} \times \mathbf{B}\,. \tag{5.227}$$

In an inhomogeneous field there is a resulting net force in the direction of decreasing field if $\mathbf{m}$ is antiparallel to $\mathbf{B}$; diamagnetic objects are repelled by magnetic fields. The governing equation is

$$\mathbf{F} = \operatorname{grad}\,(\mathbf{m} \cdot \mathbf{B})\,. \tag{5.228}$$

To levitate an object, the gravity $mg = \rho V g$ must be balanced by the force

$$F = |\operatorname{grad}\,(\mathbf{m} \cdot \mathbf{B})|\,. \tag{5.229}$$

ρ is the material density, V the volumetric extent, and g the earth gravitational constant. If the distortion of the ambient field $B = |\mathbf{B}|$ by the object is disregarded, the induced magnetic moment is given by

$$|\mathbf{m}| = \frac{\chi}{\mu_0} V B, \tag{5.230}$$

where χ is the material's susceptibility. The magnetic force results in

$$F = \frac{\chi}{\mu_0} V\,|\operatorname{grad}\,(B^2)|\,. \tag{5.231}$$

Consequently the vertical field gradient grad (B^2) required for levitation must be larger than $2\mu_0\rho g/\chi$. The force field (5.228), visualized in Figure 5.16

(right), gives rise to magnetic levitation of superconducting samples. In the case of water the magnetic susceptibility χ is -8.8×10^{-6}, and therefore field gradients of approximately 10 $\mathrm{T}^2\,\mathrm{m}^{-1}$ are required. Figure 5.16 shows a combined-function magnet [15] based on accelerator-magnet technology, which is able to create such field gradients.

References

1 Berry, M.V., Geim, A.K.: Of flying frogs and levitrons, European Journal of Physics 18, 1997

2 Bird, P.F., Friedman, M.D.: Handbook of Elliptic Integrals for Engineers and Scientists, Springer, Berlin, 1971

3 Demerdash, N.A., Nehl, T.W.: Electric machinery parameters and torques by current and energy perturbations from field computations – part 1: theory and formulation, IEEE Transactions on Energy Conversion, 1999

4 Gross, P.W., Kotiuga P.R.: Electromagnetic Theory and Computation: A Topological Approach, Cambridge University Press, Cambridge, 2003

5 Grover, F.W.: Inductance Calculations, Dover Phoenix Edition, reprint 2004 of the 1962 edition

6 Halbach, K.: Design of permanent multipole magnets with oriented rare earth cobalt material, Nuclear Instruments and Methods, 1980

7 Jackson, J.D.: Classical Electrodynamic, John Wiley & Sons, New York, 1997

8 Kellogg, O..D.: Foundations of Potential Theory, Dover, New York, 1967

9 Kurz, S.: Die numerische Behandlung elektromechanischer Systeme mit Hilfe der Kopplung der Methode der finiten Elemente und der Randelementmethode, Fortschritt-Berichte VDI, Reihe 21, Nr. 252

10 Kurz, S.: Vorlesung über Theoretische Elektrotechnik, Hamburg 2004, Lecture Notes

11 Kurz, S.: Some remarks about flux linkage and inductance, Advances in Radio Science, 2004

12 Lehner, G.: Elektromagnetische Feldtheorie für Ingenieure und Physiker, Springer, Berlin, 1990

13 Maxwell, J.C: A Treatise on Electricity and Magnetism, Dover, New York, 1954

14 Van Oosterom, A., Strackee, J.: The solid angle of a plane triangle, IEEE Transactions on Biomedical Engineering, 1983

15 Quettier, L.: Contribution methodologique a la conception de systemes supraconducteurs de levitation magnetique, These, Institut National Polytechnique de Lorraine, 2003

16 Tawada, M., Nakayama H., Satoh, K.: Special quadrupole magnets for KEKB interaction region, Proceedings of EPAC 2000

17 Wilson, M.N.: Superconducting Magnets, Oxford Science Publications, Oxford, 1983

18 Woebke, G. et al.: Precision septum halfquadrupoles for the HERA luminosity-upgrade, IEEE Transactions on Applied Superconductivity, 2000

6
Field Harmonics

Fair goes the dancing when the sitar is tuned;
Tune us the sitar neither low nor high,
and we will dance away the hearts of men.
The string overstretched breaks, the music flies,
the string overslack is dumb and the music dies;
Tune us the sitar neither low nor high.

The teaching of Buddha.

In the aperture of an accelerator magnet, free of currents and magnetized material, both magnetic scalar and vector potentials can be employed for the formulation of a boundary value problem. In two dimensions, both formulations yield a scalar Laplace equation for the magnetic scalar potential and for the z-component of the magnetic vector potential. As before, we denote the problem domain Ω and consider smooth scalar fields $\phi_{\mathrm{m}}, A_z \in \mathcal{S}(\Omega)$.

The field quality in accelerator magnets is conveniently described by a set of Fourier coefficients, known to the magnet design community as *field harmonics* or *multipole coefficients*. The method used for the calculation of field harmonics is based on finding a general solution that satisfies the Laplace equation in a suitable coordinate system. The integration constants in the general solution, obtained with the separation of variables technique, are then determined by comparison with the boundary values; in circular coordinates these boundary values are given by the radial or azimuthal field components at a given reference radius.

We will discuss three different methods for determining the field harmonics. The first is a comparison of the integration constants in the general solution with the Fourier series expansion of the field components on the domain boundary. In the case of accelerator magnets, the domain boundary is often chosen as a circle with a radius of two-thirds of the aperture radius. For the time being we shall assume that the field components are known from measurements or numerical field calculations.

Field Computation for Accelerator Magnets. Stephan Russenschuck

ISBN: 978-3-527-40769-9

For the second method, we obtain an approximation of the field harmonics from the coefficients in a suitable Taylor series expansion of the calculated (or measured) field about a symmetry axis.

By studying the theorems on harmonic functions in Section 5.3, we found that if a scalar field ϕ is harmonic in a closed domain, ϕ cannot take a maximum or minimum at any interior point of that domain. Harmonic fields are thus unable to account for the line-currents we will use for the calculation of coil fields. The field of line-currents can be found by means of the fundamental solution of the Laplace operator and the Green functions of free space; see Chapter 5. The third method of calculating field harmonics thus consists of a series expansion of the reciprocal radius in these Green functions and a comparison of the coefficients with the general solution.

We begin with 2D circular coordinates because they are the most commonly used for computation of the field in long accelerator magnets. This is perfectly in accordance with the technique of measuring magnetic fields by means of rotating pickup coils. For the case of axisymmetric (solenoidal) magnets we will examine the symmetric version of the spherical coordinates, also known as zonal coordinates. Finally, we will present the solution of a rectangular boundary value problem for the special case of a nonsaturated, iron-dominated dipole magnet.

6.1 Circular Harmonics

We will first discuss the situation in 2D circular coordinates for apertures in long accelerator magnets. A general solution that satisfies the Laplace equation, $\nabla^2 A_z = 0$, can be found by the separation of variables method.[1] For $A_z = \rho(r)\phi(\varphi)$ we obtain

$$\frac{\partial A_z}{\partial r} = \frac{\mathrm{d}\rho(r)}{\mathrm{d}r}\phi(\varphi), \tag{6.1}$$

$$\frac{\partial^2 A_z}{\partial r^2} = \frac{\mathrm{d}^2\rho(r)}{\mathrm{d}r^2}\phi(\varphi), \tag{6.2}$$

$$\frac{\partial^2 A_z}{\partial \varphi^2} = \frac{\mathrm{d}^2\phi(\varphi)}{\mathrm{d}\varphi^2}\rho(r). \tag{6.3}$$

Therefore, the Laplace equation in circular coordinates,

$$r^2\frac{\partial^2 A_z}{\partial r^2} + r\frac{\partial A_z}{\partial r} + \frac{\partial^2 A_z}{\partial \varphi^2} = 0, \tag{6.4}$$

1 First published by Jean d'Alembert (1717–1783) and taken up by Joseph Fourier himself. There exist eleven separable coordinate systems with orthogonal intersections of their coordinate surfaces.

can be rewritten in the form

$$\underbrace{\frac{1}{\rho(r)}\left(r^2\frac{\mathrm{d}^2\rho(r)}{\mathrm{d}r^2}+r\frac{\mathrm{d}\rho(r)}{\mathrm{d}r}\right)}_{n^2}=\underbrace{-\frac{1}{\phi(\varphi)}\frac{\mathrm{d}^2\phi(\varphi)}{\mathrm{d}\varphi^2}}_{n^2}. \tag{6.5}$$

Since the left-hand side of Eq. (6.5) depends only on r and the right-hand side only on φ, a separation constant n^2 can be introduced, and for the case $n \neq 0$, two ordinary differential equations are obtained:

$$r^2\frac{\mathrm{d}^2\rho(r)}{\mathrm{d}r^2}+r\frac{\mathrm{d}\rho(r)}{\mathrm{d}r}-n^2\rho(r)=0, \tag{6.6}$$

$$\frac{\mathrm{d}^2\phi(\varphi)}{\mathrm{d}\varphi^2}+n^2\phi(\varphi)=0, \tag{6.7}$$

with the solutions,

$$\rho_n(r)=\mathcal{E}_n\,r^n+\mathcal{F}_n\,r^{-n}, \tag{6.8}$$

$$\phi_n(\varphi)=\mathcal{G}_n\,\sin n\varphi+\mathcal{H}_n\,\cos n\varphi. \tag{6.9}$$

As the vector potential is single-valued, it must be a periodic function in φ with $A_z(r,0)=A_z(r,2\pi)$. The separation constant n takes integer values and therefore the general solution of the homogeneous differential equation (6.4) is

$$A_z(r,\varphi)=\sum_{n=1}^{\infty}(\mathcal{E}_n r^n+\mathcal{F}_n r^{-n})(\mathcal{G}_n\sin n\varphi+\mathcal{H}_n\cos n\varphi). \tag{6.10}$$

We can now express the field in the aperture of an accelerator magnet according to Eq. (6.10). We will define our model problem as shown in Figure 6.1, consisting of an aperture domain Ω_a and an exterior domain Ω_e, with $\partial\Omega_\mathrm{a}=\Gamma_\mathrm{a}$ and $\partial\Omega_\mathrm{e}=\Gamma_\mathrm{e}\cup\Gamma_\infty$.

Let us consider the magnet aperture as the problem domain. The condition that the flux density is finite at $r=0$ imposes $\mathcal{F}_n=0$. In the exterior domain Ω_e the field must vanish at Γ_∞, which results in $\mathcal{E}_n=0$. The introduction of new constants, $\mathcal{C}_n=\mathcal{E}_n\mathcal{G}_n$ and $\mathcal{D}_n=-\mathcal{E}_n\mathcal{H}_n$, in Eq. (6.10) yields the general solution for the vector potential in Ω_a:

$$A_z(r,\varphi)=\sum_{n=1}^{\infty}r^n(\mathcal{C}_n\sin n\varphi-\mathcal{D}_n\cos n\varphi). \tag{6.11}$$

The field components can then be expressed as

$$B_r(r,\varphi)=\frac{1}{r}\frac{\partial A_z}{\partial\varphi}=\sum_{n=1}^{\infty}nr^{n-1}(\mathcal{C}_n\cos n\varphi+\mathcal{D}_n\sin n\varphi), \tag{6.12}$$

$$B_\varphi(r,\varphi)=-\frac{\partial A_z}{\partial r}=-\sum_{n=1}^{\infty}nr^{n-1}(\mathcal{C}_n\sin n\varphi-\mathcal{D}_n\cos n\varphi), \tag{6.13}$$

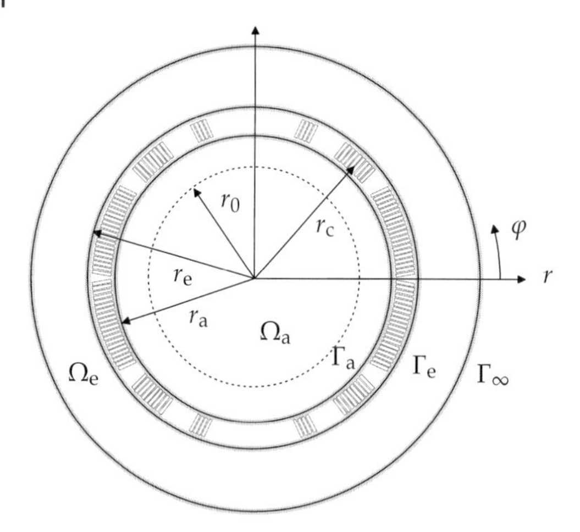

Figure 6.1 Boundary value problem in 2D circular coordinates. The aperture domain is denoted Ω_a and the exterior domain Ω_e. The boundary $\partial\Omega_a$ of the aperture domain is denoted Γ_a. The boundary of the exterior domain consists of two parts: $\partial\Omega_e = \Gamma_e \cup \Gamma_\infty$.

in Ω_a. Each value of the integer n in the solution of the Laplace equation corresponds to a specific flux distribution generated by ideal magnet geometries. The three lowest values, $n = 1,2,3$ correspond to the dipole, quadrupole, and sextupole flux density distributions.

6.1.1
Determining the Multipole Coefficients

The *multipole coefficients*, or *field harmonics* $\mathcal{C}_n$ and $\mathcal{D}_n$, are undetermined at this stage. We will discuss three different approaches to their calculation:

- Fourier series expansion of a calculated field component along a circle, or alternatively, an expansion of the voltage signal induced in a rotating measurement coil.
- In the case of an up/down symmetry, comparison of the multipole coefficients with the *Taylor coefficients*[2] of a series expansion of the (known) flux density at the horizontal median plane.
- Series expansion of the Green kernel in the potentials generated by transport and magnetization currents, and comparison of the coefficients with the multipoles.

2 Brook Taylor (1685–1731).

6.1.1.1 **Fourier Series Expansion of the Radial Field Component**

Assuming that the radial component of the magnetic flux density is measured or calculated at a reference radius $r = r_0$ as a function of the angular position φ, we obtain the Fourier series expansion of the field components

$$B_r(r_0, \varphi) = \sum_{n=1}^{\infty} (B_n(r_0) \sin n\varphi + A_n(r_0) \cos n\varphi), \tag{6.14}$$

$$B_\varphi(r_0, \varphi) = \sum_{n=1}^{\infty} (B_n(r_0) \cos n\varphi - A_n(r_0) \sin n\varphi), \tag{6.15}$$

where

$$A_n(r_0) = \frac{1}{\pi} \int_0^{2\pi} B_r(r_0, \varphi) \cos n\varphi \, \mathrm{d}\varphi, \qquad n = 1, 2, 3, \ldots, \tag{6.16}$$

$$B_n(r_0) = \frac{1}{\pi} \int_0^{2\pi} B_r(r_0, \varphi) \sin n\varphi \, \mathrm{d}\varphi, \qquad n = 1, 2, 3, \ldots. \tag{6.17}$$

Because the magnetic flux density is divergence free, $A_0 = 0$.

Remark: In computational practice, the B_r field components are numerically calculated at N discrete points in the interval $[0, 2\pi)$

$$\varphi_k = \frac{2\pi k}{N}, \qquad k = 0, 1, 2, \ldots, N-1. \tag{6.18}$$

This allows the calculation of two times N Fourier coefficients by the *discrete Fourier transform* (DFT):

$$A_n(r_0) \approx \frac{2}{N} \sum_{k=0}^{N-1} B_r(r_0, \varphi_k) \cos n\varphi_k, \tag{6.19}$$

$$B_n(r_0) \approx \frac{2}{N} \sum_{k=0}^{N-1} B_r(r_0, \varphi_k) \sin n\varphi_k. \tag{6.20}$$

The proof is deferred to Section 9.4. For our purposes sufficient accuracy is achieved by setting $N = 60$. □

Comparing the coefficients in Eqs. (6.12) and (6.14) we obtain for $r_0 < r_\mathrm{a}$:

$$A_n(r_0) = n r_0^{n-1} \mathcal{C}_n, \qquad B_n(r_0) = n r_0^{n-1} \mathcal{D}_n. \tag{6.21}$$

Thus

$$B_r(r_0,\varphi) = \sum_{n=1}^{\infty} (B_n(r_0)\sin n\varphi + A_n(r_0)\cos n\varphi)$$
$$= B_N \sum_{n=1}^{\infty} (b_n(r_0)\sin n\varphi + a_n(r_0)\cos n\varphi), \tag{6.22}$$

$$B_\varphi(r_0,\varphi) = \sum_{n=1}^{\infty} (B_n(r_0)\cos n\varphi - A_n(r_0)\sin n\varphi)$$
$$= B_N \sum_{n=1}^{\infty} (b_n(r_0)\cos n\varphi - a_n(r_0)\sin n\varphi). \tag{6.23}$$

The *normal* and *skew multipole coefficients* $B_n(r_0), A_n(r_0)$ are given in units of tesla at a reference radius r_0. For the LHC magnets, the reference radius is 17 mm, which is about 2/3 of the magnet aperture. The small $b_n(r_0)$ and $a_n(r_0)$ denote the normal and skew relative multipole coefficients related to the main field[3] $B_N(r_0)$. The $b_n(r_0)$ and $a_n(r_0)$ are dimensionless. For good field quality the *harmonic distortion factor*

$$F_d(r_0) := \sum_{n=1;n\neq N}^{K} \left(b_n^2(r_0) + a_n^2(r_0)\right) \tag{6.24}$$

must be a few units in 10^{-4} at the reference radius.

Remark 1: In some documents [6] the field harmonics are defined independently of the reference radius by

$$\mathcal{B}_n = \frac{B_n(r_0)}{r_0^{n-1}}, \tag{6.25}$$

where $[\mathcal{B}_1] = 1\,\mathrm{T}$, $[\mathcal{B}_2] = 1\,\mathrm{T\,m^{-1}}$, $[\mathcal{B}_3] = 1\,\mathrm{T\,m^{-2}}$, etc. It is also common practice, especially in American literature on the subject [3], to have the summation index run from 0 to ∞, which emphasizes the field as the fundamental quantity, rather than the potential, resulting in B_0 as the dipole field component and the removal of the powers of $n-1$ in Eq. (6.21). □

For the scaling of the multipole coefficients to any radius r inside the aperture it holds that

$$A_n(r) = \left(\frac{r}{r_0}\right)^{n-1} A_n(r_0), \qquad B_n(r) = \left(\frac{r}{r_0}\right)^{n-1} B_n(r_0). \tag{6.26}$$

3 This is B_1 for the dipole, B_2 for the quadrupole, etc.

Hence

$$b_n(r) = \frac{B_n(r)}{B_N(r)} = \frac{\left(\frac{r}{r_0}\right)^{n-1} B_n(r_0)}{\left(\frac{r}{r_0}\right)^{N-1} B_N(r_0)} = \left(\frac{r}{r_0}\right)^{n-N} b_n(r_0), \tag{6.27}$$

with a similar expression for the skew components:

$$a_n(r) = \left(\frac{r}{r_0}\right)^{n-N} a_n(r_0). \tag{6.28}$$

Thus for the field components in Ω_a,

$$B_r(r,\varphi) = \sum_{n=1}^{\infty} \left(\frac{r}{r_0}\right)^{n-1} (B_n(r_0) \sin n\varphi + A_n(r_0) \cos n\varphi)$$

$$= B_N \sum_{n=1}^{\infty} \left(\frac{r}{r_0}\right)^{n-N} (b_n(r_0) \sin n\varphi + a_n(r_0) \cos n\varphi), \tag{6.29}$$

$$B_\varphi(r,\varphi) = \sum_{n=1}^{\infty} \left(\frac{r}{r_0}\right)^{n-1} (B_n(r_0) \cos n\varphi - A_n(r_0) \sin n\varphi)$$

$$= B_N \sum_{n=1}^{\infty} \left(\frac{r}{r_0}\right)^{n-N} (b_n(r_0) \cos n\varphi - a_n(r_0) \sin n\varphi). \tag{6.30}$$

For nested corrector magnets as shown in Figure B.7, which are designed to produce any combination of horizontal and vertical fields, it is useful to rewrite Eq. (6.22) in the amplitude-phase notation

$$B_r(r_0,\varphi) = \sum_{n=1}^{\infty} F_n(r_0) \sin(n\varphi + \psi_n), \tag{6.31}$$

where the amplitude is

$$F_n(r_0) = \sqrt{A_n^2(r_0) + B_n^2(r_0)} \tag{6.32}$$

and the phase angle of the $2n$-pole term,

$$\psi_n = \begin{cases} \arctan \frac{A_n}{B_n} & \text{if } B_n \geq 0, \\ \arctan \frac{A_n}{B_n} + \pi & \text{if } B_n < 0. \end{cases} \tag{6.33}$$

In numerical field computation, it can be useful to perform a Fourier analysis of the vector potential on the reference radius, thus avoiding the calculation of the flux density by means of differential quotients. Fourier series expansion of the magnetic vector potential at a reference radius yields

$$A_z(r_0,\varphi) = \sum_{n=1}^{\infty} (\mathcal{F}_n(r_0) \cos n\varphi + \mathcal{E}_n(r_0) \sin n\varphi). \tag{6.34}$$

Using Eqs. (6.12) and (6.21), we obtain

$$B_n(r_0) = \frac{-n\,\mathcal{F}_n}{r_0}, \qquad A_n(r_0) = \frac{n\,\mathcal{E}_n}{r_0}. \tag{6.35}$$

Hence for a dipole,

$$A_z(r,\varphi) = -B_1 r_0 \sum_{n=1}^{\infty} \frac{1}{n}\left(\frac{r}{r_0}\right)^{n-1} (b_n(r_0)\cos n\varphi - a_n(r_0)\sin n\varphi). \tag{6.36}$$

The expression of field quality using the multipole coefficients is perfectly in accordance with magnetic measurements using *harmonic coils*, where the periodic variation in flux linkage in rotating coils is analyzed by means of a Fourier series expansion.

Consider a tangential coil as sketched in Figure 6.2 (left), rotating in the aperture of a magnet. For $\varphi = \omega t + \Theta$, where ω is the angular velocity, the flux linkage through coil 1 at time t is given by[4]

$$\begin{aligned}\Phi(t) &= N\ell \int_{\varphi-\delta/2}^{\varphi+\delta/2} B_r(r_0,\varphi) r_0 \mathrm{d}\varphi \\ &= \sum_{n=1}^{\infty} S_n^{\text{tan}} \left[B_n(r_0)\sin(n\omega t + n\Theta) + A_n(r_0)\cos(n\omega t + n\Theta)\right],\end{aligned} \tag{6.37}$$

where

$$S_n^{\text{tan}} := \frac{2N\ell r_0}{n} \sin\left(\frac{n\delta}{2}\right) \tag{6.38}$$

is known as the *coil sensitivity factor*. N is the number of turns in the rotating coil, r_0 the measurement coil radius, ℓ the length, δ the opening angle, and Θ the positioning angle at $t = 0$. The voltage signal at time t is consequently

$$\begin{aligned}U(t) &= -\frac{\mathrm{d}\Phi}{\mathrm{d}t} \\ &= \sum_{n=1}^{\infty} n\omega S_n^{\text{tan}} \left[-B_n(r_0)\cos(n\omega t + n\Theta) + A_n(r_0)\sin(n\omega t + n\Theta)\right].\end{aligned} \tag{6.39}$$

The geometric parameters of the measurement coil result in a constant sensitivity factor which can be calculated and calibrated. The field harmonics can be obtained by means of the Fourier transform of the voltage signal.[5] Note

4 Using the relations $\cos(x+y) = \cos x \cos y - \sin x \sin y$ and $\sin(x+y) = \sin x \cos y + \cos x \sin y$.

5 In practice, it is more complicated as the angular velocity is difficult to control to the required precision. Thus the voltage is integrated over fixed time intervals, which corresponds to measuring the flux at discrete angles. The multipoles are computed with the discrete Fourier series expansion.

that tangential coils have a "blind eye" for the multipole of order n, if $2\pi/n$ is in the range of the opening angle of the measurement coil. For example, consider an opening angle δ of 20° and a multipole order of $n_1 = 18$. No signal is induced because $\sin(n_1\delta/2) = 0$. On the other hand, the same measurement coil has maximal sensitivity for the multipole of order $n_2 = 9$ as in this case $\sin(n_2\delta/2) = 1$. This is why several coils of different opening angles are used to measure the dominant harmonics. This process is know as *bucking* [2].

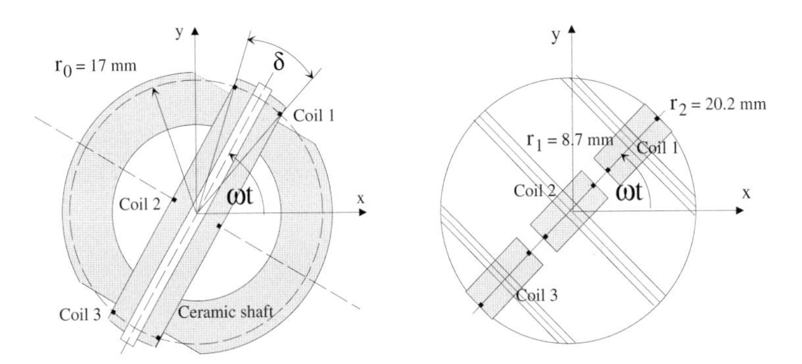

Figure 6.2 Left: Cross section of the long ceramic measuring shaft for the LHC magnets with three tangential coils centered and aligned by use of ceramic pins. Right: Radial coil assembly.

One may use also radial coil assemblies as sketched in Figure 6.2 (right). Using expression (6.30), we can calculate the flux linkage and the induced voltage in a radial measurement coil:

$$\begin{aligned}\Phi(t) &= N\ell \int_{r_1}^{r_2} B_\varphi(r,\varphi)\mathrm{d}r \\ &= \sum_{n=1}^{\infty} S_n^{\mathrm{rad}} \cdot [B_n(r_0)\cos(n\omega t + n\Theta) - A_n(r_0)\sin(n\omega t + n\Theta)], \qquad (6.40)\end{aligned}$$

where r_1 and r_2 are the coil inner and outer radii, ℓ is the length, and Θ is the positioning angle at $t = 0$. The coil sensitivity factor for the radial coil reads

$$S_n^{\mathrm{rad}} := \frac{2N\ell r_0}{n}\left[\left(\frac{r_2}{r_0}\right)^n - \left(\frac{r_1}{r_0}\right)^n\right]. \qquad (6.41)$$

The voltage signal is therefore

$$U(t) = \sum_{n=1}^{\infty} n\omega S_n^{\mathrm{rad}}\left[B_n(r_0)\sin(n\omega t + n\Theta) + A_n(r_0)\cos(n\omega t + n\Theta)\right]. \qquad (6.42)$$

Note the phase difference with respect to Eq. (6.39) for the tangential coil.

6.1.1.2 Dipole, Quadrupole, and Sextupole Field Distribution

For the dipole field ($n = 1$) it yields

$$B_r(r, \varphi) = B_1 \cos \varphi + A_1 \sin \varphi, \tag{6.43}$$

$$B_\varphi(r, \varphi) = B_1 \cos \varphi - A_1 \sin \varphi, \tag{6.44}$$

and

$$B_x(x, y) = A_1, \qquad B_y(x, y) = B_1. \tag{6.45}$$

This is the homogeneous field displayed in Figure 6.3 for the (negative) normal dipole with $A_1 = 0$ and $B_1 < 0$. By definition, the direction of the magnetic flux density is directed from the magnetic north pole to the south pole as shown in Figure 6.3 (left).

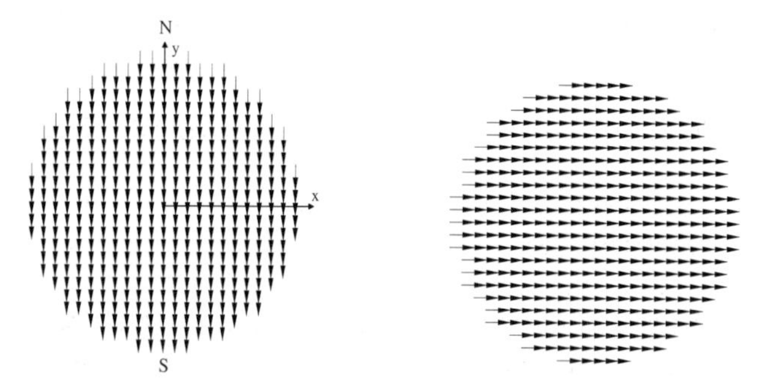

Figure 6.3 Magnetic flux density and force distribution on charged particles in the aperture of an ideal dipole. Left: Magnetic flux density of the normal dipole ($B_x = 0$, B_y constant, negative). Right: x-component of the electromagnetic force field acting on protons traveling into the positive z-direction.

Remark: There is an important remark to be made on polarity definitions of magnetic fields (see also Chapter 12) and in particular on the use of a compass to check magnet polarities. Based on the early use of permanent magnets in compasses, the two poles of the magnet were named N and S, with N being the pole that pointed north. When it was understood that opposite poles attract, the N pole of the compass needle remained defined as the one attracted to the north pole. Thus the north pole of the earth is physically a magnetic south pole. □

For the quadrupole, $n = 2$, presented in Figure 6.4, we obtain from Eqs. (6.29) and (6.30)

$$B_r(r,\varphi) = \frac{r}{r_0}\left(B_2(r_0)\sin 2\varphi + A_2(r_0)\cos 2\varphi\right), \tag{6.46}$$

$$B_\varphi(r,\varphi) = \frac{r}{r_0}\left(B_2(r_0)\cos 2\varphi - A_2(r_0)\sin 2\varphi\right), \tag{6.47}$$

and

$$B_x(x,y) = \frac{1}{r_0}\left(B_2(r_0)y + A_2(r_0)x\right), \tag{6.48}$$

$$B_y(x,y) = \frac{1}{r_0}\left(-A_2(r_0)y + B_2(r_0)x\right). \tag{6.49}$$

The amplitudes of the horizontal and vertical components vary linearly with the distance from the origin and thus the field components in a normal quadrupole ($A_2 = 0$) can be expressed as

$$B_x(x,y) = gy, \qquad B_y(x,y) = gx, \tag{6.50}$$

where g is the *gradient* expressed in $\mathrm{T\,m^{-1}}$. With zero magnetic flux density at the origin, the quadrupole field distribution provides linear focusing of the charged particles. Figure 6.4 shows the field and force distribution inside the aperture of an ideal normal quadrupole; $A_2 = 0$, $B_2 < 0$. While this quadrupole is defocusing in the horizontal direction, it is focusing in the vertical direction (with a restoring force that rises linearly with the displacement in the y-direction). A quadrupole that focuses horizontally is called a focusing or *F-type quadrupole*. Consequently the field displayed on the left acts as a defocusing or *D-type quadrupole*.

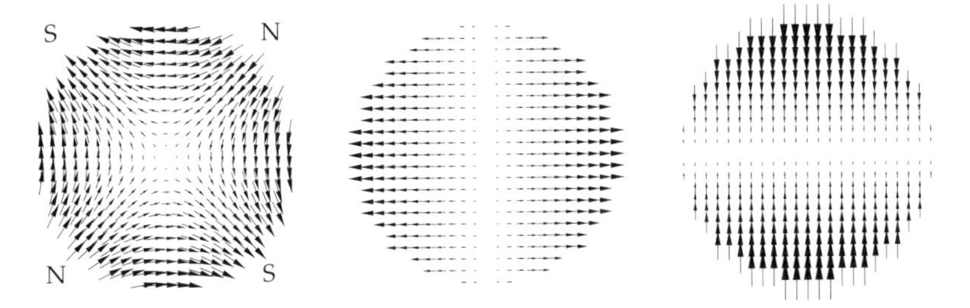

Figure 6.4 Left: Magnetic flux density of the normal quadrupole ($B_y = gx$, $B_x = gy$, gradient g negative). Middle: x-component of the electromagnetic force field on a proton beam parallel to the z-axis into the positive z-direction. Right: y-component of the force field.

For a normal quadrupole, $A_2 = 0$, we can calculate the field in a displaced coordinate system where $x' = x - x_{\mathrm{d}}$, resulting in

$$B_y(x') = \frac{1}{r_0}\left(B_2(r_0)x' + B_2(r_0)x_{\mathrm{d}}\right). \tag{6.51}$$

For a displaced quadrupole, the field contains the constant term $B_2 x_{\mathrm{d}}$ as it does for a dipole field. This effect is called *feed-down* and will be discussed in detail in Section 9.7.1.

Repeating the exercise for the case of the sextupole (n=3) yields

$$B_r(r,\varphi) = \left(\frac{r}{r_0}\right)^2 \left(B_3(r_0)\sin 3\varphi + A_3(r_0)\cos 3\varphi\right), \tag{6.52}$$

$$B_\varphi(r,\varphi) = \left(\frac{r}{r_0}\right)^2 \left(B_3(r_0)\cos 3\varphi - A_3(r_0)\sin 3\varphi\right), \tag{6.53}$$

and

$$B_x(x,y) = \frac{1}{r_0^2}\left(A_3(r_0)\,(x^2 - y^2) + 2B_3(r_0)\,xy\right), \tag{6.54}$$

$$B_y(x,y) = \frac{1}{r_0^2}\left(-2A_3(r_0)\,xy + B_3(r_0)\,(x^2 - y^2)\right), \tag{6.55}$$

which is represented in Figure 6.5 for the normal sextupole; $A_3 = 0$ and $B_3 < 0$.

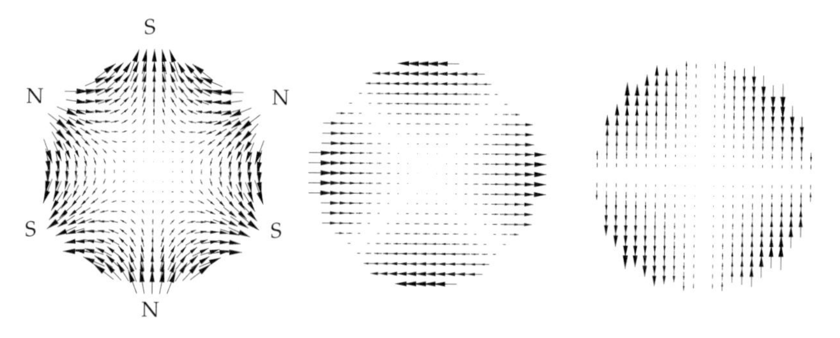

Figure 6.5 Left: Magnetic flux density of the normal sextupole. Middle: x-component of the electromagnetic force field acting on a beam parallel to the z-axis into the positive z-direction. Right: y-component of the electromagnetic force field.

6.1.1.3 Expanding the Green Function

The potential function $\phi(\mathbf{r}) = \ln(r/r_{\mathrm{ref}})$ with an arbitrary reference radius r_{ref} obeys the Laplace equation in a domain excluding the central axis.

$$\nabla^2\left(\ln\left(\frac{r}{r_{\mathrm{ref}}}\right)\right) = \frac{1}{r}\frac{\partial}{\partial r}\left(r\frac{\partial}{\partial r}\ln\left(\frac{r}{r_{\mathrm{ref}}}\right)\right) = 0. \tag{6.56}$$

Let the source point be denoted $\mathbf{r}' = (r_\mathrm{c}, \varphi_\mathrm{c})$ and the field point $\mathbf{r} = (r, \varphi)$ as shown in Figure 6.6. Also the potential

$$\phi(\mathbf{r}) = \ln\left(\frac{|\mathbf{r} - \mathbf{r}'|}{r_\mathrm{ref}}\right) \tag{6.57}$$

is harmonic in a domain excluding the source point $\mathbf{r}'$ as it constitutes merely a shift of the source point from the central axis to $\mathbf{r}'$. The multipole coefficients may therefore be determined by expanding the Green function into a power series and comparing the coefficients with the general solution of the Laplace equation.

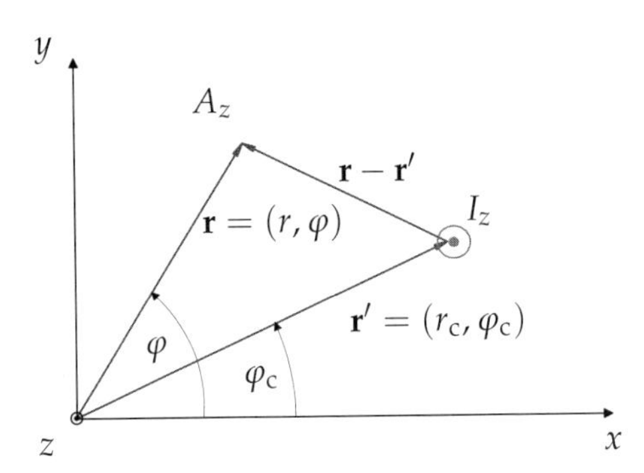

Figure 6.6 Coordinate system for the calculation of the potential of a line-current.

Field inside the aperture Let the distance between the source point and the field point be expressed by the law of cosines:[6]

$$\begin{aligned}|\mathbf{r} - \mathbf{r}'|^2 &= r_\mathrm{c}^2 + r^2 - 2\,\mathbf{r}' \cdot \mathbf{r} = r_\mathrm{c}^2 + r^2 - 2 r_\mathrm{c} r \cos(\varphi - \varphi_\mathrm{c}) \\ &= r_\mathrm{c}^2 \left(1 - \frac{r}{r_\mathrm{c}} e^{i(\varphi - \varphi_\mathrm{c})}\right)\left(1 - \frac{r}{r_\mathrm{c}} e^{-i(\varphi - \varphi_\mathrm{c})}\right).\end{aligned} \tag{6.58}$$

Hence

$$\begin{aligned}\ln\left(\frac{|\mathbf{r} - \mathbf{r}'|}{r_\mathrm{ref}}\right) = \ln\left(\frac{r_\mathrm{c}}{r_\mathrm{ref}}\right) + \frac{1}{2}\ln\left(1 - \frac{r}{r_\mathrm{c}} e^{i(\varphi - \varphi_\mathrm{c})}\right) \\ + \frac{1}{2}\ln\left(1 - \frac{r}{r_\mathrm{c}} e^{-i(\varphi - \varphi_\mathrm{c})}\right).\end{aligned} \tag{6.59}$$

Using the Taylor series expansion of $\ln(1 - z)$, which gives for $z \in \mathbb{C}, |z| < 1$:

$$\ln(1 - z) = -\sum_{n=1}^{\infty} \frac{1}{n} z^n, \tag{6.60}$$

6 Note that $e^{iz} = \cos z + i \sin z$ and $\cos z = (e^{iz} + e^{-iz})/2$ for $z \in \mathbb{C}$.

the vector potential of a line-current

$$A_z(\mathbf{r}) = -\frac{\mu_0 I}{2\pi} \ln\left(\frac{|\mathbf{r}-\mathbf{r}'|}{r_{\text{ref}}}\right) \tag{6.61}$$

can be rewritten as

$$A_z(r,\varphi) = -\frac{\mu_0 I}{2\pi} \ln\left(\frac{r_c}{r_{\text{ref}}}\right) + \frac{\mu_0 I}{2\pi} \sum_{n=1}^{\infty} \frac{1}{n} \left(\frac{r}{r_c}\right)^n \cos n(\varphi - \varphi_c) \tag{6.62}$$

for $r < r_c$. The components of the magnetic flux density are given by

$$\begin{aligned} B_r(r,\varphi) &= \frac{1}{r}\frac{\partial A_z}{\partial \varphi} = -\frac{\mu_0 I}{2\pi r_c} \sum_{n=1}^{\infty} \left(\frac{r}{r_c}\right)^{n-1} \sin n(\varphi - \varphi_c) \\ &= -\frac{\mu_0 I}{2\pi r_c} \sum_{n=1}^{\infty} \left(\frac{r}{r_c}\right)^{n-1} (\sin n\varphi \cos n\varphi_c - \cos n\varphi \sin n\varphi_c)\,, \end{aligned} \tag{6.63}$$

$$\begin{aligned} B_\varphi(r,\varphi) &= -\frac{\partial A_z}{\partial r} = -\frac{\mu_0 I}{2\pi r_c} \sum_{n=1}^{\infty} \left(\frac{r}{r_c}\right)^{n-1} \cos n(\varphi - \varphi_c) \\ &= -\frac{\mu_0 I}{2\pi r_c} \sum_{n=1}^{\infty} \left(\frac{r}{r_c}\right)^{n-1} (\cos n\varphi \cos n\varphi_c + \sin n\varphi \sin n\varphi_c)\,, \end{aligned} \tag{6.64}$$

for $r < r_c$. Comparison of the coefficients with Eq. (6.22) yields for the reference radius $r_0 < r_a$:

$$B_n(r_0) = -\frac{\mu_0 I}{2\pi r_c} \left(\frac{r_0}{r_c}\right)^{n-1} \cos n\varphi_c, \qquad A_n(r_0) = \frac{\mu_0 I}{2\pi r_c} \left(\frac{r_0}{r_c}\right)^{n-1} \sin n\varphi_c. \tag{6.65}$$

This is an important result as it allows us to calculate the harmonic content of a field generated by a number of line-currents by adding the terms in Eqs. (6.65). The results are applied to coil dominated magnets in Chapter 8.

Field outside the coil For the field outside the coil, we rewrite the cosine law (6.58) as

$$|\mathbf{r}-\mathbf{r}'|^2 = r^2 \left(1 - \frac{r_c}{r} e^{i(\varphi-\varphi_c)}\right) \left(1 - \frac{r_c}{r} e^{-i(\varphi-\varphi_c)}\right). \tag{6.66}$$

Therefore,

$$\begin{aligned} \ln\left(\frac{|\mathbf{r}-\mathbf{r}'|}{r_{\text{ref}}}\right) &= \ln\left(\frac{r}{r_{\text{ref}}}\right) + \frac{1}{2}\ln\left(1 - \frac{r_c}{r} e^{i(\varphi-\varphi_c)}\right) \\ &\quad + \frac{1}{2}\ln\left(1 - \frac{r_c}{r} e^{-i(\varphi-\varphi_c)}\right). \end{aligned} \tag{6.67}$$

Employing the Taylor series expansion (6.60) we obtain

$$A_z(r,\varphi) = -\frac{\mu_0 I}{2\pi}\ln\left(\frac{r}{r_{\text{ref}}}\right) + \frac{\mu_0 I}{2\pi}\sum_{n=1}^{\infty}\frac{1}{n}\left(\frac{r_c}{r}\right)^n \cos n(\varphi-\varphi_c) \tag{6.68}$$

for $r > r_c$. The components of the magnetic flux density are thus

$$\begin{aligned} B_r(r,\varphi) &= -\frac{\mu_0 I}{2\pi r_c}\sum_{n=1}^{\infty}\left(\frac{r_c}{r}\right)^{n+1}\sin n(\varphi-\varphi_c) \\ &= -\frac{\mu_0 I}{2\pi r_c}\sum_{n=1}^{\infty}\left(\frac{r_c}{r}\right)^{n+1}(\sin n\varphi\,\cos n\varphi_c - \cos n\varphi\,\sin n\varphi_c)\,, \end{aligned} \tag{6.69}$$

$$\begin{aligned} B_\varphi(r,\varphi) &= \frac{\mu_0 I}{2\pi r} + \frac{\mu_0 I}{2\pi r_c}\sum_{n=1}^{\infty}\left(\frac{r_c}{r}\right)^{n+1}\cos n(\varphi-\varphi_c) \\ &= \frac{\mu_0 I}{2\pi r} + \frac{\mu_0 I}{2\pi r_c}\sum_{n=1}^{\infty}\left(\frac{r_c}{r}\right)^{n+1}(\cos n\varphi\,\cos n\varphi_c + \sin n\varphi\,\sin n\varphi_c)\,, \end{aligned} \tag{6.70}$$

for $r > r_c$. Comparison of the coefficients with Eq. (6.22) at a reference radius $r_0 > r_e$ yields

$$B_n(r_0) = -\frac{\mu_0 I}{2\pi r_c}\left(\frac{r_c}{r_0}\right)^{n+1}\cos n\varphi_c, \qquad A_n(r_0) = \frac{\mu_0 I}{2\pi r_c}\left(\frac{r_c}{r_0}\right)^{n+1}\sin n\varphi_c. \tag{6.71}$$

Note that Eqs. (6.29) and (6.30) cannot be used to reconstitute the vector field of the magnetic flux density outside the coil. From Eq. (6.69) and the corresponding equation for the φ-component we get for $r_0 > r_e$:

$$B_r(r,\varphi) = \sum_{n=1}^{\infty}\left(\frac{r}{r_0}\right)^{n-1}(B_n(r_0)\sin n\varphi + A_n(r_0)\cos n\varphi), \tag{6.72}$$

$$B_\varphi(r,\varphi) = \sum_{n=1}^{\infty}\left(\frac{r}{r_0}\right)^{n-1}(-B_n(r_0)\cos n\varphi + A_n(r_0)\sin n\varphi). \tag{6.73}$$

6.1.1.4 **Field and Potential of a Strand**

Using the series expansions (6.62) and (6.68) for the field of a line-current we can calculate the flux density and magnetic vector potential of a strand of perfect circular shape of radius r_s, cross-sectional area a_s, homogeneous current density J, and total current I. We derive for $r > r_s$ outside the strand:

$$\begin{aligned} A_z(r,\varphi) &= \frac{-\mu_0}{2\pi}\int_{a_s} \ln\frac{|\mathbf{r}-\mathbf{r}'|}{r_{\text{ref}}} J\mathrm{d}a' \\ &= \frac{-\mu_0 J}{2\pi}\int_0^{r_s}\int_0^{2\pi}\left[\ln\left(\frac{r}{r_{\text{ref}}}\right)+\sum_{n=1}^{\infty}\frac{1}{n}\left(\frac{r_c}{r}\right)^n \cos n(\varphi-\varphi_c)\right] r_c \mathrm{d}\varphi_c \mathrm{d}r_c \\ &= \frac{-\mu_0 J}{2\pi}\int_0^{r_s}\ln\left(\frac{r}{r_{\text{ref}}}\right) 2\pi\, r_c \mathrm{d}r_c = \frac{-\mu_0 I}{2\pi}\ln\left(\frac{r}{r_{\text{ref}}}\right) \end{aligned} \tag{6.74}$$

and

$$B_\varphi(r,\varphi) = -\frac{\partial A_z}{\partial r} = \frac{\mu_0 I}{2\pi r}\,. \tag{6.75}$$

Inside the strand, $r < r_s$, we obtain

$$\begin{aligned} A_z(r,\varphi) &= \frac{-\mu_0 J}{2\pi}\int_r^{r_s}\int_0^{2\pi}\left[\ln\left(\frac{r_c}{r_{\text{ref}}}\right)+\sum_{n=1}^{\infty}\frac{1}{n}\left(\frac{r}{r_c}\right)^n \cos n(\varphi-\varphi_c)\right] r_c \mathrm{d}\varphi_c \mathrm{d}r_c \\ &\quad \frac{-\mu_0 J}{2\pi}\int_0^{r}\int_0^{2\pi}\left[\ln\left(\frac{r}{r_{\text{ref}}}\right)+\sum_{n=1}^{\infty}\frac{1}{n}\left(\frac{r_c}{r}\right)^n \cos n(\varphi-\varphi_c)\right] r_c \mathrm{d}\varphi_c \mathrm{d}r_c \\ &= -\mu_0 J\left[\int_r^{r_s}\ln\left(\frac{r_c}{r_{\text{ref}}}\right) r_c \mathrm{d}r_c + \int_0^{r}\ln\left(\frac{r}{r_{\text{ref}}}\right) r_c \mathrm{d}r_c\right] \\ &= \frac{\mu_0 I}{4\pi}\left[1-\left(\frac{r}{r_s}\right)^2 - 2\ln\left(\frac{r_s}{r_{\text{ref}}}\right)\right] \end{aligned} \tag{6.76}$$

and

$$B_\varphi(r,\varphi) = \frac{\mu_0 I r}{2\pi r_s^2}\,. \tag{6.77}$$

6.1.1.5 Taylor Series Expansion of the Field in the Median Plane

In iron-dominated magnets with air gaps of large width-to-height ratio, it may not be convenient to use rotating coils for field quality measurements. For design and optimization, it is required to perform a number of harmonic analyses on circular domains displaced from the air gap's center. Another possible approach is to calculate the relative field error $(B_y - B_0)/B_0$ along the horizontal median plane of the air gap. This corresponds to field quality measurements with a *Hall probe*[7] moved along the horizontal median plane. Assuming

7 Edwin Hall (1855–1938).

field symmetry about the median plane (there are no skew multipole components) we can expand the field in the median plane ($y = 0$) about an axis at x_0 [1]

$$f(x) = \sum_{n=0}^{\infty} \frac{1}{n!}(x - x_0)^n f^{(n)}(x_0), \tag{6.78}$$

where $f^{(n)}$ is the nth derivative of the function at x_0

$$f^{(n)}(x_0) := \left. \frac{\mathrm{d}^n f(x)}{\mathrm{d}x^n} \right|_{x=x_0}. \tag{6.79}$$

For the special case of $x_0 = 0$, Eq. (6.78) is called a *Maclaurin series*.[8] The y-component of the magnetic flux density on the median plane can therefore be expanded as

$$B_y(x) = B_0 + \left. \frac{\mathrm{d}B_y}{\mathrm{d}x} \right|_{x=y=0} x + \cdots + \frac{1}{n!} \left. \frac{\mathrm{d}^n B_y}{\mathrm{d}x^n} \right|_{x=y=0} x^n + \cdots, \tag{6.80}$$

where the constant term B_0 corresponds to the dipole, $\frac{\mathrm{d}B_y}{\mathrm{d}x} x$ to the quadrupole, and so on. The expansion (6.80) can be normalized to the *magnetic rigidity*[9] of the beam:

$$B_y(x) = B_0 R \left[\frac{1}{R} + \frac{1}{B_0 R} \left. \frac{\mathrm{d}B_y}{\mathrm{d}x} \right|_{x=y=0} x + \cdots + \frac{1}{B_0 R} \frac{1}{n!} \left. \frac{\mathrm{d}^n B_y}{\mathrm{d}x^n} \right|_{x=y=0} x^n + \cdots \right]. \tag{6.81}$$

In this context it is customary to name the term

$$k := \frac{Q}{p_0} \left. \frac{\mathrm{d}B_y}{\mathrm{d}x} \right|_{x=y=0} = \frac{1}{B_0 R} \left. \frac{\mathrm{d}B_y}{\mathrm{d}x} \right|_{x=y=0} \tag{6.82}$$

the *normalized gradient*. The polarity convention is discussed in Chapter 12. The relation between the multipole coefficients and those in the Maclaurin series reads

$$b_n = \frac{r^{n-1}}{B_N} \frac{1}{(n-1)!} \left. \frac{\mathrm{d}^{n-1} B_y}{\mathrm{d}x^{n-1}} \right|_{x=y=0}. \tag{6.83}$$

6.1.2
Magnetic Shielding; Permeable Cylindrical Shell in a Uniform Field

Consider a cylindrical shell of a field-independent permeability $\mu = \mu_r \mu_0$ exposed to a uniform magnetic field, as shown in Figure 6.7. The reduction of

8 Colin Maclaurin (1698–1746).
9 Definition in Chapter 11.

field inside the shell is known as *magnetic shielding*. Since there are no free currents, both the magnetic scalar and magnetic vector potentials can be used to solve the field problem.

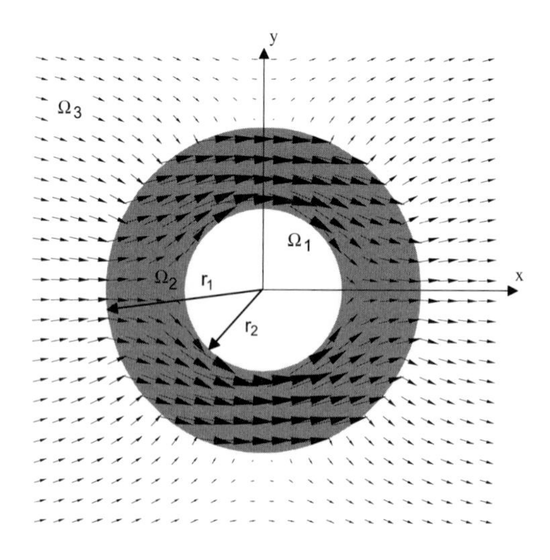

Figure 6.7 Permeable cylindrical shell in a uniform applied field.

The original homogeneous magnetic field is perpendicular to the generators of the cylinder; it points into the direction of $\mathbf{e}_x$. The magnetic scalar potential of this original (source) field can be expressed in circular coordinates as

$$\phi_{\mathrm{m}} = -H_0\, r\, \cos\varphi\,. \tag{6.84}$$

We denote the outer domain Ω_3, the permeable shell Ω_2, and the interior Ω_1. The general solution in Ω_3 has the form

$$\phi_{\mathrm{m3}} = -H_0\, r\, \cos\varphi + \sum_{n=1}^{\infty} \mathcal{A}_n r^{-n} \cos n\varphi \tag{6.85}$$

because of the symmetry about the x-axis. Inside the shell we have

$$\phi_{\mathrm{m2}} = \sum_{n=1}^{\infty} \left(\mathcal{B}_n\, r^n + \mathcal{C}_n r^{-n}\right) \cos n\varphi\,. \tag{6.86}$$

In the interior air domain Ω_1 we set

$$\phi_{\mathrm{m1}} = \sum_{n=1}^{\infty} \mathcal{D}_n r^n \cos n\varphi\,, \tag{6.87}$$

because the potential on the axis must remain finite.

The interface conditions at Γ_{12} $(r = r_1)$ and Γ_{23} $(r = r_2)$ yield four systems of equations for the unknowns $\mathcal{A}_n$, $\mathcal{B}_n$, $\mathcal{C}_n$, and $\mathcal{D}_n$. In terms of the magnetic scalar potential the continuity conditions for B_r and H_φ at the interface are expressed as

$$\frac{\partial \phi_{\mathrm{m1}}}{\partial \varphi}(r_1) = \frac{\partial \phi_{\mathrm{m2}}}{\partial \varphi}(r_1), \qquad \frac{\partial \phi_{\mathrm{m1}}}{\partial r}(r_1) = \mu_{\mathrm{r}} \frac{\partial \phi_{\mathrm{m2}}}{\partial r}(r_1), \tag{6.88}$$

and

$$\frac{\partial \phi_{\mathrm{m2}}}{\partial \varphi}(r_2) = \frac{\partial \phi_{\mathrm{m3}}}{\partial \varphi}(r_2), \qquad \mu_{\mathrm{r}} \frac{\partial \phi_{\mathrm{m2}}}{\partial r}(r_2) = \frac{\partial \phi_{\mathrm{m3}}}{\partial r}(r_2). \tag{6.89}$$

All coefficients with $n \neq 1$ vanish; a shell of constant permeability cannot generate higher-order multipole field components when it is exposed to a homogeneous field. The coefficients with index 1 are determined by solving the linear equation system

$$\begin{pmatrix} 0 & -r_1 & -r_1^2 & r_1 \\ -r_2^{-1} & r_2 & r_2^{-1} & 0 \\ 0 & -\mu_r & \mu_r r_1^{-2} & 1 \\ r_2^{-2} & \mu_r & -\mu_r r_2^{-2} & 0 \end{pmatrix} \begin{pmatrix} \mathcal{A}_1 \\ \mathcal{B}_1 \\ \mathcal{C}_1 \\ \mathcal{D}_1 \end{pmatrix} = \begin{pmatrix} 0 \\ -H_0 r_2 \\ 0 \\ -H_0 \end{pmatrix}. \tag{6.90}$$

The constant $\mathcal{D}_1$ that determines the field strength inside the shell is

$$\mathcal{D}_1 = \frac{-4 H_0 \mu_r}{(\mu_r + 1)^2 - (\mu_r - 1)^2 \left(\frac{r_1}{r_2}\right)^2}. \tag{6.91}$$

For $r_1 = r_2$ the factor is $-H_0$, as expected. If the applied field is of multipole order N, a similar calculation shows that

$$\mathcal{D}_N = \frac{-4 H_0 \mu_r}{(\mu_r + 1)^2 - (\mu_r - 1)^2 \left(\frac{r_1}{r_2}\right)^{2N}}. \tag{6.92}$$

The reduced field in Ω_3 can be calculated from Eq. (6.85) and the constant

$$\mathcal{A}_1 = \frac{H_0 \left(-1 + \mu_r^2\right) r_2^2 \left(r_1^2 - r_2^2\right)}{(\mu_r - 1)^2 r_1^2 - (\mu_r + 1)^2 r_2^2}. \tag{6.93}$$

Example: Consider as typical values $r_1 = 25$ mm and $r_2 = 26.7$ mm for the cold-bore tube in the LHC main dipoles and quadrupoles. The tube is made from austenitic steel with a relative permeability of 1.01. We find $(H - H_0)/H_0 = -0.0305 \times 10^{-4}$. □

6.1.3
Integrated Multipoles in Accelerator Magnets

For the calculation of the integrated multipole content in the magnet extremities, no analytical equation has been found. The coefficients $B_n(r_0, z)$ and

$A_n(r_0, z)$ calculated or measured at some longitudinal position z, henceforth called 3D field harmonics, can still be derived by means of the Fourier series expansion of the radial field component $B_r(r_0, \varphi, z)$.

The scaling laws (6.26) and (6.27) derived for field harmonics of the 2D field may **not**, however, be applied to the 3D field harmonics, as can be seen from Figure 6.8. The solid curve shows the relative sextupole as a function of the longitudinal position in the coil end region, calculated at $r_1 = 17$ mm. The wrong dotted curve shows the result obtained from a calculation of the relative sextupole at $r_0 = 10$ mm and a scaling to 17 mm by means of Eq. (6.27).

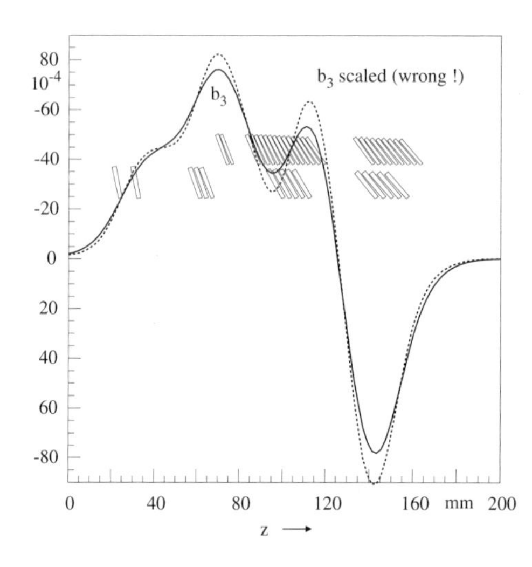

Figure 6.8 Relative multipole component $b_3(r_0, z)$ as a function of the longitudinal position z. Solid: Calculated at r_0 = 17 mm reference radius, Dashed: Sextupole component at 17 mm reference radius scaled from 10 mm using the 2D scaling laws (wrong). For the geometry of the coil ends and their cut planes see Figure 19.3.

As it is sufficient to calculate the integrated transverse multipole coefficients,[10] the local errors presented in Figure 6.8 cancel out. This observation triggers the following proposition.

Proposition: The scalar potential in the magnet aperture satisfies the Laplace equation

$$\nabla^2 \phi_\mathrm{m}(x, y, z) = \frac{\partial^2 \phi_\mathrm{m}(x, y, z)}{\partial x^2} + \frac{\partial^2 \phi_\mathrm{m}(x, y, z)}{\partial y^2} + \frac{\partial^2 \phi_\mathrm{m}(x, y, z)}{\partial z^2} = 0\,. \tag{6.94}$$

We define $\overline{\phi}_{\mathrm{m}(x,y)}$ by

$$\overline{\phi}_\mathrm{m}(x, y) := \int_{-z_0}^{z_0} \phi_\mathrm{m}(x, y, z)\mathrm{d}z. \tag{6.95}$$

10 The effect of the z-component of the magnetic field on the LHC beam can be disregarded since the magnets are short with respect to the betatron wavelength.

It follows that

$$\nabla^2 \overline{\phi}_{\mathrm{m}}(x,y) = \frac{\partial^2 \overline{\phi}_{\mathrm{m}}(x,y)}{\partial x^2} + \frac{\partial^2 \overline{\phi}_{\mathrm{m}}(x,y)}{\partial y^2} = 0, \tag{6.96}$$

if the magnet is symmetric with respect to the center, or the integration path is extended far enough outside the magnet so that the field has dropped to zero. Consequently, the scaling laws (6.26) and (6.27) for two dimensions can be applied to the integrated multipoles derived from $\overline{\phi}_{\mathrm{m}}$.

Proof. We must show that

$$\begin{aligned} \frac{\partial^2 \overline{\phi}_{\mathrm{m}}(x,y)}{\partial x^2} + \frac{\partial^2 \overline{\phi}_{\mathrm{m}}(x,y)}{\partial y^2} &= \int_{-z_0}^{z_0} \left(\frac{\partial^2 \phi_{\mathrm{m}}}{\partial x^2} + \frac{\partial^2 \phi_{\mathrm{m}}}{\partial y^2} \right) \mathrm{d}z \\ &= \int_{-z_0}^{z_0} \left(- \frac{\partial^2 \phi_{\mathrm{m}}}{\partial z^2} \right) \mathrm{d}z = - \left. \frac{\partial \phi_{\mathrm{m}}}{\partial z} \right|_{-z_0}^{z_0} \\ &= H_z(-z_0) - H_z(z_0) \stackrel{!}{=} 0 \,. \end{aligned} \tag{6.97}$$

This requirement is fulfilled for symmetric magnets where $H_z(z_0) = -H_z(-z_0)$, or in the case that the longitudinal field has dropped to zero. □

The *magnetic length* or *effective length* for a $2N$-pole magnet is defined by

$$l_{\mathrm{mag}}(I) := \frac{1}{B_N^{\infty}(I)} \int_{-\infty}^{\infty} B_N(I,z) \, \mathrm{d}z \tag{6.98}$$

where $B_N^{\infty}(I)$ is the 2D solution of the main field component. Magnetic saturation caused by end effects results in a small decrease in magnetic length with increasing excitation current. The average relative field harmonics as a function of the excitation current I can be calculated by integrating the $B_n(I,z)$ and $A_n(I,z)$ harmonics along the z-axis and dividing by $l_{\mathrm{mag}} \, B_N^{\infty}$.

6.2 Spherical Harmonics

The Laplace equation in spherical coordinates (R, ϑ, φ) provides a multipole expression for devices with azimuthal symmetry (solenoids). It reads

$$\nabla^2 \phi_{\mathrm{m}} = \frac{1}{R^2} \frac{\partial}{\partial R} \left(R^2 \frac{\partial \phi_{\mathrm{m}}}{\partial R} \right) + \frac{1}{R^2 \sin \vartheta} \frac{\partial}{\partial \vartheta} \left(\sin \vartheta \frac{\partial \phi_{\mathrm{m}}}{\partial \vartheta} \right) + \frac{1}{R^2 \sin^2 \vartheta} \frac{\partial^2 \phi_{\mathrm{m}}}{\partial \varphi^2} = 0. \tag{6.99}$$

For axial symmetry $\partial\phi_{\rm m}/\partial\varphi = 0$. In accordance with the standard mathematical texts [4] we substitute $x = \cos\vartheta$ for $x \in [-1,1]$ and recall that $\mathrm{d}x = -\sin\vartheta\,\mathrm{d}\vartheta$. Equation (6.99) can be rewritten as

$$\nabla^2\phi_{\rm m}(R,x) = \frac{\partial}{\partial R}\left(R^2\frac{\partial\phi_{\rm m}}{\partial R}\right) + \frac{\partial}{\partial x}\left[\left(1-x^2\right)\frac{\partial\phi_{\rm m}}{\partial x}\right] = 0\,. \tag{6.100}$$

Applying again the separation of variables method with

$$\phi_{\rm m}(R,x) = \rho(R)\xi(x) \tag{6.101}$$

gives

$$\underbrace{\frac{1}{\rho(R)}\frac{\partial}{\partial R}\left(R^2\frac{\partial\rho(R)}{\partial R}\right)}_{k} + \underbrace{\frac{1}{\xi(x)}\frac{\partial}{\partial x}\left[\left(1-x^2\right)\frac{\partial\xi(x)}{\partial x}\right]}_{-k} = 0, \tag{6.102}$$

that decomposes into two ordinary differential equations

$$\frac{\mathrm{d}}{\mathrm{d}R}\left(R^2\frac{\mathrm{d}\rho(R)}{\mathrm{d}R}\right) - k\rho(R) = 0\,, \tag{6.103}$$

$$\frac{\mathrm{d}}{\mathrm{d}x}\left[\left(1-x^2\right)\frac{\mathrm{d}\xi(x)}{\mathrm{d}x}\right] + k\xi(x) = 0\,. \tag{6.104}$$

The general solution of the *Euler differential equation* (6.103) is given by

$$\rho_n(R) = \mathcal{E}_n R^n + \mathcal{F}_n R^{-n-1} \tag{6.105}$$

for[11] $k = n(n+1)$. Equation (6.104) becomes

$$\frac{\mathrm{d}}{\mathrm{d}x}\left[\left(1-x^2\right)\frac{\mathrm{d}\xi(x)}{\mathrm{d}x}\right] + n(n+1)\xi(x) = 0\,, \tag{6.106}$$

which is the ordinary *Legendre differential equation* [5]. It has the general solution

$$\xi_n(x) = \mathcal{G}_n P_n(x) + \mathcal{H}_n Q_n(x) \tag{6.107}$$

were $P_n(x)$ and $Q_n(x)$ are the nth-order *Legendre polynomials* of the first and the second kind.

If the axis of the azimuthal symmetry is included in the problem domain, $\mathcal{F}_n$ in Eq. (6.105) and $\mathcal{H}_n$ in Eq. (6.107) must be zero.[12] Consequently, the general

11 This can be seen by inserting Eq. (6.105) into Eq. (6.103) and solving for k.

12 Note that $Q_n(x)$ have poles at $x = 0$.

solution for the magnetic scalar potential inside a sphere, free of any current sources, is given by

$$\phi_{\mathrm{m}}(R,x) = \sum_{n=0}^{\infty} \mathcal{A}_n R^n P_n(x)\,. \tag{6.108}$$

In this context, the functions $P_n(\cos\vartheta)$ are called *zonal spherical functions* or *zonal harmonics*.

Outside of a sphere containing all sources we obtain

$$\phi_{\mathrm{m}}(R,x) = \sum_{n=0}^{\infty} \mathcal{B}_n R^{-n-1} P_n(x)\,. \tag{6.109}$$

The lower-order Legendre polynomials read

$$P_0(x) = 1\,, \tag{6.110}$$

$$P_1(x) = x\,, \tag{6.111}$$

$$P_2(x) = \frac{1}{2}\left(3x^2 - 1\right)\,, \tag{6.112}$$

$$P_3(x) = \frac{1}{2}\left(5x^3 - 3x\right)\,, \tag{6.113}$$

$$P_4(x) = \frac{1}{8}\left(35x^4 - 30x^2 + 3\right)\,, \tag{6.114}$$

and in general

$$P_n(x) = \sum_{m=0}^{M} (-1)^m \frac{(2n-2m)!}{2^n r!(n-m)!(n-2r)!} x^{n-2m} \tag{6.115}$$

where $M = n/2$ if n is an even number and $M = (n-1)/2$ if n is odd. A compact way to write $P_n(x)$ is known as the *Rodriguez formula*,[13]

$$P_n(x) = \frac{1}{2^n n!}\left[\frac{\mathrm{d}^n}{\mathrm{d}x^n}\left(x^2-1\right)^n\right]\,. \tag{6.116}$$

In terms of $\cos\vartheta$ we obtain

$$P_0(\cos\vartheta) = 1\,, \tag{6.117}$$

$$P_1(\cos\vartheta) = \cos\vartheta\,, \tag{6.118}$$

$$P_2(\cos\vartheta) = \frac{1}{4}(3\cos 2\vartheta + 1)\,, \tag{6.119}$$

$$P_3(\cos\vartheta) = \frac{1}{8}(5\cos 3\vartheta + 3\cos\vartheta)\,, \tag{6.120}$$

$$P_4(\cos\vartheta) = \frac{1}{64}(35\cos 4\vartheta + 20\cos 2\vartheta + 9)\,. \tag{6.121}$$

13 Benjamin Rodriguez (1795–1851).

The Legendre polynomials obey the recurrence relations

$$(n+1)P_{n+1}(x) + nP_{n-1}(x) = (2n+1)xP_n(x)\,, \tag{6.122}$$

$$\frac{\mathrm{d}P_{n+1}(x)}{\mathrm{d}x} - \frac{\mathrm{d}P_{n-1}(x)}{\mathrm{d}x} = (2n+1)P_n(x)\,, \tag{6.123}$$

$$(x^2-1)\frac{\mathrm{d}P_n(x)}{\mathrm{d}x} - nxP_n(x) = -nP_{n-1}(x)\,. \tag{6.124}$$

The $P_n(x)$ establish an orthogonal function system on the interval $[-1,1]$ by

$$\int_{-1}^{+1} P_m(x)P_n(x)\mathrm{d}x = 0, \qquad m \neq n \tag{6.125}$$

and

$$\int_{-1}^{+1} (P_n(x))^2\mathrm{d}x = \frac{2}{2n+1}\,, \qquad n = 0,1,2,\dots\,. \tag{6.126}$$

For differentials of the Legendre polynomials, one finds [4]

$$\frac{\mathrm{d}P_n(x)}{\mathrm{d}x} = \frac{n}{x^2-1}\left(xP_n(x) - P_{n-1}(x)\right)\,. \tag{6.127}$$

From the magnetic scalar potential, we can calculate the magnetic field expressed in spherical coordinates as

$$\mathbf{H} = -\,\mathrm{grad}\,\phi_\mathrm{m} = -\frac{\partial\phi_\mathrm{m}}{\partial R}\,\mathbf{e}_R - \frac{1}{R}\frac{\partial\phi_\mathrm{m}}{\partial\vartheta}\,, \mathbf{e}_\vartheta \tag{6.128}$$

and consequently for any point inside a sphere, not containing any current sources,

$$B_R = -\mu_0\sum_{n=0}^{\infty}\frac{\partial}{\partial R}\left(\mathcal{A}_n R^n P_n(\cos\vartheta)\right) = -\mu_0\sum_{n=1}^{\infty}\mathcal{A}_n n R^{n-1}P_n(\cos\vartheta)\,, \tag{6.129}$$

$$B_\vartheta = -\mu_0\sum_{n=0}^{\infty}\frac{1}{R}\frac{\partial}{\partial\vartheta}\left(\mathcal{A}_n R^n P_n(\cos\vartheta)\right) = \mu_0\sum_{n=1}^{\infty}\mathcal{A}_n R^{n-1}\,P_n^1(\cos\vartheta)\,. \tag{6.130}$$

where P_n^1 are the *associated Legendre polynomials* of order n and degree 1:

$$P_n^1(\cos\vartheta) = \sin\vartheta\frac{\mathrm{d}P_n(\cos\vartheta)}{\mathrm{d}\cos\vartheta} = -\frac{\mathrm{d}P_n(\cos\vartheta)}{\mathrm{d}\vartheta}\,. \tag{6.131}$$

Thus the lower-order associated Legendre polynomials are in terms of $\cos\vartheta$:

$$P_0^1(\cos\vartheta) = 1\,, \tag{6.132}$$

$$P_1^1(\cos\vartheta) = \sin\vartheta\,, \tag{6.133}$$

$$P_2^1(\cos\vartheta) = 3\sin\vartheta\cos\vartheta\,, \tag{6.134}$$

$$P_3^1(\cos\vartheta) = \frac{3}{2}\sin\vartheta(5\cos^2\vartheta - 1)\,, \tag{6.135}$$

$$P_4^1(\cos\vartheta) = \frac{5}{2}\sin\vartheta(7\cos^3\vartheta - 3\cos\vartheta)\,. \tag{6.136}$$

Generally, the associated Legendre polynomials of order m are defined for $x \in [-1, 1]$ by

$$P_n^m(x) := (-1)^m(1-x^2)^{\frac{m}{2}}\frac{\mathrm{d}^m P_n(x)}{\mathrm{d}x^m}\,, \tag{6.137}$$

or employing the Rodriguez formula (6.116):

$$P_n^m(x) = \frac{(-1)^m}{2^n n!}(1-x^2)^{\frac{m}{2}}\frac{\mathrm{d}^{m+n}}{\mathrm{d}x^{m+n}}(x^2-1)^n\,. \tag{6.138}$$

The B_z and B_r components of the magnetic flux density can be expressed in terms of R and ϑ as

$$B_z(R,\vartheta) = B_R\cos\vartheta - B_\vartheta\sin\vartheta\,, \tag{6.139}$$

$$B_r(R,\vartheta) = B_R\sin\vartheta + B_\vartheta\cos\vartheta\,. \tag{6.140}$$

Using the recurrence relations we derive

$$B_z(R,\vartheta) = -\mu_0\sum_{n=1}^{\infty}\mathcal{A}_n n R^{n-1}P_{n-1}(\cos\vartheta)\,, \tag{6.141}$$

$$B_r(R,\vartheta) = \mu_0\sum_{n=1}^{\infty}\mathcal{A}_n R^{n-1}P_{n-1}^1(\cos\vartheta)\,. \tag{6.142}$$

For the first three zonal multipole coefficients we have for $n = 1$,

$$B_R(R,\vartheta) = -\mu_0\mathcal{A}_1\cos\vartheta\,, \qquad B_\vartheta(R,\vartheta) = \mu_0\mathcal{A}_1\sin\vartheta\,, \tag{6.143}$$

$$B_z(R,\vartheta) = -\mu_0\mathcal{A}_1\,, \qquad B_r(R,\vartheta) = 0\,, \tag{6.144}$$

for $n = 2$,

$$B_R(R,\vartheta) = -\mu_0\frac{1}{2}R\mathcal{A}_2(3\cos 2\vartheta + 1)\,, \tag{6.145}$$

$$B_\vartheta(R,\vartheta) = \mu_0 3R\mathcal{A}_2\sin\vartheta\cos\vartheta\,, \tag{6.146}$$

$$B_z(R,\vartheta) = -\mu_0\mathcal{A}_2 2R\cos\vartheta\,, \tag{6.147}$$

$$B_r(R,\vartheta) = \mu_0\mathcal{A}_2 R\sin\vartheta\,, \tag{6.148}$$

and for $n = 3$,

$$B_R(R, \vartheta) = -\mu_0 \frac{3}{8} R^2 \mathcal{A}_3 (5 \cos 3\vartheta + 3 \cos \vartheta) , \tag{6.149}$$

$$B_\vartheta(R, \vartheta) = \mu_0 \frac{3}{2} R^2 \mathcal{A}_3 \sin \vartheta (5 \cos^2 \vartheta - 1) , \tag{6.150}$$

$$B_z(R, \vartheta) = -\mu_0 \mathcal{A}_3 \frac{3}{4} R^2 (3 \cos 2\vartheta + 1) , \tag{6.151}$$

$$B_r(R, \vartheta) = \mu_0 \mathcal{A}_3 R^2 3 \sin \vartheta \cos \vartheta . \tag{6.152}$$

These vector fields are displayed for positive $\mathcal{A}_n$ in Figure 6.9.

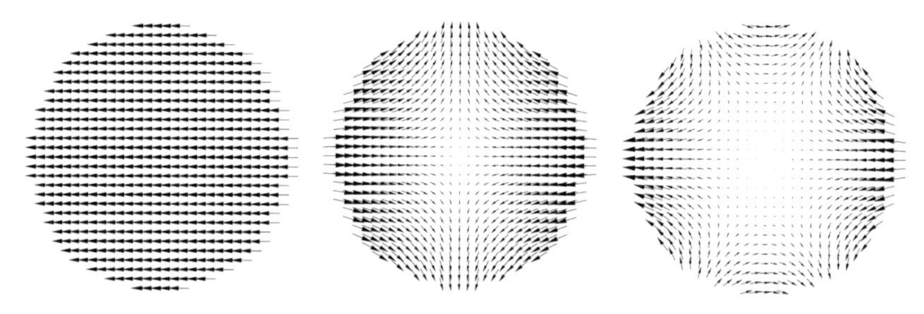

Figure 6.9 Field distribution for the first three zonal harmonics. Left: $n = 1$. Middle: $n = 2$. Right: $n = 3$.

6.2.1 Legendre Series Expression for the Vector Potential

For azimuthal symmetry the vector potential has only one component $\mathbf{A} = A(R, \vartheta)\mathbf{e}_\varphi$ and the curl–curl expression is given by

$$\text{curl curl}\, A = \frac{\partial^2 A}{\partial R^2} + \frac{2}{R}\frac{\partial A}{\partial R} + \frac{\cot \vartheta}{R^2}\frac{\partial A}{\partial \vartheta} + \frac{1}{R^2}\frac{\partial^2 A}{\partial \vartheta^2} - \frac{A}{R^2 \sin^2 \vartheta} . \tag{6.153}$$

But

$$\begin{aligned} \mu_0 \frac{\partial}{\partial \vartheta} \nabla^2 \phi_\text{m} &= \mu_0 \frac{\partial}{\partial \vartheta} \left(\frac{\partial^2 \phi_\text{m}}{\partial R^2} + \frac{2}{R}\frac{\partial \phi_\text{m}}{\partial R} + \frac{\cot \vartheta}{R^2}\frac{\partial \phi_\text{m}}{\partial \vartheta} + \frac{1}{R^2}\frac{\partial^2 \phi_\text{m}}{\partial \vartheta^2} \right) \\ &= \frac{\partial^2}{\partial R^2} \left(\mu_0 \frac{\partial \phi_\text{m}}{\partial \vartheta} \right) + \frac{2}{R}\frac{\partial}{\partial R} \left(\mu_0 \frac{\partial \phi_\text{m}}{\partial \vartheta} \right) + \frac{\cot \vartheta}{R^2}\frac{\partial}{\partial \vartheta} \left(\mu_0 \frac{\partial \phi_\text{m}}{\partial \vartheta} \right) \\ &\quad + \frac{1}{R^2}\frac{\partial^2}{\partial \vartheta^2} \left(\mu_0 \frac{\partial \phi_\text{m}}{\partial \vartheta} \right) - \frac{1}{R^2 \sin^2 \vartheta} \left(\mu_0 \frac{\partial \phi_\text{m}}{\partial \vartheta} \right) , \end{aligned} \tag{6.154}$$

which establishes

$$A_\varphi(R,\vartheta) = \sum_{n=1}^{\infty} \mu_0 \mathcal{A}_n R^n P_n^1(\cos\vartheta)\,, \tag{6.155}$$

because for $A = \mu_0 \frac{\partial \phi_{\mathrm{m}}}{\partial \vartheta}$ Eqs. (6.153) and (6.154) are identical.

6.2.2
Determining the Zonal Harmonics

6.2.2.1 Legendre Series Expansion

The multipole coefficients $\mathcal{A}_n$ are not determined at this stage. They are defined by the boundary conditions at some reference radius R_0 or can be calculated from the Legendre series expansion of the numerically calculated radial component of the magnetic flux density at a reference radius R_0. Formally

$$B_R(R_0,\vartheta) = \sum_{n=1}^{\infty} A_n P_n(\cos\vartheta)\,, \tag{6.156}$$

where

$$A_n(R_0) = \frac{2n+1}{2} \int_0^\pi B_R(R_0,\vartheta) P_n(\cos\vartheta) \sin\vartheta \,\mathrm{d}\vartheta\,. \tag{6.157}$$

Comparing the coefficients in Eqs. (6.129) and (6.156) yields

$$\mathcal{A}_n = \frac{-A_n(R_0)}{\mu_0 n R_0^{n-1}}\,. \tag{6.158}$$

6.2.2.2 Taylor Series Expansion

Particularly simple expressions result for the calculation of the multipole coefficients if we expand the field into a Maclaurin series about the origin:

$$B_z(z) = \sum_{n=1}^{\infty} \frac{1}{n!} a_n z^n\,, \qquad z < R_0\,, \tag{6.159}$$

where $a_n = \left.\frac{\mathrm{d}^n B_z}{\mathrm{d}z^n}\right|_{r=z=0}$. The solution (6.141) reduces for $\cos\vartheta = 1$ and $P_n(1) = 1$ to

$$B_z = -\mu_0 \sum_{n=1}^{\infty} \mathcal{A}_n n z^{n-1} \tag{6.160}$$

and therefore

$$\mathcal{A}_n = \frac{-a_n z}{\mu_0 n n!}\,. \tag{6.161}$$

The components of the magnetic flux density at an arbitrary point inside the sphere $R < R_0$ can be calculated with Eqs. (6.141) and (6.142).

6.2.2.3 **Expanding the Green Function**

The aim is now to express the magnetic vector potential of a ring current in zonal harmonics. We first expand the reciprocal distance

$$\frac{1}{|\mathbf{r}-\mathbf{r}'|} = \frac{1}{\sqrt{|\mathbf{r}|^2 + |\mathbf{r}'|^2 - 2\mathbf{r}\cdot\mathbf{r}'}} = \frac{1}{\sqrt{|\mathbf{r}|^2 + |\mathbf{r}'|^2 - 2|\mathbf{r}||\mathbf{r}'|\cos\alpha}}, \tag{6.162}$$

into a series of Legendre polynomials. In Eq. (6.162) the position vectors are given by

$$\mathbf{r} = |\mathbf{r}|(\sin\vartheta\cos\varphi\,\mathbf{e}_x + \sin\vartheta\sin\varphi\,\mathbf{e}_y + \cos\vartheta\,\mathbf{e}_z)\,, \tag{6.163}$$

$$\mathbf{r}' = |\mathbf{r}'|(\sin\vartheta_c\cos\varphi_c\,\mathbf{e}_x + \sin\vartheta_c\sin\varphi_c\,\mathbf{e}_y + \cos\vartheta_c\,\mathbf{e}_z)\,. \tag{6.164}$$

According to Figure 6.10 it yields

$$\mathbf{r}\cdot\mathbf{r}' = |\mathbf{r}||\mathbf{r}'|(\cos\vartheta\cos\vartheta_c + \sin\vartheta\sin\vartheta_c\cos(\varphi-\varphi_c)) = |\mathbf{r}||\mathbf{r}'|\cos\alpha\,. \tag{6.165}$$

Because of azimuthal symmetry we may set $\varphi = 0$. We list without proof the useful relation [4]

$$\frac{1}{\sqrt{1+t^2-2tx}} = \sum_{n=0}^{\infty} t^n P_n(x) \tag{6.166}$$

from which follows directly

$$\frac{1}{\sqrt{|\mathbf{r}|^2 + |\mathbf{r}'|^2 - 2|\mathbf{r}||\mathbf{r}'|\cos\alpha}} = \frac{1}{|\mathbf{r}'|}\sum_{n=0}^{\infty}\left(\frac{|\mathbf{r}|}{|\mathbf{r}'|}\right)^n P_n(\cos\alpha) \tag{6.167}$$

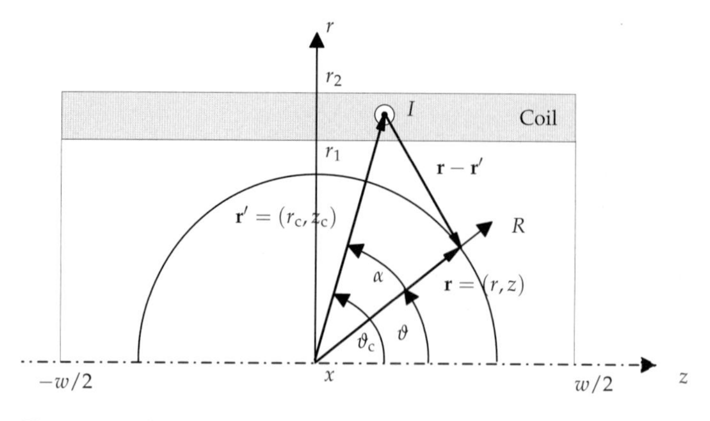

Figure 6.10 Coordinate system for the calculation of field, potential, and zonal harmonics in a solenoid.

for $|\mathbf{r}'| > |\mathbf{r}|$. In addition, we make use of the spherical harmonic addition theorem

$$P_n(\cos\alpha) = P_n(\cos\vartheta\cos\vartheta_c + \sin\vartheta\sin\vartheta_c\cos\varphi_c)$$
$$= P_n(\cos\vartheta)P_n(\cos\vartheta_c) + 2\sum_{m=1}^{n}\frac{(n-m)!}{(n+m)!}P_n^m(\cos\vartheta)P_n^m(\cos\vartheta_c)\cos m\varphi_c\,. \tag{6.168}$$

For $|\mathbf{r}'|^2 = r_c^2 + z_c^2$, $|\mathbf{r}|^2 = r^2 + z^2$ (see Figure 6.10) and the relations $\cos\vartheta = z/|\mathbf{r}|$ and $\sin\vartheta = r/|\mathbf{r}|$ we calculate for $|\mathbf{r}'| > |\mathbf{r}|$ the φ-component of the vector potential according to Eq. (5.55). Modified for $z_c \neq 0$ it yields with the expression (6.167)

$$A_\varphi = \frac{\mu_0 I r_c}{2\pi}\int_0^\pi \frac{\cos\varphi_c \mathrm{d}\varphi_c}{\sqrt{r^2 + r_c^2 + (z - z_c)^2 - 2 r r_c\cos\varphi_c}}$$
$$= \frac{\mu_0 I r_c}{2\pi}\int_0^\pi \frac{\cos\varphi_c \mathrm{d}\varphi_c}{\sqrt{|\mathbf{r}|^2 + |\mathbf{r}'|^2 - 2|\mathbf{r}||\mathbf{r}'|(\cos\vartheta\cos\vartheta_c + \sin\vartheta\sin\vartheta_c\cos\varphi_c)}}$$
$$= \frac{\mu_0 I r_c}{2}\frac{1}{|\mathbf{r}'|}\sum_{n=1}^{\infty}\left(\frac{|\mathbf{r}|}{|\mathbf{r}'|}\right)^n \frac{(n-1)!}{(n+1)!}P_n^1(\cos\vartheta)P_n^1(\cos\vartheta_c)\,. \tag{6.169}$$

Comparison of the coefficients with Eq. (6.155) yields for the contribution of one single line-current loop of radius r_c positioned at $\mathbf{r} = (R_c, \vartheta_c)$ to the nth zonal harmonic

$$\mathcal{A}_n = \frac{I r_c}{2}\frac{1}{R_c^{n+1}}\frac{1}{n(n+1)}P_n^1(\cos\vartheta_c)\,. \tag{6.170}$$

Applications to the optimal design of solenoids are presented in Chapter 8.

6.3 Separation in Cartesian Coordinates

Finally we will present a solution of a boundary value problem in Cartesian coordinates suited to the optimization of iron-dominated dipoles, when symmetry about the magnet center and a vanishing tangential field component at the (nonsaturated) iron pole can be assumed. For simplicity, we start again with the magnetic scalar potential and write

$$\phi_m = X(x)Y(y)\,. \tag{6.171}$$

It follows that

$$\frac{\partial^2\phi_m}{\partial x^2} = \frac{\partial^2}{\partial x^2}(X(x)Y(y)) = Y(y)\frac{\mathrm{d}^2 X(x)}{\mathrm{d}x^2}\,. \tag{6.172}$$

Repeating the exercise for the y-derivative and substituting the results into the two-dimensional Laplace equation yields

$$\underbrace{\frac{1}{X(x)}\frac{\mathrm{d}^2X(x)}{\mathrm{d}x^2}}_{p^2}+\underbrace{\frac{1}{Y(y)}\frac{\mathrm{d}^2Y(y)}{\mathrm{d}y^2}}_{-p^2}=0\,. \tag{6.173}$$

Employing the separation of variables method it yields the two ordinary differential equations

$$\frac{\mathrm{d}^2X(x)}{\mathrm{d}x^2}-p^2X(x)=0\,, \tag{6.174}$$

$$\frac{\mathrm{d}^2Y(y)}{\mathrm{d}y^2}+p^2Y(y)=0\,, \tag{6.175}$$

with the general solutions

$$X_0(x)=\mathcal{A}_0+\mathcal{B}_0x\,,\qquad Y_0(y)=\mathcal{C}_0+\mathcal{D}_0y\,, \tag{6.176}$$

for $p=0$ and

$$X_p(x)=\mathcal{A}_p\cosh px+\mathcal{B}_p\sinh px\,, \tag{6.177}$$

$$Y_p(y)=\mathcal{C}_p\cos py+\mathcal{D}_p\sin py\,, \tag{6.178}$$

for $p\neq 0$. The general solution for the Laplace equation is thus given by

$$\phi_{\mathrm{m}}(x,y)=\sum_{p=0}^{\infty}X_p(x)Y_p(y)\,, \tag{6.179}$$

with the two sets of unknowns $\mathcal{A}_p$, $\mathcal{B}_p$, and $\mathcal{C}_p$, $\mathcal{D}_p$. Imposing symmetry conditions about the magnet center yields

$$B_y(-x,y)=B_y(x,y)\quad\rightarrow\quad X_p(x)=X_p(-x) \tag{6.180}$$

and

$$B_y(x,-y)=B_y(x,y)\quad\rightarrow\quad \left.\frac{\mathrm{d}Y_p(y)}{\mathrm{d}y}\right|_{y=-y_0}=\left.\frac{\mathrm{d}Y_p(y)}{\mathrm{d}y}\right|_{y=y_0}. \tag{6.181}$$

The symmetry conditions can be fulfilled only for $\mathcal{B}_p=0$ and $\mathcal{C}_p=0$. The x-component of the magnetic flux density can be written as

$$B_x=\sum_{p=1}^{\infty}\mathcal{U}_p\sinh px\sin py\,, \tag{6.182}$$

where $\mathcal{U}_p = -p\mathcal{A}_p\mathcal{D}_p$. Assuming that $B_x(x, \pm h) = 0$, which is true for non-saturated iron poles at $y = h$, yields $p = (n\pi)/h$. The field components at any position inside the air-gap can be calculated from

$$B_x(x, y) = \sum_{n=1}^{\infty} \mathcal{U}_n \sinh\left(\frac{n\pi}{h}x\right) \sin\left(\frac{n\pi}{h}y\right), \tag{6.183}$$

$$B_y(x, y) = U_0 + \sum_{n=1}^{\infty} \mathcal{U}_n \cosh\left(\frac{n\pi}{h}x\right) \cos\left(\frac{n\pi}{h}y\right). \tag{6.184}$$

The eigenfunctions for $n = 1$ and $n = 2$ are displayed in Figure 6.11. The coefficients can be determined by a Fourier analysis of $B_x(x_0, y)$ at the boundary of the problem domain. Pole-shim optimization may then be performed numerically by minimizing $\mathcal{U}_n$ for $n \neq 0$.

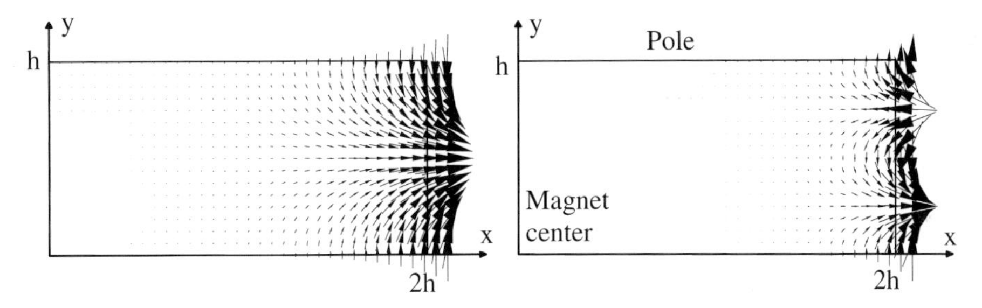

Figure 6.11 Vector field from the first and second higher multipole component in Cartesian coordinates. The boundary value problem assumes symmetry about the magnet center and vanishing x-component at the iron pole.

References

1 Bryant, P.J.: Basic theory of magnetic measurements, CAS, CERN Accelerator School on Magnetic Measurement and Alignment, CERN 92-05, Geneva, 1992

2 Jain, A.K.: Harmonic coils, CERN Accelerator School on Measurement and Alignment of Accelerator and Detector Magnets, CERN Yellow Report 98-05

3 Gupta, R., Chow, K., Dietderich, D., Gourlay, S., Millos, G., McInturff, A., Scanlan, R., Ramberger, S., Russenschuck, S.: A high field magnet design for a future hadron collider, IEEE Transactions on Applied Superconductivity, 1999

4 Hobson, E.W.: The Theory of Spherical and Ellipsoidal Harmonics, Chelsea, New York, 1955

5 Jackson, J.D.: Classical Electrodynamics, John Wiley & Sons, New York, 1997

6 The LHC study group, The Large Hadron Collider, Conceptual Design, CERN/AC/95-05

7 Moon, P., Spencer, D.E.: Field Theory Handbook: Including Coordinate Systems, Differential Equations and Their Solutions, Springer, Berlin, 1961

7
Iron-Dominated Magnets

You have broken a wooden yoke,
but in its place you will get a yoke of iron.

Jeremiah 28:13.

According to the classification given in Section 1.2, we may distinguish iron-dominated and coil-dominated magnets. This distinction depends on the ratio of the *source field* contributed by the coil to the *reduced field* from the iron magnetization. Iron-dominated magnets with coils made of copper or aluminum conductors are denoted normal-conducting magnets, while iron-dominated magnets with superconducting excitation coils are known as *superferric*. Though it is common practice to call normal-conducting magnets resistive or conventional, this is not followed here, since neither is resistivity a design criterion for normal-conducting magnets nor are superconducting magnets "unconventional."

If we disregard iron saturation and stray fields, we can achieve the conceptual design of iron-dominated magnets by applying Ampère's law and the magnetic flux conservation law in order to derive magnetic circuit equations. This back-of-the-envelope approach is also referred to as one-dimensional field calculation. Its advantage is that it allows optimal dimensioning of permanent magnets in an iron yoke; it's limitations will be seen when we compare its results with numerical calculations using the CERN field computation program ROXIE.

At the end of this chapter, we will present a rudimentary treatment of water cooling and the dimensioning of cooling manifolds.

Field Computation for Accelerator Magnets. Stephan Russenschuck

ISBN: 978-3-527-40769-9

7.1
C-Shaped Dipole

Consider the magnetic circuit shown in Figure 7.1 (left), consisting of a C-shaped iron yoke (long with respect to the width of the cross section and thus treated as a 2D field problem) with two coils around it.[1] We assume that the stray field around the air gap is small and consequently the magnetic flux through any section of the yoke and across the air gap is constant. Applying Ampère's law we may write,

$$\int_{\partial \widetilde{\mathscr{A}}} \mathbf{H} \cdot \mathrm{d}\mathbf{r} = \int_{\partial \widetilde{\mathscr{A}}} \mathbf{H} \cdot \mathbf{t} \, \mathrm{d}s = \int_{\widetilde{\mathscr{A}}} \mathbf{J} \cdot \mathbf{n} \, \mathrm{d}a \,, \tag{7.1}$$

where $\mathbf{t}$ is the field of tangent vectors to the boundary $\partial \widetilde{\mathscr{A}}$ of the surface $\widetilde{\mathscr{A}}$, and $\mathbf{n}$ is the normal vector to that surface. Assuming a piecewise-constant field in the magnetic circuit, that is, a constant H_{iron} along the iron path $\mathscr{S}_{\text{iron}}$ and a constant H_0 across the air gap $\mathscr{S}_0$, the integration of Eq. (7.1) yields

$$\int_{\partial \widetilde{\mathscr{A}}} \mathbf{H} \cdot \mathrm{d}\mathbf{r} = \int_{\widetilde{\mathscr{A}}} \mathbf{J} \cdot \mathbf{n} \, \mathrm{d}a \,,$$

$$\int_{\mathscr{S}_{\text{iron}}} \mathbf{H} \cdot \mathrm{d}\mathbf{r} + \int_{\mathscr{S}_0} \mathbf{H} \cdot \mathrm{d}\mathbf{r} = \int_{\widetilde{\mathscr{A}}_{\text{coil}}} \mathbf{J} \cdot \mathbf{n} \, \mathrm{d}a \,,$$

$$H_{\text{iron}} \, s_{\text{iron}} + H_0 \, s_0 = N I \,,$$

$$\frac{1}{\mu_0 \mu_{\mathrm{r}}} B_{\text{iron}} \, s_{\text{iron}} + \frac{1}{\mu_0} B_0 \, s_0 = N I \,, \tag{7.2}$$

where N is the total number of turns in the coils. The lengths of the integration paths inside the iron yoke and in the air gap are denoted s_{iron} and s_0. Assuming that the relative permeability of the yoke is very high, $\mu_r \gg 1$, we obtain the simple expression

$$B_0 = \frac{\mu_0 N I}{s_0} \,. \tag{7.3}$$

Disregarding all stray fields, one can also derive a simple equation for the inductance in a normal-conducting dipole magnet. As the excitation coil links the entire magnetic flux $\Phi = B_0 a_0 = \mu_0 N I \, (a_0 / s_0)$ exactly N times, the inductance can be calculated as

$$L = \frac{N \Phi}{I} = \mu_0 N^2 \frac{a_0}{s_0} \,. \tag{7.4}$$

Once iron saturation sets in this equation is no longer valid and numerical methods must be employed.

1 For computation of the magnetomotive force (ampere-turns) we refer to Figure 5.1.

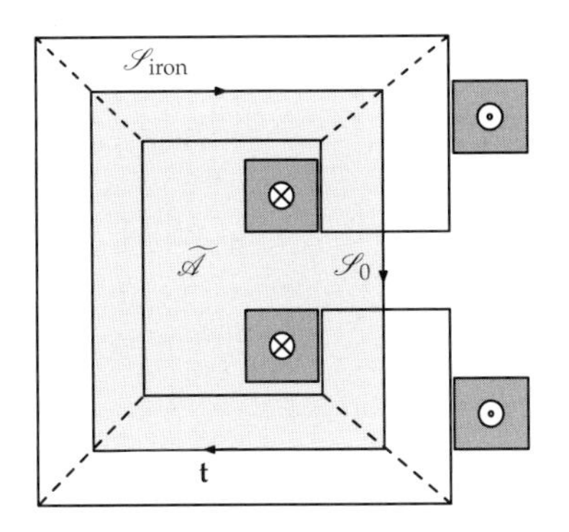

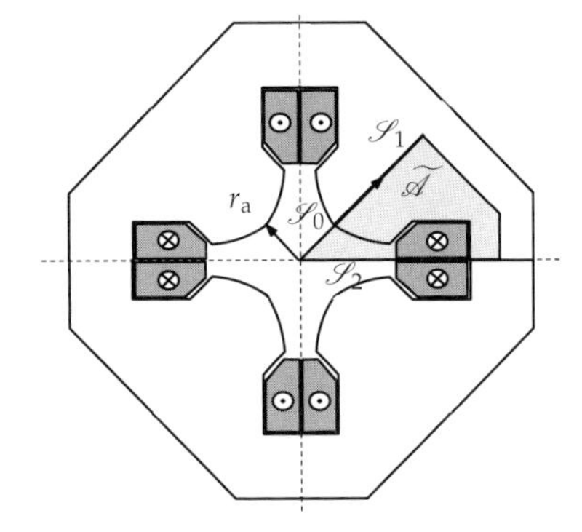

Figure 7.1 Magnetic circuit (2D) of a normal-conducting dipole magnet (left) and a quadrupole magnet (right). Notice that the sense of integration along ∂a yields a right-handed screw with the normal vector to the surface.

7.2 Quadrupole

In the case of the quadrupole magnet, with four coils of N turns each, we can split up the integration path along $\partial\widetilde{\mathscr{A}}$ in a segment from the origin to the pole ($\mathscr{S}_0$), along an arbitrary path through the iron yoke ($\mathscr{S}_1$), and back along the x-axis ($\mathscr{S}_2$); see Figure 7.1 (right). Hence

$$\int_{\partial\widetilde{\mathscr{A}}} \mathbf{H}\cdot \mathrm{d}\mathbf{r} = \int_{\mathscr{S}_0} \mathbf{H}_0 \cdot \mathrm{d}\mathbf{r} + \int_{\mathscr{S}_1} \mathbf{H}_1 \cdot \mathrm{d}\mathbf{r} + \int_{\mathscr{S}_2} \mathbf{H}_2 \cdot \mathrm{d}\mathbf{r} = N\,I\,. \tag{7.5}$$

The field in a quadrupole is defined by its gradient g, which yields

$$B_x = gy, \qquad B_y = gx; \tag{7.6}$$

see Eq. (6.50). Therefore, the magnitude of the field along the integration path $\mathscr{S}_0$ is

$$H = \frac{g}{\mu_0}\sqrt{x^2+y^2} = \frac{g}{\mu_0}r. \tag{7.7}$$

Along the x-axis ($\mathscr{S}_2$) the field integral is zero because $\mathbf{H}\cdot\mathbf{t} = 0$. With the assumption that the relative permeability of the yoke is very high, $\mu_r \gg 1$, the magnetic field will be low and the magnetomotive force along the path $\mathscr{S}_1$ can be neglected. Thus we obtain

$$\int_0^{r_\mathrm{a}} H\,\mathrm{d}r = \frac{g}{\mu_0}\int_0^{r_\mathrm{a}} r\,\mathrm{d}r = \frac{g}{\mu_0}\frac{r_\mathrm{a}^2}{2} = N\,I, \tag{7.8}$$

or

$$g = \frac{2\mu_0 N I}{r_\mathrm{a}^2}. \tag{7.9}$$

Notice that for a given NI the field decreases linearly with the size of the dipole's air gap, whereas the gradient in a quadrupole magnet is inversely proportional to the square of the aperture radius r_a.

7.3
Ohmic Losses in Dipole and Quadrupole Coils

It is instructive to calculate the ohmic losses in dipole and quadrupole coils as a function of the aperture size for constant air-gap flux density and gradient:

$$W_{\text{coil}} = RI^2 = \frac{l\, NI^2}{\varkappa a_{\text{cond}}} = \frac{l}{\varkappa \lambda a_{\text{coil}}} (NI)^2 , \tag{7.10}$$

where l is the average turn-length, a_{cond} is the cross section of the single conductor, λ is the coil packing factor, and a_{coil} the coil cross-sectional area. Employing Eqs. (7.3) and (7.9), it is easy to show that the losses scale with s_0^2 for the dipole and with r_a^4 for the quadrupole.

7.4
Magnetic Circuit with Varying Yoke Width

From Ampère's law we derive for a magnetic circuit of $n+1$ branches,

$$\sum_{i=0}^{n} H_i s_i = N\, I \tag{7.11}$$

where s_0 is the size of the air gap and s_i, $i = 1, \ldots, n$ are the mean lengths of the yoke sections. H_0 is the air-gap field and H_i the field in the ith yoke section. Because of the continuity of B_n, the flux conservation law, and the assumption that stray fields can be neglected, this yields

$$H_i = \frac{B_i}{\mu_i} = \frac{\Phi}{a_i\, \mu_i} , \tag{7.12}$$

where the a_i are the surface areas $a_i = w_i \cdot l_i$ of the longitudinal yoke sections $\mathscr{A}_i$ and **not** the transverse cross-sectional area of the laminations; see Figure 7.2 (left). This is the reason we used a tilde in Eq. (7.1) for the surface appearing in Ampère's law. As can be seen in Figure 7.2, the surface $\widetilde{\mathscr{A}}$ is outer oriented by the passing direction of the current. The outer orientation of the boundary of $\widetilde{\mathscr{A}}$ induces the inner orientation of $\mathscr{A}_i$. For the entire magnetic circuit we find

$$\Phi \sum_{i=0}^{n} \frac{s_i}{a_i\, \mu_i} = N\, I = V_{\text{m}} , \tag{7.13}$$

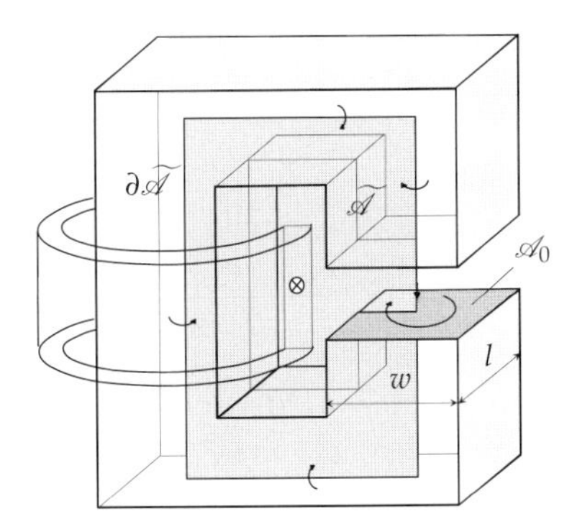

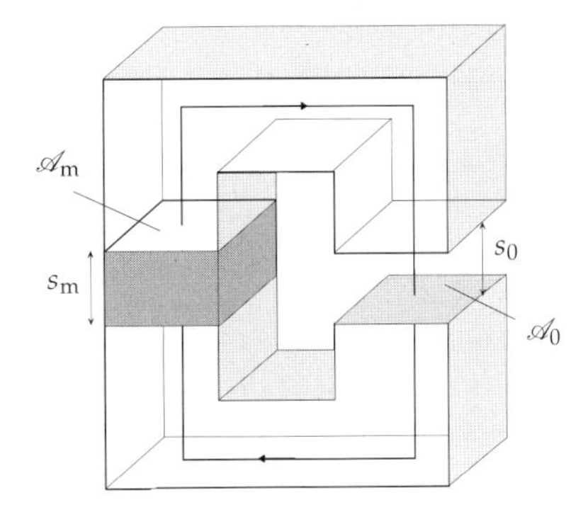

Figure 7.2 Left: C-shaped dipole with excitation coil and varying yoke width. Compare the orientation of the surfaces of integration of field strength and flux density. The surfaces are dual in the sense that the outer-oriented surface $\widetilde{\mathscr{A}}$ for the integration of the field strength induces the inner orientation of the surface $\mathscr{A}_0$. Right: Dipole with permanent magnet excitation.

which is formally identical with Ohm's law $I \sum_{i=0}^{n} \frac{s_i}{a_i \varkappa_i} = U$. Consequently,

$$R_\mathrm{m} := \frac{V_\mathrm{m}}{\Phi} = \sum_{i=0}^{n} \frac{s_i}{a_i \, \mu_i} \tag{7.14}$$

is called the *magnetic resistance*, or *reluctance*, of the magnetic circuit. The physical unit is $[R_\mathrm{m}] = \mathrm{H}^{-1} = 1\,\mathrm{A\,V^{-1}\,s^{-1}}$. The magnetic resistance is nonlinear in most cases, and thus the result in Eq. (7.14) is an approximation assuming a constant μ_i along the entire path segment $\mathscr{S}_i$. It follows that

$$N\,I = \Phi \left(\frac{s_0}{a_0 \, \mu_0} + \sum_{i=1}^{n} \frac{s_i}{a_i \, \mu_i} \right) = \Phi \frac{s_0}{a_0 \, \mu_0} \left(1 + \sum_{i=1}^{n} \frac{\mu_0}{\mu_i} \frac{s_i}{s_0} \frac{a_0}{a_i} \right) . \tag{7.15}$$

In [11] the dimensionless factor,

$$\eta := \left(1 + \sum_{i=1}^{n} \frac{\mu_0}{\mu_i} \frac{s_i}{s_0} \frac{a_0}{a_i} \right)^{-1} , \tag{7.16}$$

is called the *magnet efficiency*. Obviously, for $\mu_i \gg \mu_0$ it yields the same result as in Eq. (7.3) and thus the magnet efficiency is one. It can be seen also from Eq. (7.15) that with a large air gap s_0 the magnetic resistance of the circuit is stabilized against temperature variations in μ_i.

The back-of-the-envelope dimensioning of a magnetic circuit can now be formalized as follows:

1. For a given air-gap flux density B_0 calculate the flux as $\Phi = B_0 a_0$ and the air-gap field as $H_0 = B_0 / \mu_0$. A correction for the leakage flux can be made by adding $2 s_0 (l + w)$ to the pole surface a_0.

2. Determine $B_i = \Phi / a_i$.
3. Estimate H_i from the $B(H)$ curve of the yoke laminations.
4. Calculate the necessary excitation current from $NI = \sum_i H_i s_i$.

Results are easily obtained if the yoke width remains constant, $a_i = a_0$, $i = 1, \ldots, n$, and if the relative permeability all along the yoke sections remains constant, that is, $\mu_i = \mu_0 \mu_r$, $i = 1, \ldots, n$. Writing $s_{\text{iron}} := \sum_{i=1}^{n} s_i$ we obtain for $\mu_i \gg \mu_0$:

$$B_0 = \frac{NI}{\frac{s_{\text{iron}}}{\mu_0 \mu_r} + \frac{s_0}{\mu_0}} \approx \frac{NI \mu_0}{s_0} \left(1 - \frac{s_{\text{iron}}}{s_0 \mu_r} \right) . \tag{7.17}$$

7.5 Branched Circuits

Figure 7.3 shows the design of the superconducting undulator for LHC beam diagnostics [8]. This is a magnetic circuit in which the coils are coupled by a branched iron yoke, featuring air gaps of different lengths in order to compensate for field errors due to stray fields. As before, we disregard all stray fields and the magnetic resistance of the iron yoke. This allows a rough estimation of the magnetic flux density in the air gaps, because the iron poles are highly saturated; see Figure 7.3.

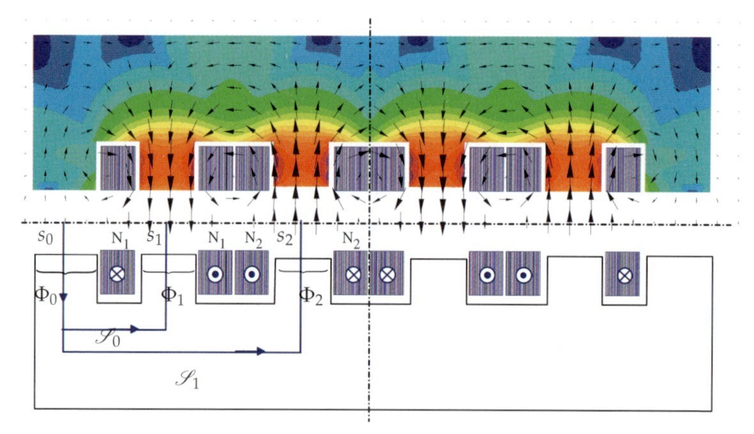

Figure 7.3 Superconducting undulator for the LHC as an example of a branched circuit. Disregarding both the magnetic resistance in the iron yoke and stray fields leads to a rough estimation of the air-gap field.

With the symmetry of the device in mind we now apply Ampère's law to the paths $\mathscr{S}_0$, $\mathscr{S}_1$ shown in Figure 7.3 (bottom):

$$H_0 s_0 + H_1 s_1 = -N_1 I \,, \tag{7.18}$$

$$H_0 s_0 + H_2 s_2 = N_2 I \,, \tag{7.19}$$

where s_0, s_1, s_2 are the half-gap lengths ($s_0 = s_1$). From the flux conservation law it follows that

$$\Phi_0 = B_0 a_0 = \mu_0 H_0 a_0 \,, \tag{7.20}$$

$$\Phi_1 = B_1 a_1 = \mu_0 H_1 a_1 \,, \tag{7.21}$$

$$\Phi_2 = B_2 a_2 = \mu_0 H_2 a_2 \,, \tag{7.22}$$

$$\Phi_0 = \Phi_1 + \Phi_2 \,. \tag{7.23}$$

The two sets of equations yield a linear equation system for the unknowns $\Phi_{0,1,2}$ and $H_{0,1,2}$.

7.6 Ideal Pole Shapes of Iron-Dominated Magnets

From Section 3.3 we recall that the gradient of a scalar potential is perpendicular to the surface of the equipotential. With the field entering highly permeable materials in normal direction to the surface (cf. Section 4.5), the iso-surfaces of the total magnetic scalar potential define the pole shapes of normal-conducting magnets. As in 2D both the z-component of the vector potential and the magnetic scalar potential satisfy the Laplace equation, so we already have the solution by analogy to the results in Chapter 6:

$$\phi_{\mathrm{m}} = \mathcal{C}_1 x + \mathcal{D}_1 y. \tag{7.24}$$

Consequently, $\mathcal{C}_1 = 0$, $\mathcal{D}_1 \neq 0$ yields a vertical (normal) dipole field and $\mathcal{C}_1 \neq 0$, $\mathcal{D}_1 = 0$ a horizontal (skew) dipole field. The equipotential surfaces are parallel to the x- or y-axis depending on the values of $\mathcal{C}$ and $\mathcal{D}$. For the quadrupole we obtain

$$\phi_{\mathrm{m}} = \mathcal{C}_2(x^2 - y^2) + 2\mathcal{D}_2 xy \,, \tag{7.25}$$

with $\mathcal{C}_2 = 0$ yielding a normal quadrupole field and with $\mathcal{D}_2 = 0$ giving a skew-quadrupole field (rotated clockwise by $\pi/4$). The quadrupole field is generated by lines of equipotential having the hyperbolic form shown in Figure 7.4 (left). For the $\mathcal{C}_2 = 0$ case, the asymptotes are the two major axes.

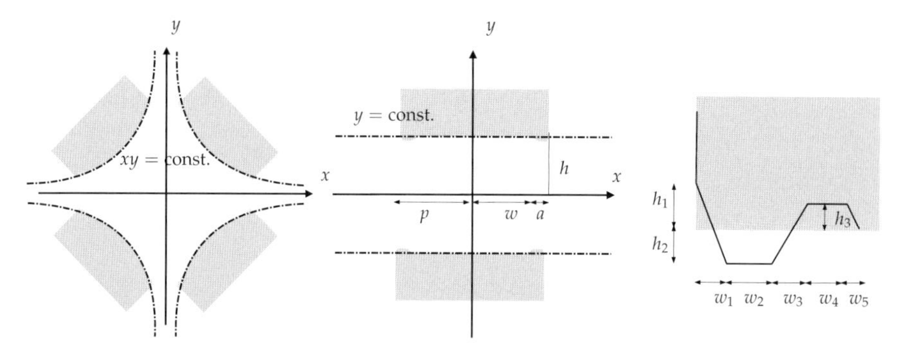

Figure 7.4 Left: Ideal and technical pole shape of a normal-conducting quadrupole. Middle: Ideal and technical pole shape of a normal-conducting dipole with the geometrical data needed for the definition of the relative pole-overhang factor. Right: Design parameters for a shim.

In practice, however, the magnets have a finite pole width due to the necessity for a magnetic flux return yoke and space for the coil. To ensure good field quality, small shims are added at the outer ends of each pole. Figure 7.4 shows the pole shape of a normal-conducting dipole and quadrupole magnet, with magnetic shims. The middle illustration also gives the geometrical data for the definition of the *relative pole overhang* by

$$\lambda_{\mathrm{p}} := \frac{a}{h} = \frac{w-h}{h}\,, \tag{7.26}$$

where h is the half-gap size, w is the half-width of the good-field zone, and a is the pole overhang. The pole half-width p is given by $p = w + a = w + \lambda_{\mathrm{p}} h$.

Figure 7.5 also shows the field distortion $Q := (B_y - B_0)/B_0$ ($[Q] = 1_{\mathrm{U}}$) as a function of the distance x from the magnet center, and for different relative pole-overhang factors λ_{p}. The half-gap and the half-width of the good-field zone are both 30 mm, such that the λ_{p} correspond to 70, 100, and 130 mm pole half-widths, respectively. The results are compared to an optimized design with shims and $\lambda_{\mathrm{p}} = 2.333$, which corresponds to a pole half-width of 100 mm.

Empirical rules for the necessary relative pole overhang as a function of the required field quality Q in the good-field zone are given in [11]:

$$\lambda_{\mathrm{p}}^{\mathrm{s}} = -0.14 \ln|Q| - 0.25\,, \tag{7.27}$$

$$\lambda_{\mathrm{p}} = -0.36 \ln|Q| - 0.90\,, \tag{7.28}$$

where $\lambda_{\mathrm{p}}^{\mathrm{s}}$ is the relative pole overhang for the pole with optimized shim.

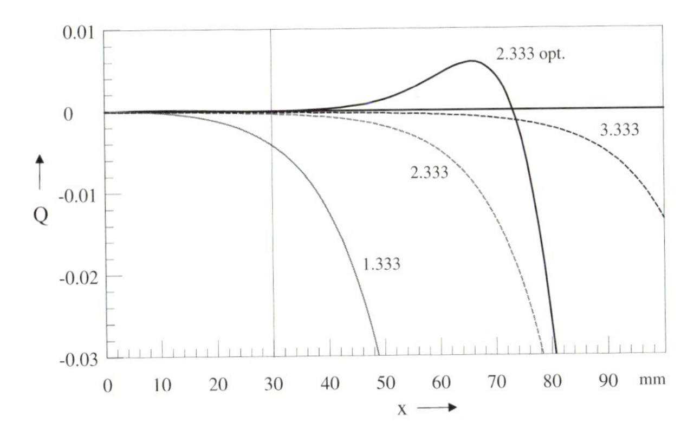

Figure 7.5 Field distortion $Q := (B_y - B_0)/B_0$ for different relative pole-overhang factors λ_P, compared to an optimized design with shims. The half-gap size is 30 mm, such that the λ_P correspond to 70, 100, and 130 mm pole half-widths, respectively. For the design with shim p = 100 mm.

7.6.1 Shimming

Figure 7.6 (left) shows the cross section of the LEP dipole with iso-surfaces of constant vector potential. Figure 7.6 (right) shows the magnitude of the magnetic flux density in the iron yoke of the LEP main quadrupole. The field quality in the dipole was improved by adding shims on the pole face. In the case of the quadrupole, the pole face is defined as a combination of a hyperbola, a straight section, and an arc. The points at which the segments are connected were found in an optimization process that not only considered the multipole

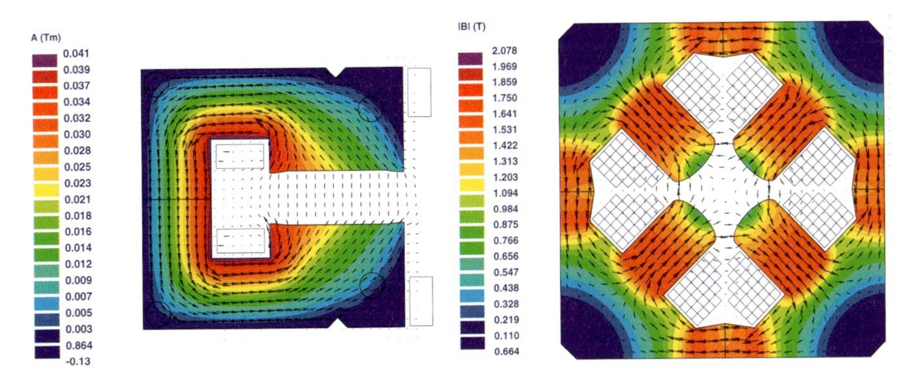

Figure 7.6 Left: Cross section of the LEP dipole magnet with iso-surfaces of constant vector potential. Right: LEP quadrupole cross section with magnetic field magnitude.

components in the cross section but also provided a partial compensation for the field errors in the end-region of the magnet [3].

Figure 7.4 (right) shows the eight design variables for a pole-shim optimization. The results are given in Figure 7.7 for a C-shaped magnet optimized for maximum field quality at nominal excitation of 1.25 T and for a minimum variation in field errors as a function of the excitation from 0.25 to 1.25 T. The relative pole overhang λ_P is 1.2. In the case shown in Figure 7.7 (left) the achieved relative multipole content at the reference radius of 35 mm is $b_2 = -0.08$ and $b_3 = -0.82$ units, while the variation from 0.25 to 1.25 T is $\Delta b_2 = 8.9$ and $\Delta b_3 = 33.8$ units.

Figure 7.7 Optimized pole shims for a C-shaped magnet with small relative pole overhang of 1.2. Left: Optimization for maximum field quality at nominal excitation. Right: Optimization for minimum variation in the multipole field errors versus excitation.

The variation in multipoles can be reduced for the design shown in Figure 7.7 (right). The results are 2.6 and 7.5 units for the variations in the quadrupole and the sextupole components, respectively. However, at nominal excitation the field quality is worse: $b_2 = -13.8$ and $b_3 = -24.2$ units. The example shows that technical optimization problems usually show multiple conflicting objectives. More on these so-called *vector optimization problems* is presented in Chapter 20.

7.7
Rogowski Profiles

As can be seen in Figure 7.9, the magnetic flux lines converge toward sharp edges, giving rise to local maxima of the field strength, limited only by the nonuniform iron saturation in the pole piece. This edge effect was discussed in connection with electrostatic field problems by Maxwell himself [6]. Setting

$$\psi := \frac{\phi}{\phi_0}\pi\,, \qquad A : \frac{a}{\pi}\,, \tag{7.29}$$

the field pattern for the case of a thin conducting electrode placed parallel to the conducting xz-plane is described by

$$x = A\left(\varphi + e^{\varphi}\cos\psi\right)\,, \qquad y = A\left(\psi + e^{\varphi}\sin\psi\right)\,, \tag{7.30}$$

where $\psi \in [0, \pi]$, $\varphi \in \mathbb{R}$, and a is the distance between the electrode and the conducting plane. Equipotentials and field lines can be found by setting ψ = const. and φ = const., respectively. The potential at any point in the problem domain is denoted ϕ, while the potential at the electrode is ϕ_0. The Eqs. (7.30) give the coordinates as a function of the potential and the parameter φ. For $\psi = 0$, and therefore for $\phi = 0$ and $y = 0$, the x-coordinate takes any value between $-\infty$ and $+\infty$. The xz-plane is thus an equipotential with $\phi = 0$. For $\psi = \pi$ it yields $y = A\pi = a$ and $x = A(\varphi - e^{\varphi})$. The differential quotient $\mathrm{d}x/\mathrm{d}\varphi = A(1 - e^{\varphi})$ is zero at $\varphi = 0$. This corresponds to $x_{\max} = -A$. The inner surface of the electrode is traced by negative values of φ, while the outer surface is traced by positive values of φ. The electrode is consequently at constant potential $\psi = \pi$, which corresponds to $\phi = \phi_0$. It is evident from Figure 7.8 (left) that near the edge of the finite electrode the field is much higher than between the plates.

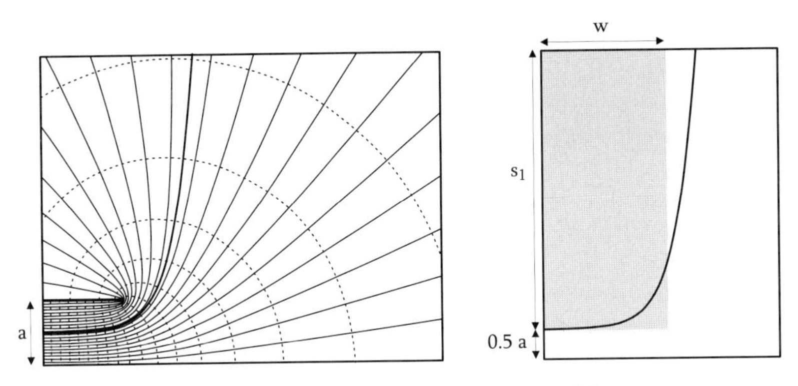

Figure 7.8 Left: Equipotentials (solid) and field lines (dashed) for an electrode at a distance a from the conducting half-plane. The equipotential $\psi_0 = \pi/2$ (thick line) is the Rogowski profile. Right: Rogowski profile for a pole piece of width $2w$ and an air gap size of a.

Following Rogowski [10] we derive from Eqs. (7.30)

$$\mathrm{d}x = -Ae^{\varphi}\sin\psi\,\mathrm{d}\psi\,, \qquad \mathrm{d}y = A\left(1 + e^{\varphi}\cos\psi\right)\,\mathrm{d}\psi\,. \tag{7.31}$$

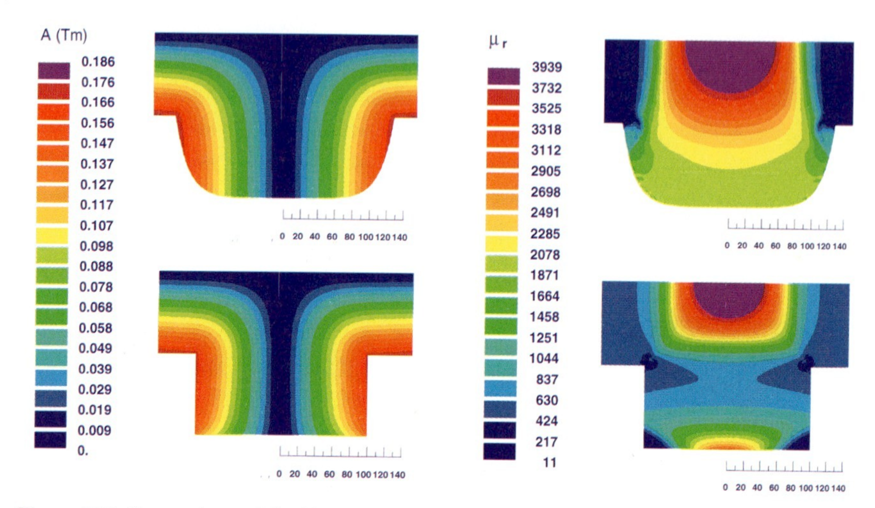

Figure 7.9 Comparison of the Rogowski pole profile with a rectangular pole. Left: Magnetic vector potential. Magnetic flux lines converge toward sharp edges. Right: Relative permeability. Notice the uniform saturation and homogeneous flux distribution for the Rogowski profile.

Thus, using Eqs. (7.29), we can derive for the differential line element along the field line φ = const.:

$$\mathrm{d}s = \sqrt{\mathrm{d}x^2 + \mathrm{d}y^2} = A\sqrt{1 + e^{2\varphi} + 2e^{\varphi}\cos\psi}\,\mathrm{d}\psi$$
$$= \frac{a}{\phi_0}\sqrt{1 + e^{2\varphi} + 2e^{\varphi}\cos\psi}\,\mathrm{d}\phi\,. \tag{7.32}$$

The modulus of the field strength in the problem domain is $E = \mathrm{d}\phi/\mathrm{d}s$. If we define $E_0 := \phi_0/a$ as the field strength in the uniform region between the electrode and the plane at distance a, it follows that

$$\frac{E}{E_0} = \frac{1}{\sqrt{1 + e^{2\varphi} + 2e^{\varphi}\cos\psi}}\,. \tag{7.33}$$

The argument of the square-root takes its minimum at $e^{\varphi} = -\cos\psi$ and is equal to $1 - \cos^2\psi = \sin^2\psi$. The maximum of the field strength along the equipotential surface ψ_0 is therefore

$$\left(\frac{E}{E_0}\right)_{\max} = \frac{1}{\sin\psi_0}\,. \tag{7.34}$$

For $\psi_0 = \pi/2$ there will be no field enhancement along the equipotential surface. Its minimum distance to the plane at $y = 0$ is

$$\frac{a}{\pi}\psi_0 = \frac{a}{2}\,. \tag{7.35}$$

For this situation, Eqs. (7.30) reduce to the so-called 90° *Rogowski profile*:

$$x = A\varphi , \qquad y = A\left(\frac{\pi}{2} + e^{\varphi}\right) . \tag{7.36}$$

A constraint for practical applications results from the fact that the electrode extends to infinity. We thus construct the space curve for the Rogowski profile of a dipole-magnet pole according to Figure 7.8 (right):

$$\mathbf{r}(\varphi) = \frac{a}{\pi}\,(\varphi + w)\,\mathbf{e}_x + \frac{a}{\pi}\left(\frac{1}{2}\pi + e^{\varphi}\right)\mathbf{e}_y , \tag{7.37}$$

where $2w$ is the pole width, a is the air-gap size, and the parameter range of φ is restricted to

$$\varphi \in \left[-\frac{w}{a}\pi,\, \ln\left(\frac{\pi s_1}{a}\right)\right] . \tag{7.38}$$

For a pole piece with a Rogowski profile, the magnetic flux density is proportional to φ and thus also to x on the pole face. This yields a uniform magnetization inside the iron pole, as illustrated in Figure 7.9 (right). Consequently, the air-gap field is less affected by iron saturation compared to a rectangular pole.

The radius of curvature derived from Eq. (7.37) by means of Eq. (3.33) is given by

$$R = \frac{\left|\frac{\mathrm{d}\mathbf{r}}{\mathrm{d}\varphi}\right|^3}{\left|\frac{\mathrm{d}\mathbf{r}}{\mathrm{d}\varphi} \times \frac{\mathrm{d}^2\mathbf{r}}{\mathrm{d}\varphi^2}\right|} = \frac{a}{\pi}\,\frac{\sqrt{1+e^{2\varphi}}^{\,3}}{e^{\varphi}} , \tag{7.39}$$

which has its minimum at $\varphi = -0.5 \ln 2$:

$$R_{\text{min}} = \frac{3\sqrt{3}}{2}\,\frac{a}{\pi} \approx 0.83\,a . \tag{7.40}$$

Thus a simple approximation for the Rogowski profile can be obtained [2] by tapering the pole piece and rounding it off with a circle of $R = 0.83\,a$.

Shaping the longitudinal profile of magnets accordingly will also limit magnetic length variations caused by local saturation in the magnet ends [9].

7.8 Combined-Function Magnets

For combined-function dipole/quadrupole magnets, the pole shape is given by the hyperbolic arc needed for the quadrupole component, together with a displacement of the aperture by x_{d} to create the dipole component by means of the feed-down effect. The displacement can be calculated from Eq. (6.51).

The combined dipole/quadrupole magnet for the CERN PS ring is shown in Figure 7.10. The magnet has a field gradient of 5 $\mathrm{T\,m^{-1}}$ and a dipole field of 1.5 T at nominal excitation of 6000 A. The pole shape is determined by a displacement x_d of 0.33 m with respect to the center of the ideal quadrupole.

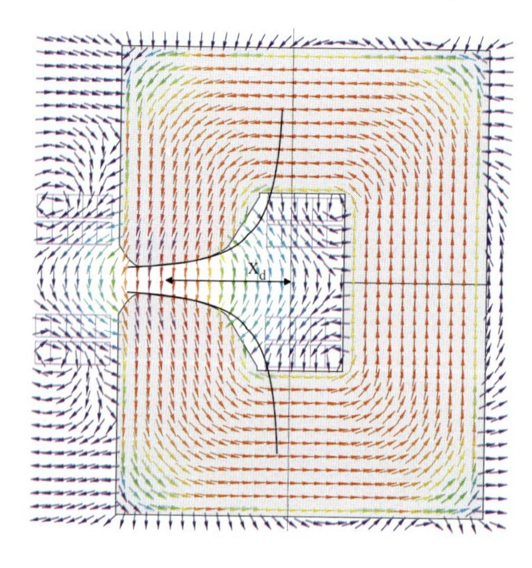

Figure 7.10 Cross section of the combined-function, dipole/quadrupole magnet for the CERN PS ring having the hyperbolic pole shape of an ideal quadrupole. Notice the color scaling of the field icons. The beam axis is displaced by x_d from the center of the quadrupole.

7.9 Permanent Magnet Excitation

For a magnetic circuit with permanent magnet excitation as shown in Figure 7.2 (right), Ampère's law yields

$$H_0\, s_0 + H_\mathrm{m}\, s_\mathrm{m} = 0\,, \tag{7.41}$$

if, again, the iron yoke is assumed to be infinitely permeable and all stray fields are neglected. The integration path within the permanent magnet is denoted s_m. For the pole surface a_0 and the magnet surface a_m we obtain

$$B_\mathrm{m} a_\mathrm{m} = B_0 a_0 = \mu_0 H_0 a_0\,. \tag{7.42}$$

From Eq. (7.41) it follows that

$$H_0 s_0 = -H_\mathrm{m} s_\mathrm{m}\,,$$

$$\frac{1}{\mu_0} B_\mathrm{m} \frac{a_\mathrm{m}}{a_0} s_0 = -H_\mathrm{m} s_\mathrm{m}\,,$$

$$B_\mathrm{m} = -\mu_0 \frac{s_\mathrm{m}}{s_0} \frac{a_0}{a_\mathrm{m}} H_\mathrm{m}\,. \tag{7.43}$$

Note that in the permanent magnet, the field and magnetic flux density have opposite direction. Result (7.43) gives rise to the definition of the *permeance coefficient*,

$$P := \frac{B_m}{\mu_0 H_m} = -\frac{s_m}{s_0}\frac{a_0}{a_m}. \tag{7.44}$$

The permeance coefficient becomes zero for $s_0 \gg s_m$ (the open circuit) and $-\infty$ for $s_m \gg s_0$ (the short circuit). The case of $a_m > a_0$ is usually referred to as the *flux concentration mode*. The permeance coefficient defines the working point on the *demagnetization curve*, which is the branch of the permanent magnet's hysteresis curve in the second quadrant. Thus H_m is called the *demagnetization field*. The line of negative slope

$$s := \frac{B_m}{H_m} = \mu_0 P = -\mu_0 \frac{s_m}{s_0}\frac{a_0}{a_m} = -\tan\alpha \tag{7.45}$$

in the $B(H)$ diagram (Figure 7.13) is known as the *load line* of the circuit.

The calculation of the magnet's working point involves an additional relationship between B and H, which is provided by the demagnetization characteristic of the magnet. The intercept of the load line with the demagnetization curve defines the working point. The vector relationship between **B**, **H**, and **M** will actually vary within the material. It should therefore be noted that in this simplified treatment we assume a scalar constitutive equation in the form

$$B_m = \mu_0(H_m + M). \tag{7.46}$$

Figure 7.11 shows the results of 2D numerical simulations of a magnetically short-circuited samarium cobalt magnet $SmCo_5$ with a remanent flux density of 0.9 T (right), as well as a circuit where $s_0 = 2s_m$ (left). Figure 7.12 shows open circuits for both rectangular and cylindrical $SmCo_5$ magnets. The load lines for the configurations with rectangular magnets are shown in Figure 7.13. The maximum slope of the load lines is determined by the magnetic resistance of the iron yoke, while an open circuit is determined by the shape of the permanent magnet. It is only for cylinders (2D) and ellipsoids (3D), that the demagnetization field, and thus the open circuit conditions, can be calculated analytically.

From Eqs. (7.41) and (7.42) we derive

$$B_m a_m s_m = -\mu_0 H_0 a_0 \frac{H_0 s_0}{H_m}, \tag{7.47}$$

and therefore,

$$H_0 = \sqrt{\frac{-B_m H_m a_m s_m}{\mu_0 a_0 s_0}} = \sqrt{\frac{-B_m H_m V_m}{\mu_0 V_0}}. \tag{7.48}$$

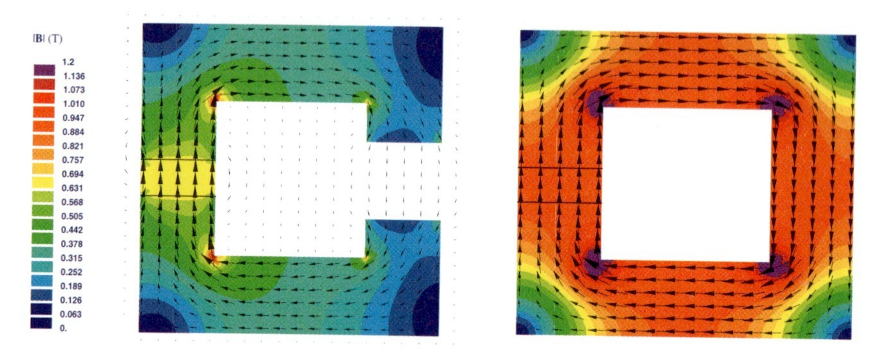

Figure 7.11 Left: A magnetic circuit with an air gap of twice the length as the thickness of the permanent magnet, $s_0 = 2s_m$. Right: Short circuited permanent magnet. B_r = 0.9 T.

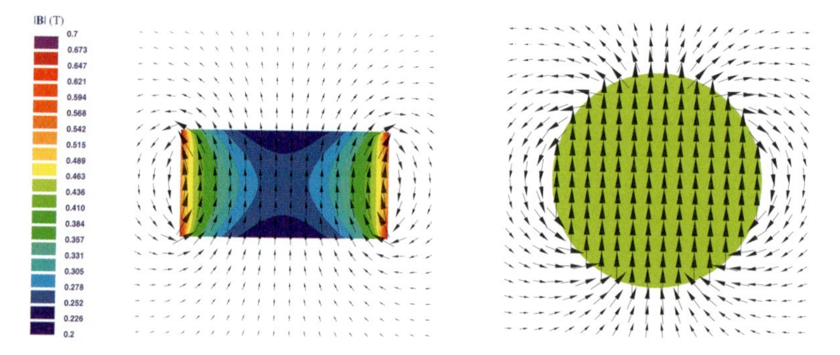

Figure 7.12 Open circuited, rectangular permanent magnet (left) and cylindrical permanent magnet (right). Notice that the demagnetization field is uniform only in the cylindrical specimen. The remanent flux density is 0.9 T in both cases.

For a given aperture volume V_0 and required air gap flux density, the magnet volume V_m can be minimized by dimensioning the circuit such that $B_m H_m$ is maximum.

The maximum product $(BH)_{max}$ is thus a good measure of quality in permanent-magnet material. As this quantity has the physical dimension of energy density, it is sometimes said that it is best to operate the permanent magnet at its energy maximum, which is misleading. This misconception results from the error of looking at the electromagnetic subsystem alone, while neglecting mechanical effects (magnetostriction) and contributions such as exchange energy (electronic spin alignment), and anisotropy energy that depend on the crystalline structure of the permanent magnet. One can say, however, that $\int_V \frac{1}{2}(BH)_{max} dV$ is the maximum magnetic energy in the air gap when the air-gap size is identical to the magnet size.

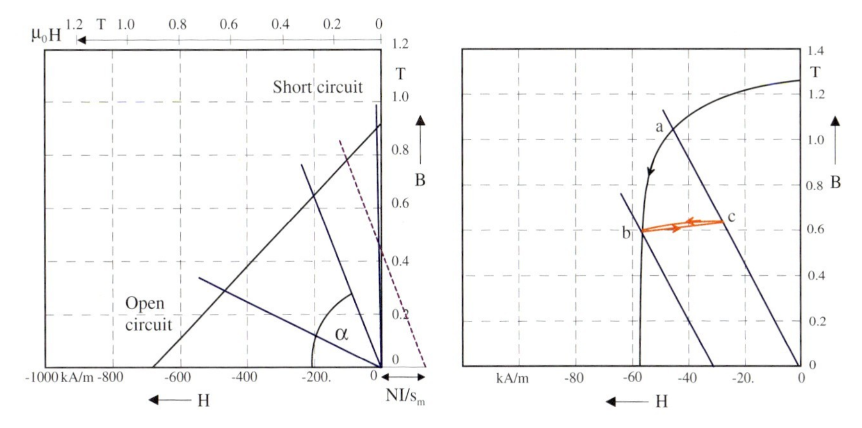

Figure 7.13 Left: Load lines for the three cases shown in Figures 7.11 and 7.12. The dashed load line corresponds to an additional excitation coil. Right: Recoil loop for dynamic operation of a magnetic circuit with a permanent magnet of nonlinear demagnetization.

For the use of permanent magnet material in accelerator magnets, demagnetization due to irradiation and thermal fluctuations must be taken into account [7]. As no permanent damage to the crystalline structure of the magnet occurs, it is possible after irradiation to magnetize the material to the nominal level.

Disregarding the leakage flux is only a rough treatment of the field problem, in particular for magnetic flux densities exceeding 1 T and for large air-gaps as shown in Figure 7.14 (left). The leakage flux is proportional to the magnetic potential difference, that is, the magnetomotive force $V_{\mathrm{m}} = \int \mathbf{H} \cdot \mathrm{d}\mathbf{r}$ between the poles; see Figure 7.15 (left). The design shown in Figures 7.14 (left) and 7.15 (left) is therefore weak, as there are large domains at high potentials. A configuration with the permanent magnets placed at the air gap has considerably less leakage flux, as can be seen in Figures 7.14 (right).

Again consider the magnetic circuit as shown in Figure 7.2 (right), but with an additional excitation coil around the iron yoke and powered such that the magnetic flux density in the circuit is enhanced for a positive current. We derive from Ampère's law

$$H_0\, s_0 + H_{\mathrm{m}}\, s_{\mathrm{m}} = N\, I\,. \tag{7.49}$$

Neglecting leakage flux, the modified load line can be calculated from

$$B_{\mathrm{m}} = -\mu_0 \frac{s_{\mathrm{m}}}{s_0} \frac{a_0}{a_{\mathrm{m}}} \left(H_{\mathrm{m}} - \frac{NI}{s_{\mathrm{m}}} \right)\,; \tag{7.50}$$

compare this to the result of Eq. (7.43). The slope of the load line still depends only on the circuit dimensions, but the line is shifted to the right, as can be

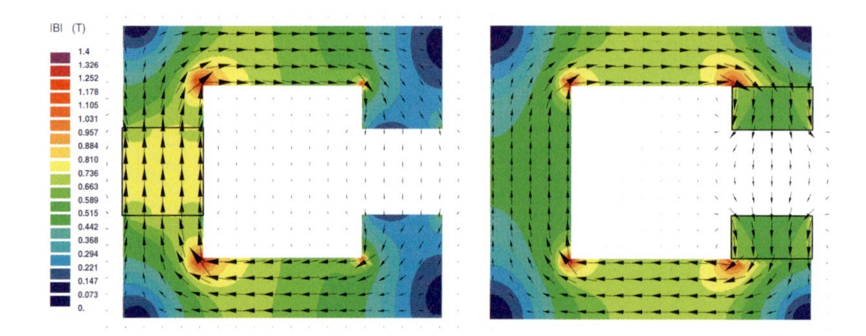

Figure 7.14 C-shaped magnet with permanent magnet excitation. For the permanent magnet brought to the air gap (right), the stray field is considerably reduced, while the flux density in the air gap is increased.

Figure 7.15 C-shaped magnet with permanent magnet excitation. The magnetic scalar potential ϕ_m is displayed for the two magnet variants presented in Figure 7.14.

seen in Figure 7.13 (left). As the flux density is enhanced in the circuit, the magnitude of the demagnetization field H_{m} is reduced.

Whereas the magnetization of the rare earth material is basically constant, equal to the saturation magnetization, AlNiCo material shows a "knee" below which irreversible demagnetization takes place. Working points below the knee might result from temperature excursions, from externally applied demagnetization fields or from changing air-gap sizes in magnetic actuator circuits.

Consider a magnet operating at point a on the demagnetization curve shown in Figure 7.13 (right). Operating conditions resulting in a change of the load line's slope or position may shift the working point from point a to b. When the amplitude of the demagnetization field is reduced, the working point will move along minor loops known as *recoil loops* toward point c. Subsequent dynamic operation with the same load will cause the magnet to operate between points b and c on the minor loop.

7.10 Cooling of Normal-Conducting Magnets

A comprehensive treatment of the thermodynamics of magnet cooling exceeds the scope of this book. We will nevertheless present some empirical laws for turbulent flow that hold for a *Reynolds number*[2] between 4000 and 100 000. The Reynolds number is the quotient of dynamic pressure and shear stress [5]:

$$\mathrm{Re} = \frac{v_\mathrm{s} D}{\nu}, \tag{7.51}$$

where $[\mathrm{Re}] = 1_U$, D is the pipe diameter, and v_s the mean fluid velocity. ν is the *kinematic fluid viscosity* defined by the dynamic viscosity[3] μ, divided by the mass density ϱ of the incompressible fluid,

$$\nu := \frac{\mu}{\varrho}, \tag{7.52}$$

where $[\nu] = 1\ \mathrm{m^2\,s^{-1}}$, $[\mu] = 1\ \mathrm{kg\,m^{-1}s^{-1}}$, $[\varrho] = 1\ \mathrm{kg\,m^{-3}}$. For water at 310 K the parameters are $\varrho = 993\ \mathrm{kg\,m^{-3}}$ and $\nu = 6.982 \times 10^{-7}\mathrm{m^2\,s^{-1}}$. Henceforth we will disregard the temperature-dependent variation in the kinematic viscosity along the cooling manifold.

Reynolds numbers below 2000 indicate laminar flow with reduced heat transfer. Reynolds numbers higher than 3000 indicate the turbulent regime, characterized by stochastic property changes and strong variations in pressure and velocity in space and time. However, an upper technical limit is given by a Reynolds number of 100 000, since very turbulent flow leads to pipe erosion-corrosion. The transitional flow regime with Reynolds numbers between 2000 and 3000 is usually avoided as in this case the pressure drop is difficult to calculate.

Cooling efficiency depends on the speed of the cooling medium and on the heat transfer at the tube boundary. The pressure drop in the pipe is proportional to ϱv_s^2 and can be calculated from

$$\Delta p = \lambda_\mathrm{f} \frac{l}{D} \frac{\varrho v_\mathrm{s}^2}{2}, \tag{7.53}$$

where the dimensionless friction factor λ_f is itself a function of the surface roughness. l is the total length of the cooling tube. For the calculation of the friction factor we use the *Blasius law*:[4]

$$\lambda_\mathrm{f} = \frac{0.3164}{\sqrt[4]{\mathrm{Re}}}, \tag{7.54}$$

2 Osborne Reynolds (1842–1912).

3 The dynamic viscosity is the ratio of the pressure exerted on the moving surface of a fluid to the velocity gradient toward a stationary boundary plate.

4 Heinrich Blasius (1883–1970).

which is sufficiently correct for round, smooth pipes and for Reynolds numbers in the range between 4000 and 100 000. For noncircular tubes, the hydraulic diameter can be introduced by

$$D_\mathrm{h} := \frac{4a_\mathrm{c}}{c}\,, \tag{7.55}$$

where a_c is the area of the tube cross section and c denotes the passage's wetted perimeter.

Aiming at a relation between the water velocity and pressure drop, we combine Eqs. (7.53) with (7.54) and (7.51), and consider the material parameters for water at 310 K:

$$v_\mathrm{s}^2 = \frac{2\Delta p}{\varrho}\frac{D}{l}\frac{1}{0.3164}\left(\frac{v_\mathrm{s}D}{\nu}\right)^{1/4},$$

$$v_\mathrm{s}^{7/4} = \frac{2}{0.3164\,\varrho\,\nu^{1/4}}\frac{\Delta p}{l}D^{5/4},$$

$$v_\mathrm{s} = 0.421\left(\frac{\Delta p}{l}\right)^{4/7}D^{5/7}\,. \tag{7.56}$$

Equation (7.56) is displayed in Figure 7.16. The evacuated heat is

$$\mathrm{d}Q = Ua_c\,\Delta T\mathrm{d}t\,, \tag{7.57}$$

where $[Q] = 1\,\mathrm{J}$. U is the heat transfer coefficient, $[U] = 1\,\mathrm{W\,m^{-2}\,K^{-1}}$, determined by

$$U = C_\mathrm{p}v_\mathrm{s}\varrho\,. \tag{7.58}$$

ΔT is the temperature difference between the inlet and outlet. The heat capacity of water is $C_\mathrm{p} = 4179\,\mathrm{J\,kg^{-1}\,K^{-1}}$. From Eqs. (7.57) and (7.58) we finally obtain

$$\frac{\mathrm{d}Q}{\mathrm{d}t} = C_\mathrm{p}v_\mathrm{s}\varrho a_c\,\Delta T\,. \tag{7.59}$$

For the heat transfer through the pipe's surface layer, two typical cases should be distinguished: constant temperature and constant heat dissipation. The latter is valid for cooled conductors:

$$q_\mathrm{w} = (T_\mathrm{w} - T_\mathrm{b})\mathrm{Nu}\frac{k}{D}\,, \tag{7.60}$$

where k is the heat conductivity, T_b the bulk temperature, T_w the wall temperature, and Nu the local *Nusselt number*[5] for fully developed pipe flow. The dimensionless Nusselt number is the ratio of convective to conductive heat transfer in normal direction to the boundary.

5 Wilhelm Nusselt (1882–1957).

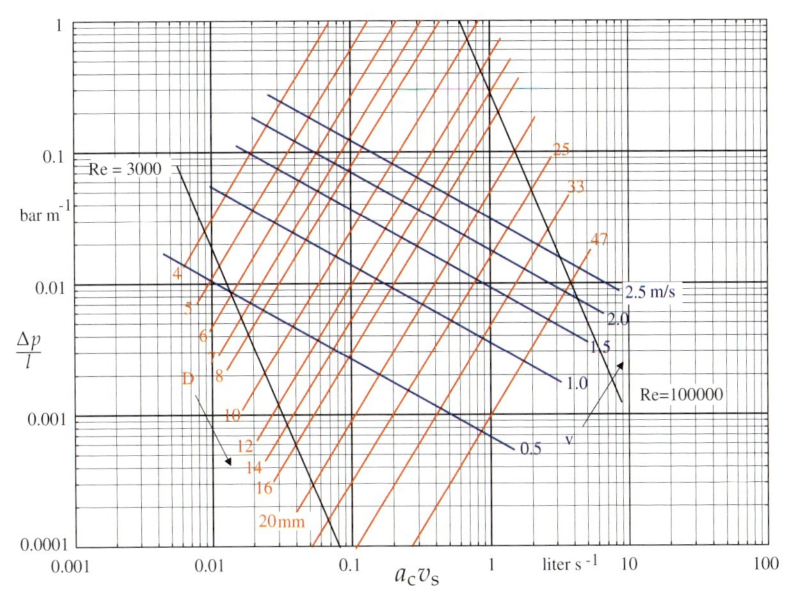

Figure 7.16 Pressure drop per unit length of tube, as a function of flow rate and diameter D of the cooling channel [12].

With a number of assumptions, namely

- developed, turbulent flow,
- smooth pipes of constant diameter and no effects from bends,
- constant heat flow along the pipe,

a feasibility study can be performed easily using the following procedure:

1. For a given cross section and length of the conductor, calculate the total ohmic loss in the coil from Eq. (7.10).
2. With the heat balance equation, Eq. (7.59), calculate the required flow rate $v_s a_c$ of water needed to evacuate the total ohmic loss: As a rule of thumb, $1\,\mathrm{l\,s^{-1}}$ evacuates $4.15\,\mathrm{kJ\,s^{-1}}$ for $\Delta T = 1\,°\mathrm{C}$. For an inlet temperature of 27 °C and an outlet temperature of 51 °C ($\Delta T = 23.9\,°\mathrm{C}$), 100 kW requires a flow rate of $1\,\mathrm{l\,s^{-1}}$.
3. For a given flow rate and cooling-hole diameter, check by means of Figure 7.16 the pressure drop per unit length of tube. Determine the corresponding Reynolds number.
4. Check that the Reynolds number is in the range of 3000–100 000. This is the turbulent flow regime for which the Blasius law holds.
5. Check that the total pressure drop in the cooling manifold of the accelerator magnet is smaller than 4–5 bar. In experimental magnets the limit for the pressure drop is around 20 bar.

6. If the limit is not obeyed, increase the cooling hole and return to (1). If the size of the conductor cannot be increased, consider multiple cooling circuits. Return to (5).

Example: The normal-conducting separation dipole MBW for the LHC cleaning insertion, shown in Figure B.14 (left), is an H-type magnet with an aperture of 52 mm and a nominal current of 720 A in two coils. Each coil is made of three "pancakes" of 14 windings. The total number of turns per magnet is 84. The pancake windings are made from rectangular, hollow copper conductors with a cooling-hole diameter of 8 mm. The conductors are insulated with fiberglass tape and impregnated with epoxy resin.

- The air-gap flux density is calculated for a nominal current of 720 A from

$$B_0 = \frac{\mu_0 N I}{s_0} = \frac{4\pi \times 10^{-7} \cdot 84 \cdot 720}{0.052}\,\mathrm{T} = 1.4\,\mathrm{T}\,. \tag{7.61}$$

- The magnetic length is 3.4 m. Including the coil ends, the total length of the conductor is 395 m per coil.
- For a conductor cross section of 15 × 18 mm and a cooling-hole diameter of 8 mm, the ohmic resistance of one coil can be calculated as

$$R = \frac{l}{\varkappa a_{\mathrm{cond}}} = \frac{395}{5.9 \times 10^7 \cdot 219.7 \times 10^{-6}}\,\Omega = 30.5\,\mathrm{m}\Omega\,. \tag{7.62}$$

- The resulting voltage drop over one coil is 22 V.
- The dissipated power per coil at nominal current is 15.8 kW. This power can be evacuated by a water flow rate of $3.8\,\mathrm{l\,s^{-1}}$ and a ΔT of 1 °C. For a ΔT of 25 °C we need $0.15\,\mathrm{l\,s^{-1}}$. To limit the pressure drop across the cooling manifold, a separate cooling circuit is installed for each pancake winding. With three circuits, the flow rate per cooling circuit is maintained at $0.05\,\mathrm{l\,s^{-1}}$.
- From the Blasius law, we obtain for a cooling hole of 8 mm diameter a pressure drop of 0.022 bar per meter. This results in a pressure drop of 2.8 bar in each circuit, which is within the technical limits. □

References

1 Bettoni, S.: Design and integration of superconducting undulators for the LHC beam diagnostics, Universita' degli studi di Milano, 2006

2 Braams, C.M.: Edge effect in charged-particle analyzing magnets, Nuclear Instruments and Methods, 1964

3 LEP Design Report: CERN-LEP 84-01, 1984

4 LHC Design Report, Vol. 1, The LHC main ring, CERN-2004-003, 2004

5 Lienhard, J.H. (IV), Lienhard, J.H. (V): A Heat Transfer Textbook, Third Edition, Phlogiston Press, USA, 2004

6 Maxwell, J.C.: A treatise on Electricity & Magnetism, Dover, New York, 1954, Reprint of the 1891 edition.

7 Okuda, S., Ohashi, K., Kobayashi, N.: Effects of Electron Beam and γ-ray Irradiation on the Magnetic Flux of Nd–Fe–B and Sm–Co Permanent Magnets

8 Ponce, L., Jung, R., Méot, F.: LHC Proton Beam Diagnostics using Synchrotron Radiation, CERN-2004-007, AB-Department, CERN, 2004

9 Ostiguy, J.-F.: Longitudinal Profile and Effective Length of a Conventional Dipole Magnet, IEEE Transactions, 1993

10 Rogowski, W.: Die elektrische Festigkeit am Rande des Plattenkondensators, Archiv für Elektrotechnik, 1923

11 Tanabe, J. T.: Iron Dominated Electromagnets: Design, Fabrication, Assembly and Measurements, World Scientific, Singapore, 2005

12 Taylor, T.: Private communication, CERN, 2003

8
Coil-Dominated Magnets

In superconducting accelerator magnets with central fields well above one tesla, the current distribution in the coils dominates the field quality. The design process thus starts with the optimal shape design of the superconducting coil.

It is reasonable to focus on the fields generated by line-currents, since any current distribution over an arbitrary cross section can be approximated by a number of line-currents distributed within the coil cross section. Moreover, superconducting magnets for accelerators are wound from Rutherford cables, ribbon type conductors, or braids made of strands with a size of about 1 mm. Thus the coil is well modeled if one considers a single line-current at each strand position. By representing keystoned Rutherford cables by two layers of line-currents, the grading of the current density due to the different compaction on the cable's narrow and wide sides is automatically modeled. A transversal cross section of a block of two Rutherford cables, with different compaction at the narrow (left) and wide sides, is shown in Figure 8.1.

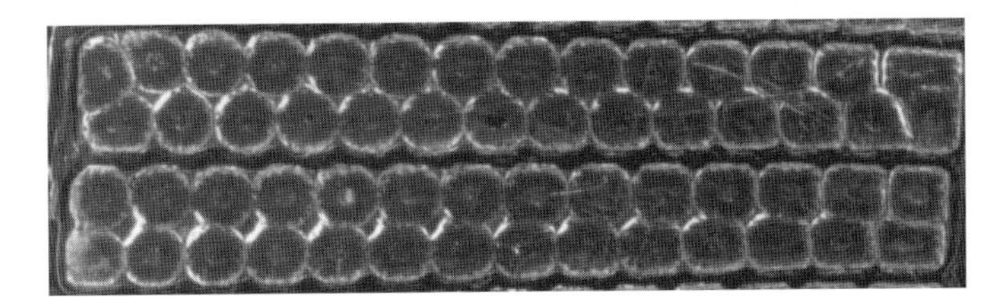

Figure 8.1 Transversal cross section of a block of cables in the inner layer of a dipole model magnet showing two Rutherford cables composed of superconducting strands. Note the increasing size of the gaps between the strands toward the outer diameter of the coil (left-hand side of the picture).

We will use the results derived in Chapter 5 to calculate the coil field in accelerator magnets and solenoids, and to derive ideal current distributions as starting points for coil field optimization. In the case of accelerator magnets, the concepts of intersecting circles and ellipses, the cos φ current distribution, and the shell type magnet cross section can facilitate the conceptual design. For solenoids, we will derive the Fabry factors [12], which are geometrical

Field Computation for Accelerator Magnets. Stephan Russenschuck

ISBN: 978-3-527-40769-9

factors for the optimum aspect ratio of solenoidal coils, and we will treat the field quality by means of zonal harmonics. Results of numerical calculations obtained with the CERN field computation program ROXIE are shown as illustrations.

Magnet design can be performed to a large extent under the assumption that the overall current density in the superconducting (copper stabilized) wire is about 400 times higher than in household wiring ($\leq 1.5\,\mathrm{A\,mm^{-2}}$) and around 100 times higher than in water-cooled conductors ($\leq 10\,\mathrm{A\,mm^{-2}}$) of normal-conducting magnets, electrical machines, and drives. In reality, however, the performance of the magnet is limited by the position of the working point with respect to the critical surface of the superconductor. We will thus calculate the field and temperature margins at nominal excitation.

8.1 Accelerator Magnets

In circular accelerators, transverse magnetic fields guide and focus the beams of charged particles. As the beam cross section and the magnet aperture is small compared to the radius of the accelerator, it is a reasonable simplification to use a 2D field analysis for magnet design. Three-dimensional effects from the coil ends can be treated separately and often compensated by a slight adjustment of the coil's cross section.

Another simplification can be arrived at by estimating the effect of the iron magnetization on the aperture field by means of the image-current method explained in Section 5.9.2. The image current at the position

$$\mathbf{r}^* = \frac{r_\mathrm{y}^2}{|\mathbf{r}'|^2}\mathbf{r}', \tag{8.1}$$

must be scaled according to

$$I^* = \lambda_\mu I := \frac{\mu_\mathrm{r} - 1}{\mu_\mathrm{r} + 1} I. \tag{8.2}$$

The normal and skew multipole coefficients at the reference radius r_0 inside the magnet's aperture are given for a set of K line currents at the positions $(r_{\mathrm{c},k}, \varphi_{\mathrm{c},k})$ carrying current I_k by

$$B_n(r_0) = -\sum_{k=1}^{K} \frac{\mu_0 I_k}{2\pi} \frac{r_0^{n-1}}{r_{\mathrm{c},k}^n} \left(1 + \lambda_\mu \left(\frac{r_{\mathrm{c},k}}{r_\mathrm{y}}\right)^{2n}\right) \cos n\varphi_{\mathrm{c},k}, \tag{8.3}$$

$$A_n(r_0) = \sum_{k=1}^{K} \frac{\mu_0 I_k}{2\pi} \frac{r_0^{n-1}}{r_{\mathrm{c},k}^n} \left(1 + \lambda_\mu \left(\frac{r_{\mathrm{c},k}}{r_\mathrm{y}}\right)^{2n}\right) \sin n\varphi_{\mathrm{c},k}, \tag{8.4}$$

for $r_0 < r_a$, cf. Eq. (6.65) and Figure 6.1. The second term in the parenthesis is due to the image currents. Some important conclusions can be drawn:

1. For a coil without an iron yoke the field errors scale with $1/r^n$ where n is the order of the multipole and r is the mid radius of the coil. It is clear, however, that an increase in coil aperture causes a linear drop in dipole field. Other limitations of the coil size are the beam separation distance, the electromagnetic forces and yoke size, as well as the stored energy, which results in an increase of the hot-spot temperature during a quench.
2. Symmetry conditions reveal that some of the multipole components vanish. Because of $\sin\varphi = -\sin\varphi$, an up/down symmetry in a dipole magnet implies that $A_n = 0$. If there is an additional left/right symmetry, only the odd $B_1, B_3, B_5, B_7, \ldots$ components remain because of $\cos(\pi - \varphi) = -\cos\varphi$. In general, $n = (2m+1)N$ for $m = 0, 1, 2, \ldots$, where $N = 1$ for the dipole, $N = 2$ for the quadrupole, etc. With the symmetry conditions applied, the $B_n \neq 0$ are termed the *allowed multipoles*.
3. The ratio of the image currents B_N^{imag} to the main field component $B_N + B_N^{\text{imag}}$ for a nonsaturated yoke with $\mu_r \gg 1$ is

$$\frac{B_N^{\text{imag}}}{B_N + B_N^{\text{imag}}} \approx \left(1 + \left(\frac{r_y}{r}\right)^{2N}\right)^{-1} . \tag{8.5}$$

   For the main dipoles, with a mean coil radius of $r = 43.5$ mm and a yoke radius of $r_y = 89$ mm, we obtain for the B_1 component a 19% contribution from the yoke, whereas for the B_5 component the influence of the yoke is only 0.07%. It is therefore appropriate to optimize for higher-order harmonics first using analytical field calculation, and to include the effect of iron saturation on the lower-order multipoles only at a later stage.

8.1.1 Generation of Pure Multipole Fields

We will now aim at finding a current distribution for the required two-dimensional field in the aperture, for example, a pure multipole field of a given order. The efficiency of the optimization process for coil-dominated magnets can be enhanced by analytical solutions, employing Eqs. (8.3) and (8.4), in the conceptual design phase.

Shell with cos $m\varphi_c$ current density Consider a current shell[1] $r_a \leq r_c \leq r_e$ with a current density that varies with the azimuthal angle φ_c, $J(\varphi_c) = J_E \cos m\varphi_c$. Here J_E denotes the maximum engineering current density in the coil, as explained in Section 1.3. Extending Eq. (8.3) from line-currents to arbitrary current distributions, the B_n components are obtained by

$$B_n(r_0) = \int_{r_a}^{r_e} \int_0^{2\pi} -\frac{\mu_0 J_E r_0^{n-1}}{2\pi r_c^n} \left(1 + \lambda_\mu \left(\frac{r_c}{r_y}\right)^{2n}\right) \cos m\varphi_c \cos n\varphi_c \, r_c \mathrm{d}\varphi_c \, \mathrm{d}r_c \quad (8.6)$$

for $r_0 < r_a$. Employing the orthogonality condition for trigonometric functions, $\int_0^{2\pi} \cos m\varphi_c \cos n\varphi_c \mathrm{d}\varphi_c = \pi\delta_{m,n}(m, n \neq 0)$, it follows that the current shell produces a pure $2m$-polar field. In the case of the dipole ($m = n = 1$) this yields for $r_0 < r_a$,

$$B_1(r_0) = -\frac{\mu_0 J_E}{2} \left((r_e - r_a) + \lambda_\mu \frac{1}{r_y^2} \frac{1}{3} \left(r_e^3 - r_a^3\right) \right) . \quad (8.7)$$

For the quadrupole shell ($m = n = 2$) as shown in Figure 8.2 (right) we obtain for $r_0 < r_a$,

$$B_2(r_0) = -\frac{\mu_0 J_E r_0}{2} \left(\ln\left(\frac{r_e}{r_a}\right) + \lambda_\mu \frac{1}{r_y^4} \frac{1}{4} \left(r_e^4 - r_a^4\right) \right) . \quad (8.8)$$

Because $\int_0^{2\pi} \cos m\varphi_c \sin n\varphi_c \, \mathrm{d}\varphi_c = 0, \forall m \neq n$, all A_n components vanish. Shells that have $\cos\varphi_c$- and $\cos 2\varphi_c$-dependent current density are displayed in Figure 8.2.

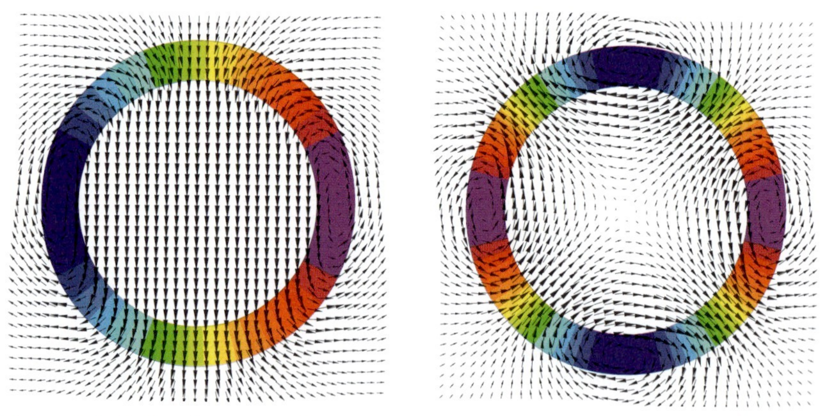

Figure 8.2 Shells with $\cos\varphi_c$– (left) and $\cos 2\varphi_c$–dependent current density (right).

1 The subscripts a and e stand for aperture and external, respectively.

The main field component inside the aperture of a shell dipole without iron yoke is given by the simple equation

$$B_1(r_0) = -\frac{\mu_0 J_E}{2}(r_e - r_a), \qquad r_0 < r_a. \tag{8.9}$$

Outside the dipole coil we find

$$B_1(r_0) = -\frac{\mu_0 J_E}{6r_0^2}(r_e^3 - r_a^3), \qquad r_a < r_0. \tag{8.10}$$

Inserting Eq. (8.9) into Eq. (6.45) yields a magnetic flux density inside the aperture of

$$\mathbf{B}(r_0, \varphi) = -\frac{\mu_0 J_E}{2}(r_e - r_a)\,\mathbf{e}_y, \qquad r_0 < r_a. \tag{8.11}$$

Inserting Eq. (8.9) into Eqs. (6.72) and (6.73) and calculating[2] the Cartesian field components for $r_0 > r_e$ yields

$$\begin{aligned}\mathbf{B}(r_0, \varphi) &= -\frac{\mu_0 J_E}{6r_0^2}\left(r_e^3 - r_a^3\right)\left((2\sin\varphi\cos\varphi)\,\mathbf{e}_x + \left(\sin^2\varphi - \cos^2\varphi\right)\,\mathbf{e}_y\right),\\ &= -\frac{\mu_0 J_E}{6r_0^2}\left(r_e^3 - r_a^3\right)\left(\sin(2\varphi)\,\mathbf{e}_x - \cos(2\varphi)\,\mathbf{e}_y\right). \end{aligned} \tag{8.12}$$

For field points located inside the current shell, that is, at an arbitrary radius r_0 with $r_a < r_0 < r_e$, we can combine Eq. (8.11) for the inner field of the outer shell (r_0 to r_e) with Eq. (8.12) for the outer field of the inner shell (r_a to r_0) and obtain[3] for $r_a < r_0 < r_e$,

$$\mathbf{B}(r_0, \varphi) = -\frac{\mu_0 J_E}{2}(r_e - r_0)\,\mathbf{e}_y - \frac{\mu_0 J_E}{6r_0^2}\left(r_0^3 - r_a^3\right)\left(\sin(2\varphi)\,\mathbf{e}_x - \cos(2\varphi)\,\mathbf{e}_y\right). \tag{8.13}$$

Let B denote the modulus of the dipole field in the aperture. We can now relate Eq. (8.13) to the critical current density J_c of the superconductor and the critical field B_{c2}, respectively:

$$B = \frac{\mu_0}{2}\lambda_{tot}\, J_c\,(r_e - r_a) = \frac{\mu_0}{2}\lambda_{tot}\, d\,(B_{c2} - B)\,(r_e - r_a), \tag{8.14}$$

where we use the scaling law, [28]

$$J_c(B) = d\,(B_{c2} - B) \tag{8.15}$$

2 Using the relations between the unit vectors in Cartesian and cylindrical coordinates; see Table 3.2.

3 Using the relations $\cos^2\varphi + \sin^2\varphi = 1$, $\sin 2\varphi = 2\sin\varphi\cos\varphi$ and $\cos 2\varphi = \cos^2\varphi - \sin^2\varphi$.

for $B < B_{c2}$ and a constant[4] negative slope d in the high-field region above 3 T

$$d := -\left.\frac{\mathrm{d}J_c}{\mathrm{d}B}\right|_{B_{c2}} . \tag{8.16}$$

The total superconductor filling factor λ_{tot}, given for the LHC cables in Table 1.3, is approximately 0.3. The slope d is 500–600 $\mathrm{A\,mm^{-2}\,T^{-1}}$. Equation (8.15) yields an estimate of the coil thickness for a desired flux density in the magnet aperture. The coil thickness approaches infinity when $B \to B_{c2}$.

Example: Consider the LHC main dipoles, with $d = 520\ \mathrm{A\,mm^{-2}\,T^{-1}}$, $B_{c2} = 13.6$ T, $\lambda_{tot} = 0.26$, and a short sample field[5] of 9.76 T. Equation (8.15) yields a coil thickness of 29.9 mm. The LHC main dipole coil is indeed made of two 15.4-mm-thick coil layers. □

Cylindrical current shell with constant current density Shells with $\cos m\varphi_c$-dependence of the current density are technically impossible to realize with superconducting cable. A step-wise approximation can be given, however, by a number of concentric shells of constant current density. A geometry for a dipole magnet containing three nested shells (symmetric about the x- and the y-axes) is shown in Figure 8.3 (left).

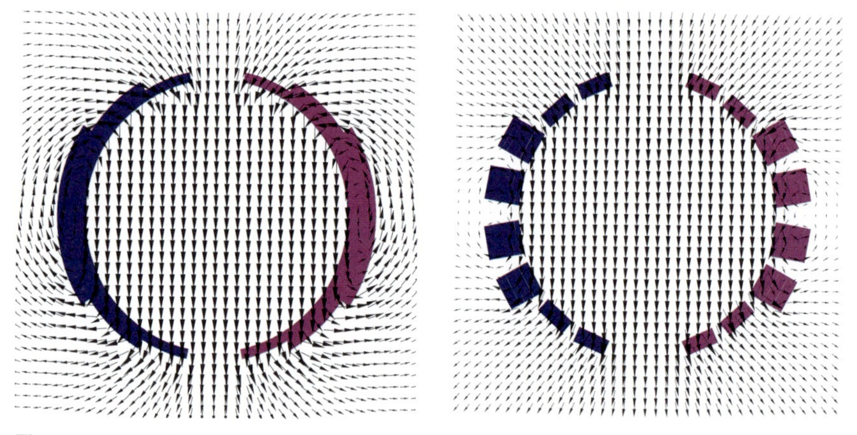

Figure 8.3 Left: Three nested cylindrical current shells with constant current density, which allow the minimization of three higher-order multipole components, here $b_3 = b_5 = b_7 = 0$. Right: Coil block arrangement with cables placed in grooves stamped into the austenitic steel collars [21].

4 This is a valid approximation for the critical surface of Nb–Ti at constant temperature. It is much less accurate for Nb_3Sn and other superconducting materials; see Chapter 16.

5 The short sample field is defined as the theoretical quench field if there is no degradation due to mechanical instabilities.

Consider a shell segment with $r_a < r_c < r_e$ and $-\varphi_s < \varphi_c < \varphi_s$ carrying constant current density J_E. Let us add a second shell segment of constant current density $-J_E$ on the interval $\pi - \varphi_s < \varphi_c < \pi + \varphi_s$. Since $\cos n\varphi_c = \cos(-n\varphi_c)$, $\sin n\varphi_c = -\sin(-n\varphi_c)$, and

$$\cos n\varphi_c - \cos n(\pi + \varphi_c) = \begin{cases} 2\cos n\varphi_c & \text{for } n = 1, 3, 5, \ldots \\ 0 & \text{for } n = 2, 4, 6, \ldots \end{cases}, \tag{8.17}$$

it follows that only odd-numbered, normal multipoles are present:

$$B_n(r_0) = \int_{r_a}^{r_e} 2\int_{-\varphi_s}^{\varphi_s} -\frac{\mu_0 J_E r_0^{n-1}}{2\pi r_c^n}\left(1 + \lambda_\mu \left(\frac{r_c}{r_y}\right)^{2n}\right) \cos n\varphi_c \, r_c \mathrm{d}\varphi_c \, \mathrm{d}r_c, \tag{8.18}$$

which for the dipole component yields

$$B_1(r_0) = \frac{-2\mu_0 J_E}{\pi}\left((r_e - r_a) + \lambda_\mu \frac{1}{r_y^2}\frac{r_e^3 - r_a^3}{3}\right) \sin n\varphi_s \,. \tag{8.19}$$

For the calculation of all higher-order multipole components ($n = 3, 5, \ldots$) we will disregard the influence of the iron yoke and make an approximation for a thin shell:

$$\int_{r_a}^{r_e} \left(\frac{r_0}{r_c}\right)^{n-1} \mathrm{d}r \approx \left(\frac{r_0}{r_m}\right)^{n-1} \Delta r_s \,, \tag{8.20}$$

where $r_m := 0.5(r_a + r_e)$ is the mean radius of the current shell and $\Delta r_s := r_e - r_a$ is its thickness. We then obtain the simple expression,

$$B_n(r_0) = -\frac{2\mu_0 J_E \Delta r_s}{\pi n}\left(\frac{r_0}{r_m}\right)^{n-1} \sin n\varphi_s \,. \tag{8.21}$$

Example: For a sector with $\varphi_s = 60°$, $r_0 = 17$ mm, and $r_m = 35.5$ mm, the first higher-order multipole is b_5, for which we obtain

$$b_5(r_0) = \frac{B_5(r_0)}{B_1(r_0)} = \frac{1}{5}\left(\frac{r_0}{r_m}\right)^4 \frac{\sin 300°}{\sin 60°} \approx -105 \times 10^{-4} \,, \tag{8.22}$$

which would be unacceptable for the LHC operation. With n symmetrical layers, however, the field quality up to the order of $2n + 1$ can be optimized; $b_3 = b_5 = \cdots = b_{2n+1} = 0$ can be achieved.[6] □

For a shell with quadrupolar symmetry we obtain in a similar way,

$$B_2(r_0) = \frac{-\mu_0 J_E r_0}{\pi}\left(\ln\left(\frac{r_e}{r_a}\right) + \lambda_\mu \frac{1}{r_y^4}\frac{r_e^4 - r_a^4}{4}\right) \sin n\varphi_s \,. \tag{8.23}$$

6 The optimization problem involves the solving of a nonlinear equation system for the B_n. Thus numerical optimization methods must be used. They are presented in Chapter 20.

Intersecting circles For the Cartesian components of the magnetic field inside a round conductor centered at the origin we found the following relations in Section 9.14:

$$H_x(x,y) = -\frac{1}{2} J_E\, y\,, \qquad H_y(x,y) = \frac{1}{2} J_E\, x\,. \tag{8.24}$$

We will now show that in the current-free center of two intersecting circles of opposite current density a homogeneous field is generated [26]. Consider the two intersecting circles shown in Figure 8.4, shifted by some distance c, with the local coordinate systems $(x_1 = x + c/2,\ y_1 = y)$ and $(x_2 = x - c/2,\ y_2 = y)$. Let us assume constant engineering current density $J_E\,\mathbf{e}_z$ in circle 1 (shifted to the left) and opposite current density $-J_E\,\mathbf{e}_z$ in circle 2 (shifted to the right). It follows that

$$B_x = B_{x,1} + B_{x,2} = -\mu_0 \frac{1}{2} J_E\,(y_1 - y_2) = 0, \tag{8.25}$$

$$B_y = B_y^{(1)} + B_{y,2} = \mu_0 \frac{1}{2} J_E \left(\left(x + \frac{c}{2} \right) - \left(x - \frac{c}{2} \right) \right) = \mu_0 \frac{1}{2} J_E\, c\,. \tag{8.26}$$

The dipole field scales linearly with the mid-thickness of the coil and its maximum engineering current density, but it does not depend on the size of the aperture. However, the amount of superconductor scales with the size of aperture, as we will show below. At the horizontal median plane, the coil thickness of the intersecting circles is the same as for the $\cos\varphi_c$ shell:

$$c = \frac{2}{\mu_0} \frac{B_y}{J_E} = \frac{2}{\mu_0} \frac{B_y}{\lambda_{tot} d\,(B_{c2} - B_y)}\,. \tag{8.27}$$

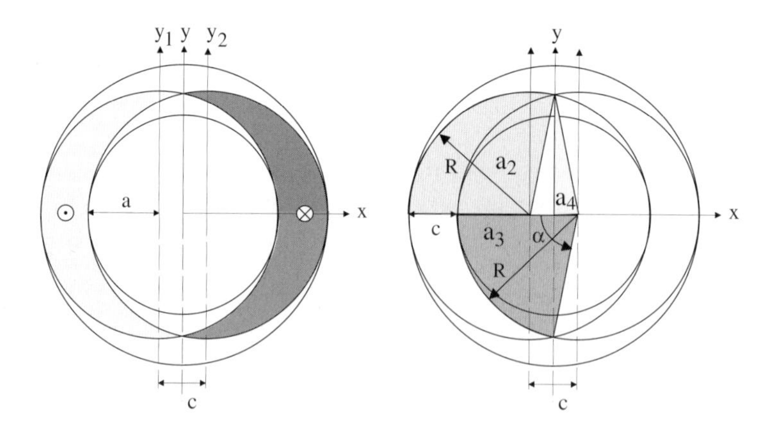

Figure 8.4 Left: Intersecting circles of opposite current density that create an ideal dipole field inside the aperture. Right: Parameters for the calculation of the coil surface represented by intersecting circles.

Note that $c \to \infty$ when B_y approaches B_{c2}. Thus the key to higher magnetic flux densities is an increased critical field of the superconductor, and not an increased coil thickness.

Example: Consider the LHC dipole coil with a central field of $B_0 = 8.33$ T and two coil layers, each 15.4 mm in width. The necessary engineering current density is 430 A mm^{-2} according to Eq. (8.26). This result must be compared to the cross-sectional area of the insulated cable, which is 32 mm^2 for the cable in the inner layer and 27 mm^2 for the cable in the outer layer. With 11 850 A in the cables this yields engineering current densities of 368 and 439 A mm^{-2}, respectively. □

The area of the coil cross section, modeled by intersecting circles, can be calculated from the geometrical parameters [30] shown in Figure 8.4 (right):

$$a = 4\,(a_2 - a_3 + 2a_4)\,, \tag{8.28}$$

where

$$a_2 = \frac{\pi - \alpha}{2} R^2, \qquad a_3 = \frac{\alpha}{2} R^2, \qquad a_4 = \frac{1}{2} c\sqrt{R^2 - c^2}\,. \tag{8.29}$$

Expressing the angle α in terms of the quantities R and c yields

$$a = 4\left(R^2\left(\frac{\pi}{2} - \arccos\left(\frac{c}{R}\right)\right) + c\sqrt{R^2 - c^2}\right). \tag{8.30}$$

With the aperture radius $r_a = R - c/2$ and the intersection c given by Eq. (8.27), we can express the surface of the coil cross section as a function of the critical field B_{c2}, the current degradation d, the filling factor λ_{tot}, the aperture radius r_a, and the magnetic flux density B_y.

Example: Consider, once more, the LHC dipole coil with a short sample field of $B_y = 8.76$ T. Using Eq. (8.27) and typical values of $d = 520$ A mm^{-2} T^{-1}, $B_{c2} = 13.6$ T, $\lambda_{tot} = 0.26$ we obtain $c = 29.9$ mm. Eq. (8.30) yields $a = 10\,790$ mm^2 for an aperture radius of $r_a = 28$ mm. The LHC main dipole coil has a copper cross section of 9128 mm^2 and a superconductor cross section of 4988 mm^2. □

Intersecting ellipses The components of the magnetic field within an elliptical conductor centered at the origin, with semiaxes a and b carrying constant engineering current density J_E, is given by[7]

$$H_x(x, y) = -J_E \frac{a}{a+b} y\,, \qquad H_y(x, y) = J_E \frac{b}{a+b} x\,. \tag{8.31}$$

7 The most elegant proof of these equations is carried out using complex potentials. It can be found in Section 9.14.

Consider ellipse 1, centered at $x = -x_0$ with semiaxes a_1 and b_1 carrying the engineering current density J_E, and ellipse 2, centered at $x = x_0$ with semiaxes a_2 and b_2 carrying the engineering current density $-J_E$. This results in

$$H_{x,1} = -J_E \frac{a_1}{a_1 + b_1} y, \qquad H_{y,1} = J_E \frac{b_1}{a_1 + b_1} (x + x_0), \tag{8.32}$$

$$H_{x,2} = J_E \frac{a_2}{a_2 + b_2} y, \qquad H_{y,2} = -J_E \frac{b_2}{a_2 + b_2} (x - x_0), \tag{8.33}$$

We can now study three different cases:

1. The dipole configuration as shown in Figure 8.5 (left) with $a_1 = a_2 = a$, $b_1 = b_2 = b$, and $x_0 = c/2$.

$$B_x = 0, \qquad B_y = \mu_0 J_E \, c \frac{b}{a + b}. \tag{8.34}$$

2. The quadrupole configuration as shown in Figure 8.5 (middle), with $a_1 = b_2 = a$, $b_1 = a_2 = b$, and $x_0 = 0$.

$$B_x = \mu_0 J_E \frac{b - a}{a + b} y, \qquad B_y = \mu_0 J_E \frac{b - a}{a + b} x. \tag{8.35}$$

3. The combined dipole/quadrupole configuration as shown in Figure 8.5 (right). In this case $a_1 = b_2 = a$, $b_1 = a_2 = b$. Coil 1 is centered at $-x_0$ and coil 2 at x_0. This yields

$$B_x = \mu_0 J_E \frac{b - a}{a + b} y, \qquad B_y = \mu_0 J_E \left(x_0 + \frac{b - a}{a + b} x \right). \tag{8.36}$$

These results have two implications: First, they pave the way for the design of coil-block magnets of constant current density, and second, the intersecting circles can be used as a model for reproducing shielding-current densities in hard superconductors, which are exposed to a variable applied field.

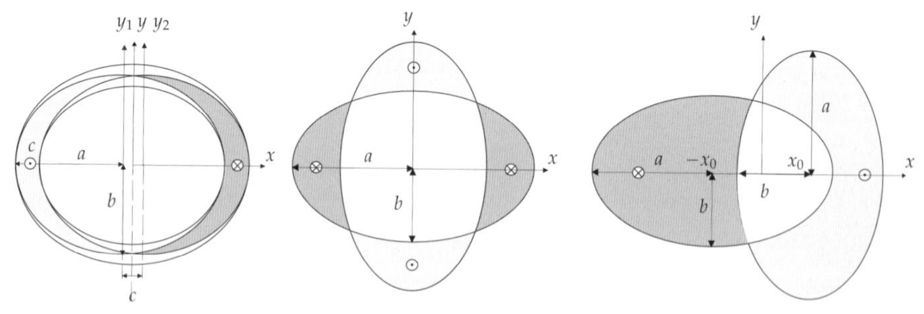

Figure 8.5 Intersecting ellipses of opposite current density that create an ideal dipole field inside the aperture (left), quadrupole field (middle), and constant gradient field (right).

Coil-block geometries The coils do not usually consist of perfectly cylindrical shells because the cables themselves are either rectangular or trapezoidal with an insufficient *keystone* angle to allow for perfect sector geometries.[8] As the cables are connected in series and consequently carry the same current, the shells are subdivided into coil blocks, separated by wedges. This gives the needed degrees of freedom to approximate the ideal current distribution. The field generated by this coil layout can be calculated by means of the line-current approximation for the superconducting cable. The field distribution is optimized with the methods explained in Chapter 20. Coil-block geometries for dipoles and quadrupoles having two layers of coil blocks are shown in Figure 8.6. Whereas the same cable is used in the inner and outer layers of

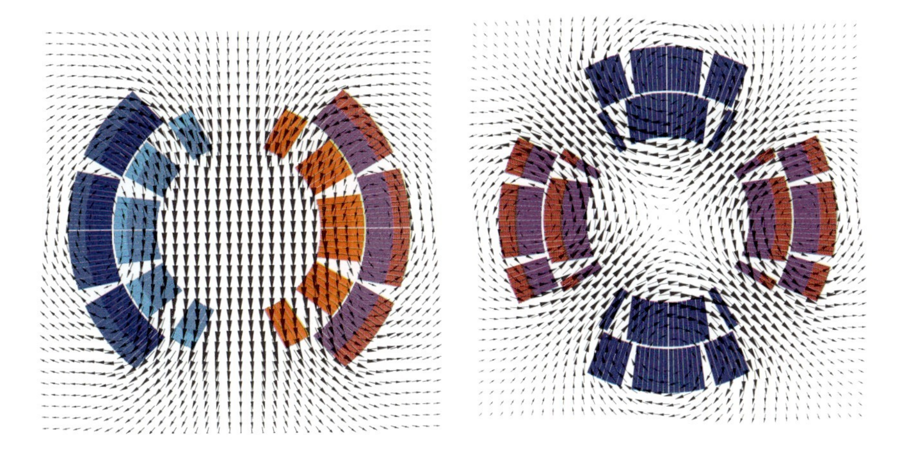

Figure 8.6 Coil-block arrangements made of Rutherford cable with grading of the current density due to the keystoning of the cable. LHC dipole model coil (left) with five coil blocks per coil and LHC main quadrupole coil cross section (right).

the LHC main quadrupole coils, the LHC main dipole coils are wound from two kinds of cables, allowing a *current density grading* with a smaller cable and hence a higher current density in the outer-layer cable, which is exposed to a lower magnetic field.

A different kind of coil-block arrangement was proposed by Patoux [21], where the (nonkeystoned) cables are placed in grooves stamped into the austenitic steel collars. A layout for a 56 mm aperture is shown in Figure 8.3 (right). The disadvantage of this design, however, is the lack of adequate control of the prestress during the cool-down of the magnet.

Figure 8.7 shows the contribution of the strand currents in a superconducting dipole coil to the B_3, B_5, B_7, and B_9 field components as a visualization of

8 If the keystone angle was large enough for perfect sector geometries, the current density would vary with $J_E(r_a - r_e)/(2r)$ because of the different compaction on the cable's narrow and wide sides.

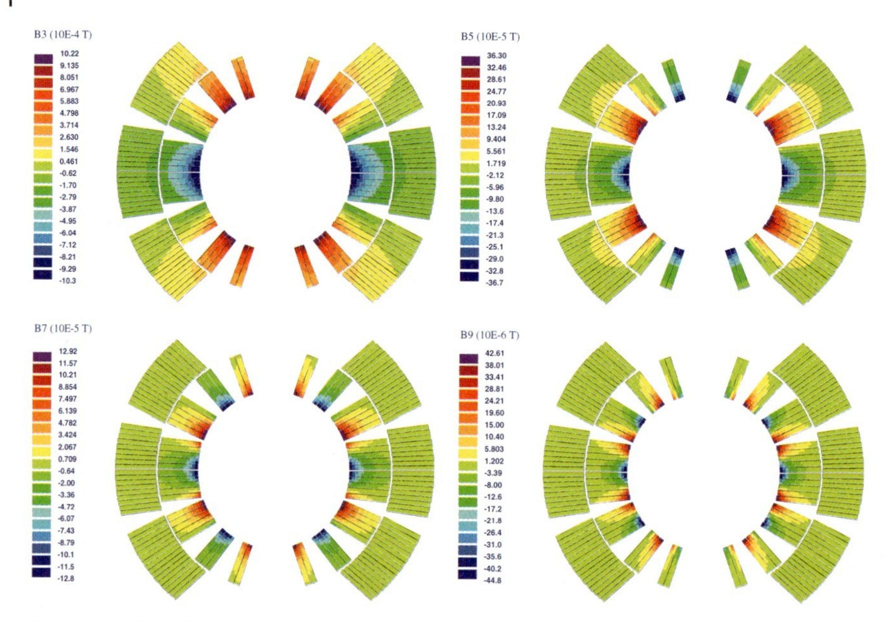

Figure 8.7 Visualization of Eq. (6.65): Contribution of the strand current to the field components B_3 (top, left), B_5 (top, right), B_7 (bottom, left), and B_9 (bottom, right) at $r_0 = 17$ mm.

Eq. (8.3). The field errors scale with r^{-n}, where n is the order of the multipole and r is the mid-radius of the coil. In a good magnet design, the positive and negative contributions to the field harmonics cancel out. It can be seen from Figure 8.7 (top left), that shims at the horizontal median plane and poles have a large effect on the sextupole coefficient. Moreover, B_9 is insensitive to the inclination of the coil blocks. Because the coverage on the aperture radius is high, B_9 is determined by the number of cables per coil block in the inner layer and thus by the winding topology of the coil.

For certain symmetry conditions in the magnet, some of the multipole components vanish: For an up/down symmetry in a dipole magnet no A_n terms are present. If there is an additional left/right symmetry, only the odd $B_1, B_3, B_5, B_7, \ldots$ components remain. In this context they are referred to as the *allowed multipoles*. The higher-order allowed multipoles are minimized in the design process. Other multipoles are avoided by the use of appropriately designed tooling for coil winding and curing so that the symmetry in the assembled magnet is preserved as much as possible. Some residual values of A_n and even-numbered B_n persist as a result of manufacturing tolerances. These residuals are often called *nonallowed multipole coefficients*.

8.1.2
Sensitivity to Coil-Block Positioning Errors

By differentiating Eq. (8.3) with respect to r_c and φ_c, we can calculate the sensitivity of the multipoles to coil-block displacements. For one line-current and $\mu_r \gg 1$, that is $\lambda_\mu = 1$, we obtain

$$\frac{\partial B_n(r_0)}{\partial \varphi_c} = -\frac{\mu_0 I_k}{2\pi} \frac{n r_0^{n-1}}{r_c^n} \left(1 + \left(\frac{r_c}{r_y}\right)^{2n}\right) \sin n\varphi_c, \tag{8.37}$$

$$\frac{\partial B_n(r_0)}{\partial r_c} = \frac{\mu_0 I_k}{2\pi} \frac{n r_0^{n-1}}{r_c^{n+1}} \left(1 - \left(\frac{r_c}{r_y}\right)^{2n}\right) \cos n\varphi_c. \tag{8.38}$$

It can be seen here that the sensitivity to azimuthal displacements is an order of magnitude higher than for radial displacements. The magnetization of the iron yoke, represented by the image currents, enhances the sensitivity to azimuthal displacements and diminishes the effect of radial displacements.

8.1.3
Force Distribution

Figure 8.8 (left) shows the magnetic forces, according to Eq. (5.193), on the strands of an LHC model dipole magnet at nominal excitation.

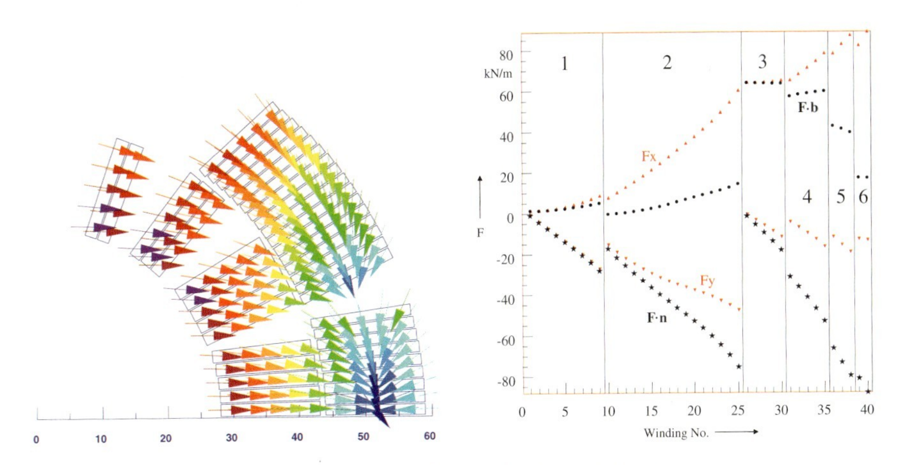

Figure 8.8 Left: Force distribution in an LHC main dipole model coil. Right: Force components (per unit length) on each conductor at nominal current. See Figure 8.9 for cable numbering and local frame.

Figure 8.9 (right) shows the force components in the local coordinate system of the cable. Notice that the radial magnetic flux component, and consequently the force in normal direction to the cable's broad face, is relatively uniform in each cable. The binormal component shows nearly linear rise within each cable. The binormal force is defined as positive in the outward direction and the normal component is defined as positive away from the pole. Numerical values are given in Figure 8.8 (right).

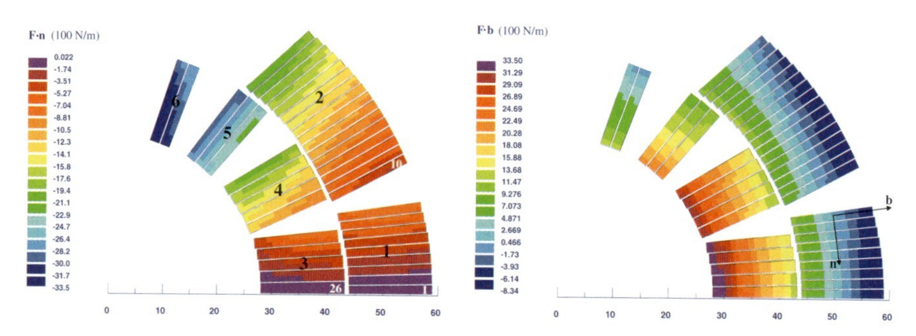

Figure 8.9 Components of the electromagnetic forces (per unit length) in the local coordinates of the cable. Left: Normal to the broad face, $\mathbf{F} \cdot \mathbf{n}$. Right: Binormal to the broad face, $\mathbf{F} \cdot \mathbf{b}$. Notice that the force normal to the cable's broad face is relatively uniform in each cable, while the binormal component shows a nearly linear rise within each cable.

8.1.4
Margins in the LHC Main Dipole

For the LHC main dipole coil, Figure 8.10 (left) shows the magnitude of the engineering current density, which varies across the cable because of the different compaction on its wide and narrow edges. The difference in current density between the inner- and the outer-layer cables is clearly visible. Figure 8.10 (right) shows the modulus of the magnetic flux density. Figure 8.11 (left) shows the margin on the *load line* for an operation temperature of 1.9 K. The load line is a curve in the parameter space $V \subset \mathbb{R}^3_+$, which is defined by

$$\mathscr{S} : [0, I_\mathrm{q}] \to V : I \mapsto \mathbf{x}(I) \,, \qquad \mathbf{x}(I) = (J_\mathrm{s}(I), B_\mathrm{p}(I), T) \,. \tag{8.39}$$

This is in general not a straight line because of the iron saturation. The parameter I is the excitation current in the interval from zero to the quench current I_q, J_s the current density in the superconducting fraction of the strands, B_p the peak field in the coil, and T the operation temperature. The quench current defines the point where the load line crosses the critical surface. The margin $M_\mathrm{L}(I)$ on the load line can thus be calculated from

$$M_\mathrm{L}(I) = \frac{s(I_\mathrm{q}) - s(I)}{s(I_\mathrm{q})} \,, \tag{8.40}$$

where s is the arc length on the load line. Figure 8.11 (right) shows the temperature margin $M_\mathrm{T}(I)$, which is defined as the difference between the opera-

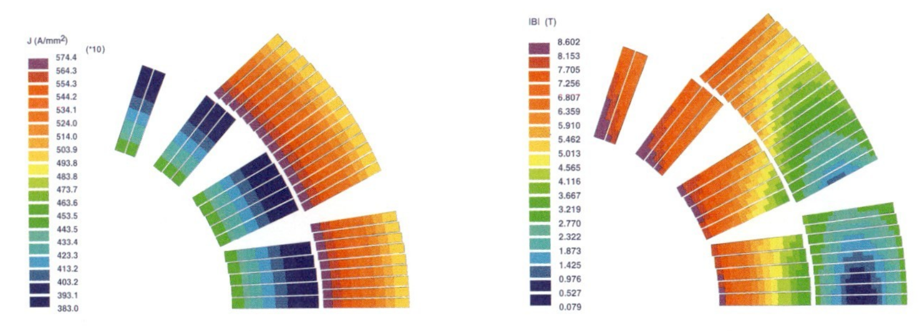

Figure 8.10 Left: Modulus of the engineering current density. Right: Modulus of the magnetic flux density. All figures for the LHC main dipole at nominal operation: 8.33 T, 11 850 A.

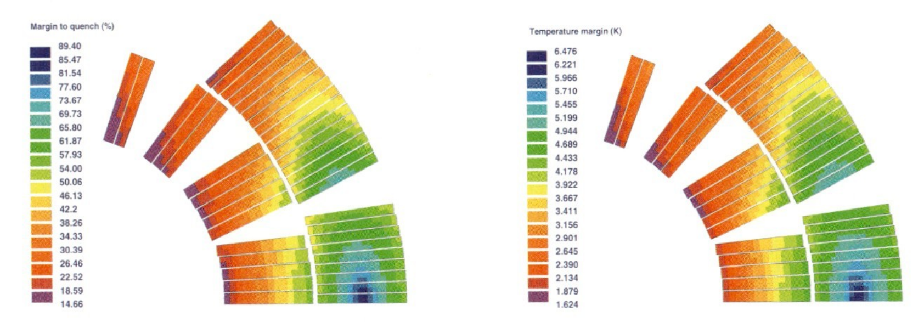

Figure 8.11 Left: Margin on the load line. Right: Temperature margin, i.e., the difference between the bath temperature and the critical temperature at operation field and current. All figures for nominal operation: 8.33 T, 11 850 A.

tional temperature T and the critical temperature $T_c(B_p(I), J_s(I))$. The critical surface is modeled by the empirical expression (1.6).

Remark: For a strand of perfectly circular shape (radius r_s, cross-sectional area a_s, uniform current density J, total current I), Eq. (6.77) yields

$$B_\varphi(r, \varphi) = \frac{\mu_0 I r}{2\pi r_s^2} . \tag{8.41}$$

The nominal current in the LHC main dipoles is 11 850 A. The outer-layer cable has 36 strands of radius 0.4125 mm. The inner-layer cable has 28 strands of radius 0.5325 mm. The maximum self-field in the strands is 0.16 T, both in the inner and outer layers.

For the operational margin of the LHC magnets the self-field in the strands must not be neglected. However, the critical surface modeling is usually performed by compensating for the self-field in the cable but not in the strand. Disregarding the strand self-field in field computations is thus consistent with the measurement method. Nevertheless, the magnet designer is advised to verify the conditions under which the measurements of the strand's critical current density are performed. □

Another aspect to be addressed in superconducting magnet design is the margin for beam-induced heat loads, which can be calculated with *Monte Carlo programs* such as CASIM [31] and FLUKA [2]. A detailed treatment of the time constants of heat pulses expected from hadronic showers in the LHC can be found in [15].

For very short heat pulses, during which the heat has insufficient time to diffuse transversally between the inner and the outer edges of the cable, it is the *volumetric enthalpy reserve* that is relevant for the quench behavior of the magnet. For heat pulses too short for significant heat conduction through the insulation but long enough to allow heat transfer into the confined helium,[9] it is the *average heat reserve* we must calculate.

The volumetric enthalpy reserve of the superconducting wires is defined by

$$\Delta h := \int_{T_\mathrm{b}}^{T_\mathrm{c}(J,B)} \rho\, c_\mathrm{p}(T)\, \mathrm{d}T\,, \tag{8.42}$$

where $[h] = 1\,\mathrm{J\,m^{-3}}$. T_b is the bath temperature of the coolant and $T_\mathrm{c}(J,B)$ critical temperature as a function of current density and applied magnetic flux density. Data for the volumetric heat capacity (VHC) $\rho c_\mathrm{p}(T)$ of copper and superconductors are provided in Annex A.4. The data must be weighted with the area ratio of copper to superconductor in the strand. Figure 8.12 (left) shows the volumetric enthalpy reserve for the coil windings of the LHC main dipole.

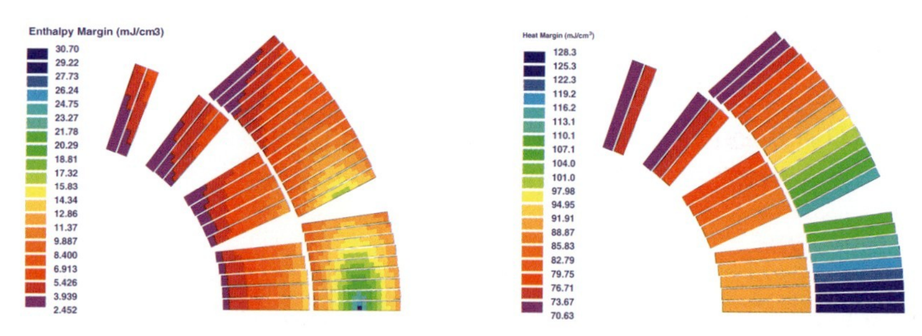

Figure 8.12 Left: Enthalpy margin of the strands. Right: Average heat reserve of the coil windings. Both at nominal excitation (11 850 A).

Taking the confined helium into account, it is no longer appropriate to speak of enthalpy, which is a thermodynamic potential at constant pressure. For helium above the lambda point, c_p and c_v differ considerably, and additional assumptions about the mass density of the confined helium must be accounted

9 The confined helium occupies the void fraction in the Rutherford type cables.

for.[10] For this weighted VCH of conductor and helium, the integral value of Eq. (8.42) is called the average heat reserve, shown in Figure 8.12 (right) for the coil windings of the LHC main dipole.

8.2 Combined-Function Magnets and the Unipolar Current Dipole

Combined-function magnets are easy to realize by nesting thin coils of different multipole order. Figure 8.13 (left) shows a four-layer, combined-function corrector for the RHIC accelerator [27]. It comprises, from inside out, a decapole, octupole, quadrupole, and dipole. Each of the inner three structures consists of a double-layer racetrack coil wound on a flat, flexible substrate, which is then bent on a support mandrel.

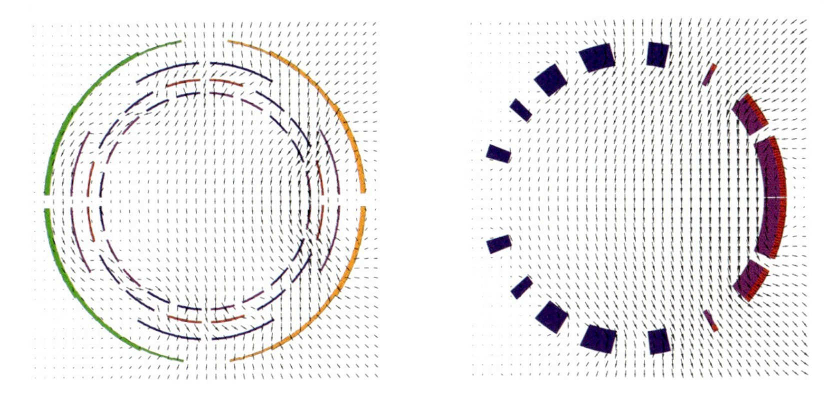

Figure 8.13 Left: Nested, combined-function corrector magnet for RHIC [27]. Right: Combined dipole/quadrupole magnet with a single coil and an opening at the horizontal median plane [25]. Blue to green color range corresponds to negative current density. Yellow to magenta range represents positive current densities.

The disadvantage of nested structures is the local field enhancement in the coils, which depends on the powering scheme. The correctors are therefore designed to operate at no more than 30% of their quench margin on the load line.

Another possibility for realizing a combined-function dipole/quadrupole magnet in a single layer coil is to open an asymmetric gap at the horizontal median plane. Figure 8.13 (right) shows a magnet with a dipole field of 2.6 T and a quadrupole field of 19 $\mathrm{T\,m^{-1}}$ designed for the beam line of the J-PARC neutrino experiment at KEK [25].

10 This will be treated in detail in Annex A.4.

A configuration with twelve individually powered superconducting racetrack coils around a cylindrical iron yoke is proposed in [17]. Any combination of correction fields up to the sextupole field component can be excited. A combined dipole/skew-quadrupole configuration is shown in Figure 8.14 (right). Stray fields can be reduced by an additional ferromagnetic shielding ring [17].

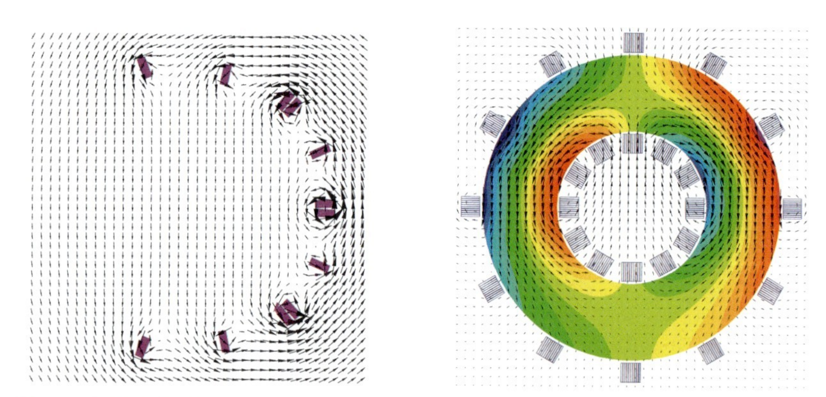

Figure 8.14 Left: Unipolar current dipole [9]. Right: Iron yoke with 12 individually powered coils, configured to produce a combined dipole/skew-quadrupole magnet [17].

Placing N coil blocks with unipolar current direction at different angular and radial positions yields enough degrees of freedom to suppress field errors up to order $2N$, where N is the number of coil blocks above the x-axis [9]. An example is shown in Figure 8.14 (left). Notice the large field outside the coil, which reduces the quench margin of the cables.

8.3
Rectangular Block-Coil Structures

Figure 8.15 shows the four-layer quadrupole design for the LHC insertion quadrupole (MQY), together with two block-coil alternatives [19]. The block-coil magnet (bottom) is made of racetrack coils, which are easier to wind and assemble than the $\cos 2\varphi_c$ coils. The design in Figure 8.15 (top right) has advantages for double-aperture magnets where the two straight sections of the racetrack coils are in different apertures and therefore the minimum bending radius of the coil is determined by the beam-separation distance. Both block-coil alternatives require a considerably larger amount of superconductor for a given gradient and aperture.

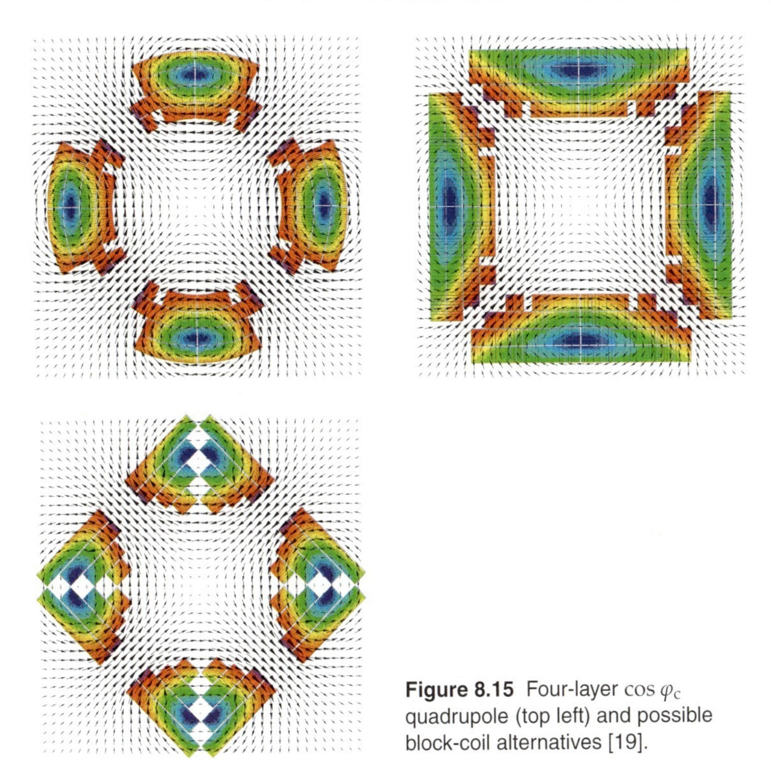

Figure 8.15 Four-layer $\cos \varphi_c$ quadrupole (top left) and possible block-coil alternatives [19].

8.4 Field Enhancement in Coil Ends of Accelerator Magnets

The performance of a superconducting magnet is limited by the maximum field in the coil. For the dipole we define the *field enhancement* as the ratio of the coil's peak field to the main field in the aperture. The field enhancement must be kept close to one in order to optimize the operational margin to quench; see also Section 1.4 and Figures 8.10 and 8.11. In the LHC dipole magnets the field enhancement B_p^{2D}/B_1 is 1.037 and $B_p^{3D}/B_1 = 1.077$. Figure 8.16 (top left) shows a standard coil configuration for a decapole corrector magnet wound from twenty coils in two layers. The additional field enhancement in the coil end, defined by

$$\lambda_E := \frac{B_p^{3D}}{B_p^{2D}}, \tag{8.43}$$

is 1.199 for this configuration. A more economical solution for the construction of multipole corrector-magnets is the winding of a $2N$-pole from only N coils. The example of the decapole built from only five coils per layer is shown in Figure 8.16 (top right). However, the additional field enhancement is then on the order of $\lambda_E = 1.5$. Figure 8.16 (bottom) shows a configuration where the second coil layer with inverted polarity is rotated by one pole pitch. In this case the field enhancement is $\lambda_E = 1.23$, close to that of the 10-coil construction.

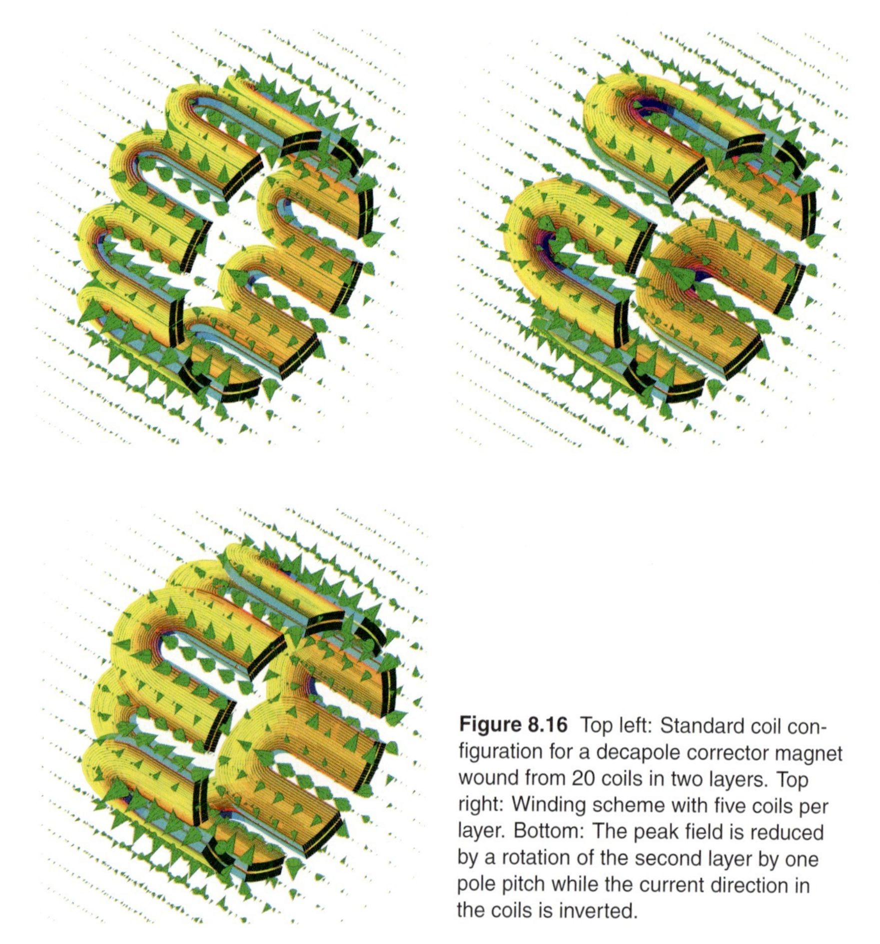

Figure 8.16 Top left: Standard coil configuration for a decapole corrector magnet wound from 20 coils in two layers. Top right: Winding scheme with five coils per layer. Bottom: The peak field is reduced by a rotation of the second layer by one pole pitch while the current direction in the coils is inverted.

8.5 Magnetic Force Distribution in the LHC Dipole Coil Ends

The design of the 3D coil geometry, shown in Figure 8.17, is driven by the objectives of maximizing the radius of curvature in the coil end, applying as little *hard-way* strain as possible to the cable (resulting from the unavoidable bend over the cable's narrow side), optimizing the multipole content of the integrated field, and limiting the peak-field enhancement. The coil end must be carefully designed, as it cannot be mechanically confined as well as can the

Figure 8.17 Magnetic force density in the LHC dipole coil-end. Left: Outer layer. Right: Inner layer. On the cable's narrow side, the colors (bright yellow to magenta) represent -15.3–29.1 MPa (positive in binormal direction $\mathbf{b}$). Nominal current of 11 800 A.

straight section. The ends have often been the limiting factor for the quench performance; see Chapter 19.

With the appropriate spacing between the coil blocks, the field quality and peak-field enhancement can be optimized. Figure 8.17 shows the coil end together with the electromagnetic force density in the normal and binormal direction to the broad face of the cable.

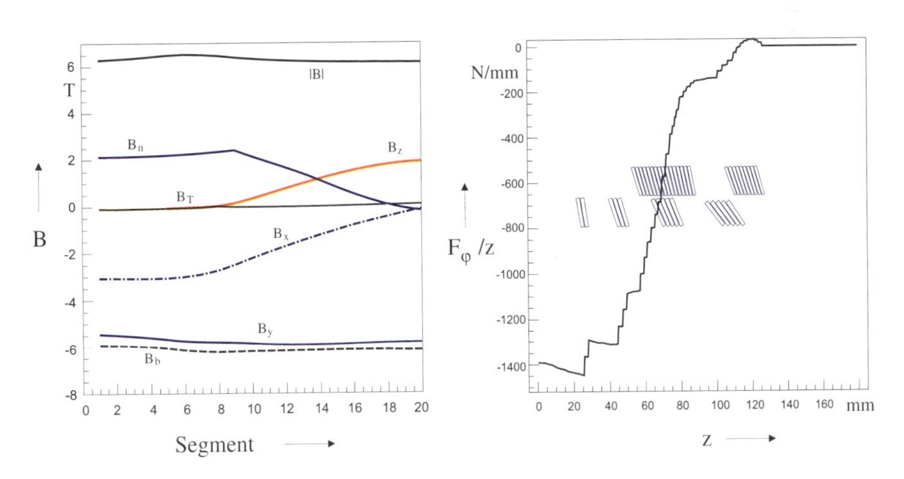

Figure 8.18 Left: Components of the magnetic flux density in the pole turn of the outer layer; see Figure 8.17 (left). Segment 1 is the onset of the coil in the xy-plane; segment 20 is located at the nose in the yz-plane. Right: Azimuthal electromagnetic force at nominal current on the coil end as a function of the z-coordinate. Nominal current is 11 800 A.

Figure 8.18 (right) shows the cross section of the coil end in the yz-plane and the electromagnetic forces in the azimuthal direction. The size of the shims placed at the horizontal median plane before collaring the end must be chosen such that the resulting prestress prevents an unloading due to the electromagnetic forces. Unloading at full excitation leads to conductor movements, resulting in so-called *disturbance quenches* well below the magnet's short sample field.

8.6 Nested Helices

Meyer [23] and Goodzeit [13] propose a dipole magnet made from nested helices. The windings of the helices are tilted at an angle α (see Figure 8.19) with respect to the coil axis. It is reported [13] that these magnets provide for a perfect field quality, ease of manufacturing, and the high flexibility to construct curved magnets as well as those with elliptical apertures.

Each coil has an axial field component with a superimposed dipole field. With two coils of opposite tilt angles and polarities, the axial fields cancel out, while the dipole field components are superimposed. The resulting field is displayed in Figure 8.20 (left).

The tilted elliptical helix can be described by a space curve $\mathscr{S} : I \to E_3 : \varphi_c \mapsto \mathbf{r}(\varphi_c)$:

$$\mathbf{r} = a \cos \varphi_c \mathbf{e}_x + b \sin \varphi_c \mathbf{e}_y + \left(b \sin \varphi_c \tan \alpha + p \frac{\varphi_c}{2\pi} \right) \mathbf{e}_z, \tag{8.44}$$

for $\varphi_c \in [0, 2N\pi]$, the pitch length p, the tilt angle α, and the two ellipse semi-axes a and b.

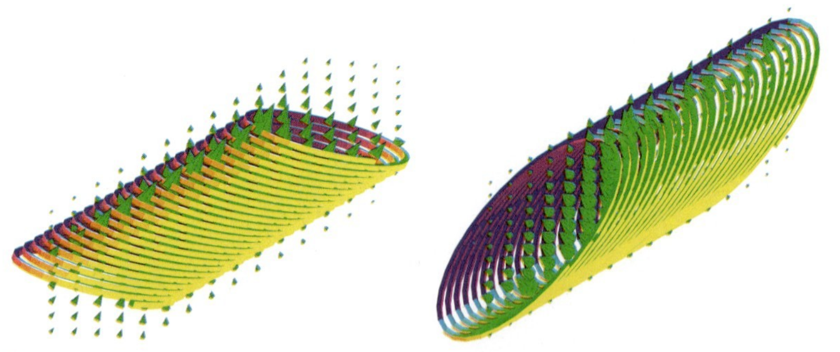

Figure 8.19 Two helices with opposite tilt angle and different polarity resulting in opposite solenoidal fields and the same B_y field components. The z-component of current density is represented by the color scheme on the conductor surfaces.

The velocity vector to this space curve is given by

$$\mathbf{v} = -a \sin \varphi_c \, \mathbf{e}_x + b \cos \varphi_c \, \mathbf{e}_y + \left(b \tan \alpha \, \cos \varphi_c + \frac{p}{2\pi} \right) \mathbf{e}_z \,, \tag{8.45}$$

(see Section 3.1.1) so that the current density in z-direction is proportional to $\cos \varphi_c$, as desired for the creation of a perfect dipole field. The principle can also be applied to higher-order multipoles. An example of a quadrupole coil is shown in Figure 8.20 (right).

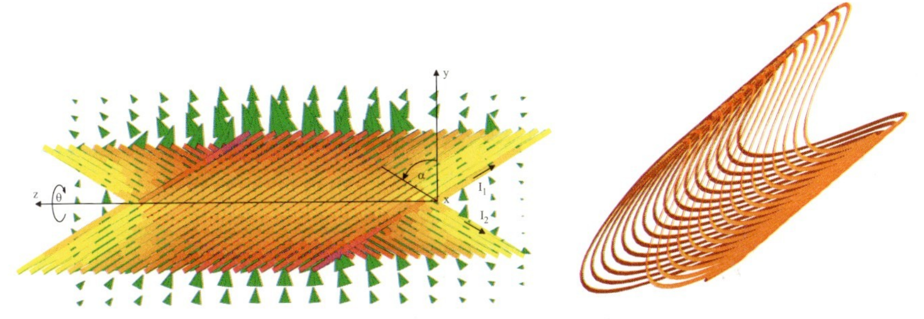

Figure 8.20 Left: Nested helices creating a pure dipole field due to the $\cos \varphi_c$-dependence of the current density J_z. Right: Changing the periodicity of the inclination results in higher-order multipole magnets, here a quadrupole.

As explained in Section 8.4, the performance of a superconducting magnet is limited by the local field enhancement in the coil. The disadvantage of the nested helices is that the inner coil is exposed to a peak field composed of the dipole and the solenoidal field components of the outer coil. The field enhancement for the short magnet displayed in Figure 8.20 (left) is 24%. The field enhancement can be reduced, however, by nesting two or more double-helices [13].

8.7 Solenoids

We turn now to magnets with axial symmetry used in medical imaging (MRI), energy storage, and particle-physics experiments. If iron yokes are present, their main function is to reduce the stray fields. The stray fields can, however, also be actively reduced by compensation coils in the end region of the main solenoid.

8.7.1 Helmholtz and Maxwell Coils

A uniform field or gradient can be generated on the symmetry axis of a split-coil arrangement, known as *Helmholtz* and *Maxwell coils*. Consider two concentric line-current loops of radius r_c at position $z = z_c$ and $z = -z_c$ with currents of the same polarity. Using the result of Eq. (5.71) we obtain for the axial field:

$$B_z = \frac{\mu_0 I}{2}\left(\frac{r_c^2}{\sqrt{r_c^2 + (z+z_c)^2}^3} + \frac{r_c^2}{\sqrt{r_c^2 + (z-z_c)^2}^3}\right). \tag{8.46}$$

Differentiating with respect to z yields

$$\frac{\mathrm{d}B_z}{\mathrm{d}z} = \frac{-3\mu_0 I r_c^2}{2}\left(\frac{z+z_c}{\sqrt{r_c^2 + (z+z_c)^2}^5} + \frac{z-z_c}{\sqrt{r_c^2 + (z+z_c)^2}^5}\right), \tag{8.47}$$

which is zero at $z = 0$ for any coil radius r_c. The second derivative is given by

$$\frac{\mathrm{d}^2B_z}{\mathrm{d}z^2} = \frac{-3\mu_0 I r_c^2}{2}\left(\frac{r_c^2 - 4(z+z_c)^2}{\sqrt{r_c^2 + (z+z_c)^2}^7} + \frac{r_c^2 - 4(z-z_c)^2}{\sqrt{r_c^2 + (z+z_c)^2}^7}\right), \tag{8.48}$$

which is zero at $z = 0$ only if $z_c = r_c/2$. Thus the spacing between the two coils must be identical to the coil radius, as shown in Figure 8.21 (left).

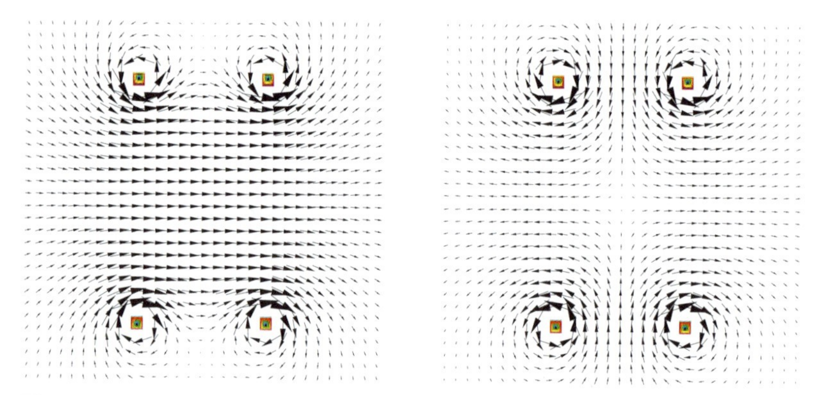

Figure 8.21 Left: Helmholtz coil with same current direction in both coils. Right: Maxwell coil with opposing currents in both coils.

If we reverse the current in the left coil, the sign of the first term in the brackets of Eq. (8.47) is reversed and the magnetic flux density at $z = 0$ is still zero. But for the gradient we obtain at the origin

$$\left.\frac{\mathrm{d}B_z}{\mathrm{d}z}\right|_{r=z=0} = \frac{3\mu_0 I r_c^2}{2}\frac{2z_c}{\sqrt{r_c^2 + (z_c)^2}^5}. \tag{8.49}$$

The magnetic flux density in the Maxwell coil is displayed in Figure 8.21 (right).

8.7.2 Fabry Factors

Constant current density Following [4], Eq. (5.71) can be used to calculate the field in the center of a thick solenoid of constant engineering current density J_E and a rectangular cross section having inner radius r_1, outer radius r_2, and length w, as shown in Figure 6.10:

$$B_z(0,0) = \int_{r_1}^{r_2} \int_{-\frac{w}{2}}^{\frac{w}{2}} \frac{\mu_0 J_E}{2} \frac{r_c^2}{\sqrt{r_c^2 + z_c^2}^3} \, dz_c dr_c$$

$$= \mu_0 \, J_E \, \frac{w}{2} \ln \left(\frac{r_2 + \sqrt{r_2^2 + (\frac{w}{2})^2}}{r_1 + \sqrt{r_1^2 + (\frac{w}{2})^2}} \right) . \qquad (8.50)$$

Introducing the dimensionless parameters $\alpha := r_2/r_1$ and $\beta := w/2r_1$ yields

$$B_z(0,0) = \mu_0 \, J_E \, r_1 \, F_S(\alpha, \beta) \, , \qquad (8.51)$$

where the last term is the *shape function*,[11]

$$F_S(\alpha, \beta) := \beta \ln \left(\frac{\alpha + \sqrt{\alpha^2 + \beta^2}}{1 + \sqrt{1 + \beta^2}} \right) , \qquad (8.52)$$

which is known to the magnet design community as the *Fabry factor* [12, 24]. Let the aperture radius r_1 and the central field be specified. The engineering current density is determined by the superconductor properties and the coil packing factor. Hence the numerical value of the shape function $F_S(\alpha, \beta)$ will automatically be given, and the design problem reduced to choosing the optimal values of α and β for a minimum coil winding volume

$$V_{\text{coil}} = \pi(r_2^2 - r_1^2)w = 2\pi r_1^3(\alpha^2 - 1)\beta =: r_1^3 \, v_{\text{coil}} \, , \qquad (8.53)$$

where v_{coil} is the dimensionless volume factor.

For normal-conducting (water-cooled) solenoids, the main objective is to minimize the ohmic losses. The current density may be expressed with the parameters α and β as

$$J_E = \frac{NI}{a} = \frac{NI}{w(r_2 - r_1)} = \frac{NI}{2\beta(\alpha - 1)r_1^2} , \qquad (8.54)$$

where N is the number of turns, and a the cross-sectional area of the coil. It follows from Eq. (8.51) that

$$B_z(0,0) = \frac{\mu_0 NI}{2\beta(\alpha - 1)r_1} F_S(\alpha, \beta) = \frac{\mu_0 NI}{2(\alpha - 1)r_1} \ln \left(\frac{\alpha + \sqrt{\alpha^2 + \beta^2}}{1 + \sqrt{1 + \beta^2}} \right) . \qquad (8.55)$$

11 Note that in some references [6] this factor contains μ_0.

The total length of the conductor is

$$l_t = \pi(r_1 + r_2)N = \pi r_1(\alpha + 1)N \tag{8.56}$$

and its cross-sectional area

$$a = \frac{\lambda(r_2 - r_1)w}{N} = \frac{\lambda(\alpha - 1)r_1^2 2\beta}{N}. \tag{8.57}$$

The resistivity can be calculated from

$$R = \frac{l_t}{\varkappa a} = \frac{N^2}{\lambda \varkappa r_1} \frac{\pi}{2\beta} \frac{\alpha + 1}{\alpha - 1}. \tag{8.58}$$

The ohmic losses of the winding amount to

$$W = RI^2 = \frac{(NI)^2}{\lambda \varkappa r_1} \frac{\pi}{2\beta} \frac{\alpha + 1}{\alpha - 1} = \frac{J^2}{\varkappa} \lambda 2\pi r_1^3 \beta(\alpha^2 - 1) = \frac{J^2}{\varkappa} \lambda V_{\text{coil}}. \tag{8.59}$$

Bringing together Eqs. (8.55) and (8.59), the expression relating the ohmic losses to the axial field reads

$$B_z(0,0) = \mu_0 G(\alpha, \beta) \sqrt{\frac{W \lambda \varkappa}{r_1}} = \mu_0 \frac{F_S(\alpha, \beta)}{\sqrt{v_{\text{coil}}}} \sqrt{\frac{W \lambda \varkappa}{r_1}}, \tag{8.60}$$

where

$$G(\alpha, \beta) = \frac{F_S(\alpha, \beta)}{\sqrt{2\pi\beta(\alpha^2 - 1)}} = \sqrt{\frac{\beta}{2\pi(\alpha^2 - 1)}} \ln\left(\frac{\alpha + \sqrt{\alpha^2 + \beta^2}}{1 + \sqrt{1 + \beta^2}}\right). \tag{8.61}$$

Figure 8.22 (left) displays the lines of constant F_S (solid) and constant volume factor v_{coil} (dashed) as a function of the parameters α and β. Designs B and C are inefficient, while A achieves the same central field with less coil volume. Figure 8.22 (right) shows the factor $G(\alpha, \beta)$ relating the ohmic losses to the central field.

The function $G(\alpha, \beta)$ has a maximum of 0.1426 at the point $(\alpha, \beta) = (3.095, 1.862)$. The most efficient design is thus achieved with a coil aspect ratio of approximately 0.6.

In practical superconducting solenoid design, the field quality and the peak field to main field ratio must be taken into account. The calculation of these quantities as a function of the parameters α and β must be done numerically. Results are shown in Figure 8.23.

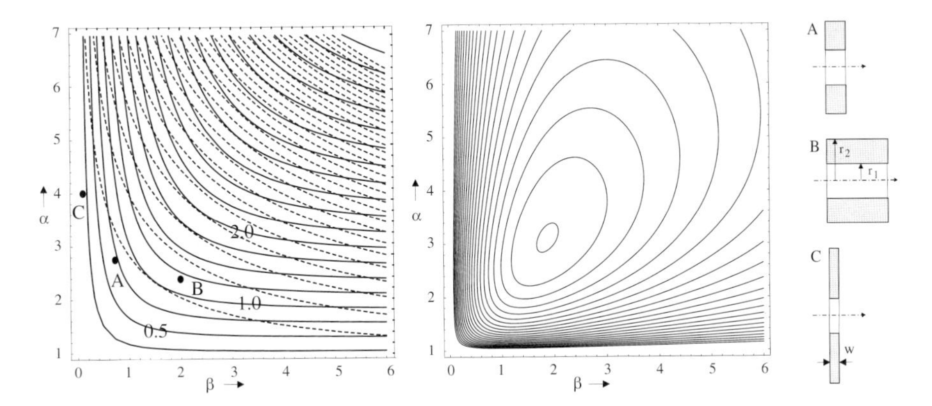

Figure 8.22 Left: The lines (solid) of constant Fabry factors F as a function of the parameters $\alpha = r_1/r_2$ and $\beta = w/2r_1$. Dashed lines represent the coil volume factor v_{coil}. Notice the three different coil shapes A–C. Right: Factor G, relating the ohmic losses to the on-axis field in the center of the solenoid.

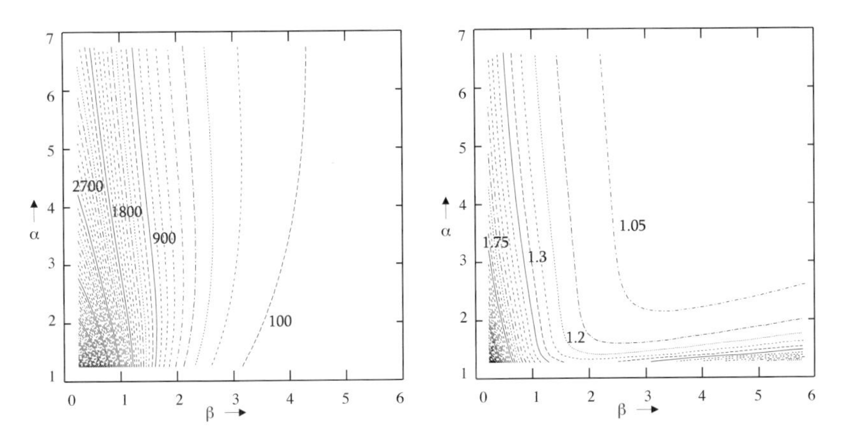

Figure 8.23 Left: Field quality (third zonal harmonic); see Section 6.2 in a thick solenoid as a function of the shape parameters α and β. Right: Peak field to main field ratio in the solenoid.

The central field in very long solenoids can be calculated from Eq. (8.50) for $w \longrightarrow \infty$:

$$
\begin{aligned}
B_z(0,0) &= \lim_{w\to\infty} \mu_0 J_{\text{E}} \frac{w}{2} \ln \left(\frac{r_2 + \sqrt{r_2^2 + (\frac{w}{2})^2}}{r_1 + \sqrt{r_1^2 + (\frac{w}{2})^2}} \right) \\
&= \lim_{w\to\infty} \mu_0 J_{\text{E}} \frac{w}{2} \ln \left(\frac{r_2 + \frac{w}{2}}{r_1 + \frac{w}{2}} \right) \\
&= \mu_0 J_{\text{E}} (r_2 - r_1) = \frac{\mu_0 N I}{w}. \qquad (8.62)
\end{aligned}
$$

This result yields a particularly easy formula for the inductance of a long thin solenoid, inside which the magnetic flux can be assumed to be constant. This flux is linked to exactly N coil windings and thus

$$L = \frac{N\Phi}{I} = \frac{NB_z a}{I} = \frac{\mu_0 N^2 a}{w}, \tag{8.63}$$

where a is the cross-sectional area of the coil and w its total length.

Bitter magnets All the above equations were derived for the case of a constant current density in the coil cross section. In the normal-conducting *Bitter magnets* [3], however, the current density in the coil is inversely proportional to the radius,

$$J = J_{\text{max}} \frac{r_1}{r_c}, \tag{8.64}$$

where J_{max} is the maximum overall current density at the inner radius r_1 of the coil. The Bitter magnet is constructed as a stack of annular copper plates with cooling holes and with an angular slit. Each plate is insulated, except for a segment that makes pressure contact with the next plate's uninsulated segment. In this way the current is forced to flow in a helical path from one end of the stack to the other; see Figure 8.24. The maximum current density is determined by

$$J_{\text{max}} = \frac{E_\varphi}{\rho} = \frac{U}{2\pi\rho r_1}, \tag{8.65}$$

where U is the voltage across the slit and ρ is the resistivity of copper ($\rho_{\text{Cu}} \approx 2 \times 10^{-8}\,\Omega\,\text{m}$).

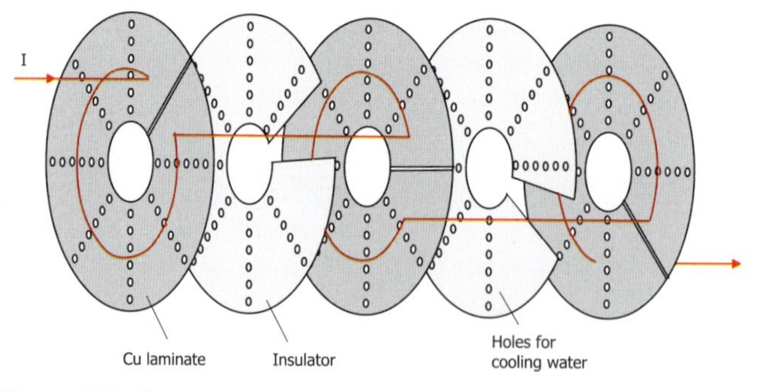

Figure 8.24 Construction principle of the Bitter magnet. Each copper plate is insulated, except for a segment that allows pressure contact to the next plate. This way the current flows in a helical path from one end of the stack to the other.

The central field is given by

$$B_z(0,0) = \int_{r_1}^{r_2} \int_{-\frac{w}{2}}^{\frac{w}{2}} \frac{\mu_0 J_{\max} r_1}{2} \frac{r_c}{\sqrt{r_c^2 + z_c^2}^3} dz_c dr_c$$

$$= \mu_0 J_{\max} r_1 \ln \left(\frac{r_2(\frac{w}{2} + \sqrt{r_1^2 + (\frac{w}{2})^2})}{r_1(\frac{w}{2} + \sqrt{r_2^2 + (\frac{w}{2})^2})} \right)$$

$$= \mu_0 J_{\max} r_1 \ln \left(\frac{\alpha\left(\beta + \sqrt{1+\beta^2}\right)}{\beta + \sqrt{\alpha^2 + \beta^2}} \right) =: \mu_0 J_{\max} r_1 F_B(\alpha, \beta), \tag{8.66}$$

where $F_B(\alpha, \beta)$ is the Fabry factor for the Bitter magnet. In this case [16], the factor $G_B(\alpha, \beta)$ is

$$G_B(\alpha, \beta) = \frac{1}{\sqrt{4\pi\beta \ln \alpha}} \ln \left(\frac{\alpha\left(\beta + \sqrt{1+\beta^2}\right)}{\beta + \sqrt{\alpha^2 + \beta^2}} \right). \tag{8.67}$$

The maximum value of $G_B(\alpha, \beta)$ is 0.1665 at $(\alpha, \beta) = (6.423, 2.146)$.

8.7.3 Off-Axis Fields

Off-axis field calculation for solenoids involves the solving of the complete elliptic integrals, as was shown in Section 5.6. Approximations for the complete elliptic integrals can be found by series expansion and integration of the components of Eqs. (5.65) and (5.66).

Example: The Compact Muon Solenoid (CMS) is one of the general-purpose detectors for the LHC, comprising a 4 T superconducting solenoid with a magnetic length of 12.5 m and a free inner-bore diameter of 6 m; see Figure 8.25. The large coil radius is necessary for the resolution of particle momenta.

The particle momenta p are determined by the measurement of the particle trajectories in the solenoidal magnetic field B. The momentum resolution is proportional to the deflection angle and sagitta of the particle trajectory[12] and therefore

$$\frac{\Delta p}{p} \propto \frac{p}{Bl^2}, \tag{8.68}$$

where l is the length of the chord. At the CMS muons travel radially by $l = 3$ m in the 4 T field resulting in $Bl^2 = 36$ T m^2. Outside the solenoid, the return

12 See Eq. (11.29) in Chapter 11.

Figure 8.25 The dimensions of the CMS detector can be appreciated from this photo taken during the installation of the cryomagnet. The vacuum tank of the magnet houses the superconducting coil with a free bore diameter of 6 m and a total length of 12.5 m.

flux yields another 5 T m^2 for particle tracking. The coil of the CMS solenoid is wound from a 32-strand cable, which is co-extruded inside a high-purity aluminum matrix and welded to two aluminum alloy stabilizers. The stabilizers are needed for magnet protection and mechanical stability [18]. The coil is epoxy impregnated after winding and indirectly cooled with two-phase helium at 4.5 K. The superconducting solenoid has a length of 13 m, and an inner diameter of 5.9 m. The total weight of the coldmass is 225 tons. The stored energy is 2.7 GJ at a nominal current of 19 500 A. The overall dimensions of the yoke are a length of 21.6 m and an outer diameter of 14.6 m, giving a total iron mass of around 11 500 tons.[13]

The excitation field is calculated with the semianalytical line-current method described above. The vector potential rA_φ and the flux density in the iron yoke (shown in Figure 8.26) are determined using the BEM–FEM coupling method presented in Chapter 15. The magnet is clearly coil-dominated, as the excitation field is 3.69 T and the iron yoke contributes only 8.4% to the total field in the magnet center. □

13 This is approximately the same mass as the Eiffel tower.

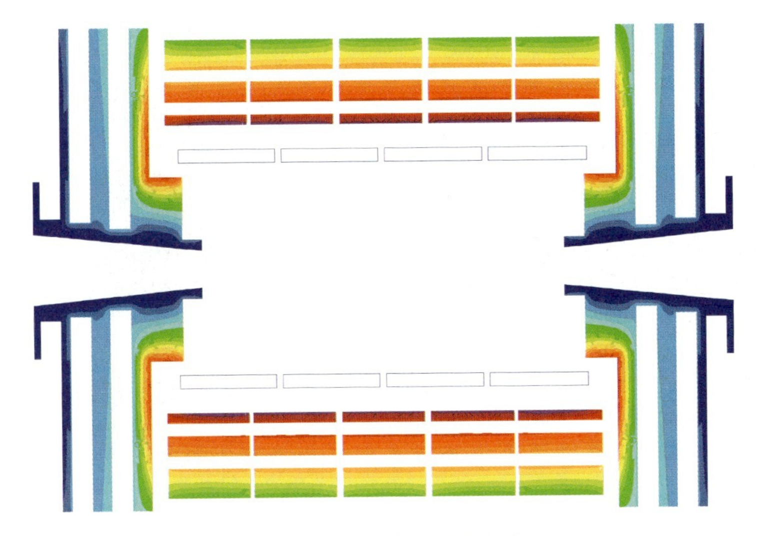

Figure 8.26 Magnetic vector potential rA_φ in the CMS yoke, calculated for axisymmetric conditions and an open boundary.

Remark 1: We have shown in Section 4.8 that in 2D, the field lines are of constant magnetic vector potential. The flux tubes between two of these field lines carry constant magnetic flux Φ. For rotationally symmetric fields, the magnetic flux through a surface parallel to the $r\varphi$-plane is calculated by means of Stoke's theorem:

$$\Phi = \int_{\mathcal{A}} \mathbf{B} \cdot \mathrm{d}\mathbf{a} = \int_{\mathcal{A}} \operatorname{curl} \mathbf{A} \cdot \mathrm{d}\mathbf{a} = \int_{\partial\mathcal{A}} \mathbf{A} \cdot \mathrm{d}\mathbf{r} = \int_0^{2\pi} A_\varphi r \mathrm{d}\varphi = 2\pi r A_\varphi \,. \quad (8.69)$$

The quantity rA_φ is displayed in the rz-plane. □

Remark 2: Although the maximum magnetic flux density in both superconducting accelerator and detector magnets is limited by the achievable engineering current density in the coil, the resulting field optimization problems are quite different. In accelerator magnets, the objective is a maximum field in a small aperture. For detector-magnet design, the objective is to generate a field on a large volume, as a large field volume is more important than high field strength for the precision measurements of the particle trajectory. The particle momentum is determined by the measurement of the sagitta S and the opening angle α of the particle trajectory in the field. The sagitta of the trajectory is approximately $QBL^2/8p$; for a proof see Chapter 11. □

8.7.4
Zonal Harmonics

Employing the result of Section 6.2.2.3, we can express the contribution of the kth line-current loop to the nth zonal harmonic as

$$\mathcal{A}_{n,k} = \frac{I r_{c,k}}{2} \frac{1}{R_{c,k}^{n+1}} \frac{1}{n(n+1)} P_n^1(\cos\vartheta_{c,k}), \qquad (8.70)$$

where $r_{c,k}$ is the radius of the loop at position $R_{c,k}, \vartheta_{c,k}$.

The results for the third zonal harmonic are displayed in Figure 8.27. The methods of optimizing the field quality are evident from the color representation of the zonal harmonic. A better field quality is obtained by single- or double-layer split-coil arrangements, as well as the so-called *notches* shown in Figures 8.27 and 8.28. The obvious method, already known from the Fabry factors, is to make the solenoid longer and thus add negative contributions (blue in Figures 8.27 and 8.28). This is of course the least cost-effective solution.

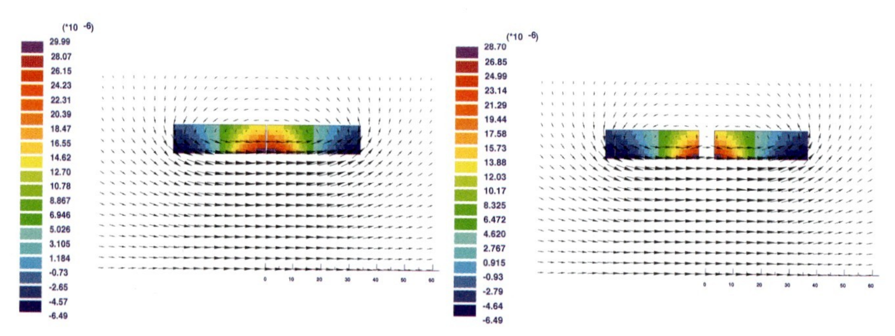

Figure 8.27 Left: Contribution to the third zonal harmonic of the line-current loops in a thick solenoid (reference radius $R_0 = 30$ mm). Right: Improvement of the field quality by splitting the coil at the symmetry plane. The optimization included the fifth zonal harmonic in an objective weighting function. Notice the different scales of the color representation.

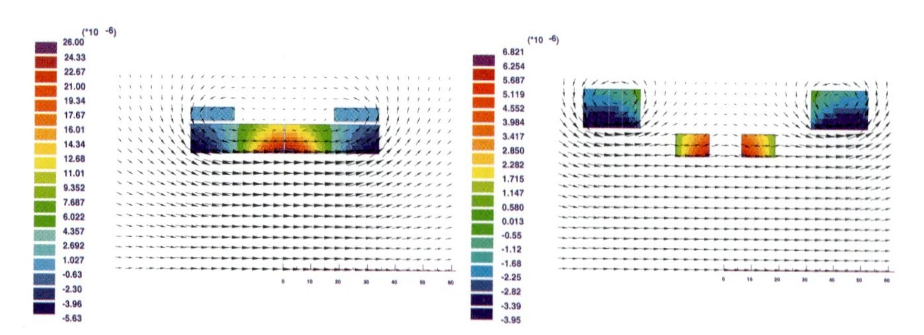

Figure 8.28 Concepts for optimization of the field quality in thick solenoids. Left: Notches. Right: Double-layer split-coil arrangement.

References

1 Beth, R.A.: Analytic design of superconducting multipolar magnets, Proceedings of the 1968 Summer Study on Superconducting Devices and Accelerators, Brookhaven National Laboratory, 1968

2 Ballarini, F. et al.: Nuclear models in FLUKA: Present capabilities, open problems and future improvements, International Conference on Nuclear Data for Science and Technology, 2004

3 Bitter, F.: The design of powerful electromagnets, part II: the magnetizing coil, Review of Scientific Instruments, 1936

4 Boom, F.W., Livingstone, R.S.: Proceedings of IRE, March 1962

5 Bottura, L.: A Practical Fit for the Critical Surface of NbTi, 16th International Conference on Magnet Technology (MT16), Florida, USA, 1999

6 Brechna, H.: Superconducting Magnet Systems, Springer, Berlin, 1973

7 Coupland, J.H.: Dipole, quadrupole and higher-order fields from simple coils, Nuclear Instruments and Methods, 1970

8 Coupland J.H. et al.: Very high field synchrotron magnets with iron yokes, Nuclear Instruments and Methods, 1973

9 Crow, J.T.: Unipolar-current Dipoles and other optimum reduced-symmetry multipoles, Journal of Applied Physics, 1997

10 CMS, The Compact Muon Solenoid, Technical Proposal, CERN/LHCC 94-38, 1994

11 Dahl, P. et al.: Superconducting magnet models for ISABELLE, Proceedings of the Particle Accelerator Conference, San Francisco, USA, 1973

12 Fabry, C.: Sur le champ magnetique au centre d'une bobine cylindrique et la construction de bobines galvonometres, Eclairage Electrique, 1898

13 Goodzeit, C.L., Ball, M.J., Meinke, R.B.: The double-helix dipole – a novel approach to accelerator magnet design, IEEE Transactions on Applied Superconductivity, 2003

14 Halbach, K.: Fields and first order perturbation effects in two-dimensional conductor dominated magnets, Nuclear Instruments and Methods, 1970

15 Jeanneret, J.B., Leroy, D., Oberli, L., Trenkler, T.: Quench levels and transient beam losses in LHC magnets, LHC project report 44, CERN, 1996

16 Kratz, R., Wyder, P.: Principles of Pulsed Magnet Design, Springer, Berlin, 2002

17 Kashikhin, V.S.: Novel design for a superconducting multipole corrector magnet, IEEE Transactions on Applied Superconductivity, 2005

18 Kircher F. et al.: Status report on the CMS superconducting solenoid for LHC, IEEE Transactions on Applied Superconductivity, 1999

19 Mokhov N.V., Kashikhin, V.V., Monville, M.E., Ferracin, P., Sabbi, G.L.: Energy deposition studies of block-coil quadrupoles for the LHC luminosity upgrade, Proceedings of PAC07, MOPAS027, 2007

20 Morgan, G.H.: Two-dimensional, uniform current density, air-core coil configurations for the production of specified magnetic fields, IEEE Transactions on Nuclear Science, 1969

21 Patoux, A., Perot, J., Rifflet, J.M.: Test of new accelerator superconducting dipoles suitable for high precision field, IEEE Transactions on Nuclear Science, 1983

22 Meß, K.H., Schmüser, P., Wolff S.: Superconducting Accelerator Magnets, World Scientific, Singapore, 1996

23 Meyer, D.I., Flasck, R.: A new configuration for a dipole magnet for use in high energy physics applications, Nuclear Instruments and Methods 80, 1970

24 Montgomery, D.B.: Solenoid Magnet Design, Wiley-Interscience, New York, 1969

25 Nakamoto, T.: Development of a prototype of superconducting combined function magnet for the 50 GeV proton beam line for the J-PARC neutrino experiment, IEEE Transactions on Applied Superconductivity, 2005

26 Rabi, I.I.: A method of producing uniform magnetic fields, Review of Scientific Instruments, 1934

27 RHIC Design Manual, 2000, http://www.bnl.gov/cad/

28 Rossi, L., Todesco, E.: Electromagnetic design of superconducting quadrupoles,

Physical Review Special Topics – Accelerators and Beams, 2006

29 Sabbi, G.: Status of Nb_3Sn accelerator R&D, IEEE Transactions on Applied Superconductivity, 2002

30 Schwerg, N., Völlinger, C.: Estimation of the Required Amount of Superconductors for High-field Accelerator Dipole Magnets, AT-MAS Technical Note, CERN, 2006.

31 Van Ginneken, A.: CASIM, Program to Simulate Transport of Hadronic Cascades in Bulk Matter, NECSCO742/01, 1975

32 Wilson, M.N.: Superconducting Magnets, Oxford Science Publications, Oxford, 1983

9
Complex Analysis Methods for Magnet Design

The imaginary number is a fine and wonderful resource of the human spirit, almost an amphibian between being and not being.

Gottfried Wilhelm Leibniz (1646–1716).

An elegant way to calculate two-dimensional fields in the aperture of superconducting magnets is by the use of complex potentials.[1] This was a standard method in the design process of the early superconducting magnet projects. The papers by Beth [2, 3] have for a long time been the main references for magnet designers.

Although numerical methods have largely replaced complex analysis in magnet design, the definition of field quality in accelerator magnets is often based on complex field components. The effect of feed-down, for example, can be much more easily understood in the framework of complex analysis.

The field of intersecting ellipses, which can be calculated by means of Cauchy's integral formula, provides an ideal current distribution of multipolar magnets and serves as the basis for a phenomenological model of persistent currents.

A remark on notation: Throughout this chapter, complex numbers are written as $z = x + iy$ and complex conjugates as $\overline{z} = x - iy$ in accordance with most mathematical texts. We will denote the field point $z = x + iy$ or $z_0 = x_0 + iy_0$, corresponding to the position vector $\mathbf{r}$ in $\mathbb{R}^2$, and write the source point as $z_c = x_c + iy_c$, which is denoted $\mathbf{r}'$ in $\mathbb{R}^2$.

1 There exists a generalization to higher dimensions: quaternions and the Clifford algebra [5, 9].

Field Computation for Accelerator Magnets. Stephan Russenschuck

ISBN: 978-3-527-40769-9

9.1
The Field of Complex Numbers

The complex field $\mathbb{C}$ is the set of ordered pairs of real numbers $(x, y) \in \mathbb{R}^2$ with addition and multiplication defined by

$$(x_1, y_1) + (x_2, y_2) = (x_1 + x_2, y_1 + y_2)\,, \tag{9.1}$$

$$(x_1, y_1)(x_2, y_2) = (x_1 x_2 - y_1 y_2, x_1 y_2 + y_1 x_2)\,. \tag{9.2}$$

The zero element is given by $(0, 0)$ and hence the additive inverse of (x, y) is $(-x, -y)$. The multiplicative identity is $(1, 0)$. The multiplicative inverse is derived from $(x_1, y_1)(x_2, y_2) = (1, 0)$, with the solution

$$x_2 = \frac{x_1}{x_1^2 + y_1^2}\,, \qquad y_2 = \frac{-x_2}{x_1^2 + y_1^2}\,. \tag{9.3}$$

Because $(x_1, 0) + (x_2, 0) = (x_1 + x_2, 0)$ and $(x_1, 0)(x_2, 0) = (x_1 x_2, 0)$, the complex numbers of the form $(x, 0)$ are isomorphic with the set of real numbers. In the same way, $(0, 1)$ is the square root of -1 because $(0, 1)(0, 1) = (-1, 0) = -1$ and $(0, 1)$ is denoted i. $\mathbb{C}$ is thus generated by adding i to $\mathbb{R}$ and is closed under addition and multiplication. Any complex number can therefore be written in the form

$$(x, y) = (x, 0) + (0, y) = x + iy =: z\,. \tag{9.4}$$

It is easily proved that the set of complex numbers $z = x + iy \in \mathbb{C}$ forms a vector space that obeys the rules of addition and scalar multiplication

$$z_1 + z_2 = x_1 + x_2 + i(y_1 + y_2), \qquad \lambda z = \lambda x + i\lambda y\,. \tag{9.5}$$

The set of complex numbers form an inner product space by means of the Euclidean scalar product $\langle x, y \rangle := x_1 x_2 + y_1 y_2$. For $x, y \in \mathbb{R}$ and $z = x + iy$ we define the real part of the complex number as $\mathrm{Re}\{z\} := x$, the imaginary part as $\mathrm{Im}\{z\} := y$, the complex conjugate as $\bar{z} := x - iy$, and the modulus as $|z| := \sqrt{x^2 + y^2}$. This yields the identities

$$\mathrm{Re}\{z\} = \frac{1}{2}(z + \bar{z})\,, \qquad \mathrm{Im}\{z\} = \frac{1}{2i}(z - \bar{z})\,, \tag{9.6}$$

and

$$|z|^2 = z\bar{z}\,, \qquad \frac{1}{z} = \frac{\bar{z}}{|z|^2}\,. \tag{9.7}$$

Moreover, $\overline{z_1 + z_2} = \overline{z_1} + \overline{z_2}$ and $\overline{z_1\, z_2} = \overline{z_1}\,\overline{z_2}$.

The assignment $\varphi \mapsto (\cos\varphi, \sin\varphi)$ defines a 2π-periodic mapping of the real line into the unit circle in $\mathbb{R}^2$. In complex notation, this reads $\varphi \mapsto \cos\varphi + i\sin\varphi$. Every nonzero complex number can be written also in the form

$$z = x + iy = r(\cos\varphi + i\sin\varphi) = r\,e^{i\varphi} \tag{9.8}$$

with

$$r = |z| = \sqrt{x^2 + y^2}, \qquad \tan\varphi = \frac{y}{x}, \tag{9.9}$$

where r is the *modulus* of z ($\mathrm{mod}\{z\}$) and φ the *argument* of z ($\arg\{z\}$), with the latter only determined to modulo 2π. It can be normalized by insisting that $\arg\{z\} \in (-\pi, \pi]$. If $z_1 = r_1\,e^{i\varphi_1}$ and $z_2 = r_2\,e^{i\varphi_2}$, then $z_1 z_2 = r_1 r_2\,e^{i(\varphi_1+\varphi_2)}$ which yields $\arg\{z_1 z_2\} = \arg\{z_1\} + \arg\{z_2\}$. In particular, the multiplication by i is a rotation by $\pi/2$ in the complex plane: $iz = re^{i(\varphi+\pi/2)}$.

9.2 Holomorphic Functions and the Cauchy–Riemann Equations

We can now study functions $f : \Omega \to \mathbb{C}$ on an open subset $\Omega \subset \mathbb{C}$,

$$f(z) = f(x + iy) = u(x, y) + iv(x, y), \tag{9.10}$$

where u and v are real-valued functions $u, v : \Omega \to \mathbb{R}$. A function f of a complex variable z is called *complex analytic* or *holomorphic* at z_0 if f is differentiable on a neighborhood[2] of z_0, that is, there exists a unique limit[3]

$$f^{(1)}(z_0) := \left.\frac{\mathrm{d}f}{\mathrm{d}z}\right|_{z_0} = \lim_{\Delta z \to 0} \frac{f(z_0 + \Delta z) - f(z_0)}{\Delta z} \tag{9.11}$$

that is independent of the direction of Δz. For $\Delta z \to 0$ along the real axis $\Delta z = \Delta x$, and for $\Delta x \in \mathbb{R}$, this yields

$$f^{(1)}(z_0) = \lim_{\Delta x \to 0} \frac{f(x_0 + \Delta x, y_0) - f(x_0, y_0)}{\Delta x} = \frac{\partial f}{\partial x}. \tag{9.12}$$

For $\Delta z \to 0$ along the imaginary axis $\Delta z = i\Delta y$ and $\Delta y \in \mathbb{R}$:

$$f^{(1)}(z_0) = \lim_{\Delta y \to 0} \frac{f(x_0, y_0 + \Delta y) - f(x_0, y_0)}{i\Delta y} = \frac{\partial f}{\partial y}\frac{1}{i}. \tag{9.13}$$

2 A holomorphic function on $X \subset \Omega$ is complex differentiable on X but the converse is not true.

3 We use the general notation $f^{(m)} := \mathrm{d}^m f/\mathrm{d}z^m$, where $f^{(1)} = \mathrm{d}f/\mathrm{d}z$ and $f^{(0)} = f$.

For the derivatives to be independent of the direction of Δz it is therefore required that $i\,\partial f/\partial x = \partial f/\partial y$. Setting $f = u + iv$ yields

$$i\frac{\partial u}{\partial x} - \frac{\partial v}{\partial x} = \frac{\partial u}{\partial y} + i\frac{\partial v}{\partial y} \tag{9.14}$$

and hence

$$\frac{\partial u}{\partial x} = \frac{\partial v}{\partial y}, \qquad \frac{\partial u}{\partial y} = -\frac{\partial v}{\partial x}, \tag{9.15}$$

which are known as the *Cauchy–Riemann equations*[4] at z_0. The function is thus holomorphic at z_0 if the Cauchy–Riemann equations are satisfied. In particular, the real and imaginary part u and v are Fréchet-differentiable at z_0, as explained in Section 3.2.

Remark: Let us identify the complex plane with $\mathbb{R}^2$ by means of

$$\mathbb{R}^2 \xrightarrow{\cong} \mathbb{C} : \mathbf{r} = (x, y) \mapsto x + iy \tag{9.16}$$

and consequently a complex function f as a mapping $f : U \to \mathbb{R}^2$ for $U \subset \mathbb{R}^2$. The function f is differentiable at $\mathbf{r} \in U$ when there exists one and only one linear mapping $T : \mathbb{R}^2 \to \mathbb{R}^2 : (x, y) \mapsto (u, v)$ for which

$$f(\mathbf{r} + t\mathbf{v}) = f(\mathbf{r}) + T(t\mathbf{v}) + R(\mathbf{r}, t\mathbf{v}) \tag{9.17}$$

and $\lim_{t\to 0} R(\mathbf{r}, t\mathbf{v})/t = 0$. The linear mapping $T(t\mathbf{v})$ is called the differential of f at $\mathbf{r}$, denoted $\mathrm{d}f|_{\mathbf{r}}$, and represented by the Jacobi matrix,

$$\mathrm{d}f = [J] = \begin{pmatrix} \frac{\partial u}{\partial x} & \frac{\partial u}{\partial y} \\ \frac{\partial v}{\partial x} & \frac{\partial v}{\partial y} \end{pmatrix}. \tag{9.18}$$

The differential is the best linear approximation for f at $\mathbf{r}$. The Cauchy–Riemann equations state that the Jacobi matrix has the form

$$\begin{pmatrix} a & -b \\ b & a \end{pmatrix} = r\begin{pmatrix} \frac{a}{r} & \frac{-b}{r} \\ \frac{b}{r} & \frac{a}{r} \end{pmatrix} = r\begin{pmatrix} \cos\varphi & -\sin\varphi \\ \sin\varphi & \cos\varphi \end{pmatrix}, \tag{9.19}$$

where $r = \sqrt{a^2 + b^2}$. The mapping T defines a local stretch and rotation while preserving angles; it is therefore called *conformal* or an "amplitwist" in [7]. □

Employing the complex conjugate and the identities (9.6) we find

$$\frac{\partial x}{\partial z} = \frac{\partial x}{\partial \bar{z}} = \frac{1}{2}, \qquad \frac{\partial y}{\partial z} = -\frac{i}{2}, \qquad \frac{\partial y}{\partial \bar{z}} = \frac{i}{2}. \tag{9.20}$$

4 Augustin Cauchy (1789–1857), Bernhard Riemann (1826–1866).

Using the chain rule this yields

$$\frac{df}{dz} = \frac{\partial f}{\partial x}\frac{\partial x}{\partial z} + \frac{\partial f}{\partial y}\frac{\partial y}{\partial z} = \frac{1}{2}\left(\frac{\partial f}{\partial x} - i\frac{\partial f}{\partial y}\right) \tag{9.21}$$

$$\frac{df}{d\bar{z}} = \frac{\partial f}{\partial x}\frac{\partial x}{\partial \bar{z}} + \frac{\partial f}{\partial y}\frac{\partial y}{\partial \bar{z}} = \frac{1}{2}\left(\frac{\partial f}{\partial x} + i\frac{\partial f}{\partial y}\right) . \tag{9.22}$$

Because

$$\frac{\partial f}{\partial x} = \frac{\partial u}{\partial x} + i\frac{\partial v}{\partial x} , \qquad \frac{\partial f}{\partial y} = \frac{\partial u}{\partial y} + i\frac{\partial v}{\partial y} , \tag{9.23}$$

the Cauchy–Riemann equations can be expressed as

$$\frac{df}{d\bar{z}} = 0 , \tag{9.24}$$

which is equivalent to $df/dz = \partial f/\partial z$. Holomorphic functions consequently depend only on z and not on $\bar{z}$. No imaging on the real axis $f : \Omega \to \mathbb{C} : z \mapsto \bar{z}$ is therefore allowed.

9.3 Power Series

A function $f : \Omega \to \mathbb{C} : z \mapsto f(z)$ for $\Omega = \{z | |z| < \rho\}$ and $\rho \in (0, \infty)$, expressed as the power series

$$f(z) = \sum_{n=0}^{\infty} c_n z^n \tag{9.25}$$

is holomorphic, and its derivative can be given element-wise:

$$f^{(1)}(z) = \sum_{n=1}^{\infty} n c_n z^{n-1} . \tag{9.26}$$

If the power series (9.25) has a nonzero radius of convergence, it follows that

$$c_n = \frac{f^{(n)}(0)}{n!} . \tag{9.27}$$

Proof. By definition $f(0) = c_0$. Repeated differentiation of the power series term-by-term yields

$$f^{(n)}(z) = n! c_n + (n+1)!\, c_{n+1}\, z + \frac{(n+2)!}{2!}\, c_{n+2}\, z^2 + \cdots \tag{9.28}$$

and the assertion (9.27) is fulfilled for $z = 0$. □

Functions that are holomorphic at all $z \in \mathbb{C}$ are said to be *entire* functions. The exponential function and the trigonometric functions defined by

$$e^z := \sum_{n=0}^{\infty} \frac{z^n}{n!} \tag{9.29}$$

$$\sin z := \sum_{n=0}^{\infty} (-1)^n \frac{z^{(2n+1)}}{(2n+1)!} \tag{9.30}$$

$$\cos z := \sum_{n=0}^{\infty} (-1)^n \frac{z^{2n}}{(2n)!} \tag{9.31}$$

are examples of entire functions. They can be seen also as *entire extensions*, or *analytic continuations*, of the corresponding real functions: In the case of the exponential function of z it is easy to verify that

$$f(z) = e^z = e^x \cos y + i e^x \sin y \tag{9.32}$$

satisfies the properties $f(z_1 + z_2) = f(z_1) f(z_2)$ and $f(x) = e^x$ for $x \in \mathbb{R}$, as expected. In the case of the trigonometric functions, we recall that

$$e^{ix} = \cos x + i \sin x\,, \qquad e^{-ix} = \cos x - i \sin x\,, \tag{9.33}$$

for $x \in \mathbb{R}$, so that

$$\sin x = \frac{1}{2i}(e^{ix} - e^{-ix})\,, \qquad \cos x = \frac{1}{2}(e^{ix} + e^{-ix})\,, \tag{9.34}$$

and by entire extensions

$$\sin z = \frac{1}{2i}(e^{iz} - e^{-iz})\,, \qquad \cos z = \frac{1}{2}(e^{iz} + e^{-iz})\,. \tag{9.35}$$

Hence the famous *Euler formula*,[5]

$$e^{iz} = \cos z + i \sin z \tag{9.36}$$

for $z \in \mathbb{C}$. The modulus of a trigonometric function of a complex variable is, however, no longer bounded by 1.

If $w \neq 0$, the most general solution of $e^z = w$ is $z = \log|w| + i \arg\{w\} + i 2\pi n$ for $n \in \mathbb{Z}$. If $\Omega \in \mathbb{C}$ is an open set and $f : \Omega \to \mathbb{C}$ is a continuous function with $e^{f(z)} = z$, f is called a branch of the logarithm on Ω. The principle path is the logarithm on $\mathbb{C} \setminus \mathbb{R}_-$.

5 This result was first discovered by Euler around 1740 for the real number Θ, that is, $e^{i\Theta} = \cos\Theta + i \sin\Theta$, from which follows the notation $z = re^{i\Theta}$, where r is called the modulus and Θ the argument of z. This gives rise to the simple expression for the multiplication of complex numbers: $z_1 z_2 = (r_1 e^{i\Theta_1})(r_2 e^{i\Theta_2}) = r_1 r_2 e^{i(\Theta_1 + \Theta_2)}$.

9.4 The Complex Form of the Discrete Fourier Transform

We are now in a position to prove the equations for the discrete Fourier transform (DFT) described in Section 6.1.1.1:

$$B_r(r_0,\varphi) = \frac{1}{2}A_0 + \sum_{n=1}^{\infty}\left(B_n(r_0)\sin n\varphi + A_n(r_0)\cos n\varphi\right), \tag{9.37}$$

where the field harmonics

$$A_n(r_0) = \frac{2}{N}\sum_{k=0}^{N-1} B_r(r_0,\varphi_k)\cos n\varphi_k, \tag{9.38}$$

$$B_n(r_0) = \frac{2}{N}\sum_{k=0}^{N-1} B_r(r_0,\varphi_k)\sin n\varphi_k\,, \tag{9.39}$$

for $n = 1,\ldots,N$ are calculated from the numerically calculated radial field components B_r at N discrete points

$$\varphi_k = \frac{2\pi k}{N}, \qquad k = 0,1,2,\ldots,N-1\,, \tag{9.40}$$

in the interval $[0,2\pi)$ at a reference radius r_0 within the aperture of the magnet. Omitting the notation of the radius-dependence of B_r and using the identities (9.33) we obtain the complex form of the Fourier series,

$$\frac{1}{2}A_0 + \sum_{n=1}^{\infty}\left(A_n\cos n\varphi + B_n\sin n\varphi\right) = \sum_{-\infty}^{\infty} D_n e^{in\varphi}\,, \tag{9.41}$$

where[6] $D_0 = \frac{1}{2}A_0 = \frac{1}{2\pi}\int_0^{2\pi} B_r(\varphi)\mathrm{d}\varphi = 0$ for all divergence-free fields and

$$\begin{aligned} D_n &:= \frac{1}{2}(A_n - iB_n) \\ &= \frac{1}{2\pi}\int_0^{2\pi} B_r(\varphi)\,(\cos n\varphi - i\sin n\varphi)\mathrm{d}\varphi = \frac{1}{2\pi}\int_0^{2\pi} B_r(\varphi)\,e^{-in\varphi}\mathrm{d}\varphi \end{aligned} \tag{9.42}$$

for $n > 0$. The coefficients are now also defined for negative indices by $D_{-n} = \frac{1}{2}(A_{-n} + iB_{-n})$, that is, $D_n = \overline{D}_{-n}$. Therefore the real coefficients are

$$A_n = D_n + D_{-n} = 2\,\mathrm{Re}\{D_n\}\,, \tag{9.43}$$

$$B_n = i(D_n - D_{-n}) = -2\,\mathrm{Im}\{D_n\}\,. \tag{9.44}$$

Because of the symmetry of D_n we need only take positive indices into account.

6 It is a common practice to denote the complex Fourier coefficients C_n but this would clash with the conventions in Section 9.7.1.

Theorem 9.1 *The complex form of the DFT is given by* $\mathcal{F} : \mathbb{R}^N \to \mathbb{C}^N : (B_r(\varphi_0), \ldots, B_r(\varphi_{N-1})) \mapsto (D_0, \ldots, D_{N-1})$ *with*

$$D_n = \frac{1}{N} \sum_{k=0}^{N-1} B_r(\varphi_k) e^{-in\varphi_k} = \frac{1}{N} \sum_{k=0}^{N-1} B_r(\varphi_k) e^{-\frac{i2\pi nk}{N}}, \tag{9.45}$$

and its inverse $\mathcal{F}^{-1}$ *given by*

$$B_r(\varphi_k) = \sum_{n=0}^{N-1} D_n e^{in\varphi_k}. \tag{9.46}$$

Proof. Equation (9.45), a Riemann sum for the *Euler–Fourier integral* (9.42), is now proved by substituting Eq. (9.45) into Eq. (9.46). We derive

$$B_r(\varphi_j) = \sum_{n=0}^{N-1} \frac{1}{N} \sum_{k=0}^{N-1} B_r(\varphi_k) e^{-in\varphi_k} e^{in\varphi_j} = \frac{1}{N} \sum_{k=0}^{N-1} B_r(\varphi_k) \sum_{n=0}^{N-1} \left(e^{\frac{2\pi i(j-k)}{N}} \right)^n. \tag{9.47}$$

For $j = k$ this yields

$$\sum_{n=0}^{N-1} \left(e^{\frac{2\pi i(j-k)}{N}} \right)^n = \sum_{n=0}^{N-1} 1 = N. \tag{9.48}$$

For all other values of $j - k$ we use the formula

$$1 + r + r^2 + \cdots + r^{N-1} = \frac{1 - r^N}{1 - r} \tag{9.49}$$

and obtain

$$\sum_{n=0}^{N-1} \left(e^{\frac{2\pi i(j-k)}{N}} \right)^n = \frac{1 - \left(e^{\frac{2\pi i(j-k)}{N}} \right)^N}{1 - e^{\frac{2\pi i(j-k)}{N}}} = \frac{1 - e^{2\pi i(j-k)}}{1 - e^{\frac{2\pi i(j-k)}{N}}} = 0 \tag{9.50}$$

because $j - k$ is an integer and $|j - k| < N$, from which follows $-1 < \frac{j-k}{N} < 1$. Consequently,

$$B_r(\varphi_j) = \sum_{n=0}^{N-1} \frac{1}{N} \sum_{k=0}^{N-1} B_r(\varphi_k) e^{-in\varphi_k} e^{in\varphi_j} = \frac{1}{N} \sum_{k=0}^{N-1} B_r(\varphi_k) N \delta_{jk} = B_r(\varphi_j). \tag{9.51}$$

□

Equation (9.46) can be written also in the matrix form as $\{B\} = [F]\{D\}$ where $\{D\} = (D_0, \ldots, D_{N-1})^T$ and $\{B\} = (B_r(\varphi_0), \ldots, B_r(\varphi_{N-1}))^T$. This is a linear transformation $\mathcal{F} : \mathbb{C}^N \to \mathbb{C}^N$ given by the *Fourier matrix*

$$[F] = \begin{pmatrix} 1 & 1 & 1 & \ldots & 1 \\ 1 & z & z^2 & \ldots & z^{N-1} \\ 1 & z^2 & z^4 & \ldots & z^{2(N-1)} \\ \vdots & \vdots & \vdots & & \vdots \\ 1 & z^{N-1} & z^{2(N-1)} & \ldots & z^{(N-1)^2} \end{pmatrix} \tag{9.52}$$

with the coefficients $z^k = \exp[2k\pi i/N]$. The commutative property expressed by

$$[F][\overline{F}] = [\overline{F}][F] = N[I], \tag{9.53}$$

where $[I]$ is the identity matrix, means that the inverse of the Fourier matrix is identical to its conjugate. Equation (9.45) may therefore be written as

$$\{D\} = \frac{1}{N}[\overline{F}]\{B\}. \tag{9.54}$$

9.5 Complex Potentials

Not all smooth functions $u(x, y)$ and $v(x, y)$ qualify as real or imaginary parts of holomorphic functions. As a consequence of the Cauchy–Riemann conditions we have the following theorem:

Theorem 9.2 *Real and imaginary parts of a holomorphic function are harmonic functions.*

Proof. If $f(z) = f(x, y) = u(x, y) + iv(x, y)$ is holomorphic, the Cauchy–Riemann equations yield

$$\nabla^2 u = \frac{\partial}{\partial x}\left(\frac{\partial u}{\partial x}\right) + \frac{\partial}{\partial y}\left(\frac{\partial u}{\partial y}\right) = \frac{\partial}{\partial x}\left(\frac{\partial v}{\partial y}\right) + \frac{\partial}{\partial y}\left(-\frac{\partial v}{\partial x}\right) = 0, \tag{9.55}$$

$$\nabla^2 v = \frac{\partial}{\partial x}\left(\frac{\partial v}{\partial x}\right) + \frac{\partial}{\partial y}\left(\frac{\partial v}{\partial y}\right) = \frac{\partial}{\partial x}\left(-\frac{\partial u}{\partial y}\right) + \frac{\partial}{\partial y}\left(\frac{\partial u}{\partial x}\right) = 0. \tag{9.56}$$

□

In the same way it can be shown that the real and imaginary parts of the conjugate of a holomorphic function are harmonic, while the conjugate itself is not holomorphic.

In two-dimensional regions Ω that are current-free and simply-connected, both the magnetic scalar potential as well as the z-component[7] of the magnetic vector potential can be used to solve Maxwell's equations:

$$\mathbf{H} = -\operatorname{grad}\phi = -\frac{\partial \phi}{\partial x}\mathbf{e}_x - \frac{\partial \phi}{\partial y}\mathbf{e}_y, \tag{9.57}$$

$$\mathbf{B} = \operatorname{curl}(\mathbf{e}_z A_z) = \frac{\partial A_z}{\partial y}\mathbf{e}_x - \frac{\partial A_z}{\partial x}\mathbf{e}_y. \tag{9.58}$$

This implies,

$$\frac{\partial A_z}{\partial y} = -\mu_0 \frac{\partial \phi}{\partial x} \quad \text{and} \quad \frac{\partial A_z}{\partial x} = \mu_0 \frac{\partial \phi}{\partial y}, \tag{9.59}$$

which are the Cauchy–Riemann differential equations of the holomorphic function:

$$w(z) := u(x,y) + iv(x,y) = A_z(x,y) + i\mu_0\phi(x,y). \tag{9.60}$$

The expression (9.60) is known as the *complex potential*. The real part is also called the *potential function*, and the imaginary part is known as the *stream function*, by analogy to fluid flow. Derivatives of w are also holomorphic[8] functions on Ω:

$$-\frac{dw}{dz} = -\frac{\partial A_z}{\partial x} - i\mu_0 \frac{\partial \phi}{\partial x} = i\frac{\partial A_z}{\partial y} - \mu_0 \frac{\partial \phi}{\partial y} = B_y(x,y) + iB_x(x,y) =: B(z). \tag{9.61}$$

It follows directly from Eq. (9.59) that the Laplace equation holds for both A_z and ϕ; the potentials are harmonic functions of z.

9.6
Conformal Mappings

As $B_y(x,y) + iB_x(x,y)$ is proportional to z^{n-1}, the complex potential $w(z) = A_z(x,y) + i\mu_0\phi(x,y)$ is proportional to z^n. If a function f and its inverse are holomorphic on U and V, respectively, then f defines a bijective *conformal mapping*[9] $f : U \to V : z \mapsto f(z)$ locally preserving angles and orientation.

7 This is the spatial z that has no relation to the complex number.
8 We emphasize that $B_x(x,y) + iB_y(x,y)$ is **not** holomorphic on Ω.
9 The special case where the upper half-plane is mapped to a polygon is known as the Schwarz–Christoffel transformation. Elwin Bruno Christoffel (1829–1900), Hermann Amandus Schwarz (1843–1921).

Examples:

1. The mapping $f : \Omega \to \Omega : z \mapsto z^2$ for $\Omega = \mathbb{C} \setminus 0$ yields

$$u(x,y) = \text{Re}\{f(z)\} = x^2 - y^2, \tag{9.62}$$

$$v(x,y) = \text{Im}\{f(z)\} = 2xy. \tag{9.63}$$

As we are interested in the pole faces of an ideal quadrupole, we plot in the xy-plane equipotentials

$$u_i = x^2 - y^2, \qquad v_j = 2xy, \tag{9.64}$$

which are mapped to the coordinate lines in the uv-plane according to Figure 9.1 (bottom, left). The hyperbolas of constant u and v, intersecting at right angles, represent the flux lines and pole faces of an ideal quadrupole.

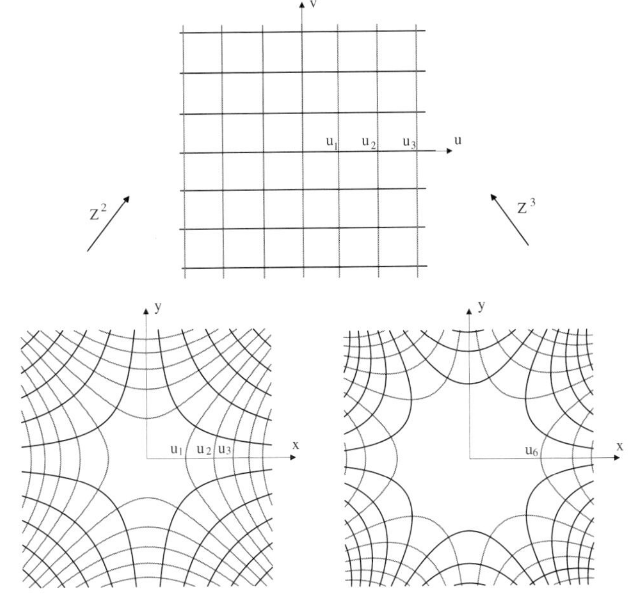

Figure 9.1 Left: Equipotentials of u and v for the function $w = f(z) = z^2$ (quadrupole flux lines and potential). Right: Equipotentials of u and v for the function $w = f(z) = z^3$ (sextupole flux lines and potential).

2. Figure 9.1 (bottom, right) shows the flux lines and pole faces for an ideal sextupole, that is, equipotentials of u and v for the mapping $f : \Omega \to \Omega : z \mapsto z^3$ for $\Omega = \mathbb{C} \setminus 0$:

$$u_i = x^3 - 3y^2x, \qquad v_j = 3yx^2 - y^3. \tag{9.65}$$

3. As a third example we will show that the mapping $f : \Omega \to \Omega : z \mapsto \log(z/z_{\text{ref}})$ for $\Omega = \mathbb{C} \setminus 0$, where $z_{\text{ref}} \neq 0$ denotes an arbitrary reference point, can be used to describe the field and potential of a line-current. For $z = x + iy = re^{i\Theta}$ and $z_{\text{ref}} = r_{\text{a}}e^{i\Theta_{\text{a}}}$ we obtain

$$w = \log(re^{i\Theta}) + \log(r_{\text{a}}e^{i\Theta_{\text{a}}}) = \ln \frac{r}{r_{\text{a}}} + i(\Theta + \Theta_{\text{a}}) = u + iv \,. \tag{9.66}$$

Circles concentric to and rays directed from the origin are mapped to the coordinate lines in the uv-plane:

$$r_{\text{a}}^2 e^{2u} = x^2 + y^2 = r^2 \tag{9.67}$$

and $\Theta = \arctan(y/x) = v - \Theta_{\text{a}}$, from which it follows that $y = x \tan(v - \Theta_{\text{a}})$. □

9.7 Complex Representation of Field Quality in Accelerator Magnets

The solution in Cartesian coordinates is needed for the calculation of particle motions in the horizontal and vertical plane of an accelerator. It can be obtained from the transformation

$$B_x = B_r \cos\varphi - B_\varphi \sin\varphi, \qquad B_y = B_r \sin\varphi + B_\varphi \cos\varphi, \tag{9.68}$$

which reads in complex notation

$$B_y + iB_x = (B_\varphi + iB_r)e^{-i\varphi} \,. \tag{9.69}$$

For $z = x + iy$ this yields

$$\begin{aligned} B_y + iB_x &= \sum_{n=1}^{\infty} (B_n(r_0) + i\,A_n(r_0)) \left(\frac{r}{r_0}\right)^{n-1} e^{i(n-1)\varphi} \\ &= \sum_{n=1}^{\infty} (B_n(r_0) + i\,A_n(r_0)) \left(\frac{z}{r_0}\right)^{n-1} \\ &= B_N \sum_{n=1}^{\infty} (b_n(r_0) + i\,a_n(r_0)) \left(\frac{z}{r_0}\right)^{n-1} , \end{aligned} \tag{9.70}$$

which in some documents is referred to as the definition of the field harmonics [11].

9.7.1 Feed-Down

An interesting example for the use of the complex field representation is the calculation of the feed-down effect due to an off-centering of a magnetic measurement coil with respect to the magnet axis, or the misalignment of magnets in the accelerator tunnel.

The transformation law for the field harmonics $C_n \mapsto C'_n$ with $C_n := B_n + i\,A_n$ can be derived for a translation of the reference frame (the center of the measurement coil) $z \to z'$, $z' := z - z_\mathrm{d}$ with $z = x + iy$, $z_\mathrm{d} = x_\mathrm{d} + iy_\mathrm{d}$ as follows: The field components in both coordinate systems (shown in Figure 9.2) must be identical, and therefore $B_y + iB_x = B_{y'} + iB_{x'}$. It follows that

$$\sum_{n=1}^{\infty} C_n \left(\frac{z}{r_0}\right)^{n-1} \overset{!}{=} \sum_{n=1}^{\infty} C'_n \left(\frac{z'}{r_0}\right)^{n-1}, \tag{9.71}$$

and the transformation law for the field harmonics is

$$C'_n = \sum_{k=n}^{\infty} C_k \binom{k-1}{n-1} \left(\frac{z_\mathrm{d}}{r_0}\right)^{k-n}. \tag{9.72}$$

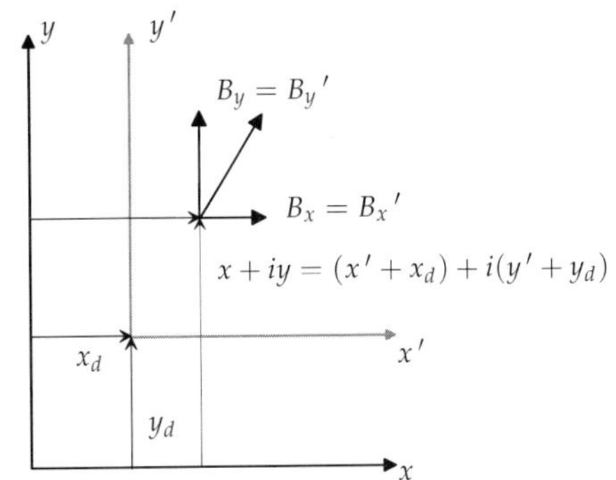

Figure 9.2 Displacement of the reference frame.

Proof. Using the *binomial series*[10]

$$\left(z_\mathrm{d} + z'\right)^{n-1} = \sum_{k=1}^{n} \binom{n-1}{k-1} \left(z'\right)^{k-1} {z_\mathrm{d}}^{n-k}, \tag{9.73}$$

the left-hand side of Eq. (9.71) becomes

$$\begin{aligned}\sum_{n=1}^{\infty} C_n \left(\frac{z}{r_0}\right)^{n-1} &= \sum_{n=1}^{\infty} \frac{C_n}{{r_0}^{n-1}} \left(z' + z_\mathrm{d}\right)^{n-1} \\ &= \sum_{n=1}^{\infty} \frac{C_n}{{r_0}^{n-1}} \sum_{k=1}^{n} \binom{n-1}{k-1} \left(z'\right)^{k-1} {z_\mathrm{d}}^{n-k}\end{aligned}$$

10 Remember that $\binom{n}{p} = \frac{n!}{p!(n-p)!}$ for $0 \le p \le n$ and zero for $0 \le n < p$.

$$= \sum_{k=1}^{\infty} \left[\sum_{n=1}^{n} \frac{C_n}{r_0^{n-1}} \binom{n-1}{k-1} z_d^{n-k} \right] (z')^{k-1}$$

$$= \sum_{k=1}^{\infty} \left[\sum_{n=k}^{\infty} C_n \binom{n-1}{k-1} \left(\frac{z_d}{r_0}\right)^{n-k} \right] \left(\frac{z'}{r_0}\right)^{k-1}$$

$$= \sum_{n=1}^{\infty} \left[\sum_{k=n}^{\infty} C_k \binom{k-1}{n-1} \left(\frac{z_d}{r_0}\right)^{k-n} \right] \left(\frac{z'}{r_0}\right)^{n-1} . \tag{9.74}$$

Comparing the coefficients of $(z'/r_0)^{n-1}$ leads to the transformation law for the complex field harmonics, Eq. (9.72). □

In particular,

$$C_2' = C_2 + 2\,C_3 \left(\frac{z_d}{r_0}\right) + 3\,C_4 \left(\frac{z_d}{r_0}\right)^2 + \cdots , \tag{9.75}$$

which shows that an off-centering of the measurement coil creates a quadrupole field component from the sextupole component C_3 of the coil field. The transformation (9.72) is therefore referred to as *feed-down*. Notice the slower convergence for the higher order multipoles, for example,

$$C_{10}' = C_{10} + 10\,C_{11} \left(\frac{z_d}{r_0}\right) + 55\,C_{12} \left(\frac{z_d}{r_0}\right)^2 + \cdots . \tag{9.76}$$

Equation (9.72) may also be used to determine the center of the measurement coil, for example, by the elimination of the measured dipole component in a quadrupole magnet. In dipole magnets, the B_{11} component (resulting from limitations in the coil design) is quite insensitive to manufacturing errors, while B_{10} is nearly zero. If C_{11}, C_{12}, and C_{13} at a reference radius r_0 are known with sufficient accuracy, the displacement z_d can be calculated from the truncated series (9.76). This holds only for small displacements where $z_d \ll r_0$. For example, in the case of $z_d/r_0 = 0.1$, the series converges slowly and more than two higher-order multipoles must be taken into account.

A simple expression is obtained if the first-order feed-down effect is considered alone:

$$B_n' \approx \frac{n}{r_0}(B_{n+1}\,x_d - A_{n+1}\,y_d), \qquad A_n' \approx \frac{n}{r_0}(B_{n+1}\,x_d + A_{n+1}\,y_d). \tag{9.77}$$

For a dipole with a sextupole component and displacement only along the x-axis, as shown in Figure 9.3 (left), this yields

$$B_2' \approx \frac{2}{r_0} B_3 x_d . \tag{9.78}$$

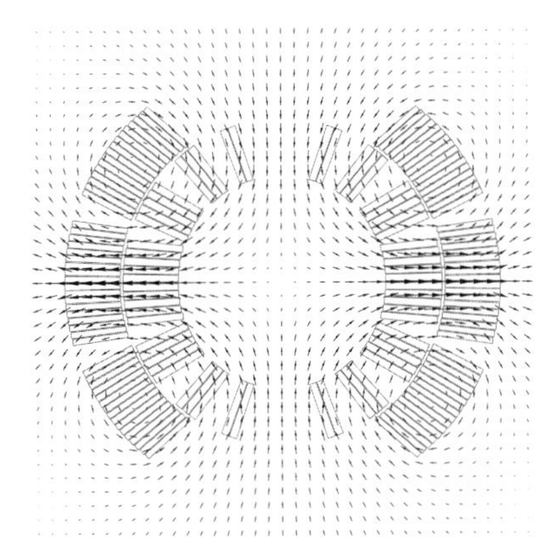

Figure 9.3 Left: Field of an ideal dipole with a sextupole field superimposed. Notice how the field in the displaced aperture (circle) resembles a negative dipole field with a quadrupole field superimposed (the negative B_y field component increases with the distance from the center). Right: LHC dipole coil powered as a quadrupole to measure the magnetic axis of the magnet by means of feed-down on the dipole component.

Notice how the field in the displaced aperture in Figure 9.3 (left) resembles a quadrupole field with the y-component of $\mathbf{B}$ increasing with the distance from the magnet center.

The feed-down effect can also be used to determine the magnetic axis of the dipole by powering the coil in a quadrupole configuration, shown in Figure 9.3 (right), and by measuring the feed-down of the quadrupole on the measured dipole field component, a quantity that should be zero [6].

Remark: What we have introduced here is the mathematical concept of an *analytic* or better, a *holomorphic continuation*: Let $f_1(z)$ be a holomorphic function in Ω_1 and $f_2(z)$ be a holomorphic continuation in an overlapping domain Ω_2, and $\Omega_1 \cap \Omega_2 \neq \emptyset$. Let $\mathscr{S}$ be a piecewise-smooth path in $\Omega_1 \cup \Omega_2$ from z to z', then $f(z')$ can be obtained using the Taylor series expansion of f_2. Its coefficients are obtained from stepwise continuation, that is, a repeated application of the feed-down equation (9.72) for a succession of overlapping disks with centers along $\mathscr{S}$. The results for two paths $\mathscr{S}_1$ and $\mathscr{S}_2$ are identical if the paths belong to the same homotopy class. This would in principle allow the calculation of the field quality in areas larger than either the circular domain covered by the measurement coil or the domain inside the reference radius used in numerical field calculation. We will not go into further detail, as numerical errors in the calculation of the multipole coefficients (in particular for the higher-order terms) make this method ill advised for practical applications. □

As a consequence of the feed-down effect, the transversal displacement of the magnetic dipole axis with respect to the actual closed orbit of the beam must be limited to ± 0.1 mm for both x_d and y_d systematic, and 0.5 mm for both σ_x and σ_y random (RMS). The quadrupole misalignment must not exceed 0.36 mm (RMS) including the survey errors at installation and the effects induced by ground motion. Quadrupole feed-down of the sextupole spool-piece correctors connected to the main dipoles imposes a random sextupole-misalignment of less than 0.5 mm (RMS) with respect to the dipole. The achievable alignment tolerances of the spool-piece correctors also define the limits for the systematic sextupole and octupole components in the main dipoles at injection energy.

9.7.2
Reference Frame Rotation

Consider the reference frame rotated by the angle φ, as shown in Figure 9.4 (left), described by the transformation $z' = r\,e^{i\Theta'} = r\,e^{i(\Theta-\varphi)} = ze^{-i\varphi}$. For the field components we derive

$$B_{y'} + iB_{x'} = (B_y + iB_x)\,e^{i\varphi} \tag{9.79}$$

and require

$$\sum_{n=1}^{\infty} C'_n \left(\frac{z'}{r_0}\right)^{n-1} = \sum_{n=1}^{\infty} C'_n \left(\frac{z\,e^{-i\varphi}}{r_0}\right)^{n-1} \stackrel{!}{=} \sum_{n=1}^{\infty} C_n \left(\frac{z}{r_0}\right)^{n-1} e^{i\varphi}\,, \tag{9.80}$$

from which follows

$$C'_n = C_n\,e^{in\varphi}\,. \tag{9.81}$$

In particular for a magnet turned upside-down (rotated by π), we obtain $C'_n = (-1)^n C_n$ and consequently

$$B'_n = (-1)^n B_n\,, \qquad A'_n = (-1)^n A_n\,. \tag{9.82}$$

Substituting $B_n = B_\mathrm{N} b_n$ and $A_n = B_\mathrm{N} a_n$, using the relations (9.82) as well as $B'_\mathrm{N} = (-1)^N B_\mathrm{N}$, we obtain for the relative multipoles

$$b'_n = (-1)^{n-N} b_n\,, \quad \text{and} \quad a'_n = (-1)^{n-N} a_n\,. \tag{9.83}$$

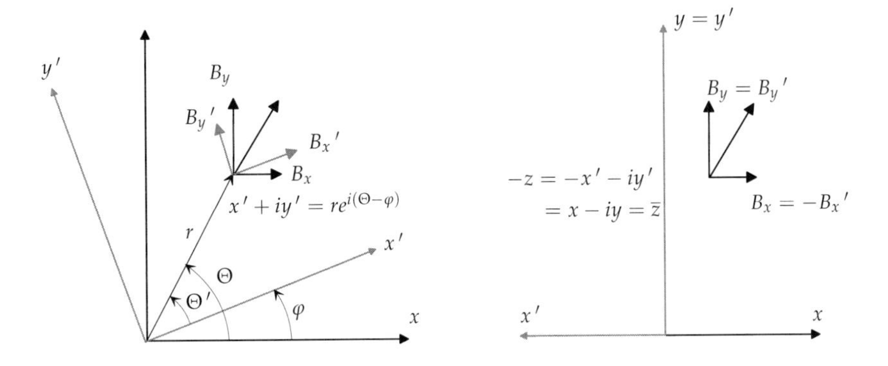

Figure 9.4 Left: Rotation of the reference frame. Right: Reflection about the y-axis.

9.7.3 Reflection about the Vertical Axis

Consider the reference frame imaged about the y-axis, as shown in Figure 9.4 (right), that is, $z' = x' + iy' = -(x - iy) = -\overline{z}$. For the field components, we obtain $B_{y'} + iB_{x'} = B_y - iB_x$ and require

$$\sum_{n=1}^{\infty} C'_n \left(\frac{z'}{r_0}\right)^{n-1} = \sum_{n=1}^{\infty} C'_n (-1)^{n-1} \left(\frac{\overline{z}}{r_0}\right)^{n-1} \stackrel{!}{=} \sum_{n=1}^{\infty} \overline{C}_n \left(\frac{\overline{z}}{r_0}\right)^{n-1}, \tag{9.84}$$

where $\overline{C}_n := B_n - iA_n$. From Eq. (9.84) it follows $C'_n = (-1)^{n-1}\overline{C}_n$ or

$$B'_n = (-1)^{n-1} B_n, \qquad A'_n = (-1)^n A_n. \tag{9.85}$$

This is an important result for the scaling of multipoles when the position of the magnet in the accelerator tunnel is turned with respect to the "normal" installation direction defined in Chapter 12. How do the relative normal and relative skew-multipole components transform? Substituting $B_n = B_N b_n$ and $A_n = B_N a_n$, using the relations in Eq. (9.85) as well as the relation $B'_N = (-1)^{N-1} B_N$, we obtain for the normal multipole magnets

$$b'_n = (-1)^{n-N} b_n \qquad a'_n = (-1)^{n-N+1} a_n, \tag{9.86}$$

and for the skew-multipole magnets ($B_n = A_N b_n$, $A_n = A_N a_n$):

$$b'_n = (-1)^{n-1-N} b_n \qquad a'_n = (-1)^{n-N} a_n. \tag{9.87}$$

9.8 Complex Integration

First we shall study the integration of entire functions on a smooth path $\mathscr{S}$ given in the parametric form $z(t) = x(t) + iy(t), t \in [t_1, t_2]$. Expressing $f(z)$ as $u(x,y) + iv(x,y)$ yields

$$\int_{\mathscr{S}} f(z)\mathrm{d}z = \int_{t_1}^{t_2} f(z(t))\frac{\partial z}{\partial t}\mathrm{d}t = \int_{t_1}^{t_2} (u + iv)\left(\frac{\partial x}{\partial t} + i\frac{\partial y}{\partial t}\right)\mathrm{d}t$$
$$= \int_{t_1}^{t_2}\left(u\frac{\partial x}{\partial t} - v\frac{\partial y}{\partial t}\right)\mathrm{d}t + i\int_{t_1}^{t_2}\left(u\frac{\partial y}{\partial t} + v\frac{\partial x}{\partial t}\right)\mathrm{d}t$$
$$= \int_{\mathscr{S}} (u\mathrm{d}x - v\mathrm{d}y) + i\int_{\mathscr{S}} (u\mathrm{d}y + v\mathrm{d}x)\,. \tag{9.88}$$

Example: Consider the path integral on a closed circle centered at the origin, $\mathscr{C} : [0, 2\pi] \to \Omega : \varphi \mapsto z(\varphi) = z_0 + re^{i\varphi}$. The complex arc element is $\mathrm{d}z = ire^{i\varphi}\mathrm{d}\varphi$ and $(z - z_0) = re^{i\varphi}$. For $n \in \mathbb{Z}$ and $r > 0$,

$$\int_{\mathscr{C}} (z - z_0)^n \mathrm{d}z = \int_0^{2\pi} (re^{i\varphi})^n ire^{i\varphi}\mathrm{d}\varphi$$
$$= \int_0^{2\pi} ir^{n+1}(\cos(n+1)\varphi + i\sin(n+1)\varphi)\mathrm{d}\varphi\,, \tag{9.89}$$

from which follows

$$\int_{\mathscr{C}} (z - z_0)^n \mathrm{d}z = \begin{cases} 2\pi i & \text{for } n = -1 \\ 0 & \text{for } n \neq -1 \end{cases} \tag{9.90}$$

□

For line integrals of entire functions this yields

$$\int_{-\mathscr{S}} f(z)\mathrm{d}z = -\int_{\mathscr{S}} f(z)\mathrm{d}z \tag{9.91}$$

if the path is traced out in opposite direction. For the change of the curve's parametric representation it holds that

$$\int_{\mathscr{S}} f(z)\mathrm{d}z = \int_{\mathscr{S}\circ\varphi} f(z)\mathrm{d}z, \tag{9.92}$$

where $\varphi : [u_1, u_2] \to [t_1, t_2]$ defines a C^1 change of parametric representation of the path. If the path is only piecewise smooth, we can redefine it as

$$\int_{\mathscr{S}} f(z)\mathrm{d}z := \sum_{i=1}^{k} \int_{\mathscr{S}_i} f(z)\mathrm{d}z\,. \tag{9.93}$$

If f can be written as the derivative of an entire function F for $F^{(1)}(z) = f(z)$, then

$$\int_{\mathscr{S}} f(z)\mathrm{d}z = F(z(t_2)) - F(z(t_1)), \tag{9.94}$$

which yields the closed-curve theorem for entire functions,

$$\int_{\mathscr{C}} f(z)\mathrm{d}z = 0, \tag{9.95}$$

for any closed curve $\mathscr{C}$. Note that here $\mathscr{C}$ is not necessarily a Jordan curve, which means that it may be self-intersecting.

9.8.1 Cauchy's Theorem and the Integral Formula

The closed-curve theorem for entire functions, Eq. (9.94), can be generalized to the following theorem.

Theorem 9.3 *(Cauchy) Suppose $\Omega \subset \mathbb{C}$ is simply-connected and open, and $f : \Omega \to \mathbb{C}$ is holomorphic. Hence*

$$\int_{\partial\Omega} f(z)\mathrm{d}z = 0. \tag{9.96}$$

Employing the terminology of Section 2.8, we can reformulate Cauchy's theorem that $\int_{\mathscr{S}} f(z)\mathrm{d}z = 0$ holds if $\mathscr{S}$ is zero-homotopic.

Proof. We invoke Green's theorem in the plane (see Section 3.10.2) applied to the first integral of Eq. (9.88)

$$\int_{\partial\Omega} (u\mathrm{d}x - v\mathrm{d}y) = \int_{\Omega} \left(-\frac{\partial v}{\partial x} - \frac{\partial u}{\partial y} \right) \mathrm{d}x\mathrm{d}y = 0, \tag{9.97}$$

as the Cauchy–Riemann equations hold in Ω. In the same manner the second integral of Eq. (9.88) can be shown to vanish in Ω. □

Any holomorphic function can be expressed in the neighborhood of a point $z_0 \in \Omega$ as

$$f(z) = f(z_0) + f^{(1)}(z)(z - z_0) + g(z) \tag{9.98}$$

where $f^{(1)}(z)$ is again holomorphic and

$$\lim_{z \to z_0} \frac{g(z)}{|z - z_0|} = 0. \tag{9.99}$$

Therefore, taking the path integral on a closed circle $\mathscr{C}$ centered at z_0 yields

$$\int_{\mathscr{C}} \frac{f(z)}{z - z_0}\mathrm{d}z = \int_{\mathscr{C}} \left(\frac{f(z_0)}{z - z_0} + f^{(1)}(z) \right) \mathrm{d}z = \int_{\mathscr{C}} \frac{f(z_0)}{z - z_0}\mathrm{d}z = 2\pi i f(z_0). \tag{9.100}$$

If now Ω can be contracted to the circle $\mathscr{C}$ without passing z_0, that is, there exists a homotopy H from $\partial\Omega$ to $\mathscr{C}$, the result is the *Cauchy integral formula*

$$f(z_0) = \frac{1}{2\pi i}\int_{\partial\Omega}\frac{f(z)}{z-z_0}\mathrm{d}z\,. \tag{9.101}$$

More generally, let $\mathscr{C}$ be any smooth curve that is null-homotopic in Ω, and let f be holomorphic on Ω. It follows that

$$n(\mathscr{C};z_0)f(z_0) = \frac{1}{2\pi i}\int_{\mathscr{C}}\frac{f(z)}{z-z_0}\mathrm{d}z\,, \qquad z_0 \in \Omega\setminus\mathscr{C}\,, \tag{9.102}$$

where $n(\mathscr{C};z_0)$ is the *winding number* of $\mathscr{C}$ about z_0 defined by

$$n(\mathscr{C};z_0) := \frac{1}{2\pi i}\int_{\mathscr{C}}\frac{1}{z-z_0}\mathrm{d}z\,. \tag{9.103}$$

Equation (9.102) is also known as the *winding number version* of the Cauchy integral formula. If $\mathscr{C}_1$ and $\mathscr{C}_2$ are two paths homotopic in $\Omega\setminus z_0$ this yields $n(\mathscr{C}_1;z_0) = n(\mathscr{C}_2;z_0)$. If $\mathscr{C}$ is zero-homotopic in Ω, on which f is holomorphic with zeros at $z_1, z_2, \ldots, z_N$, then

$$\frac{1}{2\pi i}\int_{\mathscr{C}}\frac{f^{(1)}(z)}{f(z)}\mathrm{d}z = \sum_{n=1}^{N} n(\mathscr{C};z_n) \tag{9.104}$$

because we can write $f(z) = (z-z_1)(z-z_2)\cdots(z-z_N)g(z)$, where g is holomorphic and nonzero on Ω, and therefore

$$\frac{f^{(1)}(z)}{f(z)} = \frac{1}{z-z_1} + \frac{1}{z-z_2} + \cdots + \frac{1}{z-z_N} + \frac{g^{(1)}(z)}{g(z)}\,. \tag{9.105}$$

9.8.2 Properties of Holomorphic Functions

We will now recall some theorems of complex analysis which are known from the treatment of harmonic functions in Section 5.3.

Theorem 9.4 *Let $f : \Omega \to \mathbb{C}$ be holomorphic on Ω and let $\{z \,|\, |z-z_0| < r\} \subset \Omega$ define a disk $\mathscr{D}$ of radius r centered at z_0. Then the average value of f on the boundary $\partial\mathscr{D}$ of the disk be its value at the center:*

$$f(z_0) = \frac{1}{2\pi}\int_0^{2\pi} f(z_0 + re^{it})\mathrm{d}t\,. \tag{9.106}$$

Proof. This is a nearly trivial consequence of the Cauchy integral formula,

$$f(z_0) = \frac{1}{2\pi i}\int_{\partial\mathscr{D}}\frac{f(z)}{z-z_0}\mathrm{d}z = \frac{1}{2\pi i}\int_0^{2\pi}\frac{f(z_0+re^{it})}{re^{it}}rie^{it}\mathrm{d}t\,. \tag{9.107}$$

□

Theorem 9.5 *For $f : \Omega \to \mathbb{C}$ holomorphic and $z_0 \in \Omega$ there exists exactly one expansion $f(z) = \sum_{n=1}^{\infty} c_n (z - z_0)^n$, where*

$$c_n = \frac{1}{2\pi i} \int_{\partial \mathcal{D}} \frac{f(z)}{(z - z_0)^{n+1}} \mathrm{d}z \tag{9.108}$$

and $\partial \mathcal{D}$ is again the boundary of the disk $\mathcal{D} \subset \Omega$.

Proof. Without a loss of generality we can set $z_0 = 0$. From the Cauchy integral formula and $|z| < r$, we derive

$$\begin{aligned} f(z) &= \frac{1}{2\pi i} \int_{\partial \mathcal{D}} \frac{f(z_c)}{z_c - z} \mathrm{d}z_c = \frac{1}{2\pi i} \int_{\partial \mathcal{D}} \frac{f(z_c)}{z_c} \frac{1}{1 - \frac{z}{z_c}} \mathrm{d}z_c \\ &= \frac{1}{2\pi i} \int_{\partial \mathcal{D}} \sum_{n=1}^{\infty} \frac{f(z_c)}{z_c} \left(\frac{z}{z_c} \right)^n \mathrm{d}z_c = \sum_{n=1}^{\infty} \left[\frac{1}{2\pi i} \int_{\partial \mathcal{D}} \frac{f(z_c)}{z_c^{n+1}} \mathrm{d}z_c \right] z^n . \end{aligned} \tag{9.109}$$

□

Let $|f(z)| < M$ for all $z \in \partial \mathcal{D}$, then it follows from Eq. (9.108) that

$$|c_n| \leq \frac{1}{2\pi} 2\pi r \frac{M}{r^{n+1}} = \frac{M}{r^n}, \tag{9.110}$$

which is known as the *Cauchy estimate* for the Taylor coefficients. From Eqs. (9.108) and (9.27), we obtain the generalized Cauchy integral formula

$$f^{(n)}(z_0) = \frac{n!}{2\pi i} \int_{\partial \Omega} \frac{f(z)}{(z - z_0)^{n+1}} \mathrm{d}z . \tag{9.111}$$

Theorem 9.6 *(Goursat)*[11] *A holomorphic function is infinitely complex differentiable.*

This is a corollary of the Taylor expansion theorem because of the differentiability of convergent power series.

Theorem 9.7 *(Liouville) A bounded entire function is constant.*

Proof. Let $f(z) \leq M$ for all $z \in \mathbb{C}$. Hence f can be expressed everywhere as a convergent power series with $|c_n| \leq M/r^n$ for each $r > 0$. Therefore all c_n, $n = 1, 2, \ldots$ must be zero and consequently $f(z) = c_0$. □

The importance of this theorem becomes more obvious if we express it in a different way: If our holomorphic function has no singularities at all, it is constant everywhere and therefore is of little use for magnet design. As a direct consequence we can state without proof:

Theorem 9.8 *(Fundamental theorem of algebra) Every polynomial of grade ≥ 1 has at least one zero in $\mathbb{C}$.*

11 Edouard Goursat (1858–1936).

9.8.3
The Residual Theorem

The *residual theorem* states that if a complex function is meromorphic on Ω (holomorphic on Ω except at a finite number of isolated poles) and holomorphic on the boundary $\partial\Omega$ of Ω it then holds that

$$\int_{\partial\Omega} f(z)\mathrm{d}z = 2\pi i \sum_n \mathrm{Res}(f; z_n) , \tag{9.112}$$

where the $\mathrm{Res}(f; z_n)$ are the residuals of the poles of the function $f(z)$ at the points z_n. This theorem can be derived from the Cauchy integral formula by a homologous deformation of the integration path that forms a simple loop as defined in Section 3.1. For the special case shown in Figure 9.5 it holds that

$$\begin{aligned}\int_{\mathscr{C}} f(z)\mathrm{d}z &= \int_{\mathscr{C}_1} f(z)\mathrm{d}z + \int_{\mathscr{C}_2} f(z)\mathrm{d}z + \int_{\mathscr{C}_3} f(z)\mathrm{d}z \\ &= 2\pi i \left(\mathrm{Res}(f; z_1) + \mathrm{Res}(f; z_2) + \mathrm{Res}(f; z_3)\right) .\end{aligned} \tag{9.113}$$

The circles $\mathscr{C}_{1,2,3}$ around the poles are given by $|z - z_n| = \epsilon$, $\epsilon > 0$, $n = 1, 2, 3$.

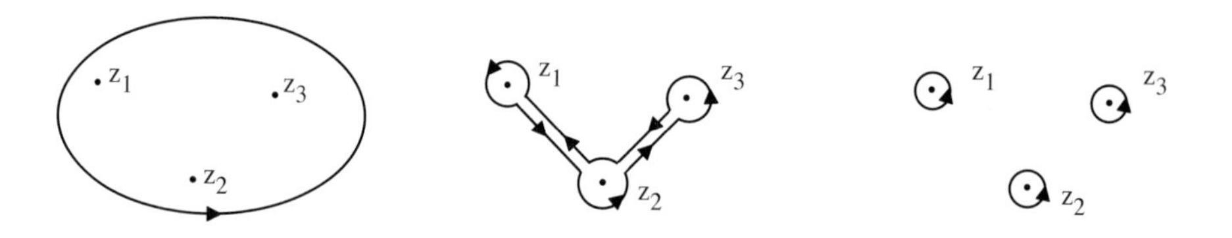

Figure 9.5 Homologous deformation of a simple loop $\mathscr{C}$ around three poles.

If f has a simple pole at z_0, then it can be expressed as $f(z) = \frac{g(z)}{z - z_0} + h(z)$ where $g(z)$ and $h(z)$ are holomorphic in a neighborhood of z_0, and

$$\mathrm{Res}(f; z_0) = \lim_{z \to z_0} (z - z_0) f(z) = g(z) . \tag{9.114}$$

Therefore Cauchy's integral formula can be interpreted as a special case of the residual theorem where f has a simple pole. If f has a double pole at z_0 it follows that

$$f(z) = \frac{g(z)}{(z - z_0)^2} + h(z) = \frac{g(z_0)}{(z - z_0)^2} + \frac{g^{(1)}(z_0)}{(z - z_0)} + \cdots + h(z), \tag{9.115}$$

and therefore $\mathrm{Res}(f; z_0) = g^{(1)}(z)$. For a pole of order n we accordingly obtain

$$\mathrm{Res}(f; z_0) = \frac{1}{n - 1} g^{(n-1)}(z_0) . \tag{9.116}$$

9.9 The Field and Potential of a Line Current

In what follows we denote the field point as $z = x + iy = re^{i\varphi}$ and the source point as $z_c = x_c + iy_c = r_c e^{i\varphi_c}$. In accordance with this convention we can rewrite the Cauchy integral formula with z instead of z_0 for the field point in the interior of Ω, and z_c instead of z for the points on the loop $\partial\Omega$:

$$f(z) = \frac{1}{2\pi i} \int_{\partial\Omega} \frac{f(z_c)}{z_c - z} dz_c . \tag{9.117}$$

The complex potential due to a line-current at the source point z_c reads

$$w(z) = -\frac{\mu_0 I}{2\pi} \log\left(\frac{z - z_c}{z_{\text{ref}}}\right) , \tag{9.118}$$

where $z_{\text{ref}} \neq 0$ is an arbitrary complex reference point. The complex magnetic flux density is then

$$-\frac{dw}{dz} = B_y(z) + iB_x(z) = -\frac{\mu_0 I}{2\pi} \frac{1}{z_c - z} , \tag{9.119}$$

which yields the well-known equations for the magnetic field components

$$B_y(z) = -\frac{\mu_0 I}{2\pi} \frac{x_c - x}{(x_c - x)^2 + (y_c - y)^2} , \tag{9.120}$$

$$B_x(z) = \frac{\mu_0 I}{2\pi} \frac{y_c - y}{(x_c - x)^2 + (y_c - y)^2} . \tag{9.121}$$

Remark: We discussed in Section 2.13 that arguments of trigonometric functions, exponential functions, and logarithms must have identity dimension due to the fact that the Taylor expansion of these functions must be dimensionally homogeneous. This is the reason an arbitrary reference point is always defined. Using the potential function for the calculation of the magnetic energy does not depend on the choice of reference point, as long as the sum of all currents is zero: Consider a set of K line-currents. We then obtain

$$w(z) = -\frac{\mu_0}{2\pi} \sum_{k=1}^{K} I_k \log\left(\frac{z - z_{c,k}}{z_{\text{ref}}}\right) = -\frac{\mu_0}{2\pi} \sum_{k=1}^{K} I_k \log(z - z_{c,k}) + C \tag{9.122}$$

where

$$C = \frac{\mu_0}{2\pi} \sum_{k=1}^{K} I_k \log(z_{\text{ref}}) = 0 , \tag{9.123}$$

if the sum of all currents is zero. □

Equation (9.118) allows an elegant notation of the potential from a line-current at position z in the aperture of a rectangular iron yoke (sides a and b) of block-coil magnets. Use of the image-current method for infinitely permeable boundaries yields the line-current distribution as shown in Figure 9.6. Hence,

$$w(z) = -\frac{\mu_0 I}{2\pi} \log \prod_{m,n=-\infty}^{\infty} \left[\left(\frac{z + \overline{z}_c + z_{mn}}{z_{ref} + \overline{z}_c + z_{mn}} \right) \left(\frac{z - \overline{z}_c + z_{mn}}{z_{ref} - \overline{z}_c + z_{mn}} \right) \left(\frac{z + z_c + z_{mn}}{z_{ref} + z_c + z_{mn}} \right) \left(\frac{z - z_c + z_{mn}}{z_{ref} - z_c + z_{mn}} \right) \right] \tag{9.124}$$

where $z_{mn} = 2ma + i\,2nb$.

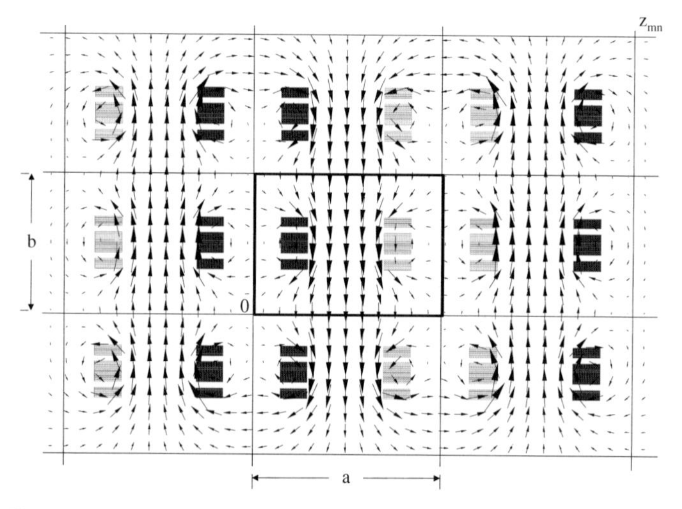

Figure 9.6 Image-current method applied to a rectangular iron yoke. The pattern consists of an infinite number of patches in the x- and y-directions (with alternating polarity in the x-direction and unchanged polarity in the y-direction).

9.9.1
Series Expansion of the Line-Current Field

For $|z| < |z_c|$ inside the circular aperture of the magnet carrying no line-currents we expand

$$\frac{1}{z_c - z} = \frac{1}{z_c(1 - \frac{z}{z_c})} = \frac{1}{z_c} \sum_{n=1}^{\infty} \left(\frac{z}{z_c} \right)^{n-1} = \sum_{n=1}^{\infty} \frac{z^{n-1}}{z_c^n} . \tag{9.125}$$

Substituting the result into Eq. (9.119) yields

$$B_y(z) + iB_x(z) = -\frac{\mu_0 I}{2\pi} \sum_{n=1}^{\infty} \frac{z^{n-1}}{z_c^n} . \tag{9.126}$$

From Eq. (9.70) we derive the multipole coefficients of the field generated by the line current by comparing the coefficients:

$$C_n(r_0) = B_n(r_0) + iA_n(r_0) = -\frac{\mu_0 I}{2\pi}\frac{r_0^{n-1}}{z_c^n}, \tag{9.127}$$

a result that is identical to Eq. (6.65).

9.9.2
Circular Sector Windings

For a circular sector winding we can write for $z_c = r_c e^{i\varphi_c}$, $r_a \leq r_c \leq r_e$, $\varphi_1 \leq \varphi_c \leq \varphi_2$ and the constant current density J, we derive

$$\begin{aligned} C_n(r_0) &= -\frac{\mu_0}{2\pi} r_0^{n-1} \int_{r_a}^{r_e} \int_{\varphi_1}^{\varphi_2} J r_c^{-n} e^{-in\varphi_c} r_c \, dr_c \, d\varphi_c \\ &= -\frac{\mu_0 i}{2\pi}\frac{J r_0^{n-1}}{n(2-n)}\left(r_e^{2-n} - r_a^{2-n}\right)\left(e^{-in\varphi_2} - e^{-in\varphi_1}\right), \end{aligned} \tag{9.128}$$

for $n \neq 2$ and

$$C_2(r_0) = -\frac{\mu_0 i}{4\pi} J r_0 \ln\frac{r_e}{r_a}\left(e^{-i2\varphi_2} - e^{-i2\varphi_1}\right). \tag{9.129}$$

For a sector, symmetric about the x-axis with $\varphi_1 = -\varphi_{sec}$, $\varphi_2 = \varphi_{sec}$, using Eq. (9.34) we obtain

$$C_n(r_0) = -\frac{\mu_0}{\pi}\frac{J r_0^{n-1}}{n(2-n)}\left(r_e^{2-n} - r_a^{2-n}\right)\sin n\varphi_{sec}, \tag{9.130}$$

for $n \neq 2$ and

$$C_2(r_0) = -\frac{\mu_0}{2\pi} J r_0 \ln\frac{r_e}{r_a}\sin 2\varphi_{sec}. \tag{9.131}$$

9.10
Multipoles Generated by a Magnetic Dipole Moment

We will now calculate the field harmonics generated by a magnetic dipole moment. For the line-current doublet located at z_c (see Figure 9.7) we can calculate from Eq. (9.119),

$$B_y(z) + iB_x(z) = -\frac{\mu_0 I}{2\pi}\left(\frac{1}{z_2 - z} - \frac{1}{z_1 - z}\right) = -\frac{\mu_0 I}{2\pi}\frac{z_1 - z_2}{(z_2 - z)(z_1 - z)}. \tag{9.132}$$

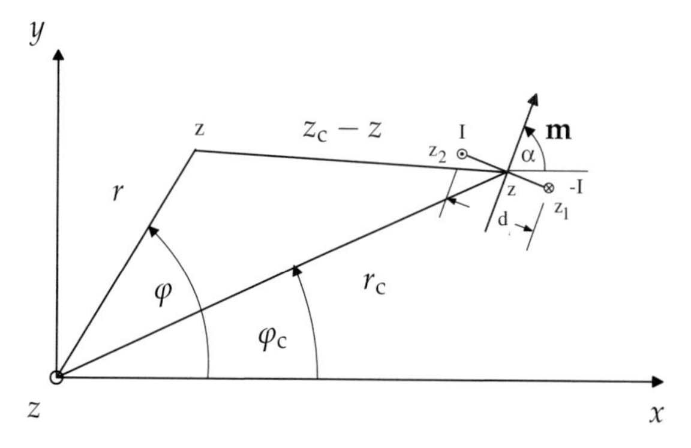

Figure 9.7 Geometric relations for the calculation of the multipole content from a magnetic moment at z.

Now if we consider that $z_1 - z_2 = -ide^{i\alpha}$ and approximate the distances from the source points z_1 and z_2 to the field point z by $z_c - z$, this yields

$$B_y(z) + iB_x(z) = \frac{i\mu_0 I}{2\pi} \frac{de^{i\alpha}}{(z_c - z)^2} = \frac{i\mu_0 I}{2\pi} \frac{de^{i\alpha}}{z_c^2(1 - \frac{z}{z_c})^2} = \frac{i\mu_0 m}{2\pi} \frac{n}{z_c^2} \sum_{n=1}^{\infty} \left(\frac{z}{z_c}\right)^{n-1}, \tag{9.133}$$

where we let $d \to 0$ and $I \to \infty$ so that the product $dI = m$ remains finite. Employing Eq. (9.127) we finally obtain

$$C_n = m \frac{i\mu_0 n}{2\pi} \frac{e^{i\alpha}}{z_c^2} \left(\frac{r}{z_c}\right)^{n-1} = m \frac{i\mu_0 n}{2\pi r_c^2} \left(\frac{r}{r_c}\right)^{n-1} e^{i(\alpha-(n+1)\varphi_c)}. \tag{9.134}$$

9.11
Beth's Current-Sheet Theorem

We will now study the case of functions that are holomorphic on Ω except for a cut of finite length according to Figure 9.8. Without a loss of generality we place this cut on the real axis, where x_1 and x_2 denote its lower and upper bounds, respectively. By homologous deformation of the loop $\mathscr{C}$ into $\mathscr{C}_1 + \mathscr{C}_2$ we obtain

$$\int_{\mathscr{C}} f(z)\mathrm{d}z = 0 = \int_{\mathscr{C}_1} f(z)\mathrm{d}z + \int_{\mathscr{C}_2} f(z)\mathrm{d}z. \tag{9.135}$$

Thus

$$\begin{aligned}\int_{\mathscr{C}_1} f(z)\mathrm{d}z &= -\int_{\mathscr{C}_2} f(z)\mathrm{d}z \\ &= -\lim_{\epsilon\to 0}\left(\int_{x_2}^{x_1} f(x-i\epsilon)\mathrm{d}x + \int_{x_1}^{x_2} f(x+i\epsilon)\mathrm{d}x\right) \\ &= -\lim_{\epsilon\to 0}\int_{x_1}^{x_2}[f(x+i\epsilon)-f(x-i\epsilon)]\mathrm{d}x = -\int_{x_1}^{x_2}[f_\mathrm{l}-f_\mathrm{r}]\mathrm{d}x\,,\end{aligned} \tag{9.136}$$

where f_l and f_r denote the limits of the holomorphic function left and right of the cut (seen when traveling in positive direction). From the expression for the magnetic flux density, Eq. (9.119), we can write more generally

$$\int_{\mathscr{C}_1} B(z)\mathrm{d}z = -\mu_0 iI = -\int_{z_1}^{z_2}(B_\mathrm{l}-B_\mathrm{r})\,\mathrm{d}z\,, \tag{9.137}$$

and for smooth distributions of I along the cut:

$$\mu_0 i\frac{\mathrm{d}I}{\mathrm{d}z} = B_\mathrm{l}-B_\mathrm{r}\,. \tag{9.138}$$

This is the *Cauchy principle value* extended to the complex plane, well known to the magnet design community as *Beth's current-sheet theorem* [2].

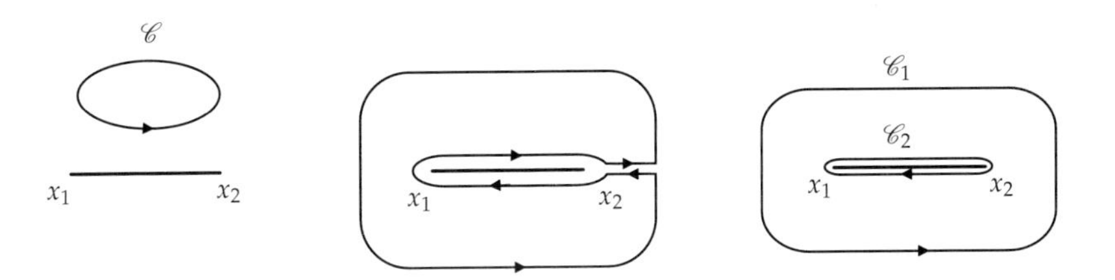

Figure 9.8 Homologous deformation of a loop $\mathscr{C}$ around a cut of finite length.

Now consider a surface-current on a cylinder centered at the origin, with parametric representation $z_\mathrm{c} = r_\mathrm{c}\,e^{i\varphi_\mathrm{c}}$, $\varphi_\mathrm{c} \in [-\pi,\pi]$. The surface-current[12] $I_\alpha(\varphi_\mathrm{c})$ is considered to be 2π-periodic and can therefore be expressed as

$$I_\alpha(\varphi_\mathrm{c}) = \sum_{n=1}^{\infty} c_n\cos(n\varphi_\mathrm{c})\,. \tag{9.139}$$

12 $[I_\alpha] = 1\,\mathrm{A\,m^{-1}}$.

Taking φ_c counter-clockwise from the x-axis we denote the magnetic flux density on the left and right sides of the current sheet by B_{in} and B_{out}, respectively. Employing the chain rule and Eq. (9.34), we obtain

$$B_{in}(z) - B_{out}(z) = \mu_0 i \frac{dI_\alpha}{dz} = \mu_0 i \frac{dI_\alpha}{d\varphi_c}\frac{d\varphi_c}{dz} = -\mu_0 i \frac{1}{iz}\sum_{n=1}^{\infty} n\, c_n \sin(n\varphi_c)$$

$$= -\mu_0 \frac{1}{z}\sum_{n=1}^{\infty} \frac{n\, c_n}{2}\left[\left(\frac{z}{r_c}\right)^n + \left(\frac{r_c}{z}\right)^n\right]. \tag{9.140}$$

Because the potential must remain finite both for $|z| \to 0$ and $|z| \to \infty$ this yields

$$B_{out}(z) = \frac{\mu_0}{2}\sum_{n=1}^{\infty} n c_n \frac{r_c^n}{z^{n+1}}, \qquad B_{in}(z) = -\frac{\mu_0}{2}\sum_{n=1}^{\infty} n c_n \frac{z^{n-1}}{r_c^n}. \tag{9.141}$$

If we now request a constant magnetic flux density within the aperture of the magnet,

$$B_{in}(z) = B_y + iB_x = -B_0. \tag{9.142}$$

All coefficients c_n for $n \neq 1$ must be zero and the solution is for $c_1 = 2B_0 r_c/\mu_0$:

$$I_\alpha(\varphi_c) = \frac{2B_0 r_c}{\mu_0}\cos\varphi_c. \tag{9.143}$$

Using Eq. (9.141), the flux density outside the current sheet can be calculated from

$$B_{out}(z) = B_0\left(\frac{r_c}{z}\right)^2. \tag{9.144}$$

Example: Consider the LHC dipole coil (without iron yoke) with a central field of B_{in} = 8.33 T and a mean coil radius of r = 43.5 mm. Integrating Eq. (9.143) from $-\pi/2$ to $\pi/2$ we calculate a total current of 1.8×10^6A. The numerical calculation yields 1.1×10^6 A. □

9.12
Electromagnetic Forces on the Dipole Coil

The force per unit length acting on a line-current I exposed to an external magnetic flux density can be written in components as

$$F_x = -I\,B_y, \qquad F_y = I\,B_x, \tag{9.145}$$

which gives rise to the definition of the complex force at the source point z_c,

$$F(z_c) := F_y(z_c) - iF_x(z_c) = -iIB(z_c). \tag{9.146}$$

Following[13] Beth [4], the magnetic force exerted on a line-current can be calculated from the total field:

$$B_{\text{tot}}(z) = B(z) + \frac{\mu_0 I}{2\pi} \frac{1}{z - z_c} . \tag{9.147}$$

It follows that

$$B_{\text{tot}}^2(z) = B^2(z) + \frac{\mu_0 I}{\pi} \frac{B(z)}{z - z_c} + \left(\frac{\mu_0 I}{2\pi}\right)^2 \frac{1}{(z - z_c)^2} \tag{9.148}$$

has the residual $(\mu_0 I/\pi)B(z_c)$. With the residual theorem it follows for the loop $\mathscr{C}$, which encloses the line-current I located at z_c,

$$\int_{\mathscr{C}} B_{\text{tot}}^2 \mathrm{d}z = 2\pi i \frac{\mu_0 I}{\pi} B(z_c) = -2\mu_0 F(z_c) . \tag{9.149}$$

The technique presented here is known as the *contour integral technique* [1]. Using Eq. (9.149) and letting the loop $\mathscr{C}$ collapse onto the circular current shell carrying the total current I yields, together with Eqs. (9.142) and (9.144), the force per unit length,

$$F = F_y + iF_x = -\int_{\mathscr{C}} \frac{B_{\text{tot}}^2}{2\mu_0} \mathrm{d}z_c . \tag{9.150}$$

Therefore, using $z_c = r_c e^{i\varphi_c}$ and $\mathrm{d}z_c = ir_c e^{i\varphi_c} \mathrm{d}\varphi_c$, we obtain

$$\begin{aligned} F &= \int_{\mathscr{C}} \frac{B_{\text{in}}^2}{2\mu_0} \mathrm{d}z_c - \int_{\mathscr{C}} \frac{B_{\text{out}}^2}{2\mu_0} \mathrm{d}z_c = \frac{B_0^2}{2\mu_0} \int_0^{\frac{\pi}{2}} \left[1 - \left(\frac{r_c}{z_c}\right)^4\right] \mathrm{d}z_c \\ &= \frac{B_0^2}{2\mu_0} \int_0^{\frac{\pi}{2}} (1 - e^{-4i\varphi_c}) ir_c e^{i\varphi_c} \mathrm{d}\varphi_c , \end{aligned} \tag{9.151}$$

resulting in the expressions for the two components

$$F_x = \frac{B_0^2}{2\mu_0} \frac{4r_c}{3} , \qquad F_y = -\frac{B_0^2}{2\mu_0} \frac{4r_c}{3} . \tag{9.152}$$

Example: For the LHC dipole coil (without iron yoke) with a central field of B_{in} = 8.33 T and a mean coil radius of r_c = 43.5 mm we have a total bursting force of $2F_x$ per unit length, which gives 3.2×10^6 N m^{-1}. The numerical calculation yields $2F_x = 2.8 \times 10^6$ N m^{-1} and $2F_y = -2.3 \times 10^6$ N m^{-1}. □

13 We use SI units and a slightly different notation.

9.13
The Field of a Polygonal Conductor

The complex potential at a field point z for a conductor of cross section Ω can be expressed as

$$w(z) = -\int_{\Omega} \frac{\mu_0 J(z_c)}{2\pi} \log\left(\frac{z - z_c}{z_{\mathrm{ref}}}\right) \mathrm{d}a\,, \tag{9.153}$$

where $J(z_c)$ is the current density in the conductor cross section. From Eq. (9.97), for $u = 0$,

$$\int_{\partial\Omega} v \mathrm{d}y_c = \int_{\Omega} \left(\frac{\partial v}{\partial x_c}\right) \mathrm{d}x_c \mathrm{d}y_c\,. \tag{9.154}$$

Substituting

$$R := \frac{z - z_c}{z_{\mathrm{ref}}}\,, \qquad v := R \log R\,, \tag{9.155}$$

into Eq. (9.154), the complex potential in Eq. (9.153) can be rewritten as

$$w(z) = -\int_{\partial\Omega} \frac{\mu_0 J(z_c)}{2\pi} R \log R \,\mathrm{d}y_c\,, \tag{9.156}$$

as $\int_{\partial\Omega} \mathrm{d}y_c = 0$. Consequently, for a uniform current density,

$$-\frac{\mathrm{d}w}{\mathrm{d}z} = B_y(z) + iB_x(z) = \frac{\mu_0 J}{2\pi} \int_{\partial\Omega} \log R \,\mathrm{d}y_c\,. \tag{9.157}$$

Now consider a polygonal conductor as shown in Figure 9.9 (left) with N vertices, $z_n = x_n + iy_n$, where the sense of vertex ordering and the current direction form a right-handed screw with the surface-normal vector. On the edge from z_{n-1} to z_n, the differential $\mathrm{d}y_c$ is proportional to $\mathrm{d}z_c$:

$$\mathrm{d}y_c = \frac{y_n - y_{n-1}}{z_n - z_{n-1}} \mathrm{d}z_c\,. \tag{9.158}$$

In Figure 9.9 (left) we find [3] the slope

$$s_n := \frac{y_n - y_{n-1}}{z_n - z_{n-1}} = \frac{a \sin\theta_n}{a e^{i\theta_n}} = e^{-i\theta_n} \sin\theta_n\,. \tag{9.159}$$

As z and z_{ref} are constant during integration, we have $\mathrm{d}z_c = -z_{\mathrm{ref}} \mathrm{d}R$, and consequently Eq. (9.157) can easily be solved on the computer, taking into account that

$$B_y(z) + iB_x(z) = \frac{\mu_0 J}{2\pi} \sum_{n=1}^{N} s_n \left(Q_n - Q_{n-1}\right), \tag{9.160}$$

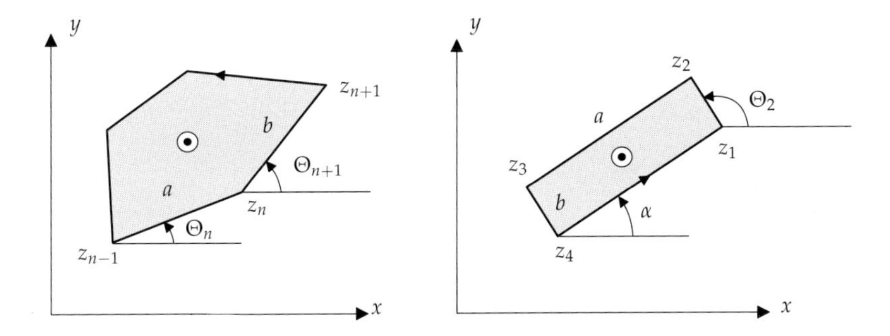

Figure 9.9 Left: Polygonal conductor with N vertices. Right: Rectangular conductor and definition of the inclination angle.

where

$$Q_n := (z - z_n) \log \frac{z - z_n}{z_{\text{ref}}} . \tag{9.161}$$

For a polygon with N vertices we derive

$$\begin{aligned} s_{n+1} - s_n &= e^{-i\theta_{n+1}} \sin \theta_{n+1} - e^{-i\theta_n} \sin \theta_n \\ &= e^{-i\theta_{n+1}} \frac{e^{i\theta_{n+1}} - e^{-i\theta_{n+1}}}{2i} - e^{-i\theta_n} \frac{e^{i\theta_n} - e^{-i\theta_n}}{2i} \\ &= \frac{-1}{2i} e^{-i(\theta_{n+1}+\theta_n)} \left(e^{i(\theta_{n+1}-\theta_n)} - e^{-i(\theta_{n+1}-\theta_n)} \right) \\ &= e^{-i(\theta_{n+1}+\theta_n)} \sin(\theta_{n+1} - \theta_n) . \end{aligned} \tag{9.162}$$

An easy equation is obtained for the flux density generated by a rectangular conductor shown in Figure 9.9 (right). In this case $\theta_n = \alpha + (n-1)\pi/2$ for $n = 1, \ldots, 4$, where α is the inclination angle from the x-axis.

$$s_{n+1} - s_n = e^{-i((2n-1)\frac{\pi}{2}+2\alpha)} \sin \frac{\pi}{2} = (-1)^{n-1} i e^{-i2\alpha} . \tag{9.163}$$

It follows that

$$B_y(z) + iB_x(z) = \frac{\mu_0 J}{2\pi} i e^{-i2\alpha} (Q_1 - Q_2 + Q_3 - Q_4) . \tag{9.164}$$

In a similar way, the complex potential can be calculated for a rectangular conductor:

$$A_z(z) + i\mu_0\phi(z) = -\frac{\mu_0 J}{2\pi} i e^{-i2\alpha} (T_1 - T_2 + T_3 - T_4) , \tag{9.165}$$

where

$$T_n := \frac{1}{2}(z - z_n)^2 \log \frac{z - z_n}{z_{\text{ref}}} . \tag{9.166}$$

9.14
Magnetic Flux Density Inside Elliptical Conductors

For the calculation of the magnetic flux density inside domains of uniform current density J, Beth [2] uses the complex potential

$$w(z) = H - \frac{1}{2} J \bar{z} \tag{9.167}$$

for $H = H_y + iH_x$. Inside the current-carrying conductor H is not holomorphic though the complex potential $w(z)$ is.

Proof. We can write Eq. (9.167) as

$$\begin{aligned} w(z) &= H - \frac{1}{2} J \bar{z} = H_y + iH_x - \frac{1}{2} J(x - iy) \\ &= \underbrace{H_y - \frac{1}{2} Jx}_{u(x,y)} + i \underbrace{(H_x + \frac{1}{2} Jy)}_{v(x,y)} . \end{aligned} \tag{9.168}$$

Calculating the partial derivatives of the real-valued functions $u(x, y)$ and $v(x, y)$ yields

$$\frac{\partial u(x,y)}{\partial x} = \frac{\partial}{\partial x} \left(H_y(x,y) - \frac{1}{2} Jx \right) = \frac{\partial H_y}{\partial x} - \frac{1}{2} J , \tag{9.169}$$

$$\frac{\partial v(x,y)}{\partial y} = \frac{\partial}{\partial y} \left(H_x(x,y) + \frac{1}{2} Jy \right) = \frac{\partial H_x}{\partial y} + \frac{1}{2} J , \tag{9.170}$$

$$\frac{\partial u(x,y)}{\partial y} = \frac{\partial}{\partial y} \left(H_y(x,y) - \frac{1}{2} Jx \right) = \frac{\partial H_y}{\partial y} , \tag{9.171}$$

$$\frac{\partial v(x,y)}{\partial x} = \frac{\partial}{\partial x} \left(H_x(x,y) + \frac{1}{2} Jy \right) = \frac{\partial H_x}{\partial x} . \tag{9.172}$$

Thus we derive from the Cauchy–Riemann equations:

$$\frac{\partial H_y}{\partial x} - \frac{\partial H_x}{\partial y} = J , \qquad \frac{\partial H_y}{\partial y} + \frac{\partial H_x}{\partial x} = 0 , \tag{9.173}$$

which are Maxwell's equations $\operatorname{curl} \mathbf{H} = \mathbf{J}$ and $\operatorname{div} \mathbf{H} = 0$ for the 2D case, valid in regions Ω with constant permeability and constant current density. Therefore, the Cauchy–Riemann equations are fulfilled for $w(z)$. □

Let a and b be the semiaxes of an infinitely long, elliptical conductor carrying a uniform current density J. Let the points on the boundary $\partial\Omega$ of this conductor be denoted $z_c = x + iy$ and the field point z. The domain inside the conductor is denoted Ω and the domain outside $\mathbb{C} \setminus \Omega$. Using the complex potentials

$$w_{\text{in}}(z_c) = H_{\text{in}}(z_c) - \frac{1}{2} J \bar{z}_c, \qquad w_{\text{out}}(z_c) = H_{\text{out}}(z_c), \tag{9.174}$$

yields

$$w_{\text{in}}(z) = \frac{1}{2\pi i} \int_{\partial\Omega} \frac{w_{\text{in}}(z_c)}{z_c - z} \mathrm{d}z_c = \begin{cases} H_{\text{in}}(z) - \frac{1}{2} J \bar{z} & \text{if } z \in \Omega \\ 0 & \text{if } z \in \mathbb{C} \setminus \Omega \end{cases}, \tag{9.175}$$

$$w_{\text{out}}(z) = -\frac{1}{2\pi i} \int_{\partial\Omega} \frac{w_{\text{out}}(z_c)}{z_c - z} \mathrm{d}z_c = \begin{cases} H_{\text{out}}(z) & \text{if } z \in \mathbb{C} \setminus \Omega \\ 0 & \text{if } z \in \Omega \end{cases}. \tag{9.176}$$

Both $w_{\text{out}}(z)$ and $w_{\text{in}}(z)$ must generate the same magnetic field $H(z_c)$ on the boundary, that is, $H_{\text{in}}(z_c) = H_{\text{out}}(z_c)$. It follows that

$$w_{\text{in}}(z_c) - w_{\text{out}}(z_c) = -\frac{1}{2} J \bar{z}_c \quad \text{on } \partial\Omega. \tag{9.177}$$

For the sum of (9.175) and (9.176), we obtain

$$\begin{aligned} w(z) &= w_{\text{in}}(z) + w_{\text{out}}(z) \\ &= \frac{1}{2\pi i} \int_{\partial\Omega} \frac{w_{\text{in}}(z_c)}{z_c - z} \mathrm{d}z_c - \frac{1}{2\pi i} \int_{\partial\Omega} \frac{w_{\text{out}}(z_c)}{z_c - z} \mathrm{d}z_c \end{aligned} \tag{9.178}$$

and therefore

$$w(z) = \frac{Ji}{4\pi} \int_{\partial\Omega} \frac{\bar{z}_c}{z_c - z} \mathrm{d}z_c = \begin{cases} H_{\text{in}}(z) - \frac{1}{2} J \bar{z} & \text{if } z \in \Omega \\ H_{\text{out}}(z), & \text{if } z \in \mathbb{C} \setminus \Omega \end{cases}. \tag{9.179}$$

The parametric form of the elliptic boundary for $0 \le \theta < 2\pi$ is given by

$$\begin{aligned} z_c &= a \cos\theta + ib \sin\theta = a\, \frac{e^{i\theta} + e^{-i\theta}}{2} + ib\, \frac{e^{i\theta} - e^{-i\theta}}{2i} \\ &= \frac{a+b}{2} e^{i\theta} + \frac{a-b}{2} e^{-i\theta} = rt + \delta t^{-1}, \end{aligned} \tag{9.180}$$

where we used the definitions

$$t := e^{i\theta}, \qquad r := \frac{a+b}{2}, \qquad \delta := \frac{a-b}{2}. \tag{9.181}$$

The semiaxes of the ellipse are $a = r + \delta$ and $b = r - \delta$. Furthermore,

$$\bar{z}_c = rt^{-1} + \delta t, \qquad \frac{\mathrm{d}z_c}{\mathrm{d}t} = r - \delta t^{-2}, \tag{9.182}$$

and therefore the generalized Cauchy integral, Eq. (9.179), can be expressed as

$$w(z) = \frac{Ji}{4\pi}\int_{\partial\Omega}\frac{\overline{z}_c}{z_c - z}\mathrm{d}z_c = \frac{Ji}{4\pi}\int_{\mathscr{C}}\frac{(rt^{-1}+\delta t)(r-\delta t^{-2})}{rt+\delta t^{-1}-z}\mathrm{d}t$$
$$= \frac{Ji}{4\pi}\int_{\mathscr{C}}\underbrace{\frac{(r+\delta t^2)(rt^2-\delta)}{t^2(rt^2-zt+\delta)}}_{S(t)}\mathrm{d}t\,, \tag{9.183}$$

where the loop $\mathscr{C}$ is the unit circle $t = e^{i\theta}, 0 \le \theta < 2\pi$. $S(t)$ is meromorphic inside that circle with poles at $t = 0$ and $t_{2,3} = \frac{1}{2r}\left(z \pm \sqrt{z^2 - 4r\delta}\right)$. $S(t)$ can be separated into partial fractions

$$S(t) = \delta + \frac{D_1}{t} + \frac{E_1}{t^2} + \frac{D_2}{t-t_2} + \frac{D_3}{t-t_3}\,, \tag{9.184}$$

where the constants are

$$D_1 = -\frac{rz}{\delta}\,, \tag{9.185}$$

$$E_1 = -r\,, \tag{9.186}$$

$$D_2 = \frac{1}{2r\delta}\left((r^2+\delta^2)z - (r^2-\delta^2)\sqrt{z^2-4r\delta}\right)\,, \tag{9.187}$$

$$D_3 = \frac{1}{2r\delta}\left((r^2+\delta^2)z + (r^2-\delta^2)\sqrt{z^2-4r\delta}\right)\,. \tag{9.188}$$

According to the residual theorem, the Cauchy integral is equal to the sum of all enclosed residuals according to

$$\int_{\mathscr{C}} S(t)\mathrm{d}t = 2\pi i \sum_n \mathrm{Res}(S;t_n). \tag{9.189}$$

The residual $\mathrm{Res}(S;t_n)$ are the coefficients of each pole of first order enclosed by the line integral, $\mathrm{Res}(S;t_n) = D_n$. The Cauchy integral can now be calculated for $z \in \Omega$ from

$$H_{\mathrm{in}}(z) = \frac{Ji}{4\pi}\int_{\mathscr{C}} S(t)\,\mathrm{d}t + \frac{1}{2}J\overline{z} = -\frac{J}{2}(D_1 + D_3 + D_4) + \frac{1}{2}J\overline{z} \tag{9.190}$$

and therefore,

$$H_{\mathrm{in}}(z) = \frac{J}{2}\left(\frac{rz}{\delta} - \frac{r^2+\delta^2}{2r\delta}z + \overline{z}\right) = \frac{J}{2}\left(\frac{a+b}{a-b}z - \frac{2(a^2+b^2)}{a^2-b^2}z + \overline{z}\right)$$
$$= -\frac{J}{2}\left(\frac{a-b}{2(a+b)}z + \overline{z}\right) = \frac{J}{a+b}(bx - iay)\,. \tag{9.191}$$

The components of the magnetic flux density within an elliptical conductor of constant current density is consequently given by

$$H_x(z) = -J\frac{a}{a+b}y\,, \qquad H_y(z) = J\frac{b}{a+b}x\,. \tag{9.192}$$

For the round conductor $a = b$ this yields the well-known relations

$$H_x(z) = -\frac{1}{2}Jy\,, \qquad H_y(z) = \frac{1}{2}Jx\,. \tag{9.193}$$

References

1 Bak, J., Newman, D.J.: Complex Analysis, Springer, Berlin, 1997

2 Beth, R. A.: An integral formula for two-dimensional fields, Journal of Applied Physics, 1967

3 Beth, R. A.: Complex representation and computation of two-dimensional fields, Journal of Applied Physics, 1966

4 Beth, R. A.: Currents and coil forces as contour integrals in two-dimensional magnetic fields, Journal of Applied Physics, 1969

5 Gürlebeck, K., Sprössig W.: Quaternionic and Clifford Calculus for Physicists and Engineers, John Wiley & Sons, New York, 1998

6 Jain, A. K.: Harmonic Coils, CERN Accelerator School on Measurement and Alignment of Accelerator and Detector Magnets, CERN Yellow Report 98-05, 1998

7 Needham, T.: Visual Complex Analysis, Claredon Press, Oxford, 1997

8 Nethe, A., Stahlmann, H.-D. (editors): Einführung in die Feldtheorie, Verlag Dr. Köster, Berlin, 2003

9 Penrose, R.: The Road to Reality: A Complete Guide to the Laws of the Universe, Jonathan Cape, London, UK, 2004

10 Perin, R.: Calculation of Magnetic Field in a Cylindrical Geometry Produced by Sector or Layer Windings, ISR-MA Internal Note, CERN, 1973

11 The LHC Study Group, The Large Hadron Collider, Conceptual Design, CERN/AC/95-05, 1995

10
Field Diffusion

In order to limit the induced voltages in the circuits of the LHC main dipoles, which contain 154 series-connected magnets, the rise time of the excitation current is set at approximately 1200 s, which corresponds to a ramp rate of 6 $\mathrm{mT\,s^{-1}}$; see Figure 1.13. Another reason for the slow ramp rate is to limit time-transient field errors in the magnets. For such low ramp rates, the induced eddy currents in the cold bore and the beam screen are insignificant. Nevertheless, we will study their analytical calculation in conductive domains in the context of a quenching magnet (with decay rates of up to 40 $\mathrm{T\,s^{-1}}$), and fast-cycling accelerator projects. By means of the analytical calculations, we will also derive scaling laws for very thin layers of conductive material that are difficult to model with finite elements.

10.1
Time Constants and Penetration Depths

Consider a conductive permeable domain as shown in Figure 10.1 (left). The domain will later be assigned the material properties of the coating of the LHC beam screen, consisting of pure copper[1] at cryogenic temperatures, exposed to an applied field of 8.33 T. A varying magnetic field induces an electric field and leads to an eddy-current distribution according to the local form of Ohm's law, $\mathbf{J} = \varkappa \mathbf{E}$. According to Lenz's rule, the field due to the eddy currents opposes the applied magnetic field. Let the boundary surface of the domain be exposed to a spatially constant but time-varying magnetic flux density in the y-direction, as shown in Figure 10.1 (left):

$$B_y(0,t) = \hat{B}_1 \cos \omega t = \hat{B}_1 \operatorname{Re}\left\{ e^{i\omega t} \right\} \tag{10.1}$$

We will derive a steady-state solution of the one-dimensional version of the diffusion equation (4.195):

$$\frac{\partial^2 B_y(x,t)}{\partial x^2} = \mu \varkappa \frac{\partial B_y(x,t)}{\partial t} \tag{10.2}$$

1 The material properties at cryogenic temperatures are discussed in Appendix A.

Field Computation for Accelerator Magnets. Stephan Russenschuck

ISBN: 978-3-527-40769-9

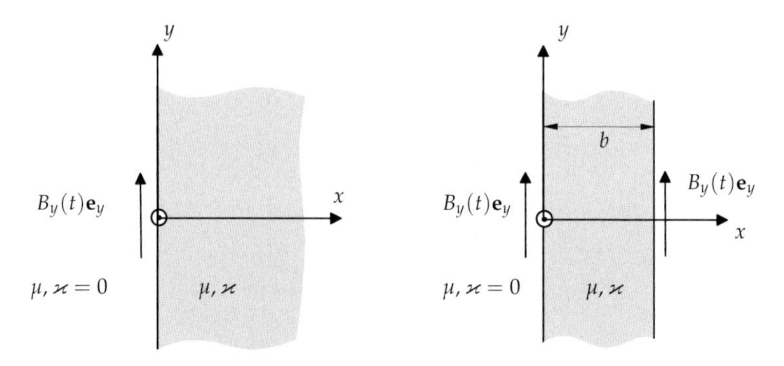

Figure 10.1 Left: Conductive half-space in the presence of a time-transient parallel magnetic field. Right: Conductive slab in a time variant, homogeneous, parallel field.

for $x \geq 0$, with the boundary condition (10.1) at $x = 0$ and the condition that $B_y(\infty, t)$ remains finite. The interface conditions, $\mathbf{n} \times [\![\mathbf{H}]\!]_{12} = \mathbf{0}$ and $\mathbf{n} \cdot [\![\mathbf{B}]\!]_{12} = 0$, and the linearity of the diffusion equation imply that the field in the conductive domain has only a y-component. A trial solution of the form

$$B_y(x, t) = \hat{B}_1 \, e^{i(\omega t - kx)}, \tag{10.3}$$

inserted into Eq. (10.2) yields $k^2 = -i\omega\varkappa\mu$, and therefore,

$$k = \frac{1 - i}{\delta} \quad \text{with} \quad \delta = \sqrt{\frac{2}{\omega\varkappa\mu}} = \frac{1}{\sqrt{\pi f \varkappa\mu}}. \tag{10.4}$$

The other formal solution $k = -(1 - i)/\delta$ is disregarded as it yields an infinite field for $x \to \infty$. The parameter δ has the dimension of length, which depends on the frequency f and the material properties $\varkappa$ and μ. The parameter δ is called the *skin depth* or *penetration depth*. At $x = \delta$, the amplitude has fallen to $1/e \approx 0.368$ times the boundary value $\hat{B}_1$. Considering the boundary condition (10.1) and taking the real part of the trial solution results in

$$B_y(x, t) = \hat{B}_1 \, \mathrm{Re}\left\{ e^{-\frac{x}{\delta} + i\left(\omega t - \frac{x}{\delta}\right)} \right\} = \hat{B}_1 \, e^{-\frac{x}{\delta}} \cos\left(\omega t - \frac{x}{\delta}\right). \tag{10.5}$$

Using the relation $\cos\varphi - \sin\varphi = \sqrt{2}\cos\left(\varphi + \frac{\pi}{4}\right)$, we can now formulate the induced current density in the conductive domain as

$$\begin{aligned} J_z(x, t) &= \frac{1}{\mu}\frac{\partial B_y(x, t)}{\partial x} = \frac{1}{\mu}\hat{B}_1 \, \mathrm{Re}\left\{ -\frac{1 + i}{\delta} e^{-\frac{x}{\delta} + i\left(\omega t - \frac{x}{\delta}\right)} \right\} \\ &= -\frac{1}{\mu}\hat{B}_1 \frac{\sqrt{2}}{\delta} e^{-\frac{x}{\delta}} \cos\left(\omega t - \frac{x}{\delta} + \frac{\pi}{4}\right), \end{aligned} \tag{10.6}$$

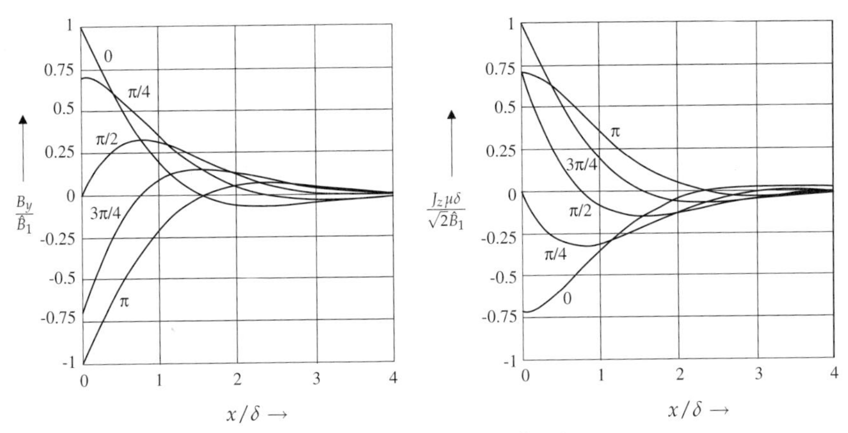

Figure 10.2 Magnetic flux density (left) and eddy-current density (right) for a half-period of a sinusoidal field, applied to a semiinfinite, conductive domain.

Figure 10.2 shows the results for the flux and the eddy-current densities.

10.2 The Laplace Transform

In this section, we will present the *Laplace transformation* as a means to solve initial value problems in field diffusion by a transformation into the complex half-plane. We will limit ourselves to problems described by ordinary linear differential equations in (x, t) with constant coefficients and to initial conditions at $t = 0$. In order to find the response to a given input, it is often easier to use the Laplace transformation, which yields algebraic equations in $s := \sigma + i\omega$, $\sigma > \sigma_0$, and then to perform an inverse Laplace transformation to the rational functions in s. This procedure is known as *Heaviside calculus*.

For didactic purposes, we begin with the definition of the Fourier transformation: A function $f : \mathbb{R} \rightarrow \mathbb{R}$ is called *Fourier-transformable* if the integral

$$\begin{aligned} \mathcal{F}\{f(t)\} &:= \int_{-\infty}^{+\infty} f(t) e^{-i\omega t} \mathrm{d}t \\ &= \int_{-\infty}^{+\infty} f(t) \cos \omega t \mathrm{d}t - i \int_{-\infty}^{+\infty} f(t) \sin \omega t \mathrm{d}t \,, \end{aligned} \tag{10.7}$$

exists as in improper Riemann integral for all $\omega \in \mathbb{R}$. The function $\tilde{f}(s) := \mathcal{F}\{f(t)\}$ is called the Fourier transform of $f(t)$. This definition resembles the complex version of the Fourier series expansion, except that the domain of f is not bounded to the length of one period and we deal with $\omega \in \mathbb{R}$ instead of $n \in \mathbb{Z}$. However, the Fourier transform of the *unit step function* or *Heaviside function* defined by[2]

$$u(t) := \begin{cases} 0 & \text{for} \quad t < 0, \\ 1 & \text{for} \quad t \geq 0, \end{cases} \tag{10.8}$$

cannot exist, for the equation

$$\begin{aligned} \mathcal{F}\{u(t)\} &:= \int_0^{+\infty} 1 \cdot e^{-i\omega t} \mathrm{d}t = \lim_{R \to \infty} \int_0^R e^{-i\omega t} \mathrm{d}t \\ &= \frac{1}{-i\omega} \lim_{R \to \infty} \left(e^{-i\omega t} \Big|_0^R \right) \end{aligned} \tag{10.9}$$

diverges because the limit $\lim_{R\to\infty} e^{-i\omega R}$ does not exist. In solving field diffusion problems, we are confronted with excitation functions of the form $e^{\alpha}t$, $\sin \omega t$, $\cos \omega t$, as well as the unit step function, which do not have Fourier transforms. Considering that the condition $f(t) = 0$ for $t < 0$ is often imposed due to the closing or opening of a switch at $t = 0$, convergence can be guaranteed by a modified excitation function

$$f^*(t) := \begin{cases} 0 & \text{for} \quad t < 0, \\ e^{-\sigma t} f(t) & \text{for} \quad t \geq 0, \end{cases} \tag{10.10}$$

for $\sigma > 0$. The Fourier transform of $f^*(t)$ is then given by

$$\mathcal{F}\{f^*(t)\} := \int_{-\infty}^{+\infty} f^*(t) e^{-i\omega t} \mathrm{d}t = \int_0^{+\infty} f(t) e^{-(\sigma + i\omega)t} \, \mathrm{d}t \,. \tag{10.11}$$

For $s := \sigma + i\omega$ this yields the *unilateral Laplace transform*, denoted $\mathcal{L}\{f(t)\}$. A *causal*, piecewise-smooth function $f(t)$ with a real argument $t \in \mathbb{R}^+$ in the *time domain* is transformed into a function $\tilde{f}(s) := \mathcal{L}\{f(t)\}$ in the *s-domain*. As this transformation is bijective for most of the practical relevant cases, transformation tables can be used to find simple algebraic relationships in the s-domain.

The unilateral Laplace transform of a causal function $f : \mathbb{R}^+ \to \mathbb{R}$ is given by

$$\tilde{f}(s) := \mathcal{L}\{f(t)\} = \int_0^{\infty} f(t) e^{-st} \mathrm{d}t \tag{10.12}$$

for $s \in \mathbb{C}$. The existence of the improper integral (10.12) is guaranteed for at least piecewise-continuous functions,

$$|f(t)| \leq M e^{\sigma_0 t}, \qquad t \in \mathbb{R}^+, \sigma_0 \in \mathbb{R}, \tag{10.13}$$

2 The unit step function is a special case of a causal function $f(t) = 0$ for $t < 0$.

for which the Laplace transform $\tilde{f}(s)$ exists for all $s \in \mathbb{C}$ with $\mathrm{Re}\{s\} \geq \sigma_0$ and is absolutely convergent. The function in t is then said to be of exponential order σ_0.

The most important property of the Laplace transformation for solving initial value problems is the correspondence of the time derivative in the time domain to a multiplication by s in the s-domain. By way of the integration-by-parts rule, we obtain

$$\begin{aligned} \mathcal{L}\{f(t)\} &= \int_0^\infty f(t) e^{-st} \mathrm{d}t = \left[\frac{-1}{s} f(t) e^{-st} \right]_0^\infty - \int_0^\infty \frac{-1}{s} e^{-st} \frac{\mathrm{d}f}{\mathrm{d}t} \mathrm{d}t \\ &= \frac{1}{s} f(0) + \frac{1}{s} \mathcal{L}\left\{ \frac{\mathrm{d}f}{\mathrm{d}t} \right\}, \end{aligned} \tag{10.14}$$

and therefore

$$\mathcal{L}\left\{ \frac{\mathrm{d}f}{\mathrm{d}t} \right\} = s\,\mathcal{L}\{f(t)\} - f(0)\,. \tag{10.15}$$

Differentiation in time results in a factor s, while a time integration results in a factor $1/s$ in the s-domain. For the convolution

$$g(t) = \int_0^t f_1(\tau) f_2(t-\tau) \mathrm{d}\tau = \int_0^t f_2(\tau) f_1(t-\tau) \mathrm{d}\tau, \tag{10.16}$$

we obtain

$$\mathcal{L}\{g(t)\} = \tilde{g}(s) = \tilde{f}_1(s)\,\tilde{f}_2(s)\,. \tag{10.17}$$

The linearity of the Laplace transformation follows directly from the linearity of the integration, and consequently for $a, b \in \mathbb{R}$ it yields

$$\mathcal{L}\{af(t) + bg(t)\} = a\mathcal{L}\{f(t)\} + b\mathcal{L}\{g(t)\}, \tag{10.18}$$

in the domain where $\mathcal{L}\{f(t)\}$ and $\mathcal{L}\{g(t)\}$ both exist. Additional properties of the Laplace transformation for functions $f(t)$, $g(t)$ and their Laplace transforms $\tilde{f}(t)$, $\tilde{g}(t)$ are given in Table 10.1.

From Eq. (10.12) follows the Laplace transform of the unit step function:

$$\mathcal{L}\{u(t)\} = \int_0^\infty e^{-st} \mathrm{d}t = \lim_{R\to\infty} \left(\left. \frac{e^{-st}}{-s} \right|_0^R \right) = \lim_{R\to\infty} \left(\frac{e^{-sR}}{-s} + \frac{1}{s} \right) = \frac{1}{s}, \tag{10.19}$$

for $\mathrm{Re}\{s\} > 0$. For $f(t) = e^{at}$ with $a \in \mathbb{R}$ it holds that for $\mathrm{Re}\{s\} > a$,

$$\mathcal{L}\{e^{at}\} = \int_0^\infty e^{-st} e^{a} t \mathrm{d}t = \int_0^\infty e^{-(s-a)t} \mathrm{d}t = \frac{1}{s-a}\,. \tag{10.20}$$

Table 10.1 Some properties of the Laplace transformation for functions $f(t), g(t)$ and their Laplace transforms $\tilde{f}(s) = \mathcal{L}\{f(s)\}, \tilde{g}(s) = \mathcal{L}\{g(s)\}$.

	Time domain	s-domain
Linearity	$af(t) + bg(t)$	$a\tilde{f}(s) + b\tilde{g}(s)$
Differentiation in s	$tf(t)$	$-\frac{\mathrm{d}}{\mathrm{d}s}\tilde{f}(s)$
	$t^n f(t)$	$(-1)^n \frac{\mathrm{d}^n}{\mathrm{d}s^n}\tilde{f}(s)$
Differentiation in t	$\frac{\mathrm{d}}{\mathrm{d}t}f(t)$	$s\tilde{f}(s) - f(0^-)$
	$\frac{\mathrm{d}^2}{\mathrm{d}t^2}f(t)$	$s^2\tilde{f}(s) - sf(0^-) - \frac{\mathrm{d}}{\mathrm{d}t}f(0^-)$
Integration in s	$\frac{f(t)}{t}$	$\int_s^\infty \tilde{f}(\sigma)\mathrm{d}\sigma$
Integration in t	$\int_0^t f(\tau)\mathrm{d}\tau$	$\frac{1}{s}\tilde{f}(s)$
Scaling	$f(at)$	$\frac{1}{\lvert a\rvert}\tilde{f}\left(\frac{s}{a}\right)$
Shift in s-domain	$e^{at}f(t)$	$\tilde{f}(s-a)$

The Laplace transform of the ramp $f(t) = t, t \geq 0$ yields

$$\mathcal{L}\{t\} = \int_0^\infty te^{-st}\mathrm{d}t = \frac{-te^{-st}}{s}\bigg|_0^\infty + \frac{1}{s}\int_0^\infty e^{-st}\mathrm{d}t = \frac{1}{s}\mathcal{L}\{u(t)\} = \frac{1}{s^2} \tag{10.21}$$

for $\mathrm{Re}\{s\} > 0$. Pairs of functions $f(t)$ and their Laplace transforms $\tilde{f}(s)$ are given in Table 10.2. As a direct consequence of the linearity, expressed by Eq. (10.18), it can be easily shown that

$$\begin{aligned}\mathcal{L}\{\cosh \omega t\} &= \frac{1}{2}\left(\mathcal{L}\{e^{\omega t}\} + \mathcal{L}\{e^{-\omega t}\}\right)\\ &= \frac{1}{2}\left(\frac{1}{s-\omega} + \frac{1}{s+\omega}\right) = \frac{s}{s^2-\omega^2}\end{aligned} \tag{10.22}$$

and

$$\mathcal{L}\{\sum_{n=0}^{n} c_n t^n\} = \sum_{n=0}^{n} c_n \mathcal{L}\{t^n\} = \sum_{n=0}^{n} \frac{c_n n!}{s^{n+1}}\,. \tag{10.23}$$

Having restricted ourselves to causal functions of exponential order σ_0, which are piecewise-continuous on $[0,\infty]$, guarantees that there exists a unique inverse Laplace transform $\mathcal{L}^{-1}\{\tilde{f}(s)\} = f(t)$, a result known as *Lerch's theorem*[3] [3].

3 Mathias Lerch (1860–1922).

Table 10.2 Pairs of functions $f(t)$ and $\tilde{f}(s)$.

Time domain	s-domain	Convergence
$\delta(t-\tau)$	$e^{-\tau s}$	
$\delta(t)$	1	
$u(t)$	$\frac{1}{s}$	$\mathrm{Re}\{s\} > 0$
$u(t-\tau)$	$\frac{e^{-\tau s}}{s}$	$\mathrm{Re}\{s\} > 0$
$t\,u(t)$	$\frac{1}{s^2}$	$\mathrm{Re}\{s\} > 0$
$e^{-at}\,u(t)$	$\frac{1}{s+a}$	$\mathrm{Re}\{s\} > -a$
$\sin(\omega t)\,u(t)$	$\frac{\omega}{s^2+\omega^2}$	$\mathrm{Re}\{s\} > 0$
$\cos(\omega t)\,u(t)$	$\frac{s}{s^2+\omega^2}$	$\mathrm{Re}\{s\} > 0$

For pairs of functions not listed in Table 10.2 the inverse Laplace transform can be obtained by means of the *Fourier–Mellin integral* [4]

$$\mathcal{L}^{-1}\{\tilde{f}(s)\} = \frac{1}{2\pi i}\int_{\gamma-i\infty}^{\gamma+i\infty} \tilde{f}(s)e^{st}\mathrm{d}s\,, \tag{10.24}$$

where γ is defined as a real number such that the contour path of integration is located in the domain of convergence of $\tilde{f}(s)$. The vertical line at γ is known as the *Bromwich line*.[5] However, the inversion formula (10.24) is almost never used in practice, because it requires a working knowledge of the integration of complex functions on contours in $\mathbb{C}$. Since $\tilde{f}(s)$ is analytic in $\mathrm{Re}\{s\} > \sigma_0$, all the poles of $\tilde{f}(s)$ must lie to the left-hand side of the *Bromwich line*. For a sufficiently large radius R, the countour Γ_R surrounds all the poles of $\tilde{f}(s)$ at z_n. Thus from the residual theorem we can derive

$$\mathcal{L}^{-1}\{\tilde{f}(s)\} = \frac{1}{2\pi i}\int_{\Gamma_R} \tilde{f}(s)e^{st}\mathrm{d}s = \sum_n \mathrm{Res}(F; z_n)\,. \tag{10.25}$$

Example: Consider the lumped-circuit analog of a diffusion problem, that is, the current decay in a circuit containing a series-connected resistor R and inductance L. Applying Kirchhoff's voltage law,[6] the circuit equation for the current yields

$$RI(t) + L\frac{\mathrm{d}I(t)}{\mathrm{d}t} = 0\,, \tag{10.26}$$

4 Hjalmar Mellin (1854–1933).
5 Thomas John l'Anson Bromwich (1875–1929).
6 See Section 17.1.4.

and thus in the s-domain

$$R\tilde{I}(s) + L(s\tilde{I}(s) - I(0)) = 0\,, \tag{10.27}$$

where $\tilde{I}(s) := \mathcal{L}\{I(t)\}$ and $I(0)$ is the initial current impressed by a short-circuited current source or by induction. Solving for $I(s)$ yields

$$\tilde{I}(s) = I(0)\frac{1}{s + \frac{R}{L}} \tag{10.28}$$

from which it follows that

$$I(t) = \mathcal{L}^{-1}\{\tilde{I}(s)\} = I(0)\,e^{-\frac{R}{L}t} = I(0)\,e^{-\frac{t}{\tau}}\,, \tag{10.29}$$

where $\tau = L/R$ is the time constant. Be aware that for the power dissipation in transient cases, $P(s) \neq R(s)I^2(s)$, even for constant $R(s)$ because of the nonlinearity of Joule's law: $RI^2(s) = R[\int_0^\infty e^{-st}I(t)\mathrm{d}t]^2 \neq \int_0^\infty e^{-st}RI^2(t)\mathrm{d}t$.
□

10.3 Conductive Slab in a Time-Transient Applied Field

By means of the Laplace transformation we will now study the eddy-current distribution in a conductive slab exposed to a time-varying field. Consider the conductive slab shown in Figure 10.1 (right). With the normalized time defined[7] by

$$\tau := \frac{t}{T_\mathrm{d}} = \frac{t}{\mu\varkappa b^2} \tag{10.30}$$

and the normalized coordinate

$$\xi := \frac{x}{b}\,, \qquad 0 \leq \xi \leq 1\,, \tag{10.31}$$

where b is the slab thickness, the diffusion equation becomes

$$\frac{\partial^2 B_y(\xi,\tau)}{\partial \xi^2} = \frac{\partial B_y(\xi,\tau)}{\partial \tau}\,. \tag{10.32}$$

The initial and boundary conditions are given by

$$B_y(\xi,0) = 0, \qquad B_y(0,\tau) = B_y(1,\tau) = f(\tau)\,. \tag{10.33}$$

7 For a conductor of typical size b and diffusion time T_d, an approximation of the diffusion equation for the magnetic flux density reads $B/b^2 \approx \mu\varkappa\frac{B}{T_\mathrm{d}}$ and therefore $T_\mathrm{d} \approx \mu\varkappa b^2$.

This boundary value problem can be solved by using the Laplace transform of Eq. (10.32):

$$\frac{\partial^2 \tilde{B}_y(\xi, s)}{\partial \xi^2} = s\tilde{B}_y(\xi, s), \tag{10.34}$$

where $\tilde{B}(\xi, s) := \mathcal{L}\{B(\xi, \tau)\}, \tau > 0$. The general solution of Eq. (10.34) in the s-domain is given by

$$\tilde{B}_y(\xi, s) = C_1 e^{-\sqrt{s}\xi} + C_2 e^{\sqrt{s}\xi}. \tag{10.35}$$

The transforms of the boundary conditions $\tilde{B}_y(0, s) = \tilde{B}_y(1, s) = \tilde{f}(s)$ define the parameters C_1 and C_2:

$$\tilde{B}_y(0, s) = C_1 + C_2 = \tilde{f}(s)\,, \tag{10.36}$$

$$\tilde{B}_y(1, s) = C_1 e^{-\sqrt{s}} + C_2 e^{\sqrt{s}} = \tilde{f}(s)\,. \tag{10.37}$$

Solving for C_1 and C_2 yields

$$C_1 = \tilde{f}(s)\,\frac{1 - e^{\sqrt{s}}}{e^{-\sqrt{s}} - e^{\sqrt{s}}}, \qquad C_2 = \tilde{f}(s)\,\frac{e^{-\sqrt{s}} - 1}{e^{-\sqrt{s}} - e^{\sqrt{s}}}. \tag{10.38}$$

Eq. (10.35) can be rewritten as

$$\tilde{B}_y(\xi, s) = \tilde{f}(s)\,\frac{e^{-\sqrt{s}\xi} - e^{\sqrt{s}(1-\xi)} - e^{\sqrt{s}\xi} + e^{-\sqrt{s}(1-\xi)}}{e^{-\sqrt{s}} - e^{\sqrt{s}}}. \tag{10.39}$$

This can be further simplified, by substituting $\frac{1}{2}\left(e^x - e^{-x}\right) = \sinh x$, to

$$\tilde{B}_y(\xi, s) = \tilde{f}(s)\frac{\sinh\left(\sqrt{s}(1-\xi)\right) + \sinh\left(\sqrt{s}\xi\right)}{\sinh\sqrt{s}}. \tag{10.40}$$

The Laplace transform of an excitation function, $\tilde{f}(s) := \mathcal{L}\{f(\tau)\}$, can now be inserted into Eq. (10.40) and the result be inverse transformed into the time domain.

10.3.1 The Step-Excitation Function

Let us apply to the slab surface a magnetic field of the excitation function

$$f(\tau) = B_1\, u(\tau), \tag{10.41}$$

where $u(\tau)$ is the unit step function. Its Laplace transform is given by

$$\tilde{f}(s) = B_1\,\frac{1}{s}\,. \tag{10.42}$$

The inverse Laplace transformation can be performed by summing up all residuals of Eq. (10.40) for the Laplace transform of the excitation function, Eq. (10.42), multiplied by $e^{s\tau}$:

$$B_y(\xi,\tau) = \sum \text{Res}\,\left(\tilde{B}_y(\xi,s)\,e^{s\tau};s\right). \tag{10.43}$$

Equation (10.40) has poles at $s_0 = 0$ and $s_n = -(n\pi)^2$ for $n \geq 1$. The residual at the pole $s_0 = 0$ can be calculated from

$$\text{Res}\,\left(\tilde{B}_y(\xi,s)\,e^{s\tau};s_0\right) = \lim_{s\to 0}\left(B_1\,\frac{\sinh\left(\sqrt{s}(1-\xi)\right)+\sinh\left(\sqrt{s}\xi\right)}{\sinh\sqrt{s}}\,e^{s\tau}\right), \tag{10.44}$$

where the limit value is found by employing the *l'Hôpital rule*,[8]

$$\lim_{x\to x_0}\frac{g(x)}{h(x)} = \lim_{x\to x_0}\frac{g'(x)}{h'(x)}. \tag{10.45}$$

This produces the result

$$\begin{aligned}\text{Res}\,\left(\tilde{B}_y(\xi,s)\,e^{s\tau};s_0\right) &= \lim_{s\to 0}\left(B_1\,\frac{(1-\xi)\cosh\left(\sqrt{s}(1-\xi)\right)+\xi\cosh\left(\sqrt{s}\xi\right)}{\cosh\sqrt{s}}\right)\\ &= B_1(1-\xi+\xi) = B_1.\end{aligned} \tag{10.46}$$

The other n residuals give, for $s_n = -(n\pi)^2$,

$$\begin{aligned}&\text{Res}\,\left(\tilde{B}_y(\xi,s)\,e^{s\tau};s_n\right) =\\ &\lim_{s\to s_n} B_1\left(\frac{1}{s}\frac{\sin\left(i\sqrt{s}(1-\xi)\right)}{\sin i\sqrt{s}}e^{s\tau}\left(s+(n\pi)^2\right)+\frac{1}{s}\frac{\sin\left(i\sqrt{s}\xi\right)}{\sin i\sqrt{s}}e^{s\tau}(s+(n\pi)^2)\right)\\ &= \frac{B_1}{-(n\pi)^2}\,f(\xi)\,e^{-(n\pi)^2\tau}\lim_{s\to s_n}\frac{s+(n\pi)^2}{\sin i\sqrt{s}},\end{aligned} \tag{10.47}$$

where $f(\xi) := \sin n\pi\xi + \sin n\pi(1-\xi)$. Applying the l'Hôpital rule to the limit yields

$$\lim_{s\to s_n}\frac{s+(n\pi)^2}{\sin i\sqrt{s}} = \lim_{s\to s_n}\frac{-i2\sqrt{s}}{\cos i\sqrt{s}} = \frac{-2n\pi}{\cos n\pi} = \frac{-2n\pi}{(-1)^n}, \tag{10.48}$$

and therefore the final result,

$$B_y(\xi,\tau) = B_1\left(1+\sum_{n=1}^{\infty}\frac{2}{n\pi}(-1)^n\,f(\xi)\,e^{-(n\pi)^2\tau}\right). \tag{10.49}$$

8 Guillaume Marquis de l'Hôpital (1661–1704).

Figure 10.3 (left) shows the distribution of the magnetic flux density inside the slab as a function of time. The eddy currents result from

$$J_z(\xi,\tau) = \frac{b}{\mu}\frac{\partial B_y(\xi,\tau)}{\partial \xi}. \tag{10.50}$$

This function is displayed in Figure 10.3 (right).

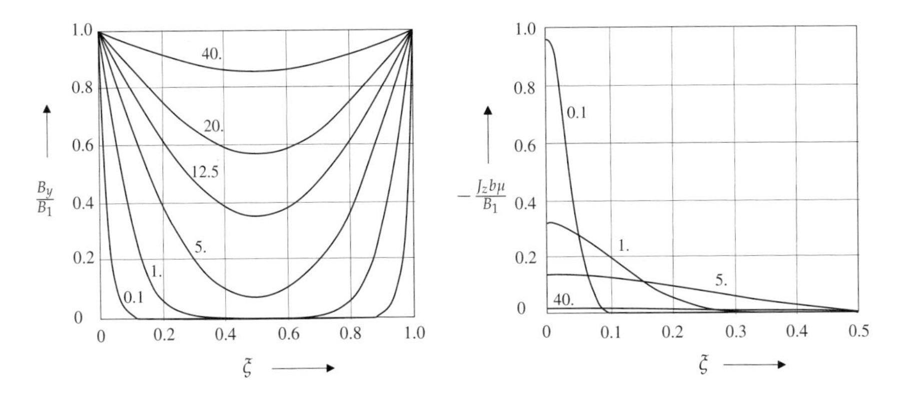

Figure 10.3 Magnetic flux density B_y (left) and eddy-current density J_z (right) as a function of ξ at 0.1–40 ms after the jump in the applied field. The material is pure copper, with $\varkappa = 14.5 \times 10^8$ S m^{-1} and $\mu \approx 1$. The width of the slab is 1 cm.

10.3.2 Linear Ramp of the Applied Field

For the calculation of the eddy currents, we begin with Eq. (10.40) and the excitation function given by

$$f(\tau) = k_B T_\mathrm{d}\, \tau \tag{10.51}$$

with its Laplace transform

$$\tilde{f}(s) = k_B T_\mathrm{d}\, \frac{1}{s^2}, \tag{10.52}$$

and the ramp rate of $k_B := \mathrm{d}B_1(t)/\mathrm{d}t$.

The residual at the double pole[9] is calculated by

$$\text{Res}\ \left(\tilde{B}_y(\xi,s)\, e^{s\tau}; s_0\right) = \lim_{s\to 0} k_B T_\mathrm{d} \frac{\mathrm{d}}{\mathrm{d}s}\left(\frac{\sinh\left(\sqrt{s}(1-\xi)\right)+\sinh\left(\sqrt{s}\xi\right)}{\sinh\sqrt{s}} e^{s\tau}\right)$$

$$= k_B T_\mathrm{d}\left(\tau + \frac{1}{2}(\xi-1)\xi\right). \tag{10.53}$$

The other n residuals are found following the same procedure as in Section 10.3.1:

$$\text{Res}\ \left(\tilde{B}_y(\xi,s)\, e^{s\tau}; s_n\right) = -\frac{2k_{B_1} T_\mathrm{d}}{(n\pi)^3}(-1)^n\, f(\xi)\, e^{-(n\pi)^2\tau}, \tag{10.54}$$

with the result

$$B_y(\xi,\tau) = k_B T_\mathrm{d}\left(\tau + \frac{1}{2}(\xi-1)\xi - \sum_{n=1}^{\infty}\frac{2}{(n\pi)^3}(-1)^n\, f(\xi)\, e^{-(n\pi)^2\tau}\right). \tag{10.55}$$

After completion of the diffusion process, we can approximate the current and field distribution merely by applying Faraday's law. For $\xi(t) \gg b/2$ we obtain $\partial B_y(x,t)/\partial t = k_B$ for all x; the field inside the slab changes at the same rate k_B in the entire cross section. The diffusion equation (4.195) reduces to

$$\frac{\partial^2 B_y(x,t)}{\partial x^2} = \varkappa\mu k_B. \tag{10.56}$$

A solution can be found by integrating twice:

$$B_y(x,t) = \varkappa\mu k_B x^2 + p(t)\, x + q(t). \tag{10.57}$$

The coefficients $p(t)$ and $q(t)$ are determined by the boundary conditions (10.33):

$$p(t) = -\varkappa\mu k_B \frac{b}{2}, \qquad q(t) = k_B\, t. \tag{10.58}$$

Faraday's law yields

$$\frac{\partial J_z(x,t)}{\partial x} = \varkappa \frac{\partial B_y(x,t)}{\partial t} = \varkappa k_B, \tag{10.59}$$

and therefore by integration,

$$J_z(x) = \varkappa(x\, k_B + c(t)). \tag{10.60}$$

9 The residual of a function $f(z)$ at a double pole at z_0 is calculated by $\lim_{z\to z_0} \frac{\mathrm{d}}{\mathrm{d}z}\left(f(z)(z-z_0)^2\right)$.

Setting $c(t) = -k_B\, b/2$, to assure that the total induced current over the cross section is zero, it follows that

$$J_z(x) = \varkappa k_B \left(x - \frac{b}{2} \right). \tag{10.61}$$

The magnetic field is given by

$$\frac{1}{\mu}\frac{\partial B_y(x,t)}{\partial x} = J_z(x,t). \tag{10.62}$$

Integration yields

$$B_y(x,t) = \varkappa\mu k_B \left(\frac{x^2}{2} - \frac{xb}{2} \right) + q(t). \tag{10.63}$$

After completion of the diffusion process, $\xi(t) \gg b/2$, the approximate equation (10.63) corresponds to the first term in Eq. (10.55) because the sum is zero for $\tau \rightarrow \infty$.

10.3.3 Sinusoidal Excitation

Now let us apply a sinusoidal field of a frequency $f = \omega/2\pi$. Proceeding in the same way as above, we express ωt as $\Omega\tau = \omega\varkappa\mu b^2\tau$, and thus rewrite the excitation function as

$$f(\tau) = \hat{B}_1 \cos \Omega\, \tau\,. \tag{10.64}$$

Its Laplace transform is

$$\tilde{f}(s) = \hat{B}_1 \frac{s}{s^2 + \Omega^2}, \tag{10.65}$$

which has two conjugate poles at $s_1 = i\Omega$ and $s_2 = -i\Omega$. The residual at s_2 is given by

$$\mathrm{Res}\left(\tilde{B}_y(\xi,s)\, e^{s\tau}; -i\Omega\right) = \frac{\hat{B}_1}{2} e^{-i\Omega\tau} \frac{\sinh\left(\sqrt{-i\Omega}(1-\xi)\right) + \sinh\left(\sqrt{-i\Omega}\xi\right)}{\sinh\sqrt{-i\Omega}}. \tag{10.66}$$

Substituting $k = \pm\frac{1-i}{\delta}$ with $\delta = \sqrt{\frac{2}{\omega\varkappa\mu_0}}$ we obtain

$$\mathrm{Res}\left(\tilde{B}_y(\xi,s)\, e^{s\tau}; -i\Omega\right) = \frac{\hat{B}_1}{2} e^{-i\Omega\tau} \frac{\sinh\left(bk\left(1-\xi\right)\right) + \sinh\left(bk\xi\right)}{\sinh bk}. \tag{10.67}$$

The quotient of two uneven functions is an even function, and therefore both values for k give the same result. The other n residuals are calculated as in Section 10.3.1, with the result

$$B_y(\xi,\tau) = \hat{B}_1 \,\mathrm{Re}\left\{ e^{-i\Omega\tau} \frac{\sinh(bk(1-\xi)) + \sinh(bk\xi)}{\sinh bk} \right\} + \hat{B}_1 \sum_{n=1}^{\infty} \frac{2(n\pi)^3}{(n\pi)^4 + \Omega^2} (-1)^n f(\xi)\, e^{-(n\pi)^2\tau}. \tag{10.68}$$

10.4 Eddy Currents in the LHC Cold Bore and Beam Screen

We will now investigate eddy-current-induced field errors and resistive losses in very thin conductive layers. To this end, consider the configuration shown in Figure 10.4 (left). Let r be the radius of the field point and r_c the radius of the cylinder with wall thickness d.

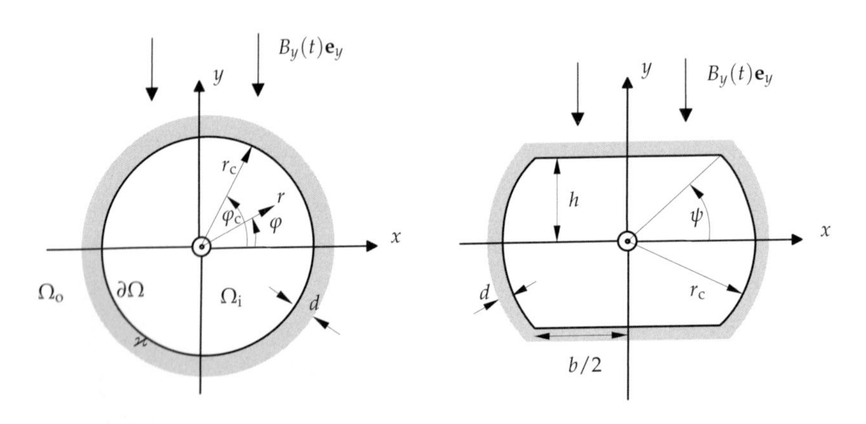

Figure 10.4 Left: Conductive cylinder in a spatially uniform, time-transient applied field. Right: Schematic representation of the LHC beam screen liner.

The cylinder is exposed to a spatially homogeneous, time-transient applied field $\mathbf{B}(t) = B_y(t)\,\mathbf{e}_y$ with the ramp rate $k_B := \mathrm{d}B_y(t)/\mathrm{d}t = -7 \times 10^{-3}\,\mathrm{T\,s^{-1}}$. For $r \gg r_c$, the magnetic flux density can be written in cylindrical coordinates as

$$\mathbf{B}(r,\varphi,t) = k_B\, t\,(\sin\varphi\,\mathbf{e}_r + \cos\varphi\,\mathbf{e}_\varphi). \tag{10.69}$$

Expressing the r-component of the magnetic flux density by the magnetic vector potential yields

$$\frac{1}{r}\frac{\partial A_z}{\partial \varphi} = k_B t \sin\varphi. \tag{10.70}$$

The z-component of the magnetic vector potential is thus given by

$$A_z = A_\infty = -k_B\, tr \cos\varphi + c(t). \tag{10.71}$$

The integration constant $c(t)$ must be chosen such that the total induced current,

$$\int_\Omega J_z\, \mathrm{d}a = \int_\Omega -\varkappa \frac{\partial A_z}{\partial t}\, \mathrm{d}a\,, \tag{10.72}$$

vanishes in the conductive domain Ω. For a very thin cylinder, $d \to 0$, the induced currents may be treated as surface-currents by introducing the *surface-conductivity*,

$$\varkappa_s := \varkappa d, \tag{10.73}$$

where $[\varkappa_s] = 1\,\Omega^{-1} = 1\,\mathrm{S}$. Let A_i, A_o, and A_∞ denote the z-component of the magnetic vector potentials inside and outside the conductive domain, and at $r \to \infty$, respectively. The interface condition for H_t is

$$\frac{\partial A_\mathrm{o}}{\partial r} - \frac{\partial A_\mathrm{i}}{\partial r} = \mu\varkappa d \frac{\partial A_z}{\partial t} \tag{10.74}$$

at $r = r_\mathrm{c}$. The vector potential is continuous at the interface of the conductive and air domains. The right-hand side of Eq. (10.74) can therefore be calculated from either A_i or A_o. Using the general solution of the Laplace equation in cylindrical coordinates for the vector potential outside the cylinder, we can express the potential as

$$A_\mathrm{o} = -k_B\, tr\cos\varphi + D(t)\frac{r_\mathrm{c}}{r}\cos\varphi, \tag{10.75}$$

where we have accounted for $\lim_{r\to\infty} A_\mathrm{o} = A_\infty$. The potential inside the cylinder is

$$A_\mathrm{i} = C(t)\frac{r}{r_\mathrm{c}}\cos\varphi. \tag{10.76}$$

The interface conditions yield

$$C(t) + \frac{1}{2}\mu\varkappa_s r_\mathrm{c}\frac{\mathrm{d}C(t)}{\mathrm{d}t} = -k_B t r_\mathrm{c}\,, \tag{10.77}$$

$$D(t) = C(t) + k_B t a\,. \tag{10.78}$$

For a time constant T_d defined by

$$T_\mathrm{d} := \frac{1}{2}\mu\varkappa_\mathrm{s} r_\mathrm{c} \tag{10.79}$$

and the initial conditions $A_\mathrm{i}(0) = 0$ and $A_\mathrm{o}(0) = A_\infty(0)$, the solution reads

$$C(t) = -k_B r_\mathrm{c}\left(t - T_\mathrm{d}\left(1 - e^{-\tau}\right)\right), \tag{10.80}$$

$$D(t) = k_B r_\mathrm{c} T_\mathrm{d}\left(1 - e^{-\tau}\right), \tag{10.81}$$

for $\tau = t/T_\mathrm{d}$. The vector potentials are

$$A_\mathrm{i} = -k_B r\cos\varphi\left(t - T_\mathrm{d}\left(1 - e^{-\tau}\right)\right), \tag{10.82}$$

$$A_\mathrm{o} = -k_B r\cos\varphi\left(t - T_\mathrm{d}\left(1 - e^{-\tau}\right)\left(\frac{r_\mathrm{c}}{r}\right)^2\right). \tag{10.83}$$

The magnetic flux density is calculated using Eq. (4.10):

$$B_\mathrm{i} = k_B\sin\varphi\left(t - T_\mathrm{d}\left(1 - e^{-\tau}\right)\right), \tag{10.84}$$

$$B_\mathrm{o} = k_B\sin\varphi\left(t - T_\mathrm{d}\left(1 - e^{-\tau}\right)\left(\frac{r_\mathrm{c}}{r}\right)^2\right), \tag{10.85}$$

where B_i and B_o denote the radial components of the flux densities inside and outside of the cylinder, respectively.

Field errors and ohmic losses from the LHC cold bore The field attenuation caused by a hollow conductive cylinder, such as the LHC cold bore, exposed to a time-transient field, is given by

$$\Delta B := B_\mathrm{i} - B_\infty = k_B\sin\varphi\left(-T_\mathrm{d}\left(1 - e^{-\tau}\right)\right), \tag{10.86}$$

which simplifies for $t \gg T_\mathrm{d}$ to

$$\Delta B_1 = -T_\mathrm{d} k_B. \tag{10.87}$$

A cylindrical conductor in a homogeneous transversal field effects only the first-order harmonic of the radial component.

The LHC cold bore is made of austenitic steel with a residual resistivity ratio (RRR) close to 1. At 1.9 K its conductivity $\varkappa$ is about 10^{-4} times that of copper. Assuming $\varkappa = 4.5 \times 10^4\ \mathrm{S\,m^{-1}}$, $k_B = -7 \times 10^{-3}\ \mathrm{T\,s^{-1}}$, $d = 1.5$ mm and $r_\mathrm{c} = 25.75$ mm we obtain $B_1 = 2.46 \times 10^{-8}$ T. The attenuation of the main field during the ramp is therefore negligible.

The ohmic losses per unit length of the hollow cylinder are

$$P = \int_\Omega \mathbf{J}\cdot\mathbf{E}\,\mathrm{d}a = \int_{\partial\Omega} \boldsymbol{\alpha}\cdot\mathbf{E}\,\mathrm{d}s, \tag{10.88}$$

where the surface-current is given by $\boldsymbol{\alpha} = \varkappa_s \mathbf{E} = -\varkappa_s \partial \mathbf{A}/\partial t$. For the hollow cylinder in a ramping homogeneous transversal field Eq. (10.88) is rewritten as

$$P = \int_0^{2\pi} \varkappa_s \, k_B^2 r_c^2 \cos^2 \varphi_c \, r_c \, \mathrm{d}\varphi_c = \pi \varkappa_s \, k_B^2 r_c^3. \tag{10.89}$$

With the materials as above, ohmic losses per unit length of the cold bore amount to 0.571 μW m^{-1}.

Field errors and ohmic losses from the beam screen The 50-µm-thick copper layer on the inner surface of the beam screen[10] is referred to as the *beam screen liner*; see Figure 10.4 (right). The pumping slots are disregarded in the following calculations. We assume that the field generated by the induced eddy currents is negligible outside the screen. The field inside the aperture is well approximated for wall thicknesses $d \ll r_c, h$. In this case, we can express the induced currents on the straight sections as surface-currents,

$$\alpha_z = x \varkappa_s k_B \tag{10.90}$$

and on the arc sections,

$$\alpha_z = \varkappa_s \, k_B h \cot \varphi_c, \qquad \text{for } \psi < \varphi_c < \pi - \psi, \tag{10.91}$$

$$\alpha_z = \varkappa_s k_B r_c \cos \varphi_c, \qquad \text{for } \pi - \psi < \varphi_c < \pi + \psi. \tag{10.92}$$

As the geometry of the beam screen is not circular, the eddy-current distribution will deviate from the $\cos \varphi_c$ current distribution, and therefore all odd normal multipoles will be excited.

From the Biot–Savart law we derived Eq. (5.44) for the 2D case, which can now be integrated over the conductive domain Ω:

$$A_z(\mathbf{r}) = \int_\Omega -\frac{\mu J_z(r_c, \varphi_c)}{2\pi} \ln\left(\frac{|\mathbf{r} - \mathbf{r}'|}{r_{\text{ref}}}\right) \mathrm{d}a, \tag{10.93}$$

for an arbitrary reference radius r_{ref}. For the calculation of the magnetic field inside the aperture, the applied and eddy-current-induced fields must be added. Equation (10.93) can be rewritten as

$$A_z(r, \varphi) = \int_\Omega -\frac{\mu J_z(r_c, \varphi_c)}{2\pi} \left(\ln\left(\frac{r_c}{r_{\text{ref}}}\right) - \sum_{n=1}^{\infty} \frac{1}{n} \left(\frac{r}{r_c}\right)^n \cos\left(n(\varphi - \varphi_c)\right) \right) \mathrm{d}a\,; \tag{10.94}$$

10 The function of the LHC beam screen is described in Section 1.7.

see Section 6.1.1.3. Hence the nth multipole component at a reference radius r_0 is given by

$$B_n = \int_\Omega -\frac{\mu J_z(r_c, \varphi_c)}{2\pi} \frac{r_0^{n-1}}{r_c^n} \cos n\varphi_c \, da \,. \tag{10.95}$$

For surface-currents, this equation yields

$$B_n = \int_{\mathscr{S}} -\frac{\mu \alpha_z(r_c, \varphi_c)}{2\pi} \frac{r_0^{n-1}}{r_c^n} \cos n\varphi_c \, ds. \tag{10.96}$$

Substituting $x = h \cot \varphi_c, r_c = h/\sin \varphi_c$ for $\psi < \varphi_c < \pi - \psi$, and $x = -h \cot \varphi_c, r_c = -h/\sin \varphi_c$ for $\pi + \psi < \varphi_c < 2\pi - \psi$, the effects of both straight sections are given by

$$B_{n,\text{str}} = -\frac{\mu \varkappa_s k_B r_0^{n-1}}{2\pi h^{n-2}} \cdot \left(\int_\psi^{\pi-\psi} \cos n\varphi_c \, \cos \varphi_c \, (\sin \varphi_c)^{n-3} \, d\varphi_c \right.$$
$$\left. - \int_{\pi+\psi}^{2\pi-\psi} (-1)^n \cos n\varphi_c \, \cos \varphi_c \, (\sin \varphi_c)^{n-3} \, d\varphi_c \right). \tag{10.97}$$

The two integrals yield the same contribution and can be solved by means of the Mathematica™ program, which applies a *Gauss quadrature* to the one-dimensional problems. For the arc sections, Eq. (10.96) yields

$$B_{n,\text{arc}}(r_0) = -\frac{\mu \varkappa_s k_B r_0^{n-1}}{2\pi r_c^{n-2}} \left(\int_{-\psi}^{\psi} \cos n\varphi_c \cos \varphi_c \, d\varphi_c \right.$$
$$\left. + \int_{\pi-\psi}^{\pi+\psi} \cos n\varphi_c \cos \varphi_c \, d\varphi_c \right). \tag{10.98}$$

For $h = 18.45$ mm, $r_c = 23.25$ mm, $\psi = 52.52°$, $d = 50$ µm, $\varkappa = 14.5 \times 10^8 \, \text{S m}^{-1}$, $k_B = -7 \times 10^{-3} \, \text{T s}^{-1}$, we obtain the results as shown in Table 10.3 for $B_n = B_{n,\text{str}} + B_{n,\text{arc}}$ at $r_0 = 17$ mm.

Table 10.3 Field harmonics due to eddy currents in the LHC beam-screen liner at a ramp-rate $k_B = -7 \times 10^{-3} \, \text{T s}^{-1}$ and a reference radius $r_0 = 17$ mm. $[B_n] = 1$ T.

n	$B_{n,\text{str}}$	$B_{n,\text{arc}}$	B_n
1	0.831	6.599	7.442
3	−1.136	0.902	−0.235
5	0.563	−0.327	0.235
7	−0.097	−0.007	−0.104

Because of the up/down symmetry, no skew components are excited. The left/right symmetry of the beam screen results in the absence of even-numbered, normal field components. The most important field error is again the attenuation of the dipole component, mainly due to the arc sections. The higher-order components are dominated by the straight sections. The field errors induced by the austenitic steel structure of the beam screen ($d = 1$ mm, $\varkappa = 14.5 \times 10^4$ S m^{-1}) are smaller by 2–3 orders of magnitude.

As in Section 10.4, we calculate the ohmic losses per unit length of the beam screen by integrating $\boldsymbol{\alpha} \cdot \mathbf{E}$ along its contour. Using Eqs. (10.90) and (10.91), we can write

$$P = 2 \int_{-\frac{b}{2}}^{\frac{b}{2}} \varkappa_S k_B^2 x^2 \, dx + 2 \int_{-\psi}^{\psi} \varkappa_S k_B^2 r_c^2 \cos^2 \varphi_c \, r_c \, d\varphi_c, \tag{10.99}$$

and finally, with ψ expressed in radians,

$$P = \varkappa_S k_B^2 \left(\frac{b^3}{6} + r_c^3 (2\psi + \sin 2\psi) \right). \tag{10.100}$$

For the same material properties as above and $b = 36.9$ mm, P is 131 μW m^{-1}. This is low compared to the estimated total heat load of 130 mW m^{-1} on the beam screen in each aperture of the LHC machine.

References

1 Beerends, R.J., ter Morsche, H.G., van den Berg, J.C., van de Vrie, E.M.: Fourier and Laplace Transform, Cambridge University Press, Cambridge, 2003

2 Davis, B.: Integral Transforms and their Applications, Springer, Berlin, 2002

3 Doetsch, G.: Introduction to the Theory and Application of the Laplace Transformation, Springer, Berlin, 1994

4 Dyke, P.P.G.: An Introduction to Laplace Transform and Fourier Series, Springer, Berlin, 1999

5 Henrici, P.: Applied and Computational Complex Analysis, Vol. 2: Special Functions, Integral Transforms, Asymptotics, Continued Fractions, John Wiley & Sons, New York, 1991

11
Elementary Beam Optics and Field Requirements

In this chapter, we will give an introduction to the transverse dynamics of charged particles in a synchrotron. The linear equations of motion will be derived from the Frenet–Serret equations[1] of the reference orbit. Requirements for the bending fields of dipoles and the focusing strength of quadrupoles will be reviewed, and the basic principles of a magnet lattice discussed. We will derive equations for the optical parameters of the beam and discuss the aperture limitations in the magnets. Finally, as a concrete example, we will present the LHC beam requirements for the field quality in the magnet elements and for the tolerances in magnet alignment.

In order to keep this chapter brief, we refer to the LHC design report [12], to the proceedings of the CERN accelerator school [2], and general books on accelerator physics by, for example, Edwards and Syphers [7], Wille [29], and Wilson [30].

11.1
The Equations of Charged Particle Motion in a Magnetic Field

A particle with charge[2] e moving with velocity $\mathbf{v}$ through an electromagnetic field is subjected to the electromagnetic force according to the *Lorentz law*[3] $\mathbf{F} = e(\mathbf{v} \times \mathbf{B} + \mathbf{E})$. The rate of change of the particle's momentum is given by

$$\frac{\mathrm{d}\mathbf{p}}{\mathrm{d}t} = m\frac{\mathrm{d}}{\mathrm{d}t}(\gamma\mathbf{v}) = m\left(\gamma\frac{\mathrm{d}\mathbf{v}}{\mathrm{d}t} + \frac{\mathrm{d}\gamma}{\mathrm{d}t}\mathbf{v}\right) = e(\mathbf{v} \times \mathbf{B} + \mathbf{E}), \tag{11.1}$$

with the *Lorentz factor*

$$\gamma = \frac{1}{\sqrt{1-\beta^2}} = \frac{1}{\sqrt{1-\left(\frac{v}{c}\right)^2}}, \tag{11.2}$$

1 See Chapter 3.
2 In dealing with electrons and protons the charge is just the electronic charge e.
3 Hendrik Antoon Lorentz (1853–1928).

Field Computation for Accelerator Magnets. Stephan Russenschuck

ISBN: 978-3-527-40769-9

the speed of light in vacuum c, and the mass m of the particle [21]. The rate of change of the particle's energy is

$$\frac{dW}{dt} = \mathbf{v} \cdot \mathbf{F} = \underbrace{e\,\mathbf{v} \cdot (\mathbf{v} \times \mathbf{B})}_{=0} + e\,\mathbf{v} \cdot \mathbf{E}. \tag{11.3}$$

No work is performed by the magnetic field, and only the electric field contributes to the change in energy. Magnetic fields serve for guiding and focusing of the particle beams. At relativistic speed, electric and magnetic fields have the same effect on the particle trajectory if $|\mathbf{E}| \equiv c|\mathbf{B}|$. A magnetic flux density of 1 T is equivalent to an electric field strength of about 3×10^8 V m^{-1}. While a magnetic flux density of 1 T can easily be achieved even with iron-dominated magnets, electric field strengths in the GV m^{-1} range are technically unfeasible. This is why magnetic fields are used for beam steering in all high-energy, circular particle accelerators.

Assuming γ is constant and there is no electric field and no particle acceleration, the force equation takes the simple form

$$\frac{d\mathbf{p}}{dt} = m\gamma \frac{d\mathbf{v}}{dt} = e\,\mathbf{v} \times \mathbf{B}\,. \tag{11.4}$$

By convention, the *reference orbit* $\mathscr{C}$, also known as the *design orbit*, is a term for the closed trajectory that a particle with reference momentum would follow in an ideal machine, with perfectly positioned magnets at nominal strength. It can be shown that a closed trajectory always exists for ideal, planar machines. Since the transverse deviations of the particle trajectories with respect to the reference orbit are typically small compared to the machine radius, we introduce a Frenet frame that follows the reference orbit $\mathbf{r}_0(s)$, parametrized with respect to the arc length s, as shown in Figure 11.1. If we assume that the reference orbit is composed of piecewise-flat curves lying either in the horizontal or vertical plane (the space curve being torsion-free), it follows that the Frenet–Serret equations, derived in Chapter 3, reduce to

$$\mathbf{T}' = \kappa \mathbf{N}\,, \qquad \mathbf{N}' = -\kappa \mathbf{T}\,, \qquad \mathbf{B}' = \mathbf{0}\,, \tag{11.5}$$

where the prime symbol denotes derivation with respect to s. Because of the singular points on the space curve, where the reference orbit changes from the horizontal to the vertical plane, we define a new right-handed frame $\{\mathbf{e}_x, \mathbf{e}_y, \mathbf{T}\}$ by $\mathbf{e}_x = -\mathbf{N}$ and $\mathbf{e}_y = \mathbf{B}$ when the orbit lies in the horizontal plane. If the curvature is positive and the particle orbits in clockwise direction seen from above the ring, the unit vector $\mathbf{e}_x$ points outward from the ring. If the orbit lies in the vertical plane we set $\mathbf{e}_x = -\mathbf{B}$ and $\mathbf{e}_y = -\mathbf{N}$. This yields a moving frame

$$\mathbf{e}_x' = \kappa_x \mathbf{T}\,, \qquad \mathbf{e}_y' = \kappa_y \mathbf{T}\,, \qquad \mathbf{T}' = -\kappa_x \mathbf{e}_x - \kappa_y \mathbf{e}_y\,, \tag{11.6}$$

for $\kappa_x(s)\kappa_y(s) = 0$. Because of misalignments, field errors, momentum deviations, and injection errors, the trajectory $\mathscr{S}$ of the individual particle does not coincide with the reference orbit $\mathscr{C}$. In the Frenet frame, a single particle trajectory can be expressed as

$$\mathbf{r}(x, y, s) = \mathbf{r}_0(s) + x(s)\,\mathbf{e}_x + y(s)\,\mathbf{e}_y\,, \tag{11.7}$$

where x and y are the *transverse coordinates* or the *fiducial coordinates*; see Figure 11.1.

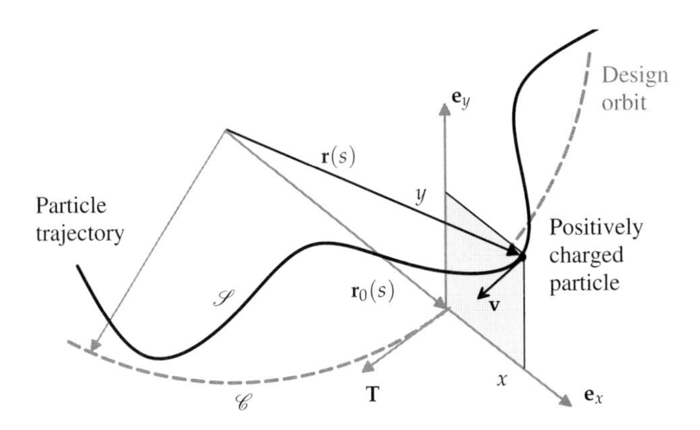

Figure 11.1 Motion of a particle on its trajectory $\mathscr{S}$, parametrized with respect to its arc-length σ, close to the ideal orbit $\mathscr{C}$ (arc-length s) shown with its local frame $(\mathbf{e}_x, \mathbf{e}_y, \mathbf{T})$.

Let σ denote the arc length on the individual particle trajectory $\mathscr{S}$. Omitting the notation of the s-dependence, the velocity of the particle is given by $v = \mathrm{d}\sigma/\mathrm{d}t$. Using

$$\frac{\mathrm{d}\mathbf{r}}{\mathrm{d}t} = \frac{\mathrm{d}\sigma}{\mathrm{d}t}\frac{\mathrm{d}\mathbf{r}}{\mathrm{d}\sigma} = v\frac{\mathrm{d}s}{\mathrm{d}\sigma}\frac{\mathrm{d}\mathbf{r}}{\mathrm{d}s} = \frac{v}{\sigma'}\mathbf{r}' \tag{11.8}$$

and

$$\frac{\mathrm{d}^2\mathbf{r}}{\mathrm{d}t^2} = \frac{v}{\sigma'}\frac{\mathrm{d}}{\mathrm{d}s}\left(\frac{v}{\sigma'}\mathbf{r}'\right) = \frac{v^2}{\sigma'^2}\left(\mathbf{r}'' - \frac{\sigma''}{\sigma'}\mathbf{r}'\right), \tag{11.9}$$

the derivation of the individual particle trajectory with respect to s yields

$$\begin{aligned}\mathbf{r}' &= \mathbf{r}_0{}' + x'\mathbf{e}_x + y'\mathbf{e}_y + x\,\mathbf{e}_x' + y\,\mathbf{e}_y' \\ &= (1 + \kappa_x x + \kappa_y y)\,\mathbf{T} + x'\mathbf{e}_x + y'\mathbf{e}_y\,,\end{aligned} \tag{11.10}$$

where we have used the relations of Eq. (11.6). In a similar way, we obtain the second derivative:

$$\mathbf{r}'' = \left(\kappa_x' x + \kappa_y' y + 2\kappa_x x' + 2\kappa_y y'\right) \mathbf{T} \\ + \left(x'' - \kappa_x(1 + \kappa_x x)\right) \mathbf{e}_x + \left(y'' - \kappa_y(1 + \kappa_y y)\right) \mathbf{e}_y \,. \tag{11.11}$$

From $|\mathbf{v}| = v = \frac{v}{\sigma'}|\mathbf{r}'|$ it yields

$$\sigma' = |\mathbf{r}'| = \sqrt{(1 + \kappa_x x + \kappa_y y)^2 + x'^2 + y'^2} \,. \tag{11.12}$$

From Eq. (11.4) we obtain $m\gamma\mathbf{r}'' = e\,\mathbf{v} \times \mathbf{B}$, and thus

$$m\gamma \frac{v^2}{\sigma'^2}\left(\mathbf{r}'' - \frac{\sigma''}{\sigma'}\mathbf{r}'\right) = \frac{ev}{\sigma'}\left(\mathbf{r}' \times \mathbf{B}\right) \,. \tag{11.13}$$

We can now express the magnetic flux density in the local coordinates as

$$\mathbf{B}(x, y, s) = B_t(x, y, s)\,\mathbf{T} + B_x(x, y, s)\,\mathbf{e}_x + B_y(x, y, s)\,\mathbf{e}_y \,. \tag{11.14}$$

Let us furthermore substitute Eqs. (11.10) and (11.11) into Eq. (11.13) and recall that $\kappa_x \kappa_y = 0$. Comparing the components on either side of Eq. (11.13) yields the equations of motion [17]:

$$x'' - \frac{\sigma''}{\sigma'}x' = \kappa_x h - \frac{e}{p}\sigma'(hB_y - y'B_t) \,, \tag{11.15}$$

$$y'' - \frac{\sigma''}{\sigma'}y' = \kappa_y h + \frac{e}{p}\sigma'(hB_x - x'B_t) \,, \tag{11.16}$$

$$\frac{\sigma''}{\sigma'} = \frac{1}{h}\left(\kappa_x' x + \kappa_y' y + 2\kappa_x x' + 2\kappa_y y'\right) - \frac{e}{p}\frac{\sigma'}{h}\left(x'B_y - y'B_x\right) \,, \tag{11.17}$$

where we have defined

$$h := 1 + \kappa_x x + \kappa_y y \,. \tag{11.18}$$

Equation (11.17) can be used to express σ''/σ' in Eqs. (11.15) and (11.16). The particle momentum is given by $p = m\gamma v$ and can be expanded around the ideal momentum p_0 by

$$p = p_0(1 + \delta) \,, \tag{11.19}$$

where the *relative momentum deviation* is defined by

$$\delta := \frac{p - p_0}{p_0} = \frac{\Delta p}{p_0} \,. \tag{11.20}$$

The road map to the linear equations of motion is now established: Insert Eq. (11.19) into Eqs. (11.15)–(11.17), express σ''/σ' in Eqs. (11.15) and (11.16) by the right-hand side of Eq. (11.17) and then make approximations: Disregard the momentum deviation ($\delta = 0$) and the tangential field component ($B_t(x, y, s) = 0$), keep only the linear terms in x, x', y, y', and use a Taylor expansion of the magnetic flux density about the reference orbit.

11.2 Magnetic Rigidity and the Bending Magnets

As the most simple case we consider planar, circular particle motion on the ideal orbit ($x = x' = 0$ and $y = y' = 0$), and therefore $h = 1, \sigma' = 1, \sigma'' = 0$. Further, we assume zero momentum deviation ($\delta = 0$) and no tangential field ($B_t = 0$). Equation (11.15) yields the simple expression

$$\kappa_x = \frac{e}{p_0} B_y(0, 0, s) . \tag{11.21}$$

Introducing the constant bending radius $R = 1/\kappa_x$, assuming constant magnetic field $B_0 := B_y(0, 0, s)$, and rearranging terms yields

$$p_0 = e\, B_0\, R . \tag{11.22}$$

It is customary to express the particle momentum as

$$p_0 = \{p_0\}_{\mathrm{GeV\,c^{-1}}} \frac{10^9 \cdot 1.602\ldots \times 10^{-19}}{2.997\ldots \times 10^8} \frac{\mathrm{V\,A\,s}}{\mathrm{m\,s^{-1}}} . \tag{11.23}$$

For the electronic charge

$$e = 1.602\ldots \times 10^{-19}\,\mathrm{A\,s} , \tag{11.24}$$

the particle momentum is thus given in units of $\mathrm{GeV\,c^{-1}}$ by

$$\{p_0\}_{\mathrm{GeV\,c^{-1}}} \approx 0.3\,\{R\}_{\mathrm{m}}\,\{B_0\}_{\mathrm{T}} . \tag{11.25}$$

The term $B_0 R$, corresponding to the ratio of momentum to charge p_0/e, is called the *magnetic rigidity* and is a measure of the beam's "stiffness" in the bending field.

Example: The magnetic rigidity of the LHC is 1500 Tm at injection (26 GeV) and 23 356 Tm at collision energy (7 TeV). □

The bending angle ϕ of a dipole is determined by the equation

$$\sin \frac{\phi}{2} = \frac{L}{2R} , \tag{11.26}$$

see Figure 11.2 (left), where L is the magnetic length of the dipole. Equation (11.26) yields an approximation for small angles ($\sin\phi \approx \phi$),

$$\phi \approx \frac{LB_0}{B_0 R}, \tag{11.27}$$

and numerically,

$$\{\phi\}_{\text{rad}} \approx 0.3\,\{L\}_{\text{m}}\,\frac{\{B_0\}_{\text{T}}}{\{p\}_{\text{GeV c}^{-1}}}. \tag{11.28}$$

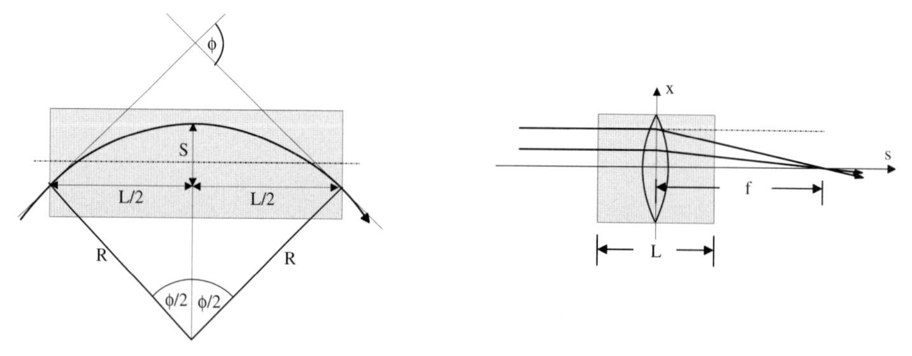

Figure 11.2 Left: The bending angle of a dipole magnet and the sagitta of the beam orbit. Right: The focal length of a thin quadrupole magnet. A ray parallel to the optical axis is bent to pass through the focal point of the lens at a distance f away.

Example: For a magnetic flux density of 8.33 T in the dipoles with a magnetic length of 14.3 m, and a particle momentum of 7 TeV c^{-1}, we calculate a bending angle of approximately 5.1 mrad. Consequently, 1232 bending magnets are needed for the LHC ring. According to Eq. (11.27), the bending radius of the dipoles is approximately 2800 m. Due to the dipole filling factor of 65% this bending radius is considerably smaller than the average machine radius of 4243 m. □

The aperture of the bending magnet must be large enough to contain the sagitta S of the beam, which is the distance between the arc's apex and the chord; see Figure 11.2 (left). The sagitta can be calculated from the equation

$$S = R\left[1 - \cos\left(\frac{\phi}{2}\right)\right] \approx \frac{R\phi^2}{8} = \frac{L^2}{8R} = \frac{eB_0 L^2}{8p}, \tag{11.29}$$

using the expansion $\cos x = 1 - \frac{x^2}{2!} + \frac{x^4}{4!} - \cdots$ for $|x| < \infty$ up to second order.

Example: In the case of LHC with an effective dipole length of 14.3 m and a bending radius of 2800 m we calculate a sagitta of 9.1 mm. □

According to Eq. (11.25), the trajectory radius of the particle increases with the particle momentum. As both the maximum field and the maximum dimensions of the magnets are limited, the magnetic field must be ramped synchronously with the particle energy. This is the principle of the synchrotron accelerator, the foundations of which were laid in the early 1940s by Szilard, Oliphant, McMillan, and Veksler [20, 26].

11.3 The Linear Equations of Motion

To study the motion of a single particle, we first make a linear approximation of the equations of motion. We set

$$\sigma' = h\sqrt{1 + \frac{x'^2}{h^2} + \frac{y'^2}{h^2}} \approx h = 1 + \kappa_x x + \kappa_y y\,, \tag{11.30}$$

$$\sigma'' \approx h' = \kappa_x' x + \kappa_x x' + \kappa_y' y + \kappa_y y'\,, \tag{11.31}$$

and thus disregard the term $\frac{\sigma''}{\sigma'}x' \approx \frac{h'}{h}x'$. Furthermore, we can approximate $(1+\delta)^{-1} \approx 1 - \delta$ for small momentum deviations and expand the magnetic flux density up to first order:

$$B_x(x, y, s) = B_x(0, 0, s) + \left.\frac{\partial B_x}{\partial x}\right|_{x=y=0} x + \left.\frac{\partial B_x}{\partial y}\right|_{x=y=0} y\,, \tag{11.32}$$

$$B_y(x, y, s) = B_y(0, 0, s) + \left.\frac{\partial B_y}{\partial x}\right|_{x=y=0} x + \left.\frac{\partial B_y}{\partial y}\right|_{x=y=0} y\,. \tag{11.33}$$

The normalized gradient k_n and the skew normalized gradient k_s are defined by

$$k_\mathrm{n} := \frac{e}{p_0}\left.\frac{\partial B_y}{\partial x}\right|_{x=y=0}\,, \qquad k_\mathrm{s} := \frac{e}{p_0}\left.\frac{\partial B_x}{\partial x}\right|_{x=y=0}\,. \tag{11.34}$$

The physical unit of the normalized gradient is $[k_{\mathrm{n,s}}] = 1\ \mathrm{m}^{-2}$. A positive value of the normalized gradient corresponds to the horizontal focusing of a positively charged particle. With the field gradient $g := \left.\frac{\partial B_y}{\partial x}\right|_{x=y=0}$, we obtain the numerical value expression

$$\{k_\mathrm{n}\}_{\mathrm{m}^{-2}} \approx 0.3 \frac{\{g\}_{\mathrm{T\,m}^{-1}}}{\{p_0\}_{\mathrm{GeV\,c}^{-1}}}\,. \tag{11.35}$$

Let us take into account that the transversal components of the flux density obey the Cauchy–Riemann equations,[4]

$$\frac{\partial B_x}{\partial x} = -\frac{\partial B_y}{\partial y}, \qquad \frac{\partial B_x}{\partial y} = \frac{\partial B_y}{\partial x}. \tag{11.36}$$

Now the equation of motion in the horizontal plane, Eq. (11.15), can be written as

$$x'' = \kappa_x h - (1-\delta)h^2 \frac{e}{p_0} B_y = \kappa_x h - (1-\delta)h^2 (\kappa_x + k_\text{n} x - k_\text{s} y). \tag{11.37}$$

We continue to make linear approximations,

$$h^2 \approx 1 + 2\kappa_x x + 2\kappa_y y, \tag{11.38}$$

$$\kappa_u h \approx \kappa_u + \kappa_u^2 u, \tag{11.39}$$

$$(1-\delta)h^2 \kappa_u \approx \kappa_u + 2\kappa_u^2 u - \kappa_u \delta, \tag{11.40}$$

$$(1-\delta)h^2 k_\text{n} u \approx k_\text{n} u, \tag{11.41}$$

$$(1-\delta)h^2 k_\text{s} u \approx k_\text{s} u, \tag{11.42}$$

where u substitutes for the x- and y-coordinates. Eq. (11.37) thus reduces to

$$x'' + \left(k_\text{n} + R_x^{-2}\right) x - k_\text{s} y = \delta R_x^{-1}. \tag{11.43}$$

In the same way, the equation of motion in the vertical plane can be derived:

$$y'' - \left(k_\text{n} - R_y^{-2}\right) y - k_\text{s} x = \delta R_y^{-1}. \tag{11.44}$$

In Eqs. (11.43) and (11.44), $R_x = \kappa_x^{-1}$ and $R_y = \kappa_y^{-1}$ are the radii of curvature in the xs- and ys-planes, respectively. Notice how the skew gradient introduces linear coupling between the equations of motion in the different planes. This has implications for the manufacturing tolerances of the magnets, the measurement of the magnetic axis, and the alignment of the quadrupoles in the tunnel.

11.4 Weak Focusing

For a circular reference orbit of radius R and a magnetic field that is independent of s, the equations of motion for a particle with design momentum are

$$x'' + \left(k_\text{n} + R^{-2}\right) x = 0, \qquad y'' - k_\text{n} y = 0, \tag{11.45}$$

4 See Chapter 9.

as long as the skew gradient k_s is zero. Both the normalized gradient and the term R^{-2} contribute to the focusing strength in the accelerator.[5] A particle traveling outside the ideal orbit will take a longer path through the bending magnet and will thus be deflected more than a particle in the ideal orbit; see Figure 11.3 (left). Notice that the R^{-2} term in the vertical plane is missing, as we assume that bending magnets deflect only in the horizontal plane.

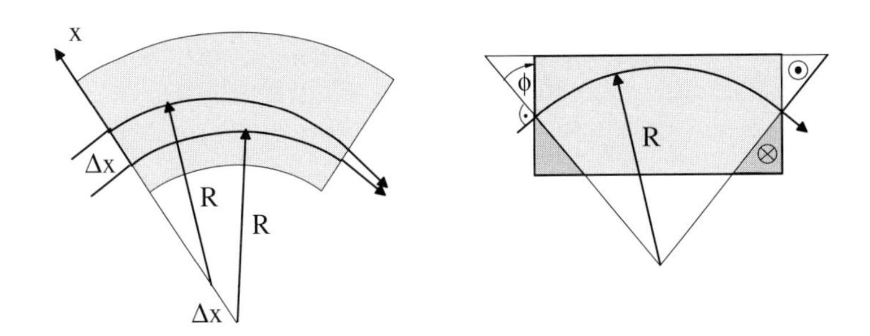

Figure 11.3 Left: Weak focusing in a bending magnet. Right: Edge focusing in a rectangular magnet. The particle in the horizontal plane "sees" two focusing quadrupoles at the entrance and exit of the magnet.

It is impossible to increase the horizontal focusing without regard to the vertical focusing. Unavoidably, focusing in one plane, for example, by constructing combined-function magnets with dipole and quadrupole field components, leads to a defocusing in the other plane. Stable motion in both planes can be achieved only if the equations have oscillatory solutions for both planes. This condition leads to the requirements

$$k_n + R^{-2} > 0\,, \qquad k_n < 0\,, \tag{11.46}$$

and thus to

$$\frac{1}{R} > -\frac{1}{B_0} g > 0\,, \tag{11.47}$$

known as the *weak focusing* condition, for reasons explained below. With the *field index n* defined by $n := -Rg/B_0$, this can simply be written as $1 > n > 0$, which is known as the *Steenbeck criterion* [25]. If the condition (11.47) is obeyed, the equation of motion in the horizontal plane has the solution

$$x(s) = x_0 \cos\left(\sqrt{k_n + R^{-2}}\, s - \varphi\right)\,, \tag{11.48}$$

where x_0 and φ are integration constants. The wavelength

$$\lambda = \frac{2\pi}{\sqrt{k_n + R^{-2}}} \tag{11.49}$$

5 It is customary to define $K := k_n + R^{-2}$ as the total focusing strength.

of this oscillation is called the *betatron oscillation* and the number of these oscillations around the machine circumference is called the *tune*

$$Q = R\sqrt{k_{\mathrm{n}} + R^{-2}}\,. \tag{11.50}$$

The stability condition for weak focusing therefore requires that $Q < 1$.

By adjusting the field index, for example, by combining edge focusing and a field gradient, it is possible to achieve net focusing in both planes. Unfortunately, the strength of the achievable focusing under these conditions is very limited. With typical values of the LHC, the weak focusing condition would allow only for small gradients in the range of 0.003 Tm^{-1}. Weak focusing leads to large beam sizes (weakly focused) and therefore costly machines with large-aperture magnets.

Below we will discuss the *strong focusing* scheme for $k_n \gg R^{-2}$ leading to symmetric equations in the horizontal and the vertical planes. For a long time, it was believed that strong focusing was not possible because of the constraint imposed by Laplace's equation. After all, one cannot construct a magnetic lens that focuses in both planes simultaneously. The strong-focusing concept arose from the realization that net positive focusing can be obtained in both planes with a focusing lens followed by a defocusing lens, provided the distance between the two lenses is set appropriately. Even though there is still a limitation in the maximum focusing with this technique, the focusing is much stronger than what can be achieved with a single magnetic lens.

11.5 Thin-Lens Approximations

The concepts of geometrical optics can be applied to the calculation of the angular deflection of a particle traversing a short quadrupole; see Figure 11.2 (right). Consider the particle entering the quadrupole at a horizontal displacement x with respect to the central axis. The horizontal displacement is assumed to be constant during the passage through the magnet. The angular displacement of the particle is found by integrating the differential equation (11.45), with the weak-focusing term neglected:

$$\int_0^L x''\,\mathrm{d}s = x'(L) - x'(0) = \tan\phi \tag{11.51}$$

and

$$\int_0^L k_{\mathrm{n}}\,x\,\mathrm{d}s = L\,k_{\mathrm{n}}\,x\,. \tag{11.52}$$

Thus we obtain

$$\tan\phi = -Lk_{\mathrm{n}}x = -\frac{LB_y}{B_0R} = -\frac{Lgx}{B_0R}\,. \tag{11.53}$$

In geometrical optics, the focal length f of a lens is related to its angle of deflection ϕ by $\tan\phi = -x/f$ (see the right-hand side of Figure 11.2), and consequently

$$f = \frac{1}{k_n L} = \frac{B_0 R}{gL}. \qquad (11.54)$$

11.6
Transfer Matrices

If there is no skew gradient in the magnetic field, and if the deflection occurs only in the horizontal plane, the equation of transverse motion of a particle with design momentum ($\delta = 0$) is given by Eq. (11.45):

$$x'' + K(s)x = 0, \qquad (11.55)$$

where $K(s) := R(s)^{-2} + k_n(s)$ combines the focusing strength of the quadrupole and the weak-focusing term of the dipole field. When $K(s)$ is periodic, Eq. (11.55) is known as *Hill's equation*,[6] which originally arose in the context of orbit calculations in astronomy.

Following Wille [29] we derive the *transfer matrices* $[M]$ for piecewise-constant $K(s)$ along a length L of the magnet element. In this case, Hill's equation (11.55) reduces to the equation of a harmonic oscillator. The solution for a horizontally defocusing quadrupole ($k_n < 0$) is

$$x(s) = A\cosh\sqrt{|k_n|}s + B\sinh\sqrt{|k_n|}s, \qquad (11.56)$$

$$x'(s) = \sqrt{|k_n|}A\sinh\sqrt{|k_n|}s + \sqrt{|k_n|}B\cosh\sqrt{|k_n|}s. \qquad (11.57)$$

The parameters A and B are determined by inserting the initial conditions $(x(s_0), x'(s_0))^T = (x_0, x'_0)^T$. It yields

$$x(s) = x_0\cosh\sqrt{|k_n|}s + \frac{x'_0}{\sqrt{|k_n|}}\sinh\sqrt{|k_n|}s, \qquad (11.58)$$

$$x'(s) = x_0\sqrt{|k_n|}\sinh\sqrt{|k_n|}s + x'_0\cosh\sqrt{|k_n|}s. \qquad (11.59)$$

6 George Hill (1838–1914).

These equations, which map the trajectory vector from the onset of a magnet to a point s within, can be rewritten in matrix form for the transport through a quadrupole of length L as

$$\begin{pmatrix} x(s_0+L) \\ x'(s_0+L) \end{pmatrix} = \begin{pmatrix} \cosh\sqrt{|k_\mathrm{n}|}L & \frac{1}{\sqrt{|k_\mathrm{n}|}}\sinh\sqrt{|k_\mathrm{n}|}L \\ \sqrt{|k_\mathrm{n}|}\sinh\sqrt{|k_\mathrm{n}|}L & \cosh\sqrt{|k_\mathrm{n}|}L \end{pmatrix} \begin{pmatrix} x(s_0) \\ x'(s_0) \end{pmatrix}, \tag{11.60}$$

or in more compact form as $\{x(s_0+L)\} = [M]_\mathrm{qd}\{x(s_0)\}$, where $[M]_\mathrm{qd} = (m_{ij}) : P_2 \to P_2 : \{x(s_0)\} \mapsto \{x(s_0+L)\}$ is the *transfer matrix*, $\{x\} \in P_2$ the *trajectory vector*, and P_2 is the 2D *phase space*.[7] In a similar manner, we can derive for the horizontally focusing quadrupole ($k_\mathrm{n} > 0$):

$$[M]_\mathrm{qf} = \begin{pmatrix} \cos\sqrt{k_\mathrm{n}}L & \frac{1}{\sqrt{k_\mathrm{n}}}\sin\sqrt{k_\mathrm{n}}L \\ -\sqrt{k_\mathrm{n}}\sin\sqrt{k_\mathrm{n}}L & \cos\sqrt{k_\mathrm{n}}L \end{pmatrix} \tag{11.61}$$

and for a field-free drift space of length L_d,

$$[M]_0 = \begin{pmatrix} 1 & L_\mathrm{d} \\ 0 & 1 \end{pmatrix}. \tag{11.62}$$

For a bending magnet of length L not containing any gradient field, we can set $K(s) = R^{-2}$ in Eq. (11.55) and obtain the transfer matrix

$$[M]_\mathrm{b} = \begin{pmatrix} \cos\frac{L}{R} & R\sin\frac{L}{R} \\ -\frac{1}{R}\sin\frac{L}{R} & \cos\frac{L}{R} \end{pmatrix}. \tag{11.63}$$

Notice the weak-focusing effect of the dipole represented by the negative component m_{21} of $[M]_\mathrm{b}$.

The additional edge-focusing effect due to the tilt angle ψ of the magnet end-plate about the s-axis, proportional to $\tan\psi/R$, is described by the transfer matrix

$$[M]_\mathrm{ef} = \begin{pmatrix} 1 & 0 \\ -\frac{\tan\psi}{R} & 1 \end{pmatrix}. \tag{11.64}$$

A schematic is shown in Figure 11.3 (right). We see that $\det[M] = 1$, which is generally true for the matrix representation of a volume-preserving, linear transformation $f : V \to V$ from a vector space V into itself. If the particle

7 See Section 2.12.

motion in both planes is described by trajectory vectors $\{x\} = (x, x', y, y')^T$ in a four-dimensional phase space P_4, the transfer matrix for the horizontally focusing quadrupole ($k > 0$) is

$$[M]_{\mathrm{QF}} = \begin{pmatrix} \cos\sqrt{k_\mathrm{n}}L & \frac{1}{\sqrt{k_\mathrm{n}}}\sin\sqrt{k_\mathrm{n}}L & 0 & 0 \\ -\sqrt{k_\mathrm{n}}\sin\sqrt{k_\mathrm{n}}L & \cos\sqrt{k_\mathrm{n}}L & 0 & 0 \\ 0 & 0 & \cosh\sqrt{|k_\mathrm{n}|}L & \frac{1}{\sqrt{|k_\mathrm{n}|}}\sinh\sqrt{|k_\mathrm{n}|}L \\ 0 & 0 & \sqrt{|k_\mathrm{n}|}\sinh\sqrt{|k_\mathrm{n}|}L & \cosh\sqrt{|k_\mathrm{n}|}L \end{pmatrix}. \tag{11.65}$$

Thin-lens approximation[8] of the transfer matrices in P_2 yields for the horizontal plane:

$$[M]_{\mathrm{qd}} = \begin{pmatrix} 1 & 0 \\ k_\mathrm{n}L & 1 \end{pmatrix}, \qquad [M]_{\mathrm{qf}} = \begin{pmatrix} 1 & 0 \\ -k_\mathrm{n}L & 1 \end{pmatrix}. \tag{11.66}$$

11.7 Strong Focusing and the FODO Cell

The alternating gradient focusing or strong focusing scheme invented by Courant an others [5] at Brookhaven and independently by Christofilos [4] uses azimuthally varying gradients and allows focusing strengths of $k_\mathrm{n} \gg R^{-2}$. For one doublet of a defocusing quadrupole and a focusing quadrupole of the same strength and length L, connected via a drift space of length L_d, the thin-lens approximation is

$$\begin{aligned} [M]_{\mathrm{qf}}[M]_0[M]_{\mathrm{qd}} &= \begin{pmatrix} 1 & 0 \\ -k_\mathrm{n}L & 1 \end{pmatrix} \begin{pmatrix} 1 & L_\mathrm{d} \\ 0 & 1 \end{pmatrix} \begin{pmatrix} 1 & 0 \\ k_\mathrm{n}L & 1 \end{pmatrix} \\ &= \begin{pmatrix} 1 + k_\mathrm{n}L\,L_\mathrm{d} & L_\mathrm{d} \\ -k_\mathrm{n}^2L^2\,L_\mathrm{d} & 1 - k_\mathrm{n}L\,L_\mathrm{d} \end{pmatrix}. \end{aligned} \tag{11.67}$$

The component m_{21} of the transfer matrix is negative and thus the assembly focuses in the horizontal plane. With Eq. (11.54) the focal length f_{fod} of the doublet is

$$\frac{1}{f_{\mathrm{fod}}} = \frac{L_\mathrm{d}}{f^2}, \tag{11.68}$$

8 $L \to 0$ and $k_n \to \infty$ while $k_n L$ remains constant.

where f is the focal length of the individual quadrupole. Thus

$$[M]_{\mathrm{qf}}[M]_0[M]_{\mathrm{qd}} = \begin{pmatrix} 1+\frac{L_\mathrm{d}}{f} & L_\mathrm{d} \\ -\frac{L_\mathrm{d}}{f^2} & 1-\frac{L_\mathrm{d}}{f} \end{pmatrix}. \tag{11.69}$$

A similar study for the four-dimensional phase space shows that FODO cells composed of horizontally focusing (F) and defocusing quadrupoles (D) with the same strength, connected by empty spaces (O) have a net focusing strength in both the horizontal and the vertical planes. The empty spaces may contain drift spaces, radio-frequency accelerator structures, or bending magnets, all having an insignificant effect on the focusing scheme [11].

Focusing in both planes occurs because particles entering parallel to the axis have a larger displacement in the focusing lens than in the defocusing lens; see Figure 11.2 (right). This assumption is true for small *beam envelopes* and for the separation d between two quadrupoles obeying $d < 2|f|$.

We will now derive a stability criterion for the entire FODO cell shown in Figure 11.4. We can calculate the transfer matrix for the half-cell (subscript hc)

$$[M]_{\mathrm{hc}} = [M]_{\mathrm{qd}/2}[M]_0[M]_{\mathrm{qf}/2} \tag{11.70}$$

from the middle of the focusing quadrupole to the middle of the defocusing quadrupole.[9] As we are dealing with thin lenses, the length of the drift space between the quadrupoles is just half the cell length, $L_\mathrm{d} = L_\mathrm{cell}/2$. Given that a half-strength quadrupole has twice the focusing length, we obtain by comparison to Eq. (11.69),

$$[M]_{\mathrm{hc}} = \begin{pmatrix} 1-\frac{L_\mathrm{cell}}{4f} & \frac{L_\mathrm{cell}}{2} \\ -\frac{L_\mathrm{cell}}{8f^2} & 1+\frac{L_\mathrm{cell}}{4f} \end{pmatrix}. \tag{11.71}$$

The matrix for the second half-cell can be obtained by inverting the sign of the focal length, with the result

$$\begin{aligned}[M]_{\mathrm{FODO}} &= \begin{pmatrix} 1+\frac{L_\mathrm{cell}}{4f} & \frac{L_\mathrm{cell}}{2} \\ -\frac{L_\mathrm{cell}}{8f^2} & 1-\frac{L_\mathrm{cell}}{4f} \end{pmatrix} \begin{pmatrix} 1-\frac{L_\mathrm{cell}}{4f} & \frac{L_\mathrm{cell}}{2} \\ -\frac{L_\mathrm{cell}}{8f^2} & 1+\frac{L_\mathrm{cell}}{4f} \end{pmatrix} \\ &= \begin{pmatrix} 1-\frac{L_\mathrm{cell}^2}{8f^2} & L_\mathrm{cell}\left(1+\frac{L_\mathrm{cell}}{4f}\right) \\ \frac{L_\mathrm{cell}^2}{16f^3}-\frac{L_\mathrm{cell}}{4f^2} & 1-\frac{L_\mathrm{cell}^2}{8f^2} \end{pmatrix}. \end{aligned} \tag{11.72}$$

9 It is, strictly speaking, a $\frac{1}{2}$FODO$\frac{1}{2}$F symmetrical cell, which results in a more compact transfer matrix.

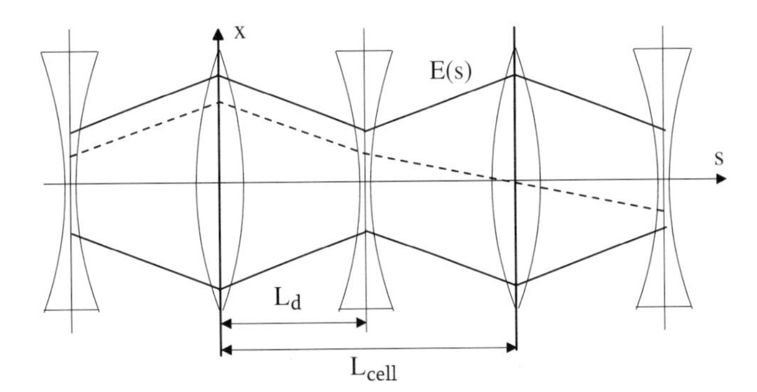

Figure 11.4 Symmetric FODO cell, with schematic particle trajectory and beam envelope $E := \sqrt{\sigma\beta(s)}$. The beta function takes its minimum at the defocusing quadrupole, ensuring the net focusing effect of the cell. We see one half an oscillation in two cells, which corresponds to 90° phase advance per cell. Notice the orbit kick due to the thin-lens approximation.

The motion is stable if the *trace*[10] of the matrix obeys the condition [1, 18]:

$$|\mathrm{tr}[M]| = \left|2 - \frac{L_{\mathrm{cell}}^2}{4f^2}\right| < 2\,. \tag{11.73}$$

The focal length of the quadrupoles must therefore be greater than half their spacing:

$$\left|\frac{L_{\mathrm{hc}}}{2f}\right| < 1\,. \tag{11.74}$$

11.8 The Beta Function, Tune, and Transverse Resonances

We will at this stage briefly recall the assumptions we have made so far:

- A torsion-free orbit with bends either in the horizontal or the vertical plane.
- No particle acceleration.
- No longitudinal magnetic field components, which implies neglecting the stray fields in the magnet end regions and the absence of solenoids.
- No skew gradients resulting from a roll misalignment of the quadrupole magnets.

10 The trace of a matrix is the sum of its diagonal elements, $\mathrm{tr}[A] = \sum_i a_{ij}$; see Section 2.12.

- No higher-order field components.
- No momentum deviation.
- A single particle only.

We will first extend the method to a composite beam with many particles. If we disregard the field of the bending magnet, $R^{-2} = 0$, Eq. (11.45) yields

$$x'' + k_{\mathrm{n}}(s)x = 0\,. \tag{11.75}$$

It should be emphasized that the normalized gradient k_{n} is a function of position s. In circular accelerators, the focusing elements are periodic after one revolution. The FODO cell sequence has an even smaller period. Consider the periodic solution that stems from a simple harmonic motion, albeit that the amplitude is not constant and the phase does not advance linearly in time and distance s:

$$x(s) = \sqrt{\epsilon}\sqrt{\beta(s)}\cos(\psi(s) - \phi)\,. \tag{11.76}$$

Dropping for the moment the explicit indication of the s-dependence yields the derivatives of the displacement function:

$$x' = \sqrt{\epsilon}\frac{\beta'}{2\sqrt{\beta}}\cos(\psi - \phi) - \sqrt{\epsilon}\sqrt{\beta}\psi'\sin(\psi - \phi)\,, \tag{11.77}$$

$$\begin{aligned} x'' = {} & \frac{\sqrt{\epsilon}}{4}\left(\frac{2\beta\beta'' - \beta'^2}{\beta^{3/2}}\right)\cos(\psi - \phi) - \sqrt{\epsilon}\frac{\beta'}{\sqrt{\beta}}\psi'\sin(\psi - \phi) \\ & - \sqrt{\epsilon}\sqrt{\beta}\psi''\sin(\psi - \phi) - \sqrt{\epsilon}\sqrt{\beta}\psi'^2\cos(\psi - \phi)\,. \end{aligned} \tag{11.78}$$

The reason for writing the amplitude function as $\sqrt{\epsilon}\sqrt{\beta(s)}$ can be explained as follows: First, for convenience let us define

$$\alpha(s) := -\frac{\beta'(s)}{2}\,, \qquad \gamma(s) := \frac{1 + \alpha^2(s)}{\beta(s)}\,. \tag{11.79}$$

The position-dependent parameters $\alpha(s), \beta(s), \gamma(s)$ are known as the Twiss parameters or the Courant–Snyder parameters [6]. The physical dimensions of these parameters are $[\alpha] = 1_{\mathrm{D}}$, $[\beta] = 1$ m, $[\gamma] = 1\ \mathrm{m}^{-1}$. From Eq. (11.76) it follows that

$$\cos(\psi(s) - \phi) = \frac{x}{\sqrt{\epsilon}\sqrt{\beta(s)}}\,. \tag{11.80}$$

Inserting this result into Eq. (11.77) yields

$$\sin(\psi(s) - \phi) = \frac{\sqrt{\beta}x'}{\sqrt{\epsilon}} + \frac{\alpha(s)x}{\sqrt{\epsilon}\sqrt{\beta(s)}}\,. \tag{11.81}$$

Because $\sin^2\varphi + \cos^2\varphi = 1$ the constant ϵ can be expressed in terms of x and x' as

$$\frac{x^2}{\beta(s)} + \left(\frac{\alpha(s)}{\sqrt{\beta(s)}}x + \sqrt{\beta(s)}x'\right)^2 = \epsilon\,. \tag{11.82}$$

From this we finally obtain

$$\gamma(s)x^2(s) + 2\alpha(s)x(s)x'(s) + \beta(s)x'^2(s) = \epsilon\,, \tag{11.83}$$

which is the equation of a general ellipse centered at the coordinate axis of the phase space.

Proof. In a rotated coordinate system $(\tilde{x}, \tilde{x}')$, where the semiaxes a and b of the ellipse correspond to the coordinate axis, the parametric form of the ellipse equation is given by $\tilde{x} = a\cos\varphi$, $\tilde{x}' = b\sin\varphi$ for $\varphi \in [0, 2\pi]$. A rotation of the ellipse into the (x, x') coordinate system by the angle ϕ yields

$$\begin{pmatrix} x \\ x' \end{pmatrix} = \begin{pmatrix} \cos\phi & \sin\phi \\ -\sin\phi & \cos\phi \end{pmatrix} \begin{pmatrix} a\cos\varphi \\ b\sin\varphi \end{pmatrix}. \tag{11.84}$$

Solving for $\sin\varphi$ and $\cos\varphi$ yields

$$a\,b = \frac{b}{a}(x\cos\phi - x'\sin\phi)^2 + \frac{a}{b}(x\sin\phi + x'\cos\phi)^2\,. \tag{11.85}$$

Equation (11.83) is now proved by inserting the definitions

$$\beta := \frac{b}{a}\sin^2\phi + \frac{a}{b}\cos^2\phi\,, \tag{11.86}$$

$$\alpha := \left(\frac{a}{b} - \frac{b}{a}\right)\cos\phi\sin\phi\,, \tag{11.87}$$

$$\gamma := \frac{b}{a}\cos^2\phi + \frac{a}{b}\sin^2\phi\,. \tag{11.88}$$

The amplitude factor ϵ, called the *transverse emittance*, is the product of the ellipse's semiaxes and thus proportional to the area of the phase ellipse. The relation between the Twiss parameters and the cardinal points of the phase ellipse is shown in Figure 11.5. The Courant–Snyder parameters of a circle of radius r are $\beta = \gamma = 1$, $\alpha = 0$, and, consequently $\epsilon = r^2$, as expected. □

In the context of beam physics, emittance preservation is a consequence of Liouville's theorem that holds for a synchrotron in the absence of acceleration and radiation. It is important to emphasize that in a multiparticle context, Liouville's theorem strictly applies to a phase space of dimension $6N$, where N

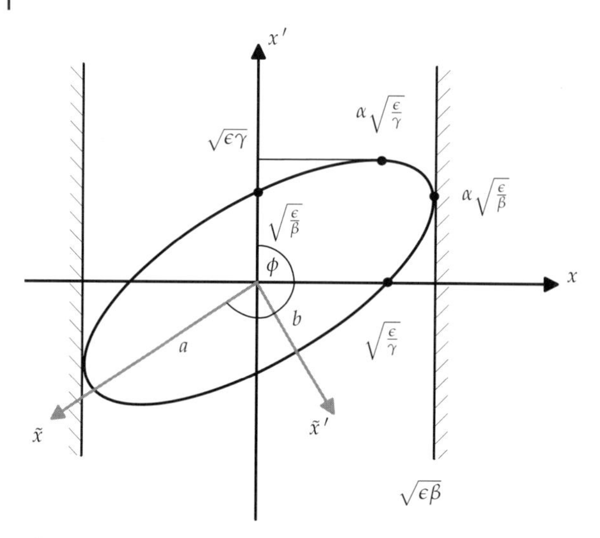

Figure 11.5 The phase-space ellipse of particle motion with the relation between the Twiss parameters and the cardinal points of the ellipse. The largest displacement in the accelerator is attained where β takes its maximum value.

is the number of particles. Conservation of the 6D phase-space volume arises when particle interactions, for example, space charge, intrabeam scattering, and wake fields, can be disregarded. The phase-space area is conserved in a 2D subspace, but only if the particle dynamics is fully decoupled from the other phase-space coordinates.

The position-dependent amplitude function $\beta(s)$, simply called the *beta function*, is affected by the beam focusing. The particles perform oscillations around the ideal orbit within a range of amplitudes bounded by the *envelope* defined by

$$E(s) := x_{\max}(s) = \sqrt{\epsilon\beta(s)}\,; \tag{11.89}$$

see Figure 11.4. The beta function thus defines the transverse size of the beam for a given emittance. It can be calculated from a given value at s_0 by matrix algebra employing the transfer matrices. Since for our purpose the emittance can be considered invariant in time, we obtain

$$\begin{aligned}\epsilon &= \gamma(s_1)x^2(s_1) + 2\alpha(s_1)x(s_1)x'(s_1) + \beta(s_1)x'^2(s_1)\\ &= \gamma(s_0)x^2(s_0) + 2\alpha(s_0)x(s_0)x'(s_0) + \beta(s_0)x'^2(s_0)\,,\end{aligned} \tag{11.90}$$

where s_1 is used as shorthand for $s_0 + L$. The mapping of the trajectory vector from s_0 to $s_1 = s_0 + L$ is given by $\{x(s_1)\} = [M]\{x(s_0)\}$, or

$$\begin{pmatrix} x(s_1) \\ x'(s_1) \end{pmatrix} = \begin{pmatrix} m_{11} & m_{12} \\ m_{21} & m_{22} \end{pmatrix} \begin{pmatrix} x(s_0) \\ x'(s_0) \end{pmatrix} . \tag{11.91}$$

Solving for $\{x(s_0)\}$ gives

$$\begin{pmatrix} x(s_0) \\ x'(s_0) \end{pmatrix} = \begin{pmatrix} m_{22} & -m_{12} \\ -m_{21} & m_{11} \end{pmatrix} \begin{pmatrix} x(s_1) \\ x'(s_1) \end{pmatrix} . \tag{11.92}$$

Inserting this result into Eq. (11.90) yields

$$\begin{aligned} \epsilon &= \gamma(s_1)\, x^2(s_1) + 2\alpha(s_1)\, x(s_1) x'(s_1) + \beta(s_1)\, x'^2(s_1) \\ &= \gamma(s_0)\, (m_{22} x(s_1) - m_{12} x'(s_1))^2 \\ &\quad + 2\alpha(s_0)\, (m_{22} x(s_1) - m_{12} x'(s_1))^2 (-m_{21} x(s_1) + m_{11} x'(s_1))^2 \\ &\quad + \beta(s_0)\, (-m_{21} x(s_1) + m_{11} x'(s_1))^2 . \end{aligned} \tag{11.93}$$

Simplifying, sorting terms, and comparing coefficients yields

$$\begin{aligned} \beta(s_1) &= m_{11}^2\, \beta(s_0) - 2\, m_{12} m_{11}\, \alpha(s_0) + m_{12}^2\, \gamma(s_0) , \\ \alpha(s_1) &= -m_{21} m_{11}^2\, \beta(s_0) + (m_{22} m_{11} + m_{12} m_{21})\, \alpha(s_0) - m_{22} m_{12}\, \gamma(s_0) , \\ \gamma(s_1) &= m_{21}^2\, \beta(s_0) - 2\, m_{22} m_{21}\, \alpha(s_0) + m_{22}^2\, \gamma(s_0) , \end{aligned} \tag{11.94}$$

or in matrix notation,

$$\begin{pmatrix} \beta(s_1) \\ \alpha(s_1) \\ \gamma(s_1) \end{pmatrix} = \begin{pmatrix} m_{11}^2 & -2 m_{11} m_{12} & m_{12}^2 \\ -m_{21} m_{21} & m_{11} m_{22} + m_{12} m_{21} & -m_{22} m_{12} \\ m_{21}^2 & -2 m_{22} m_{21} & m_{22}^2 \end{pmatrix} \begin{pmatrix} \beta(s_0) \\ \alpha(s_0) \\ \gamma(s_0) \end{pmatrix} . \tag{11.95}$$

Twiss parameters at s_0 can be transformed to any other position in the ring using a beam optics program such as MAD-X [10]. The matrix coefficients are determined by the focusing properties of the lattice elements and are just those derived in Section 11.6, for the transport of a single particle.

The matrix elements for the transformation of a circle, for which $\beta = \gamma = 1$, $\alpha = 0$, to an ellipse are $m_{11} = \sqrt{\beta}$, $m_{12} = 0$, $m_{21} = -\alpha/\sqrt{\beta}$, $m_{22} = 1/\sqrt{\beta}$. The points needed for drawing the phase-space ellipse can thus be obtained by transforming equally spaced points on a circle using the mapping (11.91) with these matrix elements [19]. For a drift space of length L_d, the matrix elements of $[M]_0$ are $m_{11} = m_{22} = 1$, $m_{12} = L_\mathrm{d}$, and $m_{21} = 0$ so that

$$\begin{pmatrix} \beta(s_0+L_\mathrm{d}) \\ \alpha(s_0+L_\mathrm{d}) \\ \gamma(s_0+L_\mathrm{d}) \end{pmatrix} = \begin{pmatrix} 1 & -2L_\mathrm{d} & L_\mathrm{d}^2 \\ 0 & 1 & -L_\mathrm{d} \\ 0 & 0 & 1 \end{pmatrix} \begin{pmatrix} \beta(s_0) \\ \alpha(s_0) \\ \gamma(s_0) \end{pmatrix}, \tag{11.96}$$

and therefore,

$$\beta(s_0 + L_\mathrm{d}) = \beta(s_0) - 2L_\mathrm{d}\alpha(s_0) + L_\mathrm{d}^2\gamma(s_0)\,. \tag{11.97}$$

Consequently, β varies parabolically in drift spaces. At $s_0 = 0$ the parameter $\alpha(0) := -\frac{\beta'(0)}{2}$ is zero, and thus for symmetrical optics it yields $\beta^* := \beta(0)$, and

$$\beta(s) = \beta^* + \frac{s^2}{\beta^*}, \tag{11.98}$$

because γ is constant and $\gamma(0) = 1/\beta^*$. A special lattice structure, called *low-beta insertion*, must be designed around the large drift spaces where the particle detectors are installed. The easiest structure to construct would be two symmetrically placed quadrupole doublets. This would result in a smaller β^* in one plane than the other, which might be desirable for limiting beam–beam effects in electron–positron colliders. To realize a beam crossing for which $\beta_x^* = \beta_y^*$, two quadrupole triplets are required.

The beta function also defines the phase of the betatron oscillations as we will now show: Inserting Eqs. (11.78) and (11.77) into the equation of motion (11.76) yields

$$\frac{\sqrt{\epsilon}}{\beta^{3/2}}\left(\frac{\beta\beta''}{2} - \frac{\beta'^2}{4} - \beta^2\psi'^2 + \beta^2 k_\mathrm{n}(s)\right)\cos(\psi - \phi) \\ - \frac{\sqrt{\epsilon}}{\sqrt{\beta}}(\beta'\psi' + \beta\psi'')\sin(\psi - \phi) = 0\,. \tag{11.99}$$

This equation can be fulfilled only if both of the following equations hold:

$$\frac{\beta\beta''}{2} - \frac{\beta'^2}{4} - \beta^2\psi'^2 + \beta^2 k_\mathrm{n}(s) = 0 \tag{11.100}$$

$$\beta'\psi' + \beta\psi'' = (\beta\psi')' = 0\,. \tag{11.101}$$

The second equation can be integrated to yield $\beta\psi'(s) = \mathcal{C}$. The integration constant $\mathcal{C}$ can be set to 1, which yields

$$\psi' = \frac{1}{\beta(s)}\,, \qquad \psi(L) = \int_{s_0}^{s_0+L} \frac{1}{\beta(s)} \mathrm{d}s\,, \tag{11.102}$$

where $\psi(L)$ is called the *betatron phase advance*. The resonant behavior of the beam depends on the betatron phase ψ over one complete revolution in the accelerator. The phase advance is inversely proportional to the beta function, which depends on the beam focusing and varies with position. Assuming an accelerator of circumference $C = NL$ with N identical cells, the phase advance per revolution is $N\psi$. Because β is never zero, the number of betatron oscillations per revolution can thus be calculated from

$$Q = \frac{1}{2\pi} \int_0^C \frac{1}{\beta(s)} \mathrm{d}s\,, \tag{11.103}$$

which is called the *Q-value* or *tune*. If the beta function was constant, Eq. (11.103) would simplify to $\beta = R/Q$, where R is the machine radius. The beta function can be interpreted as the local wavelength of the oscillation, divided by 2π.

We have seen that the elements of the transfer matrix $[M]$ can be used to determine the optical functions. In an equivalent way it is possible to express the transfer matrix in terms of the optical parameters with [29]

$$[M] = \begin{pmatrix} \sqrt{\frac{\beta(s_0+L)}{\beta(s_0)}}(\cos\psi + \alpha(s_0)\sin\psi) & \sqrt{\beta(s_0+L)\beta(s_0)}\sin\psi \\ \frac{(\alpha(s_0)-\alpha(s_0+L))\cos\psi-(1+\alpha(s_0)\alpha(s_0+L))\sin\psi}{\sqrt{\beta(s_0+L)\beta(s_0)}} & \sqrt{\frac{\beta(s_0+L)}{\beta(s_0)}}(\cos\psi - \alpha(s_0)\sin\psi) \end{pmatrix}. \tag{11.104}$$

For the periodic FODO cell ($\alpha(s_0+L) = \alpha(s_0)$, $\beta(s_0+L) = \beta(s_0)$), the transfer matrix reduces to

$$[M]_{\text{FODO}} = \begin{pmatrix} \cos\psi + \alpha(s)\sin\psi & \beta(s)\sin\psi \\ -\gamma(s)\sin\psi & \cos\psi - \alpha(s)\sin\psi \end{pmatrix}, \tag{11.105}$$

where $\mathrm{tr}[M] = 2\cos\psi$. For the calculation of β_{max} in the FODO cell, we set $\alpha = 0$ and compare the matrices (11.105) and (11.72). This yields

$$1 - \frac{L_{\text{cell}}^2}{8f^2} = \cos\psi = 1 - 2\sin^2\frac{\psi}{2} \tag{11.106}$$

and therefore

$$\sin\frac{\psi}{2} = \pm\frac{L_{\text{cell}}}{4f}\,. \tag{11.107}$$

The maximum of the beta function is

$$\beta_{\max} = \frac{L_{\text{cell}}}{\sin\psi}\left(1+\sin\frac{\psi}{2}\right). \tag{11.108}$$

For a FODO cell, the minimum $\beta_{\max}$ and thus the minimum requirements on the aperture are realized when $\psi = 76°$. In addition to beam size, chromaticity is a concern. Chromaticity increases rapidly for phase advances $> 90°$. The standard choices are usually 45, 60, and 90° because they are favorable to correction schemes. The LHC features an FODO lattice with an approximately 90° phase advance per cell in both planes. Figure 11.6 displays the beta functions in both planes for an LHC arc cell. The beta functions were calculated with the beam optics program MAD-X [10]. The horizontal beta function β_x takes its maximum at the center of the horizontally focusing quadrupole and its minimum at the defocusing quadrupoles.

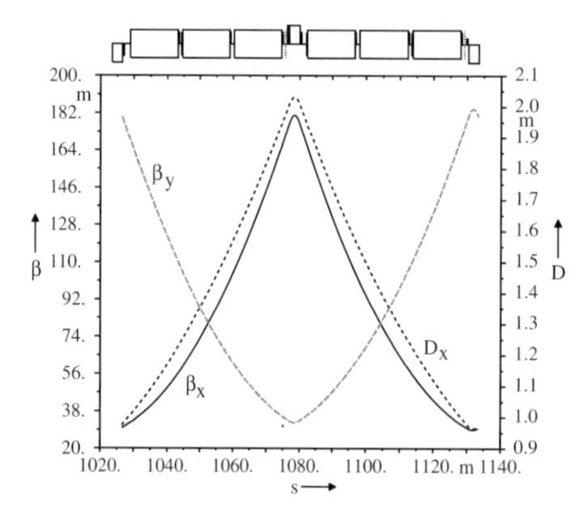

Figure 11.6 Periodic optics solution for an LHC arc cell, which is a matched, symmetric FODO cell. Notice how the beta function varies parabolically in the regions without quadrupolar focusing; cf. Eq. (11.97).

We briefly mentioned that a special insertion lattice is required around the drift spaces for the particle detectors. Low beta-function values are achieved in both planes by means of quadrupole triplets to the left and right of the interaction point. Additional quadrupoles are installed to match the optical parameters of the triplet to the parameters of the arc cells. Figure 11.7 shows how the beta function varies across the high luminosity insertion IP5 of the LHC, where the CMS detector is located [24]. A detailed view on the region of the inner triplet is shown in Figure 11.8 for the collision optics. Notice how the beta function varies within the magnet elements. As a consequence, one should not compensate for the local field errors in the magnet ends by altering the layout of the magnet cross section. Instead we must aim at minimizing the integrated multipole field errors for each end separately. Figure 11.8 (right) shows the quantity $\sqrt{\beta(s)}$, which is proportional to the beam envelope.

The minimum beta function β^* is 11 m for the LHC injection optics (version 6.5) and 0.55 m for the collision optics; compare Figure 11.7 (left) and

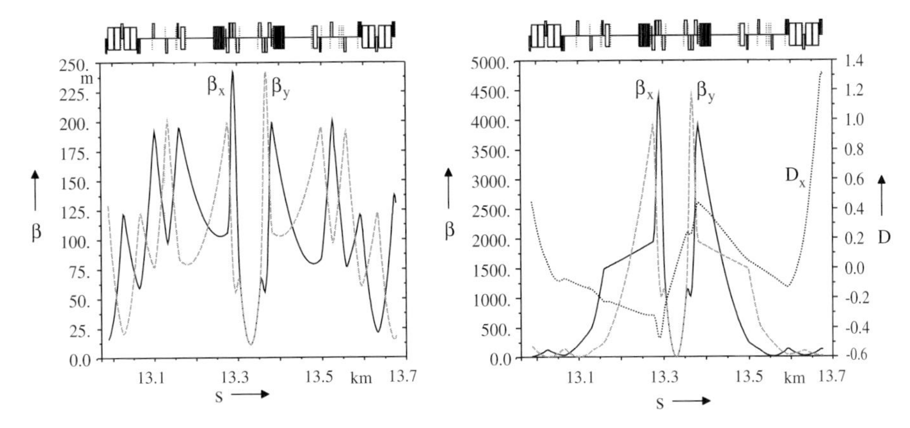

Figure 11.7 Injection optics (left) and collision optics (right) of the high luminosity insertion at IP5, the location of the CMS detector.

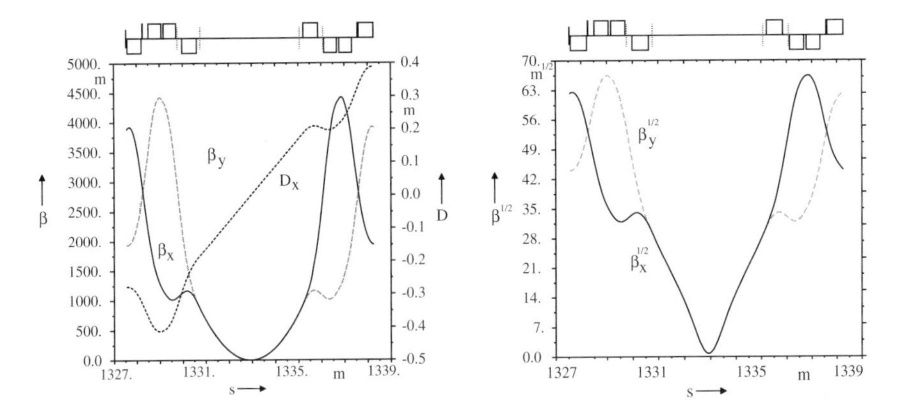

Figure 11.8 Left: Horizontal and vertical beta function in the inner-triplet region at point 5 of the LHC (collision optics). Right: $\sqrt{\beta(s)}$ that is proportional to the beam envelope [24].

(right). The optics in the insertion region must provide for the different β^* values by smooth changes in quadrupole strength from the injection to the collision energy. Hardware constraints are imposed by the common channel for both beams, which results in the same quadrupole strength for both. The small β^* value at collision results in large beta functions in the inner-triplet quadrupoles. The magnets are thus wide-aperture types with a bore of 70 mm, equipped with large diameter beam-screens. Good magnet alignment is required to limit the orbit kicks caused by feed-down.[11]

Because of the periodicity of the beta function, the tune is independent of the s-position. The pair (Q_x, Q_y) of horizontal and vertical tunes is called the

11 See Section 6.1.1.2.

working point of the machine. To avoid resonances, the tunes must avoid integer values and ratios of small integers. Resonances occur when the betatron phase takes the same value after an integral number of turns, resulting in a coherent amplitude growth. In a working accelerator, the strength of resonances depends on the nature of the field errors, as well as their spatial distribution weighted by the beam size. High-order resonances are driven by high-order, multipolar field errors and coupling resonances by skew multipoles. For example, roll misalignments of quadrupoles (in particular of the strong low-beta quadrupoles), introduce skew-quadrupole field errors, which drive second-order resonances.

The condition for transverse resonances may be expressed as

$$mQ_x + nQ_y = p, \qquad m, n, p \in \mathbb{N}, \tag{11.109}$$

where $m + n$ is the order of the resonance.

Example: For the LHC $Q_x = 64.28$ and $Q_y = 59.31$ at injection. At injection, the beta function in the arc has a maximum of 177 m and a minimum of 30 m. The beam screen has a height of 34.4 mm and a width of 44 mm. Setting a minimum aperture of 10 σ with respect to the RMS beam size, and including an allowance for magnet alignment tolerances with respect to the central beam trajectory[12] results in a peak nominal beam size of 1.2 mm. This also implies a maximum acceptable, normalized transverse beam emittance of 3.75 μm at a proton beam energy of 450 GeV. □

The choice of a working point in the betatron tune diagram is a very important step in the design of an accelerator. To study the dependence of the *dynamic aperture* on the choice of the working point, particle tracking studies were performed [16] over a wide tune range; see the dotted line in Figure 11.9 (right). The dynamic aperture is defined as the maximum initial oscillation amplitude that guarantees stable particle motion over a given number of turns. The dynamic aperture is normally expressed in multiples of the RMS beam size σ.

Figure 11.10 shows how the dynamic aperture is reduced in the region of the main resonances, in particular the third-order resonances driven by sextupoles, but also for fourth and higher orders. The nominal working point is chosen in the save region between the third- and fourth-order resonances.

12 Between 1.2 mm transverse radially for the MQX insertion quadrupoles and 2 mm for the main dipoles.

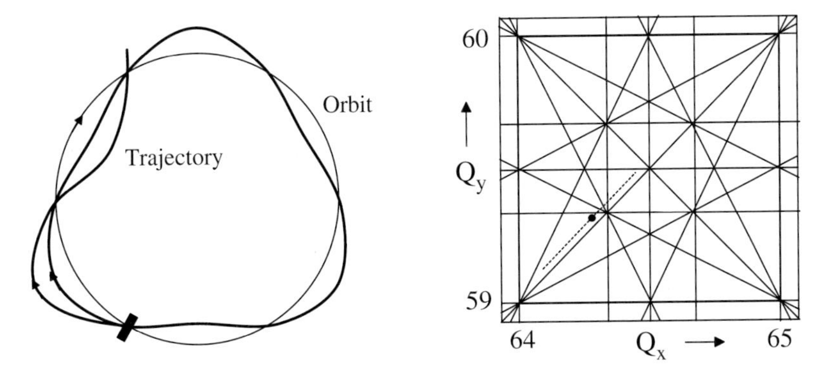

Figure 11.9 Left: Resonant excitation of betatron oscillations by a dipole field error at an integer value of the tune. Right: Transverse resonances in both planes up to order 3. Notice that higher-order multipole field errors in the magnet elements excite higher-order resonances, which are not displayed.

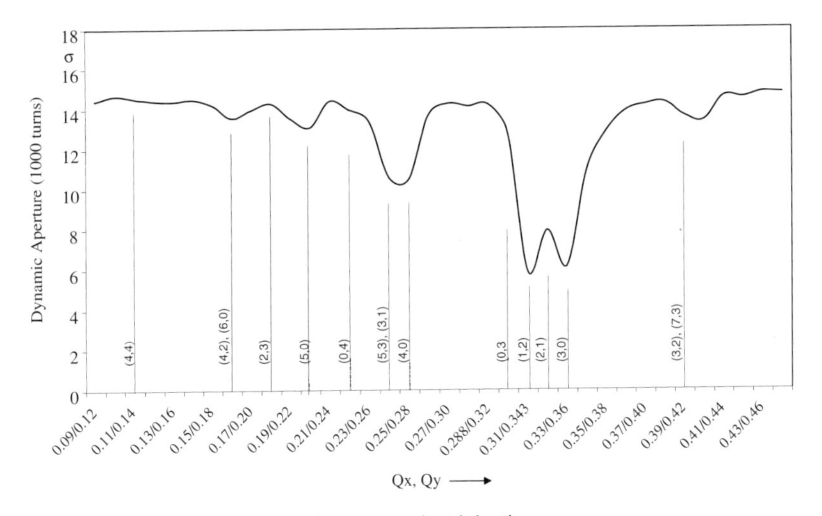

Figure 11.10 Dynamic aperture (1000 turns) at injection energy, as a function of the tune (see dotted line in Figure 11.9 (right)) using the target error-table for the main dipoles and quadrupoles. Average values over 60 random seeds are given for each pair of tunes [16].

11.9 Off-Momentum Particles

We will now return to the more general equation (11.43) of motion for off-momentum particles in the storage ring. The magnet elements induce effects that show similarities with geometrical light optics:

- In an optical prism, *dispersion* causes the spatial separation of white light into the spectral components of different wavelength (colors). This is due to the fact that the *refractive index* of the glass medium increases with shorter wavelength of the light. In the accelerator, the effect of momentum deviation on the particle's trajectory through a dipole magnet gives rise to a new equilibrium orbit that can be described by the normalized transverse coordinate known as local dispersion function. As we will show, this function can be calculated with beam-optics programs employing extended transfer matrices.
- In photographic lenses dispersion causes *chromatic aberration*, which is seen as color fringes around the image of an object. Photographic lenses can be made achromatic by combining lens elements made of glasses with different refractive indices. However, the focal length of a quadrupole magnet always varies with momentum. Thus, particles of different momentum "see" different focusing strengths in the quadrupoles and thus have different betatron oscillation frequencies. The effect can be compensated for by installing sextupole magnets in the accelerator.

11.9.1 Dispersion

Eq. (11.43) states that the effect of a momentum deviation on the particle's trajectory only is important if $R_x^{-1} \neq 0$, and consequently the equations of motion can be reduced to the case inside the bending magnets. Limiting ourselves again to the horizontal plane we must find a general solution of

$$x'' + \frac{1}{R_x^2} x = \delta \frac{1}{R_x} \tag{11.110}$$

as a sum of the complete solution of the homogeneous equation and a particular solution of the inhomogeneous equation

$$x_g(s) = x(s) + x_D(s)\,. \tag{11.111}$$

The betatron motion $x(s)$ is superimposed on a new equilibrium orbit $x_D(s)$. As δ is assumed to be constant for many revolutions in the circular accelerator, a solution for a given δ can be scaled linearly. It is therefore convenient to normalize the transverse coordinate x_D with respect to the momentum error:

$$D(s) = \frac{x_D}{\delta} = \frac{x_D}{\Delta p / p_0}\,. \tag{11.112}$$

$D(s)$ is called the local dispersion function. It is the orbit of an ideal particle ($\delta = 1$). The dispersion function must satisfy the inhomogeneous differential equation[13]

$$D''(s) + \frac{1}{R^2} D(s) = \frac{1}{R} . \tag{11.113}$$

Its general solution is

$$D(s) = A \cos \frac{s}{R} + B \sin \frac{s}{R} + R , \tag{11.114}$$

and thus

$$D'(s) = -\frac{A}{R} \sin \frac{s}{R} + \frac{B}{R} \cos \frac{s}{R} . \tag{11.115}$$

Because of the initial conditions $D(0) = D_0$ and $D'(0) = D_0'$ we obtain

$$D(s) = D_0 \cos \frac{s}{R} + D_0' R \sin \frac{s}{R} + R \left(1 - \cos \frac{s}{R}\right) , \tag{11.116}$$

$$D'(s) = -\frac{D_0}{s} \sin \frac{s}{R} + D_0' R \cos \frac{s}{R} + \sin \frac{s}{R} , \tag{11.117}$$

which yields the following solution, in matrix form, for the transport through a bending magnet of length L:

$$\begin{pmatrix} D(s_0+L) \\ D'(s_0+L) \\ 1 \end{pmatrix} = \begin{pmatrix} \cos \frac{L}{R} & R \sin \frac{L}{R} & R(1 - \cos \frac{L}{R}) \\ -\frac{1}{R} \sin \frac{L}{R} & \cos \frac{L}{R} & \sin \frac{L}{R} \\ 0 & 0 & 1 \end{pmatrix} \begin{pmatrix} D(s_0) \\ D'(s_0) \\ 1 \end{pmatrix} . \tag{11.118}$$

Hence, using Eq. (11.111) we can derive for the transverse position of the dispersive particle:

$$x_g(s) = x(s) + x_D(s) = x(s) + D(s) \frac{\Delta p}{p_0} . \tag{11.119}$$

The path length of the trajectory is consequently a function of momentum, which in turn determines the period of the revolution. In the interaction point the dispersion should be zero in order to guarantee high luminosity.

The remedy is therefore to suppress the dispersion at the exit of the arcs until the next bend on the opposite side of the interaction point. Starting with initial conditions $D(0) = D'(0) = 0$, it is necessary to build up, in what is

13 We will henceforth omit the subscript x.

called the *dispersion–suppressor region*, a number of FODO cells (two for the LHC) with a special dipole distribution (two instead of three) and individual powering of the medium-current quadrupole magnets shown in Figure B.2. The LHC requires three magnetic lengths of these MQM quadrupoles of a nominal current rating of 6 kA and lengths ranging from 2.4 to 4.8 m.

11.9.2 Chromaticity

Consider a particle beam with a certain momentum deviation that is small with respect to the design momentum. Particles with a momentum $p = p_0 + \Delta p$ traveling through a quadrupole with field gradient g are subjected to a modified focusing strength given by

$$k(p) = -\frac{e}{p}g = -\frac{e}{p_0 + \Delta p}g \approx \frac{e}{p_0}\left(1 - \frac{\Delta p}{p_0}\right)g =: k_0 - \Delta k\,. \tag{11.120}$$

As the off-momentum particle maintains its momentum for many revolutions in the circular accelerator, the particle is in fact subjected to a focusing of modified strength in all the quadrupoles, which creates a tune-shift ΔQ, proportional to the relative momentum deviation $\delta = \frac{\Delta p}{p}$. The quantity

$$Q' := p\frac{\mathrm{d}Q}{\mathrm{d}p} \tag{11.121}$$

is called the *chromaticity*. $[Q'] = 1_{\mathrm{D}}$. In large accelerators, the tune-shift can be so large that it would be impossible to keep the working point between the resonance lines in the tune diagram. The chromaticity correction is accomplished with magnets of a focusing strength that depends on the transverse coordinate, as shown in Figure 11.11. We will need the next higher component in the expansion of the transverse magnetic field:

$$\begin{aligned} B(x,0,s) &= B_0(0,0,s) + \frac{\partial B_x}{\partial x}x + \frac{1}{2}\frac{\partial^2 B_x}{\partial x^2}x^2 + \frac{1}{3}\frac{\partial^3 B_x}{\partial x^3}x^3 + \cdots \\ &=: B_0(0,0,s) + gx + \frac{1}{2}g'x^2 + \frac{1}{3}g''x^3 + \cdots\,, \end{aligned} \tag{11.122}$$

where g' and g'' are the sextupole and octupole field components, respectively. The sextupoles are installed at places where the dispersion is nonzero and the particles are sorted according to their momenta; see the schematic of Figure 11.11.

Sextupole magnets have the undesirable effect, however, of exciting third-integer and nonlinear resonances due to the nonlinear forces[14] on the particles, thereby generating anharmonic betatron oscillations.

14 See Section 6.1.1.2.

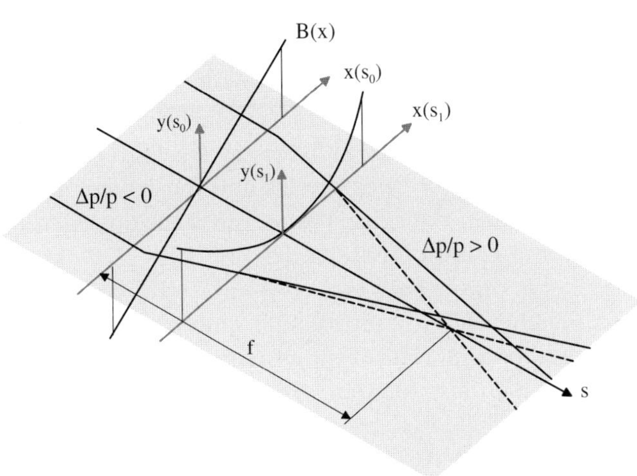

Figure 11.11 Schematic correction of chromatic aberrations in the horizontal plane. A sextupole magnet at s_1 acts as an additional focusing or defocusing element on the dispersive particles. The sextupole's focusing strength is zero on the axis, and therefore the magnet has no effect on particles with nominal momentum.

The chromaticity or lattice sextupoles (MS) and lattice skew sextupoles (MSS) for the LHC are built from the same magnet module (turned by 30° in the case of the MSS), mounted with a horizontal or vertical dipole orbit corrector MCBH(V) in a twin-aperture structure. These assemblies are installed on the downstream side of the main quadrupoles. The sextupole modules are powered in series, forming one skew-sextupole family and four sextupole families per arc and per beam. The lattice skew sextupole correctors serve to compensate for the chromatic coupling induced by the skew sextupole field errors in the dipole magnets.

The operational margins for the chromaticity are based on collective instabilities and experience with previous hadron colliders. The target chromaticity of the LHC is 2 units.

Sextupole spool-piece correctors (MCS) are single-aperture magnets installed downstream of each main dipole magnet, one for each aperture. The cross section is shown in Figure B.4. The spool-piece sextupole magnets are designed to compensate for the field-dependent, integrated sextupole component attributable to superconducting filament magnetization (persistent currents) and saturation effects in the main dipoles. The spool-piece sextupoles thus correct the impact of these field errors on the linear chromaticity.

11.10
Field Error Specifications

At injection field level, the main dipoles are the most demanding magnet elements for the machine performance. At collision energy, the high beta functions in the insertions make the triplet quadrupoles and the recombination dipoles the most critical magnets in the ring. The required tolerances in the main dipole geometry are comprehensively studied in [8].

The tolerances in the relative dipole field error (b_1) at 17 mm reference radius are zero systematic, 6.5 units in 10^{-4} uncertainty, and 8.0 units random (RMS). For the relative skew dipole (a_1) it is 6.5 units average per arc cell (systematic and uncertainty) and 8.0 units random. The different types of field errors and their sources are explained in Section 1.9. The tolerances in a_1 and b_1 are directly linked to the tolerances in the alignment of the main quadrupole magnets in the LHC, which should not exceed $\pm$ 0.36 mm [27].

Because of the symmetry in the double-aperture dipole magnets , the systematic b_2 and b_4 field errors have opposite signs in the two apertures; see Figure 12.12. Because the beam travels equal distances in the inner and outer apertures, these systematic field errors have no impact on the global beam parameters. This is untrue for uncertainty errors, all of which lead to tune spreads and are thus limited to 0.8 units in 10^{-4} for b_2 and 0.4 units for b_4.

In principle, the systematic b_3 field errors in the main dipoles can be corrected with the sextupole spool-pieces (MCS). However, quadrupolar feed-down effects induced by random misalignment of the spool pieces ($<$ 0.5 mm RMS) impose tolerances in the systematic b_3 of 10.7 units at injection and 3 units at collision, leaving a margin of 1.35 units for the sextupole corrector strength.

Vertical dispersion and linear coupling limit uncertainty errors in the skew quadrupoles to 0.9 units and the random a_2 to 1.6 units at collision energy.

These field quality requirements demanded a substantial R&D effort for the LHC main magnets, stringent quality assurance procedures, and the development of specialized electromagnetic design and optimization tools. These tools were used not only for the design phase, but also for intercepting and tracing manufacturing errors. But before we treat these specialized numerical field computation methods, we must address the worst field error of all: the wrong polarity.

References

1 Arnold, V.I.: Gewöhnliche Differentialgleichungen, Springer, Berlin, 2001

2 Brandt, D.: Proceedings of the CERN Accelerator School – Intermediate Accelerator Physics, CERN-2006-002

3 Bryant, P.J., Johnson, K.: The Principles of Circular Accelerators and Storage Rings, Cambridge University Press, Cambridge, 1993

4 Christofilos, N.C.: US Patent No. 2,736,799, 1950

5 Courant, E. D., Livingston, M. S., Snyder, H. S.: The strong-focusing synchrotron – a new high energy accelerator, Physical Review, 1952

6 Courant, E. D., Snyder, H. S.: Theory of the alternating-gradient synchrotron, Annals of Physics, 1958

7 Edwards, D.A., Syphers, M. J.: An Introduction to the Physics of High Energy Accelerators, John Wiley & Sons, New York, 1993

8 Fartoukh, S., Brüning, O.: Field quality specification for the LHC main dipole magnets, LHC Project Report 501, CERN, 2001

9 Grote, H., Iselin, F. C.: The MAD program (Methodical Accelerator Design), User's Reference Manual, CERN/SL/90-13

10 Grote, H., Schmidt, F.: MAD-X – An upgrade from MAD8, Particle Accelerator Conference, 2003

11 Holzer, B.: Lattice design in high-energy particle accelerators, Proceedings of the CERN Accelerator School – Intermediate accelerator physics, CERN-2006-002

12 LHC Design Report, Vol. 1, The LHC main ring, CERN-2004-003, 2004

13 The LHC Study Group, Large Hadron Collider, The Accelerator Project, CERN/AC/93-03, 1993

14 The LHC Study Group, The Large Hadron Collider, Conceptual Design, CERN/AC/95-05, 1995

15 ISO Standards Handbook 2, Units of Measurement, 1982

16 Jin, L., Schmidt, F.: Tune Scan Studies for the LHC at Injection Energy, LHC Project Report 377, 2000

17 Martini, M.: Transverse Beam Dynamics, Handout for the Joint University Accelerator School, Archamps, 1996

18 Michelotti, L.: Intermediate Classical Dynamics with Applications to Beam Physics, John Wiley & Sons, New York, 1995

19 Moore, R.B.: Elliptical transformations for Beam Optics, http://www.physics.mcgill.ca/~moore/Notes, 2004

20 McMillan, E.M.: Physics Review, 86, 1945

21 Okun, L.B.: The concept of mass, Physics Today, 1989

22 Rossbach, J., Schmüser, P.: Basic Course on Accelerator Physics, CERN Accelerator School, Proceedings Vol. 1, CERN 94-01, 1994

23 Sanford, J.R., Matthews, D.M. (editors): Site-Specific Conceptual Design of the Superconducting Super Collider, Superconducting Super Collider Laboratory, 1990

24 Schmidt, F.: Private communications, 2006

25 Steenbeck, M: US Patent 2.103.303, 1935

26 Veksler, V.: Journal of Physics (USSR), 1945

27 Verdier, A. (ed.): Report on the mini-workshop on the LHC alignment, LHC Project Note 247, CERN, 2001

28 Wiedemann, H.: Particle Accelerator Physics, Springer, Berlin, 1993

29 Wille, K.: The Physics of Particle Accelerators, Oxford University Press, Oxford, 2000

30 Wilson, E.J.N.: Introduction to Particle Accelerators, 1997

12
Reference Frames and Magnet Polarities

09:00 or 15:00 ?

Any error in the busbar interconnections between the magnet elements would spell serious trouble in the operation of a superconducting accelerator. Examples of wrong connections are the inversion of magnet polarity, connection of a magnet to a wrong circuit, and inversion of polarity at the level of a power converter or a current lead. Detecting such errors is very difficult after the closing of the machine, and a repair after commissioning of the accelerator extremely costly.

It was of paramount importance, to establish consistency of polarity conventions between those used in magnet construction and those specified in the electrical layout database. In particular, the definition of multipole field errors in two different reference frames (magnet measurement frame and moving frame of Beam 1 for beam physics calculations) resulted in confusion concerning the polarity of the vertical orbit correctors, the polarity of compensators in the insertion regions 1 and 8, conventions for the lattice correctors in the inner triplets, and the treatment of measured skew multipole errors in the beam physics program MAD [3].

This chapter describes how coherence between the definitions of field errors in the different reference frames and the LHC magnet polarity conventions was established. For this purpose, we will make also extensive use of the transformation rules for the multipole coefficients derived in Chapter 9.

Field Computation for Accelerator Magnets. Stephan Russenschuck

ISBN: 978-3-527-40769-9

12.1 Magnet Polarity Conventions

The polarity conventions in [5] are conceived to yield a simple identification of the polarities of the magnets installed in the ring, without reference to the different coordinate systems used for beam tracking and field measurements. The set of rules allows magnets of a given type to be manufactured and assembled without prior knowledge of their position or function in the accelerator.

The polarity of the excitation current, and consequently the optical function of the magnet, is determined by the connection of the magnet terminals. The set of rules does not follow the conventions of the beam-optics program MAD or the conventions for magnetic field computations and measurements. Appropriate transformations into the moving frame of the circulating beam or the magnet reference frame must be applied.

The conventions for the LHC magnets are summarized below:

- The reference beam called Beam 1 travels clockwise in the LHC main ring as seen from above. Beam 2 travels counter-clockwise as seen from above.
- The observer is looking in the direction Beam 1 travels, so that the center of the machine is on his right-hand side.
- The terms *upstream* and *downstream* are defined with respect to the direction of Beam 1.
- In the twin-aperture magnets (or magnet assemblies), the left aperture seen from the connection terminals is Aperture 1 and the right is Aperture 2.
- The main dipole and quadrupole magnets are installed in the tunnel with their connection terminals upstream. Hence Aperture 2 becomes the "internal" aperture and Aperture 1 the "external" aperture.
- The magnet connection terminals are denoted A and B and **not** $+$ and $-$.
- The fields and gradients are said to be positive if the current enters the A-terminal. A positive field is defined as pointing upward (deflecting Beam 1 to the inside of the ring) while a positive gradient is defined such that the vertical field increases along the outward-pointing machine radius (focusing Beam 1 in the horizontal direction). The vector fields of magnetic flux density for these cases are shown in Figures 12.1 and 12.2.
- The skew-multipole magnets of order N are tilted clockwise by an angle of $\pi/2N$ degrees, where N=1 for the dipole, N=2 for the quadrupole etc. Thus a positive skew dipole is deflecting Beam 1 downward; see Figure 12.3.
- In the case of magnets where both beams pass through a single aperture, for example, in the MQXA and MQXB magnets, Beam 1 is used to define the polarity.

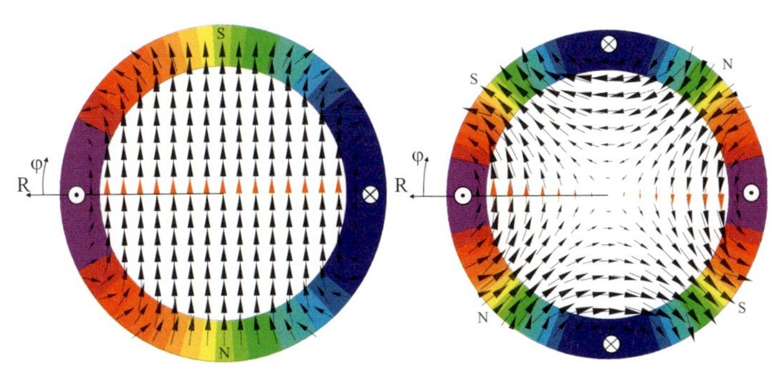

Figure 12.1 Current and field distribution in a single-aperture dipole (left) and single-aperture quadrupole for current entering the A-terminal (positive dipole field and gradient in the quadrupole). Downstream view.

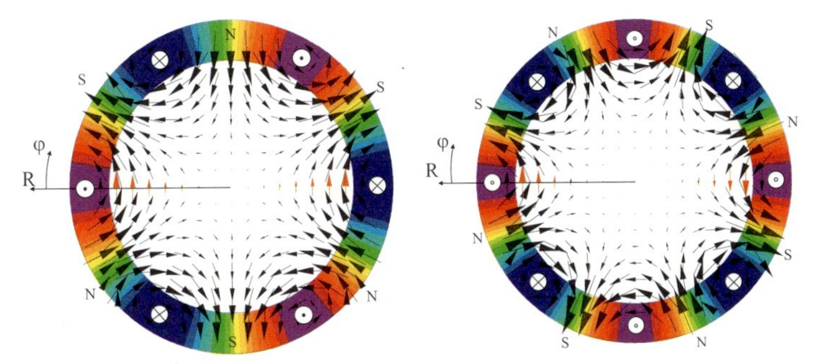

Figure 12.2 Current and field distribution in a single-aperture sextupole (left) and single-aperture octupole for current entering the A-terminal (positive sextupole and octupole fields). Notice the field distribution in the horizontal median plane (icons in red). Downstream view.

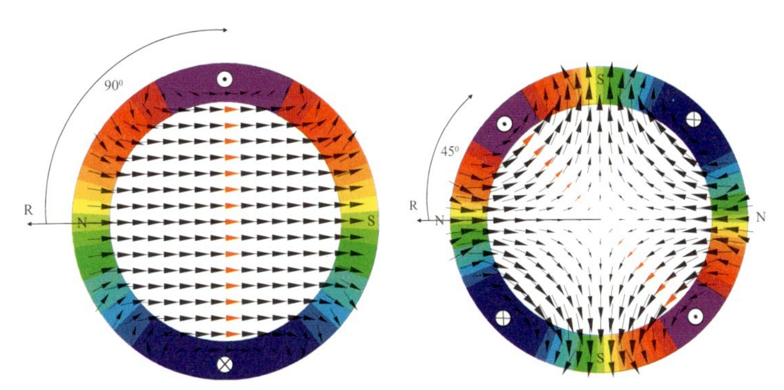

Figure 12.3 Current and field distribution in a single-aperture skew dipole (left) and single-aperture skew quadrupole for current entering the A-terminal (positive skew dipole and skew quadrupole fields). The magnets are rotated clockwise (observer looking downstream) by $\pi/2$ and $\pi/4$, respectively. A positive skew-dipole field deflects Beam 1 downward.

Note that *tilt* here defines a rotation around the longitudinal axis of the magnet; it is better referred to as *roll*. Often the term tilt is used synonymously with *pitch*, as explained in Section 19.5.

Remark: There is an easy way to determine the deflection of the proton beam: Recall that wires in current-carrying cables attract each other, while two conductors with currents flowing in opposite directions repel. Thus the proton beam is attracted by the technical current flowing in the same direction. □

12.1.1 Spool-Piece Correctors

The conventions for the polarity of corrector magnets powered from a bipolar power supply follow those of the main magnets. A positive sextupole will compensate for the persistent current effect in the main dipole.

- Current entering the A-terminal of the sextupole and decapole corrector magnets have an upward-pointing field direction in the horizontal plane; see Figure 12.2 (left) for the sextupole.
- Current entering the A-terminal of the octupole correctors results in a vertical field increasing along the outward-pointing machine radius, Figure 12.2 (right).
- In Sectors 1-2, 5-6, 6-7, and 7-8, where Beam 1 is in the external aperture and the current of the main dipole enters the A-terminal, all the corrector magnets in the external aperture are powered with current entering the A-terminal. The corrector magnets in the internal aperture have their current entering the B-terminal.

12.1.2 Twin-Aperture and Two-in-One Magnets

From here on we will make a distinction between *twin-aperture* and *two-in-one* magnets. Twin-aperture magnets are assembled with two coil pairs in one common iron yoke. For the powering of the apertures they may have one connection terminal, as in the case of the main dipole, or two connection terminals, as in the case of the main quadrupoles. We refer to two-in-one magnets, if they are assembled from magnetically and mechanically separate magnet modules in a common support structure.

For twin-aperture magnets and two-in-one magnet assemblies the following rules have been defined:

- For twin-aperture magnets such as the main dipole, and two-in-one, normal-conducting magnets with only one pair of terminals, the external aperture (Aperture 1) takes priority for the application of the rules. Consequently, if the current enters the A-terminal of the main dipole, the field direction is downward in the internal aperture (Aperture 2) and upward in the external aperture. The vector field of magnetic flux density for this case is shown in Figure 12.4.

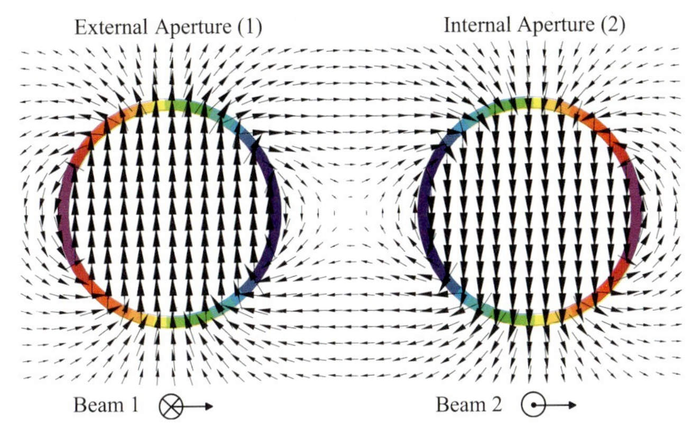

Figure 12.4 Current and field distribution in the twin-aperture main dipole (MB). Current entering the A-terminal.

- For twin-aperture magnets with two connection terminals, for example, the main quadrupoles, and for the two-in-one magnet assemblies the rules apply to both apertures independently. Therefore the field gradient is identical in both apertures of the main quadrupoles. Consequently, if the current enters the *A* terminals in the twin-aperture quadrupoles, the fields in both apertures increase along the outward-pointing machine radius. If Beam 1 is in the external aperture (as in Sectors 1-2, 5-6, 6-7,

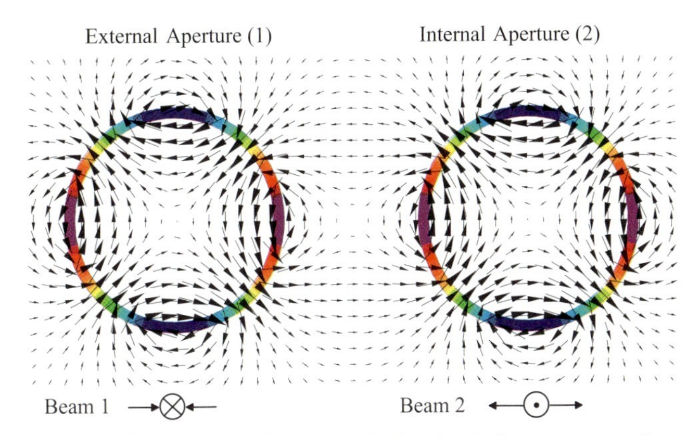

Figure 12.5 Current and field distribution in a twin-aperture main quadrupole. Current entering the A-terminals, focusing Beam 1 and defocusing Beam 2 in the horizontal direction.

7-8) it will be focused in the horizontal direction. Beam 2 (circulating counter-clockwise in these sectors) will be defocused; see Figure 12.5.

- For two-in-one quadrupoles with the same optical function in both apertures (see Figure B.13 (left) for an example) the polarity convention is as follows: If Beam 1 is in the external aperture (in Sectors 1-2, 5-6, 6-7, 7-8) both beams will be focused in the horizontal direction if the current enters the A-terminal.
- When the current enters the A-terminal in the twin-aperture separator dipoles, the field points upward both in the external and the internal apertures.

12.2 Reference Frames

In accordance with the above-mentioned rules we have avoided reference frames up to now. In Section 8.1.1 we showed that a $\cos\varphi$ current distribution generates an ideal dipole field in the aperture. Obviously, the direction of the current at $\varphi = 0$ must be the same for both the dipole and the quadrupole (and higher-order multipole) magnets having $\cos n\varphi$ current distribution. The rule whereby a positive dipole field bends Beam 1 inward and a positive quadrupole focuses Beam 1 in the horizontal direction, can only be met if the angle φ is counted as positive, as indicated in Figures 12.1 and 12.2. This corresponds to the reference frame used for particle tracking shown in Figure 12.6 (right).

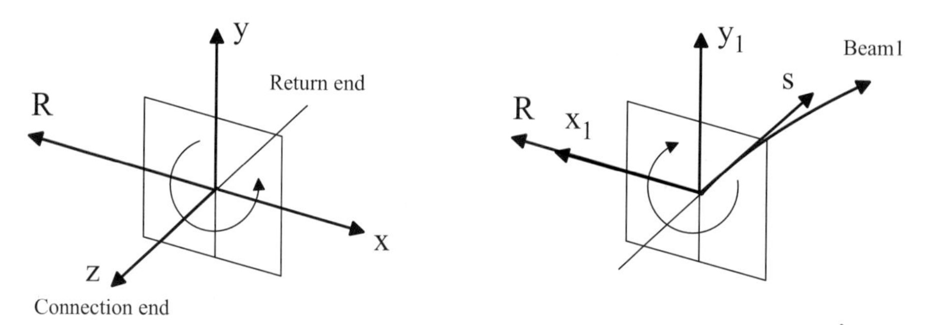

Figure 12.6 Left: Reference frame used in magnet design and measurement. The x-axis points toward the center of the machine. Right: Reference frame used for particle tracking, where the x_1-axis is pointing in the outward normal direction of the machine. Note the orientation of the plane $\mathbb{R}^2$, which we identify with the complex plane.

12.3 Multipole Expansions

12.3.1 The Magnet Frame

The Cartesian components of the magnetic flux density in the magnet aperture are combined in the complex function $B = B_y + iB_x$, which is holomorphic in $\Omega = \{z| \, |x+iy| < \rho\}$, with ρ denoting the aperture radius. B can be expanded as

$$B_y + iB_x = \sum_{n=1}^{\infty} (B_n + i\,A_n) \left(\frac{z}{r_0}\right)^{n-1} = B_N \sum_{n=1}^{\infty} (b_n + i\,a_n) \left(\frac{z}{r_0}\right)^{n-1} . \tag{12.1}$$

The normal and skew multipole coefficients $B_n(r_0), A_n(r_0)$ are given in units of tesla at a reference radius r_0 of 17 mm. The small $b_n(r_0), a_n(r_0)$ denote the relative multipole coefficients at the reference radius, which are related to the main field component $B_N(r_0)$ ($B_1(r_0)$ for the dipole, $B_2(r_0)$ for the quadrupole, etc.). The scaling of the multipoles with respect to the reference radius and the transformation under frame changes are discussed in [8] and Chapter 9.

We can identify a complex number with the point $(x,y) \in \mathbb{R}^2$ by means of

$$\mathbb{R}^2 \xrightarrow{\cong} \mathbb{C} : (x,y) \mapsto x + iy , \tag{12.2}$$

and consequently may regard a complex function f as a mapping $f : \Omega \to \mathbb{R}^2$ for $\Omega \subset \mathbb{R}^2$. Normally, the x-axis is drawn to the right. For the observer looking downstream this implies that the x-axis is pointing toward the machine center, an orientation that corresponds to the frame used by the magnet builders, hereafter referred to as the magnet frame. This makes perfect sense for numerical field computation, as the problem domain can in most cases be confined to the first quadrant. However, if we draw the x-axis pointing in outward normal direction of the machine (see Figure 12.6), the orientation of the plane is consistent with the crossing direction of Beam 1 through the plane because the ambient space is usually oriented in the sense of a right-handed screw. This seems to be the natural choice for beam optics. In order to emphasize our choice of orientation we shall write the magnetic field in the magnet frame (here with the x-axis pointing inward) as

$$\begin{aligned} B_y + iB_x &= \sum_{n=1}^{\infty} (B_n^{\text{mag}} + i\,A_n^{\text{mag}}) \left(\frac{z}{r_0}\right)^{n-1} \\ &= B_N^{\text{mag}} \sum_{n=1}^{\infty} (b_n^{\text{mag}} + i\,a_n^{\text{mag}}) \left(\frac{z}{r_0}\right)^{n-1} , \end{aligned} \tag{12.3}$$

with a small amount of additional typesetting well worth the effort. Recall that we are writing in shorthand B_n, A_n for the radius-dependent $B_n(r_0), A_n(r_0)$. From $B_\varphi + iB_r = (B_y + iB_x)e^{i\varphi}$ we obtain in the magnet frame

$$\begin{aligned} B_\varphi + iB_r &= \frac{r_0}{r} \sum_{n=1}^{\infty} (B_n^{\mathrm{mag}} + i\,A_n^{\mathrm{mag}}) \left(\frac{z}{r_0}\right)^n \\ &= B_N^{\mathrm{mag}} \frac{r_0}{r} \sum_{n=1}^{\infty} (b_n^{\mathrm{mag}} + i\,a_n^{\mathrm{mag}}) \left(\frac{z}{r_0}\right)^n , \end{aligned} \tag{12.4}$$

and the field components at any radius $r < \rho$:

$$B_r(r,\varphi) = \sum_{n=1}^{\infty} \left(\frac{r}{r_0}\right)^{n-1} (B_n^{\mathrm{mag}} \sin n\varphi + A_n^{\mathrm{mag}} \cos n\varphi), \tag{12.5}$$

$$B_\varphi(r,\varphi) = \sum_{n=1}^{\infty} \left(\frac{r}{r_0}\right)^{n-1} (B_n^{\mathrm{mag}} \cos n\varphi - A_n^{\mathrm{mag}} \sin n\varphi). \tag{12.6}$$

and

$$B_x(r,\varphi) = \sum_{n=1}^{\infty} \left(\frac{r}{r_0}\right)^{n-1} (B_n^{\mathrm{mag}} \sin(n-1)\varphi + A_n^{\mathrm{mag}} \cos(n-1)\varphi), \tag{12.7}$$

$$B_y(r,\varphi) = \sum_{n=1}^{\infty} \left(\frac{r}{r_0}\right)^{n-1} (B_n^{\mathrm{mag}} \cos(n-1)\varphi - A_n^{\mathrm{mag}} \sin(n-1)\varphi). \tag{12.8}$$

The definition (12.3), in the magnet frame shown in Figure 12.6 (left), results in the following signs for the multipole coefficients:

- B_1^{mag} is the dipole field that points into the positive y-direction, that is, a positive field bends a positively charged particle inward. This is compliant with the polarity conventions.
- A_1^{mag} is the skew dipole field pointing into the positive x-direction (inward). A positive skew dipole field bends a positively charged particle downward. This is compliant with the polarity conventions.
- A positive B_2^{mag} is a quadrupole field, which implies horizontal defocusing of Beam 1. This is a sign reversal with respect to the polarity conventions.
- A positive A_2^{mag} is a skew quadrupole, which implies defocusing of Beam 1 in the xz-plane rotated clockwise by $\pi/4$ (points in this rotated plane have coordinates $y = -x$). This is a sign reversal with respect to the polarity conventions.

12.3.2 The Local Reference Frame of Beam 1

In the local reference frame of Beam 1 the field is expanded as

$$B_{y_1} + iB_{x_1} = \sum_{n=1}^{\infty} (B_n^{\text{Beam}} + i\,A_n^{\text{Beam}}) \left(\frac{z_1}{r_0}\right)^{n-1}$$

$$= B_N^{\text{Beam}} \sum_{n=1}^{\infty} (b_n^{\text{Beam}} + i\,a_n^{\text{Beam}}) \left(\frac{z_1}{r_0}\right)^{n-1} , \tag{12.9}$$

where $z_1 = x_1 + iy_1$. This definition results in the following signs for the multipole coefficients:

- B_1^{Beam} is the dipole field pointing into the positive y_1-direction. A positive field bends a positively charged particle inward. This is compliant with the polarity conventions.
- A_1^{Beam} is the skew dipole field pointing into the positive (outward) x_1-direction. A positive skew dipole field bends a positively charged particle upward. This is a sign reversal with respect to the polarity conventions.
- A positive B_2^{Beam} is a quadrupole field, which implies horizontal focusing of Beam 1. This is compliant with the polarity conventions.
- A positive A_2^{Beam} is a skew quadrupole, which implies defocusing of Beam 1 in the x_1s-plane rotated clockwise by $\pi/4$ (points in this rotated plane have coordinates $x_1 = y_1$). This is a sign reversal with respect to the polarity conventions.

Remark: Both in the magnet frame and in the local frame of Beam 1, the mapping $B_n \mapsto A_n$ implies a rotation of the magnet in the mathematically negative (!) sense. For the observer looking downstream this is a clockwise rotation of the magnet for $B_n^{\text{mag}} \mapsto A_n^{\text{mag}}$, and a counter-clockwise rotation for $B_n^{\text{Beam}} \mapsto A_n^{\text{Beam}}$. Rossbach and Schmüser [6] expand the field as

$$B_y + iB_x = \sum_{n=1}^{\infty} (B_n - i\,A_n) \left(\frac{z}{r_0}\right)^{n-1} , \tag{12.10}$$

where the mapping $B_n \mapsto A_n$ implies a rotation of the magnet in the mathematically positive sense, which seems more natural. However, the multipole coefficients must match the calculated or measured values at a given reference radius. Hence consider B_r expressed according to Eq. (12.5). The coefficients are determined by a Fourier analysis of the voltage signal obtained from a rotating-coil measurement. If the Fourier series is written as $f(x) = \frac{1}{2}a_0 + \sum_{n=1}^{\infty} a_n \cos nx + b_n \sin nx$, it is rather the plus sign that must be invoked; yielding $B = B_n + i\,A_n$. □

12.3.3 Definition of Field Errors in the Accelerator Design Program MAD

The MAD program [3] uses a Maclaurin series expansion of the integrated field about the magnet axis:

$$\begin{aligned} B_{y_1}(x_1) &= B_0 + \left.\frac{\mathrm{d}B_{y_1}}{\mathrm{d}x_1}\right|_{x_1=y_1=0} x_1 + \cdots + \frac{1}{n!}\left.\frac{\mathrm{d}^n B_{y_1}}{\mathrm{d}x_1^n}\right|_{x_1=y_1=0} x_1^n + \cdots \\ &= \sum_{n=0}^{\infty} \frac{1}{n!} B_{n,\mathrm{n}}^{\mathrm{MAD}} x_1^n \\ &= \sum_{n=0}^{\infty} \frac{1}{n!} B\rho K_{n,\mathrm{n}} x_1^n, \end{aligned} \tag{12.11}$$

where $B\rho$ is the magnetic rigidity of the beam and the subscript n denotes the normal multipole coefficients. The same definition holds for the skew field components (denoted $B_{n,s}^{\mathrm{MAD}}$) with the reference frame rotated clockwise by $\pi/2N$ around the beam axis. Therefore the sign conventions for these multipole coefficients follow those of Section 12.3.2:

- $B_{0,\mathrm{n}}^{\mathrm{MAD}}$ is the dipole field pointing in the positive y_1-direction. A positive field bends a positively charged particle inward.
- $B_{0,\mathrm{s}}^{\mathrm{MAD}}$ is the skew dipole field pointing in the positive (outward) x_1-direction. A positive skew dipole field thus deflects Beam 1 upward.
- A positive $B_{1,\mathrm{n}}^{\mathrm{MAD}}$ is a quadrupole field that implies horizontal focusing of Beam 1.
- A positive $B_{1,\mathrm{s}}^{\mathrm{MAD}}$ is a skew quadrupole that implies defocusing of Beam 1 in the $x_1 s$-plane rotated clockwise by $\pi/4$ (points in this rotated plane have coordinates $x_1 = y_1$).

12.3.4 Transformation between the Magnet and the Beam 1 Frames

Tables 12.1 and 12.2 show the transformations for the MAD-X input, the multipole coefficients in the moving frame of Beam 1, the polarity conventions, and the multipole coefficients in the magnet frame.

It can be seen that these polarity conventions do not follow the definitions used in beam physics but are compatible with the signs in the Beam 1 frame as far as the normal multipoles are concerned. They are also **not** consistent with the MAD conventions for the skew multipoles. Recall that a clockwise rotation of the magnet element (observer looking downstream) results in the mappings $B_n^{\mathrm{mag}} \mapsto A_n^{\mathrm{mag}}$ but $B_n^{\mathrm{Beam}} \mapsto -A_n^{\mathrm{Beam}}$.

Table 12.1 Transformation table for MAD-X input, polarity convention [5], and dipole coefficients in the magnet frame[a].

MAD-X Input	HKICKER	VKICKER	$B_{0,n}^{\mathrm{MAD}}$	$B_{0,n}^{\mathrm{MAD}}$ tilt	$B_{0,s}^{\mathrm{MAD}}$	$B_{0,s}^{\mathrm{MAD}}$ tilt
$B_n^{\mathrm{Beam}}, A_n^{\mathrm{Beam}}$	$-B_1^{\mathrm{Beam}}$	A_1^{Beam}	B_1^{Beam}	$-A_1^{\mathrm{Beam}}$	A_1^{Beam}	B_1^{Beam}
Defl. Beam 1	Outward	Upward	Inward	Downward	Upward	Inward
Polarity Convention Ref. [5]	Negative dipole	Negative skew dipole	Positive dipole	Positive skew dipole	Negative skew dipole	Positive dipole
$B_n^{\mathrm{mag}}, A_n^{\mathrm{mag}}$	$-B_1^{\mathrm{mag}}$	$-A_1^{\mathrm{mag}}$	B_1^{mag}	A_1^{mag}	$-A_1^{\mathrm{mag}}$	B_1^{mag}

[a] TILT is the so-called roll angle about the longitudinal axis (a positive angle represents a clockwise rotation of the magnet). Note that in MAD-X a tilted dipole corresponds to a negative skew dipole.

Table 12.2 Transformation table for MAD-X input, polarity conventions, and quadrupole coefficients in the magnet frame[a].

MAD-X Input	$B_{2,n}^{\mathrm{MAD}}$	$B_{2,n}^{\mathrm{MAD}}$ tilt	$B_{2,s}^{\mathrm{MAD}}$	$B_{2,s}^{\mathrm{MAD}}$ tilt
$B_n^{\mathrm{Beam}}, A_n^{\mathrm{Beam}}$	B_2^{Beam}	$-A_2^{\mathrm{Beam}}$	A_2^{Beam}	B_2^{Beam}
Effect on on Beam 1	Focusing in horizontal plane	Q1 + Q2 −	Q1 − Q2 +	Focusing in horizontal plane
Polarity Conv. Ref. [5]	Positive quadrupole	Positive skew quad.	Negative skew quad.	Positive quadrupole
$B_n^{\mathrm{mag}}, A_n^{\mathrm{mag}}$	$-B_2^{\mathrm{mag}}$	$-A_2^{\mathrm{mag}}$	A_2^{mag}	$-B_2^{\mathrm{mag}}$

[a] TILT is the roll angle about the longitudinal axis (a positive angle represents a clockwise rotation of the magnet). Note that in MAD-X a tilted quadrupole corresponds to a negative skew quadrupole. The notation Q1 is shorthand for the quadrupole in Aperture 1.

The transformation laws between the multipole coefficients are

$$\frac{r_0^{n-1}}{(n-1)!} B_{n,n}^{\mathrm{MAD}} = B_n^{\mathrm{Beam}} = (-1)^{n-1} B_n^{\mathrm{mag}} , \tag{12.12}$$

$$\frac{r_0^{n-1}}{(n-1)!} B_{n,s}^{\mathrm{MAD}} = A_n^{\mathrm{Beam}} = (-1)^{n} A_n^{\mathrm{mag}} . \tag{12.13}$$

For the relative multipoles of a normal magnet we must consider $B_n = B_N b_n$, $A_n = B_N a_n$, and the relation $B'_N = (-1)^{N-1} B_N$. This yields

$$b_n^{\mathrm{Beam}} = (-1)^{n-N} b_n^{\mathrm{mag}} , \qquad a_n^{\mathrm{Beam}} = (-1)^{n-N+1} a_n^{\mathrm{mag}} . \tag{12.14}$$

For a skew magnet ($B_n = A_N b_n$, $A_n = A_N a_n$) we obtain

$$b_n^{\mathrm{Beam}} = (-1)^{n-N+1} b_n^{\mathrm{mag}} , \qquad a_n^{\mathrm{Beam}} = (-1)^{n-N} a_n^{\mathrm{mag}} . \tag{12.15}$$

The proofs can be found in Section 9.7.3.

12.4 Orbit Correctors

Orbit correctors are powered from bipolar power supplies. As can be seen from Table 12.1, a positive horizontal kick on Beam 1, deflecting outward, requires positive current into the B-terminal, while a positive kick on Beam 2, deflecting outward, requires that current enter the A-terminal. In a similar manner, a positive vertical kick on Beam 1, deflecting upward, requires positive current into the B-terminal while a positive kick on Beam 2, deflecting upward, requires current to enter the A-terminal. When both beams pass through a single aperture, Beam 1 is used to define the polarity (Beam 1 being deflected outward/upward and Beam 2 inward/downward with a positive kick).

In the electrical layout scheme a positive kick corresponds to a positive current setting on the bipolar power supply. If Beam 1 is affected, the current enters the B-terminal; when Beam 2 is affected, the current enters the A-terminal.

12.5 Position of the Connection Terminals

The *installation direction* of magnets (or magnet assemblies) in the tunnel defines the position of the magnet connection terminals, upstream or downstream. Multipole correctors, in particular, might have their connection terminals facing the opposite side with respect to the terminal of the main magnet they are attached to. Examples include the MCB in the SSS, and the MCS in the main dipole coldmass; cf. Table B.2 (column D).

The magnet's optical function changes depending on the multipole order according to

$$B_n^{\text{up}} = (-1)^{n-1} B_n^{\text{down}} , \qquad A_n^{\text{up}} = (-1)^n A_n^{\text{down}} . \tag{12.16}$$

The connection terminals are labeled such that if the current enters the A-terminal the field is indeed positive in the sense of the polarity conventions. Let us emphasize that the field quality of the magnet modules is always measured in the magnet frame and that consequently the relative, higher-order field harmonics may change sign in the magnet assembly depending on the multipole order,

$$b_n^{\text{up}} = (-1)^{n-N} b_n^{\text{down}} , \qquad a_n^{\text{up}} = (-1)^{n-N+1} a_n^{\text{down}} , \tag{12.17}$$

for the relative multipoles of a normal magnets, and

$$b_n^{\text{up}} = (-1)^{n-N+1} b_n^{\text{down}} , \qquad a_n^{\text{up}} = (-1)^{n-N} a_n^{\text{down}} \tag{12.18}$$

for skew magnets.

The normal installation direction of the magnets is given in Tables B.2 and B.3, together with the number of magnets, the operation temperature and current, the magnetic length, the inductance, and the resistance at room temperature. The kickers and experimental magnets are excluded.

The arrangements of magnet assemblies, with the position of the connection terminals, are sketched in Figures B.9–B.12. The figures define the normal position of the connection terminals.

12.6 Turned Magnets and Magnet Assemblies

For various reasons, for example, space constraints on connections and vacuum equipment, a magnet or an entire magnet assembly may be installed in the LHC tunnel reversed with respect to the normal direction, that is, rotated by π about its vertical axis. The construction and internal connections of these magnets are not changed, nor are the naming of the connection terminals. However, the magnet's optical function may change depending on the multipole order,

$$B_n^{\text{turn}} = (-1)^{n-1} B_n^{\text{norm}}\,, \qquad A_n^{\text{turn}} = (-1)^n A_n^{\text{norm}}\,. \tag{12.19}$$

In this case the polarity is changed on the warm side of the magnet, which is documented in the electrical layout database and layout drawing, where the magnet is marked with a star.

The relative, higher-order field harmonics may change sign in the magnet assembly depending on the multipole order,

$$b_n^{\text{turn}} = (-1)^{n-N} b_n^{\text{norm}}\,, \qquad a_n^{\text{turn}} = (-1)^{n-N+1} a_n^{\text{norm}}\,, \tag{12.20}$$

for the relative multipoles of a normal magnets, and

$$b_n^{\text{turn}} = (-1)^{n-N+1} b_n^{\text{norm}}\,, \qquad a_n^{\text{turn}} = (-1)^{n-N} a_n^{\text{norm}} \tag{12.21}$$

for skew magnets.

Example 1: Compensators in IR2 and IR8. The electrical layouts of the spectrometer dipole magnet compensations in IR2 and IR8 are shown in Figure 12.7. The experiments at these insertion points use spectrometer (dipole) magnets, which distort the beam trajectories. This effect is locally compensated with three orbit correctors placed in the straight sections between the interaction point and the final focusing triplet [1]. The compensators are powered according to the rules for the orbit correctors in the arc, that is, a positive kick (upward or outward) on Beam 1 is obtained by a positive setting on the

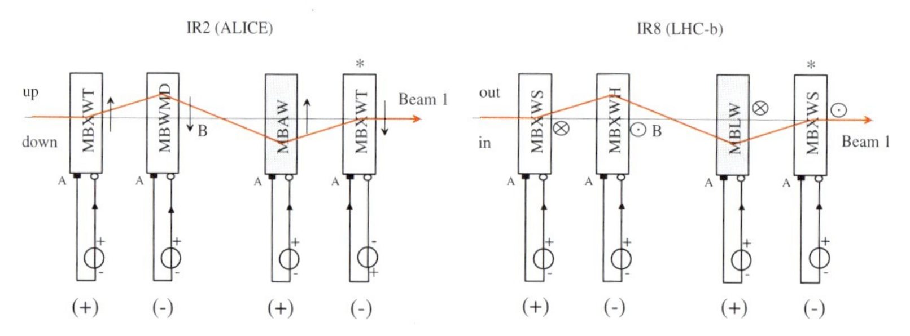

Figure 12.7 Electrical layouts of the spectrometer dipole magnet compensation scheme in IR2 (left) and IR8 (right). Note the change of polarity for the turned, vertically deflecting magnet MBXWT in IR2 while the turned, horizontally deflecting magnet MBXWS keeps its optical function.

bipolar power supply with the current entering the B-terminal of the compensators. Note the change of polarity for the turned, vertically deflecting magnet MBXWT in IR2, while the turned, horizontally deflecting magnet MBXWS keeps its optical function. □

Example 2: Polarities of inner-triplet magnets The polarities[1] of the inner-triplet quadrupoles and their adjacent lattice correctors are shown in Figure 12.8. For each magnet element the following information is provided: The optical function of the quadrupoles; the multipole order of the magnet element; stars indicating turned magnets; an indication whether or not the polarity changes when the magnet is turned; the position of the connection

1 Compliant with the polarity conventions, independent of the magnet frame and the local frame of Beam 1.

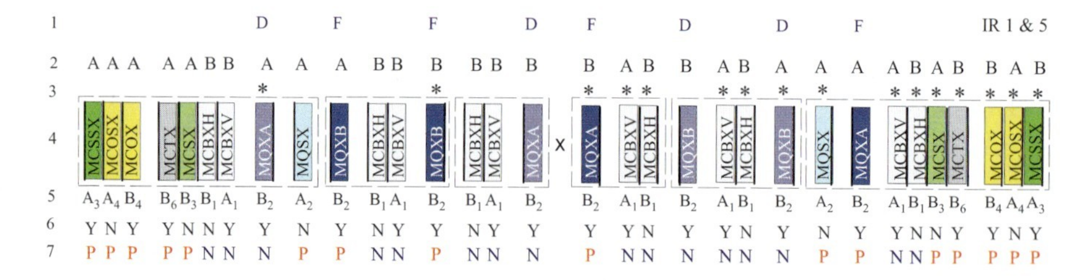

Figure 12.8 Polarity conventions for the inner triplets in IR1 and IR5. Row 1: Optical function of the quadrupole. Row 2: Current entering A- or B-terminal. Row 3: Stars indicate turned magnets. Row 4: Name of the magnet element. MCSSX, MCOSX, and MCOX are nested magnet elements in the MCSOX subassembly. Row 5: Multipole order of the magnet element. Row 6: Polarity changes when the magnet is turned; Y = yes, N = no. Row 7: Polarity of the magnet elements when the bipolar power supply has a positive setting.

terminal [2]; the terminal in which the current enters; and the magnet polarity (according to [5]) when the bipolar power supply has a positive setting.

The powering scheme of the triplet corrector elements differs with respect to the spool-piece circuits because the triplet correctors do not provide a magnet-by-magnet correction. Instead, they provide an overall kick minimization, taking into account all triplet quadrupoles, the D1 and D2 separation dipoles, and the Q4 quadrupole magnets to the left and right of the interaction points. Consequently, the magnet polarities follow the rule that a positive current entering the A-terminal implies a positive field, independent of the polarity of the quadrupole to which the correctors are attached to. □

12.7 Electrical Circuits in the LHC Machine

The main dipole circuit and the two main quadrupole circuits together require six busbars rated at 13 kA. Twenty auxiliary busbars rated at 600 A supply the spool-piece correctors in the coldmass of the main dipoles. All these busbars are joined at each interconnection plane between two cryomagnets. A flexible superconducting cable with 42 auxiliary busbars feeds the corrector magnets housed in the SSS. The busbar cable is routed through line N, located outside the main magnet coldmasses. The line N can be seen in Figure 1.15. Junctions between the line N cable segments are made at each interconnection between a cryodipole and an SSS; see Figure 1.19. The polarity of a magnet can be checked by means of voltage taps connected to the A-terminal of the magnet units. All the signals from the voltage taps are routed out of the coldmass via an instrumentation feedthrough system.

Figure 12.9 shows the electrical circuits for the main dipoles in the arc cells. The role of the two sets of switches and protection resistors will be explained in Chapter 18. According to the above rules, the dipole field in the twin-aperture dipoles is defined as positive if the current entering the A-terminal creates a downward-directed field in the internal aperture. The half-cells

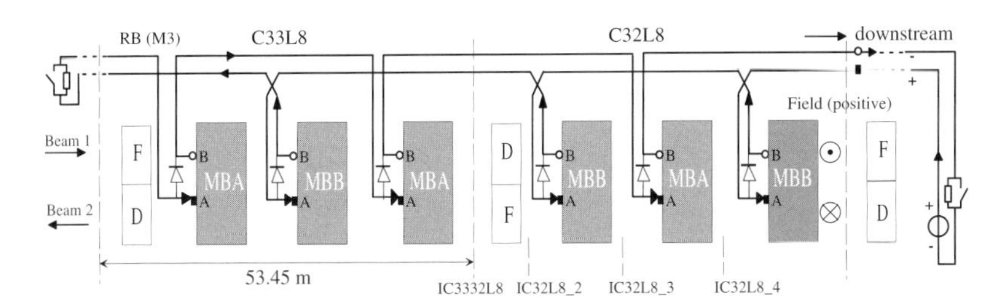

Figure 12.9 Main dipole circuit (RB).

C33L8 and C32L8 are given as an example. The anode of the protection diode is located on the left-hand side seen looking downstream. In Sector 7-8 of the LHC machine, Beam 1 is the external aperture of the twin-aperture magnets.

Figure 12.10 shows the powering scheme for the main quadrupole circuits, RQD for the focusing aperture, and RQF for the defocusing aperture. Again, the half-cells C33L8 and C32L8 are given as an example. Odd-numbered cells in Sector 7-8 have the focusing quadrupole in the external aperture (Beam 1).

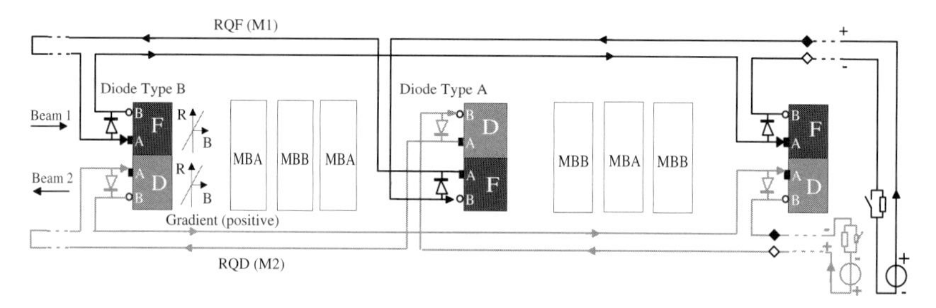

Figure 12.10 Main quadrupole circuits RQD and RQF.

Figure 12.11 shows the spool-piece corrector circuits, RCS for the sextupole spool pieces, RCD for the decapole, and RCO for the octupole spool pieces. We recall that current entering the A-terminal of the sextupole and decapole spool pieces have an upward-pointing field direction in the horizontal plane. Therefore a positive sextupole will compensate for the persistent current effects in the main dipole. As the relative sextupole and decapole components have the same sign in both apertures (see Figure 12.12) the correction requires the same setting at the bipolar power supplies of the two sextupole spool-piece circuits. The same holds for the power supplies of the decapole circuits.

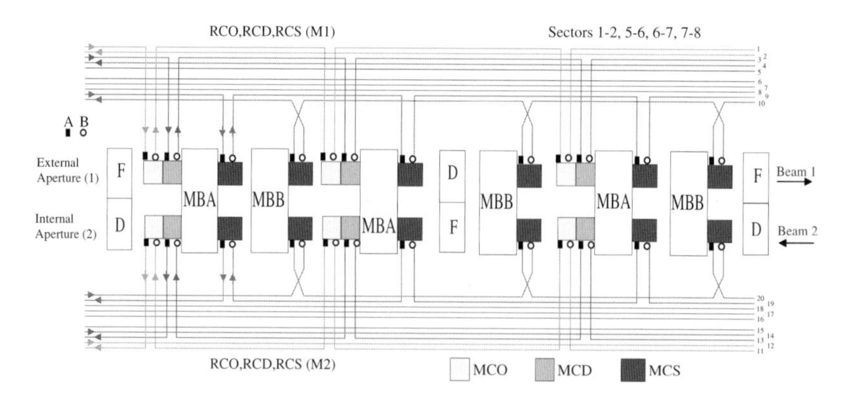

Figure 12.11 Spool-piece corrector circuits.

Current entering the A-terminal of the octupole spool pieces produces a vertical field, increasing along the outward-pointing machine radius. But as

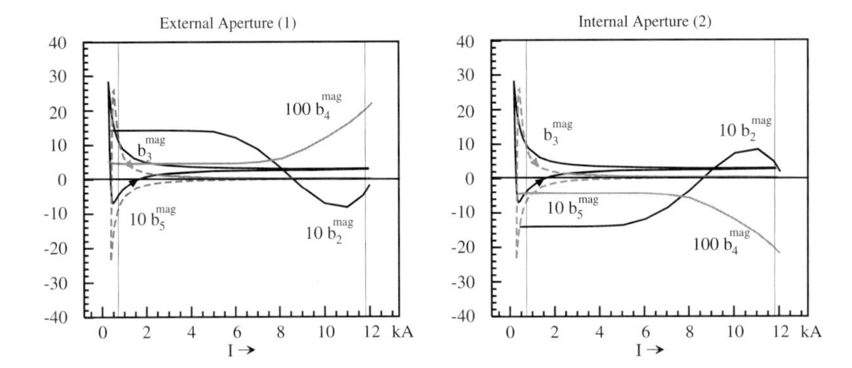

Figure 12.12 Calculated relative multipole field errors versus excitation for the two apertures of the LHC main dipole.

the relative octupole (and quadrupole) field components in the main dipole (resulting from the twin-aperture design) change sign between apertures, the power supplies also require different polarity for compensation of the octupole field component. Table 12.3 gives the settings for the sextupole (MCS), octupole (MCO), and decapole (MCD) corrector circuits required for the correction of the corresponding field errors in the main dipoles.

Table 12.3 Polarity for the sextupole (MCS), octupole (MCO), and decapole (MCD) corrector circuits for Beam 1 and Beam 2, for a correction of the corresponding field errors in the main dipoles (MB) at injection field level[a].

	External aperture (1)						Internal aperture (2)					
Circuit	b_n^{mag}	B_1	B_n^{mag}	B_n^{Beam}	T	P	b_n^{mag}	B_1	B_n^{mag}	B_n^{Beam}	T	P
RCS	–	+	–	–	A	+	–	–	+	+	B	+
RCO	+	+	+	–	A	+	–	–	+	–	B	–
RCD	+	+	+	+	A	–	+	–	–	–	B	–

[a]Beam 1 is in the external aperture as in Sectors 1-2, 5-6, 6-7, and 7-8. T = Terminal, P = Polarity. The field components B_n to be corrected refer to the field in the main dipole magnets.

References

1 Brüning O., Herr, W.: A Beam Separation and Collision Scheme for IP2 and IP8 at the LHC for Optics Version 6.1, LHC project report 367, CERN, 2004

2 CERN Drawing Repository, CDD document LHCLSX__%

3 Grote, H., Iselin, F. C.: The MAD program (Methodical Accelerator Design), User's Reference Manual, CERN/SL/90-13, 1990

4 Nielsen, L.: Short Straight Section (SSS) Types in the Arcs from Q12 to Q12, CERN-EDMS 103939

5 Proudlock, P., Russenschuck, S., Zerlauth, M.: LHC Magnet Polarities, Engineering Specification, EDMS Document Nr. 90041, CERN, 2004

6 Rossbach, J., Schmüser, P.: Basic Course on Accelerator Physics, CERN Accelerator School, Proceedings Vol. 1, CERN 94-01, 1994

7 Schmidt, F.: Private communications, CERN, 2005

8 Wolf, R.: Field Error Naming Conventions for LHC Magnets, Engineering Specification, EDMS document No. 90 250, CERN, 2001

13
Finite-Element Formulations

The purpose of computing is insight, not numbers.

R. Hamming (1915–1998).

A variety of numerical methods have been developed for the computation of electromagnetic fields. Among the best known are the finite difference (FD), the finite-element (FE), and the boundary-element (BE) methods, as well as the finite integration technique (FIT).

Numerical methods, based on the formulation of the physical laws by means of partial differential or integral equations, employ local approximations for the system variables after a suitable discretization of the problem domain.

The finite-element method (FEM) was first proposed in the 1940s and was then extensively applied to problems in structural mechanics. As general references, we look to Strang and Fix [14], Bethe [1], and Zienkiewicz and Taylor [16]. Applications to electrical engineering are treated in Silvester and Ferrari [13], and Binns, Lawrenson, and Trowbridge [2], among others.

The boundary-element method (BEM) has become popular since the 1980s [4]. As it involves fully populated matrices it is efficient in terms of computational resources for problems with a small surface-to-volume ratio. The BE method is applicable to problems for which the Green functions are known, which in general restricts the method to linear homogeneous media. It can, however, be coupled to the FE method to form a hybrid technique that is well suited to the accurate calculation of fields in accelerator magnets; see Chapter 15.

We will introduce the concepts of FE computation by means of a one-dimensional boundary value problem for which the exact solution is known. Because of its simplicity, the one-dimensional problem is ideal for demonstrating the FE method. We will then derive the weak forms of the most commonly used formulations of magnetostatic problems and eddy-current problems in two and three dimensions.

Field Computation for Accelerator Magnets. Stephan Russenschuck

ISBN: 978-3-527-40769-9

13.1
One-Dimensional Finite-Element Analysis

Consider the one-dimensional boundary value problem

$$\frac{d^2u(x)}{dx^2} = f(x), \qquad x \in \Omega \tag{13.1}$$

for $\Omega = [0,1]$, with boundary conditions of the Dirichlet type

$$u(x)|_{x=0} = u_0, \qquad u(x)|_{x=1} = u_1, \tag{13.2}$$

or the Neumann type

$$\left.\frac{du}{dx}\right|_{x=0} = q_0, \qquad \left.\frac{du}{dx}\right|_{x=1} = q_1. \tag{13.3}$$

This can be seen as a one-dimensional representation of a heat conduction problem with unit conductivity. With a constant heat source, $f(x) = c$, and homogeneous Dirichlet boundary conditions $u(x)|_{x=0} = u(x)|_{x=1} = 0$, the analytical solution of the problem can be found by adding the particular solution, $u_p(x) = cx^2/2$, to the general solution, $u_g(x) = ax + b$, of the homogeneous differential equation. The coefficients a and b are determined by the boundary conditions. The solution takes the form

$$u^*(x) = \frac{c}{2}\left(x^2 - x\right). \tag{13.4}$$

This solution is shown graphically in Figure 13.1 (left). For the numerical solution, the differential equation (13.1) will be fulfilled only approximately. The *residual*,

$$R(x) := \frac{d^2u(x)}{dx^2} - f(x), \tag{13.5}$$

is zero only for the exact solution $u(x) = u^*(x)$. A good approximation for $u^*(x)$ can be obtained if the residual error is forced to zero in a weighted, projective sense over the problem domain Ω:

$$\int_\Omega w(x)R(x)\,dx = \int_\Omega w(x)\frac{d^2u(x)}{dx^2}\,dx - \int_\Omega w(x)f(x)\,dx = 0, \tag{13.6}$$

where $w(x)$ denotes an appropriately chosen *weighting function* or *test function*. Using the integration-by-parts rule,

$$\int_a^b \phi\psi'\,dx = [\phi\psi]_a^b - \int_a^b \phi'\psi\,dx, \tag{13.7}$$

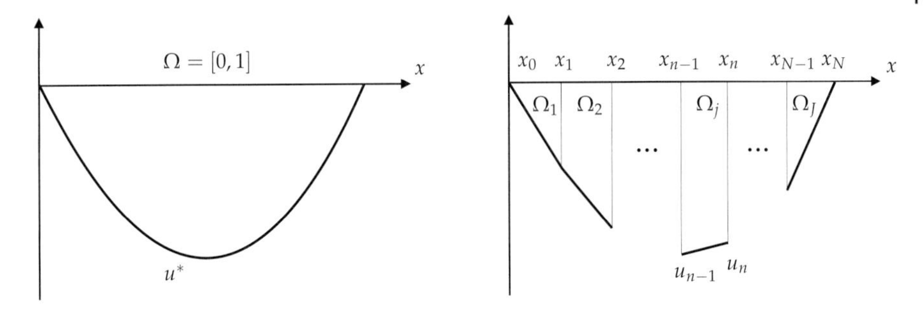

Figure 13.1 Left: Solution of the one-dimensional boundary value problem (13.1). Right: Approximation of the solution within alltogether J finite elements.

to integrate the term $\int_\Omega w(x)\frac{\mathrm{d}^2u(x)}{\mathrm{d}x^2}\,\mathrm{d}x$, yields for $\phi = w(x)$ and $\psi = \frac{\mathrm{d}u(x)}{\mathrm{d}x}$ the *weak form* of the boundary value problem:

$$-\int_\Omega \frac{\mathrm{d}w(x)}{\mathrm{d}x}\frac{\mathrm{d}u(x)}{\mathrm{d}x}\,\mathrm{d}x + \left[w(x)\frac{\mathrm{d}u(x)}{\mathrm{d}x}\right]_0^1 - \int_\Omega w(x)f(x)\,\mathrm{d}x = 0. \tag{13.8}$$

Reduced differentiability requirements for $u(x)$ are thus achieved; hence the term *weak form*. This formulation allows the approximation of the solution by linear basis functions and the accounting for Neumann-type boundary conditions, which are also known as the *natural boundary conditions*. Let us for the moment assume that these boundary conditions are homogeneous. Equation (13.8) reduces to

$$\int_\Omega \frac{\mathrm{d}w(x)}{\mathrm{d}x}\frac{\mathrm{d}u(x)}{\mathrm{d}x}\,\mathrm{d}x = -\int_\Omega w(x)f(x)\,\mathrm{d}x\,. \tag{13.9}$$

Now we partition the domain Ω into J subdomains Ω_j according to

$$\Omega = \bigcup_{j=1}^{J}\Omega_j \tag{13.10}$$

with a total of $N = J + 1$ nodes. The spatial domains Ω_j are known as the *support* of the finite elements. A finite element is characterized not only by the spatial discretization of the problem domain, but also by the number of degrees of freedom, the definition of its nodal variables, the polynomial basis of the approximation, and the interelement continuity.[1]

The potential $u(x)$ is assumed to vary linearly between the two boundary nodes of $\Omega_j = [x_{n-1}, x_n]$ and is approximated by the *basis function*,

$$u(x) = \alpha_{j1} + \alpha_{j2}x\,, \qquad x \in \Omega_j, \tag{13.11}$$

1 Using less rigorous notation, we will not distinguish between the spatial discretization and the finite element and will assign Ω_j in both cases.

shown in Figure 13.1. The approximated values u_n at the nodes of Ω_j can be expressed as

$$u_{n-1} = \alpha_{j1} + \alpha_{j2}\, x_{n-1}, \qquad u_n = \alpha_{j1} + \alpha_{j2}\, x_n\,. \tag{13.12}$$

Applying Cramer's rule to solve for α_{j1} and α_{j2} yields

$$\alpha_{j1} = \frac{x_n u_{n-1} - x_{n-1} u_n}{x_n - x_{n-1}}, \qquad \alpha_{j2} = \frac{u_n - u_{n-1}}{x_n - x_{n-1}}. \tag{13.13}$$

For any point within Ω_j, the potential $u(x)$ can be expressed as a function of the nodal values of the element:

$$u(x) = \alpha_{j1} + \alpha_{j2} x = \frac{x_n - x}{x_n - x_{n-1}} u_{n-1} + \frac{-x_{n-1} + x}{x_n - x_{n-1}} u_n. \tag{13.14}$$

The nodal values of the potential are usually referred to as *degrees of freedom*. The quotients in Eq. (13.14) are called the *local element shape functions* $N_{jk}(x)$, $k = 1, 2$, where k is the local node number in Ω_j:

$$N_{j1}(x) := \frac{x_n - x}{x_n - x_{n-1}}, \qquad N_{j2}(x) := \frac{-x_{n-1} + x}{x_n - x_{n-1}}. \tag{13.15}$$

The shape functions depend only on the spatial coordinates of the nodes. We can rewrite the weak form (13.9) for Ω_j as

$$\int_{\Omega_j} \frac{\mathrm{d}w_l(x)}{\mathrm{d}x} \sum_{k=1,2} \frac{\mathrm{d}N_{jk}(x)}{\mathrm{d}x} u_{jk}\, \mathrm{d}x = -\int_{\Omega_j} w_l(x) f(x)\, \mathrm{d}x\,, \qquad l = 1, 2. \tag{13.16}$$

Here $u_{j1} = u_{n-1}$ and $u_{j2} = u_n$ are the node potentials in the local numbering scheme of Ω_j.

The numerical methods differ in the choice of the weighting functions. *Galerkin's method*[2] uses the element shape functions $N_{jk}(x)$ as the weighting functions $w_l(x)$. This yields

$$\int_{\Omega_j} \frac{\mathrm{d}N_{jl}(x)}{\mathrm{d}x} \sum_{k=1,2} \frac{\mathrm{d}N_{jk}(x)}{\mathrm{d}x} u_{jk}\, \mathrm{d}x = -\int_{\Omega_j} N_{jl}(x) f(x)\, \mathrm{d}x\,, \qquad l = 1, 2. \tag{13.17}$$

Omitting the argument of N_{jk} results in the following linear equation system for Ω_j:

$$\begin{aligned} \int_{x_{n-1}}^{x_n} \left(\frac{\mathrm{d}N_{j1}}{\mathrm{d}x} \frac{\mathrm{d}N_{j1}}{\mathrm{d}x} u_{n-1} + \frac{\mathrm{d}N_{j1}}{\mathrm{d}x} \frac{\mathrm{d}N_{j2}}{\mathrm{d}x} u_n \right) \mathrm{d}x &= -\int_{x_{n-1}}^{x_n} N_{j1} f(x)\, \mathrm{d}x, \\ \int_{x_{n-1}}^{x_n} \left(\frac{\mathrm{d}N_{j2}}{\mathrm{d}x} \frac{\mathrm{d}N_{j1}}{\mathrm{d}x} u_{n-1} + \frac{\mathrm{d}N_{j2}}{\mathrm{d}x} \frac{\mathrm{d}N_{j2}}{\mathrm{d}x} u_n \right) \mathrm{d}x &= -\int_{x_{n-1}}^{x_n} N_{j2} f(x)\, \mathrm{d}x. \end{aligned} \tag{13.18}$$

2 Boris Galerkin (1871–1945).

The linear equation system (13.18) can be expressed in matrix form as

$$[k_j]\{u_j\} = \{f_j\}, \tag{13.19}$$

where

$$[k_j] := \int_{x_{n-1}}^{x_n} \begin{pmatrix} \frac{dN_{j1}}{dx}\frac{dN_{j1}}{dx} & \frac{dN_{j1}}{dx}\frac{dN_{j2}}{dx} \\ \frac{dN_{j2}}{dx}\frac{dN_{j1}}{dx} & \frac{dN_{j2}}{dx}\frac{dN_{j2}}{dx} \end{pmatrix} dx \tag{13.20}$$

is the symmetric *element stiffness matrix*,

$$\{u_j\} := \begin{pmatrix} u_{j1} \\ u_{j2} \end{pmatrix} = \begin{pmatrix} u_{n-1} \\ u_n \end{pmatrix} \tag{13.21}$$

the column vector of the node potentials, and

$$\{f_j\} := -\int_{x_{n-1}}^{x_n} \begin{pmatrix} N_{j1} \\ N_{j2} \end{pmatrix} f(x)dx \tag{13.22}$$

the *element force vector*. Inhomogeneous Neumann boundary conditions, for example, $\frac{du}{dx}|_{x=0} = q_0$ in Ω_1, are taken into account by the element boundary vector

$$\{t_1\} = \begin{pmatrix} N_{11}(x)|_{x=0} \\ 0 \end{pmatrix} = \begin{pmatrix} 1 \\ 0 \end{pmatrix}, \tag{13.23}$$

which is added to the linear equations system of Ω_1:

$$[k_1]\{u_1\} + q_0\{t_1\} = \{f_1\}. \tag{13.24}$$

To assemble the linear equation system for the boundary value problem, the contributions $[k_j]$ must be sorted into the overall stiffness matrix according to the global numbering scheme of the degrees of freedom.

In the final step, the Dirichlet boundary conditions must be accommodated and are imposed by writing the boundary values into the $\{u_j\}$ vector of the element. To avoid an overdetermined equation system, the corresponding rows and columns are eliminated. The boundary conditions are in this way impressed into the linear equation system and are therefore known as the *essential boundary conditions*.

Example: For the numerical solution of the boundary value problem the domain Ω is divided into four finite elements $\Omega_j, j = 1, \ldots, 4$ of equal length l, shown in Figure 13.2 (right). The stiffness matrices $[k_j]$ yield

$$[k_j] = \int_{x_{n-1}}^{x_n} \begin{pmatrix} \frac{dN_{j1}}{dx}\frac{dN_{j1}}{dx} & \frac{dN_{j1}}{dx}\frac{dN_{j2}}{dx} \\ \frac{dN_{j2}}{dx}\frac{dN_{j1}}{dx} & \frac{dN_{j2}}{dx}\frac{dN_{j2}}{dx} \end{pmatrix} dx$$

$$= \int_{x_{n-1}}^{x_n} \begin{pmatrix} \frac{1}{(x_n - x_{n-1})^2} & \frac{-1}{(x_n - x_{n-1})^2} \\ \frac{-1}{(x_n - x_{n-1})^2} & \frac{1}{(x_n - x_{n-1})^2} \end{pmatrix} dx = \begin{pmatrix} \frac{1}{l} & \frac{-1}{l} \\ \frac{-1}{l} & \frac{1}{l} \end{pmatrix}, \tag{13.25}$$

and the vectors $\{f_j\}$ are for $f(x) = c$,

$$\{f_j\} = -\int_{x_{n-1}}^{x_n} \begin{pmatrix} N_{j1} \\ N_{j2} \end{pmatrix} f(x)\, dx = -c \int_{x_{n-1}}^{x_n} \begin{pmatrix} \frac{x_n - x}{x_n - x_{n-1}} \\ \frac{-x_{n-1} + x}{x_n - x_{n-1}} \end{pmatrix} dx$$

$$= -\frac{c}{2l} \begin{pmatrix} 2x_n x - x^2 \\ -2x_{n-1}x + x^2 \end{pmatrix} \Bigg|_{x_{n-1}}^{x_n} = -\frac{c}{2l} \begin{pmatrix} (x_n - x_{n-1})^2 \\ (x_{n-1} - x_n)^2 \end{pmatrix}$$

$$= -\frac{1}{2} \begin{pmatrix} cl \\ cl \end{pmatrix}. \tag{13.26}$$

Assembling all the contributions into the overall stiffness matrix gives

$$\begin{pmatrix} \frac{1}{l} & \frac{-1}{l} & 0 & 0 & 0 \\ \frac{-1}{l} & \frac{2}{l} & \frac{-1}{l} & 0 & 0 \\ 0 & \frac{-1}{l} & \frac{2}{l} & \frac{-1}{l} & 0 \\ 0 & 0 & \frac{-1}{l} & \frac{2}{l} & \frac{-1}{l} \\ 0 & 0 & 0 & \frac{-1}{l} & \frac{1}{l} \end{pmatrix} \begin{pmatrix} u_1 \\ u_2 \\ u_3 \\ u_4 \\ u_5 \end{pmatrix} = -\frac{1}{2} \begin{pmatrix} cl \\ 2\,cl \\ 2\,cl \\ 2\,cl \\ cl \end{pmatrix}. \tag{13.27}$$

Setting $u_1 = u_5 = 0$, the first and last rows and columns in the global stiffness matrix eliminate, and the equation system reduces to

$$\begin{pmatrix} \frac{2}{l} & \frac{-1}{l} & 0 \\ \frac{-1}{l} & \frac{2}{l} & \frac{-1}{l} \\ 0 & \frac{-1}{l} & \frac{2}{l} \end{pmatrix} \begin{pmatrix} u_2 \\ u_3 \\ u_4 \end{pmatrix} = -\begin{pmatrix} cl \\ cl \\ cl \end{pmatrix}. \tag{13.28}$$

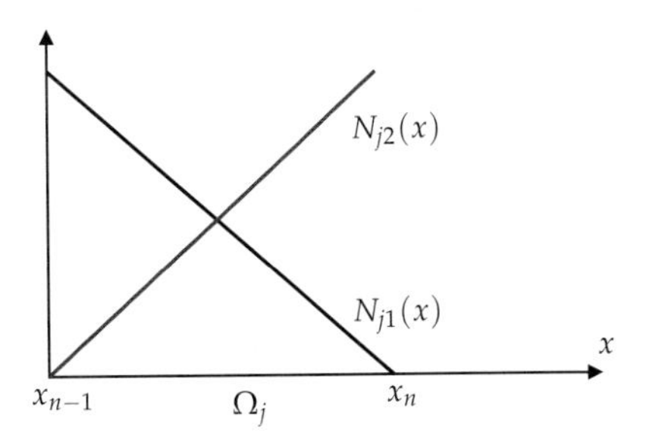

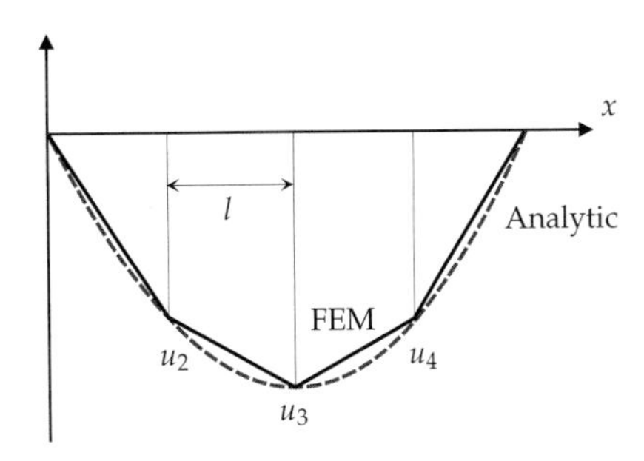

Figure 13.2 Left: Element shape functions. Right: Finite-element solution of the boundary value problem with homogeneous Dirichlet boundary conditions.

Finally, setting $c = 4$ and $l = 0.25$ yields the solution

$$\begin{pmatrix} u_2 \\ u_3 \\ u_4 \end{pmatrix} = -\begin{pmatrix} \frac{3l}{4} & \frac{l}{2} & \frac{l}{4} \\ \frac{l}{2} & l & \frac{l}{2} \\ \frac{l}{4} & \frac{l}{2} & \frac{3l}{4} \end{pmatrix} \begin{pmatrix} cl \\ cl \\ cl \end{pmatrix} = \begin{pmatrix} -0.375 \\ -0.5 \\ -0.375 \end{pmatrix}. \tag{13.29}$$

The analytical solution and its finite-element approximation are shown in Figure 13.2 (right). □

13.1.1 Quadratic Elements

One-dimensional, second-order elements, also called *quadratic elements*, have three nodes: one at each end point of the element, while the third point is placed at the element's center; see Figure 13.3. Within each element, the basis function is given by

$$u(x) = \alpha_{j1} + \alpha_{j2}x + \alpha_{j3}x^2, \quad x \in \Omega_j. \tag{13.30}$$

Employing the local numbering scheme allows the approximated values $u^{(k)}$ at the nodes of Ω_j to be expressed as

$$\begin{aligned} u_{j1} &= \alpha_{j1} + \alpha_{j2}x_1 + \alpha_{j3}x_1^2, \\ u_{j2} &= \alpha_{j1} + \alpha_{j2}x_2 + \alpha_{j3}x_2^2, \\ u_{j3} &= \alpha_{j1} + \alpha_{j2}x_3 + \alpha_{j3}x_3^2. \end{aligned} \tag{13.31}$$

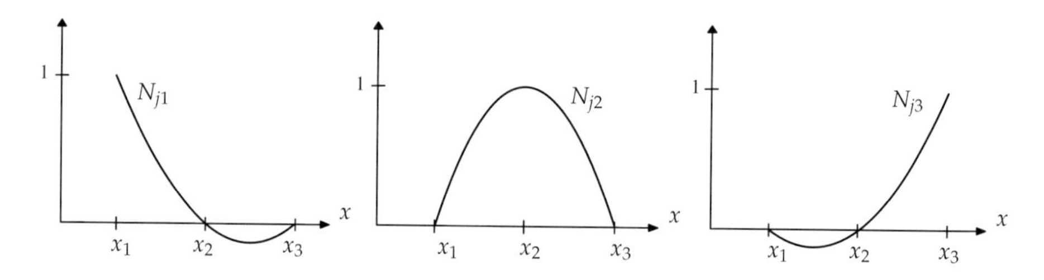

Figure 13.3 One-dimensional, quadratic element shape functions.

Applying Cramer's rule to solve for α_{j1}, α_{j2}, α_{j3} and substituting the result into Eq. (13.30) yields

$$u(x) = \sum_{k=1}^{3} N_{jk}(x) u_{jk} , \tag{13.32}$$

where

$$N_{j1}(x) = \frac{(x - x_2)(x - x_3)}{(x_1 - x_2)(x_1 - x_3)} , \qquad N_{j2}(x) = \frac{(x - x_1)(x - x_3)}{(x_2 - x_1)(x_2 - x_3)} ,$$

$$N_{j3}(x) = \frac{(x - x_1)(x - x_2)}{(x_3 - x_1)(x_3 - x_2)} . \tag{13.33}$$

The element stiffness matrix $[k_j]$ is given by

$$[k_j] = \int_{x_1}^{x_3} \begin{pmatrix} \frac{\mathrm{d}N_{j1}}{\mathrm{d}x}\frac{\mathrm{d}N_{j1}}{\mathrm{d}x} & \frac{\mathrm{d}N_{j1}}{\mathrm{d}x}\frac{\mathrm{d}N_{j2}}{\mathrm{d}x} & \frac{\mathrm{d}N_{j1}}{\mathrm{d}x}\frac{\mathrm{d}N_{j3}}{\mathrm{d}x} \\ \frac{\mathrm{d}N_{j2}}{\mathrm{d}x}\frac{\mathrm{d}N_{j1}}{\mathrm{d}x} & \frac{\mathrm{d}N_{j2}}{\mathrm{d}x}\frac{\mathrm{d}N_{j2}}{\mathrm{d}x} & \frac{\mathrm{d}N_{j2}}{\mathrm{d}x}\frac{\mathrm{d}N_{j3}}{\mathrm{d}x} \\ \frac{\mathrm{d}N_{j3}}{\mathrm{d}x}\frac{\mathrm{d}N_{j1}}{\mathrm{d}x} & \frac{\mathrm{d}N_{j3}}{\mathrm{d}x}\frac{\mathrm{d}N_{j2}}{\mathrm{d}x} & \frac{\mathrm{d}N_{j3}}{\mathrm{d}x}\frac{\mathrm{d}N_{j3}}{\mathrm{d}x} \end{pmatrix} \mathrm{d}x . \tag{13.34}$$

If the element nodes are equidistantly spaced, $x_3 - x_2 = x_2 - x_1 = l$, the element stiffness matrix takes the form

$$[k_j] = \begin{pmatrix} \frac{7}{6l} & \frac{-8}{6l} & \frac{1}{6l} \\ \frac{-8}{6l} & \frac{16}{6l} & \frac{-8}{6l} \\ \frac{1}{6l} & \frac{-8}{6l} & \frac{7}{6l} \end{pmatrix} . \tag{13.35}$$

In the special case of a constant force term, $f(x) = c$, the element's force vector is

$$\{f_j\} = -\int_{x_1}^{x_3} \begin{pmatrix} N_{j1} \\ N_{j2} \\ N_{j3} \end{pmatrix} f(x)\,\mathrm{d}x = -\frac{1}{3}c \begin{pmatrix} l \\ 4l \\ l \end{pmatrix} . \tag{13.36}$$

Consider once more the example shown in Figure 13.2 (right), this time with only two quadratic elements but the same number of degrees of freedom. For homogeneous Dirichlet boundary conditions at the nodes 1 and 5, we obtain the linear equation system

$$\begin{pmatrix} \frac{16}{6l} & \frac{-8}{6l} & 0 \\ \frac{-8}{6l} & \frac{14}{6l} & \frac{-8}{6l} \\ 0 & \frac{-8}{6l} & \frac{16}{6l} \end{pmatrix} \begin{pmatrix} u_2 \\ u_3 \\ u_4 \end{pmatrix} = -\frac{1}{3} \begin{pmatrix} 4cl \\ 2cl \\ 4cl \end{pmatrix} . \tag{13.37}$$

For $c = 4$ and $l = 0.25$, the solution of the problem is $u_2 = u_4 = -0.375$ and $u_3 = -0.5$ as before. But in spite of the same number of degrees of freedom, the approximate solution has improved by the higher-order approximation for the potential function between the nodes.

13.2 FEM with the Vector-Potential (Curl–Curl) Formulation

We now turn to FE procedures for two- and three-dimensional problem domains. Consider the elementary model problem presented in Section 4.5, consisting of the iron region Ω_i and the air region Ω_a. Let us state further that $\Omega = \Omega_a \cup \Omega_i$. The regions are connected at the interface Γ_{ai} and may contain areas with impressed current densities $\mathbf{J}$, which do not intersect the interface. Using the results of Section 4.9.2, we can formulate the boundary value problem for the magnetic vector potential $\mathbf{A} \in \mathcal{V}(\Omega)$:

$$\operatorname{curl} \frac{1}{\mu} \operatorname{curl} \mathbf{A} - \operatorname{grad} \frac{1}{\mu} \operatorname{div} \mathbf{A} = \mathbf{J} . \tag{13.38}$$

The boundary and interface conditions (4.141)–(4.143), (4.148), and (4.149) can all be expressed in terms of the magnetic vector potential:

$$\mathbf{A} \cdot \mathbf{n} = 0 \quad \text{on } \Gamma_H, \tag{13.39}$$

$$\frac{1}{\mu} \operatorname{div} \mathbf{A} = 0 \quad \text{on } \Gamma_B, \tag{13.40}$$

$$\mathbf{n} \times (\mathbf{A} \times \mathbf{n}) = \mathbf{0} \quad \text{on } \Gamma_B, \tag{13.41}$$

$$\mathbf{n} \times \left(\frac{1}{\mu} (\operatorname{curl} \mathbf{A}) \times \mathbf{n} \right) = \mathbf{0} \quad \text{on } \Gamma_H, \tag{13.42}$$

$$\left[\!\left[\frac{1}{\mu} \operatorname{div} \mathbf{A} \right]\!\right]_{\text{ai}} = 0 \quad \text{on } \Gamma_{\text{ai}}, \tag{13.43}$$

$$\mathbf{n} \times \left[\!\left[\frac{1}{\mu} \left(\operatorname{curl} \mathbf{A} \right) \right]\!\right]_{\mathrm{ai}} = \mathbf{0} \quad \text{on } \Gamma_{\mathrm{ai}}, \tag{13.44}$$

$$\left[\!\left[\mathbf{A} \right]\!\right]_{\mathrm{ai}} = \mathbf{0} \quad \text{on } \Gamma_{\mathrm{ai}}. \tag{13.45}$$

For the approximate solution of $\mathbf{A}$ in an FE mesh, we can only approximately fulfill the differential equation (13.38):

$$\operatorname{curl} \frac{1}{\mu} \operatorname{curl} \mathbf{A} - \operatorname{grad} \frac{1}{\mu} \operatorname{div} \mathbf{A} - \mathbf{J} = \mathbf{R}, \tag{13.46}$$

where $\mathbf{R} \in \mathcal{V}(\Omega)$ is the residual vector field. A linear equation system for the unknown nodal values $\mathbf{A}_n$ of the vector potential can then be obtained by minimizing the weighted residual $\mathbf{R}$ integrated over the domain Ω:

$$\int_{\Omega} \mathbf{w}_a \cdot \mathbf{R} \, \mathrm{d}V = 0 \tag{13.47}$$

for $a = 1, 2, 3$, and the vector weighting functions

$$\mathbf{w}_1 := \begin{pmatrix} w_1 \\ 0 \\ 0 \end{pmatrix}, \quad \mathbf{w}_2 := \begin{pmatrix} 0 \\ w_2 \\ 0 \end{pmatrix}, \quad \mathbf{w}_3 := \begin{pmatrix} 0 \\ 0 \\ w_3 \end{pmatrix}, \tag{13.48}$$

where w_1, w_2, w_3 are scalar functions. The vector weighting functions $\mathbf{w}_a$ must obey the homogeneous boundary conditions

$$\mathbf{w}_a \cdot \mathbf{n} = 0 \quad \text{on } \Gamma_H, \tag{13.49}$$

$$\mathbf{w}_a \times \mathbf{n} = \mathbf{0} \quad \text{on } \Gamma_B. \tag{13.50}$$

The relevance of these boundary conditions can be best explained by means of an example [8]: Consider Γ_H to be parallel with the yz-plane. The boundary conditions for the vector potential imply that $A_y = A_z = 0$. This corresponds to the condition (13.49), which takes the form $w_2 = w_3 = 0$ at the yz-plane. Similar reasoning follows for the condition (13.50).

Forcing the weighted residual (13.47) to zero yields

$$\int_{\Omega} \mathbf{w}_a \cdot \left(\operatorname{curl} \frac{1}{\mu} \operatorname{curl} \mathbf{A} - \operatorname{grad} \frac{1}{\mu} \operatorname{div} \mathbf{A} \right) \mathrm{d}V = \int_{\Omega} \mathbf{w}_a \cdot \mathbf{J} \, \mathrm{d}V \tag{13.51}$$

for $a = 1, 2, 3$.

13.2.1
The Weak Form in 3D

Applying the generalizations of the integration-by-parts rule[3]

$$\int_{\Omega}\left(\operatorname{curl}\frac{1}{\mu}\operatorname{curl}\mathbf{A}\right)\cdot\mathbf{w}_a\,dV=\int_{\Omega}\frac{1}{\mu}\operatorname{curl}\mathbf{A}\cdot\operatorname{curl}\mathbf{w}_a\,dV-\int_{\Gamma}\frac{1}{\mu}(\operatorname{curl}\mathbf{A}\times\mathbf{n})\cdot\mathbf{w}_a\,da,\quad(13.52)$$

$$\int_{\Omega}\left(-\operatorname{grad}\frac{1}{\mu}\operatorname{div}\mathbf{A}\right)\cdot\mathbf{w}_a\,dV=\int_{\Omega}\frac{1}{\mu}\operatorname{div}\mathbf{A}\operatorname{div}\mathbf{w}_a\,dV-\int_{\Gamma}\frac{1}{\mu}\operatorname{div}\mathbf{A}(\mathbf{n}\cdot\mathbf{w}_a)\,da,\quad(13.53)$$

Eq. (13.51) yields

$$\begin{aligned}&\int_{\Omega}\frac{1}{\mu}\operatorname{curl}\mathbf{A}\cdot\operatorname{curl}\mathbf{w}_a\,dV-\int_{\Gamma_H,\Gamma_B}\frac{1}{\mu}(\operatorname{curl}\mathbf{A}\times\mathbf{n})\cdot\mathbf{w}_a\,da\\&+\int_{\Omega}\frac{1}{\mu}\operatorname{div}\mathbf{A}\operatorname{div}\mathbf{w}_a\,dV-\int_{\Gamma_H,\Gamma_B}\frac{1}{\mu}\operatorname{div}\mathbf{A}(\mathbf{n}\cdot\mathbf{w}_a)\,da\\&-\int_{\Gamma_{ai}}\left(\frac{1}{\mu}\operatorname{div}\mathbf{A}_i(\mathbf{n}_i\cdot\mathbf{w}_a)+\frac{1}{\mu_0}\operatorname{div}\mathbf{A}_a(\mathbf{n}_a\cdot\mathbf{w}_a)\right)da\\&-\int_{\Gamma_{ai}}\left(\frac{1}{\mu}(\operatorname{curl}\mathbf{A}_i\times\mathbf{n}_i)+\frac{1}{\mu_0}(\operatorname{curl}\mathbf{A}_a\times\mathbf{n}_a)\right)\cdot\mathbf{w}_a\,da=\int_{\Omega}\mathbf{w}_a\cdot\mathbf{J}\,dV\end{aligned}\quad(13.54)$$

for a = 1,2,3. The boundary conditions (13.40) and (13.42)–(13.44), and the conditions (13.49) and (13.50) for the vector weighting functions $\mathbf{w}_a$, cause all the boundary integrals in Eq. (13.54) to vanish. Hence Eq. (13.54) is reduced to

$$\int_{\Omega}\frac{1}{\mu}\operatorname{curl}\mathbf{w}_a\cdot\operatorname{curl}\mathbf{A}\,dV+\int_{\Omega}\frac{1}{\mu}\operatorname{div}\mathbf{w}_a\operatorname{div}\mathbf{A}\,dV=\int_{\Omega}\mathbf{w}_a\cdot\mathbf{J}\,dV\quad(13.55)$$

for a = 1,2,3. Equation (13.55) is the weak form of the 3D boundary value problem (13.38)–(13.45). The essential boundary conditions (13.39), (13.41), and (13.45) can be enforced, once the matrix of the linear equation system is assembled.

3 Equation (13.52.) is proved by substituting $\mathbf{a}=\mathbf{w}_a$ and $\mathbf{b}=\frac{1}{\mu}\operatorname{curl}\mathbf{A}$ in Eq. (3.160). Equation (13.53) is proved by substituting $\phi=\frac{1}{\mu}\operatorname{div}\mathbf{A}$ and $\mathbf{a}=\mathbf{w}_a$ in Eq. (3.163).

13.2.2 The Weak Form in 2D

For two-dimensional field problems ($\partial/\partial z = 0$), the vector potential has only a z-component, and the Coulomb gauge is automatically fulfilled. Hence, Eq. (13.55) takes the form

$$\int_{\Omega_2} \frac{1}{\mu} \operatorname{curl} \mathbf{w}_3 \cdot \operatorname{curl} \mathbf{A}_z \, da = \int_{\Omega_2} w_3 \, J_z \, da \,. \tag{13.56}$$

Employing the relation $\operatorname{curl}(g_z \mathbf{e}_z) = \operatorname{grad} g_z \times \mathbf{e}_z$, derived from Eq. (3.69), it follows:

$$\int_{\Omega_2} \frac{1}{\mu} \operatorname{grad} w_3 \cdot \operatorname{grad} A_z \, da = \int_{\Omega_2} w_3 \, J_z \, da \,. \tag{13.57}$$

The essential boundary condition $\mathbf{n} \times (\mathbf{A} \times \mathbf{n}) = \mathbf{0}$ on Γ_B takes the form $A_z = 0$. The boundary condition $\mathbf{A} \cdot \mathbf{n} = 0$ on Γ_H is automatically fulfilled because $\mathbf{n}$ is perpendicular to $\mathbf{e}_z$.

Remark 1: The current density $\mathbf{J}$ appears on the right-hand side of the weak integral equation (13.55). Consequently, the complicated shape of the superconducting coils must be modeled in the finite-element mesh; see Figure 13.4. While this is cumbersome in 2D, it is not even possible in 3D without some simplification of the geometry. □

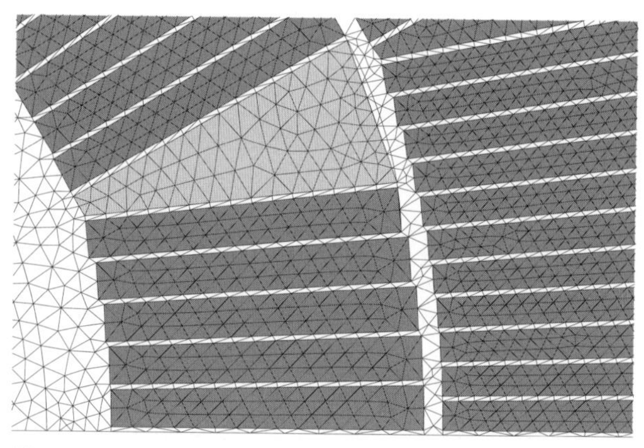

Figure 13.4 Finite-element mesh of the LHC main dipole coil. The mesh required for the accurate modeling of a superconducting coil is very dense, resulting in a large number of unknowns, in particular if the surrounding iron yoke geometry must be modeled.

Remark 2: The FE method ensures only the approximate solution of the weak form. Consequently, an error in the fulfillment of the Coulomb gauge is inevitable. The continuity condition (13.43) at the iron/air interface implies that this error is much higher in the ferromagnetic regions. There it implies a large error in the fulfillment of Ampére's law, since the penalty term $\operatorname{grad} \frac{1}{\mu} \operatorname{div} \mathbf{A}$ in Eq. (13.38) is far from zero. This is a shortcoming of the gauged formulation for 3D problems. In two dimensions the penalty term is not needed and therefore the problem does not exist. □

13.3 Complementary Formulations

We will now turn to complementary formulations that do not require the representation of the coils in the FE mesh. One approach uses the reduced vector-potential method by Bardi et al. [10]. The method also reduces the effect of the far-field boundary. Moreover, it allows the magnet designer to distinguish between the geometrical field errors caused by the coil geometry and the saturation-dependent field errors attributable to the iron yoke.

Another method is to restrict the FE domain to the iron yoke, which is assumed to be free of impressed current sources. The FE method is then coupled to the BE method by means of the continuity conditions at the iron/air interface. The FE portion of this coupling method will be derived in this section.

13.3.1 FEM with Reduced Vector-Potential Formulation

Consider again the elementary model problem of Section 4.5. The vector potential $\mathbf{A}$ is now split into two parts,

$$\mathbf{A} = \mathbf{A}_\mathrm{s} + \mathbf{A}_\mathrm{r} \tag{13.58}$$

where $\mathbf{A}_\mathrm{r}$ is the reduced vector potential due to the iron magnetization and $\mathbf{A}_\mathrm{s}$ is the source vector potential due to the currents in the air domain. This allows us to write,

$$\mathbf{B} = \mu_0 \mathbf{H}_\mathrm{s} + \operatorname{curl} \mathbf{A}_\mathrm{r}. \tag{13.59}$$

The source vector potential can be calculated from Biot–Savart-type integrals of the current distribution according to Eq. (5.29). The weak form can be derived as follows: From Eq. (4.151), replacing $\mathbf{A}$ with $\mathbf{A}_\mathrm{s} + \mathbf{A}_\mathrm{r}$, we derive

$$\operatorname{curl} \frac{1}{\mu} \operatorname{curl} (\mathbf{A}_\mathrm{r} + \mathbf{A}_\mathrm{s}) - \operatorname{grad} \frac{1}{\mu} \operatorname{div} (\mathbf{A}_\mathrm{r} + \mathbf{A}_\mathrm{s}) = \mathbf{J} \tag{13.60}$$

for $\mathbf{A}_\mathrm{r}, \mathbf{A}_\mathrm{s}, \mathbf{J} \in \mathcal{V}(\Omega)$. Equation (13.60) can be rewritten as

$$\begin{aligned}\operatorname{curl}\frac{1}{\mu}\operatorname{curl}\mathbf{A}_\mathrm{r} - \operatorname{grad}\frac{1}{\mu}\operatorname{div}\mathbf{A}_\mathrm{r} &= \mathbf{J} - \operatorname{curl}\frac{1}{\mu}\operatorname{curl}\mathbf{A}_\mathrm{s}\\ &= \operatorname{curl}\left(\mathbf{H}_\mathrm{s} - \frac{\mu_0}{\mu}\mathbf{H}_\mathrm{s}\right). \end{aligned} \tag{13.61}$$

The boundary and interface conditions yield

$$\mathbf{A}_\mathrm{r}\cdot\mathbf{n} = 0 \quad \text{on } \Gamma_H, \tag{13.62}$$

$$\frac{1}{\mu}\operatorname{div}\mathbf{A}_\mathrm{r} = 0 \quad \text{on } \Gamma_B, \tag{13.63}$$

$$\frac{1}{\mu}\operatorname{curl}\mathbf{A}_\mathrm{r}\times\mathbf{n} + \frac{\mu_0}{\mu}\mathbf{H}_\mathrm{s}\times\mathbf{n} = \mathbf{0} \quad \text{on } \Gamma_H, \tag{13.64}$$

$$\mathbf{n}\cdot\operatorname{curl}\mathbf{A}_\mathrm{r} + \mu_0\mathbf{n}\cdot\mathbf{H}_\mathrm{s} = 0 \quad \text{on } \Gamma_B, \tag{13.65}$$

$$\left[\!\left[\frac{1}{\mu}\operatorname{div}\mathbf{A}_\mathrm{r}\right]\!\right]_\mathrm{ai} = 0, \tag{13.66}$$

$$\left[\!\left[\frac{1}{\mu}\left(\operatorname{curl}\mathbf{A}_\mathrm{r}\right)\times\mathbf{n}\right]\!\right]_\mathrm{ai} = \mathbf{0}, \tag{13.67}$$

$$\left[\!\left[\mathbf{A}_\mathrm{r}\right]\!\right]_\mathrm{ai} = \mathbf{0}. \tag{13.68}$$

The left-hand side of Eq. (13.61) and the boundary conditions can be brought together by means of the identities (13.52) and (13.53), this time applied to the reduced vector potential. The right-hand side of Eq. (13.61) can be rewritten as

$$\begin{aligned}\int_\Omega \operatorname{curl}\left(\mathbf{H}_\mathrm{s} - \frac{\mu_0}{\mu}\mathbf{H}_\mathrm{s}\right)\cdot\mathbf{w}_a\,\mathrm{d}V &= \int_\Omega \left(\mathbf{H}_\mathrm{s} - \frac{\mu_0}{\mu}\mathbf{H}_\mathrm{s}\right)\cdot\operatorname{curl}\mathbf{w}_a\,\mathrm{d}V\\ &\quad - \int_\Gamma \left(\mathbf{H}_\mathrm{s}\times\mathbf{n} - \frac{\mu_0}{\mu}\mathbf{H}_\mathrm{s}\times\mathbf{n}\right)\cdot\mathbf{w}_a\,\mathrm{d}a. \end{aligned}\tag{13.69}$$

Assuming that both the reduced and the source fields obey the same boundary conditions at Γ_H and thus $\mathbf{H}_\mathrm{s}\times\mathbf{n} = \mathbf{0}$ on Γ_H, and enforcing the boundary condition (13.50) for $\mathbf{w}_a$, the weak integral form of the reduced vector potential formulation is given by

$$\begin{aligned}\int_\Omega \operatorname{curl}\mathbf{w}_a\cdot\frac{1}{\mu}\operatorname{curl}\mathbf{A}_\mathrm{r}\,\mathrm{d}V + \int_\Omega \operatorname{div}\mathbf{w}_a\cdot\frac{1}{\mu}\operatorname{div}\mathbf{A}_\mathrm{r}\,\mathrm{d}V\\ = \int_\Omega \operatorname{curl}\mathbf{w}_a\cdot\left(\mathbf{H}_\mathrm{s} - \frac{\mu_0}{\mu}\mathbf{H}_\mathrm{s}\right)\mathrm{d}V \end{aligned}\tag{13.70}$$

for $a = 1,2,3$. In the air region Ω_a, we set $\mu = \mu_0$, and therefore the right-hand side of the above equation is zero. The current density does not appear explicitly in the equation. The required source field in the iron region can be calculated by means of the Biot–Savart law.

Experience has shown that the influence of the far-field boundary on the multipole field errors is smaller when the reduced vector-potential formulation is used. The reason is that the numerically calculated reduced field accounts for only 10–30% of the total field in superconducting accelerator magnets. Although the mesh in the air region does not need to match the coil geometry, the air region must nevertheless be meshed. Figure 13.5 shows an example. Though this does not pose a problem in 2D modeling it proves troublesome in the 3D case.

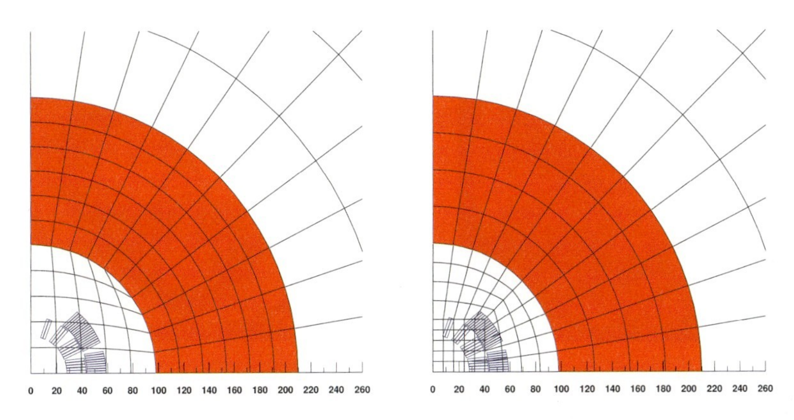

Figure 13.5 Finite-element meshes for a round iron yoke in a rectangular problem domain. When using the reduced vector-potential formulation, the coils do not need to be represented by the FE mesh. Collinear sides of elements (see left figure at the 45° plane) must be avoided as they lead to poor numerical stability.

Figure 13.6 shows the results of field calculations employing the reduced vector-potential formulation in the case of the LHC main dipole magnet and an alternative twin-aperture design [7].

It is instructive to distinguish between the magnetic flux density generated by the superconducting coil (modulus of the source field B_s) and the magnetic flux density due to the magnetization in the iron yoke (modulus of the reduced field B_r). The source field and the reduced field show left/right asymmetries in the two apertures due to the twin-aperture design, an effect that is commonly referred to as *crosstalk* between magnet apertures. We limit this crosstalk in the total field by an optimal design of the iron yoke; see Section 20.10.4.

The top figures in Figure 13.6 show the distribution of the source field, the middle figures the reduced field distribution, and the bottom figures the to-

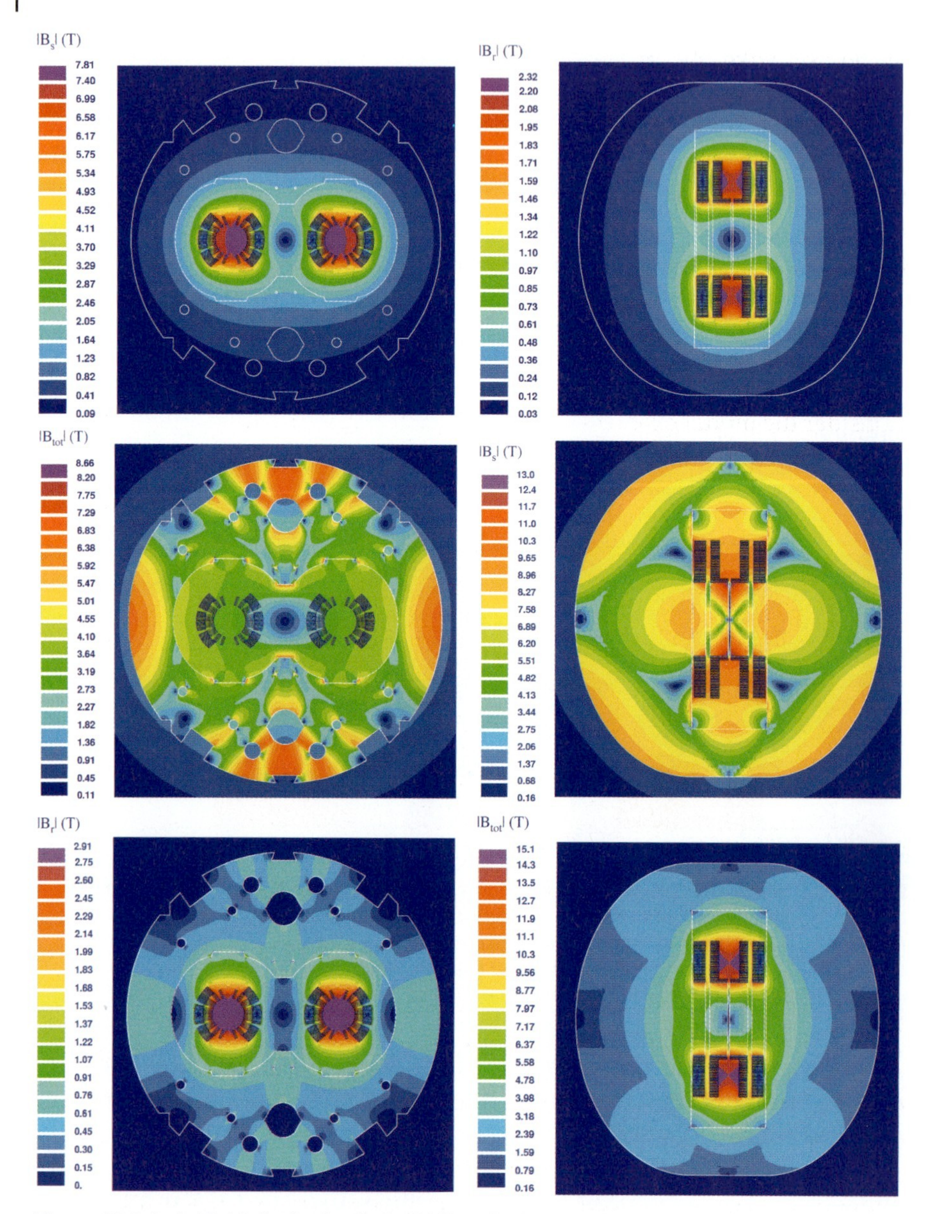

Figure 13.6 Left: Field distribution in the LHC main dipole magnet. Right: Field distribution in a twin-aperture window frame magnet. Top: Distribution of the source field $|B_\mathrm{s}|$. Middle: Reduced field $|B_\mathrm{r}|$ from the iron magnetization. Bottom: Total field distribution $|B_\mathrm{tot}|$.

tal field for nominal excitation. For the LHC main dipole magnets shown in Figure 13.6 (left), both excitational and reduced fields show a strong field gradient in each aperture due to the crosstalk. Owing to the optimized shape of the iron yoke, the two field gradients cancel each other in the total field. This results in a symmetric field distribution in the aperture with only a small quadrupole field component.

Remark: It is now easy to explain the figure on the front cover of this book. It shows the reduced field of the MQXA insertion quadrupole magnet. The function and design of this magnet are explained in Appendix B. □

13.3.2 FEM, Employing the Vector Poisson Equation

Consider the model problem shown in Figure 13.7 for a single-aperture model dipole. The model problem features both Dirichlet and Neumann boundary conditions.

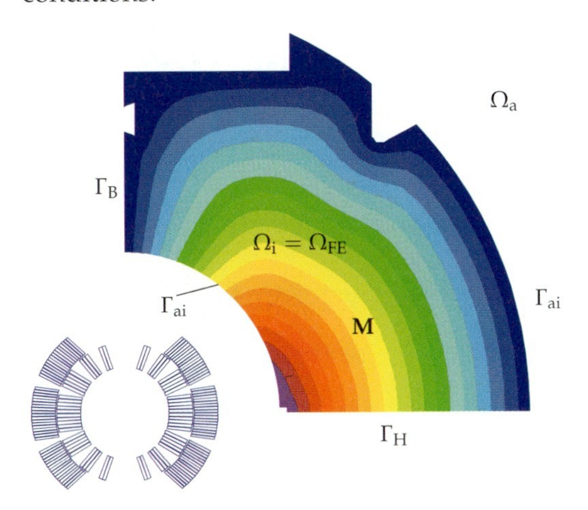

Figure 13.7 Model problem for the numerical field calculation of a superconducting single-aperture magnet. Display of the total vector potential in the iron domain. The air region Ω_a contains a number of conductor sources that do not intersect the iron region Ω_i.

The FE domain is restricted to the iron yoke denoted Ω_i. This FE formulation is the part of the BEM–FEM coupling described in Chapter 15. The air region Ω_a may contain a certain number of areas of impressed current density that do not intersect the iron boundary Γ_{ai}. The field in the FE domain is derived from a gauged vector-potential formulation. Beginning with the vector Poisson equation (4.155), the formulation of the problem is

$$-\frac{1}{\mu_0}\nabla^2\mathbf{A} = \mathbf{J} + \operatorname{curl}\mathbf{M} \qquad \text{in } \Omega_i, \tag{13.71}$$

together with the boundary and interface conditions:

$$\mathbf{A} \cdot \mathbf{n} = 0 \quad \text{on } \Gamma_H, \tag{13.72}$$

$$\frac{1}{\mu_0} \operatorname{div} \mathbf{A} = 0 \quad \text{on } \Gamma_B, \tag{13.73}$$

$$\mathbf{n} \times (\mathbf{A} \times \mathbf{n}) = \mathbf{0} \quad \text{on } \Gamma_B, \tag{13.74}$$

$$\frac{1}{\mu}(\operatorname{curl} \mathbf{A}) \times \mathbf{n} = \mathbf{0} \quad \text{on } \Gamma_H, \tag{13.75}$$

$$\left[\!\left[\frac{1}{\mu_0} \operatorname{div} \mathbf{A}_\mathrm{a} \right]\!\right]_\mathrm{ai} = 0, \tag{13.76}$$

$$\mathbf{n} \times \left[\!\left[\frac{1}{\mu} (\operatorname{curl} \mathbf{A}) \right]\!\right]_\mathrm{ai} + \mathbf{n} \times [\![\mathbf{M}]\!]_\mathrm{ai} = \mathbf{0}, \tag{13.77}$$

$$[\![\mathbf{A}]\!]_\mathrm{ai} = \mathbf{0}. \tag{13.78}$$

Equation (13.77) is the continuity condition of $H_\mathrm{ti} = H_\mathrm{ta}$ on the iron/air interface. Forcing the weighted residual to zero yields

$$-\int_{\Omega_\mathrm{i}} \frac{1}{\mu_0} \nabla^2 \mathbf{A} \cdot \mathbf{w}_a \, dV = \int_{\Omega_\mathrm{i}} (\mathbf{J} + \operatorname{curl} \mathbf{M}) \cdot \mathbf{w}_a \, dV \tag{13.79}$$

for $a = 1, 2, 3$. The weighting functions $\mathbf{w}_a$ must obey the homogeneous boundary conditions

$$\mathbf{w}_a \cdot \mathbf{n} = 0 \quad \text{on } \Gamma_H, \tag{13.80}$$

$$\mathbf{w}_a \times \mathbf{n} = \mathbf{0} \quad \text{on } \Gamma_B. \tag{13.81}$$

Green's first identity,

$$\int_{\Omega_\mathrm{i}} \nabla^2 \mathbf{A} \cdot \mathbf{w}_a dV = -\int_{\Omega_\mathrm{i}} \operatorname{grad} (\mathbf{A} \cdot \mathbf{e}_a) \cdot \operatorname{grad} w_a dV + \int_{\partial\Omega_\mathrm{i}} (\partial_{\mathbf{n}_\mathrm{i}} \mathbf{A}) \cdot \mathbf{w}_a da, \tag{13.82}$$

and the relation

$$\int_{\Omega_\mathrm{i}} \operatorname{curl} \mathbf{M} \cdot \mathbf{w}_a \, dV = \int_{\Omega_\mathrm{i}} \mathbf{M} \cdot \operatorname{curl} \mathbf{w}_a \, dV - \int_{\partial\Omega_\mathrm{i}} (\mathbf{M} \times \mathbf{n}_i) \cdot \mathbf{w}_a \, da, \tag{13.83}$$

yield the weak form of the boundary value problem,

$$\begin{aligned} &\frac{1}{\mu_0} \int_{\Omega_\mathrm{i}} \operatorname{grad} (\mathbf{A} \cdot \mathbf{e}_a) \cdot \operatorname{grad} w_a \, dV \\ &- \frac{1}{\mu_0} \int_{\Gamma_\mathrm{B}, \Gamma_\mathrm{H}, \Gamma_\mathrm{ai}} (\partial_{\mathbf{n}_\mathrm{i}} \mathbf{A} - (\mu_0 \mathbf{M} \times \mathbf{n}_\mathrm{i})) \cdot \mathbf{w}_a \, da \\ &\qquad = \int_{\Omega_\mathrm{i}} \mathbf{M} \cdot \operatorname{curl} \mathbf{w}_a \, dV + \int_{\Omega_\mathrm{i}} \mathbf{w}_a \cdot \mathbf{J} \, dV \end{aligned} \tag{13.84}$$

for $a = 1,2,3$. $\mathbf{e}_a$ are the unit vectors in the x-, y-, and z-directions. With the boundary conditions (13.72)–(13.75), and zero current density in the iron domain, Eq. (13.84) reduces to

$$\frac{1}{\mu_0}\int_{\Omega_\mathrm{i}} \operatorname{grad}(\mathbf{A}\cdot\mathbf{e}_a)\cdot\operatorname{grad} w_a \,\mathrm{d}V - \frac{1}{\mu_0}\int_{\Gamma_\mathrm{ai}} (\partial_{\mathbf{n}_\mathrm{i}}\mathbf{A} - (\mu_0\mathbf{M}\times\mathbf{n}_\mathrm{i}))\cdot\mathbf{w}_a \,\mathrm{d}a = \int_{\Omega_\mathrm{i}} \mathbf{M}\cdot\operatorname{curl}\mathbf{w}_a \,\mathrm{d}V \quad (13.85)$$

for $a = 1,2,3$. The vanishing boundary integral on Γ_H can be explained as follows: $\mathbf{A}\cdot\mathbf{n} = 0$ implies that the normal derivative $\partial_{\mathbf{n}_\mathrm{i}}\mathbf{A}$ vanishes as well. From $\mathbf{w}\cdot\mathbf{n} = 0$, it follows that $\mathbf{w}$ is orthogonal to $\mathbf{n}$. But the term $\mu_0\mathbf{M}\times\mathbf{n}$ is parallel to $\mathbf{n}$ and therefore the boundary integral vanishes. The boundary conditions on Γ_B lead to similar conclusions for the boundary integral on Γ_B. The only remaining boundary integral on Γ_ai will later serve as the coupling term between the FE domain and the boundary-element (BE) domain.

13.3.3 The A-ϕ Formulation for Eddy-Current Problems

In Section 4.10, we derived the boundary value problem for eddy-current problems in the time domain

$$\nabla^2\mathbf{A} - \varkappa\mu_0\frac{\partial\mathbf{A}}{\partial t} - \varkappa\mu_0\operatorname{grad}\phi = -\mu_0\operatorname{curl}\mathbf{M}, \quad (13.86)$$

$$\operatorname{div}\left(\varkappa\operatorname{grad}\phi + \varkappa\frac{\partial\mathbf{A}}{\partial t}\right) = 0, \quad (13.87)$$

together with the boundary and interface conditions

$$\frac{1}{\mu_0\mu_r}(\operatorname{curl}\mathbf{A})\times\mathbf{n} = \mathbf{0} \quad \text{on } \Gamma_\mathrm{H}, \quad (13.88)$$

$$\mathbf{A}\times\mathbf{n} = \mathbf{0} \quad \text{on } \Gamma_\mathrm{B}, \quad (13.89)$$

$$\left[\!\left[\frac{1}{\mu_0}(\operatorname{curl}\mathbf{A})\right]\!\right]_\mathrm{ai} + \mathbf{n}\times[\![\mathbf{M}]\!]_\mathrm{ai} = \mathbf{0}, \quad (13.90)$$

$$[\![\mathbf{A}]\!]_\mathrm{ai} = \mathbf{0}, \quad (13.91)$$

$$\left(\varkappa\operatorname{grad}\phi + \varkappa\frac{\partial\mathbf{A}}{\partial t}\right)\cdot\mathbf{n}_\mathrm{i} = 0, \quad (13.92)$$

and $\phi(\mathbf{r}_0) = \phi_0$ at an arbitrary point $\mathbf{r}_0$ in Ω_i. Applying the weighted residual method yields

$$\int_{\Omega_\mathrm{i}} \left(\nabla^2 \mathbf{A} - \varkappa\mu_0 \frac{\partial \mathbf{A}}{\partial t} - \operatorname{grad} \varkappa\mu_0\phi \right) \cdot \mathbf{w}_a \, \mathrm{d}V$$

$$= - \int_{\Omega_\mathrm{i}} \mu_0 \left(\operatorname{curl} \mathbf{M} \right) \cdot \mathbf{w}_a \, \mathrm{d}V \quad (13.93)$$

for $a = 1, 2, 3$. Moreover,

$$\int_{\Omega_\mathrm{i}} \operatorname{div} \left(\varkappa \operatorname{grad} \phi + \varkappa \frac{\partial \mathbf{A}}{\partial t} \right) \cdot \mathbf{w}_a \, \mathrm{d}V = 0 \quad (13.94)$$

for $a = 1, 2, 3$. The vector weighting-functions are again $\mathbf{w}_1 = (w_1, 0, 0)^T, \mathbf{w}_2 = (0, w_2, 0)^T, \mathbf{w}_3 = (0, 0, w_3)^T$. Employing Green's first theorem yields the weak form of the boundary value problem

$$\int_{\Omega_\mathrm{i}} \left(\operatorname{grad} (\mathbf{A} \cdot \mathbf{e}_a) \cdot \operatorname{grad} w_a + \varkappa \left(\operatorname{grad} \phi + \frac{\partial \mathbf{A}}{\partial t} \right) \cdot \mathbf{w}_a \right) \mathrm{d}V$$

$$+ \int_{\Gamma_\mathrm{ai}} \partial_{\mathbf{n}_\mathrm{a}} \mathbf{A} \cdot \mathbf{w}_a \, \mathrm{d}a = \mu_0 \int_{\Omega_\mathrm{i}} \mathbf{M} \cdot \operatorname{curl} \mathbf{w}_a \, \mathrm{d}V, \quad (13.95)$$

and

$$\int_{\Omega_\mathrm{i}} \varkappa \left(\operatorname{grad} \phi + \frac{\partial \mathbf{A}}{\partial t} \right) \cdot \operatorname{grad} w_a \, \mathrm{d}V = 0 \quad (13.96)$$

for $a = 1, 2, 3$.

References

1 Bethe, K. J.: Finite Element Procedures in Engineering Analysis, Wiley, New York, 1981

2 Binns, K. J., Lawrenson, P. J., Trowbridge, C. W.: The Analytical and Numerical Solution of Electric and Magnetic Fields, Wiley, New York, 1992

3 Braess, D.: Finite Elemente, Springer, Berlin, 1991

4 Brebbia, C. A.: The Boundary Element Method for Engineers, Pentech Press, UK, 1978

5 Collie, C. J.: Magnetic fields and potentials of linearly varying current or magnetization in a plane bounded region, Proceedings of the 1st Compumag Conference on the Computation of Electromagnetic Fields, Oxford, UK, 1976

6 Fetzer, J., Abele, S., Lehner, G.: Die Kopplung der Randelementmethode und der Methode der finiten Elemente zur Lösung dreidimensionaler elektromagnetischer Feldprobleme auf unendlichem Grundgebiet, Archiv für Elektrotechnik, 1993

7 Fetzer, J., Kurz, S., Lehner, G.: Comparison between different formulations for the solution of 3D nonlinear magnetostatic problems using BEM–FEM coupling, IEEE Transactions on Magnetics, 1996

8 Fetzer, J.: Die Lösung statischer und quasistationärer elektromagnetischer Feldprobleme mit Hilfe der Kopplung der Methode der finiten Elemente und der Randelementmethode, Fortschritt Berichte VDE, Reihe 21, VDE Verlag, 1992

9 Kurz, S., Fetzer, J., Rucker, W.M.: Coupled BEM–FEM methods for 3D field calculations with iron saturation, Proceedings of the First International ROXIE Users Meeting and Workshop, CERN, March 16–18, 1998

10 Preis, K., Biro, O., Magele, C.A., Renhart, W., Richer, K.R., Vrisk, G.: Numerical analysis of 3D magnetostatic fields, IEEE Transactions on Magnetics, 1991

11 Rain, O.: Asymptotisch optimale Randelementmethoden für die Maxwell Gleichungen, Diplomarbeit, University of the Saarland, Germany, 2001

12 Rischmüller, V., Fetzer, J., Haas, M., Kurz, S., Rucker, W.M.: Computational efficient BEM–FEM coupled analysis of 3D nonlinear eddy current problems using domain decomposition, Proceedings of the 8th International IGTE Symposium, Graz, Austria, 1998

13 Silvester P. P., Ferrari, R. L.: Finite Elements for Electrical Engineers, Cambridge University Press, Cambridge, 1996

14 Strang, G., Fix, G. J.: An Analysis of the Finite Element Method, Prentice-Hall, New Jersey, 1973

15 Steinbach, O.: Numerical Approximation Methods for Elliptic Boundary Value Problems – Finite and Boundary Elements, Springer, Berlin, 2008

16 Zienkiewicz, O. C., Taylor, R. L.: The Finite Element Methdo, Vol. 1: Basic Formulation and Linear Problems, McGraw-Hill, New York, 1989

14
Discretization

For it is unworthy of excellent men
to lose hours like slaves in the labour of calculation
which could safely be regulated to anyone else if machines were used.

Gottfried Wilhelm Leibniz (1646–1716).

When applying numerical field computation to the design of accelerator magnets and to electromagnetic devices in general, we find ourselves faced with four sources of errors.

- *Modeling errors* due to simplifications and approximations of the geometry, as well as errors in the description of material properties.
- *Discretization* or *truncation errors* when a continuous partial differential equation is approximated using the shape functions associated with discrete points, lines, or faces.
- *Residual errors* resulting from the iterative solution of the linear equation system.
- *Machine rounding errors* due to the limited word length available on the processor.

The multipole field errors calculated for the LHC main dipole at nominal excitation are shown in Figure 14.1 (left). On the right-hand side, the multipole components for a dipole coil in a perfectly round iron yoke with a bore radius of 98 mm are displayed (using semianalytical field computation with the image-current method). The higher-order, odd multipoles are insensitive to the yoke design. It can be seen from the nonallowed skew multipoles that the machine rounding errors are more pronounced in the numerical field computation. The magnitude of these rounding errors is, however, technically irrelevant.

In this chapter we will present discretization schemes useful for reducing the errors in numerical field computation. The main items are a parametric

Field Computation for Accelerator Magnets. Stephan Russenschuck

ISBN: 978-3-527-40769-9

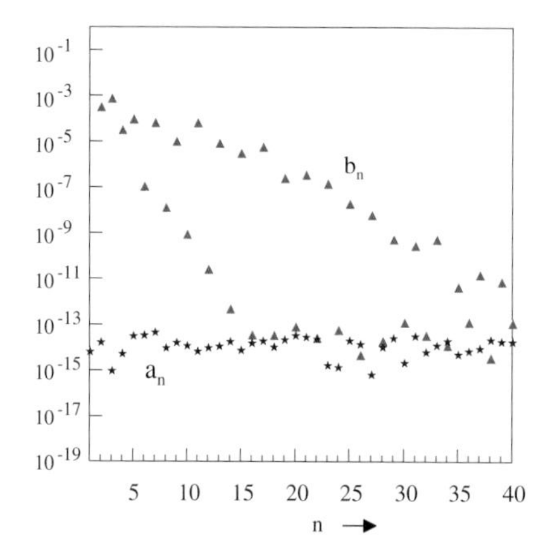

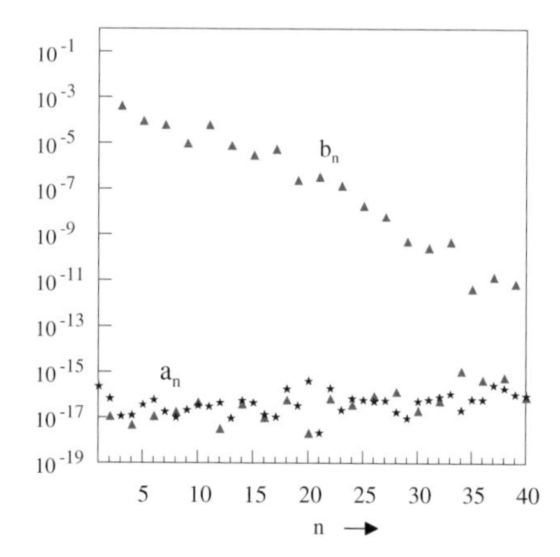

Figure 14.1 Left: Multipole field errors in the LHC main dipole. Right: Multipole field errors for an LHC dipole coil in a perfectly round iron yoke with a bore radius of 98 mm. Normal multipole errors are represented by triangles, skew multipoles by stars. Notice the even-numbered multipoles (left) resulting from the twin-aperture design of the main dipoles. The higher-order, odd-numbered normal multipoles are very insensitive to the yoke design. The machine rounding errors are slightly higher in the numerically calculated field distribution, however, at a level that is technically unimportant.

geometry modeler and mesh generator for the iron yoke. This preprocessor allows the automatic performance of sensitivity studies of geometrical tolerances. Discretization errors can be reduced by using higher-order quadrilateral elements, which also allow the accurate modeling of curved iron-yoke boundaries.

14.1 Quadrilateral Mesh Generation

Although it may seem an intuitive matter to construct a mesh, we will nonetheless begin with a formal definition. Let us consider the 2D domain Ω bounded by $\partial\Omega$ in the form of a polygon so that the domain can be partitioned into triangular or quadrilateral subdomains (cells) denoted Ω_j. A *partitioning* or *tesselation*, consisting of nodes, edges, and faces is called a *mesh* if it obeys the following conditions:

- The union of all J faces spans the entire problem domain: $\Omega = \bigcup_{j=1}^{J} \Omega_j$.
- Each edge Γ has two distinct nodes.
- Each node is associated with an edge and each edge with a surface.
- Each edge inside the domain is associated with exactly two surfaces.
- Each edge on the boundary is associated with exactly one surface.

- Two points are connected with exactly one edge.
- Two surfaces have at most one edge in common.
- Boundaries of material subdomains coincide with edges.

The order of a point $\mathscr{P}$ in the mesh is defined as the number of cells containing $\mathscr{P}$. The point is said to be *regular* if its order is 4 for quadrilateral and 6 for triangular meshes. The mesh is regular if all its nonboundary nodes are regular.

The literature includes different methods for the generation of regular, quadrilateral meshes in two-dimensional polygonal regions. Among these are quadtree-based methods, where the domain is first subdivided into squares of different sizes and subsequently the intersections of the element and material boundaries are adjusted [1] to yield uniform material filling in the elements. An alternative is the *advancing-front* technique, where elements are "paved" from the domain boundary into the domain [3].

We will present a mesh generator based on topology and domain decomposition [6]. The following extensions have been added to this mesh generator:

- Parametric input for the definition of design variables for mathematical optimization.
- Implementation of design features for the definition of material boundaries.
- Extension of the method to eight-noded quadrilateral elements.
- A morphing algorithm for optimization and sensitivity studies that avoids remeshing and changing mesh topologies.
- An extrusion of the mesh into the third dimension.

14.1.1 Parametric Modeling

Any mesh generator for finite elements requires the definition of the material contour and properties. This is in general accomplished by defining keypoints, lines, and surfaces as input data. By means of algebraic operations on user-supplied, free parameters, it is possible to define dependent parameters needed to avoid topologically impossible geometries, for example, intersecting material boundaries or negative values assigned to radii or thicknesses. Keypoints, defined by these dependent parameters, are then joined by contour segments, which include straight, circular, parabolic, and elliptic line segments, as well as bars, notches, circles, and rectangles. The mesh density on the contour segment is a user-supplied parameter. The mesh points on the contour are generated according to a user-supplied density function and a contraction parameter. Simply-connected macrodomains Ω_d (not to be confused with the finite-element domains Ω_j) of different material properties are defined by closed chains of contour segments traced in a mathematically positive sense (anticlockwise).

A macrodomain can be bounded by any number $i \geq 3$ of contour segments. The total number of mesh points and element edges along the contour must be even. Holes of any shape are defined in the same manner and then cut into one and only one macrodomain. It must be guaranteed that two holes do not intersect within a domain. The number of holes in the domain specifies its order: Simply-connected domains are *domains of order 0*; domains with one hole are of order 1, etc.

The quadrilateral mesh generator is based on the methods of topology decomposition and domain decomposition [6]. In the first step, the macrodomain is decomposed into subdomains that are topologically equivalent to a disk. Holes are eliminated by means of *cutting edges* Γ; see Figure 14.3 (left).

14.1.2
Topology Decomposition

Consider a mesh of J quadrilateral elements and an even number of edges (m_b) at its boundary. By counting the number of inner edges (m_i) and boundary edges in a mesh, we obtain $4J = m_\text{b} + 2m_\text{i}$, from which it results that m_b is even. This is the reason for requiring an odd number of mesh points between the nodes of the contour segments.

If the domain Ω is multiply-connected, the order of Ω is reduced by one, either by a cutting edge between a hole and the contour $\partial\Omega$ or by connecting two holes inside the domain. The cutting edge is then treated as a new part of the boundary. A domain of order g can thus be reduced to a topological disk by means of g cutting edges. Optimal cutting edges $\Gamma := \mathscr{P} + t(\mathscr{Q} - \mathscr{P})$, $t \in [0, 1]$ are obtained by solving the minimization problem

$$\min\{f(\Gamma)\} = \min\{\lambda_1 g_1^*(\Gamma) + \lambda_2 g_2^*(\Gamma) + \lambda_3 g_3^*(\Gamma)\}\,, \tag{14.1}$$

where

$$g_1(\Gamma) := 4\pi - \varphi_1(\mathscr{P}) - \varphi_2(\mathscr{P}) - \varphi_1(\mathscr{Q}) - \varphi_2(\mathscr{Q})\,, \tag{14.2}$$

$$\begin{aligned} g_2(\Gamma) := & \sum_{i=1}^{2} \min\left\{\varphi_i(\mathscr{P}) - \frac{\pi}{2},\, \varphi_i(\mathscr{P}) - \pi,\, \varphi_i(\mathscr{P}) - \frac{3\pi}{2}\right\} \\ & + \sum_{i=1}^{2} \min\left\{\varphi_i(\mathscr{Q}) - \frac{\pi}{2},\, \varphi_i(\mathscr{Q}) - \pi,\, \varphi_i(\mathscr{Q}) - \frac{3\pi}{2}\right\}\,, \end{aligned} \tag{14.3}$$

$$g_3(\Gamma) := L\,, \tag{14.4}$$

for all combinations of points $\mathscr{P}$ at the hole and $\mathscr{Q}$ at the contour segments. L is the length of Γ. The λ_i are weighting factors for the three objectives. These are scaled to the interval $[0,1]$ by

$$g_i^* := \frac{2}{\pi} \arctan g_i \,. \tag{14.5}$$

By default $\lambda_1 = 1$ and $\lambda_2 = \lambda_3 = 0.5$. The minimization of the objective (14.2) maximizes the inner angles $\varphi_1 + \varphi_2$ at points $\mathscr{P}$ and $\mathscr{Q}$; see Figure 14.2 (left). The minimization of (14.3) ensures that the angles between Γ and the boundary are close to multiples of the right angle. The cutting edge must be verified as *feasible*, that is, $\Gamma \in \Omega_\mathrm{d}$, $\partial\Gamma \in \partial\Omega_\mathrm{d}$, $\Gamma \cap \partial\Omega_\mathrm{d} = \{\mathscr{P}, \mathscr{Q}\}$.

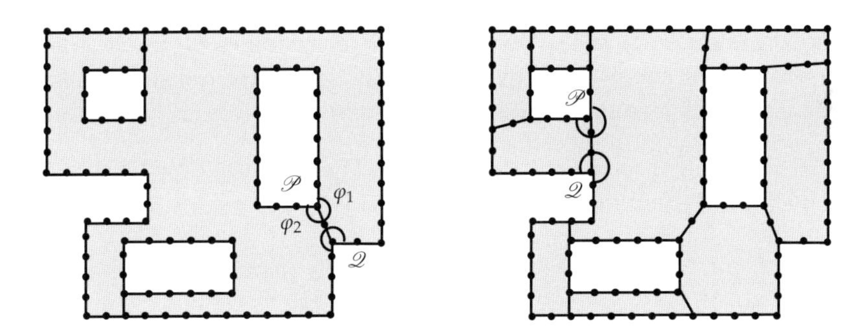

Figure 14.2 Basic principles of topology and domain decomposition.

14.1.3 Domain Decomposition

After the domain has been reduced to order zero, additional cutting edges are introduced. The shapes of the two resulting subdomains Ω_1 and Ω_2 are optimized by solving the minimization problem

$$\min\{f(\Gamma)\} = \min\left\{\lambda_1 g_1^*(\Gamma) + \lambda_2 g_2^*(\Gamma) + \lambda_3 g_3^*(\Gamma) + \lambda_4 g_4^*(\Gamma)\right\} , \tag{14.6}$$

where g_1 and g_2 are the same as in Eqs. (14.2) and (14.3), and

$$g_3(\Gamma) := \frac{L^2}{\min\{a_1, a_2\}} , \tag{14.7}$$

$$g_4(\Gamma) := \frac{L}{\min\{C_1, C_2\}} , \tag{14.8}$$

for all combinations of points $\mathscr{P}$ and $\mathscr{Q}$ on the subdomain boundary. The objective (14.7) relates the length L of the edge to the resulting surfaces a_1 and a_2. L is squared to obtain a dimensionless quantity. The objective (14.8) aims at cutting a surface of maximal circumference C for a minimum length of the cutting edge.

The domain decomposition is applied recursively to Ω_1 and Ω_2 until the remaining subdomains can be considered simple [6]. Additional points are inserted along the cutting edges according to the density of points at the subdomain boundary. A domain Ω is defined as simple if:

- The number of contour segments is less than 5.
- The domain does not contain bottlenecks; the square of the circumference related to the surface C^2/a approaches the value 4π, which is the limiting value for the circle.
- The largest inner angle of the nodes $\varphi(\mathscr{P})$ is less than π.
- The condition $m_1 + m_2 \leq m_3$ is fulfilled for triangles (m_i is the number of element edges along the ith contour segment) in order to avoid unfavorable element angles resulting from the mesh-closing algorithm.

Figure 14.3 shows the process of topological and geometrical domain decomposition for the yoke lamination of the LHC main dipole.

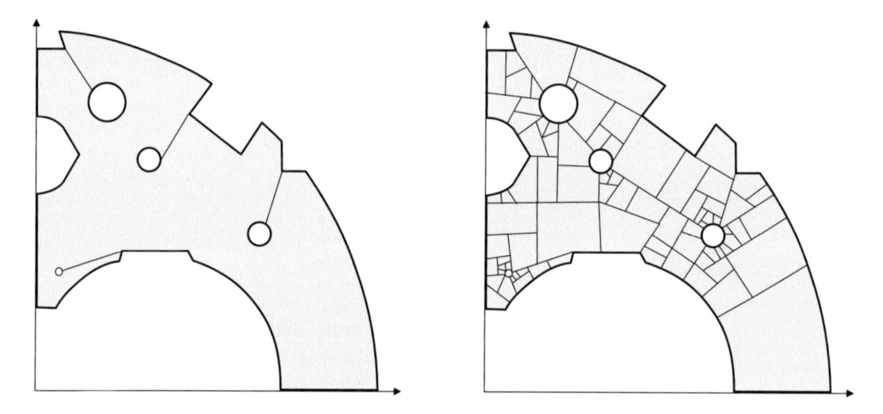

Figure 14.3 The main steps of quadrilateral mesh generation. Left: The geometry after the topology decomposition. Right: Domain decomposition into simple domains.

14.1.4 Meshing of Simple Domains

Simple domains are filled with quadrilateral elements using a modified *paving* or *advancing-front* strategy [3], wherein an area is filled from the boundary by adding full rows of quadrilateral elements; see Figure 14.4. The resulting mesh is regular; each node is connected to exactly four quadrilaterals.

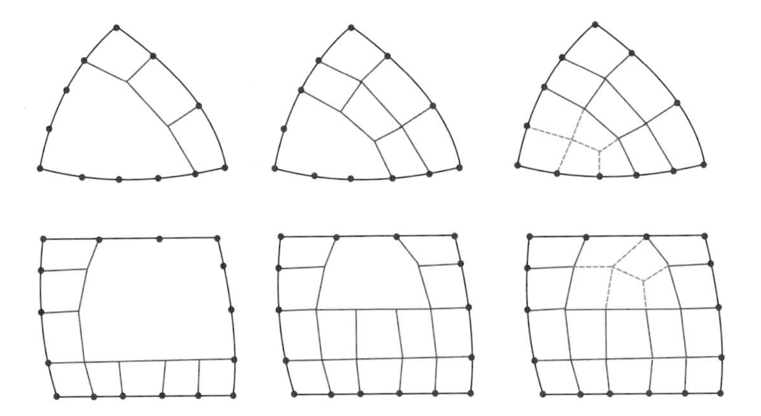

Figure 14.4 Paving and mesh closing in simple domains. The triangular domain that is to be closed has three contour segments; $m_1 = 2, m_2 = 3, m_3 = 4$. The subdomain is paved from the contour having the smallest number of element edges.

The computing time needed for the domain decomposition increases with $\mathcal{O}(m^3)$, where m denotes the number of contour points. It is therefore important to minimize the ratio C^2/a during the domain decomposition. The time needed for paving simple areas with quadrilaterals increases only with $\mathcal{O}(m^2)$ [6].

If no new row of elements can be added to the domain, the mesh must be completed by means of a set of mesh-closing rules; two are illustrated in Figure 14.4 (right).

14.1.5 Smoothing

The final step involves the application of a *smoothing* algorithm, which includes three techniques:

- *Laplace smoothing or barycentric regularization:* Enlargement of small angles by placing point $\mathcal{P}$ into the barycenter of the surrounding polygon.
- *Edge smoothing:* Reduction of large angles by calculation of the smallest distance of a point $\mathcal{P}$ to the points of the surrounding polygon. The closest neighbor shall be denoted $\mathcal{Q}$. The position of point $\mathcal{P}$ is moved by $t(\mathcal{P} - \mathcal{Q})$ for $t \in [0, 1]$.
- *Angle smoothing:* Let us denote the points with the largest and smallest inner angle as $\mathcal{P}$ and $\mathcal{Q}$, respectively. We aim at right angles in the element by a displacement of the corner points. The displacement patterns depend on the position of $\mathcal{P}$ with respect to $\mathcal{Q}$; see Figure 14.5.

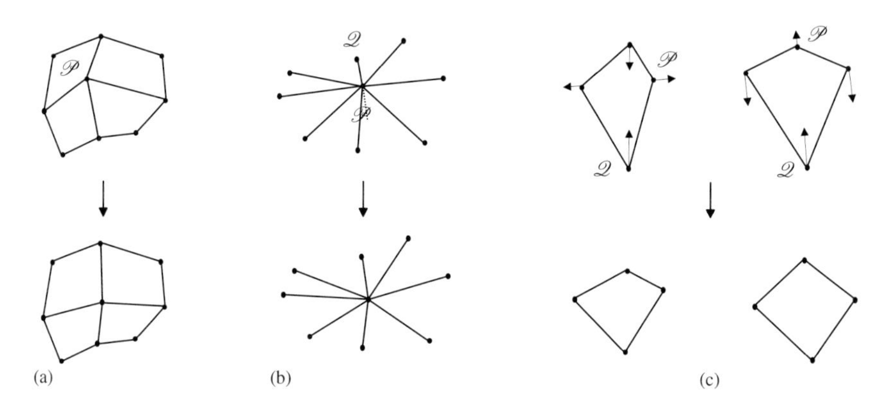

Figure 14.5 (a) Laplace smoothing. (b) Edge smoothing. (c) Angle smoothing.

The smoothing techniques are visualized in Figure 14.5. Examples for smoothed quadrilateral meshes are shown in Figure 14.6. The left-hand illustration shows the meshed iron yoke, insert, and collar of the LHC main dipole. The right-hand side shows the FE mesh of the MQM quadrupole magnet.

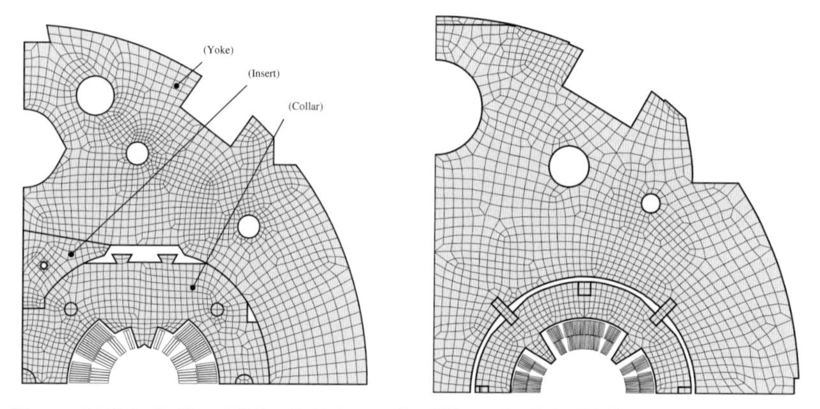

Figure 14.6 Left: Quadrilateral, higher-order FE mesh of the LHC main dipole iron yoke, insert, and austenitic steel collar. Right: Mesh for the MQM quadrupole magnet.

14.1.6
Remeshing and Morphing

When mathematical optimization and sensitivity analysis are carried out, the FE mesh must be updated for each trial solution. The obvious though computer-intensive method for realizing mesh changes is the calculation of the updated domain boundaries and the repetition of the domain decomposition. Parametric studies show that this procedure may imply changes in the mesh topology even for very small variations in the geometry. Discretization errors may thus compromise the field solutions.

An alternative to remeshing is *point-based morphing*, known from computer graphics [2,5]. This method uses the translation vectors $\mathbf{s}(\mathscr{P}_i)$ pointing from the original to the updated position of the contour nodes $\mathscr{P}_i, i = 1, \ldots, I$. The translation vector $\mathbf{s}(\mathscr{Q})$ of any mesh point $\mathscr{Q}$ inside the domain can then be calculated by

$$\mathbf{s}(\mathscr{Q}) = \sum_{i=1}^{I} \frac{\mathbf{s}(\mathscr{P}_i) \cdot w_i}{|\mathbf{r}(\mathscr{Q}) - \mathbf{r}(\mathscr{P}_i)|} \left(\sum_{i=1}^{n} \frac{w_i}{|\mathbf{r}(\mathscr{Q}) - \mathbf{r}(\mathscr{P}_i)|} \right)^{-1}, \tag{14.9}$$

where w_i are weighting factors; $w_i = 1$ by default. The advantage of morphing is that it conserves the mesh topology. Figure 14.7 demonstrates this for a magnet geometry of quadrupolar symmetry. In the upper left and lower right figures the mesh was generated for $\alpha = 20°$ and $\alpha = 27°$, respectively. The other two figures show the preservation of the mesh topology when the morphing strategy is applied.

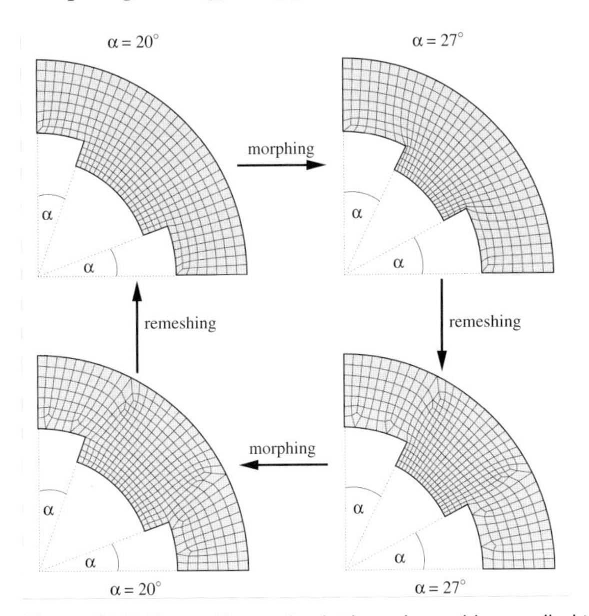

Figure 14.7 Remeshing and point-based morphing applied to a simple iron yoke for a quadrupole.

The application of the morphing strategy is particularly important for sensitivity analysis. Figure 14.8 shows meshes for the ferromagnetic insert of the LHC main dipole yoke. We will now study the sensitivity of field errors to the position of nonmagnetic tightening rods used for the preassembly of the laminations. The b_2 field component is the most sensitive field error. Figures 14.8(a) and (b) show two meshes of different topologies generated for two different rod positions.

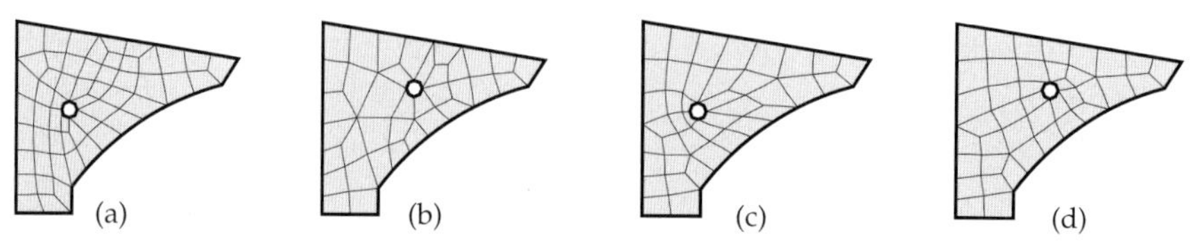

Figure 14.8 (a), (b): Meshed ferromagnetic insert for rods at $\mathscr{P}_1$ and $\mathscr{P}_2$. (c), (d): A mesh is created for a rod at $\mathscr{P}_1$, and the morphing algorithm is applied following a shift of the rod to $\mathscr{P}_2$.

Figure 14.9 shows the b_2 component as a function of the rod position, shifted along the line $\mathbf{r}(\lambda) = \mathbf{r}(\mathscr{P}_1) + \lambda(\mathbf{r}(\mathscr{P}_2) - \mathbf{r}(\mathscr{P}_1))$ for $\lambda \in [0, 1]$. It can be seen in Figure 14.9 that the quadrupole component at nominal field level is sensitive to small geometric changes in the insert. On the other hand, it is insensitive to the tightening rod position at injection field level ($I = 0.06\, I_{\text{nom}}$, $I_{\text{nom}} = 11\,800$ A) and at half the nominal excitation $I = 0.5\, I_{\text{nom}}$. At low and medium excitation, the iron insert is not saturated, and therefore the effect of the rod on the total magnetic resistance of the insert is insignificant.

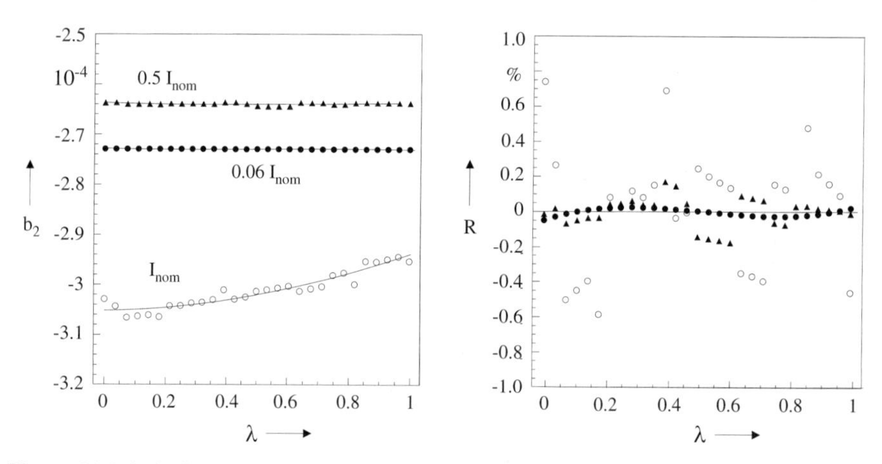

Figure 14.9 Left: Quadrupole component for three different excitation levels (0.06, 0.5, and 1 times the nominal current, $I_{\text{nom}} = 11\,800$ A), as a function of the rod position. The geometry is remeshed for each calculation. Right: Corresponding residuals.

Subtracting the calculated harmonics from those obtained by fitting with a polynomial of third order leads to the residuals

$$R := \frac{b_2 - b_2^{\mathrm{fit}}}{b_2}, \tag{14.10}$$

shown in Figure 14.9 (right). Jumps are visible at nominal field level. They can be explained by topology changes in the FE mesh and consequently by varying numerical errors in the field solution. If the rod position were a parameter in the field optimization, these discontinuities would have an impact on the convergence of the algorithms, in particular if gradient-based optimization strategies were used. Figure 14.8(d) shows the mesh generated for the rod in the lower position (c), and the application of the morphing algorithm for the displaced rod. The mesh topology remains unchanged. Figure 14.10 (left) shows the results of the field computations.

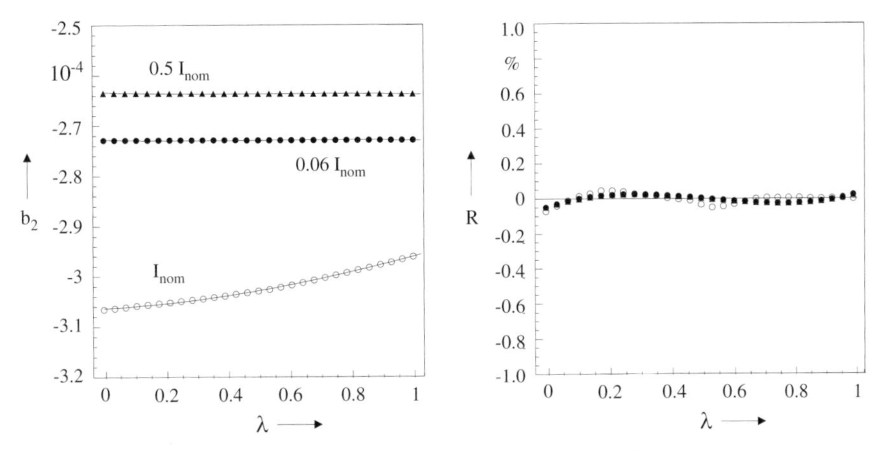

Figure 14.10 Left: Quadrupole component b_2 for three different excitation levels (see caption of Figure 14.9) as a function of the rod position. The morphing algorithm is applied for each calculation. Right: Corresponding residuals.

The residual plot in Figure 14.10 (right) confirms that the observed discontinuities result from changes in the mesh topology. The example underlines the importance of using the morphing algorithm rather than remeshing for sensitivity studies.

14.2 Finite-Element Shape Functions

With the tessellation of the problem domain accomplished, we shall study the finite-element shape functions of triangles and quadrangles in 2D, and parallelepipeds (hexahedra) for 3D meshes. These hexahedra are generated by extrusion from the 2D mesh. We will begin with the simplest, three-noded, triangular element in the 2D plane and limit ourselves to the *Lagrange elements*,[1] which have the node potentials as the degrees of freedom. The conventions for the numbering of nodes and elements in the problem domain are

- Number of finite element, $j = 1, \ldots, J$.
- Local number of nodes in jth element, $k = 1, \ldots, K$.
- Global number of all nodes, $n = 1, \ldots, N$.

14.2.1
The Linear Triangular Element, 2D

Within a particular element Ω_j the z-component of the vector potential $A(\mathbf{r}) = A_z(x, y)$ is approximated by the basis functions

$$A(\mathbf{r}) = \alpha_1 + \alpha_2 x + \alpha_3 y. \tag{14.11}$$

The basis functions are defined only on the element Ω_j; they are said to have *compact support*. The α_i are called the *generalized parameters* of the approximation. The approximated node potentials[2] A_k at the three local nodes $k = 1, 2, 3$ of Ω_j can be expressed as

$$A_1 = \alpha_1 + \alpha_2 x_1 + \alpha_3 y_1, \tag{14.12}$$

$$A_2 = \alpha_1 + \alpha_2 x_2 + \alpha_3 y_2, \tag{14.13}$$

$$A_3 = \alpha_1 + \alpha_2 x_3 + \alpha_3 y_3, \tag{14.14}$$

which can be written in the matrix form as

$$\begin{pmatrix} A_1 \\ A_2 \\ A_3 \end{pmatrix} = \begin{pmatrix} 1 & x_1 & y_1 \\ 1 & x_2 & y_2 \\ 1 & x_3 & y_3 \end{pmatrix} \begin{pmatrix} \alpha_1 \\ \alpha_2 \\ \alpha_3 \end{pmatrix}, \tag{14.15}$$

$\{A\} = [C]\{\alpha\}$ for short. The solution for $\alpha_1, \alpha_2, \alpha_3$ is obtained from $\{\alpha\} = [C]^{-1}\{A\}$. Using Cramer's rule, we obtain

$$\alpha_i = \frac{\Delta_i}{\det[C]}, \tag{14.16}$$

1 Joseph–Louis Lagrange (1736–1813).
2 The subscript z and any notation of the element number is omitted hereafter.

where Δ_i is the determinant of the matrix obtained by substituting the ith column in $[C]$ with $\{A\}$. As an example,

$$\alpha_1 = \frac{1}{\det[C]} \begin{vmatrix} A_1 & x_1 & y_1 \\ A_2 & x_2 & y_2 \\ A_3 & x_3 & y_3 \end{vmatrix}, \tag{14.17}$$

where

$$\begin{aligned} \det[C] &= x_2y_3 - x_3y_2 + x_3y_1 - x_1y_3 + x_1y_2 - x_2y_1 \\ &= (x_2y_3 - x_3y_2) + (y_2 - y_3)x_1 + (x_3 - x_2)y_1 \\ &= (x_3y_1 - x_1y_3) + (y_3 - y_1)x_2 + (x_1 - x_3)y_2 \\ &= (x_1y_2 - x_2y_1) + (y_1 - y_2)x_3 + (x_2 - x_1)y_3 . \end{aligned} \tag{14.18}$$

The determinant of $[C]$ equals two times the area of the triangle: $S = 0.5\det[C]$. Using the abbreviations

$$\begin{aligned} &a_1 := x_2y_3 - x_3y_2, \quad && a_2 := x_3y_1 - x_1y_3, \quad && a_3 := x_1y_2 - x_2y_1, \\ &b_1 := y_2 - y_3, && b_2 := y_3 - y_1, && b_3 := y_1 - y_2, \\ &c_1 := x_3 - x_2, && c_2 := x_1 - x_3, && c_3 := x_2 - x_1, \end{aligned} \tag{14.19}$$

defined by cyclic permutation of indices in a_i, b_i, and c_i, we obtain the compact notation $\det[C] = a_k + b_kx_k + c_ky_k$ for $k = 1, 2, 3$, and the solution for $\alpha_1, \alpha_2, \alpha_3$ can be written as

$$\alpha_1 = \frac{1}{2S}\,(a_1A_1 + a_2A_2 + a_3A_3), \tag{14.20}$$

$$\alpha_2 = \frac{1}{2S}\,(b_1A_1 + b_2A_2 + b_3A_3), \tag{14.21}$$

$$\alpha_3 = \frac{1}{2S}\,(c_1A_1 + c_2A_2 + c_3A_3). \tag{14.22}$$

For a general point within Ω_j, the potential can be approximated with

$$\begin{aligned} A(\mathbf{r}) &= \alpha_1 + \alpha_2x + \alpha_3y \\ &= [(a_1 + b_1x + c_1y)A_1 + (a_2 + b_2x + c_2y)A_2 + (a_3 + b_3x + c_3y)A_3]\,\frac{1}{2S} \\ &=: \sum_{k=1}^{3} N_k(\mathbf{r})A_k \end{aligned} \tag{14.23}$$

where the A_k are the unknown node potentials and N_k the local element shape functions (also called *trial* or *interpolation functions*) on Ω_j:

$$N_k(\mathbf{r}) = \frac{a_k + b_kx + c_ky}{a_k + b_kx_k + c_ky_k}, \qquad k = 1, 2, 3. \tag{14.24}$$

In the FE formulation, we find only the element shape functions but not the basis functions (14.11). The shape functions depend on the spatial coordinates of the element only; they possess the important property that

$$N_k(\mathbf{r}_k) = 1\,, \tag{14.25}$$

which results directly from $A(\mathbf{r}_k) = A_k$. If all nodal values A_k are equal, the solution must be constant: $A = A_k$. This yields a second characteristic for the shape functions,

$$\sum_{k=1}^{3} N_k(\mathbf{r}) = 1\,. \tag{14.26}$$

Because of the property (14.25), we obtain

$$N_k(\mathbf{r}_l) = \delta_{kl} = \begin{cases} 1 & \text{for } l = k \\ 0 & \text{for } l \neq k \end{cases} \tag{14.27}$$

Figure 14.11 shows the element-wise defined basis functions $A = \alpha_1 + \alpha_2 x + \alpha_3 y$, the complete linear nodal function $N_{G,n}$ of a global node (index n), and the shape function N_k in the local node (index k) of Ω_j.

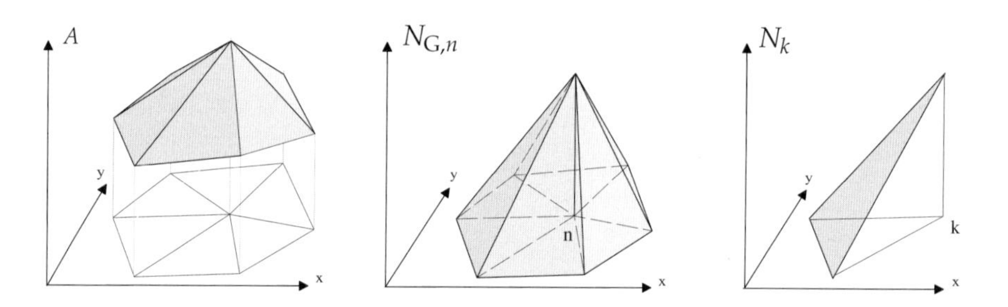

Figure 14.11 Left: Element-wise defined (linear) basis functions $A = \alpha_1 + \alpha_2 x + \alpha_3 y$. Middle: Complete (linear) nodal function $N_{G,n}$ of the global node n. Right: Shape function N_k in the local node k.

Remark: The field problem is solved by means of a potential function that must be continuous not only for the exact solution but also for the approximated solution across element boundaries (C^0 continuity). The continuity of the approximated solution is guaranteed because the basis functions and the node potentials are identical along the common boundary Γ_{ij} between two elements i and j. The tangential derivatives $\mathbf{t} \cdot \operatorname{grad} A$ of the basis functions are continuous, whereas the normal derivatives $\mathbf{n} \cdot \operatorname{grad} A$ are not. Consequently, the normal component of the magnetic flux density B_n is continuous whereas the tangential component of the field H_t may exhibit a jump discontinuity across the element boundary. H_t is said to be *continuous in the weak sense.* □

14.2.2
Barycentric Coordinates

In Section 14.2.1 the shape functions of the first-order triangular elements were derived in Cartesian (global) coordinates by the inversion of a 3×3 matrix. If a higher-order polynomial is chosen as the basis function, the derivation of the shape functions will be complicated and not by any means intuitive. It is therefore worthwhile to study ways of writing down the shape functions directly, by making use of Eqs. (14.25) and (14.26). It will turn out that Eq. (14.24) is nothing but a transformation from the Cartesian coordinates to a natural system of *barycentric coordinates*.

Consider three points $\mathscr{P}_1, \mathscr{P}_2, \mathscr{P}_3$ in E_2 with position vectors $\mathbf{r}_1, \mathbf{r}_2, \mathbf{r}_3$ that are not aligned. For each point $\mathscr{Q} \in E_2$ (position vector $\mathbf{r}$) there is one and only one set $\{\lambda_1, \lambda_2, \lambda_3\}$ of real numbers (weights) for which $\mathbf{r} = \sum_{i=1}^{3} \lambda_i \mathbf{r}_i$ and $\sum_{i=1}^{3} \lambda_i = 1$. The point $\mathscr{Q}$ is by definition the center of mass of the weighted point masses at positions $\mathscr{P}_i$, and the numbers λ_i are said to be the barycentric coordinates; see Section 2.5. Consequently, the coordinate transformation for the point $\mathscr{Q}$, expressed in terms of its barycentric coordinates, is given by

$$x = \lambda_1 x_1 + \lambda_2 x_2 + \lambda_3 x_3 \,, \tag{14.28}$$

$$y = \lambda_1 y_1 + \lambda_2 y_2 + \lambda_3 y_3 \,, \tag{14.29}$$

$$1 = \lambda_1 + \lambda_2 + \lambda_3 . \tag{14.30}$$

The last equation above indicates that only two of the three barycentric coordinates are linearly independent. Solving the equation system in $\lambda_1, \lambda_2, \lambda_3$ yields $\lambda_1 = N_1$, $\lambda_2 = N_2$, $\lambda_3 = N_3$ and proves that for linear triangular elements the shape functions are merely the triangle's barycentric coordinates. However, the relations between the barycentric coordinates and the shape functions are not always so straightforward.

Any point $\mathscr{Q}$ within the element is determined by the three barycentric coordinates which can be expressed as

$$\lambda_1 = \frac{S_{\mathscr{Q}23}}{S_{123}} \,, \qquad \lambda_2 = \frac{S_{\mathscr{Q}31}}{S_{123}} \,, \qquad \lambda_3 = \frac{S_{\mathscr{Q}12}}{S_{123}} \,, \tag{14.31}$$

where the $S_{\mathscr{Q}ij}$ are the areas of the subtriangles and S_{123} is the total area of the triangle shown in Figure 14.12. For this reason barycentric coordinates are often named *area coordinates*.[3] If the point $\mathcal{Q}$ is located on the edge 2-3, the area $S_{\mathcal{Q}23}$ is zero and $\lambda_1 = 0$; hence the edge 2-3 is the axis of $\lambda_1 = 0$.

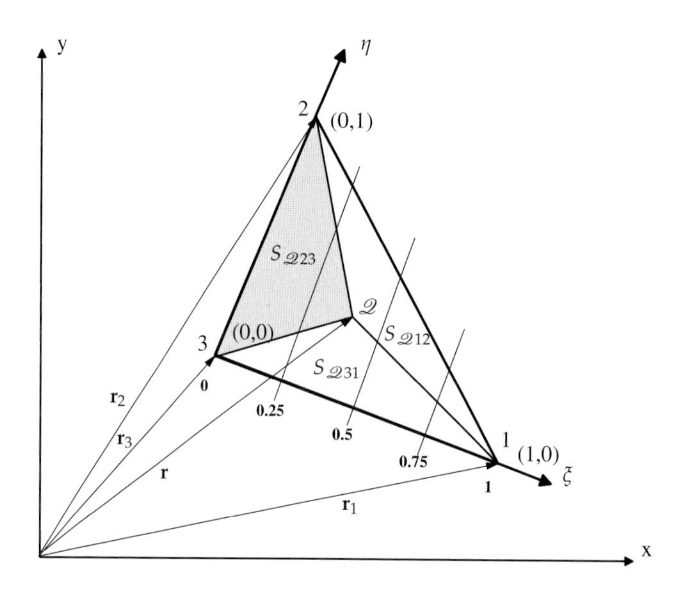

Figure 14.12 Barycentric and local coordinates in a linear triangular element.

14.2.3 Local Coordinates

In order to simplify the evaluation of the spatial derivatives of the shape functions, local coordinates (ξ, η) are introduced (see Figure 14.12) through the affine mapping

$$x = x_3 + (x_1 - x_3)\xi + (x_2 - x_3)\eta\,, \tag{14.32}$$

$$y = y_3 + (y_1 - y_3)\xi + (y_2 - y_3)\eta\,. \tag{14.33}$$

It follows that

$$N_1 = \xi, \qquad N_2 = \eta, \qquad N_3 = 1 - \xi - \eta. \tag{14.34}$$

3 This concept can easily be extended to three dimensions, e.g., for a four-noded tetrahedral element, by relating the volumes $\lambda_1 = V_{\mathcal{Q}234}/V_{1234}$.

It is a simple task to express the local coordinates in terms of the Cartesian coordinates if we resort to the barycentric coordinates:

$$\xi = \frac{2S_{\mathcal{Q}23}}{2S_{123}} = \frac{a_1 + b_1 x + c_1 y}{\det[C]}, \tag{14.35}$$

$$\eta = \frac{2S_{\mathcal{Q}31}}{2S_{123}} = \frac{a_2 + b_2 x + c_2 y}{\det[C]}, \tag{14.36}$$

with a_i, b_i as defined in Eq. (14.19). Expressed in the spatial coordinates of the element, Eqs. (14.35) and (14.36) yield

$$\xi = \frac{x_2 y_3 - x_3 y_2 + (y_2 - y_3)x + (x_3 - x_2)y}{x_2 y_3 - x_3 y_2 + x_3 y_1 - x_1 y_3 + x_1 y_2 - x_2 y_1}, \tag{14.37}$$

$$\eta = \frac{x_3 y_1 - x_1 y_3 + (y_3 - y_1)x + (x_1 - x_3)y}{x_2 y_3 - x_3 y_2 + x_3 y_1 - x_1 y_3 + x_1 y_2 - x_2 y_1}. \tag{14.38}$$

14.2.4 Mapped Elements

Triangular and tetrahedral elements require that curved domain boundaries be modeled by polygonal approximations. This has proven insufficiently accurate for accelerator magnet design. Quadrilateral, curvilinear elements are an alternative found in most commercial software packages. Quadrilaterals also avoid numerically unfavorable prisms when the geometry is extruded into the third dimension.

Higher-order, quadrilateral elements are mappings from basic rectangles in the local coordinate system onto curved elements in the Cartesian system; see Figure 14.13. The coordinates in the local frame are denoted ξ, η; the element in this frame is called the *parent element*.

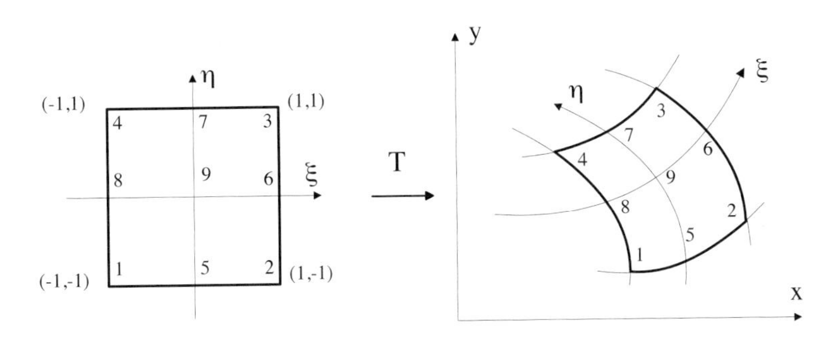

Figure 14.13 Local and global coordinate systems of a quadrilateral element.

The coordinate transformation for the nodes is often derived from the element shape functions used for the potential approximation; a concept first introduced in [8]. In the 2D case, the points with coordinates x_j, y_j coincide with the corresponding points at the element boundary. Using the definition $(\boldsymbol{\xi}) := (\xi, \eta)$ we obtain

$$A(\boldsymbol{\xi}) = \sum_{k=1}^{K} N_k(\boldsymbol{\xi}) A_k \,, \tag{14.39}$$

$$x_j(\boldsymbol{\xi}) = \sum_{k=1}^{K} N_k(\boldsymbol{\xi}) x_k \,, \tag{14.40}$$

$$y_j(\boldsymbol{\xi}) = \sum_{k=1}^{K} N_k(\boldsymbol{\xi}) y_k \,. \tag{14.41}$$

As the shape functions N_k are polynomials in ξ and η, the transformation is nonlinear, and the resulting element shapes are curvilinear as desired; see Figure 14.13. The term *isoparametric* is inspired by the fact that the same polynomial degree is used for the potential function and the shape of the element. The isoparametric elements permit the representation of curved material boundaries, as can be seen in Figure 13.5. Note, however, that a circular shape is represented by parabolic functions.

The four-noded, quadrilateral element Suitable sets of basis functions can be found by means of the *Pascal triangle* shown in Table 14.1.

Table 14.1 Pascal triangle for the construction of a set of basis functions for higher-order nodal elements.

					1					
				ξ		η				
			ξ^2		$2\xi\eta$		η^2			
		ξ^3		$3\xi^2\eta$		$3\xi\eta^2$		η^3		
	ξ^4		$4\xi^3\eta$		$6\xi^2\eta^2$		$4\xi\eta^3$		η^4	
ξ^5		$5\xi^4\eta$		$10\xi^3\eta^2$		$10\xi^2\eta^3$		$5\xi\eta^4$		η^5

For the quadrilateral (parent) element Ω_j shown in Figure 14.14 we set

$$A(\boldsymbol{\xi}) = \alpha_1 + \alpha_2\xi + \alpha_3\eta + \alpha_4\xi\eta = (1, \xi, \eta, \xi\eta)\{\alpha\} =: \{P\}^T\{\alpha\} \,, \tag{14.42}$$

which yields a linear equation system for the four coefficients. Written in matrix notation,

$$\begin{pmatrix} A_1 \\ A_2 \\ A_3 \\ A_4 \end{pmatrix} = \begin{pmatrix} 1 & \xi_1 & \eta_1 & \xi_1\eta_1 \\ 1 & \xi_2 & \eta_2 & \xi_2\eta_2 \\ 1 & \xi_3 & \eta_3 & \xi_3\eta_3 \\ 1 & \xi_4 & \eta_4 & \xi_4\eta_4 \end{pmatrix} \begin{pmatrix} \alpha_1 \\ \alpha_2 \\ \alpha_3 \\ \alpha_4 \end{pmatrix} . \tag{14.43}$$

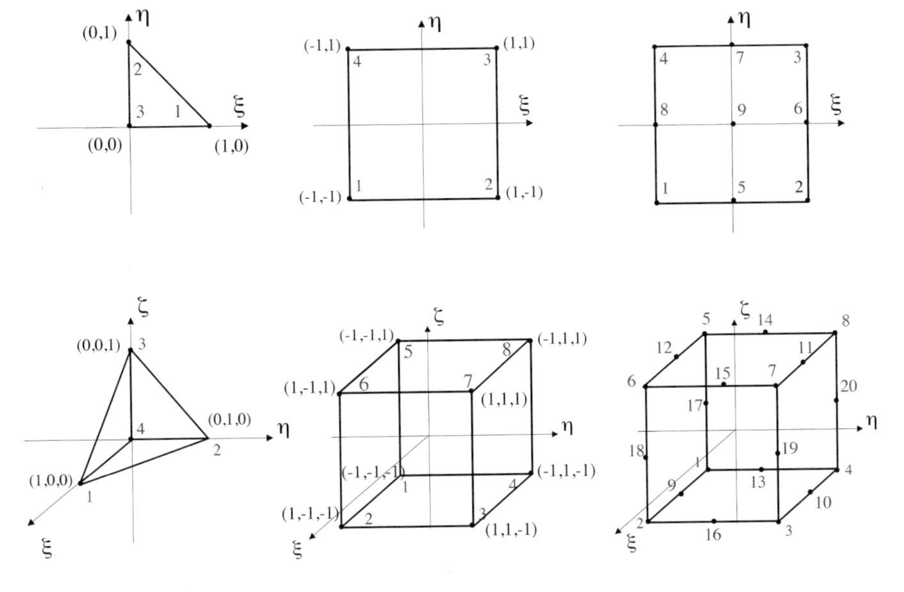

Figure 14.14 Various types of parent elements.

Inserting the coordinates of the corner points in Figure 14.14 yields

$$\begin{pmatrix} A_1 \\ A_2 \\ A_3 \\ A_4 \end{pmatrix} = \begin{pmatrix} 1 & -1 & -1 & 1 \\ 1 & 1 & -1 & -1 \\ 1 & 1 & 1 & 1 \\ 1 & -1 & 1 & -1 \end{pmatrix} \begin{pmatrix} \alpha_1 \\ \alpha_2 \\ \alpha_3 \\ \alpha_4 \end{pmatrix} \tag{14.44}$$

or the shorthand $\{A\} = [C]\{\alpha\}$. The solution is $\{\alpha\} = [C]^{-1}\{A\}$. It follows that

$$A(\boldsymbol{\xi}) = \{P\}^T\{\alpha\} = \{P\}^T[C]^{-1}\{A\} =: \{N\}^T\{A\} = \sum_{k=1}^{4} N_k(\boldsymbol{\xi})A_k\,. \tag{14.45}$$

The matrix $[C]$ is orthogonal because the scalar products of its columns are zero. Thus, its inverse is easily obtained:

$$[C]^{-1} = \frac{1}{4}[C]^T = \frac{1}{4} \begin{pmatrix} 1 & 1 & 1 & 1 \\ -1 & 1 & 1 & -1 \\ -1 & -1 & 1 & 1 \\ 1 & -1 & 1 & -1 \end{pmatrix}. \tag{14.46}$$

From $(N_1, N_2, N_3, N_4) = (1, \xi, \eta, \xi\eta)[C]^{-1}$ it follows that

$$N_1 = \frac{1}{4}(1 - \xi - \eta + \xi\eta) = \frac{1}{4}(1 - \xi)(1 - \eta), \tag{14.47}$$

$$N_2 = \frac{1}{4}(1 + \xi - \eta - \xi\eta) = \frac{1}{4}(1 + \xi)(1 - \eta), \tag{14.48}$$

$$N_3 = \frac{1}{4}(1 + \xi + \eta + \xi\eta) = \frac{1}{4}(1 + \xi)(1 + \eta), \tag{14.49}$$

$$N_4 = \frac{1}{4}(1 - \xi + \eta - \xi\eta) = \frac{1}{4}(1 - \xi)(1 + \eta). \tag{14.50}$$

The two-dimensional, linear, triangular element We can now derive the results of Section 14.2.2 in a much simpler fashion. For the element displayed in Figure 14.14 (top, left) the polynomial basis is $\{P\}^T = (1, \xi, \eta)$. Thus we obtain

$$[C]^{-1} = \begin{pmatrix} 0 & 0 & 1 \\ 1 & 0 & -1 \\ 0 & 1 & -1 \end{pmatrix}. \tag{14.51}$$

From $(N_1, N_2, N_3) = (1, \xi, \eta)[C]^{-1}$ the result of Eq. (14.34) follows directly.

The three-dimensional, four-noded, linear tetrahedron For the element displayed in Figure 14.14 (bottom, left), we obtain for the polynomial basis $\{P\}^T = (1, \xi, \eta, \zeta)$ and for the inverse of the transformation matrix:

$$[C]^{-1} = \begin{pmatrix} 0 & 0 & 0 & 1 \\ 1 & 0 & 0 & -1 \\ 0 & 1 & 0 & -1 \\ 0 & 0 & 1 & -1 \end{pmatrix}. \tag{14.52}$$

Hence the shape functions are

$$N_1 = \xi, \quad N_2 = \eta, \quad N_3 = \zeta, \quad N_4 = 1 - \xi - \eta - \zeta. \tag{14.53}$$

14.2.5 Generation of the Shape Functions

The shape functions were originally derived empirically for elements without central nodes. Thus such elements are said to be from the *serendipity class*. The most widely used higher-order elements in two dimensions are the eight-noded quadrilaterals. The polynomial basis for these elements is given by

$$\{P\}^T = (1, \xi, \eta, \xi^2, \xi\eta, \eta^2, \xi^2\eta, \xi\eta^2). \tag{14.54}$$

The element is shown in Figure 14.13 (top, right). The function approximation is exactly quadratic along the element edges. The same holds for triangles of equal complete order, that is, six nodes. Quadrilateral and six-noded triangular elements can thus be combined in a tessellation of the problem domain without loss of continuity across the inter-element boundaries.

Complete polynomial families can be found for triangular elements that render the elements *geometrically isotropic*, which means that the orientation of the element within the global coordinate system does not affect the accuracy of the approximation. However, the two-dimensional family of monomials $(1, \xi, \eta, \xi^2, \xi\eta, \eta^2, \xi^2\eta, \xi\eta^2)$ allows the modeling of quadratic functions along the ξ- and η-coordinates, and a cubic function along the line $\xi = \eta$. The element is therefore geometrically anisotropic.

Consider the rectangular element and its local coordinate system as shown in Figure 14.14. For an element containing only the corner nodes one to four, the product of linear *Lagrangian polynomials* of the form

$$\frac{1}{4}(1+\xi)(1+\eta) \tag{14.55}$$

results in a unit value at the top right corner where $\xi = \eta = 1$, and zero at all other corners. We denote the local coordinates of node k as $\xi_k, \eta_k \in \{-1, 0, 1\}$. The shape functions can then be written in the compact form

$$N_k = \frac{1}{4}(1+\xi\xi_k)(1+\eta\eta_k), \qquad k = 1, \ldots, 4. \tag{14.56}$$

Extending this idea to the incomplete quadratic element with eight nodes results in

$$N_{1,2,3,4} = \frac{1}{4}(1+\xi\xi_k)(1+\eta\eta_k)(\xi\xi_k + \eta\eta_k - 1), \tag{14.57}$$

$$N_{5,7} = \frac{1}{2}(1-\xi^2)(1+\eta\eta_k), \tag{14.58}$$

$$N_{6,8} = \frac{1}{2}(1+\xi\xi_k)(1-\eta^2). \tag{14.59}$$

If a ninth, central node is added, all terms of a complete fourth-order expansion are available. Hence, the shape function of the central node is

$$N_9 = (1-\xi^2)(1-\eta^2)\,. \tag{14.60}$$

Three-dimensional H8 and H20 elements Defining $\xi_k, \eta_k, \zeta_k \in \{-1, 0, 1\}$ as the local coordinates at node k, the shape functions of the H8 element can be expressed as

$$N_k = \frac{1}{8}(1+\xi\xi_k)(1+\eta\eta_k)(1+\zeta\zeta_k)\,. \tag{14.61}$$

For the H20 element, we obtain for the corner nodes,

$$N_{1,\dots,8} = \frac{1}{8}(1+\xi\xi_k)(1+\eta\eta_k)(1+\zeta\zeta_k)(-2+\xi\xi_k+\eta\eta_k+\zeta\zeta_k)\,, \tag{14.62}$$

and for the nodes on the sides parallel to the $\xi-$, $\eta-$, and ζ-axes,

$$N_{9,10,11,12} = \frac{1}{4}(1-\xi^2)(1+\eta\eta_k)(1+\zeta\zeta_k)\,, \tag{14.63}$$

$$N_{13,14,15,16} = \frac{1}{4}(1-\eta^2)(1+\xi\xi_k)(1+\zeta\zeta_k)\,, \tag{14.64}$$

$$N_{17,18,19,20} = \frac{1}{4}(1-\zeta^2)(1+\xi\xi_k)(1+\eta\eta_k)\,. \tag{14.65}$$

14.2.6 Transformation of Differential Operators

Finite-element procedures require the differentiation and integration of potential functions with respect to the global coordinate system. However, it is easier to calculate the derivatives with respect to ξ, η, and ζ because the shape functions are given in local coordinates. Applying the rules of partial differentiation,

$$\frac{\partial N_k}{\partial x} = \frac{\partial N_k}{\partial \xi}\frac{\partial \xi}{\partial x} + \frac{\partial N_k}{\partial \eta}\frac{\partial \eta}{\partial x}\,, \tag{14.66}$$

yields

$$\begin{pmatrix} \frac{\partial}{\partial x} \\ \frac{\partial}{\partial y} \end{pmatrix} N_k = \begin{pmatrix} \frac{\partial \xi}{\partial x} & \frac{\partial \eta}{\partial x} \\ \frac{\partial \xi}{\partial y} & \frac{\partial \eta}{\partial y} \end{pmatrix} \begin{pmatrix} \frac{\partial}{\partial \xi} \\ \frac{\partial}{\partial \eta} \end{pmatrix} N_k = [J]_{T^{-1}} \begin{pmatrix} \frac{\partial}{\partial \xi} \\ \frac{\partial}{\partial \eta} \end{pmatrix} N_k\,, \tag{14.67}$$

where $[J]_{T^{-1}}$ is the Jacobi matrix of the inverse transformation

$$T^{-1} : \mathbb{R}^2 \to \mathbb{R}^2 : (x,y) \mapsto (\xi,\eta). \tag{14.68}$$

The Jacobi matrix of the inverse transformation $[J]_{T^{-1}}$ is the inverse of the Jacobi matrix of the forward transformation $[J]_T^{-1}$ and can therefore easily be calculated:

$$[J]_{T^{-1}} = \begin{pmatrix} \frac{\partial \xi}{\partial x} & \frac{\partial \eta}{\partial x} \\ \frac{\partial \xi}{\partial y} & \frac{\partial \eta}{\partial y} \end{pmatrix} = \begin{pmatrix} \frac{\partial x}{\partial \xi} & \frac{\partial y}{\partial \xi} \\ \frac{\partial x}{\partial \eta} & \frac{\partial y}{\partial \eta} \end{pmatrix}^{-1} = [J]_T^{-1}\,, \tag{14.69}$$

where

$$[J]_T = \begin{pmatrix} \frac{\partial x}{\partial \xi} & \frac{\partial y}{\partial \xi} \\ \frac{\partial x}{\partial \eta} & \frac{\partial y}{\partial \eta} \end{pmatrix} = \begin{pmatrix} \sum_{k=1}^{K} \frac{\partial N_k}{\partial \xi} x_k & \sum_{k=1}^{K} \frac{\partial N_k}{\partial \xi} y_k \\ \sum_{k=1}^{K} \frac{\partial N_k}{\partial \eta} x_k & \sum_{k=1}^{K} \frac{\partial N_k}{\partial \eta} y_k \end{pmatrix}$$

$$= \begin{pmatrix} \frac{\partial N_1}{\partial \xi} & \frac{\partial N_2}{\partial \xi} & \cdots & \frac{\partial N_K}{\partial \xi} \\ \frac{\partial N_1}{\partial \eta} & \frac{\partial N_2}{\partial \eta} & \cdots & \frac{\partial N_K}{\partial \eta} \end{pmatrix} \begin{pmatrix} x_1 & y_1 \\ x_2 & y_2 \\ \vdots & \vdots \\ x_k & y_k \end{pmatrix} . \qquad (14.70)$$

For the special case of the linear, triangular element discussed in Section 14.2.2, this yields

$$[J]_T = \begin{pmatrix} 1 & 0 & -1 \\ 0 & 1 & -1 \end{pmatrix} \begin{pmatrix} x_1 & y_1 \\ x_2 & y_2 \\ x_3 & y_3 \end{pmatrix} = \begin{pmatrix} x_1 - x_3 & x_2 - x_3 \\ y_1 - y_3 & y_2 - y_3 \end{pmatrix} . \qquad (14.71)$$

The following relations help in the calculation of the inverse Jacobi matrix: For two dimensions

$$[J]^{-1} = \begin{pmatrix} j_{11} & j_{12} \\ j_{21} & j_{22} \end{pmatrix}^{-1} = \frac{1}{\det[J]} \begin{pmatrix} j_{22} & -j_{12} \\ -j_{21} & j_{11} \end{pmatrix} \qquad (14.72)$$

and in 3D:

$$[J]^{-1} = \begin{pmatrix} j_{11} & j_{12} & j_{13} \\ j_{21} & j_{22} & j_{23} \\ j_{31} & j_{22} & j_{33} \end{pmatrix}^{-1}$$

$$= \frac{1}{\det[J]} \begin{pmatrix} j_{22}j_{33} - j_{32}j_{23} & j_{13}j_{32} - j_{12}j_{33} & j_{12}j_{23} - j_{13}j_{22} \\ j_{31}j_{23} - j_{21}j_{33} & j_{11}j_{33} - j_{13}j_{31} & j_{21}j_{13} - j_{23}j_{11} \\ j_{21}j_{32} - j_{31}j_{22} & j_{12}j_{31} - j_{32}j_{11} & j_{11}j_{22} - j_{12}j_{21} \end{pmatrix} . \qquad (14.73)$$

Remark: Isoparametric elements, as opposed to quadrilateral elements with straight edges, need not be convex. However, the coordinate transformation and its inverse must be unique. This condition is a requirement for obtaining compatible sides of adjacent elements. Consequently, collinear edges in the mapped element, shown on the left-hand side of Figure 13.5, must be avoided. The two sides that meet in the common mesh point are orthogonal in the $\xi\eta$-plane but collinear in the xy-plane. This can only be accomplished if the Jacobian becomes singular at this point.[4] □

4 The magnitude $\det[J]_T$ denotes the local area magnification, $\mathrm{d}x\mathrm{d}y = \det[J]_T \mathrm{d}\xi \mathrm{d}\eta$, while the coefficients are a measure of the relative twisting and stretching in the different coordinate directions.

The right-hand side of Eq. (14.67) can be easily evaluated, because the shape functions N_k are given in local coordinates. Moreover, x and y are explicitly given by the coordinate mapping, so that the Jacobi matrix can be found explicitly in terms of the local coordinates.

References

1 Baehmann, P.L. et al.: Robust, geometrically based, automatic two-dimensional mesh generation, International Journal for Numerical Methods in Engineering, 1987

2 Beier, T. et al.: Feature-based image metamorphosis, Computer Graphics, 1992

3 Blacker, T.D., Stephenson M.B.: Paving: A new approach to automated quadrilateral mesh generation, International Journal for Numerical Methods in Engineering, 1991

4 Fetzer, J.: Die Lösung statischer und quasistationärer elektromagnetischer Feldprobleme mit Hilfe der Kopplung der Methode der finiten Elemente und der Randelementmethode, Fortschritt Berichte VDE, Reihe 21, VDE Verlag, 1992

5 Gröller, E.: Interactive transformation of 2D vector data, Technical University Vienna, Institute for Computer Graphics

6 Nowottny, G. D.: Netzerzeugung durch Gebietszerlegung und duale Graphenmethode, Shaker Verlag, Aachen, 1999

7 Saad, Y., Schultz, M.H.: GMRES: A generalized minimal residual algorithm for solving nonsymmetric linear systems, SIAM Journal of Scientific Statistical Computing, 1986

8 Taig I.C.: Structural analysis by the matrix displacement method. English Electric Aviation Report No. S017, 1961

9 Touzot, G., Dhatt, G.: The finite element method displayed, John Wiley & Sons, New York, 1984

10 Van der Vorst, H.A.: Bi-CGSTAB: a fast and smoothly converging variant of Bi-CG for the solution of nonsymmetric linear systems, SIAM Journal of Scientific Statistical Computing, 1992

15
Coupling of Boundary and Finite Elements

Magnets for particle accelerators have always been a key application of numerical methods in electromagnetism. In 1963, Hornsby [10] developed a code based on the finite difference (FD) method for the solving of elliptic partial differential equations and applied it to the design of accelerator magnets. Winslow [16] created the computer code TRIM (triangular mesh) with a discretization scheme based on an irregular grid of plane triangles, using a generalized finite difference scheme. He also applied a variational principle and showed that the two approaches lead to the same result. In this respect, the work can be viewed as one of the earliest examples of the finite-element method applied to the design of magnets.

The POISSON code, developed by Halbach and Holsinger [9], was the successor to the TRIM code and was still being applied to the optimization of the LHC superconducting magnets during the early stages of the project. As early as 1967, Halbach had introduced a method for optimizing coil arrangements and magnet pole shapes based on the TRIM code, an inverse field approach he consequently named MIRT [8]. In the early 1970s, a general-purpose program (GFUN) for static fields had been developed by Newman and others, which was based on the magnetization integral equation and applied to magnet design [1].

We explained in Section 13.2.2 that applying the total vector potential formulation causes the current density **J** to appear on the right-hand side of the weak forms (13.57) and (13.55). Consequently, the relatively complicated shape of the coils must be modeled in the FE-mesh. The dense mesh required for the accurate modeling of the coil, leads to a large number of unknowns, particularly if the surrounding iron yoke and vacuum vessel of the cryostat is taken into account. Simplifications of the coil geometry, on the other hand, yield inaccurate field quality estimates.

It is therefore advantageous to use methods that do not require the modeling of the coils in the finite-element mesh and which allow a distinction between the coil field and the iron magnetization. In this way modeling problems on the coils are reduced, and FEM-related numerical errors are confined to the magnetization effects in the iron yoke. The integral equation method

Field Computation for Accelerator Magnets. Stephan Russenschuck

ISBN: 978-3-527-40769-9

of GFUN would qualify but also lead to a very large, fully populated matrix of the linear equation system. A finite-element formulation using the reduced vector potential is discussed in Chapter 13.

In this chapter, we will discuss a field computation method that is particularly suited to the design and optimization of superconducting magnets for accelerators. This method, known as BEM-FEM [11], couples boundary and finite elements. The elementary model problem is shown in Figure 13.7. The magnetization of the iron parts is calculated from the source vector potential on the iron/air interface. The flux density at the field point is the superposition of the source field $\mathbf{B}_\mathrm{S}$ and the reduced field $\mathbf{B}_\mathrm{r}$ due to the iron magnetization.

The application of BEM–FEM coupling to magnet design has the following advantages:

- The coil field can be taken into account in terms of its source vector potential $\mathbf{A}_\mathrm{S}$, which is obtained from the line-currents by means of Biot–Savart–type integrals (Section 5.5) and therefore without the meshing of the coil. This is of particular importance for 3D field computations.
- BEM–FEM coupling allows for the direct computation of the reduced vector potential $\mathbf{A}_\mathrm{r}$ instead of the total vector potential $\mathbf{A}$. Consequently, errors do not influence the dominating contribution $\mathbf{A}_\mathrm{S}$ from the superconducting coil.
- Since the aperture field is calculated by integration over all the BEM elements, the local field errors in the iron yoke cancel out and the calculated multipole content is sufficiently accurate even for sparse meshes.
- The surrounding air region need not be meshed at all. This simplifies the preprocessing and avoids artificial boundary conditions at far-field boundaries. Moreover, the geometry of the permeable parts can be modified without regard to a mesh in the surrounding air region, an approach that is essential to the feature-based, parametric geometry modeling required for mathematical optimization.
- The method can be applied to both 2D and 3D field problems.

15.1 The Boundary-Element Method

Let us recall Green's first identity,

$$\int_\Omega (\operatorname{grad}\phi \cdot \operatorname{grad}\psi + \phi\nabla^2\psi)\,\mathrm{d}V = \int_\Gamma \phi \operatorname{grad}\psi \cdot \mathbf{n}\,\mathrm{d}a\,, \tag{15.1}$$

and the second,

$$\int_\Omega \left(\phi\nabla^2\psi - \psi\nabla^2\phi\right)\,\mathrm{d}V = \int_\Gamma (\phi\partial_\mathbf{n}\psi - \psi\partial_\mathbf{n}\phi)\,\mathrm{d}a\,, \tag{15.2}$$

which are 3D generalizations of the integration-by-parts rule. Green's theorems play a vital role in numerical field computation: They constitute the junction between the FEM and the BEM, as shown in Figure 15.1.

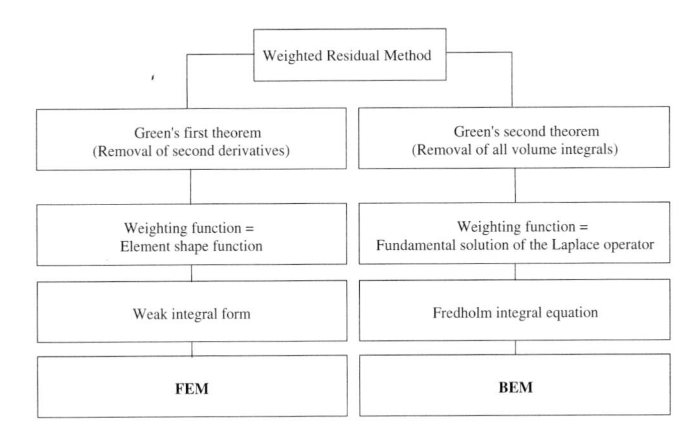

Figure 15.1 The role of Green's first and second identity for the FEM and BEM.

By definition, the BE domain Ω_a contains no iron, and therefore $\mathbf{M} = 0$ and $\mu = \mu_0$. Equation (4.155) reduces to

$$\nabla^2 \mathbf{A} = -\mu_0 \mathbf{J}, \qquad \text{in } \Omega_a. \tag{15.3}$$

If Cartesian coordinates are used, Eq. (15.3) resolves into three scalar Poisson equations. For an approximate solution of these equations the weighted residual is forced to zero. This yields

$$\int_{\Omega_a} \nabla^2 A \, w \, dV = -\int_{\Omega_a} \mu_0 J \, w \, dV, \tag{15.4}$$

where A and J are the Cartesian components of the vector potential and the current density, respectively. Applying Green's second identity yields

$$\int_{\Omega_a} A \nabla^2 w \, dV = -\int_{\Omega_a} \mu_0 J w \, dV + \int_{\Gamma} A \, \partial_{\mathbf{n}_a} w \, da - \int_{\Gamma} w \, \partial_{\mathbf{n}_a} A \, da, \tag{15.5}$$

where $\partial_{\mathbf{n}_a}$ is the directional derivative with respect to the outward unit vector $\mathbf{n}_a$, understood in the limiting sense, approaching the iron/air interface from the air domain.

In Eq. (15.5), we have already taken into account the vanishing of the boundary integrals along the far-field boundary $\Gamma_\infty^{\text{BEM}}$. The weighting function is chosen as the fundamental solution of the Laplace equation, which 3D is

$$w = u^*(\mathbf{r}, \mathbf{r}') := \frac{1}{4\pi |\mathbf{r} - \mathbf{r}'|}. \tag{15.6}$$

The normal derivative of this fundamental solution is given by

$$q^*(\mathbf{r},\mathbf{r}') := \partial_{\mathbf{n}_\mathrm{a}} u^* = -\frac{(\mathbf{r}-\mathbf{r}')\cdot\mathbf{n}_a}{4\pi|\mathbf{r}-\mathbf{r}'|^3}, \tag{15.7}$$

which can be proved using Eq. (3.110). From

$$\nabla^2 w = -\delta(|\mathbf{r}-\mathbf{r}'|) \tag{15.8}$$

and

$$\int_\Omega \delta(|\mathbf{r}-\mathbf{r}'|)\,\mathrm{d}V = 1, \tag{15.9}$$

it follows that

$$\int_\Omega A(\mathbf{r})\nabla^2 w\,\mathrm{d}V = \int_\Omega A(\mathbf{r})\delta(|\mathbf{r}-\mathbf{r}'|)\,\mathrm{d}V = A(\mathbf{r}'). \tag{15.10}$$

Substituting this result into Eq. (15.5) yields, after an exchange of $\mathbf{r}$ and $\mathbf{r}'$, the *Fredholm integral equation*[1] of the second kind:

$$\frac{\Theta}{4\pi}A(\mathbf{r}) + \int_\Gamma Q\,u^*(\mathbf{r},\mathbf{r}')\,\mathrm{d}a' + \int_\Gamma A\,q^*(\mathbf{r},\mathbf{r}')\,\mathrm{d}a' = \int_{\Omega_\mathrm{a}} \mu_0 J u^*(\mathbf{r},\mathbf{r}')\,\mathrm{d}V', \tag{15.11}$$

where $Q := -\partial_{\mathbf{n}_\mathrm{a}} A$. The only remaining domain integral on the right-hand side of Eq. (15.11) is of the Biot–Savart type for the source vector potential A_s according to Eq. (5.26). The terms

$$\boldsymbol{\alpha}(\mathbf{r}') := \frac{1}{\mu}Q(\mathbf{r}') \quad \text{and} \quad \boldsymbol{\tau}(\mathbf{r}') := \frac{1}{\mu}A(\mathbf{r}') \tag{15.12}$$

have the physical dimensions of a surface-current, $[\boldsymbol{\alpha}] = 1\ \mathrm{A\,m^{-1}}$, and a surface-density of magnetic dipoles $[\boldsymbol{\tau}] = 1$ A. The corresponding integrals in Eq. (15.11) are therefore referred to as the single- and double-layer potentials.

The result (15.11) is identical to the Kirchhoff theorem derived in Section 5.2 for field points inside the air domain. If the field point is located at the domain boundary, the single-layer potential will become singular and the double-layer potential will exhibit a jump discontinuity. The numerical evaluation of the double-layer integral in the sense of the Cauchy principle value leads to the *edge factor* $\Theta/4\pi$ where $\Theta(\mathbf{r})$ is the solid angle enclosed by the domain Ω_a in the vicinity of the field point $\mathbf{r}$; see Table 15.1.

For two-dimensional, magnetostatic problems the Fredholm equation is given by

$$\frac{\beta}{2\pi}A(\mathbf{r}) + \int_\Gamma Q\,u^*(\mathbf{r},\mathbf{r}')\,\mathrm{d}\mathbf{r}' + \int_\Gamma A\,q^*(\mathbf{r},\mathbf{r}')\,\mathrm{d}\mathbf{r}' = \int_{\Omega_\mathrm{a}} \mu_0 J u^*(\mathbf{r},\mathbf{r}')\,\mathrm{d}a', \tag{15.13}$$

1 Ivar Fredholm (1866–1927).

Table 15.1 Different solid angles Θ and edge factors for 3D domains.

Ω_a	90° Corner	90° Cone inner	Half-space	90° Cone outer
Θ	$\frac{1}{2}\pi$	$(2-\sqrt{2})\pi$	2π	$(2+\sqrt{2})\pi$
$\frac{\Theta}{4\pi}$	$\frac{1}{8}$	$\frac{2-\sqrt{2}}{4}$	$\frac{1}{2}$	$\frac{2+\sqrt{2}}{4}$

where

$$u^*(\mathbf{r},\mathbf{r}') = -\frac{1}{2\pi}\ln|\mathbf{r}-\mathbf{r}'|, \qquad q^*(\mathbf{r},\mathbf{r}') = -\frac{(\mathbf{r}-\mathbf{r}')\cdot\mathbf{n}_a}{2\pi|\mathbf{r}-\mathbf{r}'|^2}, \tag{15.14}$$

and $\beta(\mathbf{r})$ is the plane angle enclosed by the domain Ω_a in the vicinity of $\mathbf{r}$.

For 2D problems the edge factor is one, if the field point is situated within the BEM domain, and 0.5 if it is on a straight boundary segment of the domain; see Table 15.2.

Table 15.2 Different plane angles β and edge factors for 2D domains.

Ω_a	90° Wedge	Half-plane	90° Inverse wedge
β	$\frac{1}{2}\pi$	π	$\frac{3}{2}\pi$
$\frac{\beta}{2\pi}$	$\frac{1}{4}$	$\frac{1}{2}$	$\frac{3}{4}$

The components of the vector potential $\mathbf{A}$ at arbitrary points in the air domain can be computed from (15.11) or (15.13) when the components of the vector potentials $\mathbf{A}$ and their normal derivatives $\mathbf{Q}$ on the iron/air interface are known.

15.1.1
The Node Collocation Method

Let Γ denote a three-dimensional iron/air interface. For the discretization of Γ into individual boundary elements Γ_e, $e = 1, \ldots, E$, we use 8-noded quadrilateral elements, which are C^0-continuous and isoparametric. The numbering scheme for the nodes and elements is given in Table 15.3. In the 2D case, line elements with three nodes are employed at the boundary. These boundary elements must be consistent with the adjacent elements from the FE domain. Equation (15.11) can be written in discrete form as

$$C(\mathbf{r}_p)A(\mathbf{r}_p) + \sum_{e=1}^{E} \int_{\Gamma_e} Q\, u^*(\mathbf{r}, \mathbf{r}_p) \mathrm{d}a + \sum_{e=1}^{E} \int_{\Gamma_e} A\, q^*(\mathbf{r}, \mathbf{r}_p) \mathrm{d}a = A_s(\mathbf{r}_p) , \quad (15.15)$$

where $C(\mathbf{r}_p)$ are the edge factors on the nodes $p = 1, \ldots, P$, and Γ_e denotes the eth boundary element. A linear equation system can be obtained from Eq. (15.15) by successively placing the evaluation point $\mathbf{r}$ at the location of each nodal point $\mathbf{r}_p$ on the iron/air interface Γ. This procedure is known as *node collocation*. The component functions A and Q in each boundary element are expanded on Γ_e with respect to the element shape functions $N_m(\mathbf{r})$ and the nodal values A_m and Q_m, as follows:

$$A(\mathbf{r}) = \sum_{m=1}^{M} N_m(\mathbf{r}) A_m \quad \text{on } \Gamma_e , \quad (15.16)$$

$$Q(\mathbf{r}) = \sum_{m=1}^{M} N_m(\mathbf{r}) Q_m \quad \text{on } \Gamma_e , \quad (15.17)$$

where m is the local node number of the boundary element. Substituting the expressions (15.16) and (15.17) into Eq. (15.15) yields

$$C(\mathbf{r}_p)A(\mathbf{r}_p) + \sum_{e=1}^{E} \left(\sum_{m=1}^{M} \int_{\Gamma_e} N_m Q_m u^*(\mathbf{r}, \mathbf{r}_p)\, \mathrm{d}a \right)$$
$$+ \sum_{e=1}^{E} \left(\sum_{m=1}^{M} \int_{\Gamma_e} N_m A_m q^*(\mathbf{r}, \mathbf{r}_p)\, \mathrm{d}a \right) = A_s(\mathbf{r}_p) . \quad (15.18)$$

Equation (15.18) holds for all three Cartesian components of the vector potential. For one component, the following equation system is obtained:

$$[G]\{Q\} + [H]\{A\} = \{A_s\} . \quad (15.19)$$

The matrices $[G]$, resulting from the boundary integral with the kernel u^*, and $[H]$, resulting from the boundary integral with the kernel q^*, are asymmetric and fully populated. The column vector $\{A_s\}$ contains the values of the source vector potential at the nodal points $\mathbf{r}_p$ for $p = 1, 2, \ldots, P$:

$$\{A_s\} := (A_s(\mathbf{r}_1), \ldots, A_s(\mathbf{r}_P))^T . \quad (15.20)$$

Table 15.3 Conventions for the numbering of nodes and elements.

j	$1,\ldots,J$	Number of finite element.
k	$1,\ldots,K$	Local number of nodes in element Ω_j.
m	$1,\ldots,M$	Local number of nodes in the boundary element.
l	$1,\ldots,K$	Index of the shape functions in element Ω_j.
n	$1,\ldots,N$	Global number of all nodes in the mesh.
e	$1,\ldots,E$	Number of boundary elements.
p	$1,\ldots,P$	Global number of all nodes on the domain boundary.
a	1, 2, 3	x-, y-, and z-coordinates.

The equation system (15.19) alone cannot be solved because it is underdetermined. On the iron/air interface no boundary conditions are imposed and thus each node carries two or six degrees of freedom in the 2D and 3D case, respectively. A unique solution can be found by coupling of the boundary-element problem to the finite-element formulation inside the iron domain.

15.2 BEM–FEM Coupling

BEM–FEM coupling [11] combines the finite-element method inside magnetic bodies $\Omega_\mathrm{i} = \Omega_\mathrm{FE}$ with the boundary-element method in the domain outside the magnetic material $\Omega_\mathrm{a} = \Omega_\mathrm{BE}$ by means of the boundary value and the normal derivative of the vector potential on the iron/air interface.

Let us recall the weak solution of the FE part derived in Section 13.3.2:

$$\frac{1}{\mu_0}\int_{\Omega_\mathrm{i}} \operatorname{grad}(\mathbf{A}\cdot\mathbf{e}_a)\cdot\operatorname{grad} w_a \,\mathrm{d}V - \frac{1}{\mu_0}\int_\Gamma (\partial_{\mathbf{n}_\mathrm{i}}\mathbf{A} - (\mu_0\mathbf{M}\times\mathbf{n}_\mathrm{i}))\cdot\mathbf{w}_a\,\mathrm{d}a = \int_{\Omega_\mathrm{i}} \mathbf{M}\cdot\operatorname{curl}\mathbf{w}_a\,\mathrm{d}V \tag{15.21}$$

for all three Cartesian components (a = 1,2,3) of the vector potential and the iron magnetization. It is shown in [6] that the continuity condition of H_t at the iron/air interface,

$$\frac{1}{\mu_0}(\operatorname{curl}\mathbf{A}_\mathrm{i} - \mu_0\mathbf{M})\times\mathbf{n}_\mathrm{i} + \frac{1}{\mu_0}(\operatorname{curl}\mathbf{A}_\mathrm{a})\times\mathbf{n}_\mathrm{a} = \mathbf{0}\,, \tag{15.22}$$

is equivalent to

$$\partial_{\mathbf{n}_\mathrm{i}}\mathbf{A}_\mathrm{i} - (\mu_0\mathbf{M}\times\mathbf{n}_\mathrm{i}) + \partial_{\mathbf{n}_\mathrm{a}}\mathbf{A}_\mathrm{a} = \mathbf{0}\,, \tag{15.23}$$

where $\mathbf{n}_\mathrm{i}$ is the normal vector on Γ pointing outward from the FE domain Ω_i, and $\mathbf{n}_\mathrm{a}$ is the normal vector on Γ pointing from the BE domain Ω_a into the FE domain. The boundary integral term in Eq. (15.21) serves as the coupling term between the BE and FE domains, since the normal derivative of the vector potential on Γ

$$\mathbf{Q} := -\partial_{\mathbf{n}_a} \mathbf{A}_a \tag{15.24}$$

is given by the BE description. Equation (15.21) can then be rewritten as

$$\frac{1}{\mu_0} \int_{\Omega_i} \operatorname{grad}(\mathbf{A} \cdot \mathbf{e}_a) \cdot \operatorname{grad} w_a \, dV - \frac{1}{\mu_0} \int_{\Gamma} \mathbf{Q}_\Gamma \cdot \mathbf{w}_a \, da = \int_{\Omega_i} \mathbf{M} \cdot \operatorname{curl} \mathbf{w}_a \, dV \tag{15.25}$$

for $a = 1,2,3$.

We can now assemble the linear equation system for the FE domain. The iron domain Ω_i is discretized into finite elements Ω_j. C^0-continuous, isoparametric 20-node hexahedra are used in 3D, and 8-node quadrilaterals are used in 2D. The iron/air interface Γ must coincide with the element edges.

For the following discussion, we refer again to the indices listed in Table 15.3. The component functions A and Q in Ω_j that have one edge (or face) in common with the iron/air interface are expanded with respect to the element shape functions $N_k(\mathbf{r})$ and the nodal values A_k and Q_m as follows:

$$A(\mathbf{r}) = \sum_{k=1}^{K} N_k(\mathbf{r}) A_k \quad \text{on}\, \Omega_j, \tag{15.26}$$

$$Q(\mathbf{r}) = \sum_{m=1}^{M} N_m(\mathbf{r}) Q_m \quad \text{on}\, \Gamma \cap \partial\Omega_j, \tag{15.27}$$

where K is the number of nodes of the element Ω_j and M is the number of nodes at its boundary $\Gamma \cap \partial\Omega_j$.

Considering the A_z component, for example, the Galerkin method applied to the weak form yields

$$\int_{\Omega_j} \operatorname{grad}\left(\sum_{k=1}^{K} N_k A_{z,k}\right) \cdot \operatorname{grad} N_l dV - \int_{\Gamma_e} \sum_{m=1}^{M} N_m Q_{z,m} N_l da = \mu_0 \int_{\Omega_j} \left(M_x \frac{\partial N_l}{\partial y} - M_y \frac{\partial N_l}{\partial x}\right) dV, \tag{15.28}$$

for $l = 1, 2, \ldots, K$ and $\Gamma_e := \Gamma \cap \partial\Omega_j$. Similar equations hold for the other two components of the vector potential and its normal derivative, while the components of $\mathbf{M}$ are cyclicly permuted. Combining the equations for all J elements, the following linear equation system is obtained:

$$[K]\{A\} - [T]\{Q\} = \{F(\mathbf{M})\}. \tag{15.29}$$

The domain and boundary integrals in the weak form yield the stiffness matrix $[K] \in \mathbb{R}^{N\times N}$ and the boundary matrix $[T] \in \mathbb{R}^{N\times P}$, respectively. Both matrices are sparse. The boundary matrix has nonvanishing coefficients only if Ω_j has a boundary in common with Γ_{ai}. For each component, the linear equation system has N equations for the N nodal unknowns in $\{A\}$ and the P nodal unknowns in $\{Q\}$. As listed in Table 15.3, N is the total number of nodes in the finite-element mesh, and P is the total number of nodes on the domain boundary.

An overall numerical description of the field problem can be obtained by complementing the FE description (15.29) with the BE description (15.19). Multiplying Eq. (15.19) by $[G]^{-1}$ results in

$$\{Q\} = -[G]^{-1}[H]\{A\} + [G]^{-1}\{A_{\mathrm{s}}\}, \tag{15.30}$$

which can be used to eliminate $\{Q\}$ in the equation system for the FE domain:

$$\left([K] + [T][G]^{-1}[H]\right)\{A\} = \{F(\mathbf{M})\} + [T][G]^{-1}\{A_{\mathrm{s}}\}, \tag{15.31}$$

which can be written shorthand as

$$[\overline{K}]\{A\} = \{\overline{F}(A_{\mathrm{s}}, \mathbf{M})\}. \tag{15.32}$$

Equation (15.29) gives exactly the missing relationship between the Dirichlet data $\{A\}$ and the Neumann data $\{Q\}$ on the iron/air interface. It can be shown [11] that this procedure guarantees that the continuity conditions $\mathbf{n}\cdot\mathbf{B}$ and $\mathbf{n}\times\mathbf{H}$ are obeyed across Γ.

Combining the set of equations for all three components yields

$$\begin{pmatrix} [\overline{K}] & 0 & 0 \\ 0 & [\overline{K}] & 0 \\ 0 & 0 & [\overline{K}] \end{pmatrix} \begin{pmatrix} \{A_x\} \\ \{A_y\} \\ \{A_z\} \end{pmatrix} = \begin{pmatrix} \{\overline{F}_x(A_{\mathrm{s},x}, \mathbf{M})\} \\ \{\overline{F}_y(A_{\mathrm{s},y}, \mathbf{M})\} \\ \{\overline{F}_z(A_{\mathrm{s},z}, \mathbf{M})\} \end{pmatrix}. \tag{15.33}$$

15.3 BEM–FEM Coupling using the Total Scalar-Potential

In most superconducting magnets the coil can be completely embedded in a separating surface Γ_{s} so that no path outside Γ_{s} links any current. In other words, for each conductor we can find a cut surface $\mathscr{A}_{\mathrm{c}}$ such that $\partial\mathscr{A}_{\mathrm{c}}$ is the current loop and $\mathscr{A}_{\mathrm{c}}$ lies completely in the air domain Ω_{a}. The magnetic field $\mathbf{H}$ outside $\mathscr{A}_{\mathrm{c}}$ can therefore be represented by a single-valued magnetic scalar potential, $\mathbf{H} = -\operatorname{grad}\phi_{\mathrm{m}}$. Throughout the domain $\Omega_{\mathrm{i}} \cup \Omega_{\mathrm{a}} \setminus \mathscr{A}_{\mathrm{c}}$, the magnetic scalar potential ϕ_{m} is governed by

$$\operatorname{div}\mu\operatorname{grad}\phi_{\mathrm{m}} = -\rho_{\mathrm{m}}, \tag{15.34}$$

where ρ_m is the density of the fictitious magnetic charges on $\mathscr{A}_\mathrm{c}$. The current loops are thus replaced by a magnetic double layer; see Section 5.8. In order to avoid confusion with the numbering scheme for the nodes of the boundary elements we will drop the index m for the magnetic scalar potential and magnetic charges.

The coil is surrounded by the iron yoke, represented by the finite-element subdomain $\Omega_\mathrm{FE} = \Omega_\mathrm{i}$. In the air region $\Omega_\mathrm{BE} = \Omega_\mathrm{a}$, Eq. (15.34) reduces to

$$\Delta\phi = -\frac{\rho}{\mu_0}. \tag{15.35}$$

Multiplication by the Green function u^* of free space and two integrations-by-parts yield an integral representation of the scalar potential,

$$\phi = \underbrace{\frac{1}{\mu_0}\int_{\Omega_\mathrm{a}} \rho u^* \,\mathrm{d}V}_{\phi_\mathrm{s}} + \underbrace{\int_{\Gamma} \left(\phi \partial_\mathbf{n} u^* - \partial_\mathbf{n}\phi u^*\right) \mathrm{d}a}_{\phi_\mathrm{r}} . \tag{15.36}$$

The potential ϕ at an arbitrary point $\mathbf{r}$ outside the yoke consists of a source term ϕ_s and a reduced potential ϕ_r due to the iron magnetization. The source term ϕ_s is directly computed as a double-layer potential. According to Eqs. (5.90) and (5.94) the contribution of a single loop Γ carrying the current I is

$$\phi_\mathrm{s}(\mathbf{r}) = I \int_{\mathscr{A}_\mathrm{c}} \partial_{\mathbf{n}_{\mathbf{r}'}} u^*(\mathbf{r}', \mathbf{r})\, \mathrm{d}a' , \tag{15.37}$$

which is identical to the solid angle that the current loop subtends at the field point $\mathbf{r}$. If the loop is discretized into triangular patches, the solid angle can be computed analytically from geometrical data by use of Eq. (5.100).

Equation (15.36) yields a linear equation system by discretizing the iron/air interface Γ into nodal boundary elements and point-wise collocation. It is only on this interface, where the values of ϕ_s according to Eq. (15.37) are needed. The equation systems resulting from the FE and BE domains can easily be coupled because both formulations have identical Dirichlet and Neumann boundary data, ϕ and $-\mu\,\partial_\mathbf{n}\phi$, respectively. Thus, the continuity of $\mathbf{n} \times \mathbf{H}$ and $\mathbf{n} \cdot \mathbf{B}$ across the interface is ensured. We finally obtain

$$\begin{pmatrix} [K_{\Omega_\mathrm{i}\Omega_\mathrm{i}}] & [K_{\Omega_\mathrm{i}\Gamma}] & 0 \\ [K_{\Gamma\Omega_\mathrm{i}}] & [K_{\Gamma\Gamma}] & [T] \\ 0 & [H] & [G] \end{pmatrix} \begin{pmatrix} \{\phi_{\Omega_\mathrm{i}}\} \\ \{\phi_\Gamma\} \\ \{\partial_\mathbf{n}\phi\} \end{pmatrix} = \begin{pmatrix} 0 \\ 0 \\ \{\phi_\mathrm{s}\} \end{pmatrix}, \tag{15.38}$$

where $[K]$ are the FE stiffness matrices, $[T]$ is the FE boundary matrix, $[G]$ the BE matrix resulting from the single-layer integral, and $[H]$ the BE matrix resulting from the double-layer integral. The subscripts Ω_i and Γ refer to the interior and the boundary nodes of the finite elements, respectively.

The equation system (15.38) is nonlinear and must therefore be solved iteratively. Suitable methods are the $M(H)$-iteration and the Newton method. In each iteration step a linear equation system with the same structure as (15.38) must be solved. This can be done iteratively by the Krylov subspace method, where the preconditioning is based on a domain decomposition [12].

Once the solution of Eq. (15.38) has been obtained, the reduced field can be computed from Eq. (15.36) and the relation

$$\mathbf{B}_r = -\mu_0 \operatorname{grad} \phi_r . \tag{15.39}$$

According to Eq. ((15.37), ϕ_s suffers from discontinuities on the cut surface $\mathscr{A}_c$. It is therefore advantageous to make direct use of the Biot–Savart law for the coil contribution $\mathbf{B}_s$.

15.4 The M(B) Iteration

Consider the equation system

$$[K]\{A\} = \{F(A_s, \mathbf{M})\}, \tag{15.40}$$

where $[K]$ is the stiffness matrix. The force vector depends on the vector potential of the current sources A_s and on the nonlinear relationship between the magnetization M and the flux density B in the iron yoke. Probably the easiest scheme for the solution of the equation system (15.40) is the $M(B)$ iteration based on

$$\{A_{k+1}\} = [K]^{-1}\{F(A_s, \mathbf{M}_k)\}, \tag{15.41}$$

where k is the iteration index. Subtracting $\{A_k\}$ from both sides yields

$$\{\Delta A_k\} := \{A_{k+1}\} - \{A_k\} = [K]^{-1}\{R_k\}, \tag{15.42}$$

where the residual R_k is defined by

$$\{R_k\} := \{F(A_s, \mathbf{M}_k)\} - [K]\{A_k\}. \tag{15.43}$$

The convergence rate of the iteration can be improved by use of a *relaxation parameter* ω_k:

$$\{A_{k+1}\} = \{A_k\} + \omega_k\{\Delta A_k\}, \tag{15.44}$$

where $\omega_0 = 1$ and

$$\omega_k = \frac{\omega_{k-1}}{1 - \frac{\{\Delta A_k\}\,\{\Delta A_{k-1}\}}{\|\{\Delta A_{k-1}\}\|^2}} . \tag{15.45}$$

The iteration scheme can now be written in pseudocode as:

1. Set iteration index $k = 0$, and initialize vector potentials $\{A_0\} = \{0\}$.
2. For $k = 0, 1, 2, \ldots,$ until convergence Do:
3. Compute the force vector $\{F(A_s, \mathbf{M}_k)\}$ and the residual $\{R_k\}$.
4. If $\frac{\|\{R_k\}\|}{\|\{F(A_s, \mathbf{M}_k)\}\|} < \epsilon$ Goto 8.
5. Calculate the increment $\{\Delta A_k\} = [K]^{-1}\{R_k\}$.
6. Choose the relaxation parameter.
7. $\{A_{k+1}\} = \{A_k\} + \omega_k\{\Delta A_k\}$
8. End Do.

The normalized residual in step 4 is usually given in logarithmic scale with ϵ set to 50 dB. The advantages of the $M(B)$ iteration are that the stiffness matrix needs to be inverted only once if direct solvers are used, that the method is globally convergent, and that no derivative of the $M(B)$ curve is required.

15.5 Applications

We will now apply the BEM–FEM coupling method to the design of accelerator magnets. The examples demonstrate the intrinsic advantages of the method: no meshing of the coil, open boundary conditions, distinction between source field and magnetization of the iron parts, and accurate calculation of stray fields.

15.5.1 2D Calculations

The LHC coil-test facility The LHC required a major R&D effort to guarantee that the superconducting dipoles perform according to the specifications. For this reason an in-house program for the development of the superconducting dipoles was launched in 1995. The aim was to study the influence of individual coil parameters, such as the prestress in the coils, the collar material (aluminum or austenitic steel), and yoke structures, on a series of otherwise identical model dipoles. A maximum turnaround and testing efficiency was achieved with the construction of 1-m-long, single-aperture model magnets with a reusable iron yoke. Because of the reusable parts, these models were referred to as the coil-test "facility" (CTF).

The CTF is used here as an example for the 2D BEM–FEM calculation employing the magnetic vector and scalar potential formulations. The results are shown in Figure 15.2 for the z-component of the magnetic vector potential and the total magnetic scalar potential in the yoke cross section.

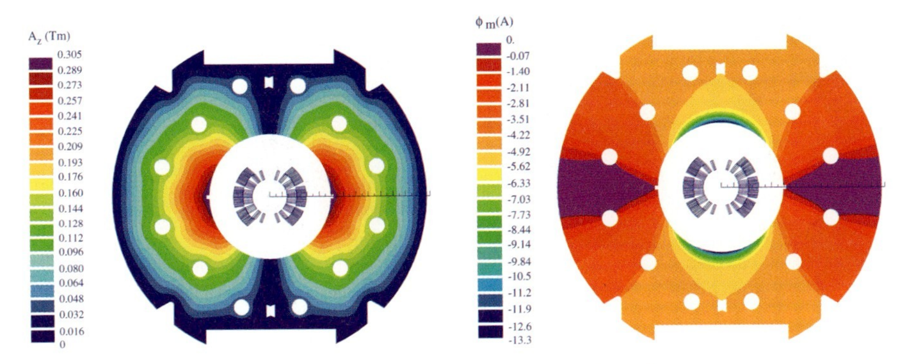

Figure 15.2 Left: The z-component of the magnetic vector potential. Right: Magnetic scalar potential.

Experience shows that the vector-potential formulation yields underestimated numerical values for the magnetic flux density while the scalar-potential formulation yields overestimated values. The convergence of the two methods for the CTF calculation is shown in Table 15.4.

Table 15.4 2D computation of the relative b_1 field error in units of 10^{-4} as a function of the mesh size[a].

Number of finite elements	60	178	449	787	2799
Total scalar potential	65.8	72.1	13.0	5.0	3.8
Vector potential	-40.5	-27.4	-7.4	-4.8	-3.8

[a] The errors were rescaled to the "true" field strength considered to be the average of the two calculations for 2799 finite elements.

Field quality in collared coils A crucial control point in the quality assurance for the LHC series-magnet production was the end of the collaring process, when the two dipole coils were assembled in the austenitic steel collars. Field-quality measurements performed at this stage revealed possible manufacturing errors at an early stage of production.

The magnetization of the steel collars ($\mu_r = 1.0025$) creates left/right asymmetries in the magnetic field when only one aperture is powered; see Figure 15.3. Powering both apertures at the same time is not possible because of the strong crosstalk between the coils. Since the BEM–FEM method neither requires the meshing of the coil nor employs far-field boundary conditions, the method is well suited for the calculation of this problem.

The systematic, relative field errors at the 17 mm reference radius in the aperture of the powered coil are given in Table 15.5. The higher-order multipoles are hardly affected by the permeability of the collars. However, the

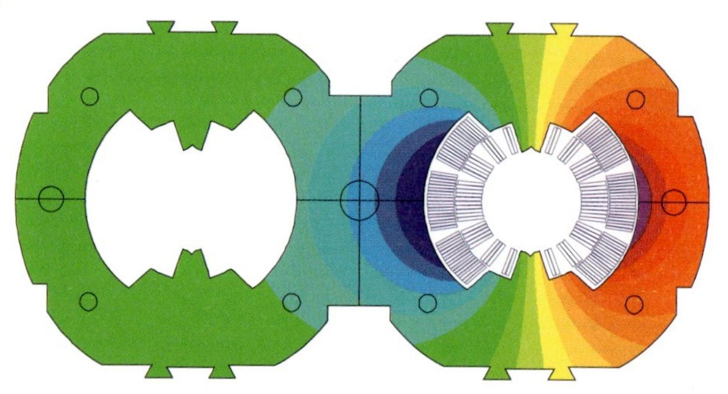

Figure 15.3 Geometric model of one dipole coil powered for room temperature measurement in the combined collar structure, assuming a constant relative permeability of $\mu_r = 1.0025$. The figure displays the magnetic flux density in the collars.

effects on the quadrupole and sextupole components cannot be disregarded and must be accounted for when inverse field computations for the tracing of manufacturing errors are performed; see Section 20.10.9.

Table 15.5 Additional field errors in a coil/collar assembly where only the coil on the right-hand side is powered (units of 10^{-4} at 17 mm)[a].

Nominal				Additional			
b_2	0.000	b_6	0.000	Δb_2	−0.239	Δb_6	0.000
b_3	3.915	b_7	0.745	Δb_3	−1.173	Δb_7	−0.058
b_4	0.000	b_8	0.000	Δb_4	−0.012	Δb_8	0.000
b_5	−1.038	b_9	0.122	Δb_5	0.305	Δb_9	0.003

[a] Nominal values for a bare coil without collar.

Magnetization of the LHC beam screen A *beam screen*, shown in Figure 1.18 (right), shields the cold bore from the synchrotron radiation emitted by the circulating proton beam. As the screen has a racetrack shape, special care was taken in the composition of the steel in order to keep the relative permeability below 1.003.

Possible field distortions caused by the beam screen are studied using BEM–FEM coupling. The calculations are performed for a 2D cross section; the effect of the pumping slots is accounted for by means of an appropriate stacking factor in the two-dimensional calculations [2]. The field errors due the magnetization of the beam screen and the cold bore are estimated by comparing field calculations with and without the screen, excited by an ideal cos φ_c current distribution. Since the cold bore has a perfectly cylindrical shape, it does not create higher-order field errors.

In order to validate the calculations, the results are compared to the measurements carried out for a prototype beam screen, made of the nonmagnetic, grade P506 austenitic steel, and centered in the cold-bore tube of a model dipole magnet. The beam screen consists of a 1-mm-thick, 300-mm-long tube of 48.5 mm diameter, with a vertical aperture of 38.9 mm, and slots at the flattened top and bottom; see Figure 1.18 (right). The internal surface of the screen is covered by a 50-μm-thick layer of copper. The nominal relative permeability μ_r of the steel used for the beam screen production is 1.003 for the screen and 1.0025 for the cooling tube.

The multipole errors due to the magnetization of the beam screen were so small that they were masked with field measurement errors stemming from the thermal cycle needed to add or remove the screen. In order to avoid these thermal cycles, the experimental setup was modified to enable the longitudinal displacement of a beam-screen sample by means of a nonmagnetic rod fastened to the measuring shaft [14]. The results of these measurements are compared to the calculations in Table 15.6.

Table 15.6 Calculations of the additional field errors in the LHC main dipoles and quadrupoles due to the magnetization of the beam screen. Comparison to the measured values of a prototype inside a dipole model, which are scaled to the nominal beam-screen thickness (all in units of 10^{-4}).

	Dipole			Quadrupole
	Calculation $\mu_r = 1.002$	Calculation $\mu_r = 1.003$	Measured	Calculation $\mu_r = 1.003$
Δb_2	0.000	0.000		0.000
Δb_3	−0.248	−0.33	−0.3	0.000
Δb_4	0.000	0.000		0.310
Δb_5	0.226	0.295	0.25	0.000
Δb_6	0.000	0.000		−0.259
Δb_7	−0.166	−0.213	−0.18	0.000
Δb_8	0.000	0.000		0.213
Δb_9	0.127	0.163		0.000

15.5.2 Saturation Effects in the Iron Yoke

Figure 15.4 shows the relative permeability (μ_r) of the iron yoke as a function of the excitation current at 4500 (left) and 10 500 A (right) for a dipole model with separate collars. It is obvious that the holes in the vertical median plane act to balance the magnetic reluctance between the two apertures and the outer region of the yoke.

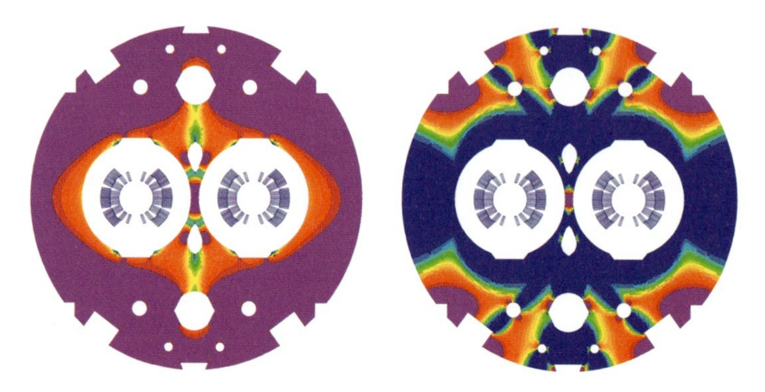

Figure 15.4 Relative permeability μ_r in the iron yoke of a model dipole with separated collars. Left: Excitation current of 4500 A. Right: 10 500 A.

Nevertheless, the crosstalk between the apertures persists and gives rise to even-numbered, normal multipoles ($b_2, b_4, \ldots$). Calculated multipole field errors of the LHC (series) dipole magnets at nominal excitation are shown in Figure 14.1 (left). Field asymmetries due to the crosstalk between apertures is more pronounced for the common-coil magnet shown in Figure 13.6 (right).

15.5.3
3D Calculations

The coil-test facility revisited The CTF has a coil length of 1.05 m, surrounded by a 402-mm-long magnetic yoke. The yoke is short in order to reduce the field in the coil ends. The geometrical model is shown in Figure 15.5. The iron saturation at nominal excitation differs considerably from that of the long magnets. A three-dimensional analysis must therefore be performed.

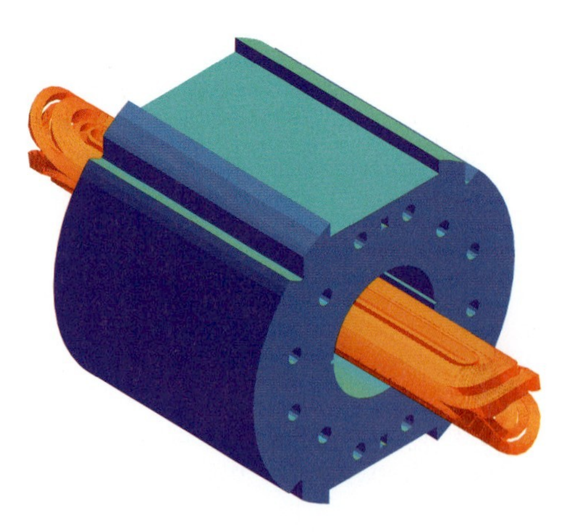

Figure 15.5 Model of the 1-m-long coil-test facility (CTF).

This example is used to illustrate the accuracy and computing cost of the two BEM–FEM formulations (vector and scalar potentials) for magnetostatic field problems. Data on the convergence of the methods are presented in Table 15.7.

Table 15.7 Performance of BEM-FEM coupling.[a]

Formulation		High field 13 000 A	Low field 3000 A		High field 13 000 A	Low field 3000 A
Allocated memory		49.7 MB				
Vector Potential $M(B)$	M(B) Steps	40	80	$b_3\ (z_1)$	1.0054	2.6827
	GMRES Steps	830	1889	$b_3\ (z_2)$	3.3111	2.6174
	Residual (dB)	-45.9	-45.1			
	T_{solv}	1.0	2.1			
	T_{integ}	18.9	18.9			
Vector Potential Newton	Newton Steps	8	9	$b_3\ (z_1)$	1.0061	2.6829
	GMRES Steps	277	2851	$b_3\ (z_2)$	3.3104	2.6178
	Residual (dB)	-46.4	-45.6			
	T_{solv}	0.46	2.5			
	T_{integ}	18.9	18.9			
Allocated memory		18.7 MB				
Scalar Potential $M(H)$	M(H) Steps	277	400	$b_3\ (z_1)$	1.5761	2.6890
	Bi-CGSTAB Steps	6223	5703	$b_3\ (z_2)$	3.7987	2.6422
	Residual (dB)	-24.6	-28.9			
	T_{solv}	1.6	1.5			
	T_{integ}	10.5	10.5			
Scalar Potential Newton	Newton Steps	13	6	$b_3\ (z_1)$	1.5958	2.6840
	Bi-CGSTAB Steps	2390	505	$b_3\ (z_2)$	3.9557	2.6325
	Residual (dB)	-49.3	-50.9			
	T_{solv}	0.52	0.19			
	T_{integ}	10.5	10.5			

[a] The normalized residual is defined by $R := 10 \log |[A]\{x\} - \{b\}|/|\{b\}|$ where $[R] = 1$ dB. Values above -40 dB indicate convergence problems. T_{solv} is the normalized CPU time for the solving of the nonlinear problem, and T_{integ} the time for calculating the Kirchhoff integrals over all the boundary elements. All CPU times are normalized to the solver time for the vector-potential formulation and $M(B)$ iteration. Relative sextupole components in the center of the magnet ($z_1 = 0$) and outside the iron yoke ($z_2 = 217$ mm) are given in units of 10^{-4} at the 17 mm reference radius.

The nonlinearity of the $B(H)$ curves is treated by an update of the magnetization ($M(B)$ iteration in the vector- and $M(H)$ in the scalar-potential case), or by the Newton method. This iteration process is usually referred to as the *outer iteration*. Preconditioned iterative solvers are applied to the solution of the linear equation system, also known as *inner iteration*. In the case of the vector-potential formulation the generalized minimum residual (GMRES) solver [13] and in the case of the scalar-potential formulation, the biconjugate gradient stabilized (Bi-CGSTAB) solver [15] are appropriate methods.

The scalar-potential formulation requires only a third of the computer memory needed for the vector potential and shows the greatest overall solver speed when combined with the Newton iteration. The advantage is smaller at nominal field. However, the time for calculating the Kirchhoff integrals is reduced and less memory allocated. This allows further mesh refinement and the calculation of more complicated cases such as twin-aperture magnets.

The results for the relative sextupole component (which is the most sensitive to the iron magnetization) in the center of the magnet and outside the iron yoke are also given in Table 15.7, as usual in units of 10^{-4} at the 17 mm reference radius. The method for the outer iteration shows only a slight effect on the sextupole component as long as the residual is below -40 dB. For small excitation currents the results are equivalent, whereas at high field the sextupole component both in the center and outside the yoke is higher if the scalar potential formulation is used. The difference is a half unit in 10^{-4}. A larger sextupole component indicates a lower saturation of the yoke and consequently a lower flux density in the aperture of the magnet. The calculated aperture field is 9.0818 T for the vector-potential formulation and 9.0699 T for the scalar-potential formulation.

Field quality in the end region of the coil-test facility The magnetic field homogeneity of the short models was systematically measured in a vertical test setup, in which the magnet was suspended inside a cryostat. The measurement of the field was performed using a "train" of five pickup coils rotating in the magnet bore.

A drawback of the short length of the CTF is that end effects also influence the magnetic field quality in the center of the magnet. The rotating pickup coils yield average multipoles over the probe length. Considering that the pickup coil in the center of the magnet has a length of 200 mm, which is half the length of the magnetic yoke, the interpretation of the measurements becomes difficult. In order to study systematic effects in the field quality, it is necessary to calculate, with high precision, the 3D multipole field errors in these magnets as a function of the z-position. Figure 15.6 shows the relative multipole components b_3, b_5, and b_7 (related to the main field B_1 of 8.24 T calculated at 11 530 A for the 2D model, at 17 mm reference radius) as a function of the z-position, where $z = 0$ is the center of the magnet and the iron yoke ends at $z = 201$ mm.

Saturation effects in short models The results of the 3D field calculation are compared with the measurements of a single-aperture model. The multipoles are computed as a function of the excitation current over a length of 550 mm

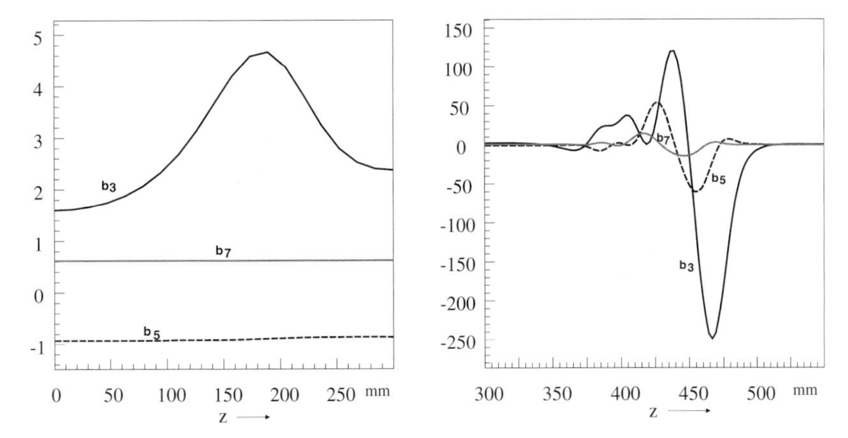

Figure 15.6 Multipole errors in the 1-m-long coil test facility as a function of the z-position. The multipoles are given in units of 10^{-4} at the 17 mm reference radius and are related to the main field B_1 of 8.24 T at 11 530 A. z = 0 is the center of the magnet and the iron yoke ends at z = 201 mm.

along the magnet bore. The dipole field $B_1(z)$ and the multipoles $b_n(z)$ are used to compute the average multipoles over the length spanned by the measurement pickup coils. The results of the measured and simulated field harmonics are given in Figure 15.7 together with the 2D approximation for the long magnet. It can be seen that the short models have globally different saturation behavior compared to the long dipoles.

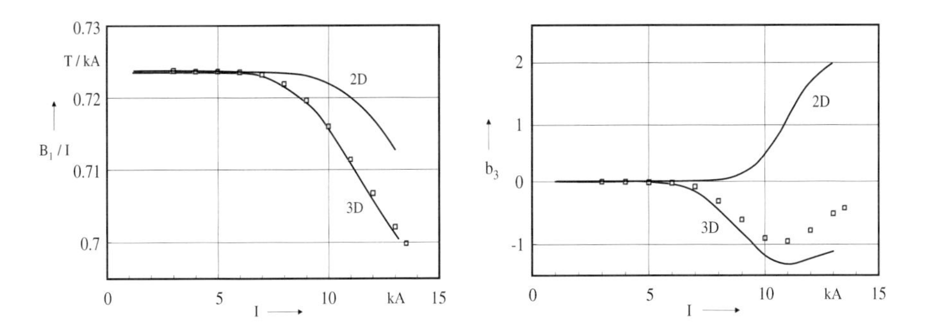

Figure 15.7 Measured and computed transfer function B_1/I in T/kA (left) and b_3 at 17 mm reference radius (right) as a function of the excitation (between injection and nominal field) averaged over the length of the measurement pickup coil (200 mm).

End fields in the dipole prototype magnets In order to reduce the peak field in the coil end and consequently to increase the quench margin in the region

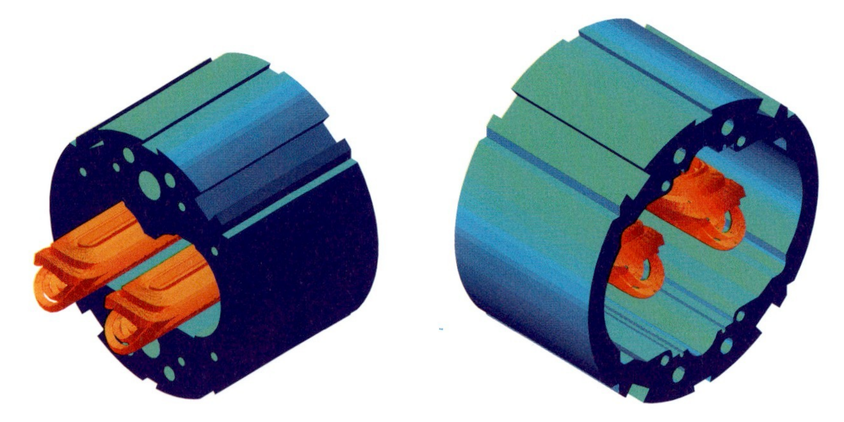

Figure 15.8 Dipole magnet extremities. Left: The LHC main dipole models until 1998. Right: The final version featuring a nonmagnetic, nested austenitic steel insert and a thin, wide-aperture iron yoke. The insert need not be modeled in the 3D field computations.

of weaker mechanical support, the magnetic iron yoke may be trimmed back at the magnet's extremity. Figure 15.8 (left) shows the trimmed back iron yoke used for the LHC dipole prototypes until 1998. Figure 15.8 (right) shows the final version, where the yoke/collar structure is made of three parts featuring a nested, austenitic steel insert. The insert need not be modeled in the 3D field computations.

The BEM–FEM coupling is used for the calculation of the end fields in the magnet models, as shown in Figure 15.8 (left). The iterative solution of the

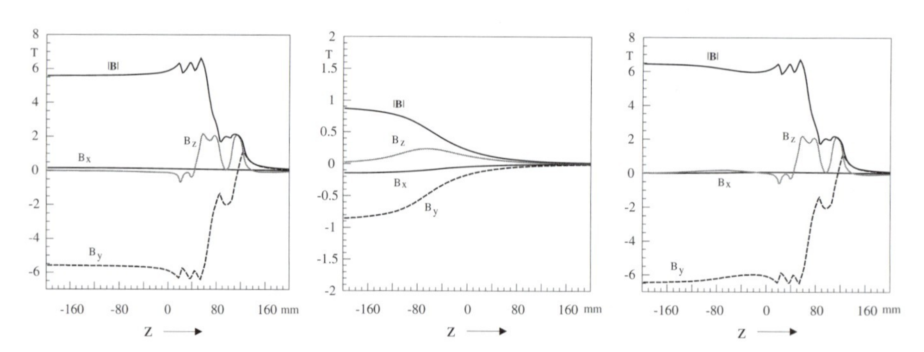

Figure 15.9 Magnetic flux density at nominal excitation along a line at $x = 97$ mm, $y = 43.6$ mm (between the inner- and the outer-layer coil) from $z = -200$ to 200 mm. The iron yoke ends at $z = -80$ mm; the onset of the coil end is at $z = 0$. Left: Coil field. Middle: Reduced field from iron magnetization. Right: Total field.

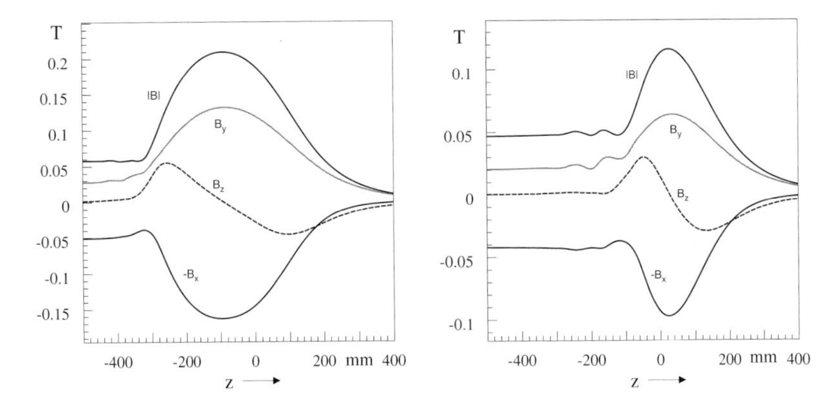

Figure 15.10 Stray field outside the dipole coldmass near the coil-end region at the position of the N-Line. Left: Connection end. Right: Return end.

linear equation system converges faster in the case of a high excitation field than in the case of the injection field at which the iron yoke is nonsaturated.

As it has already been explained, BEM–FEM coupling allows the distinction between the coil field and the reduced field attributable to the iron magnetization. Figure 15.9 shows the field components along a line in the end region (nonconnection side) of the main dipole prototype magnet, 43.6 mm above the beam axis, extending from $z = -200$ mm inside the magnet yoke to $z =$ 200 mm outside. The iron yoke ends at $z = -80$ mm; the onset of the coil end is at $z = 0$.

Figure 15.10 shows the field components along a line in the end region of the twin-aperture dipole prototype magnet (MBP2) at the position where SC cables are routed for the powering of the lattice corrector magnets. The line extends from $z = -500$ to 400 mm. The iron yoke ends at $z = 0$.

In the coil-end region the stray fields and the crosstalk between the two apertures result in a large quadrupole field component. This was the reason for adopting the nested structure as shown in Figure 15.8 (right) for the series production of the magnets.

References

1 Armstrong, A.G.A.M. , Fan, M.W., Simkin, J., Trowbridge, C.W.: Automated optimization of magnet design using the boundary integral method, IEEE Transactions on Magnetics, 1982.

2 Auchmann, B.: Field Errors due to Beam Screens, report to the Field Quality Working Group, CERN, 2005

3 Brebbia, C.A.: The Boundary Element Method for Engineers, Pentech Press, London, 1978

4 Collie, C.J.: Magnetic fields and potentials of linearly varying current or magnetization in a plane bounded region, Proceedings of the 1st Compumag Conference on the Computation of Electromagnetic Fields, Oxford, UK, 1976

5 Fetzer, J., Abele, S., Lehner, G.: Die Kopplung der Randelementmethode und der Methode der finiten Elemente zur Lösung dreidimensionaler elektromagnetischer Feldprobleme auf unendlichem Grundgebiet, Archiv für Elektrotechnik, 1993

6 Fetzer, J.: Die Lösung statischer und quasistationärer elektromagnetischer Feldprobleme mit Hilfe der Kopplung der Methode der finiten Elemente und der Randelementmethode, Fortschritt Berichte VDE, Reihe 21, VDE Verlag, 1992

7 Gupta, R., et. al: A high field magnet Design for a future hadron collider, Applied Superconductivity Conference, ASC-98, Palm Desert, CA, USA, 1998

8 Halbach, K.: A Program for inversion of system analysis and its application to the design of magnets, Proceedings of the International Conference on Magnet Technology (MT2), The Rutherford Laboratory, 1967

9 Halbach, K., Holsinger R.: Poisson user manual, Technical Report, Lawrence Berkeley Laboratory, Berkeley, 1972

10 Hornsby, J.S.: A computer program for the solution of elliptic partial differential equations, Technical Report 63-7, CERN, 1967

11 Kurz, S., Fetzer, J., Rucker, W.M.: Coupled BEM–FEM methods for 3D field calculations with iron saturation, Proceedings of the First International ROXIE users meeting and workshop, CERN, March 16–18, 1998

12 Rischmüller, V., Fetzer, J., Haas, M., Kurz, S., Rucker, W.M.: Computational efficient BEM–FEM coupled analysis of 3D nonlinear eddy current problems using domain decomposition, Proceedings of the 8th International IGTE Symposium, Graz, Austria, 1998

13 Saad, Y., Schultz, M.H.: GMRES: A generalized minimal residual algorithm for solving nonsymmetric linear systems, SIAM Journal of Scientific Statistical Computing, 1986

14 Senis. R. et al.: Preliminary test results on the Impact of the Beam Screen on the Field Harmonics, LHC-MTA-IN-2001-171, CERN, 2001

15 Van der Vorst, H.A.: Bi-CGSTAB: a fast and smoothly converging variant of Bi-CG for the solution of nonsymmetric linear systems, SIAM Journal of Scientific Statistical Computing, 1992

16 Winslow A.A.: Numerical solution of the quasi-linear Poisson equation in a nonuniform triangular mesh, Journal of Computational Physics, 1971

16
Superconductor Magnetization

The axiomatic method has many advantages over honest work.

Bertrand Russell (1872–1970).

Soon after the first liquification of helium by Kamerlingh Onnes[1] in 1908, Holst[2] measured the vanishing electrical resistance in mercury at a temperature below 4.15 K. The sudden drop in resistivity from values of some tenths of a microohm to immeasurably low values marks the *critical temperature* T_c. In pure materials the transition is usually abrupt, in the range of 0.01 K, and T_c is thus well defined. For alloys and intermetallic compounds, the critical temperature is defined as the point where the resistivity has dropped to half its extrapolated *residual resistivity*; see Figure 16.2 (left). Because T_c is influenced by the current applied to the specimen, its measurements rely on a change of magnetic permeability or the discontinuity in the material's specific heat capacity [25]. The critical temperatures of some superconducting materials are given in Table 16.1.

In 1933, Meißner and Ochsenfeld[3] discovered that superconductors cooled below T_c expel an applied magnetic flux density; they show a perfect diamagnetic behavior below a critical value B_c. In *Type I superconductors*, such as lead, mercury, and aluminum, shielding currents flow in a very thin layer on the surface of the wire. The thickness λ of this layer, called the *penetration depth*, is typically in the range of some tenths of nanometers. Once the critical field is reached, complete flux penetration occurs and the material becomes normal-conducting.

The diamagnetic property of Type I superconductors cannot be explained by classical Maxwell theory, as a perfect conductor conserves the flux rather

1 Heike Kamerlingh Onnes (1853–1926).
2 Gilles Holst (1886–1968).
3 Fritz Walther Meißner (1882–1974), Robert Ochsenfeld (1901–1993).

Field Computation for Accelerator Magnets. Stephan Russenschuck

ISBN: 978-3-527-40769-9

Table 16.1 Critical temperature and critical field of superconducting elements (Type I) and the hard superconducting alloys or compounds Nb–Ti and Nb_3Sn; sources [22, 33].

Element	Symbol	Type	Critical temperature $T_{c0}(B=0)$	Critical field $B_c(T=0)$
Unit			K	T
Aluminum	Al	1	1.175	0.011
Mercury	Hg	1	4.160	0.040–0.042
Niobium	Nb	2	8.7–8.9	0.196
Lead	Pb	1	7.22	0.080
Tin	Sn	1	3.74	0.030
Tantalum	Ta	1	4.38	0.086
Titanium	Ti	1	0.39	0.010
Alloy or intermetallic compound	Symbol	Crystal structure	Critical temperature $T_{c0}(B=0)$	Critical field $B_{c20}(T=0)$,
Niobium–titanium	Nb–Ti	A2	9.2	14.5
Niobium-3-tin (binary)	Nb_3Sn	A15	16	24
Niobium-3-tin (ternary)	Nb_3Sn	A15	18	28

than expels it.[4] In 1935, the brothers F. and H. London[5] developed the classical model of superconductivity, which enables zero resistance and perfect diamagnetism by adding

$$\operatorname{curl}\mathbf{J} = -\frac{\mathbf{B}}{\mu_0\lambda^2} \tag{16.1}$$

to the Maxwell equations.

Type I superconductors are ruled out for accelerator magnets, because their critical fields are in the range of only a few tenths of a tesla, while high current densities in an applied field of more than 4–10 T are required.

In the 1950s, Ginzburg[6] and Landau formulated a phenomenological theory that explains the increased critical magnetic field of *Type II superconductors* like niobium–titanium. Above a value B_{c1} the field can penetrate into the superconductor while the screening currents remain smaller than in the case of Type I material. This family of superconductors includes also the high temperature superconductors (HTS), such as the lanthanum-based, cuprate compounds, discovered in the 1980s.

In the *mixed state* between the lower and upper critical fields B_{c1} and B_{c2}, the penetrating flux is concentrated in *fluxoids*, or *Abrikosov vortices*,[7] parallel

4 Perfect diamagnetism requires vanishing resistivity, whereas the converse is not true.

5 Fritz London (1900–1954), Heinz London (1907–1970).

6 Vitaly Ginzburg, born 1916.

7 Alexei Abrikosov, born 1928.

to the applied field and arranged in a regular lattice. Each fluxoid, of a typical size of 0.01 to 0.1 μm, consists of a normal-conducting core penetrated by the magnetic flux,

$$\Phi_0 = \frac{h}{2e} \approx 2.07 \times 10^{-15}\,\mathrm{V\,s}, \tag{16.2}$$

where h is the Planck[8] constant $h = 6.626\ldots \times 10^{-34}$ J s, and e is the elementary charge $e = 1.602\ldots \times 10^{-19}$ C. This flux quantization was predicted in 1957 by Abrikosov's solution of the Ginsburg Landau equations, and was experimentally observed by Eßmann and Träuble [26]; see Figure 16.1 (right).

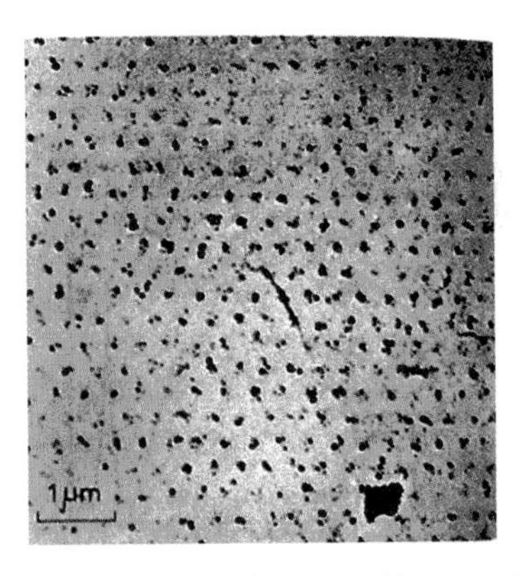

Figure 16.1 Left: Pinning centers in Nb–Ti superconductor; ©1995 IEEE [37]. The white domains are deposits of a titanium-rich phase (the so-called *α phase*). Right: The flux inside the superconductor is concentrated in fluxoids (vortices) arranged in a regular triangular lattice, which was first experimentally observed by Eßmann and Träuble [26] on the surface of a lead–indium rod cooled to 1.1 K.

Type II superconductors are actually not ideal conductors of electric current. When Lorentz forces or thermal activation cause fluxoids to move, one observes a phenomenon called *flux flow resistance*, which results in a resistive transition as shown in Figure 16.2. In addition, a regular lattice of fluxoids, and consequently a uniform flux density, is contradictory to the presence of a transport current, which requires that $\operatorname{curl}\mathbf{H}$ not be zero. The *Kim-Anderson model* accounts for temperature and field dependence of the critical current, by the existence of *flux pinning sites*.

The first commercial superconducting wires with niobium–titanium (Nb–Ti) alloy were developed at Wah Chang, Supercon, and Westinghouse in the early 1960s [6]. Normal-conducting deposits of a titanium-rich phase (the so-called α phase) serve as pinning centers for the fluxoids penetrating the Nb–Ti conductors; see Figure 16.1 (left). The pinning of fluxoids is instrumental in achieving high critical current densities. Superconductors with strong pinning

8 Max Planck (1858–1947).

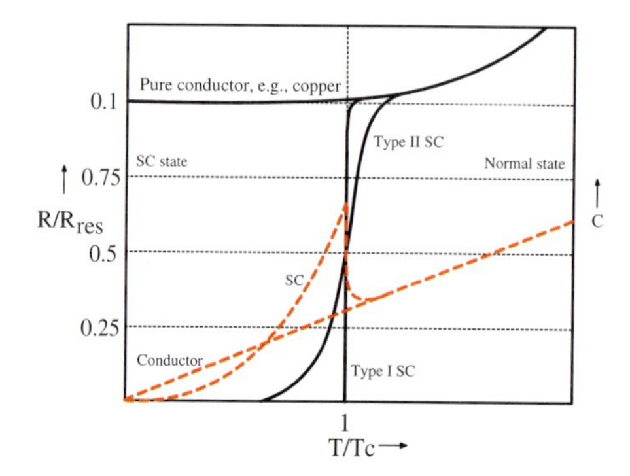

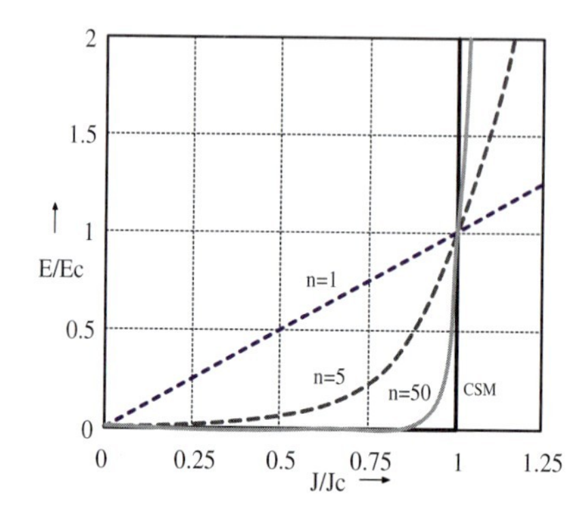

Figure 16.2 Left: The definition of the critical temperature for Type I and Type II superconductors is based on the sudden drop in resistivity. Dashed lines show the heat capacity of SC and normal-conducting material. Right: When the current density approaches J_c, the flux creep effect gives rise to an electric field that varies exponentially with n, called the *quality factor*. Ohm's law ($n = 1$) and the critical state model ($n \to \infty$) are special cases.

are called *hard superconductors*. However, flux pinning is responsible for magnetic hysteresis and thus for multipole field errors that vary with the powering cycle of the magnets.

When the fluxoids are released from the pinning centers by thermal activation, or when a current exceeds some critical value, their motion toward the low-field region induces an electric field. The phenomenological description of hard superconductors is therefore based on an electrical conductor with an $E(J)$ characteristics as shown in Figure 16.2. The $E(J)$ relation can be expressed by the power law [52],

$$\frac{E}{E_c} = \left(\frac{J}{J_c}\right)^n, \tag{16.3}$$

where $E := |\mathbf{E}|$, $J := |\mathbf{J}|$, and J_c and E_c denote the critical current density and the critical electrical field, respectively. From Eq. (16.3) we obtain in vector form

$$\mathbf{E} = E_c \left(\frac{|\mathbf{J}|}{J_c}\right)^{n-1} \frac{\mathbf{J}}{J_c}. \tag{16.4}$$

For current densities close to J_c, the flux creep effect gives rise to an electric field that varies exponentially with n, the quality factor, or *resistive transition index*. This index is as large as 50 for good multifilamentary Nb–Ti wires.[9] Measurements for LHC strands give field-dependent values of $n = 42$ at 10 T and $n = 48$ at 8 T.

9 We use the terms wire and strand synonymously, and when we speak of a conductor we mean a superconducting strand, not a cable of multiple strands.

By definition, the critical current density is reached when the electrical field attains 1 μV cm^{-1}. The nonlinear resistivity $\rho = E/J$ can be calculated from Eq. (16.3) as

$$\rho = \frac{E_c}{J_c}\left(\frac{J}{J_c}\right)^{n-1} . \tag{16.5}$$

Although a considerable effort has been made to justify the power law (16.4) theoretically, it has been validated in numerous experiments. It can therefore be taken as the constitutive equation for the hard superconductor. It also serves for the modeling of field penetration into the specimen by the process of nonlinear diffusion. However, since the resistivity of hard superconductors is nearly a step function, it has been postulated that the current density in a hard superconductor is always either zero or equal to the critical current density. The rule is known as the *critical state model* (CSM) [4]. A time-transient magnetic field induces an electric field at the surface of the conductor, which gives rise to a current density slightly above J_c, so that the resistive voltage matches the electric field. When the field sweep stops, the current in the slab decays until J_c is reached.

The definition of the critical current is not unique in the literature. The IEC standard [32] specifies two critical current I_c criteria for Nb–Ti, defined by electric fields of 1×10^{-5} V m^{-1} and 1×10^{-4} V m^{-1}, which corresponds to $\rho_{c1} = 1 \times 10^{-14}\,\Omega$ m and $\rho_{c2} = 1 \times 10^{-13}\,\Omega$ m, respectively. The n-value shall be calculated using these two I_c criteria. We might alternatively use a less common definition of I_c as the current at which thermal runaway sets in and the wire enters the normal-conducting state. The disadvantage of this lies in the fact that measurements depend on the cooling of the specimen, its cooled surface area, and surface condition.

16.1
Superconductor Magnetization

For a magnetic field applied at the strand surface, a simple way to calculate the superconductor magnetization is to employ Ampère's law, which yields a linear profile of the magnetic field within the strand. From the shielding-current distribution in the strand, it is easy to calculate the magnetic moment and its effect on the field quality in a superconducting magnet. This is the basis of the method described in the following sections. We will also require the following models and programs:

- *A macroscopical model for filament magnetization including hysteresis modeling.* Whereas these cycles are well defined for the LHC main magnets, the corrector magnets will be powered in such a way as to compensate for field errors resulting from different sources in the main magnets, such as decay and snapback, and iron saturation. The semianalytical models presented below allow for the field-dependence of the critical current density. The simple one-dimensional (or scalar) hysteresis model assumes varying field intensity but constant field direction. The two-dimensional (or vector) hysteresis model takes into account arbitrary changes of the field direction in the transverse plane and a magnetization vector not coincident with the direction of the applied field.

- *A combination of these models with numerical field computation for the calculation of the applied field.* In this way, we may account for the iron saturation and other dynamic effects, such as the interstrand coupling current described in Chapter 17. For magnets such as the LHC main dipole, where the inner radius of the iron yoke is large with respect to the outer radius of the coil, saturation effects are low, and therefore the image-current method can be used for the calculation of the persistent currents. However, some of the LHC corrector magnets have iron yokes positioned very close to the coils, and therefore saturation starts at an earlier stage. In Section 16.10.1, we discuss the partial compensation of persistent current effects by means of a ferromagnetic coil-protection sheet or ferromagnetic strips inside the cables. The study of these compensation schemes requires the use of numerical field calculation even at very low excitation levels.
- *An iteration scheme for calculating the feedback of the magnetization affecting the field distribution within the coil.* This is important at very low field levels, where the global shielding effect is relatively large and alters the local distribution of the magnetic flux density in the coil.

The advantages of this modeling approach are:

- *No meshing of the superconducting coil is needed.* The numerical treatment of the diffusion process would require the meshing of the coil down to the filament level, which is not feasible for multifilamentary wires comprising filament diameters in the micrometer range.
- *The model relies only on the measured $J_c(B)$ curve.* This model can be validated with measured multipole field errors in the magnets by adjustment of the empirical parameters used to describe the $J_c(B)$ curve.
- *The effects of different filament diameters can easily be studied.*
- *The magnetization is rate-independent and can therefore be described by the extrema of the excitation field.* Therefore, the field errors at injection (after cycling of the magnet from the nominal field to the preinjection plateau, and a subsequent up-ramp to the injection field level) can be calculated with only a few excitation steps.

The disadvantages are:

- The Meißner currents are not considered.
- The $\mathbf{E}(\mathbf{J})$ relation is idealized as a step function; consequently there is no model for dissipative effects due to flux creep.
- Explicit analytical results can be obtained only for superconductors of simple shapes such as slabs and rods of circular or elliptic cross section.

16.2 Critical Surface Modeling

In the introduction to this chapter we have identified three properties that affect the selection of superconductor material:

- *The critical temperature T_c, determined by the chemical composition and crystal structure of the material.* The critical temperature varies only by a few tenths of a kelvin in different metallurgical treatments. T_c is measured by the change of resistance, magnetic permeability, or heat capacity of the material.
- *The critical magnetic flux densities B_c and B_{c20} for Type I and Type II superconductors, respectively.* This is an intrinsic property of the material, depending on the chemical composition, the crystal structure, and atomic-scale characteristics [25]. The critical fields are usually determined by magnetization measurements.
- *The critical current density J_c is an extrinsic property, depending on metallurgical defects created by cold working and heat treatment.* J_c is measured by establishing the voltage–current characteristics, or indirectly by magnetization measurements.

The magnetization model presented in this chapter relies on the parametric representation of the critical surface $J_c(B,T)$ of the superconducting material as shown in Figure 1.9. We can compare the models to the measurements published in [8] for LHC strand 5 (see Table 1.2) and calculate the root mean square error R_{RMS} as

$$R_{\text{RMS}} := \sqrt{\frac{1}{N}\sum_{n=1}^{N}(y_n - f(x_n))^2}, \tag{16.6}$$

where the y_n are the measured data and $f(x_n)$ are the values estimated from the critical surface model. The physical unit of the error is $1\ \text{kA}\,\text{mm}^{-2}$.

For a constant operation temperature, Kim [35] proposes the empirical relation

$$J_c(B) = \frac{J_0 B_0}{B + B_0}, \tag{16.7}$$

where J_0 is the critical current density at zero field and B_0 is the magnetic flux density at which the critical current density has decreased by half. For the LHC strand 5 cooled to 1.9 K, we find $B_0 = 0.18$ T, $J_0 = 82.9\,\mathrm{kA\,mm^{-2}}$, which yields $R_{\mathrm{RMS}} = 2.1\,\mathrm{kA\,mm^{-2}}$. A modified Kim fit is proposed in [54],

$$J_c(B) = \frac{J_0 B_0}{B + B_0} + A_0 - A_1 B\,, \tag{16.8}$$

which yields $J_0 = 189\,\mathrm{kA\,mm^{-2}}$, $B_0 = 0.0375\,\mathrm{T}$, $A_0 = 8.27\,\mathrm{kA\,mm^{-2}}$ and $A_1 = -0.858\,\mathrm{kA\,mm^{-2}\,T^{-1}}$ for the above-mentioned Nb–Ti strand at 1.9 K. The error is $R_{\mathrm{RMS}} = 0.43\,\mathrm{kA\,mm^{-2}}$.

As input for the magnetization model we use the same relation [12] that was previously employed for the calculation of the working point and the temperature margin:

$$J_c(B,T) = \frac{J_c^{\mathrm{ref}} C_0 B^{\alpha-1}}{(B_{c2})^{\alpha}} (1-b)^{\beta} \left(1 - t^{1.7}\right)^{\gamma}\,, \tag{16.9}$$

where α, β, γ are fit parameters, and the normalized temperature t and field b are defined by

$$t := \frac{T}{T_{c0}} \quad \text{and} \quad b := \frac{B}{B_{c2}(T)}\,. \tag{16.10}$$

The critical field as a function of temperature is scaled as

$$B_{c2} = B_{c20} \left(1 - t^{1.7}\right)\,. \tag{16.11}$$

The upper critical field at zero temperature B_{c20} and the critical temperature T_{c0} are determined by the chemical composition of the superconducting alloy. The critical current density depends on the internal structure, influenced by special metallurgical treatment during the strand manufacture, that is, cold working and heat treatment [31].

The parameters for strand 5 are the critical current density of $J_c^{\mathrm{ref}} = 3 \times 10^9\,\mathrm{A\,m^{-2}}$ at 4.2 K and 5 T; the upper critical field at zero temperature of $B_{c20} = 14.5$ T; the critical temperature at zero flux density of $T_{c0} = 9.2$ K; the normalization constant $C_0 = 27.04$ T; and the fit parameters $\alpha = 0.57$, $\beta = 0.9$ and $\gamma = 2.32$, [12]. The error is $R_{\mathrm{RMS}} = 3.68\,\mathrm{kA\,mm^{-2}}$. The dependence of J_c on the fit parameters C_0, α, and β is shown in Figure 16.3.

A lower R_{RMS} is obtained for the fit proposed in [47]:

$$J_c(B,T) = J_c^{\mathrm{ref}}\, C_0 \left(1 + \alpha_1 e^{-\frac{B}{\delta_1}}\right) \left(1 + \alpha_2 e^{-\frac{B}{\delta_2}}\right) (1-b)^{\beta} \left(1 - t^{1.7}\right)^{\gamma}\,. \tag{16.12}$$

For $C_0 = 3.789$, $\alpha_1 = 2.48$, $\delta_1 = 0.763$ T, $\alpha_2 = 2.61$, $\delta_2 = 0.102$ T, $\beta = 1.384$, and $\gamma = 2.32$, the RMS error is $0.127\,\mathrm{kA\,mm^{-2}}$.

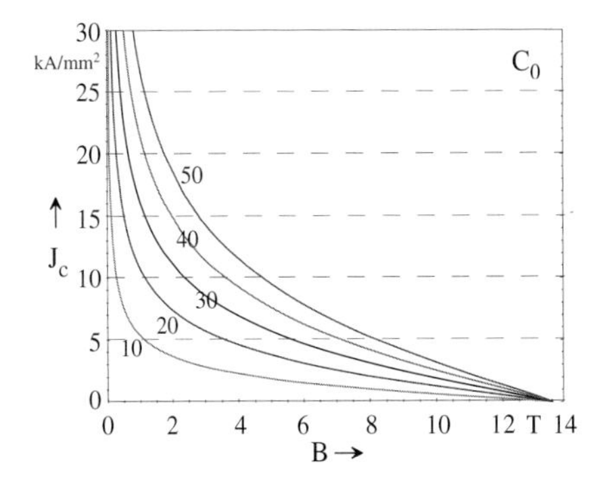

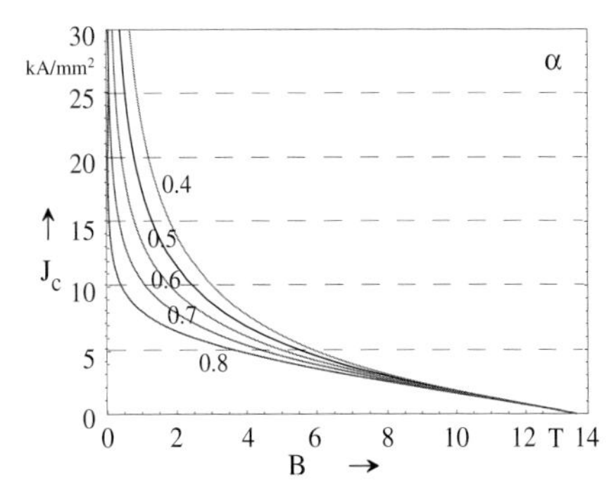

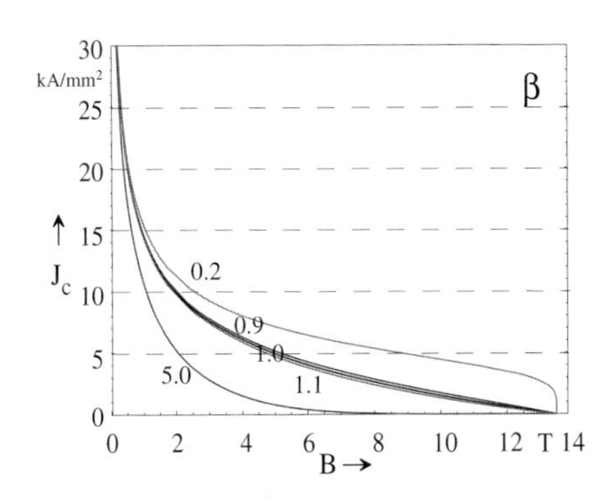

Figure 16.3 $J_c(B)$-dependence as a function of the fit parameters C_0 (top left), α (top right) and β (bottom). Notice the different shape of curve for $\beta < 1$ and $\beta > 1$.

The fit published in [50] has the advantage of remaining finite at $B = 0$:

$$J_c(B,T) = \left(C_1 + C_2 B + C_3 e^{-C_4 B} + C_5 e^{-C_6 B}\right)\left(1 - \frac{T}{T_c(B)}\right), \tag{16.13}$$

where

$$T_c(B) = T_{c0}\left(1 - \frac{B}{B_{c20}}\right)^{\alpha}. \tag{16.14}$$

For $T_{c0} = 9.2$ K, $B_{c20} = 14.5$ T, $\alpha = 0.59$, $C_1 = 10.323$ kA mm^{-2}, $C_2 = -0.7402$ kA mm^{-2}, $C_3 = 28.3793$ kA mm^{-2}, $C_4 = 1.1352$ T^{-1}, $C_5 = 107.019$ kA mm^{-2}, and $C_6 = 9.8275$ T^{-1}, the RMS error is 0.202 kA mm^{-2} for the LHC strand 5.

Multifilamentary Nb_3Sn wires have a strain-dependent critical surface. Summers [49], building on work presented in [23, 30], introduces the following empirical relation:

$$J_c(B,T,\epsilon) = C(\epsilon)\left(B_{c2}(T,\epsilon)\right)^{-\frac{1}{2}}\left(1 - t^2\right)^2 b^{-\frac{1}{2}}(1-b)^2, \tag{16.15}$$

where

$$B_{c2}(T,\epsilon) = B_{c20}(\epsilon)\left(1 - t^2\right)\left(1 - 0.31\,t^2(1 - 1.77\ln t)\right), \tag{16.16}$$

and $\epsilon := \Delta l/l$ is the strain in the specimen of length l. The normalized temperature t and the normalized flux density b are defined as

$$t := \frac{T}{T_{c0}(\epsilon)} \quad \text{and} \quad b := \frac{B}{B_{c2}(T,\epsilon)}. \tag{16.17}$$

The strain-dependent parameters are

$$C(\epsilon) := C_0\left(1 - a|\epsilon|^{1.7}\right)^{\frac{1}{2}}, \tag{16.18}$$

$$B_{c20}(\epsilon) := B_{c20m}\left(1 - a|\epsilon|^{1.7}\right), \tag{16.19}$$

$$T_{c0}(\epsilon) := T_{c0m}\left(1 - a|\epsilon|^{1.7}\right)^{\frac{1}{3}}, \tag{16.20}$$

where the values of B_{c20} and T_{c0} at zero intrinsic strain are denoted B_{c20m} (in the range of 24–28 T) and T_{c0m} (in the range of 16–19 K), respectively. C_0 is a strain-, temperature-, and field-independent parameter in the range of 12 000–40 000 A T$^{1/2}$mm^{-2}. In Eqs. (16.18) and (16.19), the dimensionless parameter a is approximately 900 for compressive strain ($\epsilon < 0$) and 1250 for tensile strain ($\epsilon > 0$). Depending on the manufacturing process, the axial compression in the strands is in the range of −0.005% to −0.4%, [22]. Equation (16.15) can be rewritten as

$$\sqrt{J_c(B,T,\epsilon)\sqrt{B}} = C_K(T,\epsilon)(1 - b), \tag{16.21}$$

where C_K is a coefficient depending on temperature and strain. For a constant strain and temperature, however, $\sqrt{J_c\sqrt{B}}$ will vary linearly in B, and therefore the measured data can be fitted by a straight line in the Kramer plot [36]. In the event that the test facility cannot achieve the flux density at which the current density will be zero, the upper critical field can thus be obtained by linear extrapolation.

Remark: We de-emphasize the choice of $J_c(B,T)$ fit because of the uncertainties in the critical current density measurement. Above a 1 T applied field, the critical current density is usually determined by the resistivity criterion according to the IEC standard [32]. These results are affected by several uncertainties including temperature fluctuations, ohmic heating in the contact resistances to the lead wires, wire movement, filament damage in the test specimen, and uncertainties in the voltage measurements on the microvolt level.

For flux density levels below 1 T, the effect of the self-field becomes important. Moreover, the power supplies are often not capable of delivering the required currents. In this case, magnetization measurements are performed and the critical current density is scaled with the critical state model. Effects of filament coupling via the copper matrix, filament twist, filament deformation, and errors in the direction of the applied field are difficult to calculate and are thus often disregarded.

No matter which fit for the critical surface is applied, it is therefore essential to validate the critical surface modeling by comparing simulated magnetization values with the raw measurement data. □

16.3 The Critical State Model

According to the critical state model (CSM) by Bean [5], a hard superconductor expels a varying applied field by generating a bipolar current distribution of the critical density J_c. This macroscopic model takes into account that the maximum current density in the conductor is directly related to the maximum pinning force per unit volume,

$$\mathbf{F}_{\mathrm{p}} = -\mathbf{J}_c \times \mathbf{B}\,. \tag{16.22}$$

The limitations of the CSM stem from ignoring the Meißner phase and reversible magnetization in Type II material. Moreover, the CSM is based on the idealization of the relation of the electrical field to the current $\mathbf{E}(\mathbf{J})$ as a step function. The CSM is also limited by the fact that the explicit solution of the Maxwell equations is possible only for simple superconductor shapes. Although the critical current density decreases with the field in all real superconductors, the original Bean model assumes a field-independent critical current density in order to simplify the mathematical treatment. Figure 16.4 shows the field and current density distribution in a superconducting slab, infinitely long in the y- and z-directions.

The slab is presumed to be initially unexposed to a magnetic field and thus in a virgin state. An applied field $\mathbf{H}_{\mathrm{a}} = H_{\mathrm{a}}\mathbf{e}_y$ creates a field inside the slab according to Ampère's law,

$$\operatorname{curl}\mathbf{H} = \frac{\partial H_y}{\partial x}\mathbf{e}_z = J_{\mathrm{c}}\mathbf{e}_z = \mathbf{J}_{\mathrm{c}}. \tag{16.23}$$

The slope of the field inside the slab is therefore equal to J_{c}, positive where J_{c} is positive and negative where J_{c} is negative.

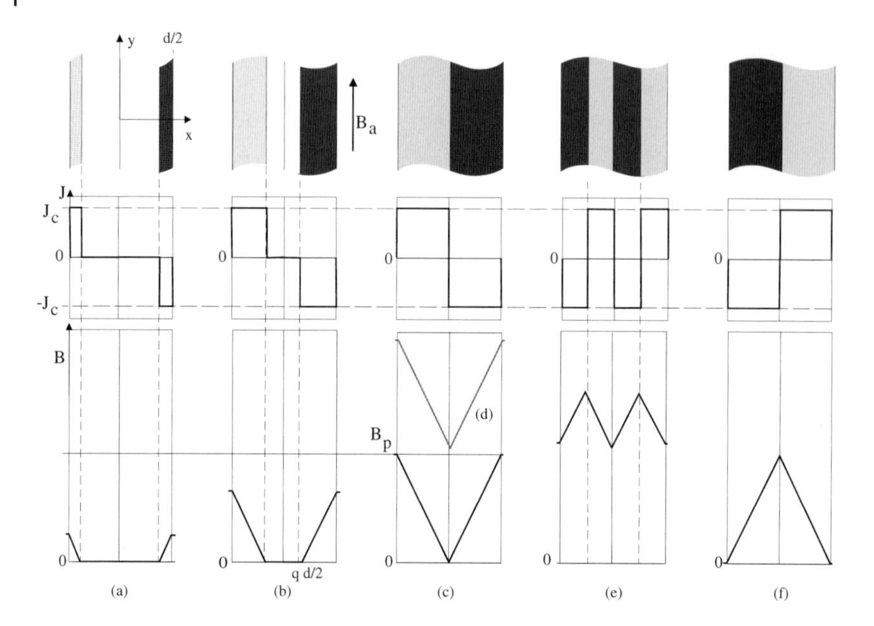

Figure 16.4 The Bean model for an infinitely long slab of superconductor. Notice the field-independent current density in the slab and consequently the constant slope of the penetrating field. (c) is the state of full penetration, (f) shows the wipe-out property.

Let $q \in [0, 1]$ denote the *relative penetration parameter*, which is zero at the surface of the filament and equals 1 at its center. For a given applied flux density $B_\mathrm{a}\mathbf{e}_y$, the field inside the slab is given by

$$B_y(q) = B_\mathrm{a} - t = B_\mathrm{a} - \mu_0 J_\mathrm{c} \frac{d}{2} q, \tag{16.24}$$

where t denotes the modulus of the shielding field. We can write the relative penetration depth as q^* and define it as the value of q at which the interior field has dropped to zero:

$$q^* := \frac{B_\mathrm{a}}{\mu_0 J_\mathrm{c}} \frac{2}{d}, \tag{16.25}$$

where d is the thickness of the slab. At $q^* = 1$ or

$$B_\mathrm{p} = \mu_0 J_\mathrm{c} \frac{d}{2}, \tag{16.26}$$

the entire slab is in the critical state and B_p is called the *penetration field*; see Figure 16.4 (c).

The critical-state model was experimentally confirmed by Coffey [21], who mapped the field distribution in a test sample of Nb–Ti by placing a Hall probe into a small gap. The field maps are presented in Figure 16.5 for different applied fields. It can be seen that the slope of the magnetic field is nearly constant but dependent on the applied field. The occurrence of both flux jumps accompanied by a temperature rise, and flux creep is visible in the field maps.

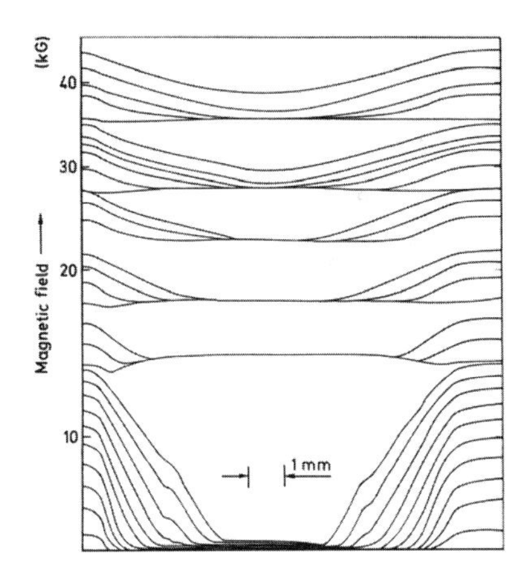

Figure 16.5 Distribution of the magnetic field in a Nb–Ti sample [21]. The field was ramped at 0.3 T/min between the field mappings. Note the deeper penetration of field before flux jumps, and the flux flow after the last two jumps.

When the applied field is reduced after it has reached its maximum B_a^{max}, the polarity of the screening current is reversed and consequently is the direction of the shielding field. The flux penetration is halted and a new shielding-current layer is induced. For $|B_a| < |B_a^{max}|$ the situation is shown in Figure 16.4 (d). The interior boundary of the new current layer propagates inwards as long as the applied field is monotonously decreased. When $|B_a^{max} - B_a| = B_p$, the inner shielding-current layer is completely *wiped out* as shown in Figure 16.4 (e). Thus the distribution of the shielding-current layers depends not only on the applied field but also on the past extremum values B_a^{max}. The $M(B)$ relationship between the superconductor magnetization and the applied field exhibits discrete memory, which is a characteristic of rate-independent hysteresis [10]. Figure 16.6 (left) shows the normalized hysteresis loop $M(B)$ obtained by applying the critical state model to the superconducting slab.

The CSM in its original form fails to model the decreasing magnetization for an applied field that is raised above the penetration field. Nor does it model that the maximum magnetization after a down-ramp does not appear at zero field, but at an applied field of about -0.1 T. This effect is referred to as the *peak-shifting* of the $M(B)$ curve. We will therefore turn to the modified magnetization model by Wilson [53] for round filaments and then present a model based on nested intersecting circles and ellipses.

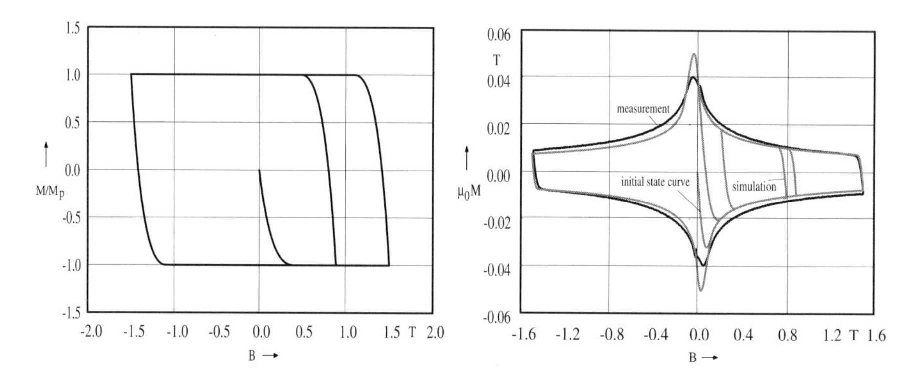

Figure 16.6 Left: Normalized hysteresis loop $M(B)$ obtained by applying the critical state model to the superconducting slab. Right: Computed magnetization curve, applying the intersecting ellipse model, compared to strand measurements [55]. Filament radius $r = 3.5\,\mu\text{m}$, filling factor $\lambda = 1/2.95$, and operation temperature $T = 1.9\,\text{K}$.

16.4
The Ellipse on a Cylinder Model

We recall from Section 8.1.1 that a $\cos\varphi_c$ current distribution in a shell creates an ideal dipole field inside its aperture. Following Wilson [53], a $\cos\varphi_c$ current distribution is approximated by a bipolar current shell with an elliptical inner boundary according to Figure 16.7 (left). The modulus of the current density in the shell is $J_c(B)$, which is spatially constant, but dependent on the applied field. The shell has the semi-major axis a, identical to the filament radius, and the semi-minor axis b. The two area elements $dxdy$ at the locations (x, y) and $(-x, y)$ contribute the quantity

$$dt = -2\frac{\mu_0 J_c}{2\pi\sqrt{x^2+y^2}}\cos\varphi_c\,dxdy \qquad (16.27)$$

to the shielding field in the center of the filament. Employing $\cos\varphi_c = \frac{x}{\sqrt{x^2+y^2}}$ we derive

$$t = -\frac{\mu_0 J_c}{\pi}\int_{-a}^{a}\left(\int_{u(y)}^{v(y)}\frac{x}{x^2+y^2}\,dx\right)dy \qquad (16.28)$$

for

$$u(y) = b\sqrt{1-\frac{y^2}{a^2}} = \frac{b}{a}\sqrt{a^2-y^2}\,, \qquad (16.29)$$

$$v(y) = \sqrt{a^2-y^2}\,. \qquad (16.30)$$

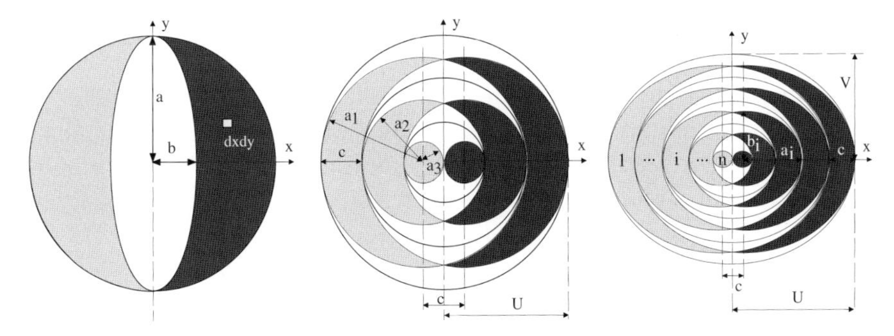

Figure 16.7 Left: Ellipse on a cylinder model. Middle: Geometry of three nested intersecting circles. Right: Geometry of nested intersecting ellipses ($n = 5$). In the semianalytical model $n \to \infty$.

Integrating Eq. (16.28) yields

$$t = -\frac{2\mu_0 J_c a}{\pi}\left(1 - \frac{b}{a}\,\frac{\arcsin\sqrt{1-\frac{b^2}{a^2}}}{\sqrt{1-\frac{b^2}{a^2}}}\right). \tag{16.31}$$

The maximum field that can be shielded from the center of the filament is reached when the current distribution resembles two half-cylinders, a situation that corresponds to $b = 0$. This gives

$$B_p = \frac{2\mu_0 J_c a}{\pi}. \tag{16.32}$$

For the calculation of the resulting magnetic moment m, we derive, from Eq. (5.79),

$$dm = -J_c\, x\, l\, dxdy \tag{16.33}$$

for a 2D area element as shown in Figure 16.7 (left). The symbol l denotes the length of the specimen. Integration over the right-hand current shell yields

$$m = -2J_c l \int_{-a}^{a}\left(\int_{u(y)}^{v(y)} x dx\right) dy = -\frac{4}{3}J_c\left(1 - \frac{b^2}{a^2}\right)a^3 l\,. \tag{16.34}$$

The magnetization, defined as the magnetic moment per unit volume, is obtained by dividing m by the volume of the filament:

$$M = \frac{m}{\pi a^2 l} = -\frac{4}{3\pi}J_c\left(1 - \frac{b^2}{a^2}\right)a\,. \tag{16.35}$$

This magnetization is also referred to as the *irreversible magnetization* in order to distinguish it from magnetization effects in soft Type II superconductors.

The magnetization can thus be reduced only when the conductor is made of a composite of very fine filaments. To avoid flux-jumping in Nb–Ti composite conductors, the filaments must be smaller than 50 μm in diameter. For the stringent conditions placed on field quality in accelerator magnets, the filament diameter is reduced to 6–10 μm. Even smaller filaments are required for the reduction of magnetization losses in magnets for rapid-cycling accelerators.

So far we have disregarded the effect of the transport current. According to the CSM, the transport current flows with critical current density inside the elliptical boundary that is free of shielding currents. Following [42], we can define the average transport current density J_t as the transport current I_t per filament, divided by the total filament cross section area πa^2. The semi-minor axis of the transport current ellipse is then given by the condition $I_t = J_t \pi a^2 = J_c \pi ab$, from which follows

$$\frac{b}{a} = \frac{J_t}{J_c} . \tag{16.36}$$

Including this result into Eq. (16.35) shows that the transport current reduces the magnetization of an otherwise saturated filament by the factor of

$$f = 1 - \left(\frac{J_t}{J_c}\right)^2 . \tag{16.37}$$

Example: Consider the cable for the LHC inner-layer dipole coil consisting of 28 strands of type 1: At injection field level the transport current in the cable is 763 A and therefore 27.25 A per strand. The strand cross-sectional area is 1.29 mm^2. Using the filling factor for the filamentary superconductor $\lambda_{SC} = 0.29$ we obtain $J_t = 72\ \mathrm{A\,mm^{-2}}$, which is small compared with the critical current density of 19 000 $\mathrm{A\,mm^{-2}}$ for Nb–Ti at injection field level. □

The standard Bean model can be refined by taking the field-dependence of the critical current density into account. Inserting Eq. (16.7) into Eq. (16.35) yields the magnetization for the fully penetrated filament ($b = 0$):

$$M_p(B_a) = -\frac{2}{3\pi} J_c(B_a) d = -\frac{2}{3\pi} \frac{J_0 B_0}{B_a + B_0} d . \tag{16.38}$$

Assuming that all the filaments in a strand are at the same level of penetration, the magnetization of an entire strand can be calculated from

$$\mathbf{M}_s = \lambda_{SC} M(B_a) \frac{\mathbf{B}_a}{|B_a|} . \tag{16.39}$$

This magnetization has the opposite direction as the applied field in the coil cross section. In Eq. (16.39), λ_{SC} is the strand filling factor defined by

$$\lambda_{SC} = \frac{1}{1+\eta} , \tag{16.40}$$

where η is the copper-to-superconductor area ratio. It is also possible first to calculate the magnetic moment for each filament and then multiply by the number of filaments in the strand. The number of solid filaments in the strands is given by

$$N_f = \frac{1}{1+\eta} \frac{r_s^2}{r_f^2}, \tag{16.41}$$

where r_s and r_f denote the strand and the filament radii, respectively. For hollow filaments this yields

$$N_f = \frac{1}{1+\eta} \frac{r_s^2}{(r_{fo}^2 - r_{fi}^2)}, \tag{16.42}$$

where r_{fo} is the outer radius and r_{fi} the inner radius of the filament.

Example: Consider the cable for the inner-layer LHC dipole coil made of strand 1: For a filling factor $\lambda_{SC} = 0.29$ and filament diameter $d = 7$ μm, the magnetic polarization $\mu_0 M$ is 10 mT at 0.535 T injection field level, where the critical current density of Nb–Ti is 19 000 A mm^{-2}. □

When the field is raised above the penetration field level, the shielding current distribution is maintained, but the filament is penetrated by the field. Hence M_p decreases proportionally with the critical current density.

Remark: As a high current density is desirable, the only way of reducing the superconductor magnetization is to reduce the filament diameter to the smallest technically feasible value. Figure 1.11 shows a micrograph of a multifilamentary strand measuring 1.065 mm in diameter with 8900 filaments of 7 μm diameter. Since the magnetization decreases with the applied field, the magnetization-induced field errors in the magnet can be reduced by increasing the beam injection energy. The *energy swing*, which is the ratio of nominal to injection beam energy, and therefore of nominal to injection dipole field level, is 15.5 for the LHC. The energy swing is 23 in the HERA, 6.5 in the Tevatron, and 8.6 in the RHIC accelerators. □

16.5 Nested Intersecting Circles and Ellipses

The shielding-current distribution in a filament can also be described as a set of nested intersecting circles or ellipses of field-dependent critical current densities. Figure 16.7 (right) shows the geometry of five nested intersecting ellipses. The semi-minor axis can be calculated as $b_i = a_i V/U$. From Section 8.1.1, we recall that the shielding field of two intersecting ellipses is constant in the aperture and given by

$$t = \mu_0 J_c c \frac{V}{U+V}, \tag{16.43}$$

where U and V are the semiaxes of the filament and c is the displacement of the intersecting ellipses with respect to each other.

Figure 16.7 (middle) shows the geometry of three nested intersecting circles. The thickness c of the current shell is identical for all circles: $c = U/n$, where U is the outer radius of the filament and n is the number of nested circles. The radius of the ith circle is

$$a_i = U - (2i-1)\frac{c}{2} \tag{16.44}$$

for $i = 1, \ldots, n$. The shielding field of each layer is given by

$$t = \frac{\mu_0}{2} J_c c \,. \tag{16.45}$$

To express the dependence of the critical current density on the applied field we use the empirical fit (16.9). For small magnetic flux densities and constant temperature, Eq. (16.9) is proportional to $B^{\alpha-1}$. With $\alpha = 0.57$ the fit can be approximated by

$$J_c(B(q)) \approx J_c(B_a) \frac{\sqrt{B_a}}{\sqrt{B(q)}}, \tag{16.46}$$

where q is again the relative penetration parameter. The uniform dipole field produced by two intersecting circles shifted by the relative distance $\Delta q = q_2 - q_1$ can therefore be expressed as

$$|\Delta t| = \frac{\mu_0 r}{2} \int_{q_1}^{q_2} J_c(B(q))\, dq, \tag{16.47}$$

where r is the filament radius and q_1 and q_2 are the relative penetration parameters corresponding to the inner and outer radii of the shielding current layer. It follows that

$$dB(q) = \xi \mu_0 \frac{r}{2} J_c(B(q))\, dq = \frac{\xi \mu_0 r\, dq}{2\sqrt{B(q)}} \mathcal{F}(B_a), \tag{16.48}$$

where

$$\mathcal{F}(B_a) := J_c(B_a)\sqrt{B_a}\,. \tag{16.49}$$

The polarity flag ξ in Eq. (16.48) is -1 in the case of ramping up and $\xi = 1$ for ramping down. During up-ramp, the orientation of the magnetic moment opposes the orientation of the applied field B_a. Equation (16.48) is a differential equation for $B(q)$, which can be solved in a closed analytical form:

$$B(q) = \left(B_\mathrm{a}^{3/2} + \frac{3}{4}\,\xi\, r\, \mathcal{F}(B_\mathrm{a}) \mu_0\, q \right)^{2/3} , \tag{16.50}$$

where the boundary condition $B(q = 0) = B_\mathrm{a}$ has been taken into account. Figure 16.8 visualizes Eq. (16.50) and $J_\mathrm{c}(B(q))$. The magnetic flux density at $q = 0$ equals the applied field B_a. The field distribution for a fully penetrated state is reached after the applied field is increased from negative field values to $B_\mathrm{a} = 0.08\,\mathrm{T}$. At $B(q) = 0$ the critical current density reaches its maximum, resulting in a steep gradient of $B(q)$. The graph of $J_\mathrm{c}(B(q))$ shows the importance of expressing J_c as a function of q rather than assuming a constant value within the filament (dotted line).

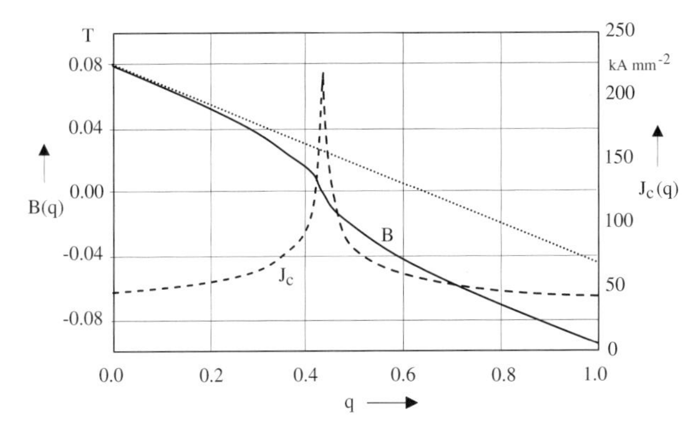

Figure 16.8 Magnetic flux density $B(q)$ as a function of the penetration parameter q (continuous line). The dashed line denotes the current density $J_\mathrm{c}(q)$. The dotted line shows the magnetic flux density for constant current density.

16.6 Hysteresis Modeling

Hysteresis is the phenomenon that causes M to lag behind B, so that the magnetization curves for increasing and decreasing fields are not the same. For M as calculated with the critical state model, the relationship is rate independent; in other words, it is invariant with respect to time scaling. At any instant t, the magnetic moment $m(t)$ depends only on $B([0, t])$ and on the history of the values attained before t. Due to the rate-independence, the output of a hysteresis operator is determined solely by the relative minima and maxima of the input variable in the time interval $[0, \tilde{t}]$, that is, by the sequence of $\{B_a^{max}(t_i)\}$ for $t_i < \tilde{t}$ at which B_a inverted its slope. We must therefore establish the memory sequence $\{q^*(t_i)\}$ for the relative penetration depths of the shielding current layers.

Here we distinguish two cases. In most circumstances it can be assumed that the applied field changes only magnitude and polarity and that the filament magnetization has the same direction as the applied field. Its polarity depends only on the history and on the ramp direction. We refer to this situation as the one-dimensional (or scalar) hysteresis problem.

In the case of nested coils assembled in one common iron yoke but powered from two or more power supplies, the direction of the applied field depends on the powering scheme. Furthermore, the filament magnetization vector will be inclined with respect to the flux density vector at what is known as the *lag angle*. We refer to this situation as the two-dimensional (or vector) hysteresis model [2].

Consider a one-dimensional field sweep of the form $\Delta B = \lambda B$, where $\lambda \in [-1, 1]$ and B is the nominal field in a direction perpendicular to the axis of the filament. This induces a shielding current layer of a relative thickness q^*, as shown in Figure 16.9 (left). The shielding current is modeled by nested intersecting circles carrying opposite current densities and shifted by the relative distance $\Delta q = q_2 - q_1$. In Figure 16.9 these nested pairs of circles have finite thickness, although the mathematical model yields a continuous shielding current and shielding field,

$$\lim_{\Delta q \to 0} \sum_{i=1}^{N} t(\Delta_i q) = \int_0^{q^*} t(q)\, dq\,, \tag{16.51}$$

where $N = \text{int}(q^* / \Delta q)$.

Figure 16.9 (left) shows the cross section of a filament after a ramp of the applied field from 0 to $B_{new} < B_p$. Each of the nested ellipses shields a fraction of the applied field from the center, thus increasing $J_c(B)$ in the inner ellipses. The shielding field $t(q)$ is a function of the flux penetration parameter. At $q = 0$ we obtain $t(0) = 0$ and $B(0) = B_a + 0$. At the inner boundary of the shielding current layer ($q = q^*$) the following continuity condition has to be obeyed:

$$B(q^*) = B_a + t(q^*) = 0\,. \tag{16.52}$$

Figure 16.9 (right) shows the case of a saturated filament ($q^* = 1$).

Figure 16.10 (left) shows the situation in which the applied field is ramped to B_{old} and subsequently reduced to B_{new}. A new layer of shielding currents is generated, leaving the remaining inner layers untouched. The field change is

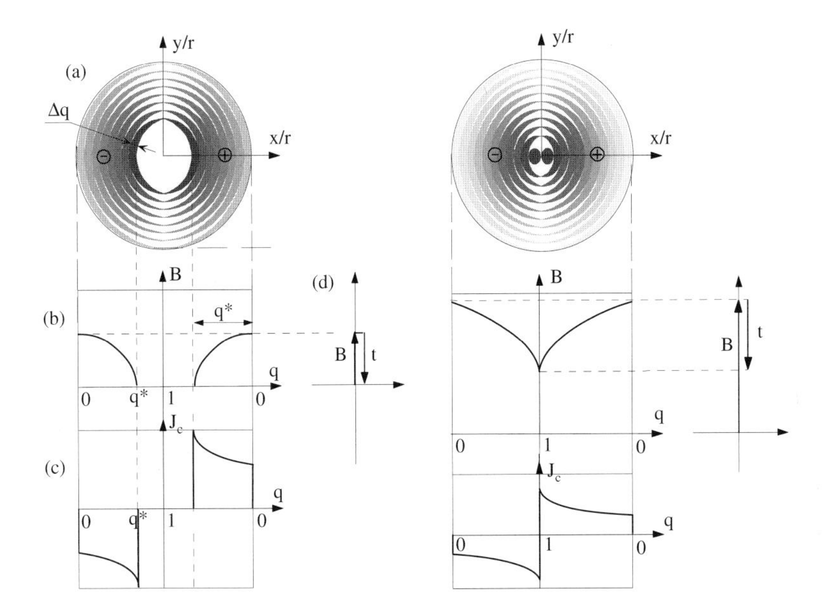

Figure 16.9 Superconducting filament exposed to a magnetic flux density B. Left: Penetration depth $q^* < 1$. (a) Circular filament modeled with layers of intersecting circles. (b) $B(q)$. (c) The shielding currents at critical density $J_c(B(q), T)$ in the filament cross section. Right: Full penetration, $q^* = 1$.

again shielded from the filament's core. The shielding field $t_{new}(q)$ must fulfill the continuity conditions on the outer ($q = 0$) and inner ($q = q^*$) boundaries of the new current layer:

$$B(0) = B_{new} + \underbrace{t_{new}(0)}_{0} = B_{new}, \tag{16.53}$$

$$B(q^*) = B_{new} + t_{new}(q^*) = B_{old} + t_{old}(q^*). \tag{16.54}$$

For given B_{old}, t_{old}, and B_{new}, the problem is reduced to the calculation of a penetration parameter q^* and corresponding shielding field t_{new} that satisfy the continuity conditions (16.53) and (16.54).

From the analytical expression for $B(q)$ inside the filament, the magnetization due to the shielding current $J_c(B(q))$ between q_i and q_{i+1} is derived, and the individual layer magnetizations are recorded in the memory sequence. For minor excitation loops, the magnetization is obtained from the superposition of n shielding current layers:

$$M = \sum_{i=1}^{n} M_i = \sum_{i=1}^{n} \int_{q_i}^{q_{i+1}} m_i(q)\, \mathrm{d}q\,, \tag{16.55}$$

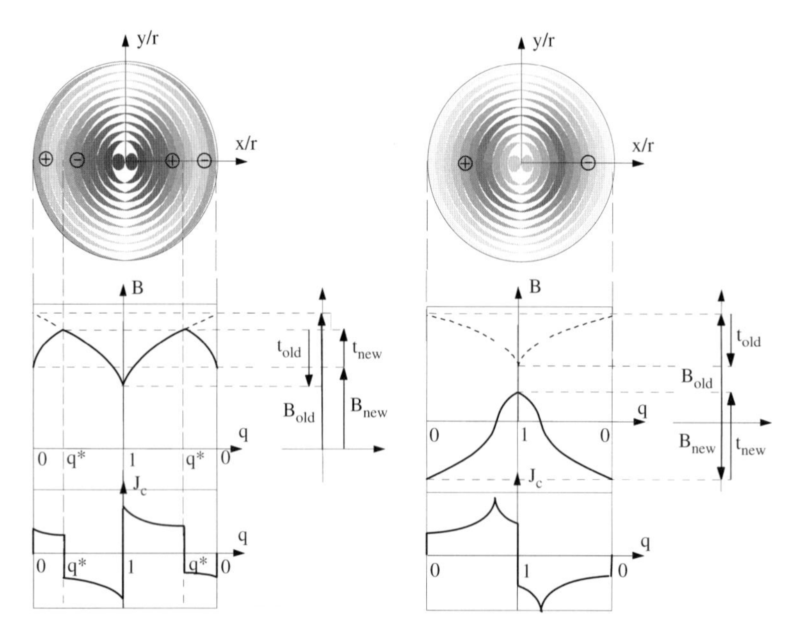

Figure 16.10 Left: The ramping down of the applied field causes the creation of a new current layer of relative thickness q^*. Right: The applied field sweep on the down-ramp exceeds the previously applied field causing complete wipe-out the previous layer ($q^* = 1$).

where $m(q)$is the *magnetization line density*,[10] $[m(q)] = 1\,\mathrm{A\,m^{-2}}$, and M_i is the magnitude of the magnetization for the ith shielding current layer. The magnetization has negative polarity if the orientation is opposite to the applied field:

$$M_i = \frac{4r\xi}{\pi} \int_{q_i}^{q_{i+1}} J_\mathrm{c}(B(q))(1-q)^2\,\mathrm{d}q = \frac{4r\xi\,\mathcal{F}}{\mu_0 \pi} \int_{q_i}^{q_{i+1}} \frac{(1-q)^2}{\sqrt{B(q)}}\,\mathrm{d}q. \tag{16.56}$$

Equation (16.56) can be solved analytically and yields, by substitution of Eq. (16.50) for $B(q)$, a closed expression for the filament magnetization:

$$M_i = \frac{4rB(q)}{5\pi\mathcal{F}^2\mu_0^2(\frac{r}{2})^3}\left[B_\mathrm{a}^3 + \xi\frac{r}{2}\mathcal{F}\mu_0\left(p\xi\frac{r}{2}\mathcal{F}\mu_0 - (q-4)\,B_\mathrm{a}^{3/2}\right)\right]\Bigg|_{q=q_i}^{q=q_{i+1}}, \tag{16.57}$$

where $p := 5 - 4q + \frac{5}{4}q^2$. The results for $B(q)$ and $m(q)$ along the virgin curve are shown in Figure 16.11 (left) for one shielding current layer from $q_1 = 0$ to $q_2(B_\mathrm{a})$. Figure 16.11 (right) shows the same quantities for a memory sequence $\{0, B_\mathrm{max}, -B_\mathrm{max}, B_\mathrm{a} > 0\}$. Since the shielding currents inside the supercon-

10 Not to be confused with the magnetic moment m where $[m] = 1\,\mathrm{A\,m^2}$.

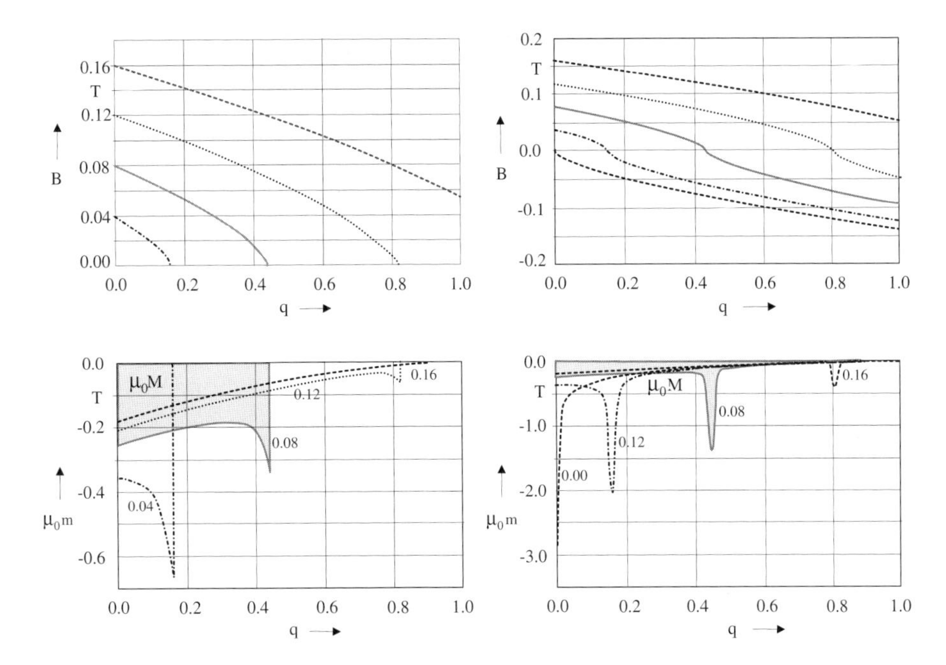

Figure 16.11 Magnetic flux density $B(q)$ and the magnetization line density $\mu_0 m(q)$ as a function of the penetration depth q. Left: For the virgin curve. Right: For the memory sequence $\{0, B_{\max}, -B_{\max}, B_a > 0\}$.

ductor persist, there is a remaining negative field $B(q)$ in the center of the filament.

With Figure 16.11 (lower right) we can also explain why the maximum magnetization does not occur at $B_a = 0$. The magnetization M is given by the integrated area under the $m(q)$ curve, which is largest for small values of $B_a \neq 0$. This characteristic behavior, known as peak-shifting, has been observed in measurements displayed in Figure 16.6.

Let B_{p1} denote the applied field where the virgin curve shows the maximum magnetization; see Figure 16.6. Since the magnetization has been calculated in a closed analytical form, Eq. (16.57), we can calculate the maximum magnetization on the virgin curve by minimizing $\mathrm{d}M(B_a)/\mathrm{d}B_a$. For a ramp from the virgin state there is only one layer of shielding currents from $q_1 = 0$ to

$$q_2 = \frac{4B_a^{3/2}}{\mu_0 \, r \, \mathcal{F}(B_a)} . \tag{16.58}$$

For $B_a = B_{p1}$ it follows that $\mathrm{d}\mathcal{F}(B_a)/\mathrm{d}B_a \approx 0$, and hence

$$B_{p1} \approx \frac{1}{2}\left(\frac{r}{2}\,\mathcal{F}(B_{p1})\,\mu_0\right)^{2/3}(15-5\sqrt{5})^{1/3}\,, \tag{16.59}$$

$$q_2 \approx \sqrt{\frac{5}{6}-\frac{5\sqrt{5}}{18}} \approx 0.46\,. \tag{16.60}$$

The recursive equation (16.59) yields a good estimate for B_{p1} after a few iterations. The maximum magnetization occurs at a penetration depth of $q_2 \approx 0.46$ and not at full penetration; see the shaded area in Figure 16.11 (lower right).

Figure 16.6 (right) shows computations of the filament magnetization according to Eq. (16.57), scaled with the strand filling factor λ_{SC}. This yields an average magnetization per unit volume of the wire, assuming that all filaments within the strand are in the same magnetic state. In Figure 16.6 the virgin curve and several minor hysteresis loops are displayed. The calculated and the measured strand magnetizations for the LHC strands are in good agreement apart from the region where the applied field is close to zero. The assumption that all filaments within one strand are in the same magnetic state is not valid for very small applied fields. In addition, the fit curve for the critical current density, described by Eq. (16.9), has a pole at $B = 0$ and therefore must be constrained to the maximum J_c^{max}.

The field to which the individual filaments are exposed depends on the strand position in the coil. Filaments in the outer layer of the coil are exposed to lower fields, as can be seen in Figure 16.12 (left). The magnitude of the superconducting filament magnetization in the coil cross section is shown in Figure 16.12 (right). Even at nominal excitation there are strands whose filaments are not fully penetrated. Figure 16.13 shows the hysteresis curves for four strands at different locations within the coil cross section. The memory sequence is $\{0,\ B_{max},\ 0.06\,B_{max},\ B_a\}$. The difference in magnetizations for the

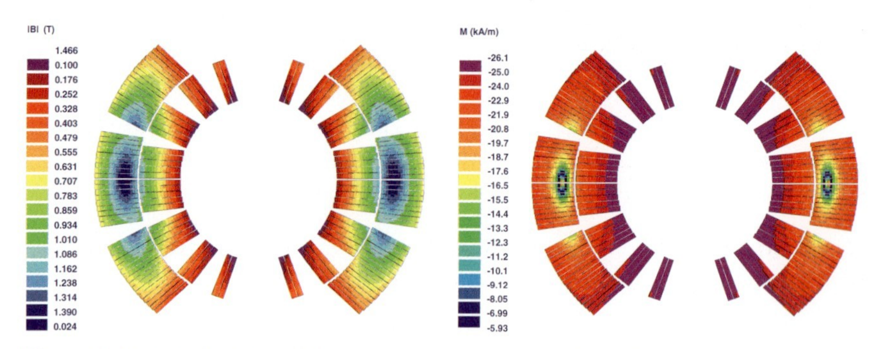

Figure 16.12 Left: Modulus of the magnetic flux density in the coil. Right: Modulus of the superconducting strand magnetization M_s. Both results are given for a ramp from zero to 1.3 T field in the aperture.

same applied field stems from the different filament sizes in the inner- and the outer-layer cables. Moreover, it can clearly be seen that $B_{\max}$ depends on the strand position in the coil.

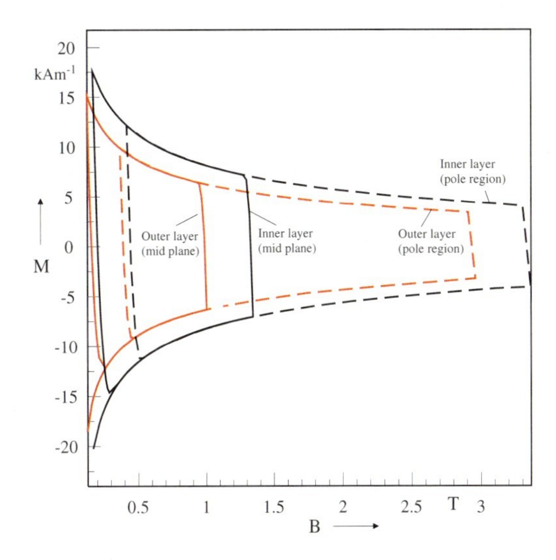

Figure 16.13 Strand magnetization M_s in the coil cross section. Notice the effect of different filament sizes in the inner- and outer-layer dipole cable.

16.7 Magnet Field Errors due to the Superconducting Filament Magnetization

The magnetic moment per unit length of a strand having cross section a is given by $\mathbf{m}/l = a\mathbf{M}$. This can be represented by a small dipole of line-currents of strength $-I_s$ and I_s, spaced by a distance s apart and located perpendicular to the applied field direction. The magnetic moment of such a dipole, $m/l = I_s\, s$, must be equal to the magnetic moment generated by the persistent currents:

$$I_s = \frac{M_s\, a}{s} = \frac{\lambda_{SC}\, M_f\, a}{s}, \tag{16.61}$$

where s can be chosen as the strand diameter. M_s is the average strand magnetization and M_f denotes the filament magnetization.

Example: For a strand surface of a = 1.29 mm^2 and a strand diameter of 1.065 mm, the formula yields 18 A at injection field level, while the transport current is 27 A in each strand. □

A more elegant and numerically precise method is to calculate the multipole harmonics directly from a series expansion of the magnetic vector potential,

$$A_z(\mathbf{r}) = \frac{\mu_0 \mathbf{m}}{2\pi} \times \operatorname{grad}_{\mathbf{r}'} \ln \left(\frac{|\mathbf{r} - \mathbf{r}'|}{r_0} \right), \tag{16.62}$$

at the field point $\mathbf{r}$ generated by a magnetic moment $\mathbf{m}$ at $\mathbf{r}'$; see Figure 16.14. Using the expansion

$$\ln \left(\frac{|\mathbf{r} - \mathbf{r}'|}{r_0} \right) = \ln \left(\frac{r_c}{r_0} \right) - \sum_{n=1}^{\infty} \frac{1}{n} \left(\frac{r_0}{r_c} \right)^n \cos(n(\varphi - \varphi_c)), \tag{16.63}$$

and the representation of the nabla operator in 2D circular coordinates,

$$\nabla = \frac{\partial}{\partial r_c} \mathbf{e}_{\mathbf{r}'} + \frac{1}{r_c} \frac{\partial}{\partial \varphi_c} \mathbf{e}_{\varphi_c}, \tag{16.64}$$

we obtain

$$\operatorname{grad}_{\mathbf{r}'} \ln \left(\frac{|\mathbf{r} - \mathbf{r}'|}{r_0} \right) = \frac{1}{r_c} \left[\left(1 + \sum_{n=1}^{\infty} \left(\frac{r_0}{r_c} \right)^n \cos(n(\varphi - \varphi_c)) \right) \mathbf{e}_{\mathbf{r}'} - \sum_{n=1}^{\infty} \left(\frac{r_0}{r_c} \right)^n \sin(n(\varphi - \varphi_c))\, \mathbf{e}_{\varphi_c} \right]. \tag{16.65}$$

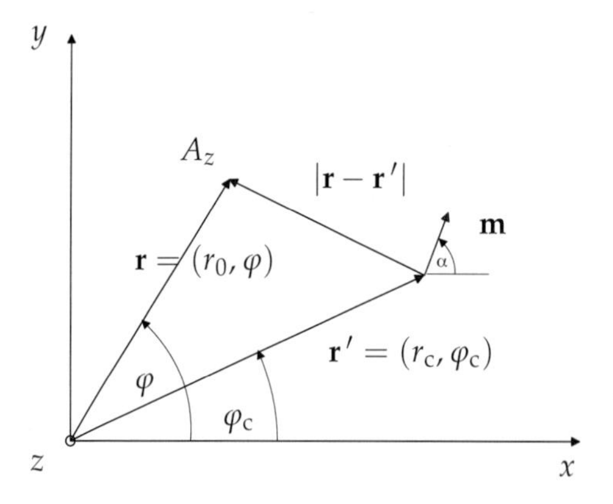

Figure 16.14 Coordinate system for the calculation of the potential of a magnetic moment.

Introducing this result into Eq. (16.62) and calculating the cross product yields

$$A_z = \frac{\mu_0}{2\pi r_c}\left[m_{\mathbf{r}'}\sum_{n=1}^{\infty}\left(\frac{r_0}{r_c}\right)^n \sin(n(\varphi-\varphi_c)) + m_{\varphi_c}\left(1+\sum_{n=1}^{\infty}\left(\frac{r_0}{r_c}\right)^n \cos(n(\varphi-\varphi_c))\right)\right]. \quad (16.66)$$

With $B_r(r_0,\varphi) = \frac{1}{r_0}\frac{\partial A_z}{\partial \varphi}$ and

$$\sin(n\varphi - n\varphi_c) = \sin n\varphi \cos n\varphi_c - \cos n\varphi \sin n\varphi_c\,, \quad (16.67)$$

$$\cos(n\varphi - n\varphi_c) = \cos n\varphi \cos n\varphi_c + \sin n\varphi \sin n\varphi_c\,, \quad (16.68)$$

it follows that

$$B_r(r_0,\varphi) = \frac{\mu_0}{2\pi r_0 r_c}\left[m_{\mathbf{r}'}\sum_{n=1}^{\infty}\left(\frac{r_0}{r_c}\right)^n n(\cos n\varphi \cos n\varphi_c + \sin n\varphi \sin n\varphi_c) - m_{\varphi_c}\sum_{n=1}^{\infty}\left(\frac{r_0}{r_c}\right)^n n(\sin n\varphi \cos n\varphi_c - \cos n\varphi \sin n\varphi_c)\right], \quad (16.69)$$

and therefore

$$B_r(r_0,\varphi) = \frac{\mu_0}{2\pi r_0 r_c}\left[\sum_{n=1}^{\infty} n\left(\frac{r_0}{r_c}\right)^n (m_{\mathbf{r}'}\cos n\varphi_c + m_{\varphi_c}\sin n\varphi_c)\cos n\varphi + \sum_{n=1}^{\infty} n\left(\frac{r_0}{r_c}\right)^n (m_{\mathbf{r}'}\sin n\varphi_c - m_{\varphi_c}\cos n\varphi_c)\sin n\varphi\right]. \quad (16.70)$$

For the multipole coefficients, we finally obtain

$$A_n = \frac{\mu_0 n}{2\pi r_c^2}\left(\frac{r_0}{r_c}\right)^{n-1}(m_{\mathbf{r}'}\cos n\varphi_c + m_{\varphi_c}\sin n\varphi_c)\,, \quad (16.71)$$

$$B_n = \frac{\mu_0 n}{2\pi r_c^2}\left(\frac{r_0}{r_c}\right)^{n-1}(m_{\mathbf{r}'}\sin n\varphi_c - m_{\varphi_c}\cos n\varphi_c)\,. \quad (16.72)$$

The contribution of the strand magnetization to the B_3 field component is displayed in Figure 16.15 (right) and compared to the contribution of the transport current (left). The color patterns show the impossibility of suppressing the persistent-current-induced multipole field error by the coil design. The

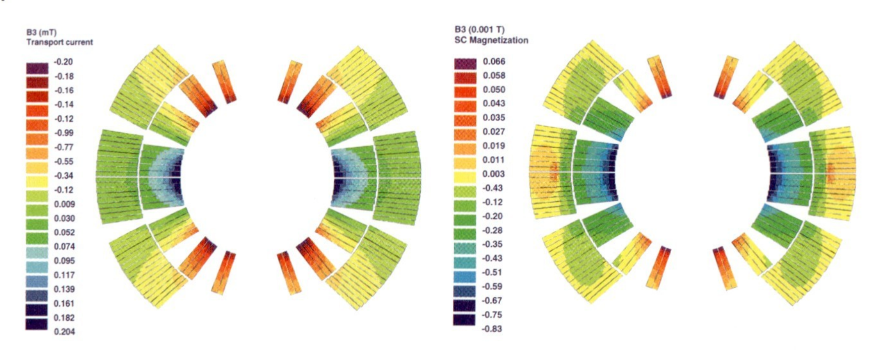

Figure 16.15 Left: Contribution of the transport current to the B_3 field component. Right: Contribution of the strand magnetization to the B_3 field component. Both calculations for 1.3 T field level ramped from zero on the virgin curve.

reason is that the field (and thus the filament magnetization) in the coil itself has no perfect cos $n\varphi_c$-dependence. It is, however, possible to compensate for the multipole field errors by means of passive turns of superconducting cable and ferromagnetic shims.

16.8 The M(B) Iteration

In order to determine the field errors due to persistent currents in a magnet, the global shielding effect of the strand magnetizations must be calculated iteratively. The need for such an iteration becomes apparent from the field plots shown in Figure 16.16.

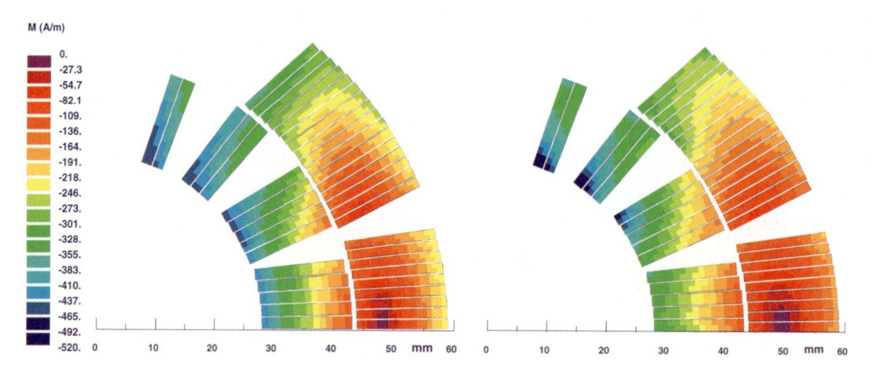

Figure 16.16 Magnetization in the coil of the LHC main dipole calculated without iteration (left) and with iteration (right).

The $M(B)$ iteration scheme is shown in Figure 16.17. For all excitational levels, the source fields in the coil and the reduced field from the iron magne-

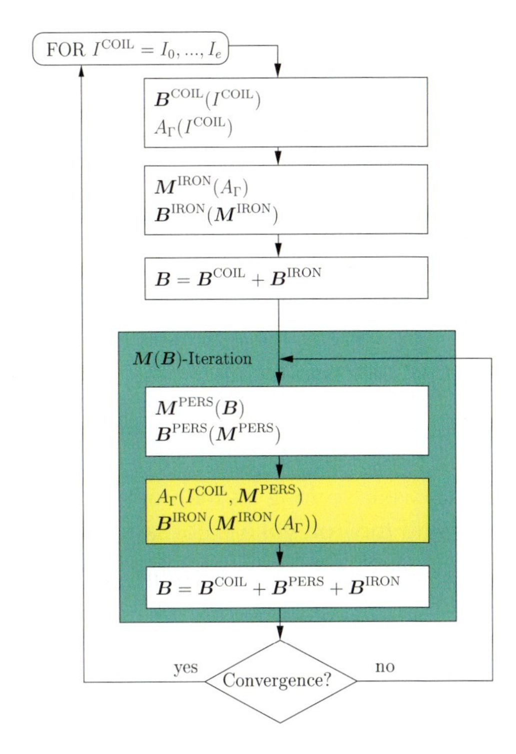

Figure 16.17 Algorithm for the computation of persistent currents with $M(B)$ iteration. Note that if the iron saturation is low, the calculation of the iron magnetization (yellow boxes) can be restricted to the first and last iteration steps.

tization are calculated, the latter by employing BEM–FEM coupling. The vector potential on the BEM–FEM boundary is updated taking into account the influence of the transport and persistent currents. The calculated iron magnetization includes the effect of the "images" of the persistent currents.

In Figure 16.17, I^{COIL} denotes the transport current in the coil and $\mathbf{B}^{\mathrm{COIL}}$ the source field at the ith strand position. A_Γ is the z-component of the magnetic vector potential on the boundary of the BEM and FEM domains. BEM-FEM coupling is used to compute $\mathbf{B}^{\mathrm{IRON}}$, which is the reduced magnetic flux density due to iron magnetization. The magnetic flux density at a strand position results from the superposition of the source (COIL) and reduced (IRON) fields. $\mathbf{M}^{\mathrm{PERS}}$ is the SC filament magnetization. The bipolar persistent currents are computed according to the CSM model. The persistent currents are added to the source currents. Iterations using updated source fields are then performed until convergence is obtained.

16.9
Software Implementation

The number of strands in a magnet, all exposed to different levels of the magnetic flux density, is relatively large. The LHC main dipole, in fact, contains 10 560 strands in all. The nested MCBX corrector magnet with decapole and dodecapole insert contains 3312 strands. We must therefore deal with a large number of function evaluations for the strand magnetization, in the nested loops of $M(B)$ iteration, excitation cycle, and, eventually, an outer loop for a mathematical optimization algorithm. The software implementation must therefore allow for different levels of sophistication in the modeling of superconducting magnets:

- *Calculation of the yoke magnetization using the image-current method, no $M(B)$ iteration*: This method works well, so long as the nonmagnetic collars have a large outer diameter and the iron yoke is not saturated. It is also required that the injection field level be relatively high, so that the global shielding effect in the coil is low.
- *Calculation of the yoke magnetization using the image-current method, iteration of the field solution*: This is necessary when the global shielding effect due to the persistent currents at low field levels influences the local field distribution in the coil.
- *Numerical field calculation for the yoke magnetization, no $M(B)$ iteration*: If the shape of the iron yoke is too irregular to be represented by an ideal cylindrical shape, the iron magnetization must be calculated by means of the BEM–FEM coupling method.
- *Numerical field calculation for the yoke magnetization, linear inner iterations*: For each excitation step a BEM–FEM calculation is carried out to yield the reduced field $\mathbf{B}_\mathrm{r}$ due to the iron magnetization. Because the persistent current effects are in the range of 10^{-3} to 10^{-4} relative to the main field, the iron saturation is assumed to be constant for each excitation step, and $\mathbf{B}_\mathrm{r}$ is kept constant during the iteration.
- *Numerical field calculation for the yoke magnetization, nonlinear inner iterations*: In this case, BEM–FEM calculations are performed at each iteration step. This is necessary for the calculation of the compensation schemes using saturating shims or ferromagnetic coil-protection sheets.

16.10
Applications to Magnet Design

It was explained in Section 16.7 that it is impossible to suppress both the geometric field errors and those due to SC filament magnetization. Still, some

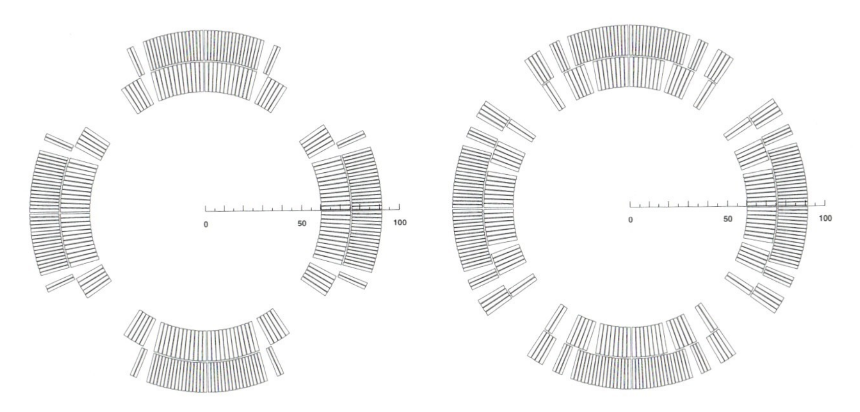

Figure 16.18 Two coil cross sections for a 120-mm-aperture quadrupole, referred to as the 4-block (left) [9] and the 6-block design (right).

differences between coil designs persist. We will discuss two variants of a wide-aperture quadrupole coil, shown in Figure 16.18. The coils are referred to as the 4-block (left) [9] and the 6-block version (right), respectively. The 6-block coil contains a low current-density block in the high-field region of the outer layer, a design feature that was developed for the LHC inner-triplet quadrupoles [44,48].

The relative multipole errors b_6 and b_{10} at a reference radius of 40 mm are given in Figure 16.19 (left). The reason for the lower persistent-current-induced field error in the 6-block coil is that both coil layers individually have

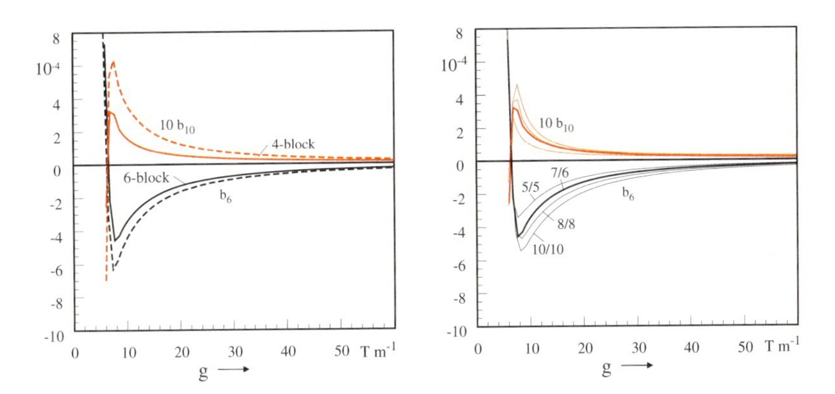

Figure 16.19 Left: Multipole field errors b_6 and b_{10} at the reference radius of 40 mm as a function of the gradient. No iron saturation has been taken into account. The memory sequence is $\{0,\ I_{\text{nom}}/2,\ 0.05 I_{\text{nom}},\ I_{\text{inj.}}\}$. Solid line is the 6-block coil, dashed line is the 4-block coil. Right: Multipole field errors in the 6-block coil as a function of the filament sizes (inner/outer layer).

minimal geometrical field errors. In the 4-block coil, a high dodecapole field error in the inner layer is compensated for by the outer layer. While this works well for the geometrical field errors, persistent current effects do not cancel because of the different field levels and filament sizes in the inner and outer layer cables. A further reduction could be obtained only with smaller filament sizes; this is shown for the 6-block coil in Figure 16.19 (right).

16.10.1 Compensation of Multipole Field Errors

The hysteresis model is now used to calculate the partial compensation of the persistent-current-induced field errors effected by ferromagnetic sheets and shims mounted at different locations in the magnet. Since persistent currents decrease with increasing applied field, the need for compensation is highest at the injection field level. Ferromagnetic sheets have the highest impact on the field quality at low fields, where the permeability is high. As the sheets saturate with increasing field levels, they qualify for persistent current compensations. However, the width of the hysteresis curve, and consequently the losses per magnetization cycle, is not reduced with these passive compensation schemes.

Ferromagnetic coil-protection sheets One possible solution, easy to implement and test, has been found for the LHC main dipoles. This entails replacing some of the *coil-protection sheets* (CPS), usually made from austenitic steel, by ferromagnetic sheets. The coil-protection sheets are placed around the coils in order to protect the coil, the ground plane insulation, and the quench heaters

Figure 16.20 A ferromagnetic coil-protection sheet mounted in the collared coils of an LHC model dipole.

against damage by the surrounding steel collar laminations. The objective is the reduction of the variation in the b_3 and b_5 components versus excitation. One can allow a constant offset of these field harmonics, as they can easily be compensated by a readjustment of the coil layout. A validation of the compensation principle was carried out by recollaring a short dipole model-magnet, as shown in Fig. 16.20.

Figure 16.21 (left) shows the first quadrant of the LHC main dipole coil with the ferromagnetic part of the coil-protection sheet. The angle α of the ferromagnetic magnetic sheet is 52°, fully covering the perimeter of the outer-layer coil.

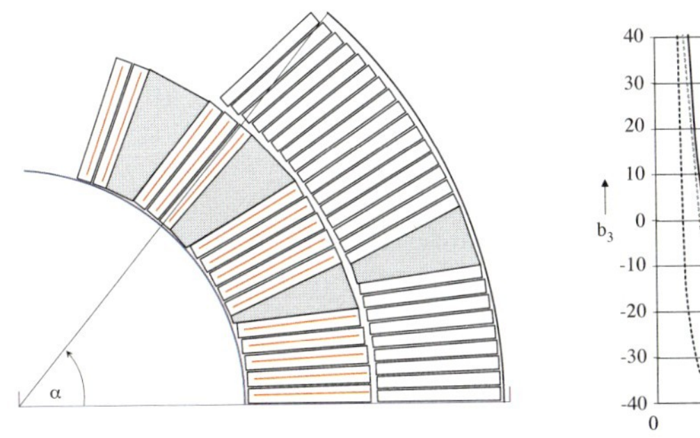

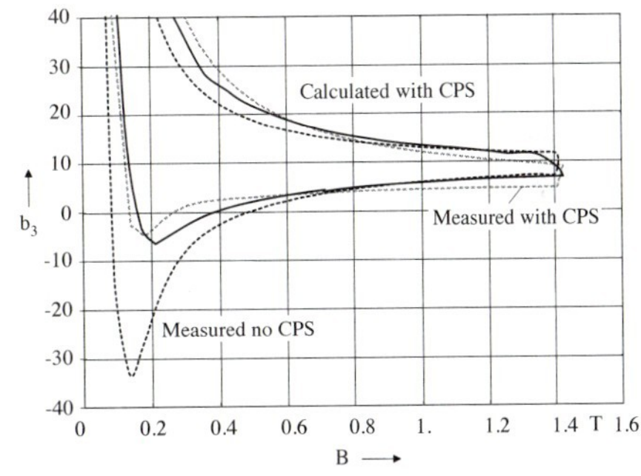

Figure 16.21 Left: Partial compensation of persistent-current-induced field errors using ferromagnetic sheets: Ferromagnetic coil-protection sheet; ferromagnetic sheet on the beam pipe; ferromagnetic strips between the two layers of strands in the Rutherford cable. Right: Multipole b_3 at 17 mm reference radius, compared with measurements of model magnets, with and without ferromagnetic coil-protection sheet.

Figure 16.21 (right) shows the measured and calculated results for the b_3 multipole variation, when the ferromagnetic coil-protection sheet is in place. An improvement in the variation of b_3 conflicts, however, with an increasing variation in the b_5 multipole. Full coverage of the outer-layer coil is therefore not the optimum solution. Different covering angles of the ferromagnetic sheet were investigated, and an angle of approximately 46° found to be the best choice for reduction of both the b_3 and b_5 multipoles. These results are displayed in Figure 16.22.

Ferromagnetic sheet on the cold bore Another effective method for partial compensation of persistent-current-induced multipoles is the addition of a thin ferromagnetic sheet on the outer radius of the cold bore; see Figure 16.20. Table 16.2 shows the calculated values for different thicknesses of such a sheet. Although the b_3 field error is compensated, the b_5 error is increased. Compared to the results obtained with the ferromagnetic coil-protection sheet, the ferromagnetic layer on the cold bore has the advantage that the circular sym-

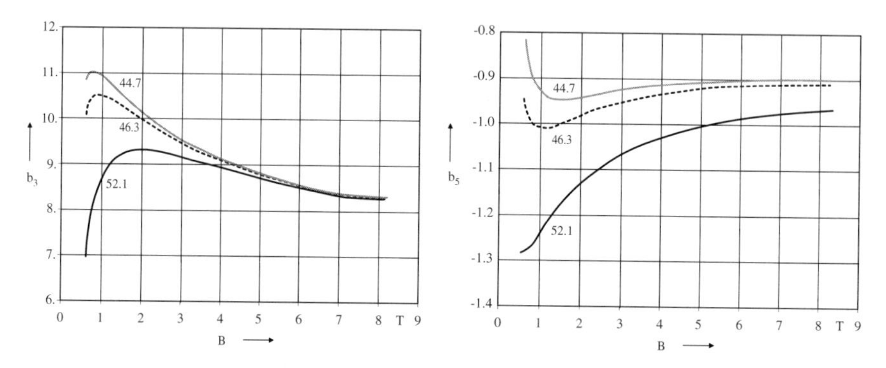

Figure 16.22 Left: The multipole b_3 versus the magnetic flux density as a function of the covering angle of the ferromagnetic portion of the coil-protection sheet installed on the superconducting coil. Right: The multipole b_5 versus the magnetic flux density.

metry of the cold bore can be exploited. In addition, the cold bore is relatively well centered in the magnet aperture, which limits skew multipole errors due to misalignment.

Table 16.2 Variation in the multipole field errors (between injection and nominal field level), including contributions from persistent currents (in units of 10^{-4}, calculated at the reference radius of 17 mm), with and without ferromagnetic sheets of various thicknesses.

	No sheet	Ferromagnetic sheet		
		40 μm	50 μm	60 μm
Δb_2	0.0001	0.034	0.005	0.002
Δb_3	5.721	1.124	0.055	1.187
Δb_4	0.002	0.050	0.018	0.035
Δb_5	0.721	1.337	1.821	2.334
Δb_6	0.000	0.005	0.011	0.011
Δb_7	0.282	0.581	0.788	1.000

Ferromagnetic shims inside the cables The feasibility of fabricating Rutherford cables with internal austenitic steel strips has been demonstrated for the rapid-cycling synchrotron project at GSI [34,54]. Austenitic steel strips reduce cross resistances in the cables and thus reduce losses due to interstrand coupling currents, which are discussed in Chapter 17. It is therefore conceivable to compensate for persistent-current-induced field errors by means of ferromagnetic strips, embedded into the winding structure of the Rutherford cable.

In the case of the LHC main dipole, for example, a 0.1-mm-thick ferromagnetic strip in the inner-layer cable would yield a strong overcompensation of the persistent current effect: an increase of b_3 to 37.4 units at injection field level and to 8.8 units at nominal field level. Ferromagnetic shims inside the cable can therefore be seen as an option for high-field dipoles wound with Nb_3Sn conductors of considerably larger filament diameter, in the range of 10–15 μm.

16.11 Nested Magnets

We will now discuss the more general case of magnets whose field in some conductors changes magnitude and direction. The hysteresis model is said to be two-dimensional and is referred to as the *vector-hysteresis model*. Starting with the situation shown in Figure 16.9 (right), a clockwise rotation and down-ramp of the magnetic flux density induces a new shielding-current layer with a thickness of q^*. The continuity equations (16.53) and (16.54) must still hold for 2D field changes. The situation is displayed in Figure 16.23 (left). The shielding vector $\mathbf{t}_{\text{new}}(q^*)$ points to $\mathbf{B}_{\text{old}} + \mathbf{t}_{\text{old}}(q^*)$. The magnetic flux density in the filament cross section is thus given by

$$\mathbf{B}(q) = \begin{cases} \mathbf{B}_{\text{new}} + \mathbf{t}_{\text{new}}(q), & 0 < q < q^* \\ \mathbf{B}_{\text{old}} + \mathbf{t}_{\text{old}}(q), & q^* < q < 1. \end{cases} \tag{16.73}$$

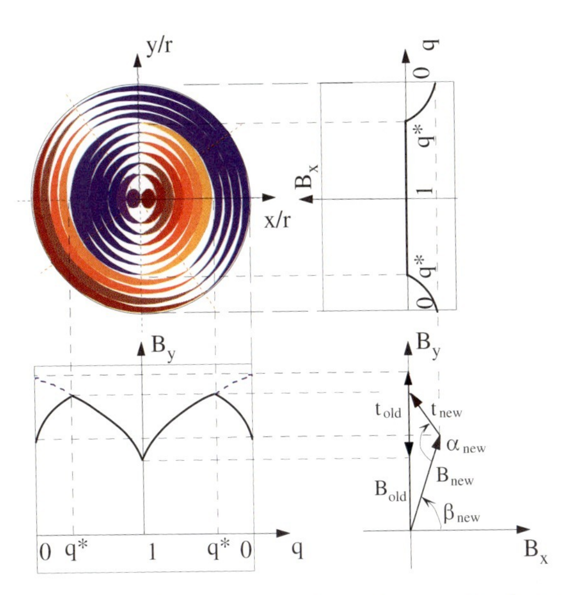

Figure 16.23 Calculation of vector hysteresis using the model of nested intersecting circles. The applied field $\mathbf{B}_{\text{new}}$ is decreased and rotated by $\beta_{\text{new}} - \beta_{\text{old}}$ with respect to the previous excitation step $\mathbf{B}_{\text{old}}$. A new current layer of thickness q^* is thus created.

From Eq. (16.47), we derive a system of differential equations for the field change within the filament:

$$\begin{pmatrix} dB_x(q) \\ dB_y(q) \end{pmatrix} = \frac{\mu_0 \mathcal{F}(B_a) r}{2\sqrt{B(q)}} \begin{pmatrix} -\cos(\alpha-\beta) \\ \sin(\alpha-\beta) \end{pmatrix} dq . \tag{16.74}$$

$\mathcal{F}(B_a)$ is defined in Eq. (16.49), and the angles are shown in Figure 16.23. The solution technique is explained in [2].

Vector-hysteresis modeling must be applied to combined function magnets that feature nested coils assembled in a common iron yoke and are powered by more than one power supply [45]. The direction of the field varies according to the powering scheme, and the filament magnetization vector will be inclined at a *lag angle* with respect to the flux-density vector.

The combined-function corrector MCBX for the LHC inner-triplet region features two nested dipole coils: The outer coil is a horizontal orbit corrector, and the inner coil is a vertical orbit corrector, which is a skew dipole with the field pointing in the horizontal direction. By powering both coils at the same time the direction of the field can be adjusted.

A hypothetical excitation cycle is now considered where the current in the outer coil is ramped to 0.1 times the nominal current and subsequently both coils are powered such that the field rotates with approximately constant magnitude. We will express the field quality in the combined-function magnet in the amplitude-phase notation according to Eq. (6.31). The multipole coefficients F_3 and F_5 are shown as a function of ψ_n in Figure 16.24. The modulus and direction of the superconductor magnetization are plotted in Figure 16.25 (left). The corresponding excitation fields are displayed on the right-hand side of the same figure.

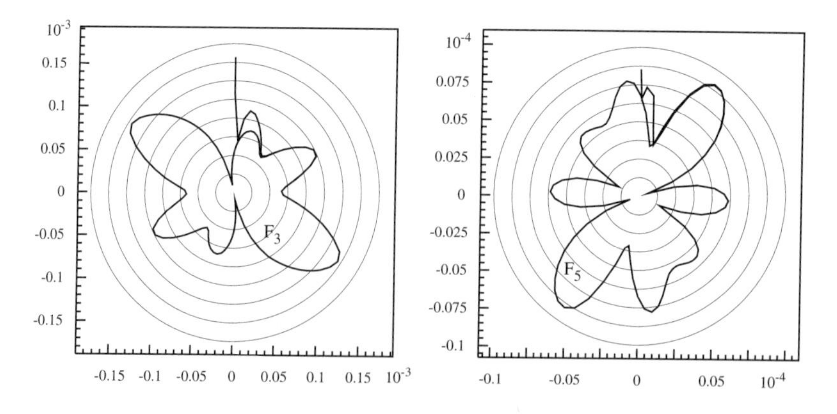

Figure 16.24 Field error $F_n = \sqrt{A_n^2 + B_n^2}$ as a function of the excitational field direction. Left: F_3. Right: F_5.

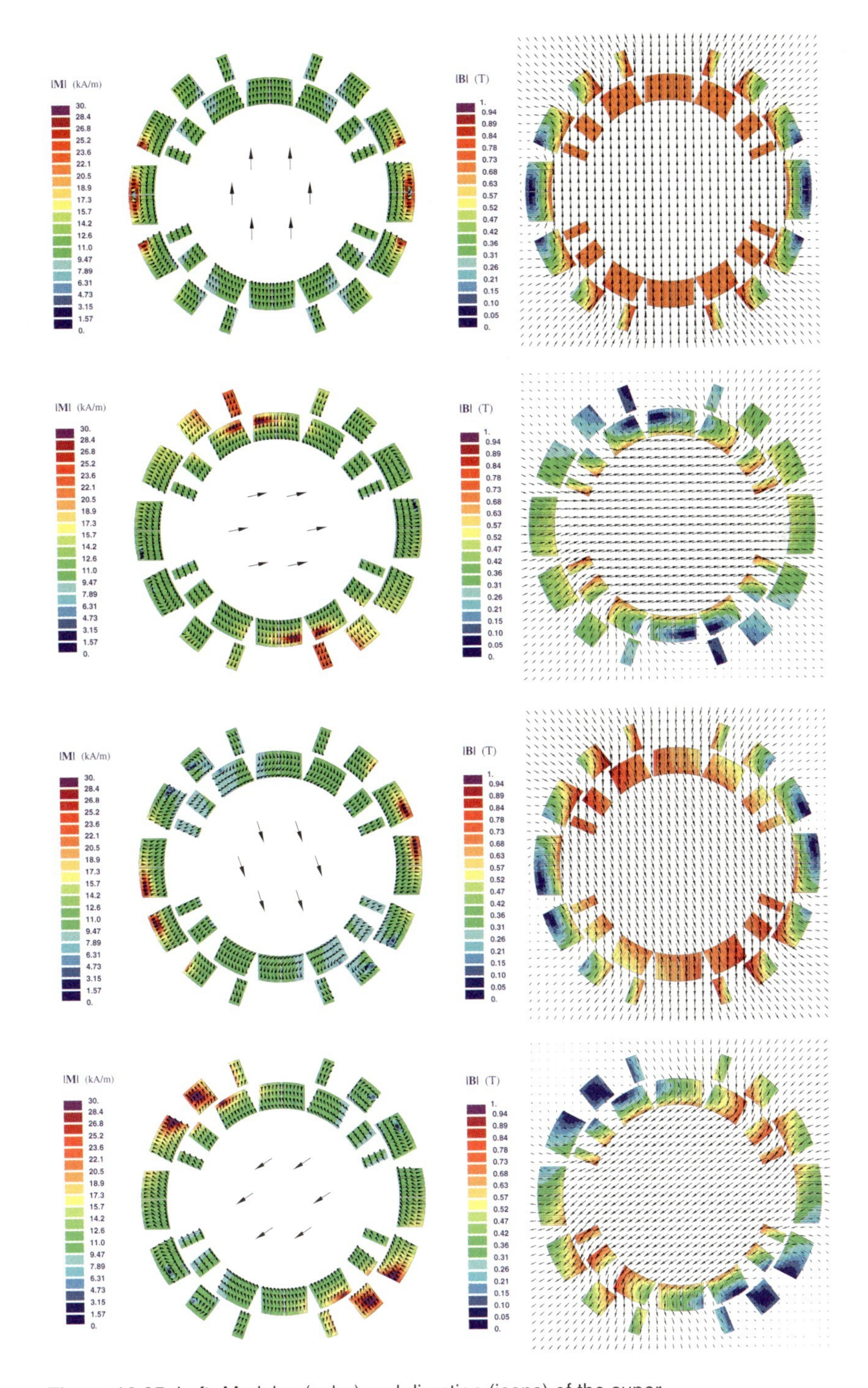

Figure 16.25 Left: Modulus (color) and direction (icons) of the superconductor magnetization. Right: Excitational field direction and flux density in the coils.

References

1 Aleksa, M., Russenschuck, S., Völlinger, C.: Magnetic field calculations including the impact of persistent currents in superconducting filaments, IEEE Transactions on Magnetics, 2002

2 Aleksa, M., Auchmann, B., Russenschuck, S., Völlinger, C.: A vector hysteresis model for superconducting filament magnetization in accelerator magnets, IEEE Transactions on Magnetics, 2004

3 Ang, Z. et al.: Magnetic performance of first low-β dipole corrector prototype, MCBX, Proceedings of the European Particle Accelerator Conference, 1998

4 Bean, C.P.: Magnetization of hard superconductors, Physical Review Letters, 1962

5 Bean, C.P.: Magnetization of high-field superconductors, Review of Modern Physics, 1964

6 Berlincourt, T. G., Hake, R. R.: Superconductivity at high magnetic fields, Physical Review, 1963

7 Beth, R. A.: An integral formula for two-dimensional fields, Journal of Applied Physics, 1967

8 Boutboul, T., Le Naour, S., Leroy, D., Oberli, L., Previtali, V.: Critical current density in superconducting Nb–Ti strands in the 100 mT to 11 T applied field range, IEEE Transactions on Applied Superconductivity, 2006

9 Borgnolutti, F., Fessia, P., Todesco, E.: Electromagnetic Design of the 120 mm Aperture Quadrupole for the LHC Phase One Ugrade, SLHC Project Report 0001, CERN, 2009

10 Bertotti, G., Mayergoyz I.: The Science of Hysteresis I-III, Elsevier, Academic Press, The Netherlands, 2006

11 Bossavit, A.: Numerical modelling of superconductors in three dimensions: a model and a finite element method, IEEE Transactions on Magnetics, 1994

12 Bottura, L.: A practical fit for the critical surface of NbTi, 16th International Conference on Magnet Technology (MT16), Florida, USA, 1999

13 Bottura, L.: $J_c(B, T, \epsilon)$ Parametrizations for the ITER Nb_3Sn Production, CERN-ITER Collaboration Report, 2008

14 Brandt, E.H.: Superconductors of finite thickness in a perpendicular magnetic field: strips and slabs, Physics Review B, 1996

15 Brechna, H.: Superconducting Magnet Systems, Springer, Berlin, 1973

16 Brück, H., Meinke, R., Müller, F., Schmüser, P.: Field Distortions from Persistent Currents in the Superconducting HERA Magnets, Zeitschrift für Physik C, Particles and Fields, 1989

17 Brück, H. et al.: Time dependent field distortions from magnetization currents in the superconducting hera magnets, Cryogenics, 1990

18 Bruzzone, P.: The index n of the voltage–current curve, in the characterization and specification of technical superconductors, Physica C, 2004

19 Carr, W.J.: AC Loss and Macroscopic Theory of Superconductors, Gordon and Breach, London, 1983

20 Clark, A. F., Ekin, J. W.: Defining critical current, IEEE Transactions on Magnetics, 1977

21 Coffey, H.T.: Distribution of magnetic fields and current in Type II superconductors, Cryogenics 7, 1967. Figure reprinted with permission from Elsevier.

22 Devred, A.: Practical Low-Temperature Superconductors for Electromagnets, CERN-Yellow Report, 2004

23 Ekin J.W.: Strain scaling law for flux pinning in practical superconductors. Part 1: Basic relationships and applications to Nb_3Sn conductors, Cryogenics V., 1980

24 Ekin J.W.: Experimental techniques for low temperature measurements, Oxford University Press, Oxford, 2006

25 Ekin, J.W.: Superconductors, in Reed R. P., Clark, A. F. (editors): Materials at Low Temperatures, American Society for Metals, OH, USA, 1983

26 Essmann, U., Träuble, H.: The direct observation of individual flux lines in type

II superconductors, Physics Letters, 24A, 1967. Photograph reprinted with permission from Elsevier.

27 Finley, D.A. et al.: Time dependent chromaticity changes in the Tevatron, Proceedings of the 12th Particle Accelerator Conference, 1987

28 Gosh, A.K., Suenaga, M.: Magnetization and critical current of tin-core multifilamentary Nb_3Sn conductors, IEEE Transactions on Magnetics, 1991

29 Green, M.A.: Residual fields in superconducting dipole and quadrupole magnets, IEEE Transactions on Nuclear Science, 1971

30 Hampshire, D.P., Jones, H., Mitchell, E.W.J.: An in-depth characterization of $(NbTa)_3Sn$ filamentary superconductor, IEEE Transactions on Magnetics, 1985

31 Hillmann H.: Fabrication technology of superconducting material, in Superconductor Materials Science, Plenum, New York, 1981

32 IEC Standard: Superconductivity, part 1: Critical current measurements, DC critical currrent of Nb–Ti composite superconductors, IEC 61788-1, 2006

33 Jensen, J.E., Tuttle, W.A., Stewart, R.B., Brechna, H., Prodell, A.G.: Selected Cryogenic Data Book, Brookhaven National Laboratory, 1980

34 Kaugerts, J.E. et al. Design of a 6 T, 1 T/s fast-ramping synchrotron magnet for GSI's planned SIS 300 accelerator, ASC2004, IEEE Transactions on Applied Superconductivity, 2005

35 Kim, Y.B., Hempstead, C.F., Strnad, A.R.: Critical currents in hard superconductors, Physical Review Letters, 1962

36 Kramer, E.J.: Scaling laws for flux pinning in hard superconductors, Journal of Applied Physics 44, 1973

37 Larbalestier, D.C., Lee P.J.: New developments in niobium titanium superconductors, Proceedings of the 1995 Particle Accelerator Conference. Figure reprinted with permission from IEEE.

38 Le Naour, S., Charifoulline, Z., Wolf, R.: The enhancement of the magnetization of twisted superconducting strands due to the distortion of the filament shape, IEEE Transactions of Applied Superconductivity, 2003

39 Lubell, M.S.: Empirical scaling formulas for critical current and critical field for commerical NbTi, IEEE Transactions on Magnetics, 1983

40 Mathematica is a trademark by Wolfram Research Inc.

41 Maslouh, M., Bouillault, F., Bossavit, A., Verite, J.-C.: From Bean's model to the H–M characteristic of a superconductor: some numerical experiments, IEEE Transactions on Applied Superconductivity, 1997

42 Meß , K.H., Schmüser, P., Wolff S.: Superconducting Accelerator Magnets, World Scientific, Singapore, 1996

43 Minervini, J.: Two-dimensional analysis of AC loss in superconductors carrying transport current, Advances in Cryogenic Engineering, 1982

44 Ostojic, R., Taylor, T.: Conceptual esign of a 70 mm aperture quadrupole for the LHC insertions, IEEE Transactions on Applied Superconductivity, 1992

45 Pekeler, M. et al.: Coupled Persistent-Current Effects in the HERA Dipoles and Beam Pipe Correction Coils, HERA Report 92-06, 1992

46 Schwerg, N., Völlinger, C., Devred, A., Leroy, D.: 2D Magnetic design and optimization of a 88-mm aperture 15 T dipole for NED, Applied Superconductivity Conference 2006, IEEE Transactions on Applied Superconductivity, 2007

47 Schwerg, N.: Determination of a fit function for the critical current density for Nb–Ti cables used in LHC main bending magnets, Technische Universität Berlin, 2005

48 Shintomi, T. et al.: Progress of the LHC Low-b quadrupole magnets for the LHC insertions, IEEE Transactions of Applied Superconductivity, 2001

49 Summers, L.T., Guinan, M.W., Miller, J.R., Hahn, P.A.: A model for the prediction of Nb_3Sn critical current as a function of field, temperature, strain, and radiation damage, IEEE Transactions on Magnetics, 1991

50 Verweij, A.: CUDI Users Manual, CERN, 2007

51 Völlinger, C., Aleksa, M., Russenschuck, S.: Calculation of persistent currents in superconducting magnets, Physics Review ST-AB, 2000

52 Walters, C.R.: Design of Multistrand Conductors for Superconducting Magnet Windings, Brookhaven National Laboratory Report, BNL 18928, 1974

53 Wilson, M.N.: Superconducting Magnets, Oxford Science Publications, Oxford, 1983

54 Wilson, M.N. et al.: Cored Rutherford cables for the GSI fast ramping synchrotron, IEEE Transactions on Applied Superconductivity, 2003

55 Wolf, R.: Persistent currents in LHC magnets, IEEE Transactions on Magnetics, 1992

17
Interstrand Coupling Currents

The coils of superconducting accelerator magnets are often wound from Rutherford type cables, which resemble the Roebel bar known in the domain of electrical machines. Two layers of fully transposed strands limit nonuniform current distributions within the cable, caused by the cable's self-field and the flux linkage between the strands. It was shown in the Rutherford Appleton Laboratory (UK) in the early 1970s that this cable could be produced, without wire or filament breakage, by rolling a hollow, twisted tube of wires into a flat keystoned cable. The cable is permeable to liquid helium, which acts as a heat sink and thus stabilizes the conductor against flux jumps and heat generated by wire motion and beam losses.

The strands contain a large number of filaments of 5-7 micron thickness embedded in a copper matrix. The uninsulated strands are in a resistive contact whose extend and location depend on the transposition-pitch length and the cable compaction. We distinguish adjacent resistances between adjacent strands of the same layer and cross resistances between transposed strands of the upper and lower layers; see Figure 1.14.

In the loops formed along the strands and across the contact resistances, eddy currents are induced by time-transient fields, for example, during the ramping or quenching of the magnets. These so-called *interstrand coupling currents* (ISCCs) are responsible for additional heat losses and magnetic field errors during ramping. During a quench they speed up the propagation of the resistive zone and thus help to protect the magnet against local overheating. This will be elaborated in Chapter 18.

The ISCCs are superimposed on intrastrand eddy currents, which flow in closed loops across the resistive copper matrix. These *interfilament coupling currents* (IFCC) can be calculated analytically and independently as long as the total current stays well below the critical current.

Interstrand coupling currents may be calculated by modeling the Rutherford cable as a linear network, employing analysis methods based on algebraic topology. Applications for the design of superconducting magnets using lumped elements are published in [3, 15, 22, 31]. Networks of distributed

Field Computation for Accelerator Magnets. Stephan Russenschuck

ISBN: 978-3-527-40769-9

elements are applied in [6,21,30]. For a comparison of these models we refer to [2].

Closed-form approximations can be derived if the cross and adjacent resistances are uniform and linear relations for the applied field distributions are assumed [11,12,34,35].

In this chapter we will present a network model weakly coupled to numerical field computation within the CERN field computation program ROXIE. The network of lumped elements is constructed according to geometrical data such as the cable position, number of strands, width, height, length, and transposition pitch.

Spatial periodic boundary conditions can be applied in order to model a long cable exposed to a longitudinally uniform field. This yields a 2D model, where end effects due to cable joints or fringe fields can be disregarded.

Because the total cable length in a magnet can be on the order of several kilometers, mesh currents with very long time constants can be excited, causing unwanted effects such as the decay of persistent currents [19] and periodic patterns of magnetic field components [9].

The coupling of network analysis with numerical field computation allows the study of the interaction between eddy currents, persistent currents, and iron saturation effects. The need for such a coupling will be demonstrated with the calculation of hysteresis effects in the fast-ramping model-magnet for the GSI-FAIR project [23].

17.1 Analysis of Linear Networks

For the calculation of interstrand coupling currents in Rutherford cables, we will employ network-analysis methods based on algebraic topology. We will limit ourselves to networks with a linear relationship between the voltage and current in its branches. Four laws are applied in network analysis:

- *Kirchhoff's current law.* If the current is confined to k thin wires connected to a single node, this law can be written as $\sum_{j=1}^{k} I_j = 0$. The sum of all currents flowing into and out of the node is always zero. The currents are counted positive if their direction is consistent with the positively oriented volume surrounding the node, that is, if they are flowing away from the node.
- *Kirchhoff's voltage law.* For a loop of k branches in a network of lumped elements, this law can be expressed as $\sum_{j=1}^{k} U_j = 0$. For any closed loop the sum of all voltages across its lumped elements is always zero. The voltages are counted positive if their direction is consistent with the orientation of the loop.

- *Ohm's law, which holds for all resistors in the network.*
- *Faraday's law of induction.*

In practical applications, the large number of equations and unknowns requires a rigorous methodology for the selection of linearly independent equations. This can be achieved in the following way:

- Establish the equations for a general linear branch as the elementary building block of the network.
- Study the topology of the network for the selection of linearly independent equations.
- Solve the matrix equations.

Remark: For the voltages we follow the norm CEI/IEC 60375 (see Figure 17.1) and the conventions in field theory, where the voltage U is a directed quantity that is positive if the scalar product $\mathbf{E} \cdot \mathbf{s}$ is positive. The voltage U is by definition directed from the high electric scalar potential (say at point a) to the low electric scalar potential at point b, and therefore $U = \int_a^b \mathbf{E} \cdot \mathrm{d}\mathbf{s} = -\int_a^b \operatorname{grad}\phi \cdot \mathrm{d}\mathbf{s} = \phi(a) - \phi(b)$. Polarity errors may occur from an overinterpretation of the *load convention* used, for example, in [10,13]. According to this load convention, the voltage arrow symbol is directed from the low to the high potential, and $U = \mathbf{E}_c \cdot \mathrm{d}\mathbf{s}$, where $\mathbf{E}_c$ is called the *back EMF*. We use the notation illustrated in Figure 17.1 (a) throughout this book. □

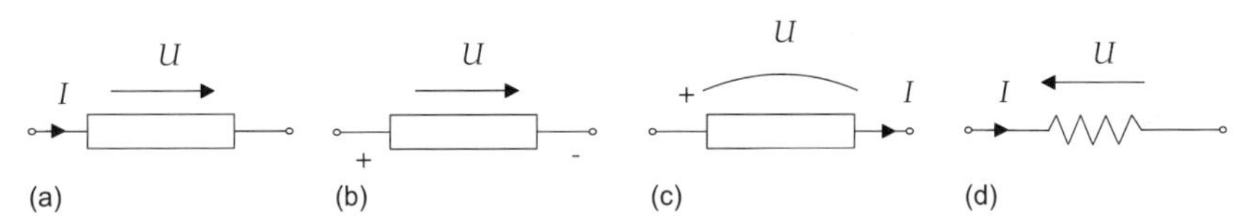

Figure 17.1 (a)–(c): Schematic representation of the reference direction of currents and the reference polarity of voltages in a resistive network according to the norm CEI/IEC 60375. (d): Load convention.

17.1.1 The Linear U(I) Relation in a Branch

The most general linear relation between the voltage U and the current I in a branch is given by

$$U = C_1 + C_2 I \,, \tag{17.1}$$

where C_1 and C_2 are constants. This relation implies that also at $I = 0$ there may exist a terminal voltage between the nodes. The general branch must thus contain a generator. The circuit that delivers the required $U(I)$ relation, known as a *voltage source*, is shown in Figure 17.2 (left). Kirchhoff's voltage law takes the following form:

$$U = U_s + R\,I, \tag{17.2}$$

which fulfills the linear relationship (17.1) for $C_1 = U_s$ and $C_2 = R$. The voltage U_s can be obtained by measuring the voltage at the open terminals, while the resistance R can be determined by measuring the current for the short-circuited branch.

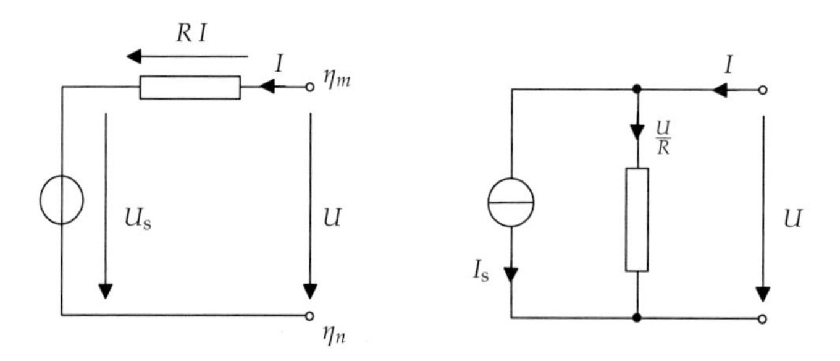

Figure 17.2 Branches with the most general, linear $U(I)$ relationship according to Eq. (17.1). Left: Voltage source. Right: Current source. Note that the load convention is used for the direction of U and I at the terminals.

A voltage source is not the only circuit that fulfills the general linear relation between U and I. An alternative is the *current source* shown in Figure 17.2 (right). Here the linear relationship between voltage and current can be written as $I = C_3 + C_4 U$, and we derive from the Kirchhoff current law:

$$I_s + GU - I = 0\,, \tag{17.3}$$

where the *conductance* is $G := 1/R$. The required relationship is obtained by setting $C_3 = I_s$ and $C_4 = G$. For short-circuited terminals we can measure $I = I_s$ while for the open branch we can measure the voltage:

$$U = U_s = \frac{I_s}{G} = I_s R\,. \tag{17.4}$$

Note that in both the cases we use the load convention, wherein the terminal voltage and current are regarded as positive in the same direction. This is convenient for treating large networks because the directions of both current and voltage can be associated with the same orientation of the branch in the *oriented graph* of the network.

Theorems dating back to the time of Helmholtz, and known as *Norton's* and *Thévenin's theorems*,[1] state that any electrical network of ideal voltage sources,

1 Edward Norton (1898–1983), Charles Thévenin (1857–1926).

ideal current sources, resistors, and one pair of connection terminals (two nodes), may be reduced to a current source or a voltage source with a single resistor only.

17.1.2
The Topology of Networks

Let us consider the network shown in Figure 17.3 (left), made of (superconducting) strands with loops formed by adjacent and cross resistances. For the study of its topology, we will draw the circuit with the oriented branches and their connections while disregarding the physical nature of the individual elements; see Figure 17.3 (right).

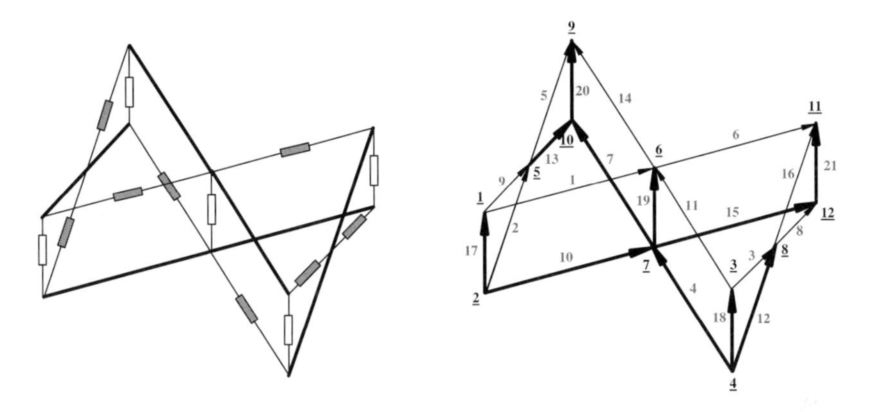

Figure 17.3 Left: Network of meshes formed by strands, adjacent (gray), and cross resistances (white); compare Figure 17.4. Right: The corresponding oriented graph of 21 branches and 12 nodes (node numbers underlined). The 11 branches of the maximal tree are drawn as bold lines. The network has 10 cotree branches (thin lines).

At this point a few formal definitions are in order: A point in the network from which one or more wires lead to different elements is called a *node* or a *vertex*. A *branch* is the oriented connection between two nodes. There are 12 nodes and 21 branches in the network shown in Figure 17.3. Branch γ_2 extends from node η_2 to node η_5, which may be expressed employing the discrete boundary operator ∂ as

$$\partial\gamma_2 = \eta_5 - \eta_2 \,. \tag{17.5}$$

A succession of branches is a *path*; its representation by a vector with coefficients indexed by branches is called a *1-chain*. A directed path determines an integer for each branch: +1 when the branch is traversed once from the start to the end point and −1 for the opposite direction. The path is obviously closed when the first and last points coincide, $\partial(\gamma_1 + \gamma_2 + \cdots + \gamma_B) = 0$ for a path of B branches. If all the branches of a path are traversed only once, the path is said to be *simple.* A simple closed path is denoted a *mesh,* for example, $\zeta = \gamma_2 - \gamma_9 - \gamma_{17}$ in Figure 17.3. A *one-dimensional complex,* also called an *oriented graph,*[2] is a collection of two sets: the set of nodes (0-cells) and branches (1-cells) together with the assignment of two distinct nodes to each branch. This definition includes the trivial case with one node and no branches but excludes branches having points in common other than their boundary points. A graph is *connected* if all nodes are connected to each other via a succession of branches. A connected graph containing no meshes is called a *tree;* a tree containing all nodes of the network is a *maximal tree.* A maximal tree for the simple network is shown in Figure 17.3 (right). The *cotree* is the set of branches not included in a chosen tree. The branches in a cotree are also referred to as *links* or *chords.* Every connected graph has a maximal tree. The procedure for finding a maximal tree consists of repeatedly cutting out branches from the meshes in a graph. A tree of N nodes has $N - 1$ tree branches. In a connected graph of N nodes and B branches there are exactly $M = B - (N - 1) = B - N + 1$ meshes.

The unknowns in a network are the branch currents and the branch voltages. In addition to the constitutive equation in the branches, we need an additional set of linearly independent equations derived from Kirchhoff's current or voltage laws. Maxwell devises two methods to accomplish this: the node-potential and the mesh-current methods. We will describe both methods in the next sections.

17.1.3
The Branch/Node Incidence Matrix and the Node-Potential Method

The *node-potential method* is based on Kirchhoff's current law, with the node potentials as the unknowns. The branch currents are expressed by means of the constitutive equations as functions of the potential difference between the adjacent nodes. From Kirchhoff's current law, we obtain the same number of equations as there are unknown variables.

2 The origin of the graph theory dates back to Euler's famous Königsberg bridges problem.

A graph composed of N nodes and B oriented branches can be topologically described by the complete branch/node *incidence matrix* $[B] := (b_{ij}) \in \mathbb{R}^{N \times B}$, with elements[3]

$$b_{ij} := \begin{cases} 1 & \text{if branch } \gamma_j \text{ exits from the node } \eta_i, \\ -1 & \text{if branch } \gamma_j \text{ enters into the node } \eta_i, \\ 0 & \text{otherwise.} \end{cases} \tag{17.6}$$

As an example, the branch/node incidence matrix for the graph of 21 branches and 12 nodes, shown in Figure 17.3, is given by

$$[B] = \begin{pmatrix} 1 & 0 & 0 & 0 & 0 & 0 & 0 & 0 & 1 & 0 & 0 & 0 & 0 & 0 & 0 & 0 & -1 & 0 & 0 & 0 & 0 \\ 0 & 1 & 0 & 0 & 0 & 0 & 0 & 0 & 0 & 1 & 0 & 0 & 0 & 0 & 0 & 0 & 1 & 0 & 0 & 0 & 0 \\ 0 & 0 & 1 & 0 & 0 & 0 & 0 & 0 & 0 & 0 & 1 & 0 & 0 & 0 & 0 & 0 & 0 & -1 & 0 & 0 & 0 \\ 0 & 0 & 0 & 1 & 0 & 0 & 0 & 0 & 0 & 0 & 0 & 1 & 0 & 0 & 0 & 0 & 0 & 1 & 0 & 0 & 0 \\ 0 & -1 & 0 & 0 & 1 & 0 & 0 & 0 & -1 & 0 & 0 & 0 & 1 & 0 & 0 & 0 & 0 & 0 & 0 & 0 & 0 \\ -1 & 0 & 0 & 0 & 0 & 1 & 0 & 0 & 0 & 0 & -1 & 0 & 0 & 1 & 0 & 0 & 0 & 0 & -1 & 0 & 0 \\ 0 & 0 & 0 & -1 & 0 & 0 & 1 & 0 & 0 & -1 & 0 & 0 & 0 & 0 & 1 & 0 & 0 & 0 & 1 & 0 & 0 \\ 0 & 0 & -1 & 0 & 0 & 0 & 0 & 1 & 0 & 0 & 0 & -1 & 0 & 0 & 0 & 1 & 0 & 0 & 0 & 0 & 0 \\ 0 & 0 & 0 & 0 & -1 & 0 & 0 & 0 & 0 & 0 & 0 & 0 & 0 & -1 & 0 & 0 & 0 & 0 & 0 & -1 & 0 \\ 0 & 0 & 0 & 0 & 0 & 0 & -1 & 0 & 0 & 0 & 0 & 0 & -1 & 0 & 0 & 0 & 0 & 0 & 0 & 1 & 0 \\ 0 & 0 & 0 & 0 & 0 & -1 & 0 & 0 & 0 & 0 & 0 & 0 & 0 & 0 & 0 & -1 & 0 & 0 & 0 & 0 & -1 \\ 0 & 0 & 0 & 0 & 0 & 0 & 0 & -1 & 0 & 0 & 0 & 0 & 0 & 0 & -1 & 0 & 0 & 0 & 0 & 0 & 1 \end{pmatrix}. \tag{17.7}$$

The sum of each column is always zero because any branch is obviously connected to exactly two nodes. The rank[4] of $[B]$ is $N - 1$, that is, there are $N - 1$ independent node potentials; one potential at the so-called reference node can therefore be set to zero.

Kirchhoff's current law asserts that because charge is conserved, the algebraic sum of the currents entering or leaving a node must be zero. In order to derive a matrix representation of this law, we construct the vector space of 0-chains, denoted C_0, as the set of vectors whose coefficients are indexed by the node numbers.[5] The vector space C_1 of 1-chains contains all vectors of which the coefficients are indexed by the branch numbers. This reflects well the colloquial meaning of the word chain as a set of connected links. Note that dim C_0 is the number of nodes and dim C_1 is the number of branches.

An important observation is that the incidence matrix $[B] \in \mathbb{R}^{N \times B}$ represents, disregarding the sign, the discrete boundary operator $\partial : C_1 \rightarrow C_0$ acting on 1-chains. As an example, consider the 1-chain consisting of branches γ_2 and γ_5 in Figure 17.3, represented by the column vector

$$\{b\} = (0,1,0,0,1,0,0,0,0,0,0,0,0,0,0,0,0,0,0,0,0)^T. \tag{17.8}$$

3 This definition of the branch/node incidence matrix is not unique in the literature. Here we attribute a 1 to the branch if it is consistently oriented with the node defined as a source.

4 The rank of a matrix is the number of independent rows.

5 Strictly speaking, these vectors are associated to the chain by isomorphism between linear spaces and the n-tuples; a formal definition can be found in [5].

Matrix multiplication $-[B]\{b\} = \{n\}$ yields

$$\{n\} = (0, -1, 0, 0, 0, 0, 0, 0, 1, 0, 0, 0)^T, \tag{17.9}$$

which is the vector representation of the boundary 0-chain consisting of nodes η_2 and η_9.

Expressing the current distribution within the network as the coefficient vector of the current 1-chain $\{I\} = (I_1, I_2, \ldots, I_B)^T$, Kirchhoff's current law can be rewritten with the discrete boundary operator as

$$[B]\{I\} = \{0\}. \tag{17.10}$$

For the first coefficient of the right-hand side we find

$$0 = I_1 + I_9 - I_{17}, \tag{17.11}$$

that is, the sum of the outgoing currents minus the sum of the entering currents to node η_1 in Figure 17.3 is zero.

The power dissipated in a branch is given by the product of the branch current and the branch voltage. For the entire network, the losses are thus calculated by $U^j I_j$ (Einstein convention), that is, by the matrix product

$$P = \{U\}^T \{I\}. \tag{17.12}$$

We observe that the coefficient vector of the branch voltages maps the coefficient vector of the current 1-chain to the set of real numbers. The voltages should therefore lie in the dual space to the vector space of 1-chains, $\mathcal{C}_1$. This dual space is called the space of *cochains*[6] and denoted $\mathcal{C}^1$. The adjoint to the boundary operator ∂ will be denoted $\mathrm{d} : \mathcal{C}^0 \to \mathcal{C}^1$ and consequently termed the *coboundary* operator.

Let a branch be labeled as γ_j with boundary points η_1 and η_2 such that $\partial\gamma_j = \eta_2 - \eta_1$. According to Kirchhoff's voltage law there exists a scalar potential ϕ at the nodes. If we associate the potentials at all nodes with a 0-cochain, such that the corresponding coefficient U^j in the branch-voltage vector $\{U\}$ is given by $U^j = \phi(\eta_1) - \phi(\eta_2)$, we find by inspection that

$$\{U\} = [B]^T\{\phi\}, \tag{17.13}$$

where $\{\phi\} \in \mathbb{R}^N$ is the node-potential vector. We can thus interpret the matrix $[B]^T$ as the representation of the discrete coboundary operator.

Using the branch/node incidence matrix $[B]$ and its transpose we can derive the matrix equation for all branch currents from the constitutive equations,

$$\{I\} = [G]\{U\} + \{I_s\}, \tag{17.14}$$

6 Note that $\dim \mathcal{C}_n = \dim \mathcal{C}^n$.

for a network containing only the current sources in the branches. In Eq. (17.14) we use the conductance matrix $[G] := \mathrm{diag}(G_j) \in \mathbb{R}^{B \times B}$ with elements $G_j = I_j / U_j$ and the current source vector $\{I_s\} := (I_{s,1}, \ldots, I_{s,B})^T \in \mathbb{R}^B$. If we employ Eq. (17.10), the linear equation system (17.14) can be solved for ϕ using matrix algebra:

$$[B]\{I\} = [B][G][B]^T\{\phi\} + [B]\{I_s\} = \{0\}, \tag{17.15}$$

which yields the following result:

$$\{\phi\} = -([B][G][B]^T)^{-1}[B]\{I_s\}. \tag{17.16}$$

17.1.4 The Mesh Matrix and the Mesh-Current Method

Maxwell's *mesh-current method* relies on the choice of an independent set of meshes, which determines the branch currents in terms of the mesh currents as unknowns. Applying Kirchhoff's voltage law to each mesh yields one equation per mesh and thus, together with the constitutive equations, as many equations as unknown variables.

In a network of N nodes and B branches we can define the maximal *tree* set of $N-1$ branches. This is a connected graph containing no meshes, but all the nodes of the network. Each cotree set is composed of all the $B-(N-1) = B-N+1$ branches in the graph that do not belong to the tree. Each cotree branch has its ends connected by a set of tree branches, since the maximal tree connects all nodes. Because the set of cotree branches contains no meshes, the connection path is unique. The cotree branch defines a mesh whose current is equal to the cotree branch current.

By closing one cotree branch at a time, a set of independent meshes can be obtained and represented by the *mesh matrix* $[M] := (m_{ij}) \in \mathbb{R}^{(B-N+1)\times B}$, which contains the elements

$$m_{ij} = \begin{cases} 1 & \text{if branch } \gamma_j \text{ belongs to mesh } \zeta_i \text{ with the same orientation,} \\ -1 & \text{if branch } \gamma_j \text{ belongs to mesh } \zeta_i \text{ with the opposite orientation,} \\ 0 & \text{otherwise.} \end{cases} \tag{17.17}$$

A mesh matrix for the graph shown in Figure 17.3 is given by

$$[M] = \begin{pmatrix}
0 & 1 & 0 & 0 & 0 & 0 & -1 & 0 & 0 & -1 & 0 & 0 & 1 & 0 & 0 & 0 & 0 & 0 & 0 & 0 & 0 \\
0 & 0 & 0 & 1 & 0 & 0 & 0 & -1 & 0 & 0 & 0 & -1 & 0 & 0 & 1 & 0 & 0 & 0 & 0 & 0 & 0 \\
0 & 0 & 0 & 0 & 0 & 0 & -1 & 0 & 1 & -1 & 0 & 0 & 1 & 0 & 0 & 0 & 1 & 0 & 0 & 0 & 0 \\
0 & 0 & 0 & -1 & 0 & 0 & 0 & 0 & 0 & 0 & 1 & 0 & 0 & 0 & 0 & 0 & 0 & 1 & -1 & 0 & 0 \\
1 & 0 & 0 & 0 & 0 & 0 & 0 & 0 & 0 & -1 & 0 & 0 & 0 & 0 & 0 & 0 & 1 & 0 & -1 & 0 & 0 \\
0 & 0 & 1 & 0 & 0 & 0 & 0 & 0 & 0 & 0 & 0 & -1 & 0 & 0 & 0 & 0 & 0 & 1 & 0 & 0 & 0 \\
0 & 0 & 0 & 0 & 1 & 0 & 0 & 0 & 0 & 0 & 0 & 0 & -1 & 0 & 0 & 0 & 0 & 0 & 0 & -1 & 0 \\
0 & 0 & 0 & 0 & 0 & 1 & 0 & 0 & 0 & 0 & 0 & 0 & 0 & 0 & -1 & 0 & 0 & 0 & 1 & 0 & -1 \\
0 & 0 & 0 & 0 & 0 & 0 & -1 & 0 & 0 & 0 & 0 & 0 & 0 & 1 & 0 & 0 & 0 & 0 & 1 & -1 & 0 \\
0 & 0 & 0 & 0 & 0 & 0 & 0 & -1 & 0 & 0 & 0 & 0 & 0 & 0 & 0 & 1 & 0 & 0 & 0 & 0 & -1
\end{pmatrix}, \tag{17.18}$$

where the set of 10 cotree branches is given by $\{1,2,3,5,6,8,9,11,14,16\}$ and the set of 11 tree branches by $\{4,7,10,12,13,15,17,18,19,20,21\}$. The first row of the mesh matrix defines the mesh ζ_1 by

$$\zeta_1 = \gamma_2 + \gamma_{13} - \gamma_7 - \gamma_{10} \,. \tag{17.19}$$

An independent set of meshes is guaranteed if each mesh contains one and only one cotree branch; this can be seen in the columns of $[M]$ corresponding to the cotree branches.

Proof. To prove the linear independence of the meshes, let ζ_i denote the mesh that includes the cotree branch γ_i. Then consider a linear combination $\sum_i c_i \zeta_i$ of meshes. As γ_i occurs only in mesh ζ_i with a coefficient of +1, the coefficient of γ_i in the sum must be c_i. Thus $\sum_i c_i \zeta_i = 0$ can only hold if all $c_i = 0$. □

All branch currents can be derived from the matrix equation $\{I\} = [M]^T\{I_\mathrm{M}\}$, where $\{I_\mathrm{M}\} \in \mathbb{R}^{B-N+1}$ is the *mesh-current vector* of currents flowing in the cotree branches. Thus the Kirchhoff laws can be written in the form

$$\{I\} = [M]^T\{I_\mathrm{M}\}, \qquad \{U\} = [B]^T\{\phi\}. \tag{17.20}$$

The constitutive equations for a network containing only voltage sources in noncoupled branches are given in matrix notation as

$$\{U\} = [R]\{I\} + \{U_\mathrm{s}\} = [R][M]^T\{I_\mathrm{M}\} + \{U_\mathrm{s}\}, \tag{17.21}$$

where $[R] := \mathrm{diag}(R_j) \in \mathbb{R}^{B\times B}$. The voltage-source vector is given by $\{U_\mathrm{s}\} := (U_{\mathrm{s},1}, \ldots, U_{\mathrm{s},B})^T \in \mathbb{R}^B$, where $U_{\mathrm{s},j}$ is the source voltage of the branch γ_j directed from node η_m to η_n, as illustrated in Figure 17.2 (left).

Kirchhoff's voltage law states that all branch voltages can be obtained from the node potentials and asserts that the voltage drop in a mesh is zero. From

$$[M]\{U\} = \{0\} \tag{17.22}$$

and Eq. (17.21), we derive the linear equation system,

$$[M][R][M]^T\{I_\mathrm{M}\} + [M]\{U_\mathrm{s}\} = \{0\}\,. \tag{17.23}$$

Its solution in $\{I_\mathrm{M}\} \in \mathbb{R}^{B-N-1}$ is given by

$$\{I_\mathrm{M}\} = -[R_\mathrm{M}]^{-1}\{U_\mathrm{M}\}, \tag{17.24}$$

and in $\{I\} \in \mathbb{R}^B$ by

$$\{I\} = -[M]^T[R_\mathrm{M}]^{-1}\{U_\mathrm{M}\}, \tag{17.25}$$

where we introduced the mesh-resistance matrix and the mesh-voltage vector,

$$[R_{\mathrm{M}}] := [M][R][M]^T, \qquad \{U_{\mathrm{M}}\} = [M]\{U_{\mathrm{s}}\}. \tag{17.26}$$

The diagonal elements of $[R_{\mathrm{M}}] \in \mathbb{R}^{M\times M}$ are the total resistances of each mesh, and the off-diagonal elements are the sum of the common resistances between different meshes. The elements of the mesh-voltage vector $\{U_{\mathrm{M}}\}$ are the sum of the voltage sources in each mesh,

$$U_{\mathrm{M},j} = \frac{\mathrm{d}}{\mathrm{d}t}\int_{\zeta_j} \mathbf{B}_{\mathrm{s}} \cdot \mathrm{d}\mathbf{a} = \frac{\mathrm{d}}{\mathrm{d}t}\int_{\partial\zeta_j} \mathbf{A}_{\mathrm{s}} \cdot \mathrm{d}\mathbf{r}, \tag{17.27}$$

where $\mathbf{A}_{\mathrm{s}}$ is the (source) magnetic vector potential.

Both vector and scalar potentials depend on the choice of the gauge. Applying the node-potential method, the electric potentials are determined up to an additive constant. Hence the potential must be set to zero at one of the nodes. Applying the mesh-current method, the branch voltages are determined up to a gradient field, which lies in the kernel[7] of the mesh matrix $[M]$.

For the entire complex, the losses can be calculated employing the matrix product (17.12), the coboundary operator (17.13), and Kirchhoff's current law (17.10):

$$P = \{U\}^T\{I\} = \left([B]^T\{\phi\}\right)^T\{I\} = \{\phi\}^T[B]\{I\} = 0. \tag{17.28}$$

This result, known as *Tellegen's theorem*,[8] states that the energy is conserved in electrical networks. The importance of this theorem lies in the fact that it contains no reference to the physical nature of the branches.

17.1.5 Transient Field Analysis

For transient field analysis, all the mutual inductances in the network must be taken into account. The constitutive equations written in matrix form yield

$$\{U\} = [R]\{I\} + [L]\frac{\mathrm{d}}{\mathrm{d}t}\{I\} + \{U_{\mathrm{s}}\}, \tag{17.29}$$

where $[L]$ is the matrix of (constant) *partial inductances*, which are the contributions of the branch to the total inductance of the mesh [17]. For the signs in Eq. (17.29) we use the load convention where RI and $L(\mathrm{d}I/\mathrm{d}t)$ have the same direction; see the remark in Section 5.11.1. Using Kirchhoff's voltage law (17.22) and the definitions (17.26), we obtain

$$[R_{\mathrm{M}}]\{I_{\mathrm{M}}\} + [L_{\mathrm{M}}]\frac{\mathrm{d}}{\mathrm{d}t}\{I_{\mathrm{M}}\} + \{U_{\mathrm{M}}\} = 0, \tag{17.30}$$

7 Let $\mathbf{E}' = \mathbf{E} + \operatorname{grad}\phi$. Then $\int_\zeta \mathbf{E}' \cdot \mathrm{d}\mathbf{r} = \int_\zeta \mathbf{E} \cdot \mathrm{d}\mathbf{r} + \int_\zeta \operatorname{grad}\phi \cdot \mathrm{d}\mathbf{r} = U_{\mathrm{M}} + 0$ as the mesh forms a closed loop.

8 Bernhard Tellegen (1900–1990).

where the diagonal elements of $[L_\mathrm{M}] = [M][L][M]^T$ are the self inductances of each mesh and the off-diagonal elements are the mutual inductances between the meshes.

For the numerical integration of Eq. (17.30) in the time domain, we use the trapezoidal approximation for the time derivative:

$$\frac{\mathrm{d}}{\mathrm{d}t}\{I\}(t^*) \approx \frac{1}{h}\left(\{I_{k+1}\} - \{I_k\}\right), \tag{17.31}$$

$$\{I\}(t^*) \approx \frac{1}{2}\left(\{I_{k+1}\} + \{I_k\}\right), \tag{17.32}$$

$$\{U_\mathrm{s}\}(t^*) \approx \frac{1}{2}\left(\{U_{\mathrm{s},k+1}\} + \{U_{\mathrm{s},k}\}\right), \tag{17.33}$$

$$\{U\}(t^*) \approx \frac{1}{2}\left(\{U_{k+1}\} + \{U_k\}\right), \tag{17.34}$$

where

$$t^* = \frac{t_{k+1} + t_k}{2}, \tag{17.35}$$

$t_k = kh$, and h is the size of the time step. The matrix equation (17.29) approximated at t^* becomes

$$\begin{aligned}\{U\}(t^*) &\approx \frac{1}{2}[R]\Big(\{I_k\}+\{I_{k+1}\}\Big)+\frac{1}{h}[L]\Big(\{I_{k+1}\}-\{I_k\}\Big)+\frac{1}{2}\Big(\{U_{\mathrm{s},k}\}+\{U_{\mathrm{s},k+1}\}\Big)\\ &= \frac{1}{2}\left(\Big([R]+\frac{2}{h}[L]\Big)\{I_{k+1}\}+\Big([R]-\frac{2}{h}[L]\Big)\{I_k\}+\{U_{\mathrm{s},k}\}+\{U_{\mathrm{s},k+1}\}\right).\end{aligned} \tag{17.36}$$

Using Eq. (17.20) for the mesh currents and applying Kirchhoff's voltage law yields

$$\begin{aligned}&[M]\left([R]+\frac{2}{h}[L]\right)[M]^T\{I_{\mathrm{M},k+1}\} =\\ &\qquad [M]\left([R]-\frac{2}{h}[L]\right)[M]^T\{I_{\mathrm{M},k}\} - [M]\{U_{\mathrm{s},k}\} - [M]\{U_{\mathrm{s},k+1}\}\end{aligned} \tag{17.37}$$

with the solution

$$\{I_{\mathrm{M},k+1}\} = -[Z_\mathrm{A}]^{-1}\Big([Z_\mathrm{B}]\{I_{\mathrm{M},k}\} + [M]\{U_{\mathrm{s},k}\} + [M]\{U_{\mathrm{s},k+1}\}\Big), \tag{17.38}$$

in which the following impedance matrices are used:

$$[Z_\mathrm{A}] := [M]\left([R]+\frac{2}{h}[L]\right)[M]^T, \quad [Z_\mathrm{B}] := [M]\left([R]-\frac{2}{h}[L]\right)[M]^T. \tag{17.39}$$

17.2 A Network Model for the Interstrand Coupling Currents

The Rutherford cable can be represented as a network of superconducting wires and lumped contact resistances between them. Figure 1.14 shows the model of a cable segment of the length of one cable transposition pitch. As shown in Section 17.1, we may analyze the network with mesh or nodal analysis. Although nodal analysis is easier to implement, the mesh analysis is applied. The main reason for this choice is the zero resistance of strands in the superconducting state, a condition that leads to numerical instabilities in the nodal analysis.

We also employ the *weak excitation* scheme, which means that we assume the currents to be sufficiently low so that the voltage drop across the superconducting strand branch, according to the power law (16.4), can be disregarded.

The network is composed of periodic segments, which we will call *bands*, of length $l_b = p_c/N_s$, where N_s is the number of strands and p_c is the length of the transposition pitch; see Figure 17.4. The topological and geometrical parameters of the cable are given in the top portion of Table 17.1. From these parameters we determine the cable dimensions and the parameters of the graph, as listed in the lower part of Table 17.1.

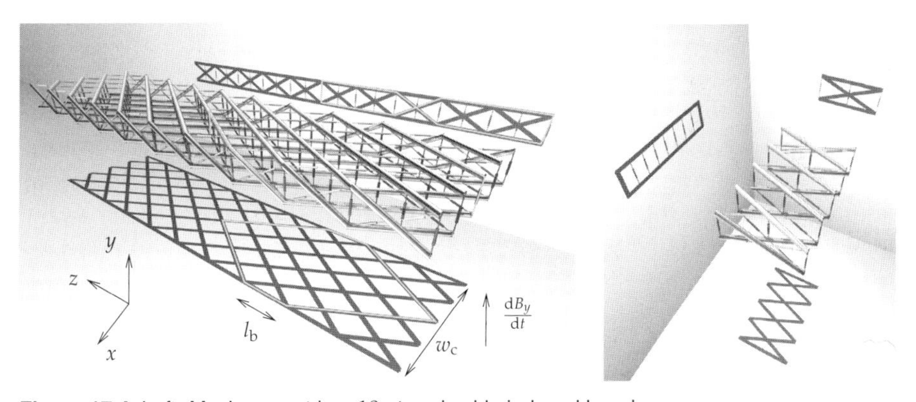

Figure 17.4 Left: Mesh current in a 10-strand cable induced by a homogeneous dB_y/dt. Right: Periodic band with 10 strands.

The contact resistances between strands are determined by the following factors:

- Strand oxidation and the surface treatment of the strands.
- Production lines of strands and cables.
- Different processes in coil fabrication, for example, curing temperatures of the coil varying between 160 and 190 °C.
- Magnet assembly procedures and the resulting prestress in the coil.

Table 17.1 Top: Cable parameters. Bottom: Dependent parameters associated with the graph of the network.

N_s	Number of strands
N_b	Number of bands
p_c	Transposition pitch
h_1, h_2	Cable height on thin and wide sides
w	Cable width
$N = 2N_sN_b + N_s$	Number of nodes
$B_s = 2N_sN_b$	Number of strand branches
$B_a = 2N_sN_b$	Number of adjacent-resistance branches
$B_c = (N_s - 1)N_b + N_s/2$	Number of cross-resistance branches
$B = B_s + B_a + B_c = (5N_s - 1)N_b + N_s/2$	Total number of branches
$M = B - N + 1 = (3N_s - 1)N_b - N_s/2 + 1$	Number of independent meshes
$l_b = p_c/N_s$	Length of a band
$l_c = N_bl_b = p_cN_b/N_s$	Length of the cable

- The size of the contact surface varying in the cable due to the different compaction in a keystoned cable.

A comprehensive discussion of these influences can be found in [31]. We distinguish the cross resistance (R_c) between the crossing strands in the upper and lower layers of the cable from the adjacent resistance (R_a) between the strands in the same layer.

The target value for the cross resistances in the LHC cables was $> 15\ \mu\Omega$. The achieved R_c values were measured on short cable samples and also derived indirectly from loss and field error measurements in the magnets [36]. Figure 17.5 shows the measurements for different *strand maps*. Each strand map corresponds to a number of unit-lengths of cable. The strand maps are selected sets of strands that guarantee uniform values for the filament magnetization in a production batch of cable. However, the resistance values vary by as much as a factor of two from batch to batch. The resistances also vary across keystoned cables due to the different compaction on the thin and thick edges, and with the longitudinal position in the magnet due to varying prestress in the coil-end regions.

For the 2D calculations, it is necessary to impose boundary conditions at the band edges. This can be implemented by expressing the periodic boundary conditions as homogeneous equations of two variables, which are added to the linear equation system.

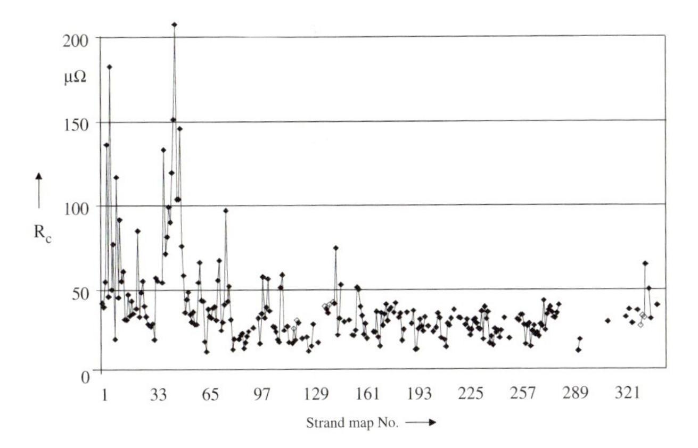

Figure 17.5 Cross resistance R_c for the LHC cable 1 used for the inner-layer coils of the main dipoles [26].

17.3
Steady-State Calculations

Consider the 2D problem of a single band with periodic boundary conditions. When the ramp of the applied field begins, the induced voltages in the meshes generate currents with different time constants. These depend on the inductive coupling between the meshes. The steady-state regime is reached for constant ramp rates after the completion of the diffusion process. In steady state, the eddy currents depend only on the contact resistances and on the flux linkages in the meshes.

The current distribution in a 10-strand cable is shown in Figure 17.6. Large currents are induced at the edge of the cable and flow in the strand branches. The currents in the cross- and adjacent-resistance branches are considerably

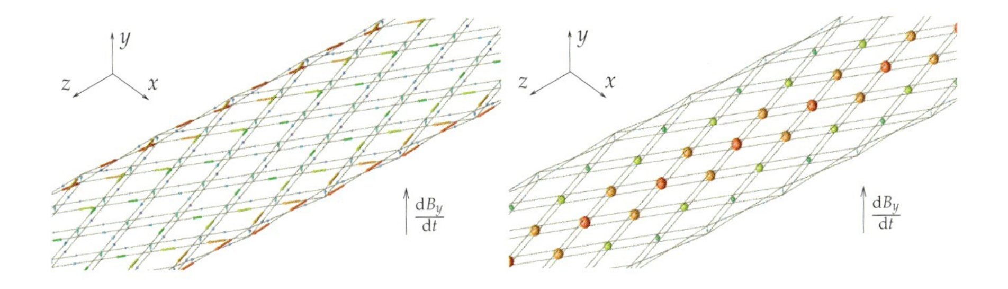

Figure 17.6 Left: Current distribution for a 10-strand cable in spatial periodic condition due to a uniform, linearly ramped magnetic field. Adjacent and cross resistance are assumed to have the same value. Right: Power distribution.

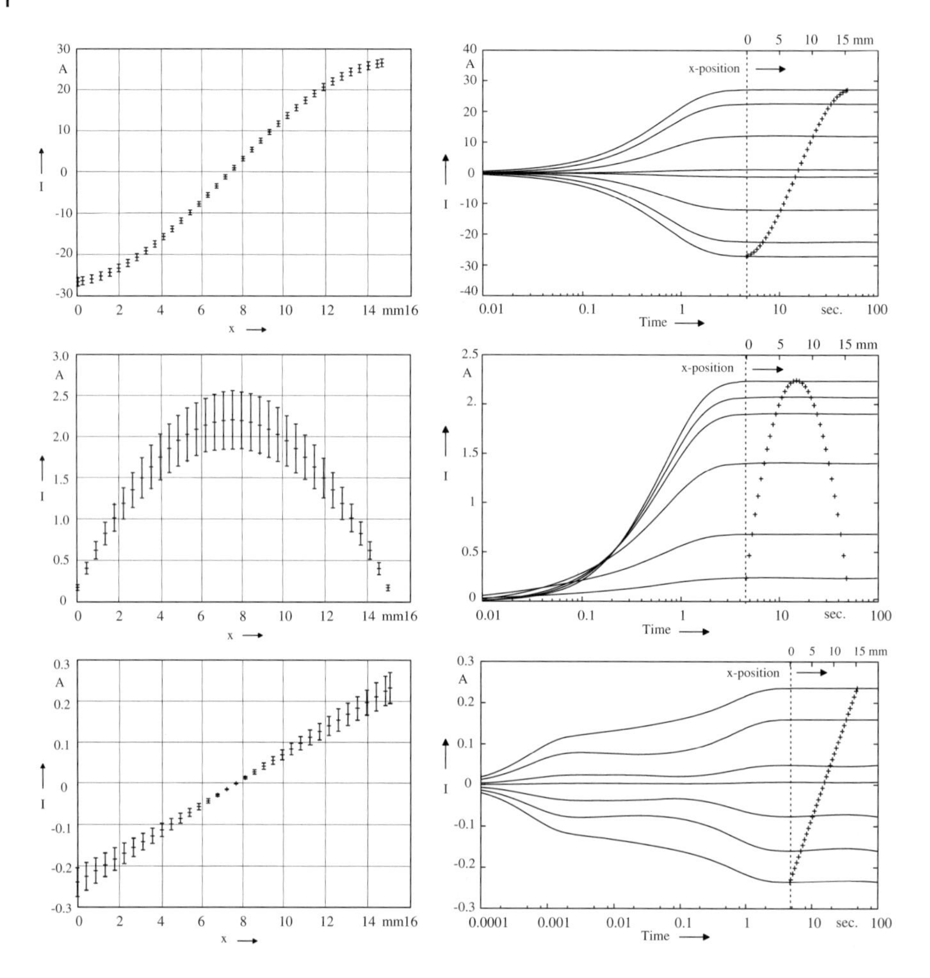

Figure 17.7 Interstrand coupling currents in the LHC cable 2 with periodic boundary conditions (2D calculation). Ramp rate $0.01\ \mathrm{T\,s^{-1}}$, cross resistance $R_c = 1\ \mu\Omega$, adjacent resistance $R_a = 1\ \mu\Omega$. Top: Currents in strand branches. Middle: Current in the cross resistance branch. Bottom: Current in the adjacent resistance branch. Left column: Steady-state calculations. The error bars show the RMS effect of random variations in the contact resistances. Right column: Time-transient calculations.

smaller. The maximum power is dissipated in the cable center, where the largest currents flow in the cross-resistance branches. The magnitude of the branch currents is shown in Figure 17.7 (left column) for a 36-strand cable of 15.1 mm width and 1.48 mm average height.

We can also calculate the effect of random variations in the contact resistances. The error bars in Figure 17.7 (left column) represent the RMS deviation from the nominal branch currents for cross and adjacent resistances varying by $\pm 50\%$ from the nominal value. The currents in the cross and adjacent resistance network branches are equally sensitive to the variation in the resistance values. The strand currents are less affected.

17.3.1 Spectral Analysis of the Solution

Because the mesh-resistance matrix $[R_\mathrm{M}] \in \mathbb{R}^{M\times M}$ in Eq. (17.24) is real, symmetric, and positive definite, it is similar to a diagonal real matrix that can be found by an *eigendecomposition*

$$[D]^{-1}[R_\mathrm{M}][D] = [R_\mathrm{D}] = \mathrm{diag}(\lambda_1, \ldots, \lambda_M) \,. \tag{17.40}$$

As $[R_\mathrm{M}][D] = [D][R_\mathrm{D}]$ and consequently $([R_\mathrm{M}]\{P\})_{\lambda_i} = \lambda_i\{P\}_{\lambda_i}$, the column vectors of $[D] \in \mathbb{R}^{M\times M}$ are the eigenvectors $\{P\}_{\lambda_i}$ of $[R_\mathrm{M}]$.

If we multiply Eq. (17.23) by $[D]^T$ and take into account that $[D][D]^T = [E]$, where $[E]$ is the identity matrix, we obtain

$$[M][R][M]^T\{I_\mathrm{M}\} + [M]\{U_\mathrm{s}\} = 0\,,$$

$$[D]^T[M][R][M]^T([D][D]^T)[M]\{I\} + [D]^T[M]\{U_\mathrm{s}\} = 0, \tag{17.41}$$

which can be rewritten taking into account the following definitions:

$$[M_\mathrm{D}] := [D]^T[M], \qquad \{I_D\} := [M_\mathrm{D}]\{I\}, \qquad \{U_\mathrm{D}\} := [M_\mathrm{D}]\{U_\mathrm{s}\}\,. \tag{17.42}$$

The result is:

$$[M_\mathrm{D}][R][M_\mathrm{D}]^T[M_\mathrm{D}]\{I\} + [M_\mathrm{D}]^T\{U_\mathrm{D}\} = 0\,, \tag{17.43}$$

where $\{I_D\}$ is the mesh-current vector, $\{U_\mathrm{D}\}$ the vector of the voltage sources in the meshes, and $[R_\mathrm{D}]$ the mesh resistance matrix. The linear equation system (17.43) has the following solution:

$$\{I\} = -[M_\mathrm{D}]^T[R_\mathrm{D}]^{-1}\{U_\mathrm{D}\}, \tag{17.44}$$

where $[R_\mathrm{D}] = [M_\mathrm{D}][R][M_\mathrm{D}]^T$. The matrix $[M_\mathrm{D}]$ is a generalized mesh matrix, fully populated with noninteger components that are each a fraction of the total induced current attributable to each branch. Because $[R_\mathrm{D}]$ is a diagonal matrix, the generalized meshes do not mutually interact. Each of the meshes represents a solution of the system when the network is excited by $\{U_\mathrm{D}\}$. In [4] these solutions are called *eigencurrents*; they can be considered as the fundamental modes of the system.

If the number of bands is an integer factor of the number of strands, the resistance mesh matrix will become singular. The dimension of the kernel is

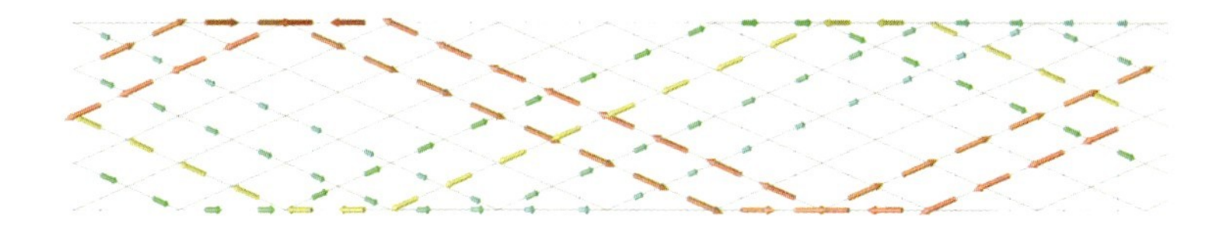

Figure 17.8 Eigencurrents corresponding to the lowest eigenvalues of the resistance mesh matrix $[R_{\mathrm{M}}]$. Ten-strand cable segment with 10 transposition pitches, periodic boundary conditions.

$N_{\mathrm{s}} - 1$. For a 10-strand cable segment, 10 transposition pitches in length, we find 9 low eigenvalues, which correspond to eigencurrents not traversing any lumped resistance. They flow only in the superconducting strands and may thus be excited without limit. These eigencurrents are shown in Figure 17.8.

17.4 Time-Transient Analysis

For the calculation of the diffusion process, the self and mutual inductances of each network branch are required. The contribution of two branches to the total inductance is called a *partial inductance* [17]. The calculation of inductances is difficult because the formulas derived in Section 5.10.1 yield inaccurate results when the size of the wire cross section is close to the branch length or to the distance between branches. The self and mutual inductances for round wire segments are approximated using the concept of the geometric mean distance (GMD); see Section 5.10.2 and Reference [17].

Mesh inductances lower than a threshold value are disregarded in order to reduce computation time:

$$L_{ij} = 0 \quad \text{if} \quad L_{ij} < \frac{\max_{i,j}\{L_{ij}\}}{\lambda} \tag{17.45}$$

for $\lambda > 0$. The smaller the factor λ, the sparser is the mesh inductance matrix. This effect was studied by computing the spectrum of the eigencurrents for a 1-m-long cable as a function of the density factor [14]. With $\lambda = 100$ this yielded errors in the range of 1%.

In Figure 17.7 (right column) the time-transient ISCCs for each type of branch are plotted for a cable with spatial periodic boundary conditions. Each line refers to a particular branch, identified by its x-coordinate within the cable. Currents in the strand and in the cross-resistance branches have time constants on the order of seconds. The time constants for the currents in the adjacent resistance are slightly smaller because the mesh areas are smaller.

Time-transient analysis without periodic boundary conditions, which involves solving a 3D diffusion problem, reveals additional dynamic effects in Rutherford cables, even though the analysis is limited to a single cable-segment because of computer hardware constraints. Figure 17.9 shows the time evolution of currents flowing in the strands at the cable edge. The segment is exposed to a uniform magnetic field with constant ramp rate, which points in the normal direction to the cable's broad face.

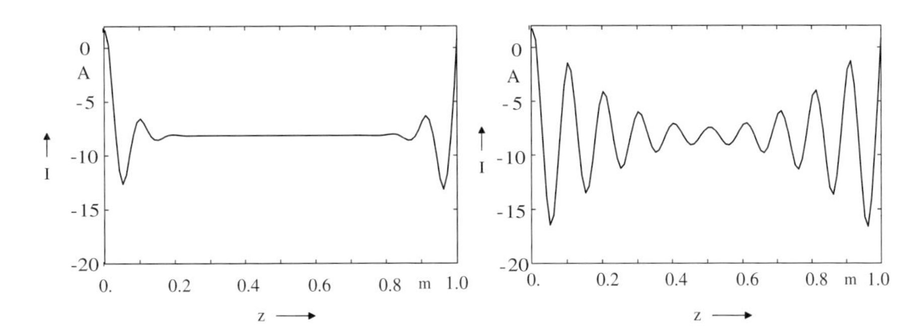

Figure 17.9 Time evolution (at 1 s left, at 10 s right) of the currents in the strands located at the cable edge $x = 0$, as a function of the longitudinal z-position. Ten-strand cable, 10 transposition pitches, no spatial periodic boundary condition. Ramp rate $0.01\ \mathrm{T\,s^{-1}}$, cross resistance $R_c = 1\ \mu\Omega$, adjacent resistance $R_a = 1\ \mu\Omega$, transposition-pitch length $p_c = 10$ cm.

Figure 17.10 shows the current distribution in the first three transposition pitches of a cable having an overall length of 10 transposition pitches. These currents, excited by the discontinuities at the cable boundaries, are referred to as *super-currents*, or *boundary-induced coupling currents* (BICC) [31]. They can also be excited by field nonuniformities in the coil-end regions. At the beginning of the field ramp, there is only one large current loop along the cable edge. The current diffuses into the center of the cable and forms smaller loops in periodic patterns.

17.4.1 Spectral Analysis of the Solution

Solving the system of differential equations (17.30) yields [4]

$$\{I_M\} = \left([E] - e^{[W]t}\right)[R_M]^{-1}\{U_s\}, \tag{17.46}$$

where $[E]$ is the identity matrix and $[W] = [L_M]^{-1}[R_M]$. If we diagonalize $[W]$, we can write Eq. (17.46) in the following form:

$$\{I_M\} = \left([E] - \sum_\lambda e^{w_\lambda t} \prod_{\lambda \neq \lambda'} \frac{[W] - w_\lambda [E]}{w_{\lambda'} - w_\lambda}\right)[R_M]^{-1}\{U_s\}, \tag{17.47}$$

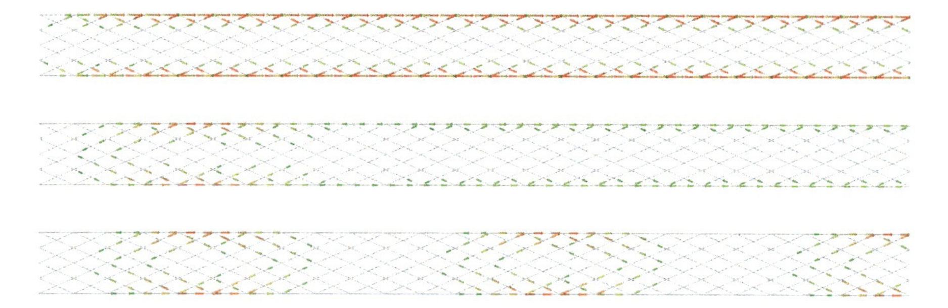

Figure 17.10 Time evolution of boundary-induced coupling currents. 10-strand cable segment, 10 transposition pitches (1 m), no spatial periodic boundary condition (3D field problem). Ramp rate $0.01\,\mathrm{T\,s^{-1}}$, $R_c = 1\,\mu\Omega$, $R_a = 1\,\mu\Omega$. Only the first three transposition pitches of the cable are displayed from left to right.

where the eigenvalues w_λ of $[W]$ represent the inverse of the time-constant spectrum of the network. In Figure 17.11 one of these eigencurrents is displayed. These modes are the cause of axial periodic field patterns in superconducting magnets, an effect first measured in the main HERA dipoles [9]. The periodicity of the measured sextupole field patterns corresponds to the transposition pitch of the cable.

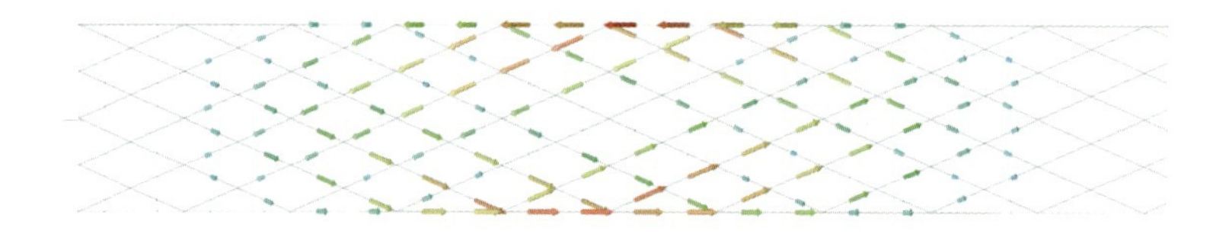

Figure 17.11 Eigencurrents corresponding to the first eigenvalue of the resistance mesh matrix $[R_{\mathrm{M}}]$. Cable model without periodic boundary conditions.

17.5 The M(B) Iteration Scheme for ISCCs

The $\mathbf{M}(\mathbf{B})$ iteration scheme according to Figure 16.17, introduced for the persistent current calculation in Section 16.8, can now be augmented to include the ramp-rate-dependent effects of interstrand coupling currents; see Figure 17.12. It is possible to calculate the global shielding effects of persistent currents and ISCCs, while taking the iron saturation into account. The need for the iteration scheme is demonstrated with the calculation of the hysteresis effects in the fast-ramping model magnet for the GSI FAIR project [23], which is discussed in Section 17.8.2.

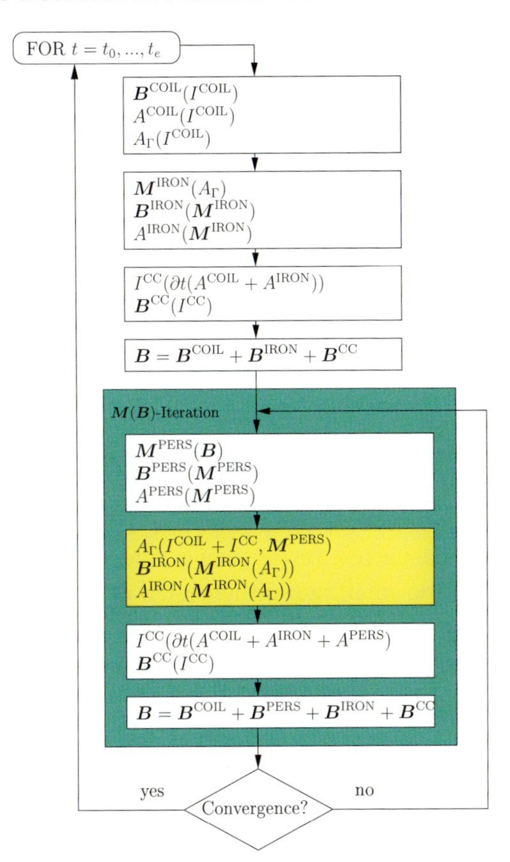

Figure 17.12 Algorithm for the computation of coupling currents and persistent currents with $M(B)$ iteration. Note that if the iron saturation is low, the calculation of the iron magnetization can be restricted to the first and last iteration step. The superscript cc stands for interstrand coupling currents (ISCC) and interfilament coupling currents (IFCC).

17.6 Approximation for the Interstrand Coupling Currents

An estimate of the ISCCs can be obtained for constant ramp rates assuming a uniform applied field in the cable's normal direction,[9] as shown in Figure 17.4. The interstrand coupling current over an entire transposition pitch can be approximated by

$$I = \frac{1}{2R_\text{c}} \frac{\text{d}\Phi}{\text{d}t} \approx \frac{3wl_\text{b}}{2R_\text{c}} \frac{\text{d}B_\text{n}}{\text{d}t}, \tag{17.48}$$

where $3wl_\text{b}$ is the area of the mesh-current loop projected onto the xz-plane. In Eq. (17.48), w is the cable width and l_b denotes the band length.

9 The local reference frame for strip surfaces is derived in Chapter 19.

Figure 8.9 (right) shows the force components in the local coordinate system of the cable. The force in the normal direction to the cable's broad face, and thus the flux density in the binormal direction, is relatively uniform. The normal component of the flux density shows a nearly linear rise within the cable. Relatively accurate results can be obtained by averaging the field components in each cable and applying the results derived in [35] for the three magnetization components in the cable's local reference frame:

$$M_{\mathrm{n}}^{\mathrm{c}} = -\frac{1}{120R_{\mathrm{c}}}\frac{\mathrm{d}B_{\mathrm{n}}}{\mathrm{d}t}\,p_{\mathrm{c}}N_{\mathrm{s}}(N_{\mathrm{s}}-1)\frac{w}{h}, \tag{17.49}$$

$$M_{\mathrm{n}}^{\mathrm{a}} = -\frac{1}{3R_{\mathrm{a}}}\frac{\mathrm{d}B_{\mathrm{n}}}{\mathrm{d}t}\,p_{\mathrm{c}}\frac{w}{h}, \tag{17.50}$$

$$M_{\mathrm{b}}^{\mathrm{a}} = -\frac{1}{8R_{\mathrm{a}}}\frac{\mathrm{d}B_{\mathrm{b}}}{\mathrm{d}t}\,p_{\mathrm{c}}\frac{h}{w}, \tag{17.51}$$

where h is the mean cable height, w the cable width, p_{c} the transposition pitch, and N_{s} the number of strands. The subscripts n and b denote the normal and binormal direction in the cable's local reference frame and the indices c and a stand for the cross and adjacent resistances, respectively. The results are derived from the induced currents in the small two-dimensional meshes seen in the projections of the cable onto the xz- and yz-planes in Figure 17.4.

17.7 Interfilament Coupling Currents

Interfilament coupling currents depend on the twist-pitch length p_{s} of the filaments, which are embedded in the copper matrix of a strand with an effective resistivity ρ_{eff}. Eddy currents are induced in meshes of up to one half a twist-pitch length. These meshes are bounded by the superconducting filaments and are closed across the resistive matrix, as illustrated in Figure 17.13.

The resulting magnetization from the eddy currents in these meshes can be calculated analytically [34] as

$$M_{\mathrm{IFCC}} = -\lambda_{\mathrm{s}}\frac{\mathrm{d}B}{\mathrm{d}t}\left(\frac{p_{\mathrm{s}}}{2\pi}\right)^2 \underbrace{\frac{1}{\rho_0+\rho_1 B}}_{\rho_{\mathrm{eff}}}. \tag{17.52}$$

In Eq. (17.52), λ_{s} is the filling factor of the twisted filaments in a strand, ρ_0 the constant part of the effective resistivity, and ρ_1 the slope of the magnetoresistive effect, $[\rho_1] = \Omega\,\mathrm{m}\,\mathrm{T}^{-1}$.

This result can be derived by considering two filaments twisted along helical paths, with radius r_0 and twist-pitch length p_{s}, around the z-axis as shown in Figure 17.13. We assume that the filaments are exposed to a uniform,

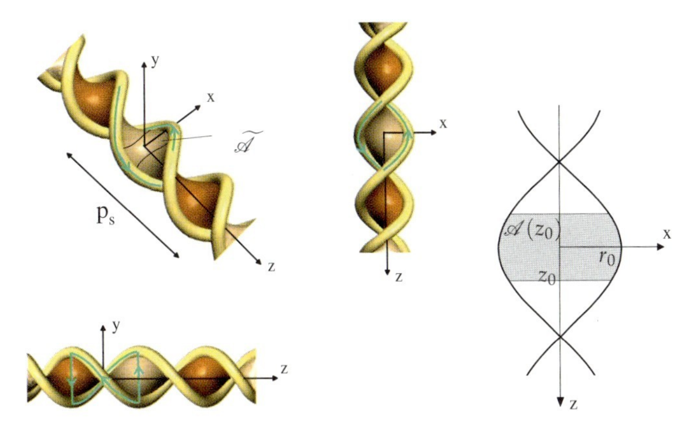

Figure 17.13 Left: Eddy-current loop between two twisted filaments in a resistive matrix. Visualization with SeifertView [29]. Right: Projection onto the xz-plane.

transversal field $\mathbf{B} = B_y\,\mathbf{e}_y$. Consider a patch $\widetilde{\mathscr{A}}(z_0)$ of the *Seifert surface*[10] and its projection $\mathscr{A}(z_0)$ onto the xz-plane, as shown in Figure 17.13. Faraday's law yields

$$\int_{\partial\widetilde{\mathscr{A}}} \mathbf{E}\cdot \mathrm{d}\mathbf{r} = -\int_{\widetilde{\mathscr{A}}} \frac{\mathrm{d}\mathbf{B}}{\mathrm{d}t}\cdot \mathrm{d}\mathbf{a} = -\int_{\mathscr{A}} \frac{\mathrm{d}B_y}{\mathrm{d}t}\mathbf{e}_y\cdot\mathbf{n}\,\mathrm{d}a$$
$$= -\int_{z_0}^{z_0}\int_{-r_0\cos(\frac{2\pi}{p_s}z)}^{r_0\cos(\frac{2\pi}{p_s}z)} \frac{\mathrm{d}B_y}{\mathrm{d}t}\,\mathrm{d}x\,\mathrm{d}z = -4\,r_0 \sin\left(\frac{2\pi}{p_s}z_0\right)\frac{p_s}{2\pi}\frac{\mathrm{d}B_y}{\mathrm{d}t}\,. \quad (17.53)$$

If the filaments are in the superconducting state, the total voltage must drop across the resistive copper matrix. Because of the symmetry, we obtain $\int_{\partial\widetilde{\mathscr{A}}} \mathbf{E}\cdot\mathrm{d}\mathbf{r} = 4(\phi(r_0,z_0) - \phi(r=0,z_0))$ and thus for a zero electrical potential at the strand axis:

$$\phi(r_0,z_0) = -\frac{p_s}{2\pi}\,r_0\,\sin\left(\frac{2\pi}{p_s}z_0\right)\frac{\mathrm{d}B_y}{\mathrm{d}t}\,. \quad (17.54)$$

If we approximate the ensemble of filaments at the radius r_0 by a thin superconducting cylinder, the potential can be expressed as a function of the φ-coordinate. For $\varphi = 2\pi z_0/p_s$, Eq. (17.54) can be rewritten as

$$\phi(\varphi) = -\frac{p_s}{2\pi}\,r_0\,\sin\varphi\,\frac{\mathrm{d}B_y}{\mathrm{d}t} = -\frac{p_s}{2\pi}\,y\,\frac{\mathrm{d}B_y}{\mathrm{d}t}\,, \quad (17.55)$$

and we obtain a uniform electric field and current density in the y-direction:

$$J_y = \frac{1}{\rho_{\mathrm{eff}}}E_y = \frac{1}{\rho_{\mathrm{eff}}}\frac{p_s}{2\pi}\frac{\mathrm{d}B_y}{\mathrm{d}t}\,. \quad (17.56)$$

10 Herbert Seifert (1907–1996).

The eddy-current loss density within the matrix material is

$$p = \frac{1}{\rho_{\text{eff}}} E_y^2 = \frac{1}{\rho_{\text{eff}}} \left(\frac{p_\text{s}}{2\pi}\right)^2 \left(\frac{\mathrm{d}B_y}{\mathrm{d}t}\right)^2 . \tag{17.57}$$

Because $\operatorname{div} \mathbf{J} = 0$, the φ-component of the surface current in the cylinder is given by

$$\alpha_\varphi(r_0, \varphi) = -J_y r_0 \cos\varphi = \frac{-1}{\rho_{\text{eff}}} \frac{p_\text{s}}{2\pi} r_0 \cos\varphi \frac{\mathrm{d}B_y}{\mathrm{d}t} . \tag{17.58}$$

The filaments in the cylinder follow helical paths with the twist pitch p_s, thus

$$\alpha_z(r_0, \varphi) = -\alpha_\varphi \frac{p_\text{s}}{2\pi r_0} . \tag{17.59}$$

Using Eq. (4.80), which relates the magnetization in a volume $\mathscr{V}$ to the surface current in $\partial\mathscr{V}$, we obtain

$$M_y = \frac{-1}{\rho_{\text{eff}}} \left(\frac{p_\text{s}}{2\pi}\right)^2 \frac{\mathrm{d}B_y}{\mathrm{d}t} . \tag{17.60}$$

Assuming a uniform filament distribution within the strand, this result can simply be scaled with the strand filling factor defined by Eq. (16.40).

17.8 Applications to Magnet Design

We will now apply the 2D transient analysis to magnet design. The first example of the LHC main dipole shows that the interstrand coupling currents enhance the field in the magnet aperture, an effect known as *field advance* [31]. It also illustrates that the main source of ISCCs is the flux linkage with the cable's broad face. It is the inner-layer dipole coil that contributes most to the transient field effects.

The second example, the GSI001 model magnet for the FAIR project, shows that the local flux distribution in the coil leads to "signatures" in the hysteresis of the multipole field errors, which cannot be explained by a mere superposition of geometrical (coil), iron saturation (yoke), and magnetization-induced field errors.

ISCCs are also the source of additional losses that become important for the rapid current decay during a quench. The methods presented in this chapter are therefore also applied to the quench simulation presented in Chapter 18.

17.8.1 Field Advance

Figure 17.14 (left) shows the magnetic flux density in the LHC main dipole coil. The cables in the inner layer are exposed to a magnetic field in nearly normal direction to their broad faces. The normal component of the flux density rises linearly within each cable, whereas the binormal component is nearly uniform; this can be seen in Figure 8.8. This is the reason for the relatively accurate results obtained from the analytical equations (17.49)–(17.51). The distribution of the ISCCs in the strand branches is shown in Figure 17.14 (right). The field generated by the ISCCs opposes the applied field in the coil but has the same direction as the source field in the aperture. The ISCCs are considerably smaller in the outer-layer coil because the applied magnetic field is smaller and tends to be parallel to the cable's broad face, thus reducing the flux linkage.

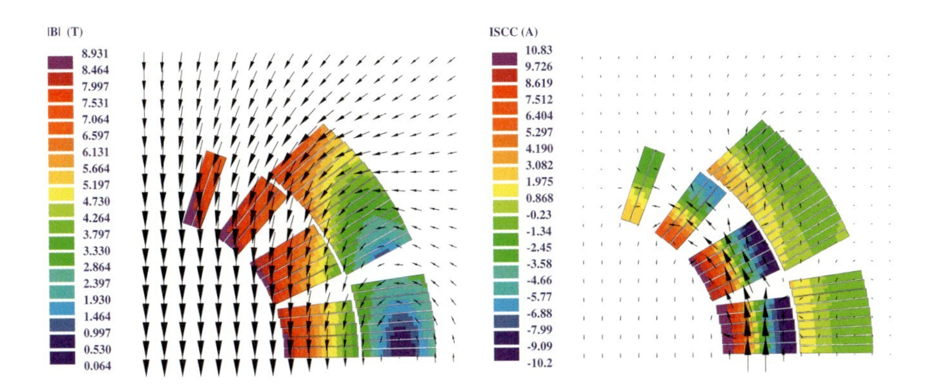

Figure 17.14 Left: Magnetic flux density in the cross section of the LHC main dipole. Right: Interstrand coupling currents for the steady-state condition with $\mathrm{d}\mathbf{B}/\mathrm{d}t = 0.094\ \mathrm{T\,s^{-1}}$. Icons represent the magnetic flux density generated by the interstrand coupling currents. Note that the icons in both plots are not to scale.

17.8.2
Rapid Cycling Magnets

A model dipole (GSI 001) was built at Brookhaven National Laboratory (BNL) for the FAIR-project (Facility for Antiproton and Ion Research) at the Gesellschaft für Schwerionenforschung (GSI), Germany [23]. It is designed for a nominal field of 4 T (7 kA) and ramp rates of up to $4\ \mathrm{T\,s^{-1}}$. The layout of the magnet cross section was based on the single-layer RHIC coil in order to take advantage of the RHIC design, existing tooling, and components. The aperture is 80 mm wide and the magnet length is approximately 1.2 m.

The aim of the GSI001 model magnet was to prove the feasibility of rapid cycling magnets and to test modifications to the cable and the insulation scheme. By placing a 25-μm-thick, 8-mm-wide austenitic steel core inside the Rutherford cables, the cross resistance in the cable was increased tenfold with respect to the RHIC cable [33]. The ramp-rate dependence of the field quality and the losses was measured at BNL [20]. The data for the transient operation were obtained with a stationary array of tangential pickup coils mounted on a cylindrical coil former. For the DC measurements the probe was used in a rotating mode.

We will use the test results of the GSI001 model magnet to validate our simulations of ramp-induced field errors. Specifically, we will analyze the contribution of persistent currents, interfilament coupling currents, and interstrand coupling currents to the aperture field distortions. For this purpose, we calculate the hysteresis of the sextupole component between the up and down ramp branch of an excitation cycle. The maximum width of the hysteresis is denoted ΔB_3.

The parameters for the calculation of the IFCCs are $\rho_0 = 9 \times 10^{-11}\,\Omega\,\mathrm{m}$, $\rho_1 = 3 \times 10^{-9}\,\Omega\,\mathrm{m}\,\mathrm{T}^{-1}$, $\lambda_S = 0.7$. The ISCC model uses the following (nominal) parameters: 30 strands per cable, twist-pitch length of 74 mm, and contact resistances of $R_\mathrm{a} = 6.4 \times 10^{-5}\,\Omega$ and $R_\mathrm{c} = 6.25 \times 10^{-2}\,\Omega$. Results are shown in Figure 17.15 (right).

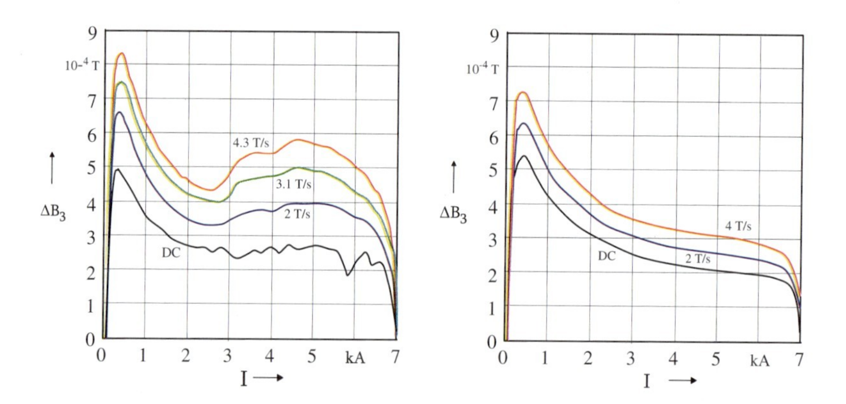

Figure 17.15 Left: Measured [20] sextupole hysteresis width at DC conditions, 2, 3.1, and 4.3 $\mathrm{T\,s^{-1}}$ ramp rates. Reference radius 25 mm. Right: Simulations assuming nominal material parameters, for ramp rates up to 4 $\mathrm{T\,s^{-1}}$.

Figure 17.15 shows a discrepancy between simulations and measurements, in particular for field levels above 3 T. For this reason we study the signature of the different magnetization effects while decreasing the adjacent resistance to $R_\mathrm{a} = 1.8 \times 10^{-5}\,\Omega$. Persistent currents (PCs) and IFCCs have similar signatures of $\Delta B_3(I)$, which is defined as the width of the sextupole-hysteresis curve (down-ramp minus up-ramp branch). Both PCs and IFCCs lessen with a decreasing magnetic field. In the case of PCs this is due to the critical current density in the superconductor, whereas in the case of the IFCCs the reason lies

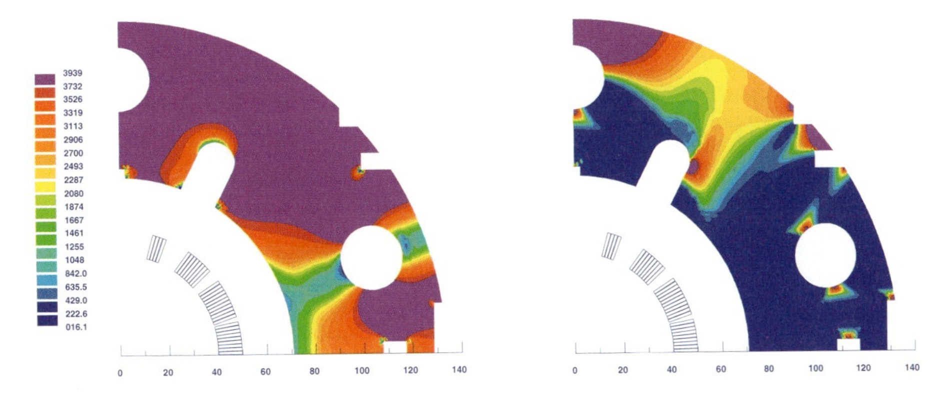

Figure 17.16 Relative permeability of the GSI001 yoke iron. Left: Low excitation (2 kA). Right: High excitation (6 kA).

in the magnetoresistance of the resistive matrix. Only the IFCCs vary linearly with the ramp rate.

Interstrand coupling currents (ISCC) have a different signature than PCs and IFCCs. The ISCC contribution to the sextupole hysteresis has an opposite sign and reduces as a function of the magnetic flux density, in particular when the iron yoke begins to saturate.[11] The cause of this is a change in field distribution in the coil cross section at this excitation level. The center of the

11 In this case, the strong saturation effects result from the nonoptimal iron yoke, which was reused from another magnet for reasons of schedule.

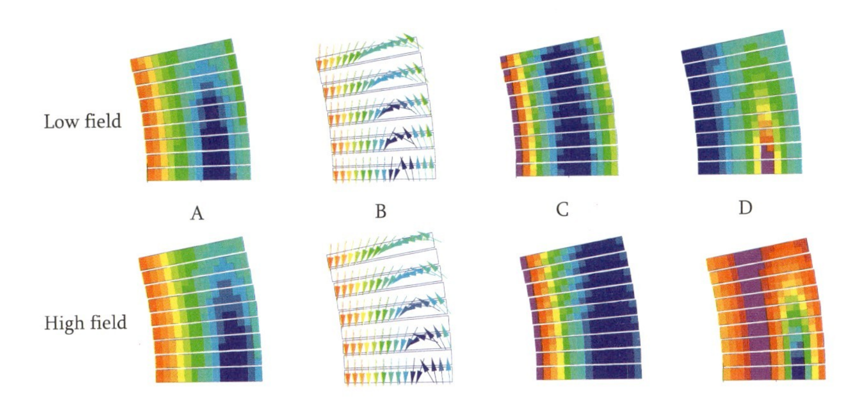

Figure 17.17 Flux distribution, interstrand coupling currents, and persistent currents in the first coil-block (counted from the horizontal median plane) of the GSI001 model magnet. Top: With 2 kA excitation current. Bottom: With 6 kA. (A) Modulus of magnetic flux density; (B) flux density vectors; (C) interstrand coupling currents; and (D) persistent currents.

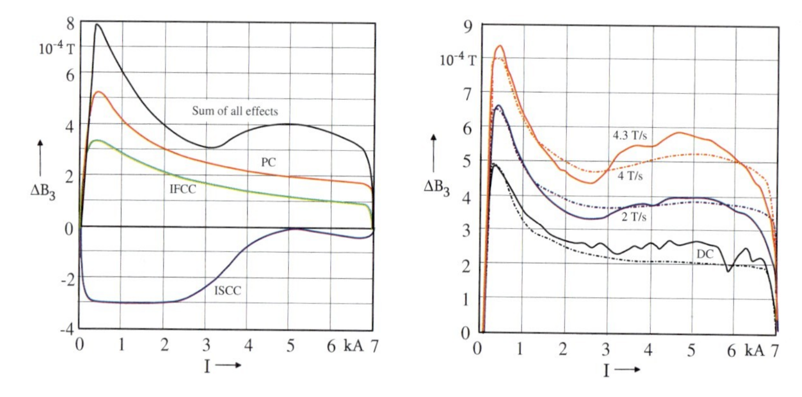

Figure 17.18 Left: Signatures of different magnetization effects in ramped magnets. Right: Measured [20] sextupole hysteresis width at DC conditions and at ramp rates of 2 $\mathrm{T\,s^{-1}}$, 3.1 $\mathrm{T\,s^{-1}}$, and 4.3 $\mathrm{T\,s^{-1}}$. Reference radius of 25 mm. Simulations with reduced, adjacent contact resistances $R_a = 1.8 \times 10^{-5}\ \Omega$.

magnetic flux-loops is displaced toward the aperture, which reduces the flux linkage through the broad side of the cables; compare Figures 17.16 and 17.17. This change of flux linkage reduces the interstrand coupling currents. Thus it is shown that the mutual interdependence of saturation effects in the iron and magnetization effects in the coil does not allow a mere superposition of the effects.

Although the simulations with adjusted material parameters reproduce the sextupole hysteresis well, the comparison of measurements and calculations of the ΔB_5 curves point to an additional, ramp-rate-dependent effect in the GSI001 model magnet. Moreover, the measurements show a ramp-rate-dependent, skew quadrupole component, which is zero in DC conditions; see Figure 17.18. This is a nonallowed multipole in the sense that it cannot be sim-

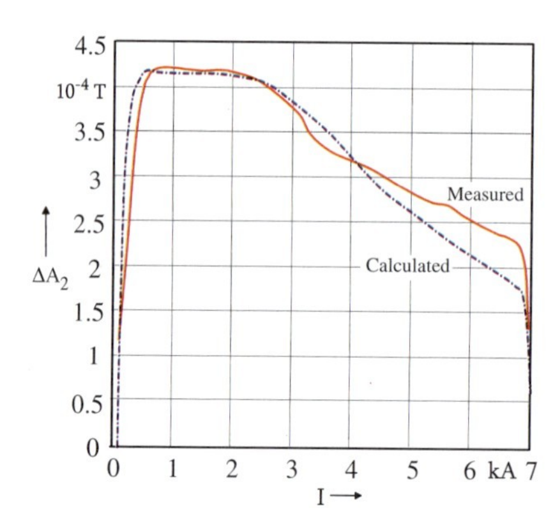

Figure 17.19 Skew quadrupole hysteresis width at 2 $\mathrm{T\,s^{-1}}$ ramp rate. $R_a = 1.8 \times 10^{-5}\ \Omega$ in the upper coil and $R_a = 1.0 \times 10^{-5}\ \Omega$ in the lower coil.

ulated with the above model, as up/down symmetry of the geometry and the material parameters is assumed.

In order to reproduce the skew quadrupole in the time-transient regime, we simulate a magnet built from coils with different properties. We assume an adjacent resistance of $R_\mathrm{a} = 1.8 \times 10^{-5}\ \Omega$ in the upper coil, and $R_\mathrm{a} = 1.0 \times 10^{-5}\ \Omega$ in the lower one. Figure 17.19 shows the ΔA_2 curve for the simulations and measurements at $2\ \mathrm{T\,s^{-1}}$.

The measured and calculated losses are summarized in Table 17.2. The remaining difference between measured and simulated losses can be explained by the iron hysteresis, which is not implemented in our model.

Table 17.2 Losses per unit length and ramp cycle in $\mathrm{J\,m^{-1}}$.

Ramp rate	Measured	Simulated nom.	Simulated adapt.
DC	40	28.5	27.5
$2\ \mathrm{T\,s^{-1}}$	83	42.5	69.3
$4\ \mathrm{T\,s^{-1}}$	126	56.5	108.7

References

1 Auchmann, B., de Maria, R., Russenschuck, S.: Calculation of field quality in fast-ramping superconducting magnets, IEEE Transactions on Applied Superconductivity, 2008

2 Akhmetov, A. A.: Compatibility of two basic models describing the A.C. loss and eddy currents in flat superconducting cables, Cryogenics, 2000

3 Akhmetov, A. A., Devred, A., Ogitsu, T.: Periodicity of crossover currents in a Rutherford-type cable subjected to a time-dependent magnetic field, Journal of Applied Physics, 1994

4 Akhmetov, A. A., Ivanov, S. S., Shchegolev, I. O.: The network approach to calculation of characteristic time constants of the flat two-layer superconducting cable, Physica C: Superconductivity, 1998

5 Bamberg, P., Sternberg, S.: A course in mathematics for students of physics, Academic Press, San Diego, 1983

6 Bottura, L., Rosso, C., Breschi, M.: A general model for thermal, hydraulic and electric analysis of superconducting cables, Cryogenics, 2000

7 Bottura, L.: A practical fit for the critical surface of NbTi, IEEE Transactions on Applied Superconductivity, 2000

8 Bottura, L., Walckiers, L., Wolf, R.: Field errors decay and snap-back in LHC model dipoles, IEEE Transactions on Applied Superconductivity, 1997

9 Brück H., Gall., D., Krzywinski, J., Meinke, R., Preißner, H., Halemeyer M, Schmüser, P., Stolzenburg, C., Stiening, R., ter Avest, D., van de Klundert, L.J.M.: Observation of a periodic pattern in the persistent-current fields in the superconducting HERA dipole magnets, Proceedings of the Particle Accelerator Conference, 1991

10 Bruhat, G.: Cours de physique générale, Masson, 1959

11 Campbell, A.M.: A.C. losses in cables of twisted multifilament superconductors, Cryogenics, 1980

12 Carr, W. J. Jr.: Conductivity, permeability, and dielectric constant in a multifilament superconductor, Journal of Applied Physics, 1975

13 Close, C.H.: The analysis of linear circuits, Harcourt College Publications, USA, 1966

14 de Maria, R.: Time Transient Effects in Superconducting Magnets, Tesi di Laurea, University "La Sapienza", Rome, 2003

15 Devred, A., Ogitsu, T.: Influence of eddy currents in superconducting particle accelerator magnets using Rutherford-type cables, CAS-CERN Accelerator School: Superconductivity in Particle Accelerators, 1995

16 Ghosh, A.K., Robins, K.E., Sampson, W.B.: The ramp rate dependence of the sextupole field in superconducting dipoles, IEEE Transactions on Magnetics, 1994

17 Grover, F.W.: Inductance Calculations, Dover Phoenix Edition, 2003 reprint of the 1962 edition

18 Gross, P.W., Kotiuga P.R.: Electromagnetic Theory and Computation: A Topological Approach, Cambridge University Press, Cambridge, 2003

19 Haverkamp, M.: Decay and Snap-Back in Superconducting Accelerator Magnets, PhD Thesis, University of Twente, The Netherlands, 2003

20 Jain, A., Ganetis, G., Gosh, A., Wing, L., Marone, A., Thomas, R., Wanderer, P.: Field quality measurements at high ramp rates in a prototype dipole for the FAIR project, IEEE Transactions on Applied Superconductivity, 2008

21 Krempasky, L., Schmidt, C.: Theory of supercurrents and their influence on field quality and stability of superconducting magnets, Journal of Applied Physics, 1995

22 Morgan, G.H.: Eddy currents in flat metal-filled superconducting braids, Journal of Applied Physics, 1973

23 Moritz, G.: Fast-pulsed SC magnets, Proceedings of the 2004 European Particle Accelerator Conference, Vienna, 2004

24 Press, W.H. et al.: Numerical recipes, Cambridge University Press, Cambridge, 2001

25 Pugnat, P., Schreiner, Th., Siemko, A.: Investigation of the periodic magnetic field modulation in LHC superconducting dipoles, IEEE Transactions on Applied Superconductivity, 2002

26 Richter, D.: Private communication, CERN, 2005

27 Saad, Y.: Iterative Methods for Sparse Linear System, PWS Publishing Company, Boston, MA, USA, 1996

28 Sampson W.B., Ghosh, A.K.: Induced axial oscillations in superconducting dipole windings, IEEE Transactions on Applied Superconductivity, 1995

29 van Wijk, J.: Seifertview, visualization of seifert surfaces, http://www.win.tue.nl/vanwijk/seifertview/, 2006

30 Turck, B.: Influence of a transverse conductance on current sharing in a two-layer superconducting cable, Cryogenics, 1974

31 Verweij, A. P.: Electrodynamics of superconducting cables in accelerator magnets, PhD Thesis, Twente University, The Netherlands, 1995

32 Verweij, A. P.: Modelling boundary-induced coupling currents in Rutherford-type cables, IEEE Transactions on Applied Superconductivity, 1997

33 Wanderer, P. et al.: Initial test of a fast-ramped superconducting model dipole for GSI's proposed SIS200 accelerator, Proceedings of the Particle Accelerator Conference, 2003

34 Wilson, M. N.: Superconducting Magnets, Oxford Science Publications, Oxford, 1983

35 Wilson, M. N.: Dipole 001 field harmonics coming from superconductor magnetization, MNW Report GSI 27, GSI Fast Pulsed Synchrotron Project, 2005

36 Wolf, R., Leroy, D. , Richter, D., Verweij, A., Walckiers, L.: Determination of interstrand contact resistance from loss and field measurements in LHC dipole prototypes and correlation with measurements on cable samples, IEEE Transactions on Applied Superconductivity, 1997

18
Quench Simulation

Give me four parameters and I can fit an elephant.
Give me five and I can wag its tail.

Attributed to C. F. Gauss, N. Bohr, Lord Kelvin,
and E. Fermi, among others.

Above certain limits on temperature, current density, and magnetic flux density, a superconductor undergoes a transition from the superconducting to the normal-conducting state. This process is known as a *resistive transition* or a *quench*. Superconducting accelerator magnets are usually operated close to their quench limits, either for economical reasons or because the application requires the highest achievable field. The LHC main dipoles, for example, operate at 88% on the load line.

It is customary to distinguish between three different types of quenches [32]:

- A *natural quench*, when the working point on the load line is moved across the critical surface by raising the excitation current and consequently the magnetic flux density at constant operation temperature.
- A *disturbance quench* can be initiated due to local heating caused by conductor movements (in particular in the coil ends with their weaker mechanical structure) or by beam losses in the accelerator. These disturbance quenches may happen with the working point well below the critical surface.
- A *training quench* is a special type of disturbance quench that results when the coil windings move slightly under the influence of the electromagnetic forces and pressure changes during quenches. This may result in successively higher fields at subsequent quenches.

Quench detection and *magnet protection* against overheating and excessive voltages during a resistive transition is an important issue in the design of superconducting magnets. It is essential to treat all involved phenomena (ther-

Field Computation for Accelerator Magnets. Stephan Russenschuck

ISBN: 978-3-527-40769-9

mal, electrical, and magnetic) as a coupled multiphysics problem. The numerical model presented in this chapter makes possible the study of the impact of different effects such as quench-back, normal-zone propagation, quench-heater performance, local field distribution, and iron saturation.

The simulation of thermal processes at cryogenic temperatures is an elaborate problem. Material properties at large temperature and pressure ranges are difficult to measure. The uncertainties and the highly nonlinear behavior of these parameters lead to ill-posed numerical problems with more model parameters than validation criteria, such as measurements of the current decay, signals at the *voltage taps*, and measurements using *quench antennas* [30].

Most quench-simulation programs are based on decoupling the electromagnetic field problem from thermal simulation. Examples are the codes QUENCH [49], DYNQUE [7], for solenoids, and QUABER, which applies a commercial network-solver to the thermal and electrical circuits [21]. The limits of these codes are reached when strong saturation effects occur in the iron yoke and the field problem can no longer be represented by lumped inductances and scaled field maps. Finite-difference (FD) and finite-element (FE) models for the calculation of quench propagation were used in [44] and [8], respectively.

Accurate modeling of the thermal processes inside Rutherford-type cables requires a large computational effort; see for example [19] for 2D and [48] for 3D models. These models have therefore not yet been applied to the simulation of entire magnets.

A *strong coupling* of field calculation, thermal simulation, and electric-circuit analysis is presented in [3] for solenoids. In this case, the thermal problem is solved on the same FE mesh as the electromagnetic problem. However, the required meshing of the coil, air domain, and iron yoke makes this approach computationally expensive for the use in the integrated design process for accelerator magnets. It is also difficult to calculate turn-to-turn voltages employing this method.

The update of the magnetic field-distribution is computationally more demanding than the calculation of a time step in the thermal model. On the other hand, the time constants of the thermal model are much shorter than those of the eddy currents. It is therefore reasonable to implement a *weak coupling* between thermal and magnetic computations, updating the magnetic fields only when the excitation current has changed significantly. But before we present this iterative method, we will briefly review the physical phenomena in a quenching magnet and give analytical expressions for a rough estimate of the size and time constants of these effects.

18.1
The Heat Balance Equation

Consider a fixed volume $\mathcal{V}$ of the coil (of uniform material properties) bounded by the closed surface $\partial\mathcal{V}$. The volumetric extent of $\mathcal{V}$ is denoted V. The rate of heat increase in $\mathcal{V}$ is equal to the rate of heat conduction across $\partial\mathcal{V}$ plus the rate of heat generation within $\mathcal{V}$. Under adiabatic conditions (no heat conduction and no cooling), we obtain the *heat-balance equation* [50]:

$$P(t) = \rho c_{\mathrm{p}}(T)\, V \frac{\mathrm{d}T}{\mathrm{d}t}, \tag{18.1}$$

where $P(t)$ is the dissipated power, and $\rho c_{\mathrm{p}}(T)$ the temperature-dependent *volumetric heat capacity* (VHC) of the medium at constant pressure, $[\rho c_{\mathrm{p}}] = \mathrm{J\,m^{-3}\,K^{-1}}$. The VHC must be averaged over the cable components, which include the superconductor, copper, insulation, and the confined helium in the voids of the cable. The dissipated power is the sum of ohmic losses in the normal-conducting zone, the eddy-current losses in the cables, and losses in the strands, as well as time-transient and continuous beam losses.

If we take into account only ohmic losses in the *normal (conducting) zone*, that is,

$$P(t) = R\, I(t)^2, \tag{18.2}$$

we obtain

$$\frac{\rho_{\mathrm{E}}(B(I), T, RRR)}{a_{\mathrm{Cu}}}\, l\, I(t)^2 = \rho c_{\mathrm{p}}(T)\, a_{\mathrm{t}}\, l \frac{\mathrm{d}T}{\mathrm{d}t}, \tag{18.3}$$

where a_{Cu} is the copper cross section of the cable, a_{t} the total[1] cross section of the cable, $\rho_{\mathrm{E}}(B, T, RRR)$ is the electrical resistivity of copper, $[\rho_{\mathrm{E}}] = 1\,\Omega\,\mathrm{m}$, and l the length of the normal zone. The resistivity of niobium–titanium in the normal-conducting state is orders of magnitude higher than the resistivity of copper; see Figure 18.2. For this reason, the resistivity of a quenched Nb–Ti conductor can be neglected in the calculations. The specific resistivity of copper has a nonlinear dependence on the magnetic field, temperature, and the *residual resistivity ratio* (RRR), which is defined[2] as the ratio of the copper's resistivity at 273 K to its resistivity at 4 K, both in the absence of magnetic fields [33]. The specific resistivity of copper, as a function of the temperature and the RRR, is plotted in Figure 18.1 (left). The field-dependence of the resistance, referred to as *magnetoresistivity*, is also a function of the RRR; see Figure 18.1 (right). The material properties needed for quench simulation are discussed in Appendix A.

1 It includes the cross sections of the filaments and the copper matrix, the confined helium, and the insulation.

2 The definition is not unique in the literature; in [9] the ratio is defined as $\rho(293\ \mathrm{K})/\rho(10\ \mathrm{K})$.

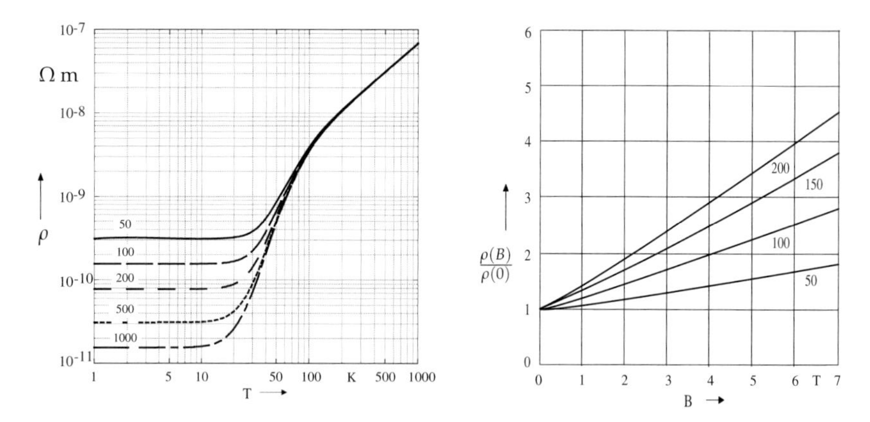

Figure 18.1 Left: The electrical resistivity of copper at zero field, as a function of temperature (T) and the residual resistivity ratio (RRR). Right: Normalized magnetoresistivity of copper at 4.5 K as a function of the RRR.

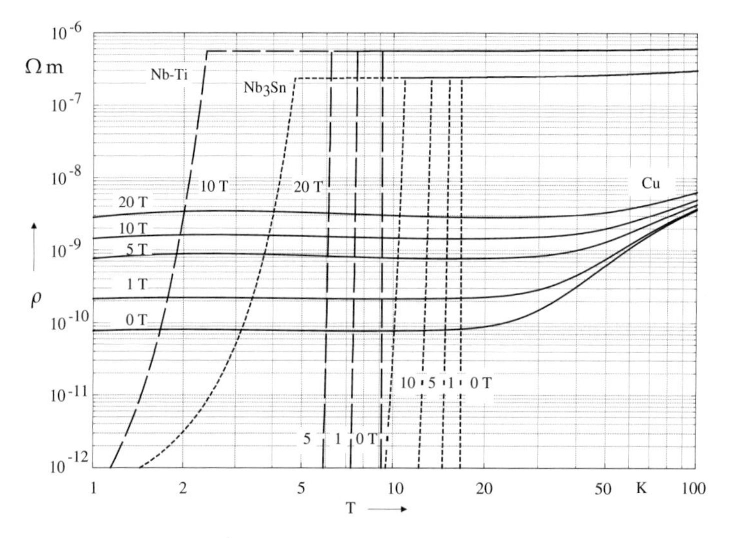

Figure 18.2 Electrical resistivity of copper and superconductors. For the SC materials a design current density of $J_c(8.4\,\mathrm{T}, 1.9\,\mathrm{K}) = 2.6 \times 10^9\,\mathrm{A\,m^{-2}}$ for Nb–Ti and $J_c(15\,\mathrm{T}, 4.2\,\mathrm{K}) = 1.4 \times 10^9\,\mathrm{A\,m^{-2}}$ for $\mathrm{Nb_3Sn}$ is assumed.

The total resistance of the quenching cable is given by

$$R = \frac{\rho_{\mathrm{E}}\, l}{a_{\mathrm{Cu}}}. \tag{18.4}$$

Rearranging Eq. (18.3) yields

$$\frac{dT}{dt} = \frac{I(t)^2 \rho_E(B(I), T, RRR)}{a_{Cu} a_t \rho c_p(T)}. \tag{18.5}$$

Suppose that the quench starts at time $t = 0$ with an initial conductor temperature T_0. Separation of variables and integration yields

$$\int_0^\infty I(t)^2 dt = a_{Cu} a_t \int_{T_0}^{T_{max}} \frac{\rho c_p(T)}{\rho_E(B_0, T, RRR)} dT. \tag{18.6}$$

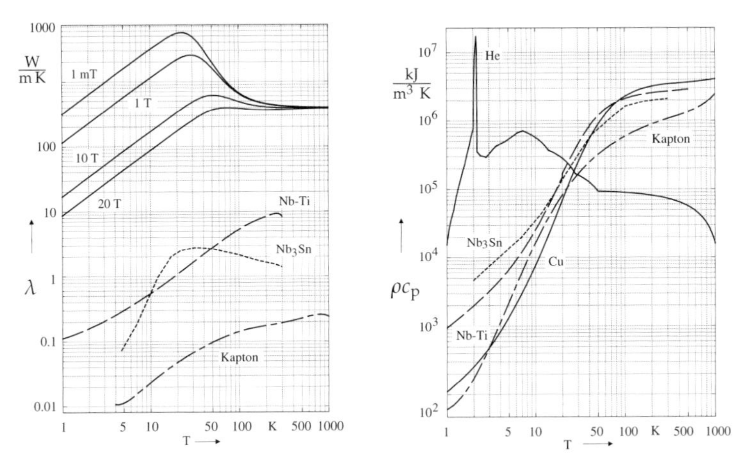

Figure 18.3 Left: Thermal conductivity of superconducting cable components as a function of temperature. Right: Temperature-dependence of the VHC. The curve for helium is scaled with the pressure-dependent mass density; see Appendix A.

The term on the left-hand side (the time integral over the square of the current) is usually expressed in units of $10^6 A^2 s$ and is termed MIIT by magnet designers. The MIITs represent the quench load and can easily be determined from the measured current decay of a quenching magnet. For a quenching LHC main dipole, the time constants are such that the result is a quench load in the range of $30 \times 10^6 A^2 s$. The term on the right-hand side of Eq. (18.6) represents the quench capacity of a specific cable under adiabatic conditions. Equation (18.6) establishes the relation between the *hot-spot temperature* T_{max} in the coil and the integral of I^2 over the duration of the quench.

More general expressions can be obtained by relating the quench load to the square of the total cross-sectional area of the cable. Following Iwasa [24], we will define the Z function as

$$Z(T_0, T_{max}) := \frac{a_{Cu}}{a_t} \int_{T_0}^{T_{max}} \frac{\rho c_p(T)}{\rho_E(B_0, T, RRR)} dT, \tag{18.7}$$

which determines the hot-spot temperature as a function of the quench load for different materials and magnetic flux densities; see Figure 18.4. Because the field changes with current and depends on the cable's position within the coil, the method yields only an approximation of the hot-spot temperature.

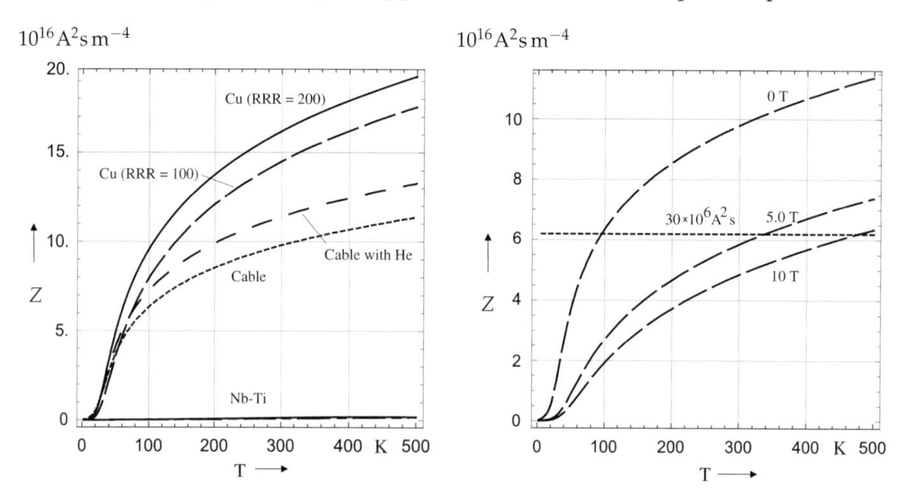

Figure 18.4 Left: Z functions of copper and components of LHC cable 2; see Table 1.3. Right: Z functions of the composite cable for three levels of applied magnetic flux density. The horizontal line represents $30 \times 10^6\ A^2\,s$ related to the squared cross-sectional area of cable 2. Notice the uncertainty in the estimation of the peak temperature due to the time and position-dependent magnetic flux density.

18.2
Electrical Network Models of Superconductors

Superconducting elements can be modeled as current-controlled voltage sources in an electrical network. Because of the nonlinear $E(J)$ characteristic of the superconductor given by Eq. (16.3), the branch currents and voltages cannot be determined by a simple inversion of the network matrix, but require an iterative method of solution. For this purpose, consider the combination of a superconducting (SC) and a resistive element (NC) in simple electrical networks shown in Figure 18.5. We will study five different network configurations:

1. *A parallel-connected resistor and superconductor driven by a current source*: The superconductor is modeled as a current-controlled voltage source $U_{SC}(I_{SC})$. The current is divided in two branch currents. The current in the superconductor determines the voltage across the terminals

and therefore also the current in the resistor. This is the *current-sharing* regime, representing a quenching copper-stabilized wire.

2. *A series-connected superconductor and resistor, driven by a current source*: The current in both elements determines the total voltage across the terminals. This is the model for a partially quenched coil.
3. *A parallel-connected resistor and superconductor driven by a voltage source*: The voltage across the terminals determines the two branch currents.
4. *A series-connected superconductor and resistor, driven by a voltage source*: This voltage divider can be used to calculate inductive voltages in superconducting loops containing resistive joints.
5. *A parallel-connected resistor and superconductor driven by a current source*: The superconductor is represented by a controlled current source, which carries at maximum the critical current I_c. The excess current to I_c flows in the resistive matrix material of the conductor. This is an approximation for a quenching wire according to Stekly [46].

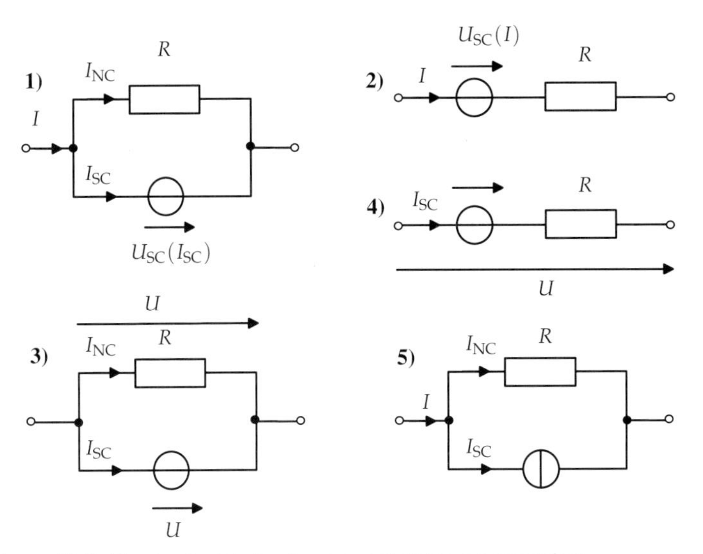

Figure 18.5 Simple electrical networks containing superconducting (SC) and normal-conducting (NC) elements. (1) Current-sharing regime of a quenching, copper-stabilized wire. (2) Voltage across a partially quenched coil. (4) Inductive voltages in a superconducting current loop with resistive joints. (5) Network describing the approximation for a quenching wire according to Stekly.

While the circuit equations for model 2, 3, and 5 can easily be solved and implemented in a quench-simulation program, the circuit equations for model 1 and 4 require linearization or iterative solution techniques.

18.3 Current Sharing

Disregarding the inductive coupling between the two branches in model 1, and using the power law,

$$U_{SC} = U_c \left(\frac{I_{SC}}{I_c} \right)^n , \tag{18.8}$$

the current in the superconductor I_{SC} can be calculated from the implicit equation,

$$R\,(I - I_{SC}) = U_c \left(\frac{I_{SC}}{I_c} \right)^n , \tag{18.9}$$

where R denotes the temperature and field-dependent resistance of the copper matrix. In Eqs. (18.8) and (18.9), the reference voltage U_c determines the critical current I_c.

The implicit equation (18.9) results in a high computational cost, especially for large n-values. Following Stekly [46], we distinguish two phases in the quench process of a superconducting composite wire. In the superconducting state, the transport current I flows entirely in the superconducting fraction of the conductor, so that $I_{SC} = I$. The temperature in the conductor increases solely from external heating and induced losses. The conductor is said to have quenched when the critical current I_c drops below the transport current. The excess current commutates into the normal conducting matrix, so that $I_{SC} + I_{NC} = I$. Ohmic heating due to I_{NC} in the matrix material increases the temperature further. The second phase of the quench is reached when the critical current has dropped to zero and the transport current is carried entirely by the matrix material, so that $I_{NC} = I$. Using this approximation, the excess current in the matrix material is given by

$$I_{NC} = \begin{cases} 0 & I_c > I, \\ I - I_c & I > I_c,\ I_c > 0, \\ I & I_c = 0. \end{cases} \tag{18.10}$$

I_c is the critical current, which depends on the cable size and parameters, the local magnetic flux density, and the temperature: $I_c = J_c(B,T)\, a_{SC}$.

Figure 18.6 (left) shows the current-sharing regime, calculated according to Eq. (18.9), in comparison to the approximation using Stekly's method and a simple step function. We note the following results:

1. *Step-function approximation*: When the applied current density in the superconductor equals the critical current density, the current commutates from the superconductor into the normal conductor (dashed line).

2. *Approximation according to Stekly*: When the current density in the superconductor exceeds the critical current density, the excess current is diverted into the copper matrix. The current density in the superconductor remains at the level of the critical current density (dotted line). The temperature at which this transition is initiated is called the current-sharing temperature T_{cs}.
3. *Iterative solution*: The current commutation from the superconductor (solid line) to the normal conductor (dash-dotted line) starts at higher temperatures compared to the Stekly approximation.

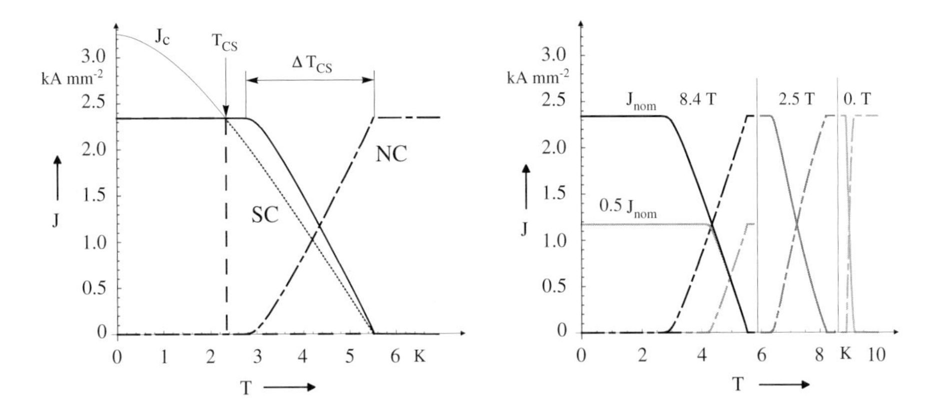

Figure 18.6 Current sharing in a Nb–Ti composite wire with a copper-to-superconductor area ratio of 1. The copper RRR is 200, and the n value is 40. Left: Different models for current sharing: iterative solution, approximation according to Stekly (dotted), and step function. The assumed current density of $2.34\,\mathrm{kA\,mm^{-1}}$ is 10% below the critical current density at $8.33\,\mathrm{T}$ and $1.9\,\mathrm{K}$. Right: Current sharing for different current and flux densities (iterative solution).

The Stekly approximation yields smaller currents in the superconductor and the ohmic losses are thus overestimated. For a small temperature rise and strong cooling, the stability of the superconductor will consequently also be underestimated. For the simulation of a quenching magnet, where the dissipated losses and the heat flow from the quench heaters exceed the cooling capability by orders of magnitude, the differences between the two models are insignificant.

The Stekly model, used henceforth for the simulation of the quench process, can be simplified with a linear approximation of the $I_c(B,T)$ relation

$$I_c(B,T) = I_c(B,T_b)\frac{T_c - T}{T_c - T_b}, \qquad T < T_c \tag{18.11}$$

where T_b is the bath temperature of the coolant, and T_c the critical temperature for the applied field B. The current-sharing temperature is identical to

T_c for zero current and drops to the bath temperature T_b when the transport current approaches the critical current; $I \rightarrow I_c(B, T_b)$. Assuming a linear relationship, the current-sharing temperature can be expressed as a function of the transport current [32]:

$$T_{cs} = T_b + (T_c - T_b)\left(1 - \frac{I}{I_c(B, T_b)}\right). \tag{18.12}$$

The ohmic losses in the copper matrix for the three temperature regimes are

$$P(T) = \begin{cases} 0 & T < T_{cs}, \\ \frac{\rho_E\, l}{a_{Cu}}(I - I_c(B, T))^2 & T_{cs} < T < T_c, \\ \frac{\rho_E\, l}{a_{Cu}} I^2 & T > T_c. \end{cases} \tag{18.13}$$

Let the temperature range ΔT_{cs} be defined as the interval between T_{cs} and the temperature at which the entire transport current flows in the resistive material. Figure 18.6 (right) shows the current-sharing regime (iterative solution) for different transport current and magnetic flux densities [42]. ΔT_{cs} decreases with field, current density, and n-value; its maximum is approximately 3 K. This justifies the use of the Stekly and step function approximations for the simulation of quenching magnets with an increase of temperature in the range of 100–300 K.

18.4 Winding Schemes and Equivalent Electrical Circuit Diagrams

As the normal zone propagates along the superconducting cable, the coil-winding scheme of the magnet with all its internal and external connections must be taken into account. As an example, we show the electrical connection of the twin-aperture LHC main dipole in Figure 18.7. Current entering the A terminal of the magnet will first flow through the lower, outer-layer coil of the inner aperture. This is the aperture seen on the right-hand side by an observer in the LHC tunnel looking downstream.[3] The current will then enter the lower-inner, upper-inner, and upper-outer coils, before it is directed to the outer aperture. Figure 18.7 also shows the *quench-heater* circuits and the outer-layer cables covered by the heaters. Only the high-field heaters (circuits YT211, YT221, YT111, YT121) are usually powered; the low-field heaters are mounted for redundancy. Figure 18.8 shows the positions of the *voltage taps* routed out of the coldmass by means of the instrumentation feedthrough system seen in Figure 1.19. The signals from the voltage taps are used for the detection of the normal zones in the magnet and the internal busbars.

3 The conventions for the positioning of the magnets in the LHC tunnel are given in Chapter 12.

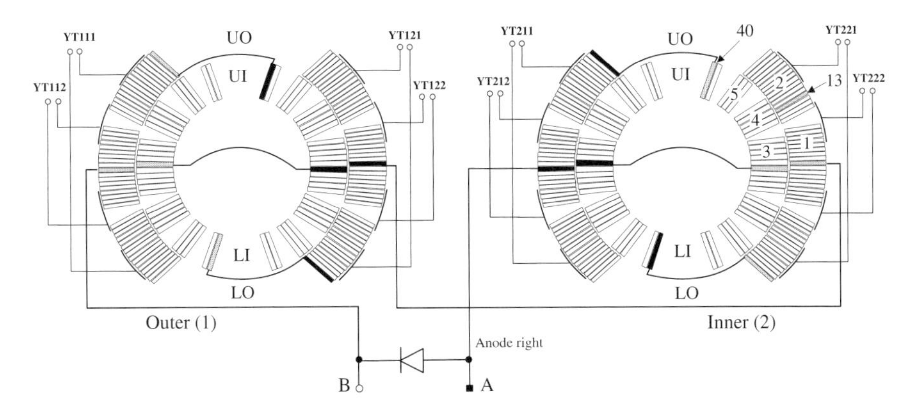

Figure 18.7 Winding scheme and internal connections in the double-aperture LHC main dipole magnet, together with the coverage of the quench heaters.

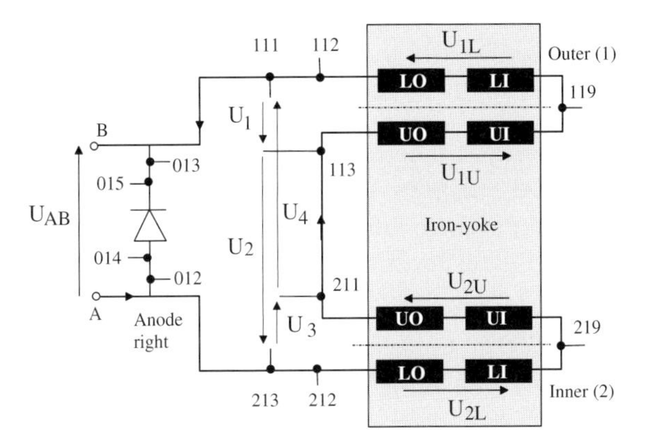

Figure 18.8 Electrical circuit diagram of the LHC main dipole with the position of the voltage taps. UO = upper-outer, LI = Lower-inner, etc.

18.5 Quench Detection

An incipient quench is detected by the resistive voltage rise across the normal zone, and must be distinguished from the inductive voltage during the ramping of the magnet. Figure 18.9 shows two different systems for accomplishing this task. The bridge detection system (left) records the voltage difference between the two apertures of the magnet. The inductive voltage is thus fully compensated. When the bridge is unbalanced, the voltage across the mid terminals is $U^* = (U_1 - U_2)/2 \neq 0$, according to Kirchhoff's voltage law. The bridge system may fail when the normal zone propagates symmetrically in the two apertures. In this case, the voltage could rise until the threshold voltage of the bypass diode is reached. An alternative quench detection system electronically compensates the inductive voltage across the magnet by means of a reference system. This reference system may be an adjacent magnet, a coil inside the magnet, or a co-wound wire within the cable. The difficulty with electronic compensation begins when iron saturation and dynamic effects result in a nonlinear or hysteretic voltage signal.

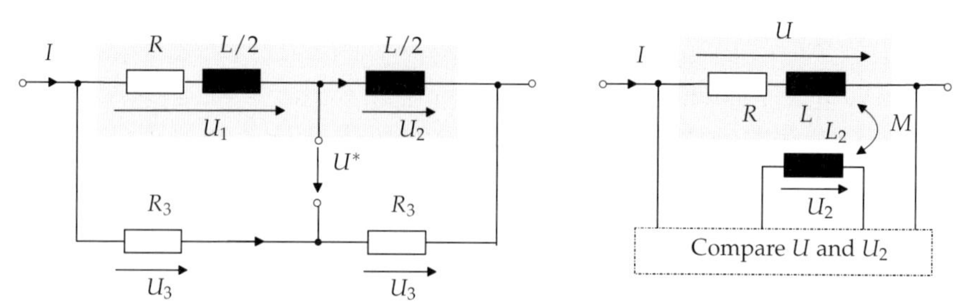

Figure 18.9 Left: Floating bridge detection system. $U^* = (U_2 - U_1)/2$. Right: Electronic voltage compensation.

The magnet is considered as quenched when the resistive voltage exceeds a threshold for a time interval called the *discrimination time span*. This slight delay is needed to exclude noise and reduce false triggers of the quench detection system. The overall quench-detection time is a crucial parameter for magnet protection as it determines a major part of the quench load, independent of any magnet protection scheme.

On a test bench it is also possible to detect a quench by measuring the aperture field change stemming from the current redistribution in the coil. Arrays of pickup coils covering the full length of the magnet allow the localization of the quench origin and the measurement of the transverse and longitudinal quench propagation velocities [43].

18.6 Magnet Protection

Magnet protection schemes can be classified in two groups. *Passive protection* schemes may include a diode or a resistor connected in parallel to the magnet, but principally they rely on a strong stabilization of the conductor such that the magnet can withstand the current decay without overheating. *Cryogenic stabilization* is established when the copper matrix is large enough to carry the entire transport current in the event of a quench. A full cryogenic stabilization guarantees that the superconductor will recover from a quench, as the ohmic heating in the matrix material will be smaller than the rate of heat dissipation into the helium bath. Full cryogenic stabilization typically requires a copper-to-superconductor area ratio between 10 and 20 [32]. Because of size constraints on accelerator magnets, we are mainly concerned with active protection schemes.

We speak of *active protection* when measures are taken to speed up the normal-zone propagation and the current decay in the magnet. The current decay is not only affected by the propagation speed of the normal zone but also strongly influenced by an external electrical circuit. Because of the high inductance of superconducting magnet circuits, the current cannot be switched off instantaneously, and therefore the power supply is short-circuited with a freewheeling diode. The current-decay rate is given by the inductance and resistance of the remaining circuit; the time constant is $\tau = L/R$.

If the magnet is series-connected in a string of magnets, the time constant can be reduced by isolating the quenching magnet from the string by means of a bypass diode. The main magnets of the LHC are bypassed, as in HERA and RHIC, by diodes operating at cryogenic temperatures. Compared to diodes at room temperature, the threshold voltage is significantly higher (6–8 V). When the voltage drop across the normal zone reaches the threshold voltage of the diode, the current in the main circuit begins to commutate into the diode. The magnet discharges independently across the diode with a current flowing in reverse direction. The maximum ramp rate of the magnet string is limited by the diode's threshold voltage. For a quench during the up-ramp phase, the switching of the diode is delayed by the inductive voltage across the magnet.

Alternatively to a bypass diode, a magnet can be protected by a parallel resistor. The size of this resistor is determined by the maximum tolerable leakage current and the resulting heat load into the cryogenic system.

Active protection often relies on quench heaters and an energy extraction system. The heaters cause a resistive transition in the covered coil windings, ensuring that the stored energy is dissipated over a larger fraction of the coil volume. The rising resistance decreases the discharge time constant and thus reduces the hot-spot temperature. A quench-heater circuit consists of the austenitic steel heater strip itself, a high-speed electromechanical circuit breaker, and an aluminum electrolytic capacitor bank. A triggered thyristor discharges the capacitor bank across the resistance of the heater.

The time between the triggering of the quench heater's power supply and the development of a quench in the covered coil winding, referred to as the *quench-heater delay*, depends on the working point of the covered coil winding, the dissipated power, the thickness of the insulation between the coil and heater strip, and the cooling conditions.

As an example, we give typical values for the LHC main dipole protection system: The threshold voltage is 0.1 V. The quench heaters are triggered after a delay of 10 ms needed for signal validation [16]. The time constant for the dissipated heater power is 37 ms. Measurements indicate that a heater-provoked quench at 1.5 kA occurs at around 80 ms after the trigger. At nominal current, the delay decreases to 35 ms [35].

For the remaining string of magnets, the current decay rate can be increased with an *energy extraction system* consisting of a high-current switch and an extraction resistor. The energy extraction system protects the bypass diode and busbar of the quenching magnet from overheating. The maximum dissipated energy in the diode determines the lowest possible resistance of the protection resistor. The maximum resistance is determined by the allowable voltage to ground in the string. The maximum current decay must be limited to reduce eddy-current-induced quenches in the magnet string. The limit is about $125\ \mathrm{A\,s^{-1}}$ for the LHC main dipole circuits at nominal excitation. This corresponds to a time constant on the order of 100 s.

Two sets of three parallel-connected 225 mΩ resistors, resulting in a total resistance of 150 mΩ, are installed at the adjacent points of each string of dipole magnets. The maximum voltage to ground is therefore ±475 V. One resistor with a volume of $3\ \mathrm{m^3}$ and a weight of 1.8 t absorbs 230 MJ during a full discharge [15]. With a maximum resistor temperature of 350 °C it takes about 2 h to cool the resistor to room temperature.

The division of the magnet into several loops decreases the circuit inductance, allowing the current in the quenching loop to decay faster. This protection scheme is known as *subdivision* and is often used for solenoidal magnets. An inductive coupling to a secondary circuit of finite resistance allows energy to be transferred away from the quenching magnet.

Figure 18.10 shows an energy extraction study, wherein the peak temperature and the fraction of the extracted to stored magnetic energy are calculated as a function of the magnet's terminal voltage [42]. The study is performed for a stand-alone, wide-aperture quadrupole magnet, shown in Figure 16.18 (left), connected to an energy extraction system with protection resistor and circuit breaker.

Figure 18.10 (left) gives the results for the magnet still superconducting at the time of the switch opening. A small protection resistance allows the slow extraction of the entire energy from the magnet. When a larger resistance is used, a quench results from eddy-current-induced losses. This effect is known as *quench-back*.[4] The peak temperature in the coil windings remains relatively low (< 100 K) because quench-back causes resistive transition of a large number of cables. The dip in the temperature curve at around 100 V can be explained by a movement of the hot spot from the inner- to the outer-layer coil.

4 Green [20] distinguishes between normal zones induced by heat transfer (thermal quench-back) and those induced by eddy-current losses (magnetic quench-back). We use the term quench-back for the latter.

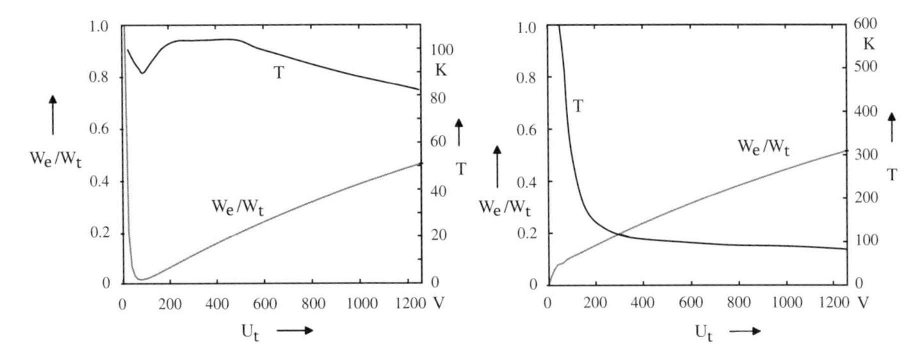

Figure 18.10 Peak temperature and ratio of extracted W_e to stored (total) magnetic energy W_t, as a function of the magnet's terminal voltage. Left: For a magnet still superconducting at the time of the switch opening. Right: For a magnet already quenching at the time of the switch opening.

A completely different behavior can be seen in Figure 18.10 (right). In the event that the magnet is already quenching at the time of the switch opening, a small protection resistance leaves the magnet essentially unprotected. The hot-spot temperature exceeds 600 K in the coil winding where the quench initiated. The ratio of extracted to stored energy grows with the protection resistance. When the current decay is sufficiently fast to provoke quench-back, the peak temperature will drop below the critical values.

18.7 Numerical Quench Simulation

A flexible quench-simulation software must include models for the following phenomena:

- Thermodynamic effects of cooling and quench propagation determined by highly nonlinear material properties.
- Electrical circuit behavior of the magnet.
- Electromagnetic behavior of the magnet, including eddy currents in the Rutherford cables and the superconductor magnetization in the filaments.
- Iron saturation.

Critical parameters for magnet design are the peak voltage and hot-spot temperature during a quench, as well as the signals at the voltage taps. In view of new challenges, such as fast-ramping magnets and high-field magnets exceeding 10 T central field, these parameters must be taken into account from the early stages of magnet design and optimization.

The field distribution in the coil is calculated by means of the coupling method between boundary and finite elements (BEM–FEM). For accelerator magnets, the numerical field computation can be often restricted to two dimensions. The rapid current decay during a quench creates losses due to the interfilament coupling currents (IFCC) and the interstrand coupling currents (ISCC); see Chapter 17. The time constants of the coupling currents are influenced by the copper resistivity, the contact resistances, and geometric cable parameters, all of which are input parameters of the simulation.

In the superconducting state, the working point lies below the critical surface. At every time step, the cable's temperature margin to the critical surface is evaluated as a function of the peak field and current density. The copper resistivity is calculated as a function of the temperature and the average magnetic flux density[5] in the cable cross section.

The dissipated power in each cable is the sum of ohmic losses due to the current in the copper matrix of the normal zone, losses due to coupling currents (IFCCs and ISCCs), the transmitted power from the quench heaters, and beam losses. The coupling-current losses can be neglected once the cable has quenched. Quench heaters are characterized by the maximum heating power and the time constant of the exponential decay. These parameters can be used to validate the model with the measured thermal coupling between the heaters and the coil.

The electromagnetic and thermal models exhibit different time constants. Moreover, field calculation is computationally more expensive than the solution of the coupled electrical and thermal network equations. Hence a weak coupling between the electromagnetic and thermal models is the most efficient method to solve the multiphysics problem.

Owing to the large number of empirical parameters to be specified, it must be noted that an important part of the simulation work is in validating the model in order to obtain a match between simulations and measurements. However, when

- all relevant phenomena have been modeled accurately,
- the material parameters are given within the range of measurement uncertainty and within physically reasonable limits, and
- the simulation results have been validated with the measured data,

we will be able to reproduce the internal states of a quenching magnet and thus determine quantities that are generally not measured due to a lack of

5 The average magnetic flux density is taken because of the transposition of the strands within the cable and the 2D approximation for the field problem.

instrumentation. Examples of such quantities are hot-spot temperature and turn-to-turn voltages. With a validated quench simulation, we will be able to study the optimal position of quench heaters, determine the most appropriate size of an extraction resistor, and perform sensitivity studies with respect to the quench origin in the coil.

18.7.1 The Thermal Model

For the numerical calculation of the quench propagation, the magnetic field distribution and the current density are assumed to be independent of the magnet's longitudinal direction. The magnetic flux density and vector potential are obtained from a 2D BEM–FEM calculation. The coil is longitudinally discretized into J sections for the thermal model. To save on computing time, the local field changes in the coil end are ignored, and the coil windings are modeled as electrically and thermally short-circuited at the ends.

The thermal model is based on the heat-balance equation

$$\rho c_{\mathrm{p}}(T,B)\frac{\partial T}{\partial t} = P + \mathrm{div}\ (\lambda(T,B)\ \mathrm{grad}\, T)\ , \tag{18.14}$$

where T is the temperature, B the modulus of the applied magnetic flux density, and P the heating power. The material parameters are the VHC ρc_{p} and the thermal conductivity λ. Equation (18.14) does not account for convective heat transfer, and therefore not for cooling by helium mass-flow. The coupling to the external circuit is accomplished by the current-dependence of the magnetic flux density. The initial condition can be written as $T(t=0,\mathbf{r}) = T_0(\mathbf{r})$, and $I(t=0) = I_0$, where $\mathbf{r}$ is the position vector of a point in the coil volume.

By using network analysis on finite volumes, Eq. (18.14) can be transformed into a system of ordinary differential equations. In the coil cross section each *coil winding*[6] is represented by one node in the network; the longitudinal discretization is a user-supplied parameter. For the calculation of the heat flow, uniform temperature and material parameters are assumed within the finite volumes.

Let $i = 1,\ldots,I$ be the index of the coil windings in the cross section and $j = 1,\ldots,J$ the index of the longitudinal subdivisions. The total number K of finite volumes (cable segments) is given by $I \cdot J$. The chain of cable segments between the connection terminals of the coil will be indexed by $k = 1,\ldots,K$. The index varies according to the winding scheme of the coils and their inter-

6 The term coil winding refers to a cable half-turn in the magnet cross section.

nal interconnections. The temperature in a cable segment is computed with the discrete form of the heat-balance equation

$$\frac{\partial T_{ij}}{\partial t} = \frac{1}{C_\mathrm{p}} \left[P_{ij} - \left(S_{ij}^{\mathrm{trans}} + S_{k(ij)}^{\mathrm{long}} + S_{ij}^{\mathrm{cool}} \right) \right], \tag{18.15}$$

where S denotes the longitudinal and transverse heat flow between adjacent cable segments, and to the coolant; P denotes the total dissipated power. The heat capacity of the element ij is given by

$$C_\mathrm{p} = V_{ij} \rho c_\mathrm{p}(T_{ij}, B_i), \tag{18.16}$$

where ρc_p is the average VHC and B_i the average flux density. The longitudinal heat flow is given by

$$S_k^{\mathrm{long}} = \frac{p_{k-1,k}}{R_{k-1,k}} (T_k - T_{k-1}) + \frac{p_{k,k+1}}{R_{k,k+1}} (T_k - T_{k+1}). \tag{18.17}$$

The heat flow between two cable segments can be modified by the parameter p to account for varying thermal conductivity, for example, due to additional copper stabilizers. There is no heat flow in and out of the connection terminals: $p_{0,1} = p_{K,K+1} = 0$. The longitudinal thermal resistance $R_{k,k+1}$ is given by

$$R_{k,k+1} = \frac{l}{a_i \lambda_\mathrm{L}(T, B_i)}, \tag{18.18}$$

where l is the segment length, and $\lambda_\mathrm{L}(T, B)$ is the longitudinal heat conductivity as a function of the mean temperature of longitudinally-connected cable segments, $T = (T_k + T_{k+1})/2$. Comparing the thermal conductivity of the materials in Figure 18.3 (left), reveals that only the copper conductivity must be accounted for. Its heat conductivity is deduced from the electrical resistivity ρ_E by the Wiedemann–Franz–Lorenz law[7] [4]

$$\lambda = \frac{L_0 T}{\rho_\mathrm{E}(T, B)}, \tag{18.19}$$

where L_0 is the Lorenz number, $L_0 = 2.44 \times 10^{-8}\,\mathrm{V^2\,K^{-2}}$. The transverse heat flow is modeled by

$$S_{ij}^{\mathrm{trans}} = \sum_{n=1, n \neq i}^{I} G_{ij,nj} (T_{ij} - T_{nj}), \tag{18.20}$$

in which the thermal conductance is given by

$$G_{ij,nj} = \frac{a_i + a_j}{2\,d(i,j)} \lambda_\mathrm{T}(T, B), \tag{18.21}$$

7 Gustav Wiedemann (1826–1899), Rudolph Franz (1827–1902), Ludvig Lorenz (1829–1891).

where $d(i, j)$ is the distance between the faces (surface areas a_i and a_j) of the cable segments, and $\lambda(T, B)$ is the thermal conductivity of the cable segments at a mean temperature $T = (T_{ij} + T_{nj})/2$ and flux density $B = (B_{ij} + B_{nj})/2$. Cooling is account for by a global node at fixed temperature T_b. The thermal conductance to this heat sink is calculated as between two coil windings,

$$S_{ij}^{\text{cool}} = G_{ij} \left(T_{ij} - T_b \right) . \tag{18.22}$$

This allows the simulation of the heat extraction under optimistic conditions, without having to consider an increase in the bath temperature or the extraction limit of the heat exchanger.

The heat transfer to liquid helium is treated in a simplified model. The thermal conductivity of phase II helium, below the lambda point exceeds the thermal conductivity of copper by several orders of magnitude. Phase I liquid helium, supercritical helium, and gaseous helium have negligible thermal conductivity. Moreover, the VHC of both phase I and II liquid helium exceeds that of copper and niobium–titanium by orders of magnitude; see Figure 18.3.

The VHC of helium is a function of pressure. We can make the following simplifying assumptions on mass and density of the confined helium.

Below the lambda-point, the heat conduction between adjacent coil windings is dominated by the helium percolating through the insulation and thus the heat conduction is limited only by the cable material. Superfluid helium looses its perfect heat flow properties above a maximum heat flow [24], which is a user-supplied parameter in the simulation. For potted coils it must be set to zero.

Above the lambda point, the thermal conductivity of helium can be disregarded. The confined helium is assumed to be in close contact with the strands and thus contributes to the overall VHC of the cable composite material. Variations in the thermal contact due to nucleate and film boiling are not taken into account. During a quench the helium is adiabatically compressed until the pressure in the cable rises to an upper limit of about 20 bar, set by the pressure release valves. Subsequently, helium is heated at constant pressure and thus the mass of the confined helium decreases proportionally. Figure 18.3 shows the VHC of the confined helium scaled from measurements published in [47].

Quench heaters are modeled as heat sources inside the coil winding by three user-supplied parameters: the effective initial heating power P_i^0, the internal delay t_i, and the discharge time constant τ. While τ is given by the heater powering circuit, the other two parameters can be used to reproduce measured heater delays. The parameters may vary among different heater circuits and are therefore assigned to the covered coil windings by

$$P_i = P_i^0 \exp \left[- \left(\frac{t - t_i}{\tau} \right) \right] , \qquad t > t_i . \tag{18.23}$$

The heater delay depends on the working point of the covered coil winding. For heater circuits of the LHC main dipoles, a delay of 30 ms and a power amplitude of $20\,\mathrm{W\,m^{-1}}$ per cable reproduce the measured heater delays at 1.5 kA and at nominal current. The heater delays of the different circuits have a tolerance of ± 5 ms [44].

18.7.2 External Electrical Circuits

The series-connected LHC main magnets are protected by high-current bypass diodes. An energy extraction system allows a faster discharge of the entire string and reduces magnet-to-magnet quench propagation. The diodes are mounted inside the liquid-helium vessel of the coldmass, thus operating at cryogenic temperatures with a maximum rise to 300 K during a quench. At standby operation, the diodes have a turn-on voltage of about 6 V. This is high enough to cope with the inductive voltage across the magnet terminals during the current ramp. When a quench has created a sufficient resistance in the magnet, the diode begins to conduct and the knee voltage drops to 1–2 V.

The external network and powering scheme can be modeled as a generic network with lumped elements, consisting of a power supply with bypass diode (also called a *crowbar*), a cold protection diode with the threshold voltage U_{th} , and an extraction resistor R_p; see Figure 18.11.

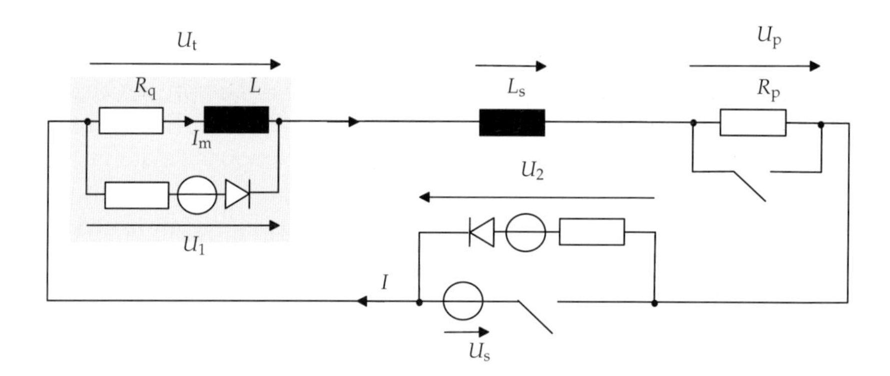

Figure 18.11 Electrical network model of a quenching magnet in a string.

The voltage/current characteristics $U_1(I)$ of the diodes is modeled by a voltage source with constant threshold voltage, a differential resistance defined by

$R^d := dU/dI$, and an ideal diode. The quenching magnet is represented by its resistance $R_q(t)$ and differential inductance,[8]

$$L := L^d(I_m) = \frac{d\Phi}{dI_m}, \tag{18.24}$$

where Φ denotes the flux linkage in the coil.

After the extraction resistor is switched in, the circuit current can be calculated by solving the differential equation

$$\frac{dI}{dt} = -\frac{(R_q + R_p)\, I + U_2}{L + L_s}, \tag{18.25}$$

where R_p is the protection resistance, U_2 the voltage across the crow-bar, and L_s the total inductance of the magnet string.

When the propagation of the normal zone causes the resistance R_q in the quenching magnet to rise, the terminal voltage increases until it reaches the threshold voltage of the cold bypass diode. The diode becomes conductive and the current commutates from the magnet into the diode branch. Then the behavior of the circuit is determined by the equations

$$\frac{dI}{dt} = -\frac{R_p I + U_1 + U_2}{L_s}, \tag{18.26}$$

$$\frac{dI_m}{dt} = -\frac{R_q I_m - U_1}{L}. \tag{18.27}$$

Stand-alone magnets, excited by their dedicated power supply, need not be isolated from the large inductance of the string - for the LHC main magnets on the test bench, the diode branch is disconnected. In this case $I_m = I$, and the network equation reduces to

$$\frac{dI}{dt} = -\frac{(R_q + R_p)\, I + U_2}{L}. \tag{18.28}$$

18.8 The Time-Stepping Algorithm

The thermal-network and electrical-network equations are coupled, nonlinear, inhomogeneous differential equations of first order and are soluble with the classical *fourth-order Runge–Kutta algorithm*. Adaptive time stepping is necessary due to the highly nonlinear material parameters, especially the VHC at cryogenic temperatures, and the normal-zone propagation in the cables.

8 See Section 5.11.1.

The calculation of the magnetic field is computationally more demanding than the evaluation of a time step in the thermal model. We therefore implement a weak coupling between thermal and magnetic computations, whereby the magnetic field is updated only when the stored energy has dropped by a user-supplied fraction (usually in the range of a few percent); see Figure 18.12.

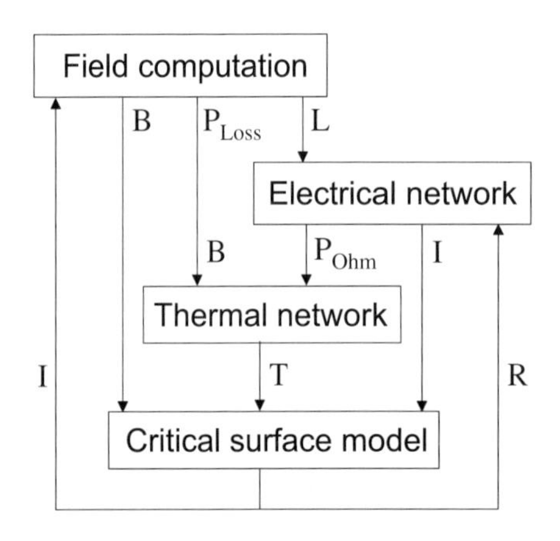

Figure 18.12 The time-stepping algorithm with a weak coupling between thermal and magnetic computations.

The induced voltage is calculated in all coil turns from the time derivative of the linked flux. To evaluate resistive voltages we can interpolate the resistivities before and after each time step. The voltage to ground of each coil winding as well as the terminal voltage are calculated by summing all internal voltages in accordance with the winding scheme.

The thermal and electrical network equations are integrated with the explicit fourth–order Runge–Kutta algorithm. The step-size is controlled by a modified Collatz method [12].

18.9 Applications

The quench-simulation technique will first be demonstrated and validated at the example of the LHC main dipole magnets. The saturation of the iron yoke has a relatively small impact on the differential inductance, which varies only by 5% between injection and nominal excitation. For thermal calculations, an average length of 14.57 m is used for each turn.

The critical current density $J_c(B, T)$ in the strands is given by the fit function (16.9). The RRR of the copper matrix is in the range of 150–250 [9]. The interstrand contact resistances of the inner- and outer-layer cables are 30 and 60 $\mu\Omega$, respectively [31]. At a ramp rate of 7.5 mT s^{-1}, ac losses of 180 mW per meter can be expected in the magnet. In the absence of induced losses, and at nominal current (11.85 kA), the longitudinal quench propagation velocities are approximately 10–20 m s^{-1} and the turn-to-turn delays about 20 ms.

We will now simulate a quench at the nominal current of 11.85 kA incipient at winding 13 (block 2 of the outer layer); see Figure 18.7. Longitudinally, the quench origin is assumed to be located in the center of the magnet.

Figure 18.13 shows the temperature margin as a function of time for each winding within the coil cross section. Three phenomena can be distinguished: quench propagation, quench-heater-provoked quenching, and quench-back. The quench is triggered at $t_0 = 0$ in the outer-layer coil. The detection voltage is reached after 14.1 ms, the quench heaters are fired after a delay of 10 ms for signal validation. These instances are marked as t_1 and t_2 in Figure 18.13. Another 37 ms later, the energy dissipated by the heaters triggers the quenches in the covered coil windings. The heater delays depend on the operational margins of the cables. At t_3, marked in Figure 18.13, the diode threshold voltage is reached and the magnet current starts to decrease. Eddy-current losses create additional heating, which leads to quench-back at around 140 ms; see t_4 in Figure 18.13. The quench propagation in the outer layer is also accelerated by the induced losses. After 260 ms all the conductors in the magnet are in the normal-conducting state.

Figure 18.14 shows the electric potential in the coil windings. Interesting patterns occur when there is a quench-heater failure; here we assume that YT211 shown in Figure 18.7 cannot be powered. The ground potential is fixed at the first cable segment with the index $k = 1$.

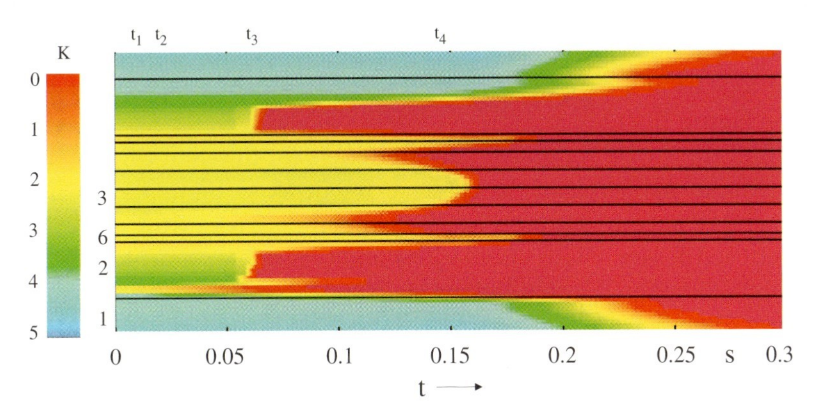

Figure 18.13 Temperature margin to quench in the cables as a function of time. The block numbers correspond to the convention in Figure 18.7. From the evolution of the temperature margin we can distinguish quench propagation, quench heater delay, and quench-back.

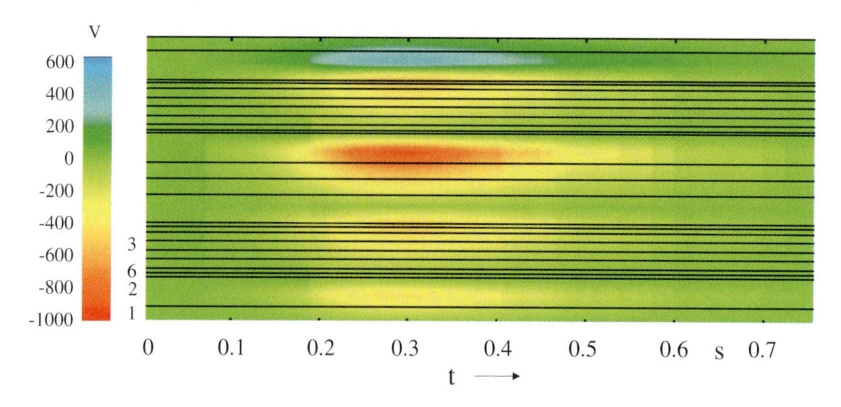

Figure 18.14 Electric potential in the coil windings as a function of time. The block numbers follow the convention in Figure 18.7.

18.9.1 Validating the Model

Current decay and voltages The simulations can now be compared with the measured current decay (Figure 18.15) and the voltage signals (Figure 18.16) during a training quench of the LHC main dipole magnet MB2381 [10]. For measurements on the test bench, the magnet is discharged over a diode, which bypasses the power converter and is mounted in an opposite direction to the magnet's bypass diode. When the power supply is switched off, the terminal voltage U_t jumps from the positive voltage over R_q to the negative forward voltage of the diode. The negative terminal voltage leads to a faster current decay than occurs in the operational conditions in the LHC tunnel. The quench is assumed to start at a current level of 12.82 kA in a pole turn (winding 40 indicated in Figure 18.7).

Figure 18.15 shows the simulated current decays for different RRR values, with and without modeling of the cable eddy currents. Simulation 1 uses RRR = 100 and disregards induced losses. Simulation 2 uses RRR = 200 and includes induced losses. Both simulations show good agreement with the measurement. Although the simulations yield similar current decay, the peak temperatures differ by 10 K. This example shows that the measurement of the current-decay curve alone is not sufficient to gauge the simulation with a large number of empirical parameters.

The measured and simulated voltages across the four poles of the twin-aperture magnet are shown in Figure 18.16. Small differences in the quench-heater delays cause asymmetric voltage distributions. The quench heaters in the numerical model were detuned by about 2 ms.

Both the measured and computed voltages show spikes during the first 100 ms. Eeach quenching cable causes a sudden increase in the resistive volt-

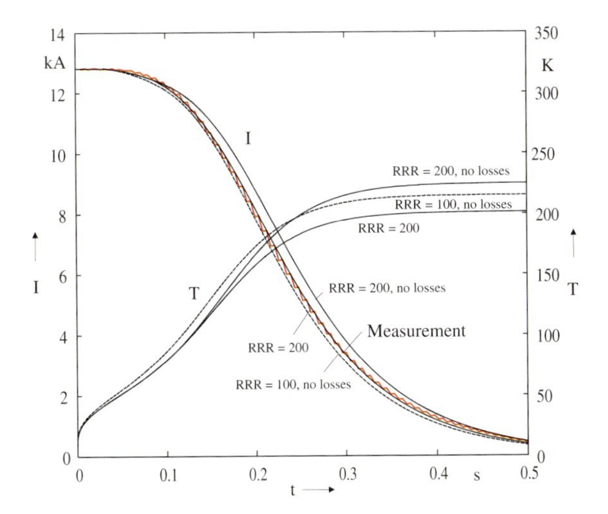

Figure 18.15 Measured current decay and simulations for different RRR values and models with and without quench-back. Both simulations (RRR = 100 no losses, and RRR = 200 with losses) show good agreement with the measurements. However, the simulated peak temperatures differ by about 10 K because quench-back helps to provide a more even dissipation of the stored energy over a larger coil volume.

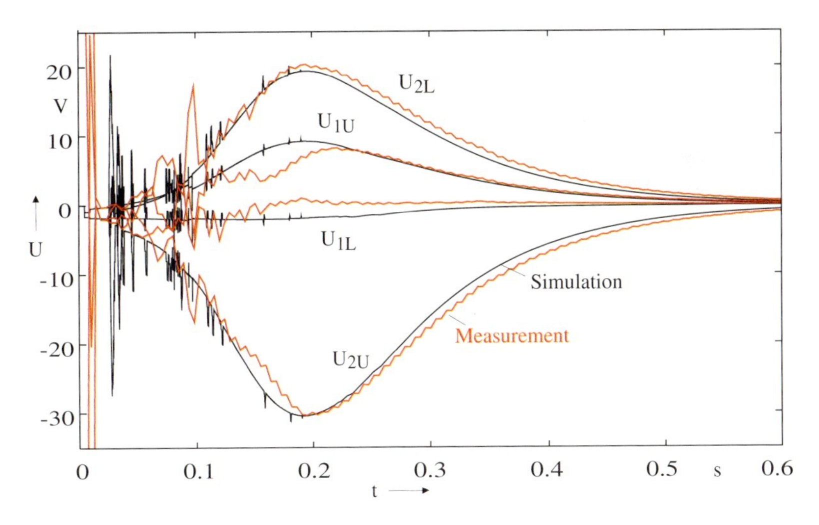

Figure 18.16 Measured and simulated pole-to-pole voltages. Naming convention according to Figure 18.8.

age. The magnet's terminal voltage, however, is clamped to the forward voltage of the diode. The increase in resistive voltage must therefore be redistributed evenly across the four poles, which results in a sudden change of all voltages. When all coil windings have quenched, the resistive voltage varies smoothly with temperature and consequently both the measured and simulated curves are smooth.

Longitudinal quench propagation The longitudinal quench propagation will now be studied for the LHC dipole corrector magnet MCBX. The magnet consists of two nested dipole coils of which only the outer-layer coil is studied. The coil is wound from a 7-strand, ribbon-type cable. The strands are connected at the magnet extremity such that the radial layers are connected in

Figure 18.17 Quench propagation in the MCBX outer-layer coil. The 3D view is longitudinally compressed by a factor of 10. The quench starts in the upper pole turn close to the end. Left: Temperature distribution after 0.1 s. Right: Temperature distribution when the current has dropped to zero. The maximum temperature is 180 K.

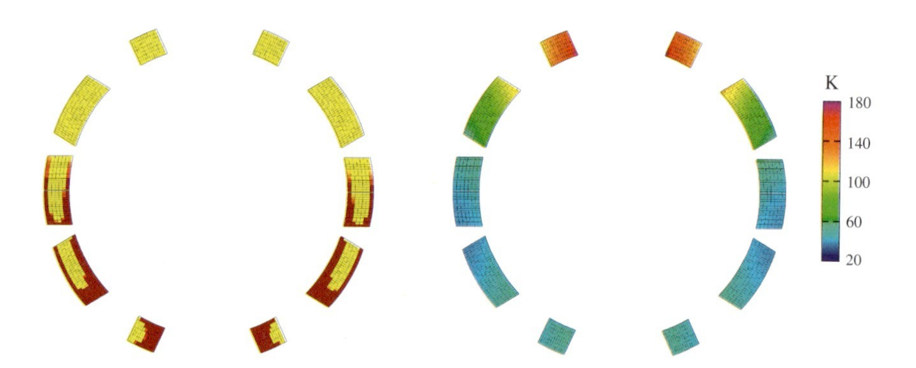

Figure 18.18 Left: Causes of the quench. Yellow: Quench triggered by normal-zone propagation. Red: Quench triggered by induced losses. Right: Peak temperature as a function of the position in the magnet cross section.

series. The coil is fully impregnated and therefore no confined helium needs to be considered. The magnet is operated at 1.9 K and the quench occurs at 734 A. For the quench simulations, the magnet protected by neither a quench heaters nor a protection resistor.

Figure 18.17 shows the quench propagation from the upper pole turn. Note that the 3D view is longitudinally compressed by the factor of 10. Simulations and measurements yield a turn-to-turn quench propagation velocity of $4\ \mathrm{m\,s^{-1}}$ and a longitudinal quench propagation velocity of $18\ \mathrm{m\,s^{-1}}$ [27].

By monitoring the integrated heat flow, induced losses, and quench heater power, the cause for the quench can be displayed for each conductor. Figure 18.18 shows that only in a few turns the quenches are triggered by induced losses. Quenches are triggered in the low field region mostly by the heat flow.

Figure 18.19 shows the maximum voltage to ground for all coil windings in the MCBX magnet.

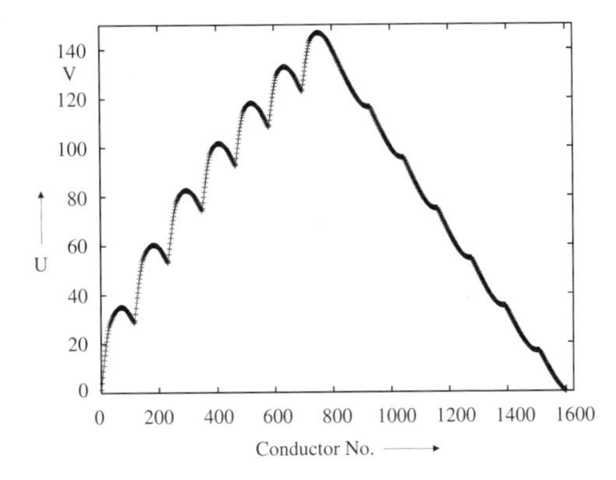

Figure 18.19 Voltage to ground of all coil windings in the MCBX magnet.

18.9.2
Fast Ramping Magnets

Ramp rates of $1\ \mathrm{T\,s^{-1}}$, corresponding to $1100\ \mathrm{A\,s^{-1}}$, are required for the SIS300 dipole magnets of the FAIR project [22]. The flat-top will last between 10 and 100 s. Considering an inductance of 25 mH for a 2.9-m-long magnet, the inductive voltage across the magnet terminals will be 27.5 V. The ring will be powered by several power converters in order to deal with the total voltage of the string. As an example, a string of 110 magnets would result in 3000 V.

Calculations show that a quench detection threshold of 1 V will be desirable during the ramp. The precision of the electronics must be at least an order of magnitude better, in order to cover unexpected phenomena such as parasitic transient effects. This causes no technological challenge at the flat-top. However, the inductive voltage rises in a short time to 27.5 V and the common mode voltage to ± 1500 V. It is especially during the acceleration and deceleration of the ramps that the required 50 dB signal-to-noise ratio[9] on a rapidly changing 84 dB common-mode background presents a challenge.

Because a magnet designed for fast-ramping operation can be ramped down faster than any SC magnet operating in present accelerators, an alternative protection scheme may be considered. It would be sufficient to detect a quench on the flat-top or at the injection plateau if the magnets could survive an undetected quench during the ramp. Bypass diodes with a high turn-on voltage would not be required and in any event would present a technological challenge. On the other hand, if state-of-the-art power converters with capacitive storage are used, even the extraction resistors and switches would be unnecessary.

The two-layer SIS300 dipole magnet [28] is now taken as an example. The input data for the quench simulations are summarized in Table 18.1. We account for conductive cooling via the polyimide insulation to the helium bath across the narrow faces of the cables. The excitation cycle is given in [28]: ramp from 1.6 to 6 T in 4.4 s, a plateau at 6 T for 11 s, followed by a down ramp to 1.6 T in 4.4 s.

Table 18.1 Data used for quench simulations based on the SIS300 dipole magnet [28] (η is the copper-to-superconductor area ratio).

Quantity	Value	Quantity	Value
Overall length	1 m	Detection threshold	0.1–1 V
Magnetic length	0.75 m	Quench heater delay	90 ms
Ramp rate	$1\,\mathrm{T\,s^{-1}}$	Extraction resistor	200 mΩ
L^d	7.5–9.7 mH	Switch delay	50 ms
Operation Temp.	4.7 K		
RRR	278	R_a	200 μΩ
η	1.38	Confined He	10% of cable voids
R_c	20 mΩ		

The temperature variation during an excitation cycle can be calculated with three different thermal diffusion models:

1. Adiabatic condition, no cooling and no confined helium, as is appropriate for potted coils.
2. Heat transfer across the cable surfaces without taking into account the confined helium.
3. Modeling of heat flow and accounting for the VHC of the confined helium.

9 Corresponding to a detection precision of 0.1 V over a signal of 27.5 V.

Figure 18.20 (left) shows the temperature variation during an excitation cycle for the three cases. The variation is an order of magnitude smaller for the wetted coils because the VHC of helium is dominant at low temperatures. In all cases the temperature decreases at the flat-top. For the adiabatic model this is due to the transverse thermal conduction. However, this is not sufficient to prevent a quench in the second excitation cycle. Henceforth we will use model 3 (with confined helium), which reproduces data published in [28].

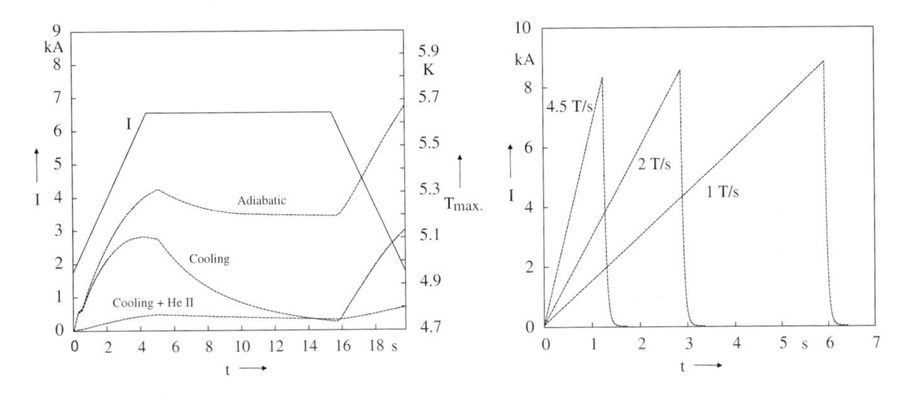

Figure 18.20 Left: Hot-spot temperature variation during an excitation cycle (no quench). Right: Quench current as a function of the ramp rate.

Ramp-rate-dependent quench limits Figure 18.20 (right) shows the excitation curves for quenches triggered near the flat-top. The magnet is ramped until a resistive voltage of 100 mV is detected. The magnet is protected with quench heaters and an extraction resistor; see Table 18.1. The ramp rate varies in the range of 1.0–4.5 T s^{-1}.

The temperature rises due to ramp-induced losses. With higher ramp rates the quench current and the peak temperature are reduced. Figure 18.21 shows the temperature margin as a function of time for each winding within the coil cross section. The results are given for a ramp rate of 4.5 T s^{-1}. The margin reduces monotonously as a consequence of increasing current and flux densities, and of the losses from the interstrand coupling currents. At $t = 1.18$ s the magnet quenches symmetrically in the second, inner-layer coil blocks seen from the poles, corresponding to block 5 in the numbering scheme shown in Figure 18.24.

The quench is detected at $t = 1.19$ s, and the quench heaters are fired at $t = 1.24$ s. At the same time, the protection resistor is switched in. The quench heaters are effective 40 ms later. Because of the fast decay of the current and flux densities, the temperature margin in the inner-layer coil increases after the

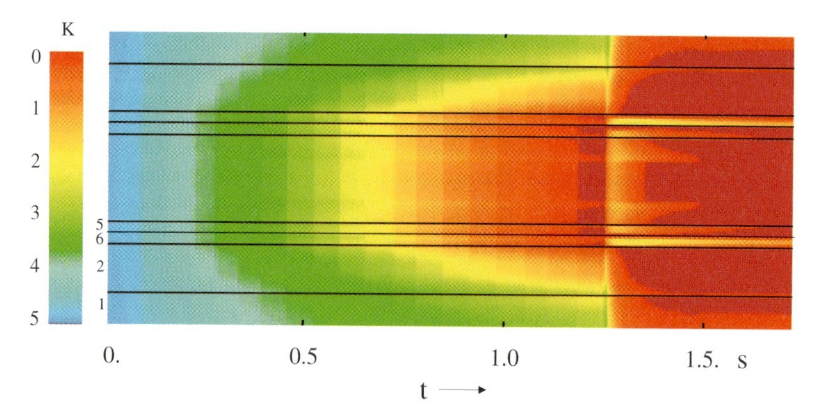

Figure 18.21 Temperature margin to quench (one pole) as a function of time for an up-ramp of 4.5 $\mathrm{T\,s^{-1}}$. The outer-layer coil blocks 1 and 2 are partially covered by quench heaters. After the opening of the extraction resistor switch, the inner-layer cables recover partially, before they are quenched as a result of cable eddy currents. Block-numbering scheme according to Figure 18.24.

extraction resistor is switched in. An additional reason for this phenomenon is the time constant of the induced eddy-current losses of about 50 ms, while the heat capacity of the confined helium results in a considerably longer thermal time constant.

The resistance built up as a result of the firing of the quench heaters creates an even faster current decay, which results in quench-back of the inner layer. Block 6 receives less-than-average induced losses because its broad side is positioned in parallel to the direction of the magnetic field.

Quench detection during the up- and down-ramp Let us now study the current decay and peak-temperature after a quench in the up-ramp phase: Quenches are triggered at 50% and 75% of the nominal field level I_{nom}. We will examine the cases in which quenches are detected at a resistive voltage of 100 mV and 1 V, and the worst-case wherein they are detected only at the flat-top. Figure 18.22 shows the current and peak temperatures for the six combinations.

The earlier the quench starts during the ramp, the higher will be the peak temperature. A detection threshold of 1 V is low enough to protect the magnet for $I_0 \geq 0.5\, I_{\mathrm{nom}}$. The delays of the protection system are longer at low excitation, since the quench propagation, and consequently the resistive voltage rise, are slower.

Passively protected magnets Finally, we will study how much copper is required in the cables for the magnet to withstand an undetected quench during

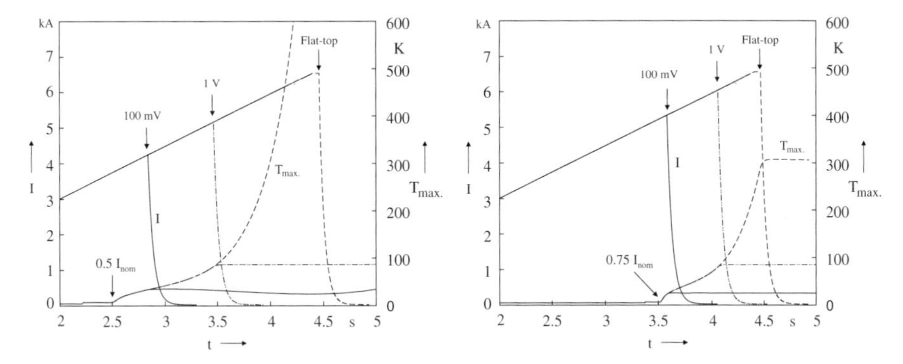

Figure 18.22 Current decay and peak-temperature for quenches triggered during the up-ramp at $I_0 = 0.5\,I_{\text{nom}}$ (left) and $0.75\,I_{\text{nom}}$ (right). Detection at the plateau (dashed line), at a resistive voltage of 1 V (dashed/dotted line), and at 100 mV (continuous).

the up- or down-ramp. Figure 18.23 (left) shows simulations for $I_0 = 0.5\,I_{\text{nom}}$ and $I_0 = 0.75\,I_{\text{nom}}$. The baseline of 100% copper is defined by the cable parameters given in Table 18.1. To reach a 150% copper content, strands are added to the cables and the copper-to-superconductor area ratio is increased such that the amount of Nb–Ti superconductor remains unchanged. For these wider cables, the coil cross section is redesigned to maintain the field quality. This can be accomplished by adding two turns to the outer layer (see Figure 18.24), which increases the inductance of the magnet by about 10%. Figure 18.23 (left) shows that the magnet withstands the quench triggered at $0.5\,I_{\text{nom}}$ but detected only at the flat-top.

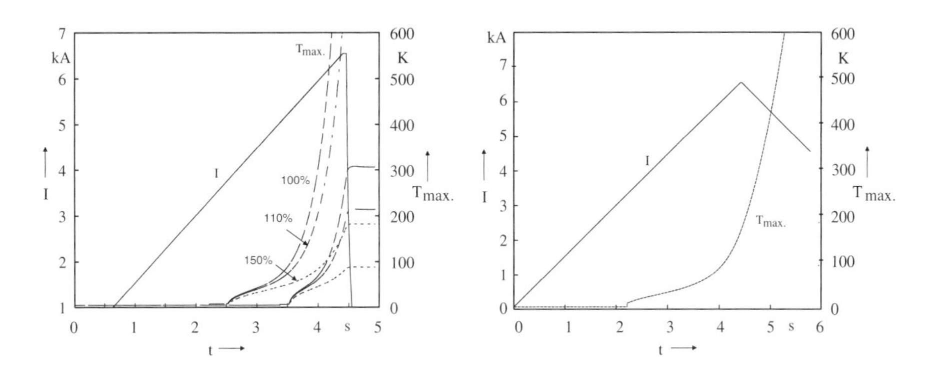

Figure 18.23 Left: Peak temperature and current evolution for quenches at $I_0 = 0.5\,I_{\text{nom}}$ and $I_0 = 0.75\,I_{\text{nom}}$. It is assumed that the quench remains undetected until the injection plateau is reached. Standard cable, and cables with 10–50% higher copper content. Right: Quench at $I_0 = 0.5\,I_{\text{nom}}$. Detection at the flat-top. The magnet is then ramped down at the standard rate.

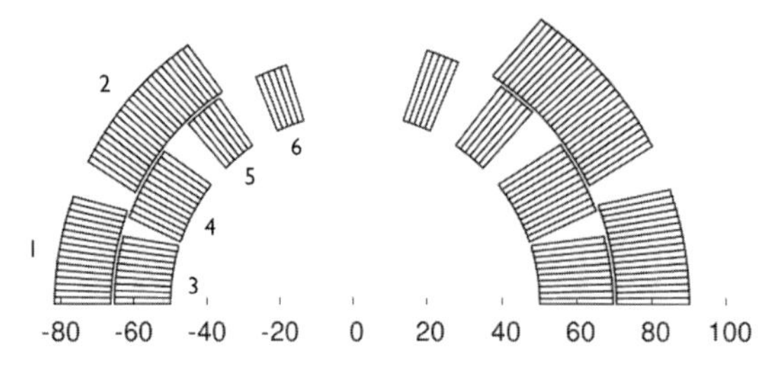

Figure 18.24 Left: The standard cross section of the SIS300 two-layer dipole. Right: The modified cross section with a wider cable and additional strands to yield a 50% increase in copper content.

A magnet protection system might appear redundant if the magnet contains enough copper to withstand a full powering cycle with a quench detected at a plateau only. However, let us consider a quench triggered at $I_0 = 0.5\,I_{\text{nom}}$ on the up-ramp. After the quench is detected at the flat-top, the magnet is ramped down with the nominal rate. Figure 18.23 (right) makes it clear that even with the wide cable (150% copper content), the magnet is still not sufficiently stabilized for this kind of operation.

References

1 Auchmann, B., de Maria, R., Kurz, S., Russenschuck, S.: 2D electromagnetic model of fast-ramping superconducting magnets, Proceedings of ICAP 2006

2 Alessandrini, M., Majkic, G., Laskaris, E. T., Salama, K.: Modeling of longitudinal and transverse quench propagation in stacks of superconducting MgB_2 wire, 2008 Applied Superconductivity Conference, 2008

3 Aird, G.J.C., Simkin J., Taylor S.C., Trowbridge, C.W., Xu, E.: Coupled transient thermal and electromagnetic finite-element simulation of quench in superconducting magnets, Proceedings of ICAP 2006

4 Brechna, H.: Superconducting Magnet Systems, Springer, Berlin, 1973

5 Calvi, M.: Quench Propagation in the LHC Superconducting Busbars, Politecnico Di Torino, Italy, 2000

6 Clark, A. F., Ekin, J. W., Radebaugh, R., Read, D. T.: The development of standards for practical superconductors, IEEE Transactions on Magnetics, 1979

7 Canali, M., Rossi, L.: Dynque: A computer code for quench simulation in adiabatic multicoil superconducting solenoids, INFN-TC-93-06, 1993

8 Caspi, S. et al.: Calculating quench propagation with Ansys, IEEE Transactions on Applied Superconductivity, 2003

9 Charifoulline, Z: Residual resistivity ratio (RRR) measurements of LHC superconducting NbTi cable strands, IEEE Transactions on Applied Superconductivity, 2006

10 Chohan, V., Veyrunes, E.: Measurement data from the training quench of magnet MB2381 at 12.82 kA, private communication, 2007

11 Chorowski, M., Lebrun, Ph., Serio, L., van Weelderen, R.: Thermohydraulics of quenches and helium recovery in the LHC prototype magnet strings, Cryogenics, 1998

12 Collatz, L.: Numerische Behandlung von Differentialgleichungen, Springer, Berlin, 1955

13 Coull, L., Hagedorn, D., Remondino, V., Rodriguez-Mateos, F.: LHC quench protection system, IEEE Transactions on Magnetics, 1994

14 Coull, L., Hagedorn, D., Krainz, G., Rodriguez-Mateos, F., Schmidt, R.: Electrodynamic behaviour of the LHC superconducting magnet string during a discharge, 5th European Particle Accelerator Conference EPAC, 1996

15 Dahlerup-Petersen, K., Kazmine, B., Popov, V., Sytchev, L., Vassiliev, L., Zubko, V.: Energy Extraction Resistors for the Main Dipole and Quadrupole Circuits of the LHC, LHC Project Report 421, 2000

16 Denz, R., Rodriguez-Mateos, F.: Detection of Resistive Transitions in LHC Superconducting Components, LHC Project Report 482, 2001

17 Denz, R.: Current measurements of the recovering MQY in Q6, private communication, 2008

18 Devred, A.: General formulas for the adiabatic propagation velocity of the normal zone, IEEE Transactions on Magnetics, 1989

19 Granieri, P., Calvi, M., Xydi, P., Baudouy, B., Bocian, D. , Bottura, L., Breschi, M., Siemko, A.: Stability analysis of the LHC cables for transient heat depositions, IEEE Transactions on Applied Superconductivity, 2008

20 Green, M. A.: Quench back in thin superconducting solenoid magnets, Cryogenics, 24, 1984.

21 Hagedorn, D., Rodriguez-Mateos, Modeling of the quenching process in complex superconducting magnet system, IEEE Transactions on Magnetics, 1992

22 Henning, W.: FAIR – an international accelerator facility for research with ions and antiprotons, Proceedings of EPAC 2004

23 Henning, W., Kurat, M.: Simulations of the quench behaviour of coated conductors with hot-spots, Proceedings of EUCAS 2007

24 Iwasa, Y.: Case Studies in Superconducting Magnets, Plenum, New York, 1994

25 Iwasa, Y.: Stability and protection in superconducting magnets – a discussion, IEEE Transactions on Applied Superconductivity, 2005

26 Kim, S.-W.: Quench simulation program for superconducting accelerator magnets, IEEE Particle Accelerator Conference, 2001

27 Karpinnen, M.: Quench propagation velocity LHC MCBX, private communication, 2008

28 Kozub, S., Bogdanov, I., Seletsky, A., Shcherbakov, P., Sytnik, V., Tkachenko, L., Zubko, V.: Final Report on the Research and Development Contract Technical Design of the SIS-300 Dipole Model, Technical Report, IHEP Protvino, 2006

29 Krainz, G.: Quench Protection and Powering in a String of Superconducting Magnets for the Large Hadron Collider, Technical University Graz, Austria, 1997

30 Leroy, D. et al.: Quench observation in LHC superconducting one meter long dipole models by field perturbation measurements, IEEE Transactions on Applied Superconductivity, 1993

31 Leroy, D.: Review of the R&D and supply of the LHC superconducting cables, IEEE Transactions on Applied Superconductivity, 2006

32 Meß, K.H., Schmüser, P., Wolff, S.: Superconducting Accelerator Magnets, World Scientific, Singapore, 1996

33 National Institute of Standards and Technology (NIST): Properties of Copper and Copper Alloys at Cryogenic Temperature, 1992

34 Pugnat, P., Siemko, A.: Review of quench performance of LHC main superconducting magnets, IEEE Transactions on Applied Superconductivity, 2007

35 Pugnat, P.: Measurement of Quench Heater Delays, private communication, 2007

36 Richter, D., Fleiter, J., Baudouy, B., Devred, A.: Evaluation of the transfer of heat from the coil of the LHC dipole magnet to helium II, IEEE Transactions on Applied Superconductivity, 2007

37 Rodriguez-Mateos, F., Pugnat, P., Sanfilippo, S., Schmidt, R., Siemko, A., Sonnemann, F., Quench Heater Experiments on the LHC Main Superconducting Magnets, LHC Project Report 418, 2000

38 Quench Heater Studies for the LHC Magnets, LHC Project Report 485, 2001

39 Schwerg, N., Auchmann B., Russenschuck, S.: Validation of a thermal electromagnetic quench model for accelerator magnets, IEEE Transactions on Applied Superconductivity, 2008

40 Schwerg, N., Auchmann B., Russenschuck, S.: Quench simulation in an integrated design environment for superconducting magnets, IEEE Transactions on Magnetics, 2008

41 Schwerg, N., Auchmann B., Russenschuck, S.: Challenges in the thermal modelling of quenches with ROXIE, IEEE Transactions on Applied Superconductivity, 2009

42 Schwerg, N.: Numerical Calculation of Transient Field Effects in Quenching Superconducting Magnets, PhD thesis, TU-Berlin, 2009

43 Siemko, A., Billan, J., Gerin, G., Leroy, D., Walckiers, L., Wolf, R.: Quench localization in the superconducting model magnets for the LHC by means of pick-up coils, IEEE Transactions on Applied Superconductivity, 1995.

44 Sonnemann, F.: Resistive transition and protection of LHC superconducting cables and magnets, PhD Rheinisch-Westfälische Technische Hochschule Aachen, 2001

45 Sonnemann, F., Calvi, M.: Quench simulation studies: Program documentation of SPQR Simulation Program for Quench Research, LHC Project Note 265, 2001

46 Stekly, Z. J. J., Zar, J. L.: Stable superconducting coils, IEEE Transactions on Nuclear Science, 1965

47 Van Sciver, S. W.: Helium Cryogenics, Plenum, New York, 1986

48 Verweij, A.: CUDI : A Model for Calculation of Electrodynamic and Thermal Behaviour of Superconducting Rutherford Cables, Departmental Report, CERN, 2006

49 Wilson, M.N.: Computer simulation of the quenching of a superconducting magnet, Rutherford High Energy Laboratory, Internal Report, RHEL/M 151, 1968

50 Wilson, M.N.: Superconducting Magnets, Oxford University Press, Oxford, 1983

19
Differential Geometry Applied to Coil-End Design

It is the pervading law of all things organic, and inorganic,
of all things physical and metaphysical,
of all things human and all things super-human,
of all true manifestations of the head, of the heart, of the soul,
that life is recognizable in its expression,
that form ever follows function. This is the law.

Louis Sullivan (1856–1924).

The coil-end design of superconducting accelerator magnets is conditioned by the objectives of minimizing the strain energy due to the winding procedure, optimizing the multipole content of the integrated field, and limiting the magnetic field enhancement. The coil ends must be carefully configured to avoid turn-to-turn short circuits, strand degradation, conductor breakage, and the tendency of turns to move toward the bore after coil curing.

As the beam pipe must not be obstructed, the coil is wound on a precision machined cylindrical tool, known as a *winding mandrel*, on which the central post and isoparametric, saddle-shaped end spacers are mounted. The coil is wound in such a way that the two narrow sides of the cables follow space curves of equal length, forming a *constant-perimeter coil end* [15] shown in Figure 19.1 (upper left). The coil ends are mechanically not as stable as the straight section and have therefore often been the zone where quenches originate.

The objectives for the coil-end optimization are:

- The minimization of the mechanical stress on the cable due to its deformation in the coil end, aiming at easy coil winding and curing. This task becomes more difficult as the bore diameter becomes smaller and the cable wider.
- The minimization of the integrated multipole coefficients along the coil end. In short corrector magnets the objective may be to compensate for

Field Computation for Accelerator Magnets. Stephan Russenschuck

ISBN: 978-3-527-40769-9

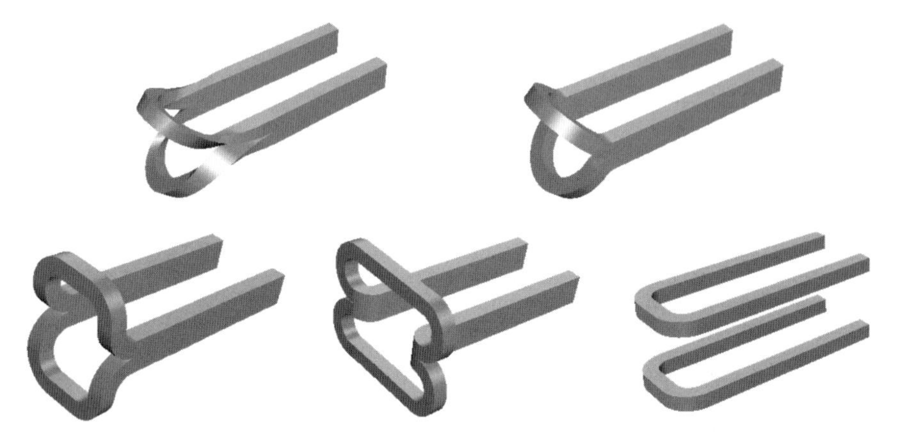

Figure 19.1 Different coil-end designs for accelerator magnets. From top left to bottom right: Iso-parametric (constant perimeter), cranked saddle, bedstead, transversally cranked saddle, racetrack.

field errors in the coil end by altering the field distribution in the straight section.

- To ensure a tight fit of the coil ends in order to avoid conductor motion during powering.
- To limit the peak-field enhancement in the coil end in order to improve the quench performance.
- To pack the windings tightly together into groups similar to the coil-blocks in the straight section in order to limit the number of end spacers and avoid inter-turn wedges.
- To guarantee tight tolerances on the coil-end geometry and maintain magnet-to-magnet consistency in the cable positioning.
- To establish an integrated design and manufacturing process with the generation of CAD data for the machining of end spacers and the validation of the design by winding tests.

Since an unstressed cable resembles at first approximation a long flat rectangle, one of its flexural rigidities is obviously much larger than the other. It is much easier to bend the cable perpendicular to its broad side than to bend it in the plane of the broad side. The two bending directions, shown in Figure 19.2, are therefore referred to as *easy-way* and *hard-way* bends. If we take one of the flexural rigidities to be infinite, that is, we do not allow any bend the hard way, then the curvature of the flat, unrolled (developed) cable is zero. Vanishing hard-way bend implies the so-called *constant-perimeter condition*[1], which is the isometry of the two opposite edges on the broad face of the cable. The

1 The converse is not true in general.

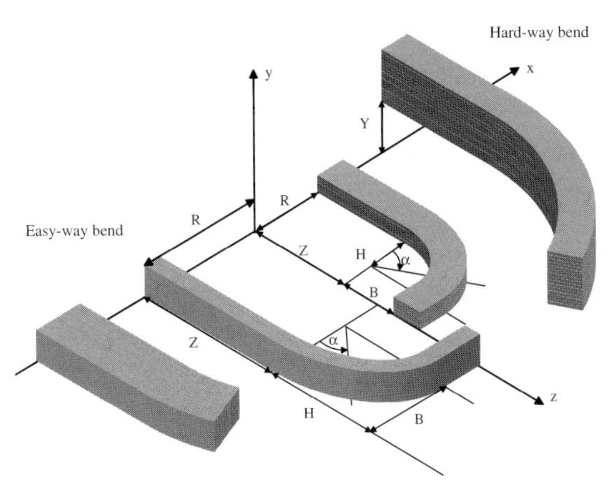

Figure 19.2 Geometrical parameters for racetrack coil ends featuring hard-way and easy-way bends of the cable.

only other coil geometry that avoids hard-way bend is the *racetrack* coil shown in Figure 19.1 (bottom right). Alternatives such as *cranked saddle* and *bedstead* coils shown in Figure 19.1 feature hard-way bends and are therefore not suited for coils made from ribbons and flat cables.

While the constant-perimeter condition guarantees a minimum overall strain in the cable, this does not hold for incremental strain at some local positions along the end. It is nonetheless possible to eliminate all local cable deformations by creating *developable* (or *ruled*) *surfaces* which can be modeled geometrically by a two-dimensional strip bent in such a way that the length of every arc in the strip is preserved. Developable surfaces in Euclidean space are surfaces that can essentially be made of a piece of paper, if we assume sufficient smoothness and thus exclude crumpling the paper. For geodesic (straight and planar) strips, it turns out that the bent strip is completely determined by the space curve of one of its edges. For our purposes, the *base curve* is defined as the edge that lies on the winding mandrel of the coil. The second edge of the strip is called the *free edge*. As the base curve often resembles an ellipse on the winding mandrel, this design principle is also known as the *ellipse on a cylinder* method.

In practice, the cable surfaces cannot precisely follow the above-mentioned developable surface. Reasons include the constraints on the interface between the straight section and the end, and the trapezoidal shape of the cable. A numerical approach for generating winding surfaces which are as close as possible to geodesic strip surfaces is presented in this chapter.

We will now apply the theory presented in Section 3.1.1, that is, the foundations of (classical) differential geometry in E_3, and the Frenet frame of space curves. As a starting point for the coil-end design, we will introduce the generalized Frenet–Serret equations for strips. Local adjustment of the twist parameters can be made in order to match the constraints which stem from the design of the coil's straight section and the aim of winding the cables in such a way that they are in perfect contact along the coil end.

The computation of the cable shapes and the end-spacer surfaces will be presented with some examples of an LHC main dipole model magnet (Figure 19.3) and the MSCB dipole orbit corrector.

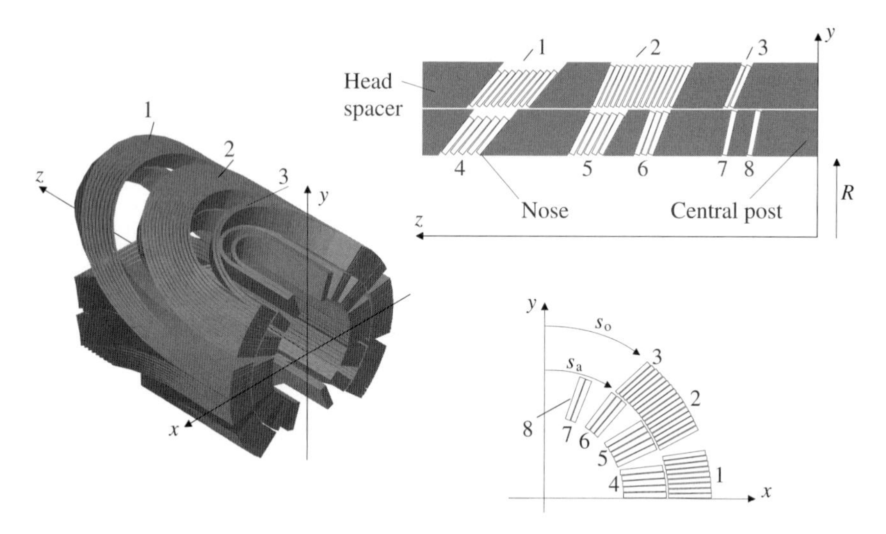

Figure 19.3 Left: Coil end of an LHC main dipole model. Right: Cross sections in the xy- and the yz-planes with the block-numbering convention, and the naming convention for the central post, nose, and head spacer.

19.1
Constant-Perimeter Coil Ends

We will now establish the winding topology with an approach that is known to magnet designers as the constant-perimeter end. The coil-winding direction is defined by the first winding on the center post when looking down on the winding mandrel. Experience has shown that the best results are obtained if left-lay cables are wound in clockwise direction, whereas right-lay cables are wound in counter-clockwise direction. It is advantageous for making internal cable joints that the two coil layers be wound in opposite direction [13]. In accordance with the cable position in the magnet cross section, the narrow edge of the keystoned cable faces the winding mandrel. The (guiding) strip surface is defined by the broad face of the cable directed toward the central post.

We assume that the base curves of the strips follow circles, ellipses, or superellipses on the surface of the winding mandrel.[2] By ensuring equal length of the two edges of the guiding strip, a geometry with minimal overall strain can be generated. The edges are discretized by polygons having segments of equal length. This allows a numerical estimation of the isometry of the bent strip.

The input parameters for the generation of a constant-perimeter end with the CERN field computation program ROXIE are the z-position at the nose[3] z_p, the semi-major axis of the ellipse on the winding mandrel, the turn's inclination angle β at the nose of the bend (in the yz-plane), and the size of the inter-turn spacers between the windings; for the definition of the parameters see Figures 19.4 and 19.5. The definition of inter-turn spacers can account for the cable's "natural" angle at the nose and for slight cable deformations (de-keystoning) that must be investigated with coil winding trails. The base curve and the free edge of the cables are determined by their position in the xy cross section, which yields the semi-minor axes a_a and a_o of the ellipses in the sz-plane shown in Figures 19.3 and 19.4.

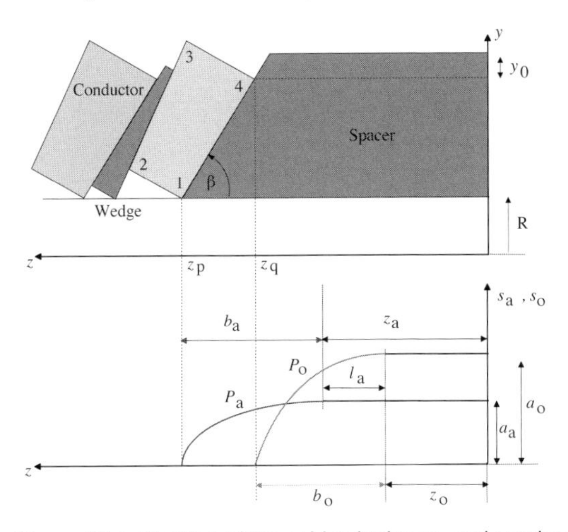

Figure 19.4 Position of the cables in the sz- and yz-planes, defining the ellipse semiaxes of the base curve and free edge of the guiding strip.

In the developed sz-plane, the base curve is described by the ellipse equation

$$\frac{s^2}{a_a^2} + \frac{z^2}{b_a^2} = 1. \tag{19.1}$$

2 In [8], the base curve lies in the developed plane on the outer radius of the coil. A shelf is attached to the end spacer to fill the empty gap between coil blocks and winding mandrel.

3 Naming convention in Figure 19.3.

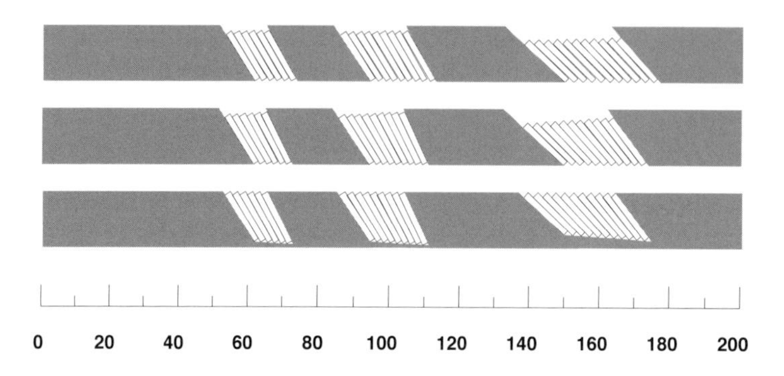

Figure 19.5 Cable positions in the yz-plane. Top: With inter-turn wedges. Middle: Without inter-turn wedges. Bottom: Cables aligned on the outer diameter of the end spacers and supported with shelfs made from the same spacer material.

Here s denotes the coordinate in the surface of the winding mandrel and does not refer to the arc length of the space curve. The perimeter of the free (outer) edge (with semiaxes a_o, b_o) is assumed to be equal to the combined length of the base curve (with semiaxes a_a, b_a) and an additional straight section l_a:

$$P_o = b_o E\left(\frac{\pi}{2}, e\right), \tag{19.2}$$

where E is the complete elliptic integral of the second kind,[4] and

$$e := \sqrt{1 - \frac{a_o^2}{b_o^2}} \tag{19.3}$$

is the eccentricity $0 \leq e < 1$ for $b > a$. Approximately, we have

$$P_o \approx \frac{1}{4}\pi(a_o + b_o)\left(1 + \frac{\lambda^2}{4} + \frac{\lambda^4}{64}\right), \qquad \lambda := \frac{a_o - b_o}{a_o + b_o}. \tag{19.4}$$

The perimeter of the base curve is calculated accordingly:

$$P_a \approx \frac{1}{4}\pi(a_a + b_a)\left(1 + \frac{\nu^2}{4} + \frac{\nu^4}{64}\right), \qquad \nu := \frac{a_a - b_a}{a_a + b_a}. \tag{19.5}$$

We further introduce a relaxation factor f_r for the perimeter of the base curve ($0.98 \leq f_r \leq 1$) to allow for some overall strain in the cable. This factor must be found by practical coil-winding trials. The semi-minor axes a_a, a_o are given by the position of the cable in the xy cross section. The ellipticity $\lambda_e := b_a/a_a$, the position of the nose z_P, and the inclination angle β are supplied by the

4 See Eq. (5.61).

designer. The straight section z_o follows directly from the geometry shown in Figure 19.4. The unknowns b_o and z_a can be calculated iteratively from the two equations

$$P_a f_r + z_a = P_o + z_o, \qquad z_a + b_a = z_p. \tag{19.6}$$

The radius of curvature of the ellipse at the onset of the bend ($z = 0$) is $R = b^2/a$. An alternative with a zero curvature at the onset of the bend is the superelliptical shape

$$\frac{s^3}{a^3} + \frac{z^3}{b^3} = 1, \tag{19.7}$$

shown in Figure 19.6 (right). The calculation of the parameters b_o and l_a is done in the same way as for the ellipse, but a polynomial approximation is used for the calculation of the perimeters.

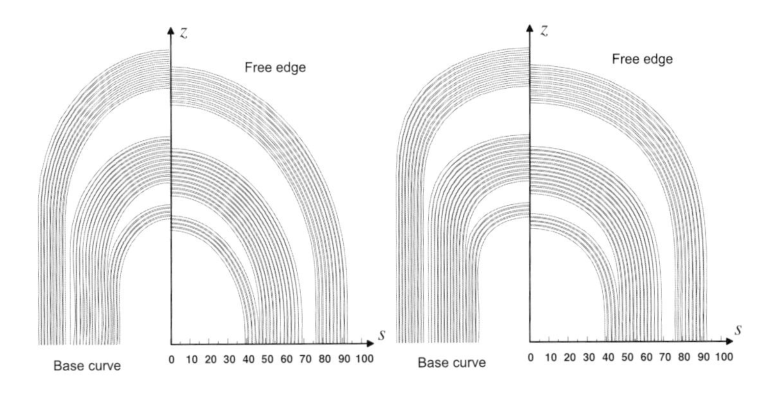

Figure 19.6 Left: Outer and inner edges of the conductors in the sz-plane (elliptical shape). Right: Conductor edges of superelliptical shape.

There is a major flow in the constant-perimeter method described in this section. The constant-perimeter condition (a global quantity) does not guarantee vanishing of the local strain. A way to minimize local hard-way bend is to apply the methods of differential geometry presented in the next section. This method is based on work described in [8,9].

19.2 Differential Geometry of the Strip Surfaces

At first approximation, the superconducting cable can be modeled purely geometrically as a two-dimensional strip to be bent, without (local or global) stretching or squeezing, into the three-dimensional space such that all arc lengths of curves on the strip are preserved. The surface of the bent strip may be unbent into the plane accordingly and is therefore called a developable surface.

In the special case of a geodesic strip, that is, a strip that is straight when developed into the plane, the strip can be modeled by the base curve and the field of Darboux vectors that define the generators of the developable surface. This result will be elaborated in the following sections.

19.2.1
The Frenet–Serret Equations for Strips

The generalized Frenet–Serret equations for strips are given by

$$\frac{\mathrm{d}\mathbf{T}}{\mathrm{d}s} = \kappa_{\mathrm{n}}\,\mathbf{n} - \kappa_{\mathrm{g}}\,\mathbf{b}, \qquad \frac{\mathrm{d}\mathbf{n}}{\mathrm{d}s} = -\kappa_{\mathrm{n}}\,\mathbf{T} + \tau\,\mathbf{b}, \qquad \frac{\mathrm{d}\mathbf{b}}{\mathrm{d}s} = \kappa_{\mathrm{g}}\,\mathbf{T} - \tau\,\mathbf{n}, \tag{19.8}$$

where κ_{g} is the geodesic curvature, κ_{n} the normal curvature, τ the torsion, and $\{\mathbf{T}, \mathbf{n}, \mathbf{b}\}$ the Frenet frame with tangent, normal, and binormal vector. The Frenet frame for strips is shown in Figure 19.7. Equations (19.8) may be written in the matrix form as

$$\begin{pmatrix} \mathbf{T}' \\ \mathbf{n}' \\ \mathbf{b}' \end{pmatrix} = \begin{pmatrix} 0 & \kappa_{\mathrm{n}} & -\kappa_{\mathrm{g}} \\ -\kappa_{\mathrm{n}} & 0 & \tau \\ \kappa_{\mathrm{g}} & -\tau & 0 \end{pmatrix} \begin{pmatrix} \mathbf{T} \\ \mathbf{n} \\ \mathbf{b} \end{pmatrix}, \tag{19.9}$$

where we have abbreviated the s-derivative with the prime, $\mathbf{T}' := \mathrm{d}\mathbf{T}/\mathrm{d}s$, a notation that we will employ henceforth.

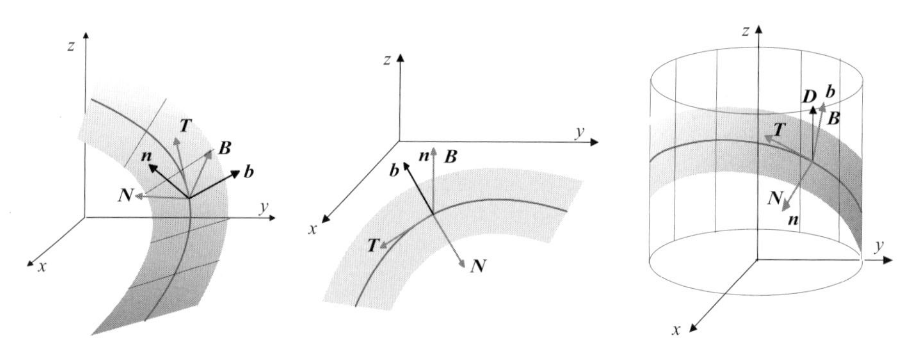

Figure 19.7 Three examples of strips. Left: General case, Middle: Developed strip ($\kappa_n = 0, \tau_g = 0$) with geodesic curvature $\kappa_g \neq 0$. Right: Strip with zero geodesic curvature. In this case the base curve (here a helix) becomes the geodesic of the strip.

Proof. Following Blaschke [6], let $\{\mathbf{a}_1, \mathbf{a}_2, \mathbf{a}_3\}$ be a moving frame with the property

$$\mathbf{a}_j \cdot \mathbf{a}_k = \delta_{jk}. \tag{19.10}$$

We can express

$$\mathbf{a}_j' = \sum_{k=1}^{3} \mathbf{a}_k \omega_{jk}, \tag{19.11}$$

as a linear combination of the moving frame. The coefficients read

$$\omega_{jk} = \mathbf{a}_j' \cdot \mathbf{a}_k. \tag{19.12}$$

Differentiating Eq. (19.10) yields

$$\mathbf{a}_j' \cdot \mathbf{a}_k + \mathbf{a}_k' \cdot \mathbf{a}_j = 0, \tag{19.13}$$

which implies that $\omega_{jk} + \omega_{kj} = 0$, and in particular $\omega_{jj} = 0$. Taking Eq. (19.11), substituting $\mathbf{a}_1 := \mathbf{T}$, $\mathbf{a}_2 := \mathbf{n}$, $\mathbf{a}_3 := \mathbf{b}$, and $\omega_{23} := \tau$, $\omega_{31} := \kappa_\mathrm{g}$, $\omega_{12} := \kappa_\mathrm{n}$, yields the generalized Frenet–Serret equations for strips. □

Given a base curve, such as an ellipse on a cylinder, and a field of frames on that base curve, the curvature parameters can be derived from Eq. (19.12):

$$\tau = \mathbf{b} \cdot \mathbf{n}' = \vartheta_\mathbf{T}', \tag{19.14}$$

$$\kappa_\mathrm{g} = \mathbf{T} \cdot \mathbf{b}' = \vartheta_\mathbf{n}', \tag{19.15}$$

$$\kappa_\mathrm{n} = \mathbf{n} \cdot \mathbf{T}' = \vartheta_\mathbf{b}'. \tag{19.16}$$

The curvature parameters τ, κ_g, and κ_n are thus the differential twist angles $\mathrm{d}\vartheta_{\mathbf{T},\mathbf{n},\mathbf{b}}$ around the Frenet frame as it is displaced on the base curve by $\mathrm{d}s$; see Figure 19.8. For the curvature of the base curve, we obtain[5]

$$\kappa = \left|\mathbf{T}'\right| = \sqrt{\kappa_\mathrm{g}^2 + \kappa_\mathrm{n}^2}, \tag{19.17}$$

and for the normal and binormal vector fields,

$$\mathbf{N} = \frac{1}{\kappa}\mathbf{T}' = \frac{-\kappa_\mathrm{g}\mathbf{b} + \kappa_\mathrm{n}\mathbf{n}}{\sqrt{\kappa_\mathrm{g}^2 + \kappa_\mathrm{n}^2}}, \tag{19.18}$$

$$\mathbf{B} = \mathbf{T} \times \mathbf{N} = \frac{\kappa_\mathrm{g}\mathbf{n} + \kappa_\mathrm{n}\mathbf{b}}{\sqrt{\kappa_\mathrm{g}^2 + \kappa_\mathrm{n}^2}}. \tag{19.19}$$

5 See Section 3.1.1.

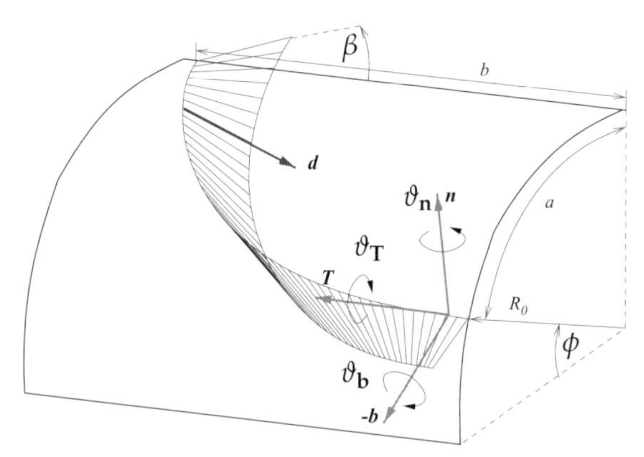

Figure 19.8 A strip bent along a base curve drawn on a cylinder of radius R_0, together with the generator lines of the developable surface.

The angle φ_1 between the principle normal $\mathbf{N}$ of the base curve and the normal vector $\mathbf{n}$ to the strip is given at some point as

$$\cos\varphi_1 = \mathbf{N}\cdot\mathbf{n} = \frac{\kappa_\mathrm{n}}{\sqrt{\kappa_\mathrm{g}^2+\kappa_\mathrm{n}^2}}\,. \tag{19.20}$$

The angle is zero in the case of vanishing geodesic curvature $\kappa_\mathrm{g} = 0$ and we find

$$\mathbf{N} = \mathbf{n}\,, \qquad \mathbf{B} = \mathbf{b}\,. \tag{19.21}$$

The set of differential equations (19.9) for this type of strip (geodesic) is then identical to the Frenet equations of the base curve, given in Eq. (3.26).

For a plane strip as shown in Figure 19.7 (middle), all tangent planes lie in the strip surface

$$\mathbf{n}' = \tau\,\mathbf{b} - \kappa_\mathrm{n}\,\mathbf{T} = 0\,, \tag{19.22}$$

and consequently $\tau = 0$ and $\kappa_\mathrm{n} = 0$. The geodesic curvature is therefore the curvature of the plane strip, and it can be shown [6] that this curvature is an invariant with respect to the bending transformation.

19.2.2
The Generators of Strips

A developable surface is generated by a one-parameter family of straight lines, the *generators* or (*rulings*) which are found to be the intersections of successive tangent planes $F(x,y,z,s)$ and $F(x,y,z,s+\mathrm{d}s)$ to the strip surface. We will now prove an important theorem:

Theorem 19.1 *The generators of a geodesic strip are defined by the Darboux vector field to the base curve.*

Proof. The tangent planes are uniquely defined by[6]

$$F(s) = (\mathbf{r} - \mathbf{r}_0) \cdot \mathbf{n}(s) = 0, \tag{19.23}$$

where $\mathbf{n}$ is a normal vector to the strip at the tip of the position vector $\mathbf{r}_0$, and $\mathbf{r}$ is another arbitrary point on the tangent plane. It thus is required that

$$F(s) = 0, \qquad \text{and} \qquad F(s + \Delta s) = 0. \tag{19.24}$$

Hence

$$F' = \lim_{\Delta s \to 0} \frac{F(s + \Delta s) - F(s)}{\Delta s} = 0. \tag{19.25}$$

A field of orthonormal frames $\{\mathbf{T}(s), \mathbf{n}(s), \mathbf{b}(s)\}$ is assigned to the base curve, where $\mathbf{T}$ is the tangent to the base curve and $\mathbf{n}$ is the normal to the strip at $\mathbf{r}_0$, as shown in Figure 19.8. Differentiating Eq. (19.23) yields

$$F' = \mathbf{r}' \cdot \mathbf{n} + (\mathbf{r} - \mathbf{r}_0) \cdot \mathbf{n}' = (\mathbf{r} - \mathbf{r}_0) \cdot (-\kappa_n \mathbf{T} + \tau \mathbf{b}) = 0, \tag{19.26}$$

where we have used the relations of Eq. (19.8). The first term vanishes because $\mathbf{T}$ is orthogonal to the normal vector. We find from Eq. (19.23) that $\mathbf{r}$ must be a linear combination of the vectors $\mathbf{T}$ and $\mathbf{b}$,

$$\mathbf{r} = \mathbf{r}_0 + k_1 \mathbf{T} + k_2 \mathbf{b}. \tag{19.27}$$

Equation (19.26) requires that $k_1/k_2 = \tau/\kappa_n$. We can write the solution as

$$\mathbf{r}(s, \lambda) = \mathbf{r}_0(s) + \lambda \, \mathbf{d}(s) \tag{19.28}$$

for a parameter λ and

$$\mathbf{d}(s) = \tau \, \mathbf{T}(s) + \kappa_n \, \mathbf{b}(s). \tag{19.29}$$

In the case of geodesic strips ($\kappa_g = 0$), the binormal vector of the strip and base curve are identical, $\mathbf{B} = \mathbf{b}$, and thus the vector field $\mathbf{d}$ is identical to the Darboux vector field $\mathbf{D}$ of the base curve. The solution of Eqs. (19.23) and (19.26) is a field of straight lines according to Eq. (19.29). They are the intersections of adjacent tangent planes to the base curve and thus span the strip surface, as required. Consequently, they will be referred to as the generators. □

6 We will omit the notation of the x, y, z coordinates.

The developable surface is thus uniquely defined by the base curve and its curvature parameters, except where the curvature of the base curve is zero, as it is the case at the transition between the straight section and the end winding ($s = 0$).

For the practical coil winding of superconducting magnets, however, the geodesic strip surface is inadequate for describing the cable surface, for the following reasons:

- The cables are wound on a cylindrical winding mandrel, and the ends must match the magnet's straight section as determined by the two-dimensional field quality optimization. The Frenet frame, however, requires that the cable meets the straight section in the radial direction. The bent strip must therefore be twisted to bring it into coincidence with the cable surface at $s = 0$.
- Bending a plane strip into a defined position consists of bending it over its successive generators; see Fig. 19.9. However, the geodesic strip generally does not guarantee the absence of intersections of generators within the cable surface. We will henceforth call this an *edge of regression* violation. The edge of regression is best explained in the discrete setting presented in the next section.
- In order to avoid placing inter-turn spacers into the coil end, the cables are wound on each other in blocks of up to 30 turns. Consequently, the tilt of the subsequent winding is defined by the shape of the previous one and thus cannot correspond to the natural inclination angle β. This effect worsens with increased keystoning of the cable.

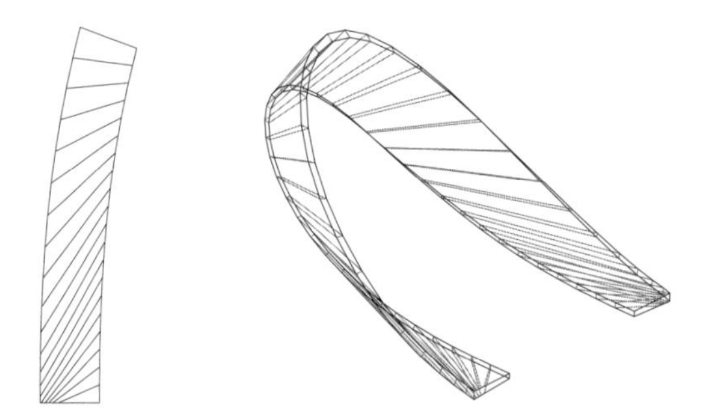

Figure 19.9 Left: Developed strip with nonvanishing geodesic curvature and edge of regression violation for the first six generators. Right: Cable bent along the generator lines.

The application of additional twist angles $\vartheta^*_{\mathrm{T}}(s)$ around the tangent vector $\mathbf{T}$ provides an additional degree of freedom to mitigate the above-mentioned problems. This inevitably results, however, in the generation of some geodesic curvature. From Eq. (19.14) it follows directly that

$$\tau^* = \tau + \frac{d\vartheta_T^*}{ds}. \tag{19.30}$$

Substituting

$$\mathbf{T}^* = \mathbf{T}, \tag{19.31}$$

$$\mathbf{n}^* = \cos\vartheta_T^* \, \mathbf{n} + \sin\vartheta_T^* \, \mathbf{b}, \tag{19.32}$$

$$\mathbf{b}^* = \cos\vartheta_T^* \, \mathbf{b} - \sin\vartheta_T^* \, \mathbf{n}, \tag{19.33}$$

into Eqs. (19.15) and (19.16) yields

$$\kappa_g^* = \cos\vartheta_T^* \, \kappa_g + \sin\vartheta_T^* \, \kappa_n, \tag{19.34}$$

$$\kappa_n^* = \cos\vartheta_T^* \, \kappa_n - \sin\vartheta_T^* \, \kappa_g. \tag{19.35}$$

Using Eqs. (19.29), (19.30) and (19.34), (19.35) , the mathematical model accounts for a start configuration on a given base curve and yields the necessary degree of freedom for the design optimization.

With the curvature parameters according to Eqs. (19.34) and (19.35), we can calculate the minimum strain energy E according to the Euler–Kirchhoff theory by

$$E = \frac{1}{2} \int_0^{s_c} \left(f_\tau \left(\tau^*(s)\right)^2 + f_n \left(\kappa_n^*(s)\right)^2 + f_g \left(\kappa_g^*(s)\right)^2 \right) \, ds, \tag{19.36}$$

where s_c is the cable length between the cross section and the nose. The simple relation (19.36), based on a linear *Hooke's law*,[7] holds in the elastic regime. The flexural rigidities f_τ, f_n, f_g are dependent on cable type. It is assumed that the cables are firmly clamped by the end spacers in such a way that their strain is not altered by the Lorentz stresses at full excitation. For

$$f_g \gg f_\tau, f_n, \tag{19.37}$$

the optimization problem of finding a minimum strain energy in a cable can be expressed as

$$\min \left\{ \int_0^{s_c} \left(\kappa_g^*(s)\right)^2 ds \right\}. \tag{19.38}$$

19.3 Discrete Theory of the Strip Surface

The calculation of the Darboux vector field can be cumbersome for complicated base curves; see the example presented in Section 3.1.1. We therefore

7 Robert Hooke (1635–1703).

avoid writing out the Darboux vector field along the superellipse that is traced out on the coil-winding mandrel. Instead we present a discrete theory of strip surfaces that allows for more general shapes of coil ends. Consider a discrete strip defined as a family of flat, quadrilateral faces embedded in the Euclidean space E_3. Except for the first and the last, each face has exactly two adjacent faces and is bounded by four edges. We distinguish between the two outer edges, one of which will become the base curve, and the edges joining adjacent faces, which become the generators of the developable surface. We will refer to these edges as *boundaries* and *joints*, respectively.

The bending of a strip is defined as the act of isometric (arc-length preserving) continuous deformation. In the discrete model, bending consists of turning the strip's faces around the joints. Figure 19.10 (left) shows two discrete strips which are geodesic and developed into the plane. Figure 19.10 (right) illustrates the bending of a nongeodesic strip shown in the middle. Obviously, the generators of a discrete strip must not intersect.

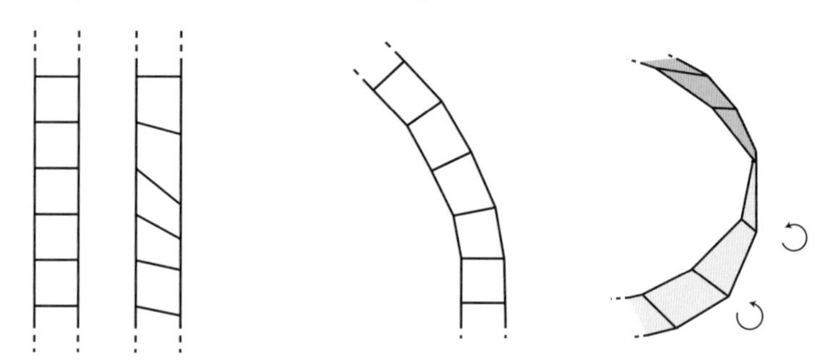

Figure 19.10 Left: Two flat straight discrete strips, differing by the position of their generators. Middle: A flat nongeodesic strip. Right: Bending of a nongeodesic strip along its generators.

The geodesic curvature of the developed strip is measured by the angles

$$\alpha^i := \angle(\mathbf{d}^i, \mathbf{T}^i) \quad \text{and} \quad \beta^i := \angle(\mathbf{d}^i, \mathbf{T}^{i+1}) \tag{19.39}$$

between the joints and boundaries, which are the generators represented by the vectors $\mathbf{d}^i$ and the tangent vectors $\mathbf{T}^i$ of neighboring base curve edges; see Figure 19.11. The notation is motivated by the fact that in continuous strip theory these would be the tangent vectors and the Darboux vectors of the strip. For $\alpha^i + \beta^i = \pi$ the strip is geodesic, although the joints may not be perpendicular to the boundaries. Recall that in general the Darboux vectors are inclined at some angle to the binormal vector. We call the measure

$$\kappa_g^i = \pi - (\alpha^i + \beta^i) \tag{19.40}$$

the discrete geodesic curvature, understood in the sense that it is the curvature integrated over the discrete strip element and thus has the physical dimension

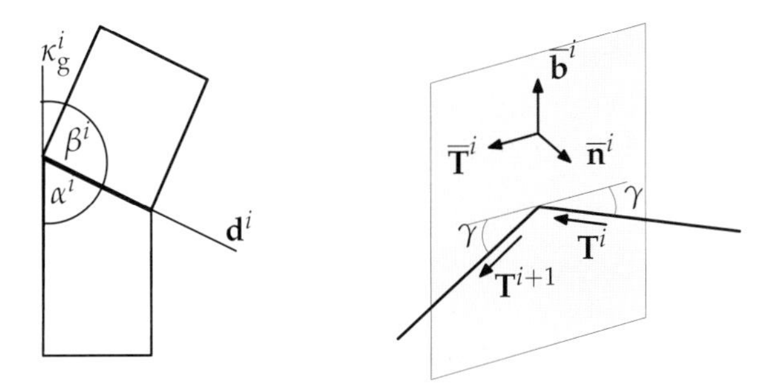

Figure 19.11 Left: The measure for the discrete geodesic curvature. Right: Tangent plane assigned to a node of the primary base curve.

of an angle; $[\kappa_g^i] = 1$ radian. This discrete geodesic curvature is an invariant with respect to the bending transformation.

The problem is now to generate a family of intersecting tangent planes to a space curve that is defined as a polygon of straight segments. Toward this aim, we introduce two discrete base curves, denoted primary and secondary base curves, as shown in Figure 19.12. The primary base curve serves for the differential quotients and the secondary base curve for the location of the Frenet frame. A mean tangent vector is defined by

$$\overline{\mathbf{T}}^i := \frac{\mathbf{T}^{i+1} + \mathbf{T}^i}{\left|\mathbf{T}^{i+1} + \mathbf{T}^i\right|}, \tag{19.41}$$

and a mean normal vector by

$$\overline{\mathbf{n}}^i := \frac{\mathbf{T}^{i+1} - \mathbf{T}^i}{\left|\mathbf{T}^{i+1} - \mathbf{T}^i\right|}. \tag{19.42}$$

The mean binormal vector is $\overline{\mathbf{b}}^i = \overline{\mathbf{n}}^i \times \overline{\mathbf{T}}^i$.

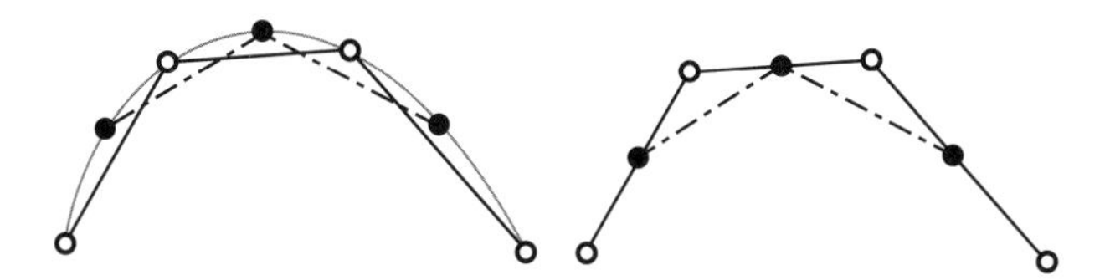

Figure 19.12 Primary (solid lines) and secondary (dashed lines) discrete base curves. Left: For a given parametric representation of the base curve, primary and secondary nodes can be located on the base curve. Right: If the base curve is defined by a polygon of straight line segments, secondary nodes are located in the barycenters of the primary edges.

The tangent vector to the secondary base curve is approximately the mean tangent vector of the adjacent edges on the primary base curve. Thus the normal vector $\overline{\mathbf{n}}^i$ and the binormal vector $\overline{\mathbf{b}}^i$ approximately represent the normal and binormal vectors of the secondary base curve; see Figure 19.13.

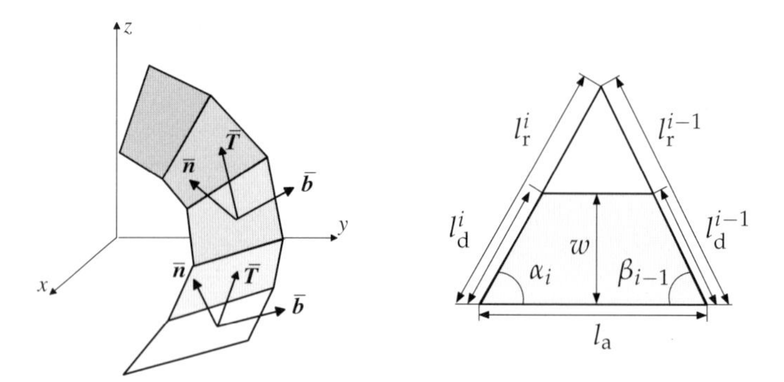

Figure 19.13 Left: Moving frame along the faces of a bent strip. Right: Intersection of two consecutive generators.

The Frenet frame to the secondary baseline thus obtained is the triad of a geodesic strip because

$$\lim_{h\to 0} \frac{(\mathbf{T}^{i+1} - \mathbf{T}^i) \cdot \mathbf{b}^i}{h} = \kappa_g \,, \tag{19.43}$$

see Eq. (19.9), and

$$(\mathbf{T}^{i+1} - \mathbf{T}^i) \cdot \mathbf{b}^i = 0 \tag{19.44}$$

by definition. In Eq. (19.43) h is a discretization density defined by $h := \sup_i(l_a^i)$ where l_a^i is the length of the ith base curve element. In the following example, we will show numerically that

$$\kappa_g = \lim_{h\to 0} \kappa_g^i \,. \tag{19.45}$$

The generators of the secondary strip are defined as the intersections of two consecutive tangent planes. The vector $\overline{\mathbf{d}}^i$ at the ith node of the secondary base curve must therefore be orthogonal to $\overline{\mathbf{n}}^i$ and $\overline{\mathbf{n}}^{i+1}$. Thus,

$$\overline{\mathbf{d}}^i = \overline{\mathbf{n}}^i \times \overline{\mathbf{n}}^{i+1}. \tag{19.46}$$

An example of the resulting bent strip is shown in Figure 19.13 (left). For $h \to 0$ the vector $\overline{\mathbf{d}}^i$ defines the same generator as the Darboux vector.

Example: For the validation of the discrete model, a geodesic strip is bent along an ellipse on a cylinder, as described in Section 19.4. The model is shown in Figure 19.14 for different discretization densities. Note that the first and last generators to complete the arc from the cross section to the yz-plane are missing. For n nodes on the base curve, only $n-2$ generators are mathematically defined. The problem can be circumvented by using the symmetry at the nose and by assuming that the strip continues as a flat, straight strip at the onset of the arc.

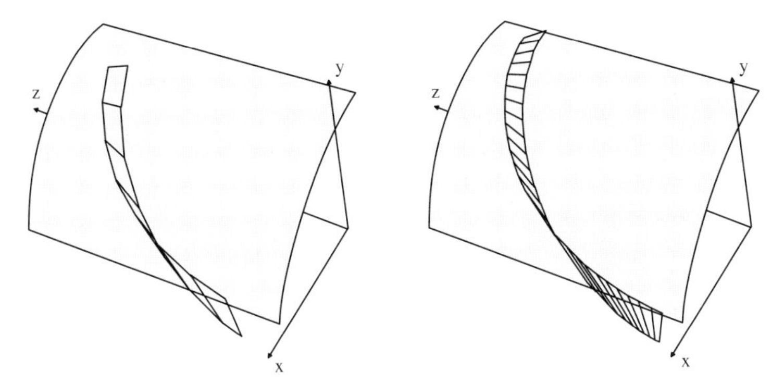

Figure 19.14 Discrete strips bent along a discrete ellipse on a cylinder. The models contain 10 (left) and 30 points on the primary base curve (right). For n nodes on the primary base curve, only $n-2$ generators are mathematically defined at the nodes of the secondary base curve.

For the numerical example shown in Figure 19.14, the measure of the discrete geodesic curvature κ_{g}^{i}, defined by Eq. (19.40), converges[8] to zero with $o(h^3)$; see Figure 19.15 (left). The generator vectors $\overline{\mathbf{d}}^{i}$ of the discrete model converge to $\lambda\mathbf{D}$ of the continuous model. Figure 19.15 (right) shows the distribution of the angle δ between the generator vectors and the Darboux vectors, which converges to zero with $o(h^2)$. □

In order to determine the length l_{d}^{i} of the generators we can calculate

$$l_{\mathrm{d}}^{i} = \frac{1}{2}\left(\frac{w}{\sin\alpha^{i}} + \frac{w}{\sin\beta^{i}}\right), \tag{19.47}$$

where the angles α^{i} and β^{i} are defined by Eq. (19.39). The width of the strip is denoted w. The family of intersections of consecutive generators is called

8 If $g : \mathbb{R} \to \mathbb{R} : x \mapsto g(x)$, then $g(x) = O(x)$ states that there exists a $K \geq 0$ such that $\left|\frac{g(x)}{x}\right| \leq K$ as $x \to 0$, i.e., $g(x)$ goes to zero at least as fast as x does. The notation $g(x) = o(x)$ means that $g(x)$ goes to zero faster than x does, that is, $K = 0$.

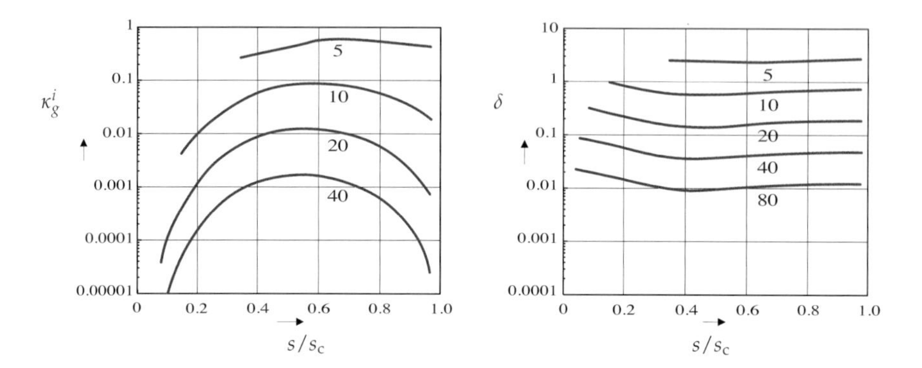

Figure 19.15 Left: Convergence of the measure for the geodesic curvature in the discrete strip. Right: The angle δ between the joints in the discrete strip and the Darboux vectors of the continuous model. The discrete models contain between 5 and 80 nodes on the primary base curve. The angles are given in degrees.

the *edge of regression*. Configuring a plane strip into a defined position consists of bending the strip over its generators; see Figure 19.9. The modulus of the Darboux vector, $\sqrt{\tau^2 + \kappa_n^2}$, denotes the differential bent angle of the strip. It follows that the generators must not intersect the surface of the cable, which would result in folding or crumbling. The intersection point of any two consecutive generators at the edge of regression, as well as their lengths l_r^i and l_r^{i-1} to the intersection point, can be calculated from geometrical data in the plane, shown in Figure 19.13 (right):

$$l_r^i \cos\alpha^i + l_r^{i-1} \cos\beta^{i-1} = l_a^i \tag{19.48}$$

and

$$l_r^i \sin\alpha^i = l_r^{i-1} \sin\beta^{i-1}, \tag{19.49}$$

where l_a^i is the length of the base curve element. We can thus verify that indeed $l_r^i > l_d^i$ and $l_r^{i-1} > l_d^{i-1}$. A violation of these criteria, which we had termed edge of regression violation in the previous section, indicates that no geodesic strip of width w can be bent onto the prescribed base curve. In order to avoid intersecting generators within the strip, additional twist of the faces around the base curve elements can be applied according to Eq. (19.30). This is also required by the nonradial position of the cable at the transition to the magnet's straight section. The maximum amount of twist must be also established by coil winding trials in order to prevent the cable from unlocking, characterized by the popping out of single strands.

19.4 Optimization of the Strip Surface

For an ellipse on a cylinder defined as the base curve of the guiding strip, the strip surface is uniquely determined by the Darboux vectors. This also yields the natural inclination angle β of the cable in the yz-plane as shown in Figure 19.16 (right).

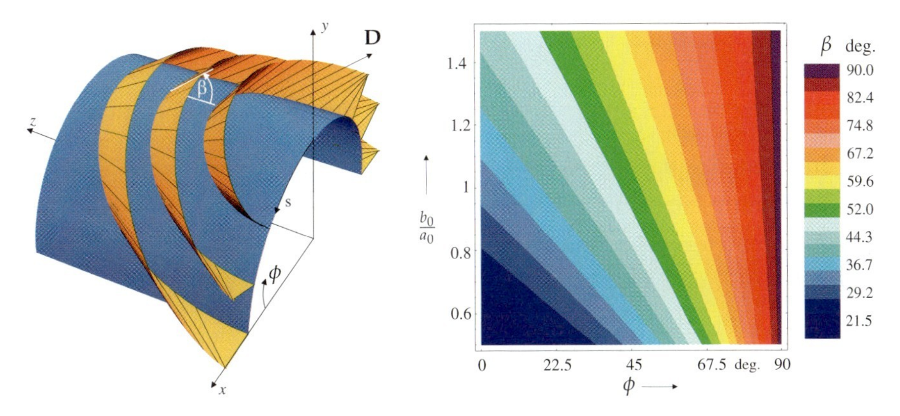

Figure 19.16 Left: Developable surfaces of the geodesic strips created by means of an ellipse on the coil-winding mandrel and generators in the direction of the Darboux vectors. Right: Natural inclination angle β of the strip surface in the yz-plane as a function of the free edge's ellipticity $\lambda_e := b_a/a_a$ and the positioning angle ϕ in the xy-plane.

Let R be the radius of the coil winding mandrel (for the inner layer of the LHC main dipole coils $R = 28$ mm), and let ϕ denote the positioning angle of the cable in the xy cross section; see Figure 19.16 (left). One of the ellipse's semiaxes is given by $a_a = R(\frac{\pi}{2} - \phi)$. The semi-major axis b_a is a design variable. For the superellipse on the cylinder the parametric representation is

$$\mathbf{r}(t) = R \sin\left(\frac{a_a}{R} \cos^{\frac{2}{n}} t\right) \mathbf{e}_x + R \cos\left(\frac{a_a}{R} \cos^{\frac{2}{n}} t\right) \mathbf{e}_y + b_a \sin^{\frac{2}{n}} t \, \mathbf{e}_z \,, \tag{19.50}$$

where t is defined in the open interval of $(0, \pi/2)$ and n is the order of the superellipse. The tangent, binormal, and Darboux vector fields, as well as curvature and torsion, can be calculated from Eqs. (3.29) and (3.30). The inclination angle β of the strip at $t = \pi/2$ is displayed in Figure 19.16 (right) for $n = 2$ as a function of the ellipticity of the base curve $\lambda_e := b_a/a_a$ and the positioning angle of the strip in the xy-plane.

With a given radius of the winding mandrel and the fixed angular position at the onset of the bend, the design parameters are:

- The ellipticity of the base curve λ_e.
- The order n of the base curve.
- The inclination angle β of the innermost turn of each coil block.
- Four knots of a cubic spline function allowing for the local adjustment of the cable torsion between the onset of the coil end and the nose, according to Eqs. (19.30), (19.34), and (19.35).

The following objectives can be considered for the design optimization:

- Integrated squared geodesic curvature of each coil block.
- Maximum curvature parameters in each coil block.
- A parameter indicating an edge of regression violation within the strip surface.

To detect an edge of regression violation, the length of the generators from the base curve to the outer edge of the cable must be determined. While this is relatively simple for the discrete case, it is a nontrivial task in the three-dimensional continuous model. Making use of two bending invariances, which are the angle between **T** and **d** and the geodesic curvature κ_g, the generator lengths can be determined by fundamental geometric operations for the unbent (developed) strip; see Figure 19.9 (left). A penalty parameter indicating the degree of the edge of regression violation is given by the integrated length of the generator vectors when intersections within the cable surface occur.

The coil-end design and optimization process is now established and is summarized as follows:

1. Optimization of the coil cross section. This determines the positioning and inclination angle of the cables and consequently the semi-minor axis a of the ellipses on the winding mandrel.
2. Estimation of the natural angle β for the innermost turn of each coil block for an ellipticity $\lambda_e = 1.2$.
3. Calculation of the Frenet frame on the base curve.
4. Matching the magnet's straight section at the onset of each winding by applying additional twist.
5. Generation of the coil block geometry (with or without inter-turn wedges).
6. Calculation of the local curvature parameters at every discretization point and in every turn in the coil blocks.

7. Numerical optimization of the strip surface with respect to the objectives described above.
8. Verification that the local geodesic curvature obeys the boundary conditions found in the winding tests. If it does not, the following measures can be applied: (a) Applying additional twist as described in Section 19.2.2. (b) Winding of thin wedges, made of polyimide foils, between the turns in the coil end. (c) Rotation of the mid-plane on the connection side; see Figure 19.17. If these measures are not sufficient, an additional end spacer must be introduced in order to reduce the number of windings in the coil blocks.
9. Optimization of the field quality by shifting the relative position of the coil blocks. For this purpose, we use the design variable z_0 described in Section 19.1. An example where the integrated sextupole field component was minimized is shown in Figure 6.8.

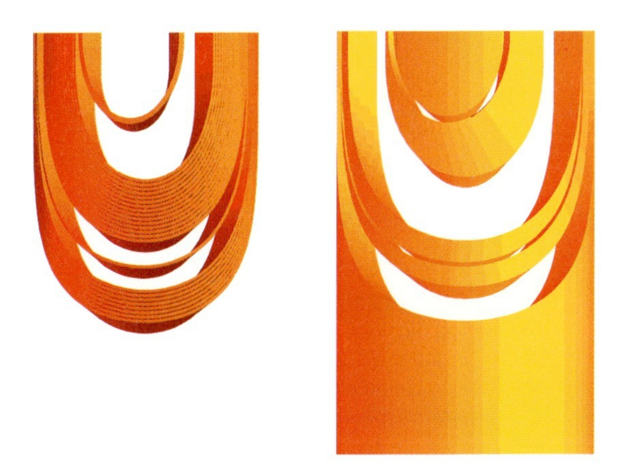

Figure 19.17 Connection side (lead end) of an LHC main dipole model coil (left) together with its end spacers (right). Notice the rotated median plane for the asymmetric inner turns.

An objective weighting function is set up for the above-mentioned objectives, while the constraint on the edge of regression violation is taken into account by means of a penalty transformation. The resulting optimization problem is solved with the deterministic optimization algorithm EXTREM discussed in Chapter 20.

19.5 Coil-End Transformations

Arbitrary arrangements of coils can be generated by applying translations and rotations, shown in Figure 19.18, in order to generate assemblies of torus coils and Helmholtz coils, or to calculate multipole feed-down effects due to magnet misalignment.

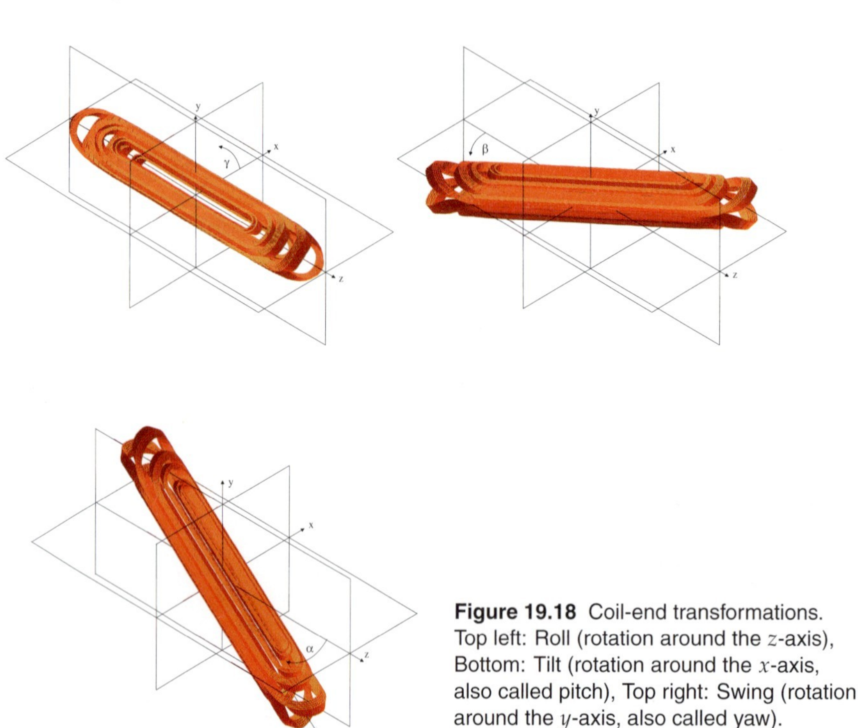

Figure 19.18 Coil-end transformations. Top left: Roll (rotation around the z-axis), Bottom: Tilt (rotation around the x-axis, also called pitch), Top right: Swing (rotation around the y-axis, also called yaw).

Every nontrivial rotation in three dimensions defines a one-dimensional linear subspace of $\mathbb{R}^3$, which is called the *rotational axis*. The rotation acts then as a two-dimensional rotation in the plane orthogonal to this axis. Right-handed screw rotations are expressed by $\{x\} \mapsto [R_i]\{x\}$, where

$$[R_1(\alpha)] = \begin{pmatrix} 1 & 0 & 0 \\ 0 & \cos\alpha & -\sin\alpha \\ 0 & \sin\alpha & \cos\alpha \end{pmatrix}, \quad [R_2(\beta)] = \begin{pmatrix} \cos\beta & 0 & \sin\beta \\ 0 & 1 & 0 \\ -\sin\beta & 0 & \cos\beta \end{pmatrix},$$

$$[R_3(\gamma)] = \begin{pmatrix} \cos\gamma & -\sin\gamma & 0 \\ \sin\gamma & \cos\gamma & 0 \\ 0 & 0 & 1 \end{pmatrix}. \tag{19.51}$$

The angles α, β, γ are the *Tait-Bryan angles*[9] used in aeronautics, representing the *tilt* (pitch), *swing* (yaw), and *roll* angles, respectively. The Euler angles, for which different conventions exist [1, 19], are equivalent to a special combination of Tait-Bryan angles. In aeronautics, the transverse x-axis runs parallel to the wings from the pilot's left to right, the longitudinal z-axis from the tail to the nose, and the vertical y-axis from the top to the bottom. The matrices are orthonormal, obeying the condition

$$[R_i]^T [R_i] = [I], \tag{19.52}$$

where $[I]$ is the unit matrix. They are length- and angle-preserving linear transformations forming the group O(3). The orthogonal matrices, which also preserve orientation, $\det[R] = 1$, are the matrices of the *special orthogonal group* SO(3). Note that the rotation matrices are not commutative, that is, SO(3) is non-Abelian, and therefore the sequence of the transformations must be respected. Another useful transformation matrix results in an imaging about a plane rolled by an angle φ:

$$[T_3(\varphi)] = \begin{pmatrix} \cos\varphi & \sin\varphi & 0 \\ \sin\varphi & -\cos\varphi & 0 \\ 0 & 0 & 1 \end{pmatrix}, \tag{19.53}$$

which is an element of the $O(3) \setminus SO(3)$ group.

19.6 Corrector Magnet Coil End with Ribbon Cables

The mathematical models and the design procedure outlined in this chapter have been validated with the coil-end design of the MQM, MQXA, and MCB magnets for the LHC. The coil end of the MCB orbit corrector was particularly difficult to design because of the large number of windings per coil block (maximum 26) and the brittle nature of the ribbon cables that does not support any geodesic curvature. The winding topology is shown in Figure 19.19. Previous designs based only on global parameters such as the constant-perimeter conditions on the inner- and outer-cable edges led to unacceptable cable lift from the winding mandrel. The effect can be understood as a redistribution of the strain energy by a deviation from the base curve, resulting in a nonzero distance between the winding mandrel and the cable's inner edge at a specific position along the end.

A first optimization aimed at a minimum geodesic curvature in the innermost winding of each block. It could be shown that in this case the strain energy of the outermost turn in a block of 26 windings is roughly 18 times higher

9 Peter Guthrie Tait (1831–1901), George Hartley Bryan (1864–1928).

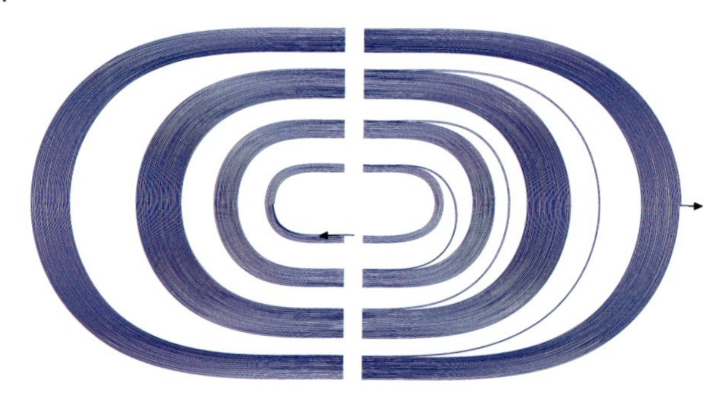

Figure 19.19 Winding topology of the MCB dipole corrector magnet. The figure shows the two outer cable edges in the developed sz-plane. Right: Connection side (lead end). Left: Nonconnection side (return end).

than in the first turn. This leads to unacceptable cable lift from the winding mandrel. For this reason the coil blocks must be clamped onto the mandrel, forcing the cable into a position that results in higher geodesic curvature and strain energy. These results are displayed in Figures 19.20 and 19.21.

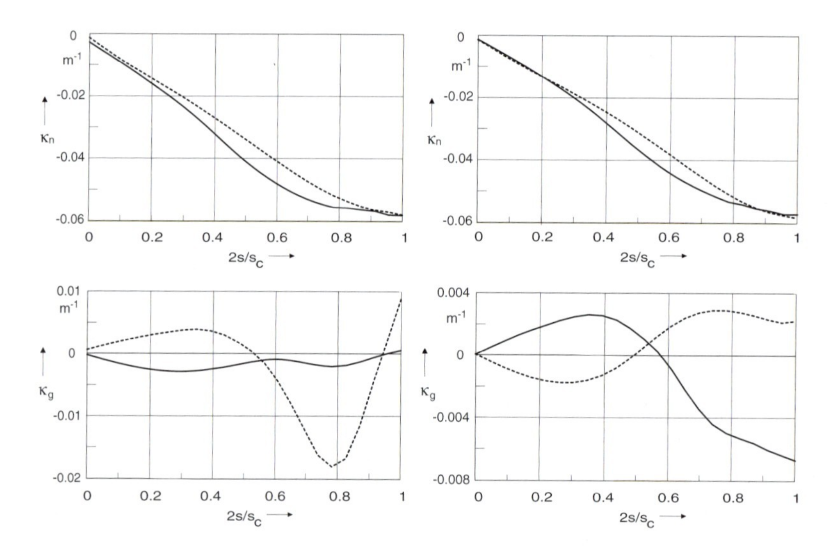

Figure 19.20 Geodesic and normal curvature as a function of the arc length in the innermost winding (solid) and the outermost winding (dashed) of the second coil-block in the MCB corrector magnet, containing 26 windings in all. Left: Result of a minimization of the geodesic curvature in the innermost winding only. Right: Optimum solution to the minimization of the total strain energy in the coil block. Notice the different scales for the geodesic curvature. s_c denotes the cable length between the cross section and the nose.

Figure 19.21 Winding trial of the MCB dipole corrector coil ends. Note the tooling for the clamping of the coil designed to prevent cable lift-off from the winding mandrel.

The minimization of the total strain energy in each coil block could balance the inevitable local geodesic curvature between the first and last turn in a coil-block. As compared to the first case, the total block strain energy could be reduced by a factor of 5. Figure 19.20 illustrates the distribution of the geodesic curvature in the innermost and outermost turns for both cases. The latter design was settled on for the series production of the magnets after a winding trial carried out in European industry.

19.7
End-Spacer Manufacturing

The CERN field computation program ROXIE contains a number of interfaces to computer aided design and manufacturing (CAD/CAM), structural analysis programs such as ANSYS [2], and commercial field computation software such as VF-OPERA [18].

A DXF[10] interface creates files for the drawing of the cross sections in the xy- and yz-planes of the magnet. Moreover, it generates the developed views of the coils and the end spacers as well as the 3D polygons for the end spacers and coils. This reduces errors and a considerable amount of drafting work. The surfaces to be machined are described by two polygons on the inner and outer diameters of the end spacers such that the straight lines joining the corresponding points on the two polygons are the generators of the developable surface. The polygons are transferred into a CAD/CAM system, for example, CATIA [10], for the calculation of the cutter movements for the CNC machining of the pieces. The spacers are machined from fiberglass-epoxy (Nema [14]

10 Data eXchange Format established by AUTODESK [5].

Grade-G11) tubes, with a five-axis CNC machine in a single pass using a cylindrical cutter. Because of the abrasive nature of the glass fibers, diamond tools must be used. (It is also possible to manufacture the pieces with a five-axis water-jet cutting machine where the direction of the water jet is given by the Darboux vectors). Figure 19.17 shows a top view of the connection-side coil and the corresponding end spacers for the outer-layer dipole coil.

As CNC milling requires costly surface modeling and part-specific tooling, a rapid prototyping method, called *solid freeform fabrication*, is applied to the manufacturing of spacers to be used for the winding tests. The *selective laser sintering* (SLS) technique uses a photoreactive polymer (nylon) powder deposited in a thin layer and scanned with a laser so that the powder particles fuse together into a cross section. The layering process is repeated until the part is completed. Accuracy and surface roughness is limited by the size of the powder grains of about 0.1 mm in diameter. The material used for the SLS process has good mechanical properties and can thus be used for winding trials; it would, however, not withstand the curing temperature of 180 °C needed to polymerize the coil with its polyimide insulation.

19.8 Splice Configurations

Figure 19.22 (left) shows two variants for the design of the splice connection between the inner and the outer coils in the lead end of the MQXA insertion quadrupole [17]. The variant on the bottom left shows a splice located outside the coil, requiring special end collars for mechanical support. The variant on the top left features an internal splice entirely contained within the outer diameter of the coil. The goal is to provide better mechanical support and to simplify the magnet assembly.

Another example of the lead-end design is the MQM quadrupole as shown in Figure 19.22 (right). The lead end is made of three sections: the layer jump, the end-spacer section, and the connection box. The 83-mm-long layer jump twists away from the inner layer until it is parallel to the outer layer, exactly one cable width away from the adjacent turn. The cable is then routed outward with constant inclination until it is aligned with the outer layer. Finally, it is then turned around the end into its final position. The lead end spacers are designed with a similar coil-block spacing as used in the return end, and consequently the peak field enhancement is very close to the value obtained for the return end. The connection between the poles is made in a connection box. An angular turn of 90° is insufficient for making a splice of 120 mm or more in arc length, required for covering at least one complete transposition pitch. Therefore, 270° turns are used for three splices in three planes, all or-

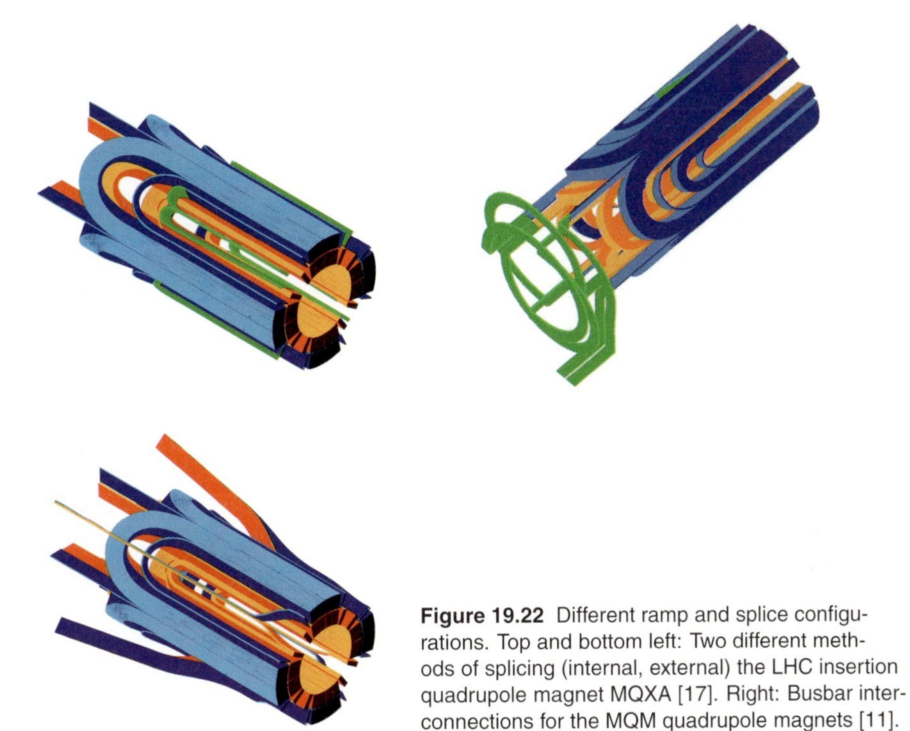

Figure 19.22 Different ramp and splice configurations. Top and bottom left: Two different methods of splicing (internal, external) the LHC insertion quadrupole magnet MQXA [17]. Right: Busbar interconnections for the MQM quadrupole magnets [11].

thogonal to the aperture axis. In addition, a fourth plane is used to bring the two leads together. The sequence of splices was chosen in such a way as to minimize their effect on the quality of the integrated field [11].

References

1 Arnold, V. I.: Mathematical Methods of Classical Mechanics, Springer, Berlin, 1989

2 Ansys Inc. software products, http://www.ansys.com

3 Auchmann, B., Russenschuck, S.: Coil end design for superconducting magnets applying differential geometry methods, IEEE Transactions on Magnetics, 2004

4 Auchmann, B., Russenschuck, S., Schwerg, N.: Discrete differential geometry applied to the coil-end design of superconducting magnets, IEEE Transactions on Applied Superconductivity, 2007

5 Autodesk: http://usa.autodesk.com

6 Blaschke, W.: Vorlesungen ueber Differentialgeometrie und geometrische Grundlagen von Einsteins Relativitaetstheorie, Reprint Chelsea Publishing Company, New York, 1967

7 Bossert, R., Cook, J., Brandt, J.: End Designs for Superconducting Magnets, Fermilab Technical Memo TM-1757, 1991

8 Cook, J.M.: An Application of Differential Geometry to SSC Magnet End Winding, SSC Note SSSL-N-720 and Fermilab Technical Memo TM-1663, 1990

9 Cook, J.M.: Strain energy minimization in SSC magnet winding, IEEE Transactions on Magnetics, 1991

10 CATIA (Computer Aided Three-dimensional Interactive Application) is a commercial CAD/CAM software suite developed by Dassault Systems and marketed by IBM.

11 Lucas, J. et al.: Performance of the final prototype of the 6-kA matching quadrupoles for the LHC insertions and status of the industrialization program, IEEE Transactions on Applied Superconductivity, 2003

12 Mathematica is a trademark by Wolfram Research Inc.

13 Morgan, G.H., Green, A., Jochen, G., Morgillo, A.: Winding mandrel design for the wide cable SSC dipole, Applied Superconductivity Conference, 1990

14 The Association of Electrical and Medical Imaging Equipment Manufacturers: http://www.nema.org

15 Rosten, H.I.: The Constant Perimeter End, Rutherford Laboratory, RL-73-096, 1973

16 Russenschuck, S. et al.: Integrated design of superconducting accelerator magnets – a case study of the main quadrupole, The European Physical Journal, Applied Physics, 1, 1998

17 Sabbi, G.: End Field Analysis of HGQ Short Models Using a BEND-ROXIE Interface, Proceedings of the First International ROXIE Users Meeting and Workshop, CERN, March 16–18, 1998

18 Vector Fields, software for electromagnetic design, http://www.vectorfields.com

19 Weisstein, E. W.: Euler Angles, form MathWorld – A Wolfram Web Resource. http://mathworld.wolfram.com/EulerAngles.html

20
Mathematical Optimization Techniques

A man once saw a butterfly struggling to emerge from the cocoon,
too slowly for his taste, so he began to blow it gently.
The warmth of his breath speeded up the process all right.
But what emerged was not a butterfly but a creature with mangled wings.

Anthony de Mello (1931–1987), One Minute Wisdom.

Mathematical optimization techniques have for decades been applied to computational electromagnetism. In 1967, Halbach [13] introduced a method for optimizing pole shapes of magnets by means of finite-element field calculation. In 1982, Armstrong et al. [3] combined optimization algorithms with the volume-integral method for the pole-profile optimization of an H-magnet. These attempts tended to be application-specific, however. Only since the late 1980s have numerical field calculation packages for both 2D and 3D applications been placed in an optimization environment. Reasons for this delay have included constraints in computing power, problems with nonsmooth objective functions arising from finite-element discretization, accuracy of the field solution, and software implementation problems.

Applying mathematical optimization techniques to numerical field computation for magnet design enforces special requirements:

- The computing time for evaluating each objective-function is in the range of minutes. The time for computing the next search point can therefore be neglected.
- As the objective functions are calculated by finite-element procedures, they are compromised by numerical errors. This makes the application of gradient methods problematic as differential quotients for the gradient approximation must be used.
- Since the objective functions are not explicitly defined, continuity, differentiability, and convexity must be assumed for the application of certain classes of optimization methods.

Field Computation for Accelerator Magnets. Stephan Russenschuck

ISBN: 978-3-527-40769-9

- Certain trial solutions might lead to physically meaningless structures and the interruption of finite-element calculations. This has to be considered during the setup of the parametric model.
- Magnet design has to deal with multiple conflicting objectives.

Let us begin with the last point. Most of the real-world optimization problems involve multiple conflicting objectives. As an example, the search for a maximum main field and a smallest possible volume of superconductor material clearly presents a contradiction. There is also a conflict between the maximum field in the aperture, obtained by bringing the iron yoke close to the coil, and small field error variations due to iron saturation.

Characteristic of these *vector optimization* problems is the appearance of an *objective conflict* in which the individual solutions for each single objective function differ and no solution exists wherein all the objectives reach their individual minimum. The procedures for solving vector-optimization problems comprise decision making, treatment of constraints, and an optimization algorithm.

- Methods for decision making, based on the optimality criterion by Pareto[1] [34], were introduced and applied to a wide range of problems in economics by Marglin [29] and Fandel [9]. They include objective-weighting, distance-function methods, and constraint formulation, among others. These methods are also referred to as *goal-programming*.
- Methods for the treatment of nonlinear constraints have been developed by Zoutdendijk [56], Fiacco and McCormick [10], and Rockafellar [43], among others, and are most often based on the optimality criterion by Karush et al. [19].
- Numerous optimization algorithms using both deterministic and stochastic elements were developed in the 1960s and covered in books by Wilde [54], Rosenbrock [44], Himmelblau [15], and Brent [5]. Since the early 1980s, triggered by the increase of computing power, researchers have focused on genetic and evolutionary algorithms. The advantages of these methods are the ability to find global optima and treat a large number of design variables simultaneously [11,16].

Since 1985 numerous papers have been published on the application of mathematical optimization to electromagnetic design, including papers by the author on the optimization of waveguide structures and on the design and optimization of permanent-magnet synchronous machines [41,45].

The large variety of optimization methods in use shows that no general method exists for solving nonlinear optimization problems in computational electromagnetism in the same way that the simplex algorithm is *the* method for linear problems. The methods have included *deterministic algorithms*,

1 Vilfredo Pareto (1848–1923).

which do not require the derivative of the objective function; *gradient methods*, including design sensitivity analysis with the adjoint state technique; and *stochastic methods*, including simulated annealing, genetic algorithms, and artificial neural networks. The performance differences between various methods of the same general type are rather small when applied to the same class of problems. Any characterization of solution methods should therefore distinguish between deterministic and stochastic algorithms. Figure 20.1 shows a classification system.

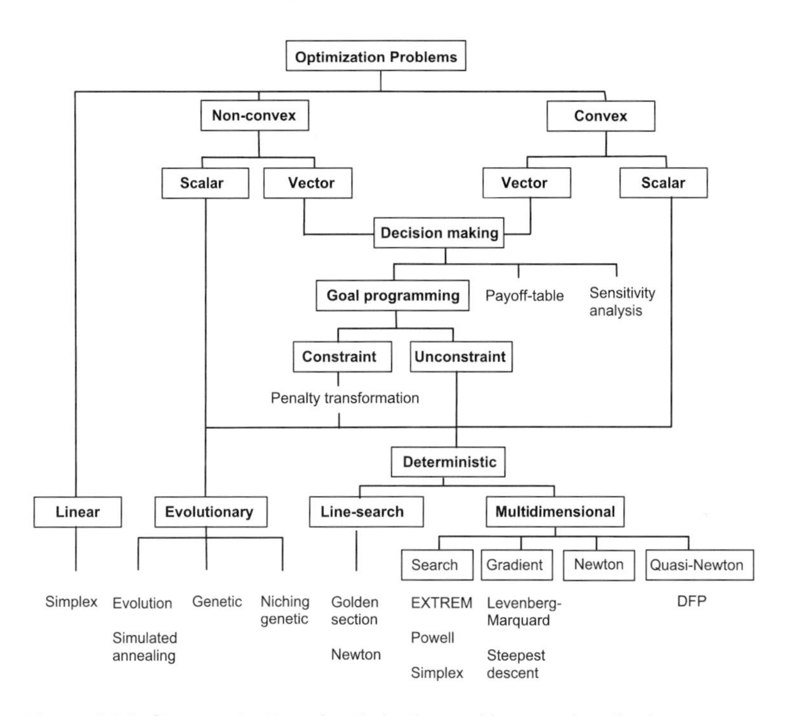

Figure 20.1 Characterization of optimization problems and methods.

20.1 Mathematical Formulation of the Optimization Problem

Optimization problems are characterized by the aim of minimizing a nonlinear *objective function* $f : X \to \mathbb{R}$ on an open domain $X \subseteq \mathbb{R}^n$ with design variables arranged in the column[2] vector $\mathbf{x} = (x_1, x_2, \ldots, x_n)^T \in X$. In magnet

2 In this chapter, we change the notation according to the conventions of matrix algebra: We denote column vectors with lower case, bold-face characters and matrices with capital bold-face characters, omitting the square brackets, e.g., $\mathbf{F} := [F] = (f_{ij})$. We also adopt the nabla notation for the gradient: $\operatorname{grad} f(\mathbf{x}) \equiv \nabla f(\mathbf{x})$.

design, this function must be calculated by means of numerical field computation and therefore sufficient smoothness must be assumed for the purpose. We are usually concerned with optimization problems of the form

$$\min\{f(\mathbf{x})\}, \tag{20.1}$$

$f : X \to \mathbb{R}$, subject to nonlinear constraints $g_i, h_j : X \to \mathbb{R}$:

$$g_i(\mathbf{x}) \leq 0, \qquad i = 1, 2, \ldots, m, \tag{20.2}$$

$$h_j(\mathbf{x}) = 0, \qquad j = 1, 2, \ldots, p, \tag{20.3}$$

where g_i is the *inequality constraint* and h_j are *equality constraint*. Each design variable x_l can take values from the interval $\Omega_l = (x_{l,\text{lower}}, x_{l,\text{upper}}) \subset \mathbb{R}$:

$$x_{l,\text{lower}} < x_l < x_{l,\text{upper}}, \qquad l = 1, 2, \ldots, n. \tag{20.4}$$

We call $x_{l,\text{lower}}$ and $x_{l,\text{upper}}$ the lower and the upper bounds of the design variables, respectively. The constraints (20.4) are also denoted *box constraints* in contrast to Eqs. (20.2) and (20.3), which are referred to as functional constraints. The box constraints can be taken into account at a later stage and are therefore omitted in the following sections. The *feasible domain* M is defined by

$$M := \{\mathbf{x} \in X | g_i(\mathbf{x}) \leq 0;\ h_j(\mathbf{x}) = 0;\ x_{l,\text{lower}} < x_l < x_{l,\text{upper}}; \\ \forall i = 1, \ldots, m;\ j = 1, \ldots, p;\ l = 1, \ldots, n\}. \tag{20.5}$$

In the case of equality constraints only, the feasible domain is a curved (hyper-) surface embedded in $\mathbb{R}^n$. A vector $\mathbf{x} \in M \subset X \subseteq \mathbb{R}^n$ is called a *feasible vector* or feasible point. The inequality constraints are said to be active at a feasible point if $g_i(\mathbf{x}) = 0$ and inactive at $\mathbf{x}$ if $g_i(\mathbf{x}) < 0$. The constraint optimization problem can also be written in the form

$$\min\{f(\mathbf{x}) | \mathbf{x} \in M\} \tag{20.6}$$

for $f : X \to \mathbb{R}$. If $M = \mathbb{R}^n$, that is, $i, j, l = 0$, the optimization problem is said to be unconstrained. A feasible vector $\mathbf{x}^* \in M$ is called a *global optimizer* if $f(\mathbf{x}^*) \leq f(\mathbf{x})$ for all $\mathbf{x} \in M$. It is called a *strict global optimizer* if $f(\mathbf{x}^*) < f(\mathbf{x})$ for all $\mathbf{x} \in M$. A feasible vector is called a *local optimizer* if $f(\mathbf{x}^*) \leq f(\mathbf{x})$ for all $\mathbf{x} \in M \cap B_\varepsilon(\mathbf{x}^*)$, where $B_\varepsilon(\mathbf{x}^*)$ is the open ϵ-ball around $\mathbf{x}^*$. The vector is a *strict local optimizer* if $f(\mathbf{x}^*) < f(\mathbf{x})$ for all $\mathbf{x} \in M \cap B_\varepsilon(\mathbf{x}^*)$.

All optimization problems can be treated as minimization problems because

$$\max\{f(\mathbf{x})\} = -\min\{-f(\mathbf{x})\}. \tag{20.7}$$

We can also impose the condition that the objective function takes only positive values:

$$\min\{f(\mathbf{x})\} = \min\{f(\mathbf{x}) + c\}, \tag{20.8}$$

where c is a positive constant. This is necessary in the case of fitness ranking for genetic algorithms. An obvious classification of optimization problems can be accomplished according to the linearity of the objective function and/or the constraints (20.2) and (20.3). In numerical field computation nonlinearity of both the objective function and the constraints must be assumed.

20.2
Optimality Criteria for Unconstrained Problems

Suppose the objective function on $X \subseteq \mathbb{R}^n$ is sufficiently smooth for our purpose; at least $f \in C^2$. The gradient of the objective function will be written as the row vector[3]

$$\nabla f(\mathbf{x}) = \left(\frac{\partial f(\mathbf{x})}{\partial x_1}, \ldots, \frac{\partial f(\mathbf{x})}{\partial x_n} \right). \tag{20.9}$$

A point $\mathbf{x} \in X$ is called a *stationary point* if

$$\nabla f(\mathbf{x}) = \mathbf{0}. \tag{20.10}$$

A necessary condition for a local minimizer $\mathbf{x}^*$ of the unconstrained optimization problem is that $\mathbf{x}^*$ be a stationary point. A sufficient condition is that the real symmetric Hesse[4] matrix $\mathbf{H} \in \mathbb{R}^{n \times n}$,

$$\mathbf{H}(\mathbf{x}^*) = \left(\frac{\partial^2 f(\mathbf{x}^*)}{\partial x_i \partial x_j} \right), \tag{20.11}$$

be a positive semidefinite matrix:

$$\mathbf{x}^T \mathbf{H}(\mathbf{x}^*) \mathbf{x} \geq 0, \qquad \mathbf{x} \in X. \tag{20.12}$$

In this case $\mathbf{x}^*$ would be a local minimizer of the unconstrained optimization problem. If the Hesse matrix is positive definite, $\mathbf{x}^*$ is a strict local minimizer.[5]

3 Note that this is often defined to be a column vector; this is a notational problem we have avoided by writing grad $f(\mathbf{x}) = \sum \partial f(\mathbf{x})/\partial x_i \, \mathbf{e}_i$ in previous chapters.
4 Otto Hesse (1811–1874).
5 In this case $\mathbf{x}^T \mathbf{H}(\mathbf{x}^*) \mathbf{x} > 0$ for all $\mathbf{x} \in X$.

Open domains $X \subseteq \mathbb{R}^n$ where for all $\mathbf{x}_1, \mathbf{x}_2 \in X$ and $\nu \in (0,1)$ the condition

$$\nu \mathbf{x}_1 + (1-\nu)\mathbf{x}_2 \in X \tag{20.13}$$

is fulfilled are called *convex*; see Figure 2.7. In particular, the intervals Ω_l are convex. The intersection of convex domains is again convex. A function $f : X \to \mathbb{R}$ on a nonempty convex domain is said to be convex if

$$f(\nu \mathbf{x}_1 + (1-\nu)\mathbf{x}_2) \leq \nu f(\mathbf{x}_1) + (1-\nu) f(\mathbf{x}_2) \tag{20.14}$$

for all $\nu \in (0,1)$ and $\mathbf{x}_1, \mathbf{x}_2 \in X$. A function is convex if the line joining two points on its graph lies nowhere below the graph. Condition (20.14) is equivalent to

$$f(\mathbf{x}_1) - f(\mathbf{x}_2) \geq \nabla f(\mathbf{x}_2)(\mathbf{x}_1 - \mathbf{x}_2) \tag{20.15}$$

for all $\mathbf{x}_1, \mathbf{x}_2 \in X$. If the Hesse matrix at $\mathbf{x}^*$ is positive definite for a convex function, then $\mathbf{x}^*$ is a strict global minimizer of the unconstrained minimization problem.

20.3 Karush–Kuhn–Tucker Conditions

The Lagrange function $\mathcal{L} : \mathbb{R}^n \times \mathbb{R}^m \times \mathbb{R}^p \to \mathbb{R}$ of the constrained optimization problem[6] (20.1)–(20.3) is defined by

$$\mathcal{L}(\mathbf{x}, \boldsymbol{\alpha}, \boldsymbol{\beta}) := f(\mathbf{x}) + \sum_{i=1}^{m} \alpha_i g_i(\mathbf{x}) + \sum_{j=1}^{p} \beta_j h_j(\mathbf{x}), \tag{20.16}$$

where the components α_i, β_j are called *Lagrange multipliers* and are arranged in the column vectors $\boldsymbol{\alpha} := (\alpha_1, \ldots, \alpha_m)^T$ and $\boldsymbol{\beta} := (\beta_1, \ldots, \beta_p)^T$. A solution of the problem

$$\min_{\mathbf{x}} \left\{ \max_{\boldsymbol{\alpha},\boldsymbol{\beta}} \{\mathcal{L}(\mathbf{x}, \boldsymbol{\alpha}, \boldsymbol{\beta})\} \right\}, \tag{20.17}$$

where for constant $\mathbf{x}$ the Lagrange function is first maximized in $\boldsymbol{\alpha}$ and $\boldsymbol{\beta}$ and then minimized in $\mathbf{x}$, is a solution of the constraint optimization problem (20.1)–(20.3) because

$$\max_{\boldsymbol{\alpha},\boldsymbol{\beta}} \{\mathcal{L}(\mathbf{x}, \boldsymbol{\alpha}, \boldsymbol{\beta})\} = \begin{cases} f(\mathbf{x}) & \text{if } g_i \leq 0, h_j = 0, i = 1, \ldots, m, j = 1, \ldots, k. \\ \infty & \text{otherwise} \end{cases} \tag{20.18}$$

6 Note that we have omitted the box constraints.

Let the objective function and constraints f, g_i, h_j be at least 1-smooth: The conditions

$$\nabla_{\mathbf{x}} \mathcal{L}(\mathbf{x}, \boldsymbol{\alpha}, \boldsymbol{\beta}) = \mathbf{0}, \quad (20.19)$$

$$g_i(\mathbf{x}) \leq 0, \quad (20.20)$$

$$\alpha_i \geq 0, \quad (20.21)$$

$$\alpha_i g_i(\mathbf{x}) = 0, \quad (20.22)$$

are called the *Karush–Kuhn–Tucker (KKT) conditions* [19, 21]. The gradient of the Lagrange function is given by

$$\nabla_{\mathbf{x}} \mathcal{L}(\mathbf{x}, \boldsymbol{\alpha}, \boldsymbol{\beta}) = \nabla f(\mathbf{x}) + \sum_{i=1}^{m} \alpha_i \nabla g_i(\mathbf{x}) + \sum_{j=1}^{p} \beta_j \nabla h_j(\mathbf{x}). \quad (20.23)$$

A vector $(\mathbf{x}^*, \boldsymbol{\alpha}^*, \boldsymbol{\beta}^*)$ that obeys the KKT conditions is called a KKT point of the optimization problem (20.1)–(20.3). If there are no constraints, the KKT conditions reduce to the necessary condition (20.10). For optimization problems where the objective function and all constraints are convex, the KKT conditions are sufficient, while for nonconvex problems they are only first-order necessary conditions. The gradients of the active inequality constraints[7] must be linearly independent for Eq. (20.19) to have a solution. A necessary condition is that the number of design variables be greater or equal to the number of constraints. A vector $\mathbf{x}^*$ is called a *regular point* if the gradients of the active constraints are linearly independent.

The KKT conditions require that the gradient of the Lagrange function $\mathcal{L}$ be zero, and that the Lagrange multipliers α_i of the active inequality constraints $g_i(\mathbf{x}^*)$ take positive values. Otherwise it would be possible to decrease the value of a constraint without increasing the objective function, which is of course not characteristic for an optimal point. Figure 20.2 gives a geometric interpretation: Point 1 is not a minimum because $-\nabla f = \alpha_1 \nabla g_1 + \alpha_3 \nabla g_3$ and $\alpha_1 < 0$. The constraint $g_1^* = 0$, causing stronger restrictions on the feasible domain, leads to point 2 as an optimum with a lower numerical value of $f(\mathbf{x})$. Thus point 1 is not an optimum. The KKT conditions are satisfied at the optimal point 3, where the negative gradient of the objective function is a positive linear combination of the gradients of the inequality constraints. A vanishing Lagrange multiplier shows that the constraint is redundant. In this case, the optimization problem is called *degenerate*. The quotient between the absolutely largest and absolutely smallest nonzero Lagrange multipliers is called *degree of degeneration*. The Lagrange multipliers are also a measure of the price to be paid when a constraint is tightened or relaxed. Pitfalls of nonconvex functions are shown in Figure 20.6.

7 That is, $g_i(\mathbf{x}) = 0$.

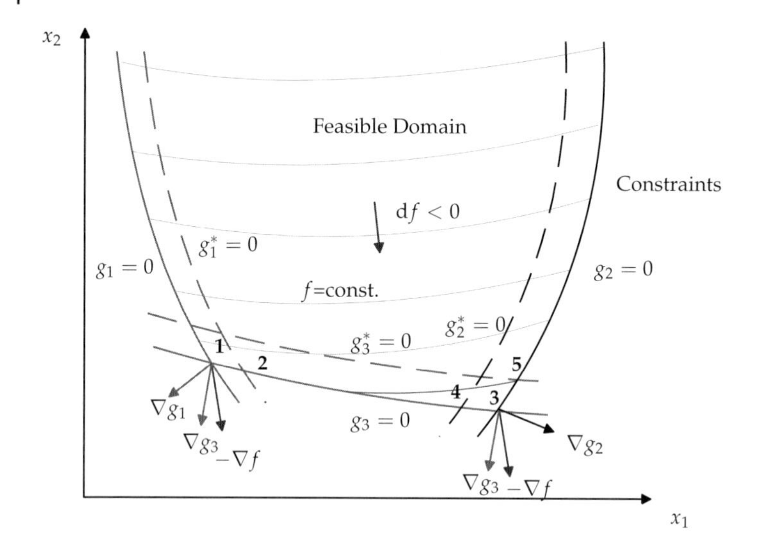

Figure 20.2 Geometrical interpretation of the KKT conditions. Point 1 is not a minimum because $-\nabla f = \alpha_1 \nabla g_1 + \alpha_3 \nabla g_3$ with $\alpha_1 < 0$. The constraint $g_1^* = 0$ causing stronger restrictions on the feasible domain leads to point 2 as an optimum with a lower numerical value of $f(\mathbf{x})$. Thus point 1 is not an optimum. The KKT conditions are satisfied in the optimal point 3.

20.4 Pareto Optimality

Real-world optimization problems involve multiple conflicting objectives that must be mutually reconciled. A vector optimization problem can be written in a standardized mathematical form as

$$\min\{\mathbf{f}(\mathbf{x})\} = \min\{f_1(\mathbf{x}), f_2(\mathbf{x}), \ldots, f_K(\mathbf{x})\}, \tag{20.24}$$

$\mathbf{f} : \mathbb{R}^n \rightarrow \mathbb{R}^K$, subject to the nonlinear constraints given in Eqs. (20.2) and (20.3). The vectors $\mathbf{f} \in \mathbb{R}^K$ are called *criterion vectors*, and the feasible set of all criterion vectors is denoted Y. For the definition of the optimal solution of the vector optimization problem we can apply the optimality criterion by Pareto, which was originally introduced for problems in economics [34,50]. A Pareto-optimal (or noninferior) solution $\mathbf{x}^*$ is given when there exists no solution $\mathbf{x}$ in the feasible domain for which

$$f_k(\mathbf{x}) \leq f_k(\mathbf{x}^*) \qquad \forall k \in [1, K], \tag{20.25}$$

$$f_k(\mathbf{x}) < f_k(\mathbf{x}^*) \qquad \text{for at least one}\, k \in [1, K]. \tag{20.26}$$

This implies that a design in which the improvement of one objective causes the degradation of at least one other objective is Pareto optimal. It is clear that this definition yields a set of solutions rather than one unique solution. The set $P(Y)$ of all Pareto-optimal solutions of the optimization problem (20.24) is called the *complete solution* or the *Pareto front* of the vector optimization problem. Figure 20.3 shows graphic interpretations of Pareto-optimal solutions for the two conflicting objectives f_1 and f_2.

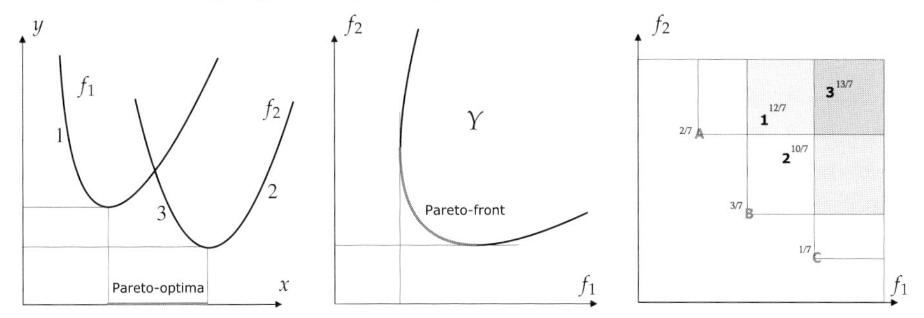

Figure 20.3 Pareto-optimal solutions. Left: Points 1 and 2 are not Pareto optimal because one objective can always be improved without degrading the other. Point 3 is a Pareto-optimal solution. Middle: Pareto front of the feasible set Y of all criterion vectors. Right: Pareto and population strength values for a set of criterion vectors. Members A, B, and C are not dominated by any other member. Member 1 is dominated by A and B, 3 by A, B, and C. The Pareto-optimal solution A dominates 1 and 3.

We express with $\mathbf{f}^* \succ \mathbf{f}$ the property that a criterion vector $\mathbf{f}^* \in \mathbb{R}^K$ is preferred to, or *strictly dominates*, another criterion $\mathbf{f} \in \mathbb{R}^K$; that is, the condition $f_k(\mathbf{x}^*) < f_k(\mathbf{x})$ is fulfilled for **all** components $f_k, k = 1, \ldots, K$. The criterion vector is weakly dominating if for all k it holds that $f_k(\mathbf{x}^*) \leq f_k(\mathbf{x})$. Two criterion vectors $\mathbf{f}_1$ and $\mathbf{f}_2$ are said to be *indifferent* if $\mathbf{f}_1$ does not dominate $\mathbf{f}_2$ and $\mathbf{f}_2$ does not dominate $\mathbf{f}_1$. Then the Pareto front is given by

$$P(Y) = \{\mathbf{f} \in Y : \{\mathbf{f}^* \in Y : \mathbf{f}^* \succ \mathbf{f}, \mathbf{f}^* \neq \mathbf{f}\} = \emptyset\} . \qquad (20.27)$$

This is the set of all nondominated solutions.

Examples: Figure 20.4 shows a real-world optimization problem with two design variables; the semi-minor axis a and semi-major axis b of the ellipse defining the inner shape of the iron yoke of a dipole magnet. Bringing the iron yoke closer to the coil increases the quench field but also increases the saturation-dependent field errors. The objective conflict is obvious. In Figure 20.4 (right), the contour plots of the two objectives are shown together with the Pareto-optimal solution set. On the Pareto front, represented by the bold line, any increase in maximum field must be paid for by an increase in field errors. Point 1 is not a Pareto optimum because in point 2 both objectives are improved.

Vector-optimization problems are also present in the pole-shim design of normal-conducting magnets. We have seen in Section 7.6.1 that with rela-

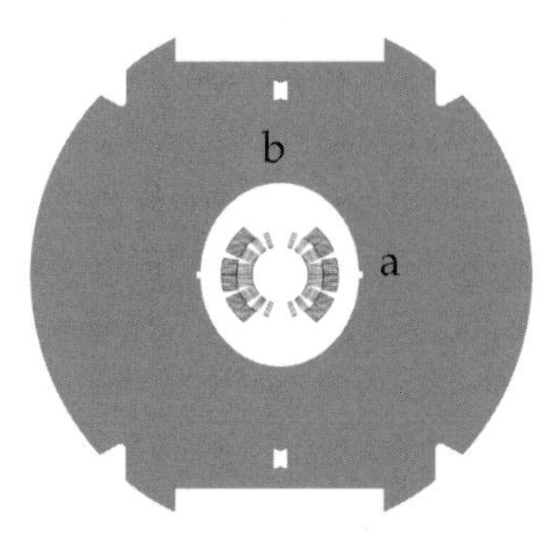

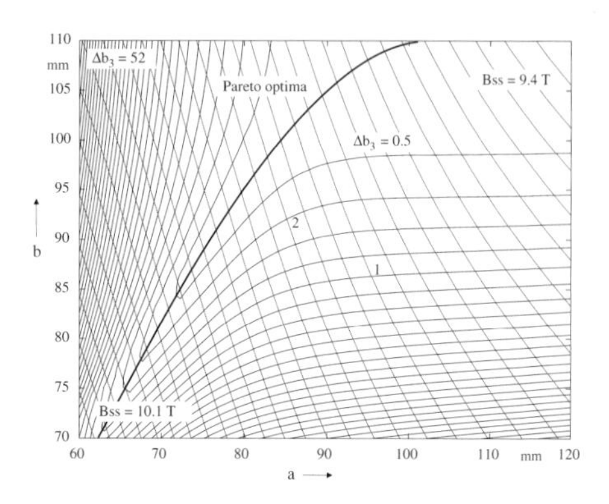

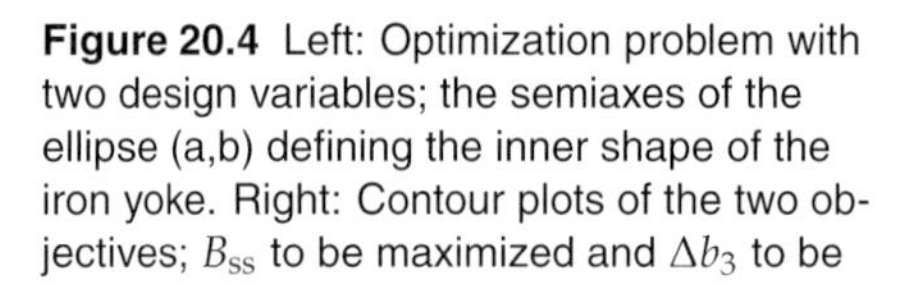

Figure 20.4 Left: Optimization problem with two design variables; the semiaxes of the ellipse (a,b) defining the inner shape of the iron yoke. Right: Contour plots of the two objectives; B_{ss} to be maximized and Δb_3 to be minimized. The Pareto-optimal solution set is indicated with the bold line. Point 1 is not a Pareto optimum because in point 2 both objectives are improved.

tively high shims it is possible to optimize the field quality for one excitation level. However, the field distribution is dependent on saturation effects in these shims. Thus for a low variation in multipole errors the shims must be made larger and thinner. Even if we consider only one excitation level, there will be a tradeoff between the quadrupole and sextupole field components. A minimum b_3 component can be achieved both with wide and short, as well as with narrow and higher shims. A minimum b_2 is obtained, however, only for relatively wide and high shims. □

20.5 Methods for Decision Making

The first issue in an optimization process is the treatment of the decision-making problem. Most important for the application of mathematical optimization algorithms is a decision-making method that guarantees a solution from the Pareto-optimal solution set. These methods are also referred to as goal programming. A comprehensive overview can be found in [7].

Some methods that we have applied to problems in computational electromagnetism are described below.

20.5.1 Goal Programming

The principle idea of goal programming is to find a substitute scalar problem for the vector-optimization problem by an appropriate transformation that guarantees a solution from the Pareto-optimal solution set. The transformation allows the incorporation of the user's preference by use of a set of weighting factors.

Objective weighting The *objective-weighting* function [19] is the sum of the weighted objectives and results in the minimization problem

$$\min \left\{ u(\mathbf{f}(\mathbf{x})) := \sum_{k=1}^{K} z_k = \sum_{k=1}^{K} t_k f_k(\mathbf{x}) \mid \mathbf{x} \in M \right\}, \tag{20.28}$$

with the weighting factors t_k representing the user's preference. For convex optimization problems, it can be proved indirectly [9] that Eq. (20.28) is a minimization problem with a unique Pareto-optimal solution:

Proof. Assume $\mathbf{x}^*$ is a minimizer of $u(\mathbf{f}(\mathbf{x}))$ but not Pareto optimal. Then there exists a solution $\mathbf{x}$ which dominates $\mathbf{x}^*$. Without loss of generality we can state that $f_1(\mathbf{x}) < f_1(\mathbf{x}^*)$ and $f_k(\mathbf{x}) \leq f_k(\mathbf{x}^*)$, for $k = 2, \ldots, K$. But $u(\mathbf{f}(\mathbf{x})) < u(\mathbf{f}(\mathbf{x}^*))$, which is a contradiction of the assumption that $u(\mathbf{f}(\mathbf{x}^*))$ is a minimum. □

The particular challenge is to find the appropriate weighting factors when the objectives have different numerical values and sensitivity. A normalization of objective function values with individual objectives of the starting point $\mathbf{x}_s$ using

$$z_k = t_k \frac{f_k(\mathbf{x})}{f_k(\mathbf{x}_s)} \tag{20.29}$$

makes the result dependent upon the initial design. Using objective weighting therefore results, in practice, in a solution process where a number of optimizations must be performed with updated weighting factors. This procedure is also referred to as the *decision-making-before-search* approach.

A geometric interpretation of objective weighting is presented in Figure 20.5 (left) for two objectives. The optimization problem (20.28) reduces to

$$\min \{u(f_1(\mathbf{x}), f_2(\mathbf{x})) = t_1 f_1(\mathbf{x}) + t_2 f_2(\mathbf{x}) \mid \mathbf{x} \in M\} \tag{20.30}$$

and can be geometrically interpreted as the problem of shifting the line

$$f_2(\mathbf{x}) = -\frac{t_1}{t_2} f_1(\mathbf{x}) + \frac{u}{t_2} \tag{20.31}$$

toward the feasible domain, as shown in Figure 20.5 (left). The objective-weighting function for the optimization problem described in Section 20.4 is given in Figure 20.7 (left). It is common in real-world applications that the minimum lies in narrow, curved "valleys" that present a challenge to the search direction and step-size adjustment schemes of optimization algorithms. It is therefore reasonable to test the performance of optimization algorithms with the Rosenbrock test function [44],

$$f(x_1, x_2) = 100 \left(x_2 - x_1^2\right)^2 + (1 - x_1)^2 . \tag{20.32}$$

This function has a global minimum at $\mathbf{x}^* = (1,1)^T$ with the objective function value $f(\mathbf{x}^*) = 0$. The minimum lies in a crescent-shaped area along the parabola $x_2 = x_1^2$.

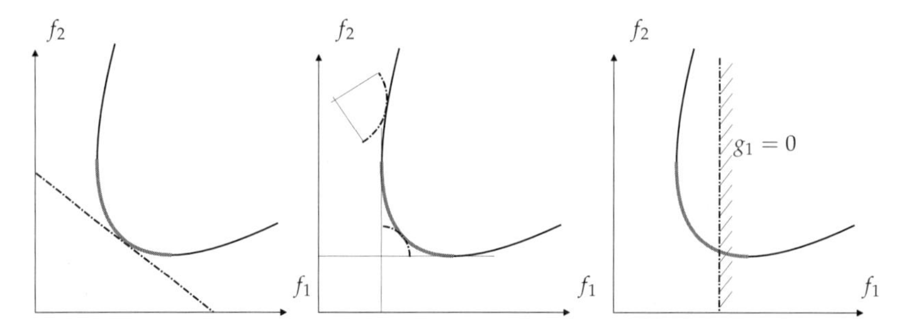

Figure 20.5 Methods for decision making. Left: Objective weighting. Line with slope $-\frac{t_1}{t_2}$. Middle: Distance function. Note that only minimizing the distance from the minimal individual solutions guarantees a Pareto-optimal solutions. Right: Constraint formulation.

Problems arise for nonconvex problems for which the Pareto-optimal solutions cannot be reached with the weighting method. Experience has shown, however, that problems in magnet design optimization are sufficiently "well behaved."

Distance function The problem of choosing appropriate weighting factors also occurs when the *distance-function* method [6] is applied. Most common is the least-squares objective function. f_k^* represents the user-supplied minimum requirement for the optimum design. In the case of inverse field computation for the tracing of manufacturing errors, f_k^* represents the measured multipole field component. The minimization problem reads

$$\min \left\{ u(\mathbf{f}(\mathbf{x})) := \|\mathbf{z}(\mathbf{x})\|_p^p = \left(\sum_{k=1}^{K} \left(t_k (f_k^*(\mathbf{x}) - f_k(\mathbf{x}))\right)^p \right)^{\frac{1}{p}} \middle| \mathbf{x} \in M, p \geq 1 \right\}, \tag{20.33}$$

which yields the least-squares problem for $p = 2$ and the min–max problem for the Tschebyscheff norm $p \to \infty$:

$$\min \left\{ \max_k |f_k^*(\mathbf{x}) - f_k(\mathbf{x})| \right\} . \tag{20.34}$$

For convex functions and for f_k^* taken as the minimal individual solutions that can be found by setting up a payoff table (Section 20.5.4), it can be proved in the same manner as for the objective weighting function that Eq. (20.33) has a unique Pareto-optimal solution; see also Figure 20.5 (middle). However, the computation time needed for finding the minimum of each individual objective is the same as for the minimization of the distance function.

The disadvantage of the application of least-squares objective functions with the Euclidean norm is the low sensitivity for residuals smaller than 1. Therefore sufficiently high weighting factors t_k must be introduced. If the one-norm $p = 1$ is applied, the disadvantage is the nondifferentiable objective function in the optimum. This is of no importance for the application of search routines but is problematic for the use of gradient methods. Figure 20.5 also shows that for arbitrarily chosen f_k^*, the solution of Eq. (20.33) is in general not a Pareto optimum. For nonconvex optimization problems the additional problem of nonglobal minima of the individual objective function arises, a combination of which could lead to solutions that are not part of the Pareto set; see Figure 20.6. The distance function for the optimization problem described in Section 20.4 is given in Figure 20.7 (right). Figure 20.8 (left) shows the distance function with the one-norm.

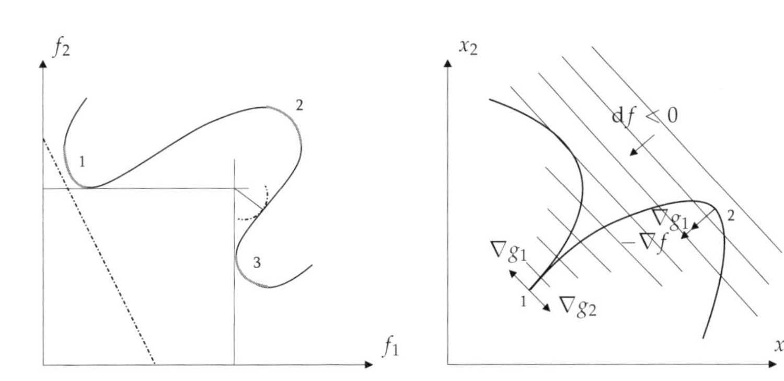

Figure 20.6 Left: For nonconvex objective functions, the local minima calculated for the individual objectives might not lead to Pareto-optimal solutions when used with the distance-function method. Right: Nonconvex constraints might satisfy the KKT conditions at some point that is not the minimum, e.g., point 2. At point 1 the gradients of the active constraints are not linearly independent. The KKT conditions cannot be verified although point 1 is indeed a minimum.

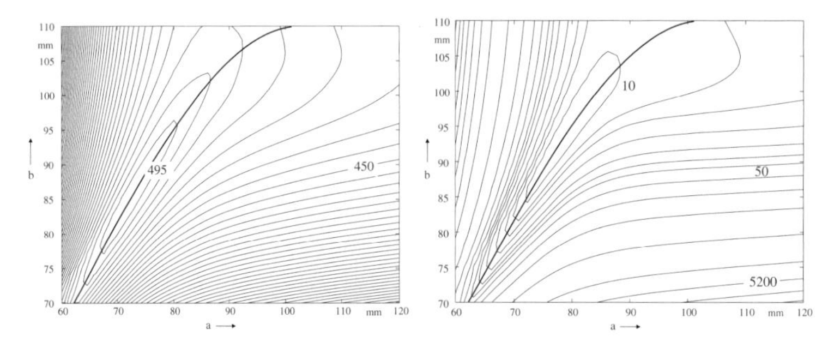

Figure 20.7 Left: Contour plot of the objective-weighting function $f(a, b) = -50B_{ss} + 3\Delta b_3$. Right: Distance function with L_2 norm $f(a, b) = 50(B_{ss} - 10.13)^2 + 3(\Delta b_3 - 0.017)^2$.

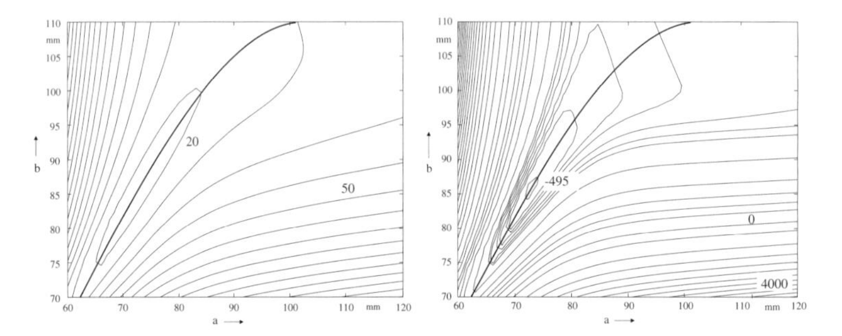

Figure 20.8 Left: Contour plot of the distance function employing the L_1 norm $f(a, b) = 50 \mid B_{ss} - 10.13 \mid +3 \mid \Delta b_3 - 0.017 \mid$; Right: Constraint formulation with penalty term for the Δb_3 constraint $f(a, b) = 50B_{ss} + 3\max\{(0, \Delta b_3 - 1)^2\}$.

20.5.2 The Pareto-Strength Algorithm

Applying goal-programming methods it is possible to find a single preference solution from the Pareto front. In particular, in combination with global optimization methods such as genetic algorithms, discussed in Section 20.9, it might be desirable to find a number of nondominated solutions on the Pareto front. The Pareto-strength algorithm [57] sorts all solutions in two groups: one containing the Pareto front and another with all solutions dominated with at least one solution. Then the strength of each Pareto-optimal solution is determined by the number of solutions it dominates:

$$S_{\mathrm{P}} := \frac{n_{\mathrm{d}}}{N + 1}, \tag{20.35}$$

where n_{d} is the number of dominated members, and N is the total number of members in the population. An example is shown in Figure 20.3 (right). Members A, B, and C are not dominated by another member. Member 1 is dominated by A and B, member 3 by A, B, and C, and so on. The Pareto-optimal solution A dominates 1 and 3 and has thus the strength $S_{\mathrm{P}} = 2/(6 + 1)$. For the known strength of the Pareto-optimal solutions, the strength of the dominated members can be calculated with

$$S_{\mathrm{m}} := 1 + \sum_{n_{\mathrm{d}}^{\mathrm{P}}} S_{\mathrm{P}}, \qquad (20.36)$$

where $n_{\mathrm{d}}^{\mathrm{P}}$ is the number of dominating Pareto solutions. As an example, member 1 in the population shown in Figure 20.3 (right) is dominated by solution A ($S_{\mathrm{P}} = 2/7$) and solution B ($S_{\mathrm{P}} = 3/7$) and therefore by $S_{\mathrm{m}} = 2/7 + 3/7 + 1$. The solutions that have the smallest strength value are declared as the most optimal. The strength value is thus a weakness rather than a strength. During an optimization procedure, all members of the Pareto front are saved for the next generation. This method is referred to as *decision-making-during-search*. Pareto sets can, however, become very large and densely populated.

20.5.3
Constraint Formulation and Sensitivity Analysis

The problem with the weighting factors can be circumvented by defining the problem in the constraint formulation [29]. Only one of the objectives is minimized, while the others are treated by additional nonlinear constraints. The resulting optimization problem reads

$$\min \{f_i(\mathbf{x})\}, \qquad (20.37)$$

subject to

$$f_k(\mathbf{x}) - r_k \leq 0, \qquad (20.38)$$

for $k = 1, \ldots, K; k \neq i$ and the constraints (20.2)–(20.4). The r_k term represents the minimum request value specified by the user for the kth objective. Combining Eq. (20.38) with Eqs. (20.2) and (20.3) while again omitting the box constraints (20.4) yields, in vector notation, $\mathbf{d} \in \mathbb{R}^{m+K-1}, \mathbf{c} \in \mathbb{R}^{p}$ and $\mathbf{g} \in \mathbb{R}^{m+K-1} : \mathbf{g} = (g_1, \ldots, g_{m+K-1})^T$ for the inequality constraints and $\mathbf{h} \in \mathbb{R}^{p} : \mathbf{h} = (h_1, \ldots, h_p)^T$ for the equality constraints:

$$\min \{f_i(\mathbf{x})\}, \qquad (20.39)$$

subject to

$$\mathbf{g}(\mathbf{x}) - \mathbf{d} \leq \mathbf{0}, \qquad (20.40)$$

$$\mathbf{h}(\mathbf{x}) - \mathbf{c} = \mathbf{0}. \qquad (20.41)$$

The separate treatment of the box constraints is described in Section 20.6. A geometric interpretation of the constraint formulation is presented in Figure 20.5 (right). This shows that care is needed in the selection of the request values to avoid the result that the feasible domain becomes empty. The method therefore requires a profound knowledge of the problem domain.

The constraint formulation can also cope with the concave part of the Pareto front; see Figures 20.5 and 20.6. The constraint formulation has the further advantage that a sensitivity analysis can be performed based on the KKT conditions at the optimum point $\mathbf{x}^*$. Let us denote the $m_t \times n$ matrix[8] composed of the row vectors $\nabla g_i(\mathbf{x}^*)$ as $\nabla \mathbf{g}$ and the $p \times n$ matrix of the rows $\nabla h_j(\mathbf{x}^*)$ as $\nabla \mathbf{h}$. The Lagrange multipliers α_i, β_j are arranged in the column vectors $\boldsymbol{\alpha} := (\alpha_1, \ldots, \alpha_{m_t})^T$ and $\boldsymbol{\beta} := (\beta_1, \ldots, \beta_p)^T$. We recall that a vector $\mathbf{x}^*$ is called a *regular point* if the gradients of the active constraints are linearly independent. Let $\mathbf{x}^*$ denote a minimum point of the constraint optimization problem (20.6) and a regular point for the constraints. Then it follows that there are vectors $\boldsymbol{\alpha}$ and $\boldsymbol{\beta}$ with $\alpha_i \geq 0$, such that

$$\nabla_{\mathbf{x}} \mathcal{L}(\mathbf{x}^*, \boldsymbol{\alpha}, \boldsymbol{\beta}) = \nabla f_i(\mathbf{x}^*) + \boldsymbol{\alpha}^T \nabla \mathbf{g}(\mathbf{x}^*) + \boldsymbol{\beta}^T \nabla \mathbf{h}(\mathbf{x}^*) = \mathbf{0}, \tag{20.42}$$

$$\mathbf{g}(\mathbf{x}^*) - \mathbf{d} = \mathbf{0}, \tag{20.43}$$

$$\mathbf{h}(\mathbf{x}^*) - \mathbf{c} = \mathbf{0}, \tag{20.44}$$

$$\alpha_i > 0. \tag{20.45}$$

By means of the corresponding Lagrange function, it can be proved that (20.39)–(20.41) is a minimization problem with a unique Pareto-optimal solution if all constraints are active. A nonactive constraint would be equivalent to a zero weight in the weighting function. The Lagrange multipliers are estimated by solving the linear equation system (20.42) by means of the variational problem

$$\min_{\boldsymbol{\alpha}, \boldsymbol{\beta}} \left\{ \left| \nabla f_i(\mathbf{x}^*) + \boldsymbol{\alpha}^T \nabla \mathbf{g}(\mathbf{x}^*) + \boldsymbol{\beta}^T \nabla \mathbf{h}(\mathbf{x}^*) \right| \right\} . \tag{20.46}$$

The Lagrange multipliers are a measure of the price that must be paid when the constraint is decreased. This is the essence of the following theorem:

Theorem 20.1 *Consider the family of optimization problems Eqs. (20.39)–(20.41) and suppose there exists a solution* $\mathbf{x}^*$ *for* $\mathbf{c} = \mathbf{0}$ *and* $\mathbf{d} = \mathbf{0}$. *Then for every* $\mathbf{c}$ *and* $\mathbf{d}$ *in a region containing* $\mathbf{0}$ *there is a vector* $\mathbf{x}(\mathbf{c}, \mathbf{d})$ *such that* $\mathbf{x}(\mathbf{0}, \mathbf{0}) = \mathbf{x}^*$ *and*

$$\nabla_{\mathbf{c}} f_i(\mathbf{x}(\mathbf{c}, \mathbf{d}))|_{\mathbf{c},\mathbf{d}=\mathbf{0}} = -\boldsymbol{\beta}^T, \tag{20.47}$$

$$\nabla_{\mathbf{d}} f_i(\mathbf{x}(\mathbf{c}, \mathbf{d}))|_{\mathbf{c},\mathbf{d}=\mathbf{0}} = -\boldsymbol{\alpha}^T. \tag{20.48}$$

8 This is the Jacobi matrix of the mapping $\mathbf{g} : X \to \mathbb{R}_{m_t}$, $X \in \mathbb{R}^n$, where we defined m_t as the total number of constraints $m_t = m + K - 1$.

Proof. In the optimum, the KKT conditions must be fulfilled:

$$\nabla_{\mathbf{x}} f_i(\mathbf{x}^*) = -\boldsymbol{\alpha}^T \nabla_{\mathbf{x}} \mathbf{g}(\mathbf{x}^*) - \boldsymbol{\beta}^T \nabla_{\mathbf{x}} \mathbf{h}(\mathbf{x}^*) \tag{20.49}$$

and both the equality and inequality constraints must be active, that is

$$\mathbf{g}(\mathbf{x}^*) = \mathbf{d}, \qquad \mathbf{h}(\mathbf{x}^*) = \mathbf{c}. \tag{20.50}$$

Employing the chain rule we obtain

$$\begin{aligned} \nabla_{\mathbf{c}} f_i(\mathbf{x}(\mathbf{c},\mathbf{d}))|_{\mathbf{c},\mathbf{d}=\mathbf{0}} &= \nabla_{\mathbf{x}} f_i(\mathbf{x}^*)\, \nabla_{\mathbf{c}} \mathbf{x}^* \\ &= -\left[\boldsymbol{\alpha}^T \nabla_{\mathbf{x}} \mathbf{g}(\mathbf{x}^*(\mathbf{c},\mathbf{d})) + \boldsymbol{\beta}^T \nabla_{\mathbf{x}} \mathbf{h}(\mathbf{x}^*)\right] \nabla_{\mathbf{c}} \mathbf{x}^*, \end{aligned} \tag{20.51}$$

$$\begin{aligned} \nabla_{\mathbf{d}} f_i(\mathbf{x}(\mathbf{c},\mathbf{d}))|_{\mathbf{c},\mathbf{d}=\mathbf{0}} &= \nabla_{\mathbf{x}} f_i(\mathbf{x}^*)\, \nabla_{\mathbf{d}} \mathbf{x}^* \\ &= -\left[\boldsymbol{\alpha}^T \nabla_{\mathbf{x}} \mathbf{g}(\mathbf{x}^*(\mathbf{c},\mathbf{d})) + \boldsymbol{\beta}^T \nabla_{\mathbf{x}} \mathbf{h}(\mathbf{x}^*)\right] \nabla_{\mathbf{d}} \mathbf{x}^*, \end{aligned} \tag{20.52}$$

and

$$\nabla_{\mathbf{c}} \left(\mathbf{g}(\mathbf{x}(\mathbf{c},\mathbf{d})) - \mathbf{d}\right)|_{\mathbf{c},\mathbf{d}=\mathbf{0}} = \nabla_{\mathbf{x}} \mathbf{g}(\mathbf{x}^*)\, \nabla_{\mathbf{c}}(\mathbf{x}^*) = \mathbf{0}, \tag{20.53}$$

$$\nabla_{\mathbf{d}} \left(\mathbf{g}(\mathbf{x}(\mathbf{c},\mathbf{d})) - \mathbf{d}\right)|_{\mathbf{c},\mathbf{d}=\mathbf{0}} = \nabla_{\mathbf{x}} \mathbf{g}(\mathbf{x}^*)\, \nabla_{\mathbf{d}}(\mathbf{x}^*) - \mathbf{I} = \mathbf{0}, \tag{20.54}$$

$$\nabla_{\mathbf{c}} \left(\mathbf{h}(\mathbf{x}(\mathbf{c},\mathbf{d})) - \mathbf{c}\right)|_{\mathbf{c},\mathbf{d}=\mathbf{0}} = \nabla_{\mathbf{x}} \mathbf{h}(\mathbf{x}^*)\, \nabla_{\mathbf{c}}(\mathbf{x}^*) - \mathbf{I} = \mathbf{0}, \tag{20.55}$$

$$\nabla_{\mathbf{d}} \left(\mathbf{h}(\mathbf{x}(\mathbf{c},\mathbf{d})) - \mathbf{c}\right)|_{\mathbf{c},\mathbf{d}=\mathbf{0}} = \nabla_{\mathbf{x}} \mathbf{h}(\mathbf{x}^*)\, \nabla_{\mathbf{d}}(\mathbf{x}^*) = \mathbf{0}, \tag{20.56}$$

where $\mathbf{I}$ is the unit matrix. Substituting (20.53)–(20.56) into Eqs. (20.51) and (20.52), the relations (20.47) are obtained. □

Because of the different numerical values of the objective function and the constraints, the Lagrange multipliers are scaled according to

$$\alpha_k^* = \alpha_k \frac{g_k(\mathbf{x}^*)}{f_i(\mathbf{x}^*)}, \qquad \beta_j^* = \beta_j \frac{h_j(\mathbf{x}^*)}{f_i(\mathbf{x}^*)}. \tag{20.57}$$

A geometric interpretation of the properties of the Lagrange multipliers is given in Figure 20.2. Moving $g_2 \to g_2^*$ does not increase the objective function appreciably (point 4), whereas moving the constraint $g_3 \to g_3^*$ with the resulting solution at point 5 leads to a stronger increase in $f(\mathbf{x})$. The numerical value of the multiplier α_2 in the equation $-\nabla f = \alpha_2 \nabla g_2 + \alpha_3 \nabla g_3$ is thus higher than that of α_3.

20.5.4 Payoff Tables

The payoff table is a tool that provides the decision maker with details of the hidden resources of the design. To create this table, K individual optimization problems are solved to find the best solution for each of the K objectives, $\mathbf{x}^i$ being the minimizer of the problem $\min\{f_i(\mathbf{x})\}$. While f_i is minimized, constraints $c_1, \ldots, c_K$ must be considered for the other objectives in order to avoid trivial results, for example, vanishing main field and therefore also vanishing field errors; see Table 20.1.

Table 20.1 Payoff table for K objectives.

	$f_1(\mathbf{x}) < c_1$	$f_2(\mathbf{x}) < c_2$	$f_3(\mathbf{x}) < c_3$	$\ldots$	$f_K(\mathbf{x}) < c_K$	Minimizer
$\min\{f_1(\mathbf{x})\}$	$f_1(\mathbf{x}^1)$	$f_2(\mathbf{x}^1)$	$f_3(\mathbf{x}^1)$	$\ldots$	$f_K(\mathbf{x}^1)$	$\mathbf{x}^1$
$\min\{f_2(\mathbf{x})\}$	$f_1(\mathbf{x}^2)$	$f_2(\mathbf{x}^2)$	$f_3(\mathbf{x}^2)$	$\ldots$	$f_K(\mathbf{x}^2)$	$\mathbf{x}^2$
$\min\{f_3(\mathbf{x})\}$	.	.	$f_3(\mathbf{x}^3)$	.	.	.
.	.	.	.	.	.	.
.	.	.	.	.	.	.
$\min\{f_K(\mathbf{x})\}$	$f_1(\mathbf{x}^K)$	$f_2(\mathbf{x}^K)$	$f_3(\mathbf{x}^K)$	$\ldots$	$f_K(\mathbf{x}^K)$	$\mathbf{x}^K$

As it will be demonstrated in the applications section, payoff tables are useful for comparing different results in their individually optimized form and at the same time for comparing their hidden resources. The best compromise solutions can be found by minimizing the distance from the infeasible *utopia solution* located on the diagonal of the payoff table. By applying different norms, for example, L_1, L_2, and L_∞, the optimal compromise solutions can be found.

20.6 Box Constraints

Since the design variables of the optimization problem can usually be varied only between upper and lower bounds, a modified objective function is applied:

$$p(\mathbf{x}) = \begin{cases} f(\mathbf{x}) & \text{no bound violated} \\ f(\mathbf{x}^*) + r(\mathbf{x}) & \text{bound violated} \end{cases}, \tag{20.58}$$

where $\mathbf{x}^* = (x_1, x_2, \ldots, x_l^*, \ldots, x_n)^T$ and $x_l^* = x_{l,\text{upper}}$ if the upper bound is violated ($x_l > x_{l,\text{upper}}$), and $x_l^* = x_{l,\text{lower}}$ if the lower bound is violated ($x_l < x_{l,\text{lower}}$). The penalty term is

$$r(\mathbf{x}) = \sum_l r_l \begin{cases} (x_l - x_{l,\text{upper}})^2 & \text{if } x_l > x_{l,\text{upper}} \\ (x_{l,\text{lower}} - x_l)^2 & \text{if } x_l < x_{l,\text{lower}} \\ 0 & \text{otherwise} \end{cases} \tag{20.59}$$

for sufficiently large penalty parameters r_l. The advantage of this procedure is that violations of the bounds are verified before a function evaluation is carried out, and impossible geometries with intersecting domain boundaries are excluded. Hence the existing algorithms for unconstrained minimization can be applied without modifications.

20.7 Treatment of Nonlinear Constraints

The necessity of treating nonlinear constraints arises with the constraint formulation of the optimization problem. One method is the transformation of the constrained problem into an (a set of) unconstrained problem(s) by means of a *penalty transformation*. The main idea behind this procedure is to add to the objective function a penalty term depending on the degree of the constraint violation. The penalty term vanishes if the constraints are satisfied. The optimization problems (20.39)–(20.41) transformed into the penalty function reads

$$\begin{aligned} z(\mathbf{x}, \mathbf{p}, \mathbf{q}) = f_i(\mathbf{x}) &+ \sum_{k=1}^{m+K-1} p_k \cdot \max{}^2\{(0., g_k(\mathbf{x}) - d_k)\} \\ &+ \sum_{j=1}^{p} q_j (h_j(\mathbf{x}) - c_j)^2. \end{aligned} \tag{20.60}$$

In order to prove the feasibility of the result, the penalty factors p_k and q_j have to tend to infinity. Large penalty factors, however, lead to ill-conditioned optimization problems. The penalty transformation thus introduces a certain "fuzziness" into the solution, as the constraints will not be exactly fulfilled.

The penalty function for the optimization problem described in Section 20.4 is given in Figure 20.8 (right).

Replacing the square terms in Eq. (20.60) with the modulus results in the *exact penalty transformation*. Here the weighting factor can be finite, $p_k > |\alpha_k|, q_j > |\beta_j|$, where α_k and β_j are the Lagrange multipliers of the constraint optimization problem. The disadvantage here is the nondifferentiability of the objective function in the optimum, the same problem that arises when applying the distance function method with the L_1 norm.

The penalty transformation and the treatment of the box constraints are illustrated in Figure 20.9.

20.8 Deterministic Optimization Algorithms

Because optimization in electromagnetism involves time-consuming, finite-element field computation, it is crucially important to find a suitable mini-

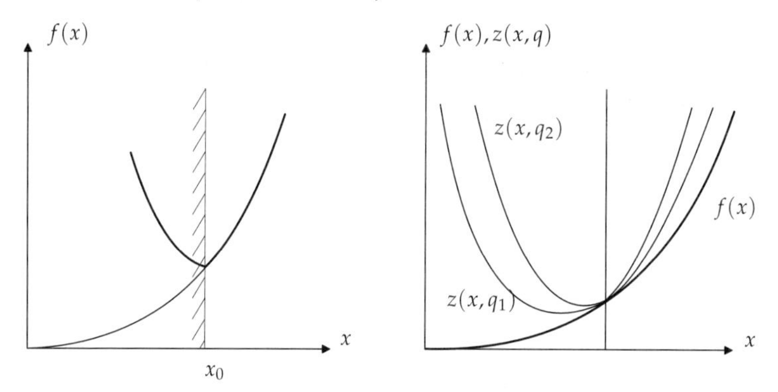

Figure 20.9 Left: Box constraint. The objective function is replaced by a penalty term added to the objective function value at the boundary. Right: Penalty transformation with different penalty parameters.

mization method corresponding to the method for decision making and to the treatment of the nonlinear constraints. For a typical optimization problem with up to 10 design variables, approximately 200 function evaluations are carried out. As an example, genetic optimization and evolution programming usually demand thousands of function evaluations and are therefore applied only in the conceptual design phase of magnets, where a number of different local optima (as opposed to a single (global) optimum) are required.

Some of the deterministic optimization algorithms frequently used for solving optimization problems in accelerator magnet design are described below.

20.8.1 Line Search

Suppose that the objective function f to be minimized is a function of a single variable only, and suppose that the optimization problem is unconstrained. The techniques employed for solving this one-dimensional problem are called *line searches* and the form the backbone of multidimensional nonlinear programming methods, since higher-dimensional problems are usually solved by a sequence of successive line searches. Starting at an initial point $\mathbf{x}_0$, the direction $\mathbf{d}$ of movement is determined and the line search is performed with

$$\min_{\mathbf{x}}\{f(\mathbf{x})\} = \min_{\lambda}\{f(\mathbf{x}_0 + \lambda\mathbf{d})\}\,. \tag{20.61}$$

The various classes of optimization routines are defined by the way in which the search direction is determined, for example, $\mathbf{d} = -\nabla f(\mathbf{x})^T$ in the steepest-descend method. By studying one-dimensional methods we will hone our tools for solving more complex problems.

20.8.1.1 The Golden Section Search

In pseudo-code the golden section search can be written as

1. Start with the intervals $[x_1, x_2]$ and $[x_2, x_3]$ of total length a between three arbitrary points, x_1, x_2, x_3.
2. Until convergence Do:
3. Divide the larger interval with the ratio $1/\tau$ to obtain x_4.
4. Calculate $f(x_4)$.
5. If $f(x_4) > f(x_2)$ then:
6. New interval is $[x_1, x_4]$
7. Else:
8. New interval is $[x_2, x_3]$
9. End if.
10. End Do.

The golden section search requires that the two possible intervals $[x_1, x_4]$ and $[x_2, x_3]$ have equal length. From Figure 20.10 we see that

$$a = \frac{a\tau}{1+\tau} + \frac{a\tau^2}{(1+\tau)^2}, \tag{20.62}$$

which is equivalent to $(1+\tau)^2 = \tau(1+\tau) + \tau^2$ and results in

$$\tau = \frac{1 \pm \sqrt{5}}{2}. \tag{20.63}$$

The positive value of $\tau = 1.618\ldots$ is the well-known *golden section*, a proportion frequently found in nature and used since antiquity in music and architecture.

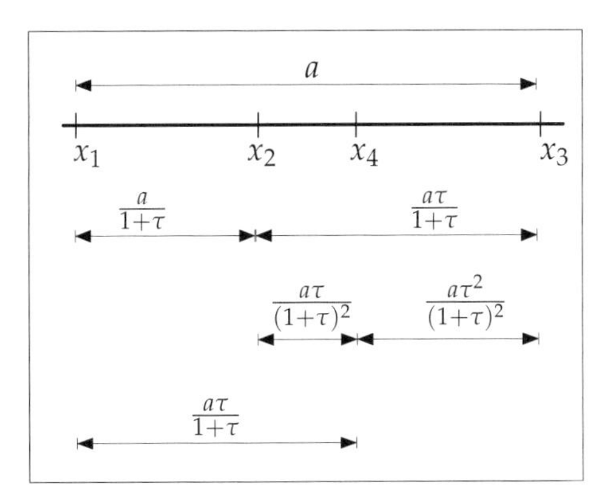

Figure 20.10 Golden section search.

20.8.1.2 **Newton's Method**

In many real-world optimization problems a sufficient smoothness of the objective function may be assumed for the purpose of solving them. In such cases we may call upon a group of optimization algorithms based on the technique of fitting a curve through a set of evaluation points in order to estimate the minimizer.

Suppose that the function f to be minimized is 2-smooth and a function of the single variable x only. Suppose further that at the point x_k it is possible to evaluate the function value $f(x_k)$ and the derivatives[9] $f'(x_k)$ and $f''(x_k)$. The truncated Taylor series

$$q(x) = f(x_k) + f'(x_k)(x - x_k) + \frac{1}{2} f''(x_k)(x - x_k)^2, \tag{20.64}$$

agrees at (x_k) with f up to second order, and therefore an estimate for x_{k+1} can be found by imposing a vanishing derivative of q:

$$q'(x_{k+1}) = f'(x_k) + f''(x_k)(x_{k+1} - x_k) = 0, \tag{20.65}$$

which results in

$$x_{k+1} = x_k - \frac{f'(x_k)}{f''(x_k)}. \tag{20.66}$$

As Newton's method does not depend on $f(x_k)$, we obtain for $g(x) := f'(x)$

$$x_{k+1} = x_k - \frac{g(x_k)}{g'(x_k)}, \tag{20.67}$$

an iterative method for solving the equation $g(x) = 0$; see Figure 20.11.

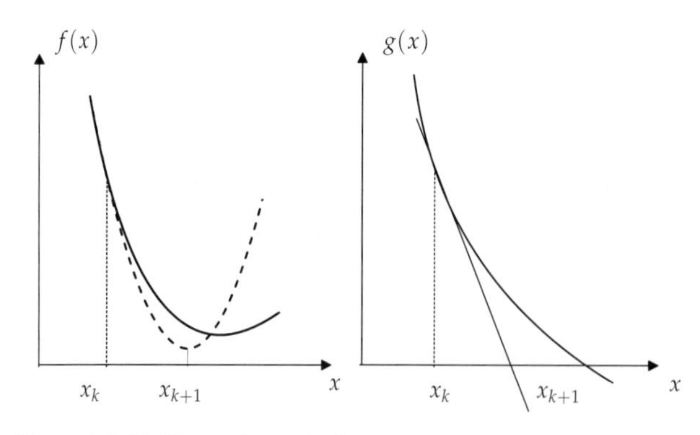

Figure 20.11 Newton's method.

9 We use the common notation $f' := \frac{df}{dx}$.

20.8.1.3 **Quadratic Fit**

The advantage of the quadratic fit is that it does not require the derivatives of the objective function. The method is used as a line search in the algorithm EXTREM, described in Section 20.8.2.2. Given x_i with corresponding function values $f(x_i)$, $i = 1, 2, 3$, the quadratic fit through these points is

$$q(x) = f(x_1)\frac{(x-x_2)(x-x_3)}{(x_1-x_2)(x_1-x_3)} + f(x_2)\frac{(x-x_1)(x-x_3)}{(x_2-x_1)(x_2-x_3)} + f(x_3)\frac{(x-x_1)(x-x_2)}{(x_3-x_1)(x_3-x_2)} \tag{20.68}$$

with the minimum

$$x = \frac{1}{2}\frac{(x_2^2-x_3^2)f(x_1)+(x_3^2-x_1^2)f(x_2)+(x_1^2-x_1^2)f(x_3)}{(x_2-x_3)f(x_1)+(x_3-x_1)f(x_2)+(x_1-x_1)f(x_3)}. \tag{20.69}$$

20.8.2
Multidimensional Search Methods

The basic algorithm for multidimensional search methods is as follows:

1. Assign $\mathbf{x}_1$.
2. For $k = 1, 2, \dots,$ until convergence Do:
3. Chose search direction $\mathbf{d}_k$.
4. Perform a line search to solve $\min_\lambda\{f(\mathbf{x}_k + \lambda\mathbf{d}_k)\}$.
5. Set $\mathbf{x}_{k+1} = \mathbf{x}_k + \lambda_{\min}\mathbf{d}_k$.
6. End Do

The problem is reduced to the question of the search direction. The classification of optimization methods depends on whether or not the derivative of the objective function is needed to specify this search direction.

20.8.2.1 **Direct Search**

The easiest choice of search directions are the coordinate directions $\mathbf{d}_k = \mathbf{e}_k$, which yield the method of *alternating variables*, also known as *coordinate descent*, Gauss–Seidel, or *direct search*, illustrated in Figure 20.12 (left). This results in a series of one-dimensional optimizations,

$$\min_{x_k}\{f(x_1, x_2, \dots, x_n)\}. \tag{20.70}$$

20.8.2.2 **EXTREM**

The optimization algorithm EXTREM by Jacob [18] is a deterministic method that does not require the derivatives of the objective function. The optimiza-

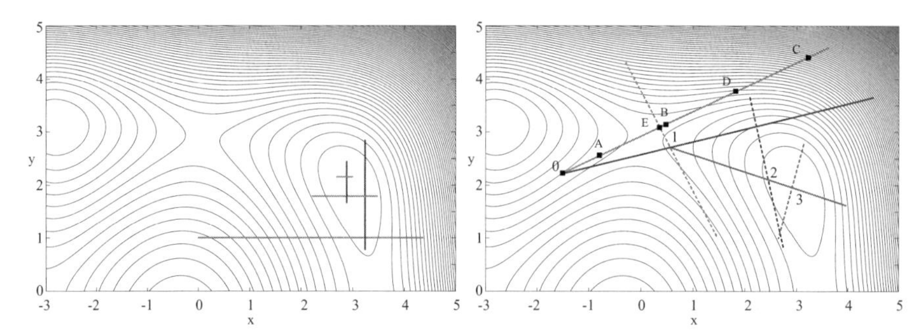

Figure 20.12 Left: Direct search. Right: Functioning of the algorithm EXTREM for the first three search steps. A–E indicate the search points in the line search along the main search direction of the initial step. The objective function is the Himmelblau function (20.71).

tion routine consists of one-dimensional minimizations in a main search direction (user-supplied) and an orthogonal direction, which is evaluated by the Erhard Schmidt orthogonalization. After these one-dimensional searches have been carried out (end of a search step), the main search direction is updated by a vector pointing from the solution of step $n-1$ to the minimum of the search step n. The user has to supply an initial step size, which is $(x_{l,\text{upper}} - x_{l,\text{lower}})/10$ by default. This user-friendly algorithm is suitable for practically all applications, including unconstrained scalar functions, distance functions, penalty functions, and augmented Lagrange functions. No parameter other than the initial step size need be supplied by the user. The principle of the algorithm is shown in Figure 20.12 (right), where the test function by Himmelblau [15] was used. This function reads

$$f(x_1, x_2) = (x_1^2 + x_2 - 11)^2 + (x_1 + x_2^2 - 7)^2 \tag{20.71}$$

and has four global optima with $f(\mathbf{x}^*) = 0$ of which only two are shown. Another local optimum is located near $\mathbf{x} = (-0.27, -0.92)^T$.

20.8.3
Gradient Methods

In Eq. (20.9), the gradient of the objective function is defined as an n-dimensional row vector $\nabla f(\mathbf{x}_k)$. We will use the notation

$$\mathbf{g}_k := \nabla f(\mathbf{x}_k)^T, \tag{20.72}$$

for the transpose of the gradient.

20.8.3.1 The Steepest Descent Method

Let the objective function $f(\mathbf{x})$ be 1-smooth in the feasible domain M. The method of the steepest descent is defined by the iteration

$$\mathbf{x}_{k+1} = \mathbf{x}_k - \lambda_k \mathbf{g}_k, \tag{20.73}$$

where λ_k is again the nonnegative scalar, minimizing $\mathbf{x}_k - \lambda \mathbf{g}_k$. The search is performed along a search direction defined by the negative gradient of the objective function at $\mathbf{x}_k$. In the case of the quadratic problem,

$$f(\mathbf{x}) = \frac{1}{2}\mathbf{x}^T\mathbf{Q}\mathbf{x} - \mathbf{x}^T\mathbf{b}, \tag{20.74}$$

where $\mathbf{Q} \in \mathbb{R}^{n \times n}$ is a positive definite and symmetric matrix, f is strictly convex. Therefore, a minimizer $\mathbf{x}^*$ can be found that satisfies the requirement that the gradient $\mathbf{g} = \mathbf{Q}\mathbf{x} - \mathbf{b}$ be zero, that is, that $\mathbf{x}^*$ is a solution of the linear equation system

$$\mathbf{Q}\mathbf{x}^* = \mathbf{b}. \tag{20.75}$$

Therefore from Eq. (20.73), we obtain $\mathbf{x}_{k+1} = \mathbf{x}_k - \lambda_k (\mathbf{Q}\mathbf{x}_k - \mathbf{b})$. Now λ_k can be calculated analytically with

$$f(\mathbf{x}_{k+1}) = f(\mathbf{x}_k - \lambda \mathbf{g}_k) = \frac{1}{2}(\mathbf{x}_k - \lambda \mathbf{g}_k)^T\mathbf{Q}(\mathbf{x}_k - \lambda \mathbf{g}_k) - (\mathbf{x}_k - \lambda \mathbf{g}_k)^T\mathbf{b} \tag{20.76}$$

derived from Eq. (20.74). Differentiating Eq. (20.76) with respect to λ we obtain

$$\lambda_k = \frac{\mathbf{g}_k^T \mathbf{g}_k}{\mathbf{g}_k^T \mathbf{Q} \mathbf{g}_k}, \tag{20.77}$$

where $\mathbf{g}_k = \mathbf{Q}\mathbf{x}_k - \mathbf{b}$. Thus from Eq. (20.73) we obtain the explicit form

$$\mathbf{x}_{k+1} = \mathbf{x}_k - \left(\frac{\mathbf{g}_k^T \mathbf{g}_k}{\mathbf{g}_k^T \mathbf{Q} \mathbf{g}_k}\right) \mathbf{g}_k . \tag{20.78}$$

20.8.3.2 Newton's Method

Newton's method is based on an approximation of the objective function $f(\mathbf{x})$ by a quadratic function, which is then exactly minimized. Using this technique in the multidimensional case, we can write the equivalent of Eq. (20.64) as

$$q(\mathbf{x}) = f(\mathbf{x}_k) + \nabla f(\mathbf{x}_k)(\mathbf{x} - \mathbf{x}_k) + \frac{1}{2}(\mathbf{x} - \mathbf{x}_k)^T\mathbf{H}(\mathbf{x}_k)(\mathbf{x} - \mathbf{x}_k), \tag{20.79}$$

where, again, $q(\mathbf{x})$ agrees with f at $\mathbf{x}_k$ up to second order. If $f \in C^2$ is strictly quadratic and the Hesse matrix is positive definite in the feasible domain, then

$$f(\mathbf{x}) > f(\mathbf{x}_k) + \nabla f(\mathbf{x}_k)(\mathbf{x} - \mathbf{x}_k) \tag{20.80}$$

and the objective function is strictly convex. An estimate for $\mathbf{x}_{k+1}$ can be found by forcing the derivative of $q(\mathbf{x})$ to vanish:

$$\nabla q(\mathbf{x})|_{\mathbf{x}_{k+1}}^T = \mathbf{0} = \mathbf{g}_k + \mathbf{H}(\mathbf{x}_k)(\mathbf{x}_{k+1} - \mathbf{x}_k). \tag{20.81}$$

It then follows that,

$$\mathbf{x}_{k+1} = \mathbf{x}_k - [\mathbf{H}(\mathbf{x}_k)]^{-1}\mathbf{g}_k. \tag{20.82}$$

However, this method has some shortcomings:

- The Hesse matrix (containing the second derivatives of the objective function) must be calculated using difference quotients because the objective function is not explicitly defined.
- The Hesse matrix must be inverted.
- Assuming that the objective function is quadratic near the solution, the method converges in one step. However, as a result of nonquadratic terms in the objective function, the Hesse matrix may not be positive definite, and Newton's method may lead to search directions diverging from the minimum.

Because of these disadvantages, the standard Newton algorithm is almost never used in practice. A more useful approach involves a modified form,

$$\mathbf{x}_{k+1} = \mathbf{x}_k - \lambda_k \mathbf{S}\mathbf{g}_k, \tag{20.83}$$

where λ is used as before to minimize f. Replacing the inverse Hessian by the unity matrix, $\mathbf{S} = \mathbf{I}$, yields the method of the steepest descent. Another approach is to take a compromise path between the two extremes, which brings us to the *Levenberg–Marquard algorithm*.

20.8.3.3 **The Levenberg–Marquard Algorithm**

This compromise between Newton's method and the method of the steepest descent is made by setting

$$\mathbf{S} = [\epsilon_k \mathbf{I} + \mathbf{H}(\mathbf{x}_k)]^{-1} \tag{20.84}$$

for a nonnegative value of ϵ_k. To assure the descent, the matrix $\mathbf{S}$ must be positive definite, a condition enforced by very large values of ϵ_k. The Levenberg–Marquard algorithm reads as follows [26]:

1. For $k = 0, 2, \ldots,$ until convergence Do:
2. Perform a Choleski factorization: $\epsilon_k \mathbf{I} + \mathbf{H}(\mathbf{x}_k) = \mathbf{G}\mathbf{G}^T$.
3. If factorization breaks down (**S** not positive definite) then:
4. Increase ϵ_k.
5. Else:
6. $\mathbf{d}_k = [\epsilon_k \mathbf{I} + \mathbf{H}(\mathbf{x}_k)]^{-1} \mathbf{g}_k$.
7. End If.
8. $\mathbf{x}_{k+1} = \mathbf{x}_k + \mathbf{d}_k$.
9. End Do.

Step 4 ensures that $\epsilon_k > -\min \alpha_i$ where α_i are the eigenvalues of $\mathbf{H}$. Because the eigenvalues of $\epsilon_k \mathbf{I} + \mathbf{H}(\mathbf{x}_k)$ are $\alpha_i + \epsilon_k$, the matrix $\mathbf{S}$ is positive definite.

The Levenberg–Marquard algorithm was originally developed for nonlinear regression problems using least-squares objective functions. The algorithm can therefore be efficiently applied to the minimization of the distance function. Let the residuals $q_i := (f_i(\mathbf{x}) - f_i^*)$, where $f_i(\mathbf{x})$ are the calculated and f_i^* are the measured values, be arranged in the vector $\mathbf{f}(\mathbf{x}) = (q_1(\mathbf{x}), q_2(\mathbf{x}), \ldots, q_m(\mathbf{x}))$. Then the objective function is given in the vector notation by

$$\min\{z(\mathbf{x})\} = \min\left\{\frac{1}{2}\sum_{i=1}^{m} q_i^2(\mathbf{x})\right\} = \min\left\{\frac{1}{2}\mathbf{f}(\mathbf{x})^T\mathbf{f}(\mathbf{x})\right\}. \tag{20.85}$$

The factor $1/2$ is included in order to avoid the factor of two in the derivatives. Using a quadratic approximation of $z(\mathbf{x})$,

$$q(\mathbf{x}) = z(\mathbf{x}_k) + \nabla z(\mathbf{x}_k)(\mathbf{x} - \mathbf{x}_k) + \frac{1}{2}(\mathbf{x} - \mathbf{x}_k)^T \mathbf{H}(\mathbf{x}_k)(\mathbf{x} - \mathbf{x}_k), \tag{20.86}$$

yields the iteration scheme

$$\mathbf{x}_{k+1} = \mathbf{x}_k - [\epsilon_k \mathbf{I} + \mathbf{H}(\mathbf{x}_k)]^{-1} \mathbf{g}_k. \tag{20.87}$$

In order to make use of the quadratic nature of the objective function we employ the Jacobi matrix $\mathbf{J}(\mathbf{x}_k)$ of $z(\mathbf{x}_k)$ to obtain

$$\nabla z(\mathbf{x}_k)^T = [\mathbf{J}(\mathbf{x}_k)]^T \mathbf{f}(\mathbf{x}_k), \tag{20.88}$$

and

$$\mathbf{H}(\mathbf{x}_k) = [\mathbf{J}(\mathbf{x}_k)]^T \mathbf{J}(\mathbf{x}_k) + \frac{\partial \mathbf{J}(\mathbf{x}_k)}{\partial \mathbf{x}_k} \mathbf{f}(\mathbf{x}_k). \tag{20.89}$$

If we neglect the term $\frac{\partial \mathbf{J}(\mathbf{x}_k)}{\partial \mathbf{x}_k}\mathbf{f}(\mathbf{x}_k)$ the iteration scheme reads

$$\mathbf{x}_{k+1} = \mathbf{x}_k - \left[\epsilon \mathbf{I} + (\mathbf{J}(\mathbf{x}_k))^T \mathbf{J}(\mathbf{x}_k)\right]^{-1} \left[\mathbf{J}(\mathbf{x}_k)\right]^T \mathbf{f}(\mathbf{x}_k). \tag{20.90}$$

The term $\epsilon \mathbf{I}$ can be regarded as an approximation of the omitted term. For $\epsilon = 0$ this method is known as the *Gauss–Newton method*; there is no need to calculate the second derivative. With a large ϵ the algorithm proceeds in a steepest-descent direction. ϵ is decreased by the optimization procedure because the neglected term becomes decreasingly important with decreasing residuals.

20.8.3.4 Conjugate Gradient (CG) Methods

Let us consider again the purely quadratic optimization problem

$$\min\{f(\mathbf{x})\} = \min\left\{\frac{1}{2}\mathbf{x}^T \mathbf{Q}\mathbf{x} - \mathbf{x}^T \mathbf{b}\right\}, \tag{20.91}$$

for a symmetric, positive definite matrix $\mathbf{Q} \in \mathbb{R}^{n\times n}$ and $\mathbf{x}, \mathbf{b} \in \mathbb{R}^n$. The global solution of this convex optimization problem is also the solution of the linear equation system

$$\mathbf{Q}\mathbf{x} = \mathbf{b}. \tag{20.92}$$

The algorithm, first published by Hestenes and Stiefel in 1952 [14] reads

1. Assign $\mathbf{x}_0$. Set $\mathbf{g}_0 = \mathbf{Q}\mathbf{x}_1 - \mathbf{b}$ and $\mathbf{d}_0 = -\mathbf{g}_0$.
2. For $k = 0, 1, 2, \ldots, n-1$ Do:
3. $\quad \alpha_k = -\dfrac{\mathbf{g}_k^T \mathbf{d}_k}{\mathbf{d}_k^T \mathbf{Q}\mathbf{d}_k}$
4. $\quad \mathbf{x}_{k+1} = \mathbf{x}_k + \alpha_k \mathbf{d}_k$
5. $\quad \beta_k = \dfrac{\mathbf{g}_{k+1}^T \mathbf{Q}\mathbf{d}_k}{\mathbf{d}_k^T \mathbf{Q}\mathbf{d}_k}$
6. $\quad \mathbf{d}_{k+1} = -\mathbf{g}_{k+1} + \beta_k \mathbf{d}_k$
7. End Do.

The gradients are calculated from $\mathbf{g}_k = \mathbf{Q}\mathbf{x}_k - \mathbf{b}$. The initial iteration step is of the steepest-descent type. The subsequent directions are linear combinations of the actual gradient and the proceeding direction vector. Here line searches are avoided. The method converges in n steps toward the global optimizer, $\mathbf{x}^*$. The iteration scheme yields a set of search directions $\{\mathbf{d}_0, \ldots, \mathbf{d}_{n-1}\}$ that are $\mathbf{Q}$-orthogonal (conjugate with respect to $\mathbf{Q}$), a property defined by

$$\langle \mathbf{d}_k, \mathbf{d}_i \rangle_{\mathbf{Q}} := \mathbf{d}_k^T \mathbf{Q}\mathbf{d}_i = 0 \tag{20.93}$$

for $i, k = 0, n-1, i \neq k$. If $\mathbf{Q}$ is symmetric and positive definite, the search directions form a set of linearly independent vectors since $\sum_{i=1}^{k} \alpha_i \mathbf{d}_i = 0$ holds only for the trivial solution[10]. For a matrix $\mathbf{Q} \in \mathbb{R}^{n \times n}$ and an arbitrary point $\mathbf{x}_0 \in \mathbb{R}^n$, the correction vector $\mathbf{x}^* - \mathbf{x}_0$ can thus be expanded as

$$\mathbf{x}^* - \mathbf{x}_0 = \alpha_0 \mathbf{d}_0 + \cdots + \alpha_{n-1} \mathbf{d}_{n-1} . \tag{20.94}$$

Multiplication by $\mathbf{Q}$ and $\mathbf{d}_i$ yields

$$\alpha_i = \frac{\mathbf{d}_i^T \mathbf{Q}(\mathbf{x}^* - \mathbf{x}_0)}{\mathbf{d}_i^T \mathbf{Q} \mathbf{d}_i} = -\frac{\mathbf{d}_i^T (\mathbf{Q}\mathbf{x}_0 - \mathbf{b})}{\mathbf{d}_i^T \mathbf{Q} \mathbf{d}_i} . \tag{20.95}$$

For the kth iteration step we obtain

$$\mathbf{x}_k - \mathbf{x}_0 = \alpha_0 \mathbf{d}_0 + \cdots + \alpha_{k-1} \mathbf{d}_{k-1} \tag{20.96}$$

and with the $\mathbf{Q}$ orthogonality

$$\mathbf{d}_k^T \mathbf{Q}(\mathbf{x}_k - \mathbf{x}_0) = \mathbf{d}_k^T \mathbf{Q} \sum_{i<k} \alpha_i \mathbf{d}_i = 0 . \tag{20.97}$$

Substituting this result into Eq. (20.95) and recalling that $\mathbf{Q}\mathbf{x}^* = \mathbf{b}$ and $\mathbf{g}_k = \mathbf{Q}\mathbf{x}_k - \mathbf{b}$, we arrive at step 3 of the algorithm:

$$\alpha_k = \frac{\mathbf{d}_k^T \mathbf{Q}(\mathbf{x}^* - \mathbf{x}_k)}{\mathbf{d}_k^T \mathbf{Q} \mathbf{d}_k} = -\frac{\mathbf{d}_k^T \mathbf{Q}(\mathbf{x}_k - \mathbf{b})}{\mathbf{d}_k^T \mathbf{Q} \mathbf{d}_k} = -\frac{\mathbf{g}_k^T \mathbf{d}_k}{\mathbf{d}_k^T \mathbf{Q} \mathbf{d}_k} . \tag{20.98}$$

A feasible way to obtain n conjugate directions might be the application of the Erhard Schmidt orthogonalization procedure to an arbitrary basis of $\mathbb{R}^n$ with respect to the inner product (20.93). Instead, the search directions are determined sequentially by calculating the gradient vector $\mathbf{g}_k$ in the kth step and adding it to a linear combination of the previous directions; see step 6 in the above algorithm. Thus a new conjugate direction is obtained. In particular, because of the linear independence of the search directions, it follows that $\mathbf{g}_k = 0$, which proves that $\mathbf{x}_k$ is a global minimizer of the quadratic optimization problem.

Now we shall prove steps 5 and 6 by induction. Consider $\mathbf{d}_0 := -\mathbf{g}_0$ and the set of $l+1$ vectors $\{\mathbf{d}_0, \ldots, \mathbf{d}_l\}$ with $\mathbf{d}_i^T \mathbf{Q} \mathbf{d}_j = 0$ for $i, j = 0, l-1, i \neq j$. Furthermore $\mathbf{g}_{l+1} \neq 0$. Employing the Erhard Schmidt orthogonalization scheme with respect to the inner product (20.93) we set

$$\mathbf{d}_{l+1} := -\mathbf{g}_{l+1} + \sum_{i=0}^{l} \beta_i^l \mathbf{d}_i . \tag{20.99}$$

The condition $\mathbf{d}_{l+1}^T \mathbf{Q} \mathbf{d}_j = 0$ for $j = 0, \ldots, l$ is fulfilled if

$$\beta_i^l = \frac{\mathbf{g}_{l+1}^T \mathbf{Q} \mathbf{d}_j}{\mathbf{d}_j^T \mathbf{Q} \mathbf{d}_j} , \tag{20.100}$$

10 Multiplication of the correction vector by $\mathbf{Q}$ and $\mathbf{d}_i$ yields $\alpha_i \mathbf{d}_i^T \mathbf{Q} \mathbf{d}_i = 0$ and therefore $\alpha_i = 0$ due to the positive definiteness of $\mathbf{Q}$.

for $j = 0, \ldots, l$. If the direction $\mathbf{d}_k$ is calculated in the same way as the $\mathbf{d}_{l+1}$ above, β_i^l will be zero for $i = 0, \ldots, l-1$, and we obtain the result by setting $\beta_l := \beta_l^l$.

The Fletcher–Reeves CG method In practical nonlinear optimization problems, $\min\{f(\mathbf{x})\}, f : \mathbb{R}^n \to \mathbb{R}$, the gradient $\mathbf{g}_k = \nabla f(\mathbf{x}_k)^T$ and the Hesse matrix $\mathbf{Q} = \mathbf{H}(\mathbf{x}_k)$ must be evaluated at each step, which is not practical for optimization in electromagnetism. If the objective function is purely quadratic the above associations are identities, whereas in the nonquadratic case the objective function is approximated in the vicinity of the search point by a quadratic function. In the nonquadratic case, the method does not converge in n steps. It is therefore necessary to terminate the conjugate gradient process after n steps and to restart the procedure with a gradient step. A method to avoid the calculation and storage of the Hesse matrix $\mathbf{H}$ is to suppress step 3 of the classical CG method and perform a line search to determine α_k. In addition, replace step 6 by

$$\mathbf{d}_{k+1} = -\mathbf{g}_{k+1} + \frac{\mathbf{g}_{k+1}^T \mathbf{g}_{k+1}}{\mathbf{g}_k^T \mathbf{g}_k} \mathbf{d}_k = -\mathbf{g}_{k+1} + \beta_k^{\mathrm{FR}} \mathbf{d}_k \,. \tag{20.101}$$

This choice becomes obvious if we consider (from the definition of α) that $\mathbf{Q}\mathbf{d}_k = \frac{1}{\alpha_k}(\mathbf{g}_{k+1} - \mathbf{g}_k)$, from which follows

$$\mathbf{g}_{k+1}^T \mathbf{Q}\mathbf{d}_k = \frac{1}{\alpha_k}(\mathbf{g}_{k+1}^T \mathbf{g}_{k+1} - \mathbf{g}_{k+1}^T \mathbf{g}_k) \,. \tag{20.102}$$

With $\mathbf{g}_{k+1}^T \mathbf{g}_k = 0$ and $-\mathbf{g}_k^T \mathbf{d}_k = \mathbf{g}_k^T \mathbf{g}_k$ this yields

$$\beta_k^{\mathrm{FR}} = \frac{\mathbf{g}_{k+1}^T \mathbf{Q}\mathbf{d}_k}{\mathbf{d}_k^T \mathbf{Q}\mathbf{d}_k} = \frac{1}{\alpha_k} \frac{\mathbf{g}_{k+1}^T \mathbf{g}_{k+1}}{\mathbf{d}_k^T \mathbf{Q}\mathbf{d}_k} = \frac{\mathbf{g}_{k+1}^T \mathbf{g}_{k+1}}{\mathbf{g}_k^T \mathbf{g}_k} \,. \tag{20.103}$$

This is the *Fletcher–Reeves CG method.*

20.8.3.5 The Davidon–Fletcher–Powell Algorithm

One of the earliest methods for the approximation of the inverse Hessian was proposed by Davidon and later refined by Fletcher and Powell. Known as a *quasi-Newton* or *variable-metric* method, its chief advantage lies in the use of an approximation of the inverse Hesse matrix instead of its true inverse, making it much less costly to calculate. This technique simultaneously generates the directions of the conjugate-gradient method and constructs the inverse Hesse matrix. It starts from the iterative process

$$\mathbf{x}_{k+1} = \mathbf{x}_k - \lambda_k \mathbf{S}_k \mathbf{g}_k, \tag{20.104}$$

where λ_k is chosen to minimize $f(\mathbf{x}_{k+1})$. The algorithm reads as follows:

1. Assign $\mathbf{x}_1$. Use any symmetric positive definite matrix $\mathbf{S}_1$.
2. For $k = 1, 2, \ldots,$ until convergence Do:
3. Set $\mathbf{d}_k = -\mathbf{S}_k \mathbf{g}_k$.
4. Solve $f(\mathbf{x}_k + \lambda \mathbf{d}_k)$ in λ and obtain $\mathbf{x}_{k+1}$ and $\mathbf{p}_k = \lambda_k \mathbf{d}_k$ and $\mathbf{g}_{k+1}$.
5. Set $\mathbf{q}_k = \mathbf{g}_{k+1} - \mathbf{g}_k$.
6. $\mathbf{S}_{k+1} = \mathbf{S}_k + \frac{\mathbf{p}_k \mathbf{p}_k^T}{\mathbf{p}_k^T \mathbf{q}_k} - \frac{\mathbf{S}_k \mathbf{q}_k \mathbf{q}_k^T \mathbf{S}_k}{\mathbf{q}_k^T \mathbf{S}_k \mathbf{q}_k}$.
7. End Do.

The derivatives of the objective function must be approximated by differential quotients which makes the convergence behavior dependent upon the accuracy of the field computation. An efficient method for the calculation of the derivatives in FE solutions has been proposed in [35] for the case of objectives that can be expressed as a function of the system variables used in the FE calculation. Even if these procedures are not available in the applied FE package, the Davidon–Fletcher–Powell algorithm is very well suited for checking the optimality conditions by means of a Lagrange-multiplier estimation, minimizing Eq. (20.46). Since the optimization parameters here are the Lagrange multipliers $\boldsymbol{\alpha}$ and $\boldsymbol{\beta}$, the derivatives can be approximated with great accuracy.

20.9 Genetic Optimization Algorithms

As can be seen in Figure 20.12, the solution found by deterministic optimization algorithms may depend on the choice of the starting point and the initial search-direction. Since we cannot exclude multiple optima in the feasible domain, the algorithms must be restarted with different initial designs. Approaches for global optimization include simulated annealing, stochastic search, evolutional optimization, and genetic algorithms. In this section we will describe the principles of genetic algorithms suited for conceptual design of superconducting magnets, in particular, their capability of handling discrete problems, which appear when the distribution of coil windings in the SC magnet cross section is optimized.

In the 1850s, Darwin[11] laid the foundations for the modern interpretation of evolution. Although organisms are capable of reproducing in an exponential way, Darwin observed that this was not the case in natural populations and concluded that some selection, or "survival of the fittest" must occur. He also observed variations between individuals of particular species. The underlying genetic basis that accounts for these variations had been formulated at the

11 Charles Darwin (1809–1882).

same time by Mendel[12] in his study of traits in pea plants, though his work remained unrecognized for more than thirty years. Whereas Darwin believed in continuous variation, Mendel described heredity as a progression in discrete steps.

Mendel's law and Darwin's theory remained for a long time unlinked. This was also to be true of genetic optimization algorithms and evolution strategies in computer science developed in parallel in the mid-1970s by Holland [16], and by Schwefel [48] and Rechenberg [40], respectively. The evolutionary strategies were based on the link between reproductive populations rather than genetic links and their binary encoding in the genetic optimization techniques. However, both methods share a common principle: A population of individuals (representing feasible technical solutions) undergoes transformations and selections based on the fitness value of each individual. The main differences in the methods are hidden on a lower level of software implementation. Genetic algorithms with richer structures than binary strings have been suggested since the early 1990s; see [30] for a comprehensive account.

In the evolutionary strategies, the representation of the individuals is given by floating point vectors $v = (\mathbf{x}, \sigma)$, where σ is a vector of standard deviations in accordance with the biological fact that smaller changes occur more often than larger ones. From a population of μ parents, *offspring* are created by adding to each component of $\mathbf{x}_i$ a Gaussian random variable with a mean zero and a standard deviation σ_i.

Facilitated by the development of computing power, evolutionary strategies and genetic algorithms have been given considerable attention in the application to problems in electromagnetic field computation [52]. With genetic algorithms, the designer is able to treat mixed continuous and integer design variables by coding them into binary strings. A genetic algorithm employs the following components:

- An encoding of the design variable vector $\mathbf{x}$ into a binary string $\mathbf{b}$.
- A means of creating an initial population (usually at random).
- An evaluation function representing the role of the environment by means of a fitness ranking of the binary string.
- Operators that alter the composition of the offspring, e.g., *crossover* and *mutation*.
- A selection scheme.
- A set of optimization parameters tuned for the technical problem at hand; chromosome length, population size, crossover rate, and mutation rate.

12 Gregor Johann Mendel (1822–1884).

20.9.1 Parameter Representation

The real and integer design variables x_k are combined into a single binary string $\mathbf{b}$, which is called a *chromosome* in accordance with the nomenclature of natural genetics. The length of the chromosome depends on the required precision; for n decimal places the domain Ω_k of x_k is cut into

$$\Delta x_k = (x_{k,\text{upper}} - x_{k,\text{lower}}) \times 10^{-n} \tag{20.105}$$

equal size ranges. Then m_k is the smallest integer such that

$$\Delta x_k \leq 2^{m_k} - 1\,. \tag{20.106}$$

The mapping from the binary string $\mathbf{b} = (b_{m_k-1} \ldots b_0)$ into a real number $x_k \in \Omega_k$ is given by $\lambda_k := \sum_{i=0}^{m_k-1} b_i \cdot 2^i$ and therefore

$$x_k = x_{k,\text{lower}} + \lambda_k \frac{x_{k,\text{upper}} - x_{k,\text{lower}}}{2^{m_k} - 1}\,. \tag{20.107}$$

Each individual is characterized by its single chromosome[13], which is represented by a binary string of length $m = \sum_{k=1}^{n} m_k$, where n is the number of design variables.

Extending the genetic metaphor, the digits in the binary string are called *genes* and the bit strings representing the design variables *alleles*. The positions of the alleles within the chromosome are called *loci*. The chromosomes of the population (which is also called the *phenotype*) form the *gene pool* or *genotype*. In what follows, we will use the terms binary string, chromosome, individual, and population member synonymously; see Figure 20.13.

It is clear that the performance of genetic algorithms is reduced when the domains of parameters are unlimited, the number of parameters is large, or high precision is required. As already suggested by Holland [17], genetic algorithms are used as a preprocessor to perform the initial search, before the solutions are fine-tuned using deterministic algorithms.

20.9.2 Gray Coding

Applying binary coding, two adjacent integers, for example, $k = 7$ where $\mathbf{b}(7) = (0111)$ and $k = 8$ where $\mathbf{b}(8) = (1000)$, differ considerably in the bit string. In this example, the Hamming[14] distance, which is defined as the

13 Note that each cell of an organism carries a certain number of chromosomes, for example, 46 in the case of homo sapiens, while in genetic optimization we employ one-chromosome individuals only.

14 Richard Hamming (1915–1998).

Phenotype	$x_1^{(1)}$	$x_2^{(1)}$	$x_3^{(1)}$	$\cdots$	$x_k^{(1)}$	Population
	$x_1^{(j)}$	$x_2^{(j)}$	$x_3^{(j)}$	$\cdots$	$x_k^{(j)}$	(Elder)
	$\downarrow$	$\downarrow$	$\downarrow$	$\downarrow$	$\downarrow$	
Gene pool	(0111)	(1001)	(1101)	$\cdots$	(1011)	
(Genotype)	(0011)	(0001)	(1110)	$\cdots$	(1101)	
	$\downarrow$	Gray coding			$\downarrow$	
Chromosomes		(010011011011...1110)				
		(001000011001...1011)				
		$\downarrow$				Recombination,
Crossover, Mutation,		Genetic Operators				Selection, Niching,
Genetic drift		$\downarrow$				Migration
Chromosomes		(010111100111...0101)				Individual, Member
		(011011001011...0010)				
	$\downarrow$	Gray decoding			$\downarrow$	
Genotype	(0110)	(1011)	(0101)	$\cdots$	(0110)	
	(0100)	(1000)	(1101)	$\cdots$	(0011)	
	$\downarrow$	$\downarrow$	$\downarrow$	$\downarrow$	$\downarrow$	
Phenotype	$x_1^{(1)*}$	$x_2^{(1)*}$	$x_3^{(1)*}$	$\cdots$	$x_k^{(1)*}$	Offspring
	$x_1^{(j)*}$	$x_2^{(j)*}$	$x_3^{(j)*}$	$\cdots$	$x_k^{(j)*}$	

Figure 20.13 Parameter encoding and decoding.

number of different bits, is 5. The Hamming distance between two binary strings can be computed with

$$HD = \sum_i b_i^{(1)} \oplus b_i^{(2)}, \tag{20.108}$$

where $\oplus$ denotes the addition modulo 2 (exclusive-or). The applied *single-distance code*, originally developed by Baudot[15] but better known as *Gray code*,[16] is a one-to-one mapping $k \to \mathbf{g}(k)$ so that the binary representations of two adjacent integers $\mathbf{g}(k)$ and $\mathbf{g}(k+1)$ differ by exactly one bit. In other words, their Hamming distance is 1 (cf. Table 20.2).

The Gray coding (binary to Gray) is achieved by a bit-shift-right and exclusive-or of the binary string of k and its shifted version (the integer of $k/2$). The binary string is truncated on the right and supplemented with a zero on the left.

$$g_i = \begin{cases} b_i & \text{if } i = 1 \\ b_{i-1} \oplus b_i & \text{if } i \geq 2 \end{cases} \tag{20.109}$$

15 Emile Baudot (1845–1903).
16 Frank Gray (1887–1969).

Table 20.2 Gray coding table[a].

Decimal	Binary		Gray	Decimal	Binary		Gray
0	(0000)	$\leftrightarrow$	(0000)	7	(0111)	$\leftrightarrow$	(0100)
1	(0001)		(0001)	8	(1000)		(1100)
2	(0010)		(0011)	9	(1001)		(1101)
3	(0011)		(0010)	10	(1010)		(1111)
4	(0100)		(0110)	11	(1011)		(1110)
5	(0101)		(0111)	12	(1100)		(1010)
6	(0110)	$\leftrightarrow$	(0101)	13	(1101)	$\leftrightarrow$	(1011)

[a] With binary coding, two neighboring integers, for example, 7 and 8, differ considerably in the bit string (5 bits) whereas the Gray encoding ensures that neighboring integers differ by one bit only.

where b_i are the bits in the binary string, g_i are the bits in the Gray code, and $i = 1$ is the most significant bit. As an example, the Gray coding of the decimal 13, $\mathbf{b}(13) = (1101)$, yields $\mathbf{g}(13) = (1101) \oplus (0110) = (1011)$. Decoding (Gray to binary) proceeds by multiple shift-right and exclusive-or operations on the shifted versions:

$$b_i = \bigoplus_{j=1}^{i} g_j \tag{20.110}$$

Decoding $\mathbf{g}(13) = (1011)$ yields $\mathbf{b}(13) = (1011) \oplus (0101) \oplus (0010) \oplus (0001) = (1101)$. Gray coding ensures that two points that are close in the representation space are also close in the problem space. Gray coding is therefore advantageous when the Hamming distance is used as a distance measure in the genetic algorithm. Gray coding also improves a mutation operator's likelihood of making small incremental improvements to the phenotype.

20.9.3 Genetic Operators

Genetic optimization algorithms are driven by the following main operators:

- *Crossover* is a recombination of two strings by breakage at a random point and reunion of the alleles. This is the underlying mechanism of sexual reproduction.
- *Mutation* is the process of an alteration in a chromosome structure. In optimization, this process avoids preliminary convergence toward a local minimum.

- *Selection* is the force behind changes in the genotype in populations through differential reproduction; less fit members of the population having a smaller mating probability.

Crossover Crossover combines the features of two parent strings to form two offspring members by swapping corresponding alleles of the parents. The single-point crossover operator can be best explained with Figure 20.14. The bits on the left-hand side of a crossing point of chromosome A are connected to the bits on the right-hand side of the crossing point within chromosome B and vice versa. Although the crossing point is chosen at random, the offspring created do not cover the entire search space, as can be seen in Figure 20.15.

Chromosome A:	(0101001|101)	(0101001) (011)	(0101001|011)
	:	×	:
Chromosome B:	(1011010|011)	(1011010) (101)	(1011010|101)

Figure 20.14 Single-point crossover of two chromosomes.

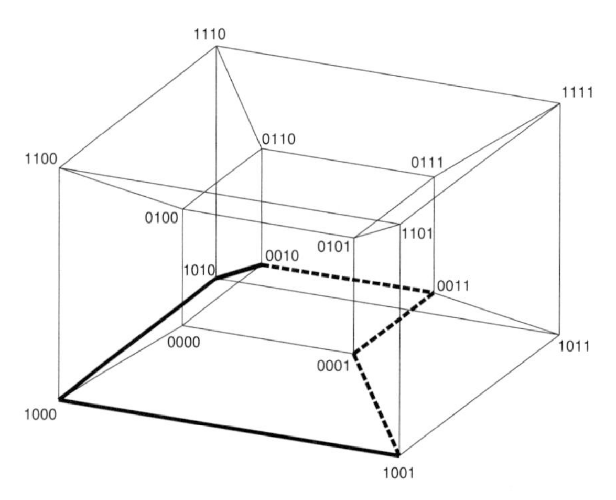

Figure 20.15 A single-point crossover of the chromosomes (0010) and (1001). Only the points on the bold lines can be reached, offspring do not cover the entire search space.

Numerous crossover schemes have been proposed in the literature, including two-point or multipoint crossover; adaptive crossover where marks in the string representation keep track of particularly beneficial crossing points; segmented crossover where the number of crossing points may vary; or uniform crossover that applies to single bits rather than alleles. See [30] for a comprehensive treatment.

Mutation The mutation operator acts with a certain probability (in our case 0.15) on a bit-by-bit basis. Each time the mutation operator is applied, a single bit is changed (mutated) from 0 to 1 or vice versa. Each bit has exactly the same chance of being mutated. The effect of this uniform mutation is twofold. On one hand it avoids preliminary convergence of the entire population toward a local minimum, and on the other hand it improves local convergence by a hill-climbing mechanism. Although these mechanisms seem mutually contradictory, they both result from the different significance of bits in the binary string, as shown in Figure 20.16. In the first example a change in the fifth bit of the genotype results in a change of the second parameter of the phenotype from 7 to 8. In the second example a change in the second bit of the genotype results in a change of the first parameter of the phenotype from 10 to 13.

	10	7	2	1	1	0	Phenotype, k
	1111	0100	0011	0001	0001	0000	Genotype, $\mathbf{g}(k)$
a)	10	8	2	1	1	0	Phenotype, k
	1111	**1**100	0011	0001	0001	0000	Genotype, $\mathbf{g}(k)$
b)	13	7	2	1	1	0	Phenotype, k
	1**0**11	0100	0011	0001	0001	0000	Genotype, $\mathbf{g}(k)$

Figure 20.16 The influence of mutation. Single bit changes have different influence on the phenotype. (a) The mutation of bit 5 results in a phenotype difference of 1. (b) The mutation of bit 2 leads to a phenotype difference of 3.

Selection In each generation, the individuals are evaluated using the objective function on the decoded sequence of variables, and a new population is selected with respect to a probability distribution based on the fitness values.

Selection methods can be classified into the following categories:

- *Fitness proportional* (or *fairy wheel*) *selection*: A fitness value that represents a probability of reproduction is assigned to each of the chromosomes so that the fit members of the population will be selected more frequently.
- *Tournament selection* acts on a group of k individuals, randomly chosen from the population, wherein the best chromosome from these k elements is copied into the next generation. The parameter k is called the *tournament size*. An increased number k increases the selective pressure.

- The preservation of the best individual may always be enforced. This is called the *elitist model*.
- *Niching methods* introduce a concept of distance according to the observation that a population spread over a geographic range will become genetically differentiated in a number of sub-populations (species) in so-called niches.

The principle of fairy wheel selection is shown in Table 20.3. The algorithm is constructed as follows:

- Evaluate the chromosome, that is, calculate the objective function value of the decoded sequence. Formally, assign $e(\mathbf{b}) = f(\mathbf{x}) + c$.
- Find the *probability of reproduction* or *fitness* for each individual:

$$p_i := \frac{e(\mathbf{b}_i)}{\sum_{j=1}^{n} e(\mathbf{b}_j)}, \tag{20.111}$$

  where n is the population size. Assuming that all objective function values are positive, we aim at maximizing the fitness. This is not a restriction, since we can always add a sufficiently large bias c.
- Calculate the *cumulative probability* q_i for each individual:

$$q_i := \sum_{j=1}^{i} p_j . \tag{20.112}$$

Table 20.3 Fairy wheel selection[a].

Index	Parent population	Objective functional value	Fitness value		Index of selected Chromosome	Child population
1	(1000111010)	0.3	0.30		3	(1001101101)
2	(1110101101)	0.4	0.22		1	(1000111010)
3	(1001101101)	0.5	0.18		4	(1011010010)
4	(1011010010)	0.8	0.11	$\rightarrow$	1	(1000111010)
5	(0111001100)	0.9	0.10		2	(1110101101)
6	(0111011010)	1.7	0.05		3	(1001101101)
7	(0011000101)	2.6	0.04		1	(1000111010)

[a] A fitness value is assigned to each individual according to the objective function value. The new generation is formed by fitness proportional reproduction. Note that the index does not impose any ranking.

- Generate a random float number $r \in [0, 1]$.
- Spin the fairy wheel: Select the ith individual if $q_{i-1} < r < q_i$.

Note that while fit individuals will probably be selected more than once, less fit members of the population also have a certain likelihood of reproduction. Over a large number of generations, however, members with the best chromosomes will always beget more copies, the average will maintain their numbers, and the less fit will die out. This is proved with the *schema theorem* [30].

Theoretical studies often assume an infinite population size and an infinite number of function evaluations, while real-world applications must deal with limited population sizes and a relatively small number of iterations. As a result, optima are frequently missed because of premature convergence. The cause of this process has been identified as *genetic drift*, in other words, the reduction of diversity due to the random propagation of alleles from parents to offspring. The search space of the crossover operator is reduced by every discarded allele and can only be regained by mutation.

Figure 20.17 shows the development of a population of 30 individuals over 500 function evaluations using a genetic algorithm with fairy wheel selection for the minimization of the analytical test function [15]

$$f(x, y) = \frac{(x^2 + y - 11)^2 + (x + y^2 - 7)^2}{2186}, \tag{20.113}$$

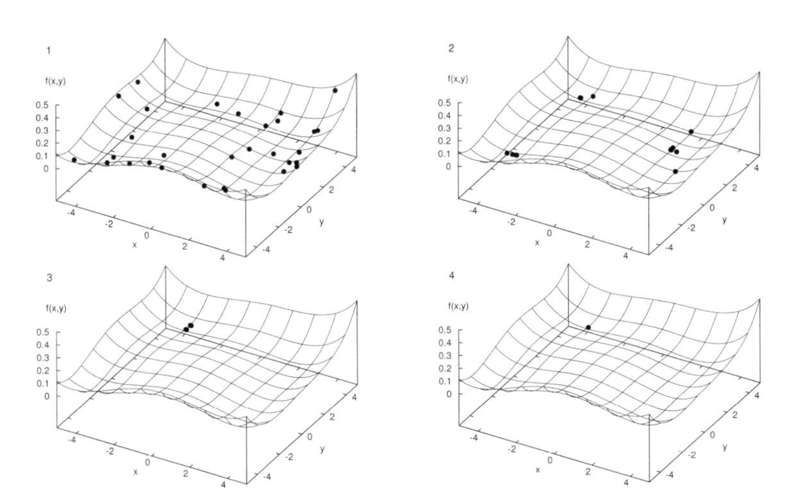

Figure 20.17 Genetic algorithm with crossover and fitness ranking. The development of a population of 30 individuals is shown over 500 function evaluations. All of the 30 individuals degenerate into a single stable solution.

which has four global minima of zero function-value at the positions (3, 2), (3.58443, −1.84813), (−2.80512, 3.13131), and (−3.77931, −3.28319). All the individuals drift into a single solution.

This property of genetic algorithms is not useful for magnet design optimization because not all the objectives in coil optimization can be mathematically formulated and included in the definition of the objective function. Therefore, a number of local optima are sought to be investigated in more detail. The most obvious method for this is iteration, with simple repetitions of the algorithms. However, since not all local optima are equally likely, methods that permit the formation of sub-populations have been developed. These niching techniques will be discussed next.

20.9.3.1 **The Niching Genetic Algorithm**

The purpose of niching at the algorithmic level is to categorize solutions by their design variables into groups, known as niches, and to evolve them in the optimization run. In order to do so, a distance measure $d(\mathbf{x}_i, \mathbf{x}_j)$ is defined, either on the phenotype with the parameter values, or on the genotype at the parameter encoding level. Niching methods may be classified as follows:

- *Sharing* methods use a function $s(d)$ that determines the neighborhood for each member in the population. The simplest type of sharing function assigns 1 to chromosomes that are close in parameter space and 0 to individuals which are further away than a certain dissimilarity threshold. A fitness value is assigned as

$$F_s(\mathbf{x}_i) := \frac{f(\mathbf{x}_i)}{\sum_{j=1}^{n} s(d(\mathbf{x}_i, \mathbf{x}_j))}, \tag{20.114}$$

  which results in a reduced fitness of individuals with a large number of neighbors. The probability of reproduction is therefore reduced in densely populated niches.

- *Crowding* methods apply a replacement mechanism based on the diversity of the offspring with respect to the individuals of the population. The offspring replace the string that has the smallest Hamming distance if its objective function value is lower and thus its fitness is higher.

- *Clearing* methods remove individuals from the population when a niche is overpopulated. Only the k best individuals survive the selection process. Individuals are assigned to a niche if their distance is lower than a dissimilarity threshold σ_s.

The crowding method, which we use for conceptual magnet design, is best explained using Table 20.4. In fitness-proportional selection, the probability of reproduction of the new string would have been rather low. However, when competition is restricted to the closest member (here with index 6), it will be selected for the next generation.

Table 20.4 Niching selection[a].

Index	Parent population	Objective functional value	Hamming distance		Index of selected chromosome	Child population
1	(1000111010)	0.3	5		1	(1001101101)
2	(1110101101)	0.4	7		2	(1000111010)
3	(1001101101)	0.5	6		3	(1011010010)
4	(1011010010)	0.8	3	→	4	(1000111010)
5	(0111001100)	0.9	3		5	(0111011010)
6	(0111011010)	1.7	2		New	(0011001010)
7	(0011000101)	2.6	4		7	(1000111010)
New	(0011001010)	1.1	0			

[a] A new member of the population is tested for its genetic similarity (Hamming distance) to the other members of the population (here chromosome with index 6). The new member replaces this closest neighbor, which has a higher objective function value and lower fitness.

Figure 20.18 shows a comparison of the classical ("royal road") genetic algorithm with fairy wheel selection, with the applied method of niching, in which for each offspring the chromosome with the smallest Hamming distance is localized and selected if its fitness is better than that of the elder.

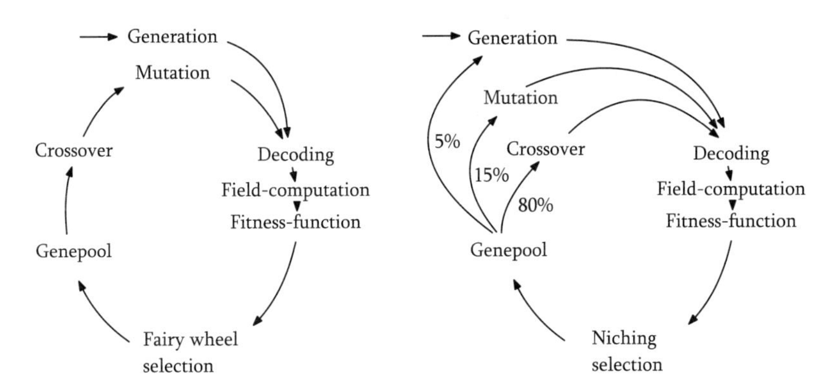

Figure 20.18 Left: Royal-road genetic algorithm. Right: Genetic algorithm with niching.

Whereas in the royal-road genetic algorithm the whole population is subject to a fitness ranking, the selection in the niching genetic algorithm is performed on the level of each individual. Selected chromosomes are immediately joined to the population. The effect of the niching method is that a number of local optima are found, which can then be analyzed in detail. Figure 20.19 shows the convergence of the niching genetic algorithm for the analytical test function (20.113).

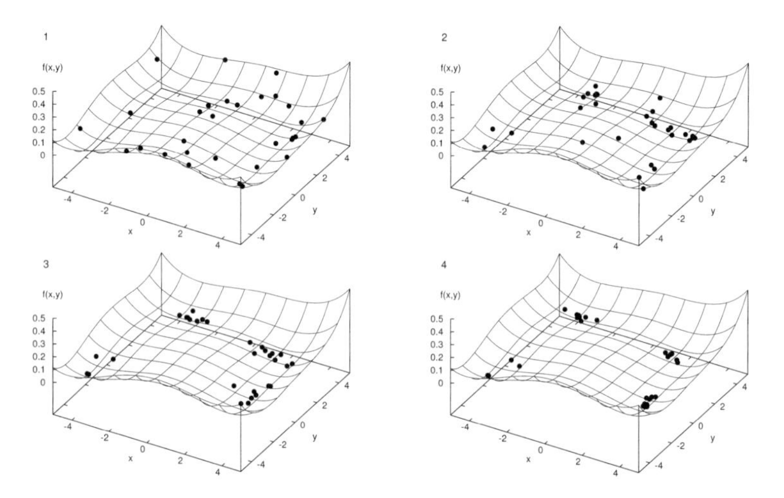

Figure 20.19 Development of a population of 30 individuals shown over 1000 function evaluations of the niching genetic algorithm. All four minima are found. Niching is therefore crucial for the conceptual design phase.

Important for overall execution time is that offspring that already exist in the population are not evaluated for their fitness again. Robustness is guaranteed by the following properties:

- No special knowledge of the feasible domain is needed. A number of local minima are obtained.
- No derivatives of the objective function are needed. Integer variables can be directly encoded.
- The population size and the number of bits for the encoding can be automatically adjusted. The convergence has proven to be relatively insensitive to the mutation, crossover, and generation rate.

20.9.4 Convergence

Genetic diversity in the population is maintained because niching decreases the selective pressure [30]. There is a trade-off between high selective pressure, which leads to premature convergence, and weak selective pressure, which leads to an ineffective search process. Therefore the main parameters (the number of iterations and the population size) must be adjusted to the optimization problem. Mahfoud [27] suggests keeping the number of individuals in a population at roughly 10 times the number of expected local optima. For the coil-block optimization problem presented in Section 20.10.1, a population size of 60 was chosen in all cases. The crossover, mutation, and generation rate were determined by a number of convergence tests. Convergence graphs are shown in Figure 20.20 for a crossover rate of 0.8, mutation rate of 0.15, and generation rate of 0.05.

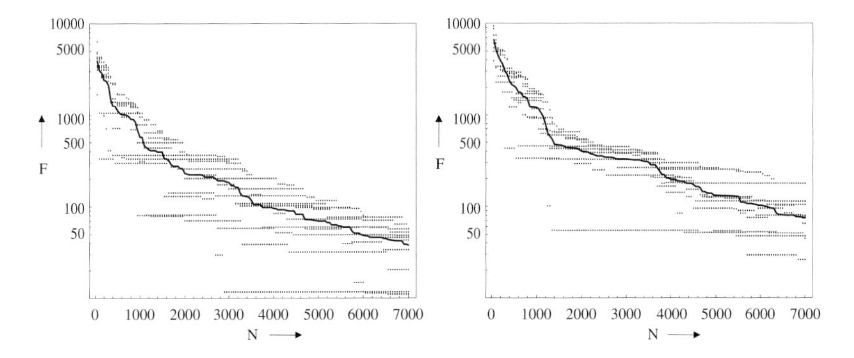

Figure 20.20 Convergence tests (N = number of function evaluations) performed on the coil optimization problem described in Section 20.10.1. Fitness values of the 10 best chromosomes and the average of the population (continuous line). Left: Default parameters: Crossover rate 0.8, mutation rate 0.15, and generation rate 0.05. Right: Crossover rate 0.35, mutation rate 0.6, and generation rate 0.05.

Increasing the mutation rate to 0.6 shows a slower initial convergence rate. After a few thousand iterations, when the population becomes stable, a higher mutation rate allows for further improvement, since alterations of single bits often correspond to small changes in the parameter space. Results for single and two-point crossover show no significant differences.

Although convergence of the algorithm can be seen in Figure 20.20, it is difficult to prove mathematically. We will present the schema theorem as an attempt [16] to prove convergence of the royal road genetic algorithm: A subset of individual strings is described by a schema **s** consisting of bit strings composed by a three-letter alphabet $A = \{0, 1, \star\}$ where 0 and 1 denote fixed bits and the asterisk $\star$ is a wild-card symbol for 0 or 1. As an example consider

$$\mathbf{b} = 1\,1\,0\,1\,0\,1\,0\,0\;0\,1,$$

$$\mathbf{s}_1 = \star\star\star\star\,0\,1\,0\,\star\star\,1,$$

$$\mathbf{s}_2 = \star\star\,0\,1\,\star\star\star\;\star\star\star.$$

The *order* $o(\mathbf{s})$ is the number of fixed positions in the schema string. The orders of string $\mathbf{s}_1$ and $\mathbf{s}_2$ are thus 4 and 2, respectively. The *defining length* $\delta(\mathbf{s})$ is the distance from the first to the last fixed position in the string and is a measure of the compactness of the information contained in the schema. For our examples $\delta(\mathbf{s}_1) = 10 - 5 = 5$ and $\delta(\mathbf{s}_1) = 4 - 3 = 1$. A schema with a single fixed position has a defining length of zero.

At an iteration step t, $m(\mathbf{s}, t)$ denotes the number of strings in the population matched by the schema $\mathbf{s}$. The set of strings matched by the schema is $\{\mathbf{b}_1, \ldots, \mathbf{b}_{m(\mathbf{s},t)}\}$. The fitness of a schema is defined by

$$p(\mathbf{s}, t) := \frac{\sum_{i=1}^{m(\mathbf{s},t)} e(\mathbf{b}_i)}{m(\mathbf{s}, t)} . \tag{20.115}$$

During the selection step, an intermediate population is created to which strings have been copied according to their fitness defined by Eq. (20.111). For the individual string matched by the schema, the probability of reproduction is $p_i := e(\mathbf{b}_i)/\sum_{j=1}^{n} e(\mathbf{b}_j)$ and the number of strings matched by the schema is $m(\mathbf{s}, t)$. Thus the number of strings in the intermediate population matched by the schema is

$$m(\mathbf{s}, t+1) = m(\mathbf{s}, t)\, n \frac{p(\mathbf{s}, t)}{\sum\limits_{i=1}^{n} e(\mathbf{b}_i)} p_o(\mathbf{s}, t) , \tag{20.116}$$

where n is the number of strings in the parent population, and $p_o(\mathbf{s}, t)$ is the probability of reproduction due to crossover and mutation. Low-order schemata are less likely to be cut by crossover operations and have a higher probability of reproduction. Schemata of large order are more likely to be changed by mutation.

Schema theorem: Short, low-order, above-average schemata receive exponentially increasing trials in subsequent generations. □

In [12] we read: "Just as a child creates magnificent fortresses through the arrangement of simple blocks of wood, so does a genetic algorithm seek near-optimal performance through the juxtaposition of short, low-order, high-performance schemata."

The number of possible schemata incorporated in a population is proportional to the number of possible bit combinations 3^l, including wild-cards, where l is the number of bits in the string and n the number of strings in a population. This allows a string to represent at most 2^l different possible schemata. The genetic algorithm evaluates those schemata in parallel, which is seen as a major advantage of genetic algorithms.

The fact that genetic algorithms are able to test design spaces of multiple minima and discrete design variables makes them appropriate in the conceptual design phase of equipment. However, genetic algorithms are computationally ineffective. Whenever the search can be restricted to a few continuous parameters, local deterministic optimization methods should be applied. Robustness is traded for speed of convergence, and the quality of the final result for the ability to work on global problems. This dilemma is known as the *no free lunch theorem* [55].

20.10 Applications

The optimization process starts by selecting the design variables and therefore requires routines for the parametric definition of the coil and iron-yoke geometry. The geometric position of the cables in the magnet cross section can be calculated from the input data shown in Figures 20.21 and 20.22:

- In the case of $\cos n\varphi$ magnets, the input data are the number of blocks, the number of cables per block, cable type (specified in a cable database), radius of the winding mandrel, and positioning and inclination angles of the blocks. The fact that the keystoning of the cables is not sufficient to allow for the construction of arc segments is fully accounted for. This effect increases with the inclination of the coil blocks with respect to the radial direction. The grading of the engineering current density within the cable is taken into account by a discretization of the cable into $N_1 \cdot N_2$ line-currents, where N_1 is the number of strands in the narrow direction, and N_2 is the number of strands in the direction of the broad side ($2 \cdot 18$ in the case of the LHC outer-layer dipole cable).
- In the case of window frame magnets, the user-supplied data comprise the number of blocks, the number of cables per block, cable type, x- and y-coordinates of the lower left corner of the block, and inclination angle with respect to the x-axis.
- In the case of beam-pipe magnets, input data are the number of blocks, the number of cables per block, the radius of the winding mandrel, the positioning angle of the first cable, and the increment angle for the subsequent cables.
- In the case of hollow conductors (cable in conduit), the geometry is created as in the above cases. An inlaying cylindrical conductor with N_1 arc segments is generated from the cable boundary. The arc segments have an inner radius such that exactly one strand is inscribed within each segment.

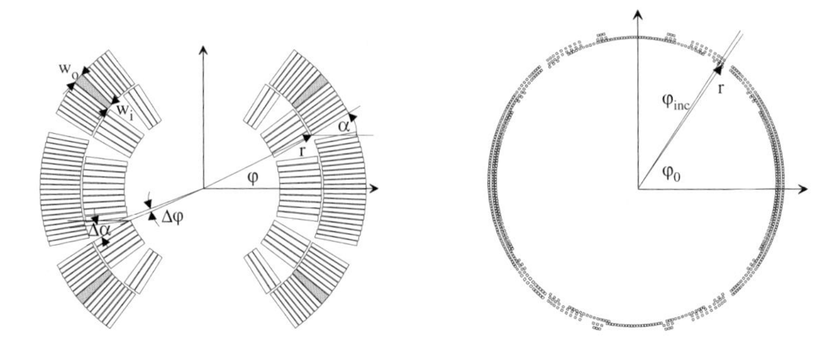

Figure 20.21 Input data for the generation of coil geometries of the $\cos \varphi$-type (left) and the beam-pipe magnets (right).

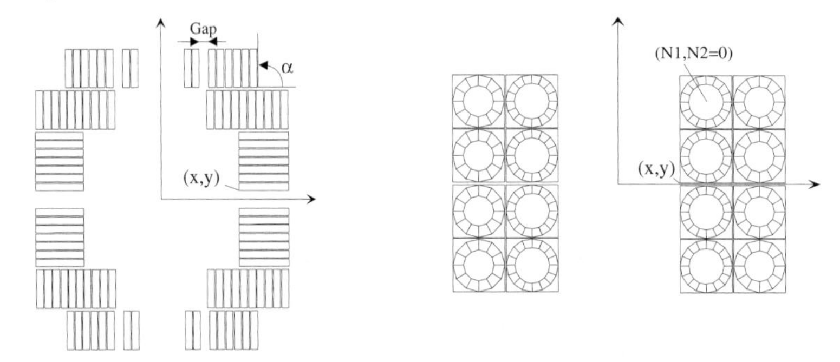

Figure 20.22 Input data for the generation of block-coil magnets (left) and hollow conductors (right).

The input parameters for the coil-end geometry are the relative longitudinal positions of each coil block, its inclination angle at the nose, the length of its straight section, and the size of inter-turn spacers between the cables. Additional twist angles can be applied in order to match the transition between the coil end and the straight section.

With the geometric modeling complete, every feature (strand, cable, block, layer) can be subjected to geometric transformations such as translation, rotation, scaling, and imaging. At the same time, constraints are defined for these operations in order to avoid penetration or physically meaningless structures. Not only the geometric properties of the magnet but also its material properties such as the number of strands, the current density in the cables and strands, filling factors, and unit price may be changed in the optimization process.

Table 20.5 shows at a glance the different steps where mathematical optimization methods have been applied to the design of the superconducting magnets and lists the typical number of design variables and function evaluations.

Table 20.5 Optimization problems in magnet design.

	Decision-making method	Optimization algorithm	Typical no. of design variables	Typical no. of function evaluations
Conceptual coil design	Objective weighting	Genetic algorithms	15 (60 Bits)	7000
Coil cross section	Distance function	EXTREM	8	250
Conceptual yoke design (material distribution)	Distance function	Genetic algorithms	170 (170 Bits)	$20\,000 \cdot 4$ FEM
Yoke cross section	Objective weighting	EXTREM	10	$100 \cdot 6$ FEM
Sensitivity analysis	Lagrange multiplier estimation	Davidon–Fletcher–Powell algorithm	10	10
	Payoff tables	EXTREM		$6 \cdot 100 \cdot 6$ FEM
3D Coil end optim.	Objective weighting	EXTREM	5	100
Inverse field calc.	Distance function	Levenberg–Marquard	50	700

20.10.1 Conceptual Coil Design with Genetic Algorithms

Genetic optimization algorithms require a much larger number of function evaluations to reach an optimum than do deterministic methods. On the other hand, they can provide the user with a number of local optima representing suitable starting points in the initial design phase.

It is important to obtain a number of alternative solutions (local optima), since not all criteria in the coil optimization can be mathematically described and accounted for in the objective function. Examples of such criteria include the following:

- The unwanted multipoles are to be minimized. However, using the goal-programming methods, the weighting factors for the field components of different sensitivity must be found in an iterative procedure.
- The local force distribution in the coil/collar structure must be optimized, as the electromagnetic forces are enormous. The force distribution depends on the coil layout, and therefore computations of coupled electromagnetic and mechanical problems are required.
- Manufacturing considerations, such as the ease of coil winding and collaring, impose geometrical constraints on the pole angle and yoke geometry. Setting too many geometrical constraints, however, results in ill-conditioned optimization problems.
- During the manufacturing process, systematic errors occur due to the applied tooling and assembly procedures. After the preseries production of the magnets, the coil-block positions must be tuned to compensate for these errors. As during the series production of the magnets,

the coil layout can no longer be changed, and the design must feature sufficient flexibility for adjustments.

On the other hand, the following goals can easily be included when using the objective-weighting function or nonlinear constraints.

- Maximum main field component.
- Minimum field harmonics, with emphasis on small b_9 and b_{11} components.
- Minimum volume of superconductor.
- A constraint on the pole angle (smaller than 80° in the case of the dipole).
- A constraint on the working point of the superconducting cables; more than 20% margin to quench along the load line.
- Minimum deviation of the cable positions from the ideal (radial) position. This objective is mathematically expressed as

$$\min \left\{ \sum_{k}^{N_\mathrm{b}} \left| \sum_{i}^{N_\mathrm{cb}} (\alpha_i - \varphi_i) \right| \right\} , \tag{20.117}$$

where N_b is the number of coil blocks, N_cb is the number of cables in the kth coil block, α_i is the inclination angle of the cable, and φ_i is its positioning angle.

The higher-order multipole components b_9 and b_{11} cannot be optimized, unless the cable distribution in the coil blocks can be altered by the algorithm. This can be seen from the distribution of field errors as a function of the strand position; see Figure 8.7. The design variables for the optimization are:

- The number of turns per coil block.
- The shape of the copper wedges between the coil blocks.
- The excitation current in the cables.

The current must be included as a design variable in the optimization. This guarantees a feasible solution not exceeding the quench limit of the superconducting cable, which depends on the local magnetic field in the coil. The quench limit is considered as an additional constraint in the objective function by means of a penalty transformation.

The design parameters are coded into a binary string as described in Section 20.9.1. The angles of the coil blocks are encoded by 4-bit strings while the number of turns per block are encoded by 3-bit strings each, thus resulting in chromosomes of 55 bits for the six-block coils. A population size of 60 chromosomes proves to be sufficient, using a crossover probability of 0.8 and a mutation rate of 0.005. The main parameters are summarized in Table 20.6.

Table 20.6 Main parameters of the genetic optimization runs.

Number of individuals	60
Chromosome length (bit)	55
Representation scheme	Gray
Reproduction method	Niching
Crossover method	Uniform
Crossover probability	0.8
Mutation rate	0.005
Termination criterion	No. of eval. > 6000

For the parameter sets resulting in 55 bit-long chromosomes typically 6000 function evaluations must be performed. Although this number is relatively high it should be noted that the design space spanning the different cable distributions comprises more than a million combinations. It is therefore impossible to explore the design space with deterministic methods.

For magnets in the 6 T range it is not necessary to use different cables in the inner and outer layer. This avoids the development of a second cable and facilitates coil winding by avoiding internal splices. Genetic algorithms were used to find 5-, 6-, and 7 block coil-configurations of a dipole magnet for the fast-pulsed synchrotron (SIS300) accelerator project at GSI in Germany [32,33]. The magnets have an aperture diameter of 100 mm. The cable is very similar to the outer-layer LHC dipole cable but with an additional steel strip between the two layers in order to increase the interstrand cross-resistance. The results shown in Figure 20.23 have all comparable field quality in the aperture. Notice that both 5-block design variants feature wedges on the horizontal median plane, in order to provide sufficient degrees of freedom for the field-quality optimization.

20.10.2
Deterministic Optimization of Coil Cross Sections

Deterministic optimization methods may be used to find a coil cross section with partial compensation for the persistent-current-induced field errors at injection. This can be done by an appropriate positioning of the coil blocks.

In the final optimization step, the angular positions of the coil blocks are adjusted, while the number of turns remains constant. The target values for the partial compensation of the persistent-current effects can be defined, which naturally leads to the distance function (least-squares) method. The aim is to compensate for 75% of the persistent-current multipoles at injection field level. The b_3 component is compensated by only four units because of the limited strength of the sextupole spool-piece corrector magnets.

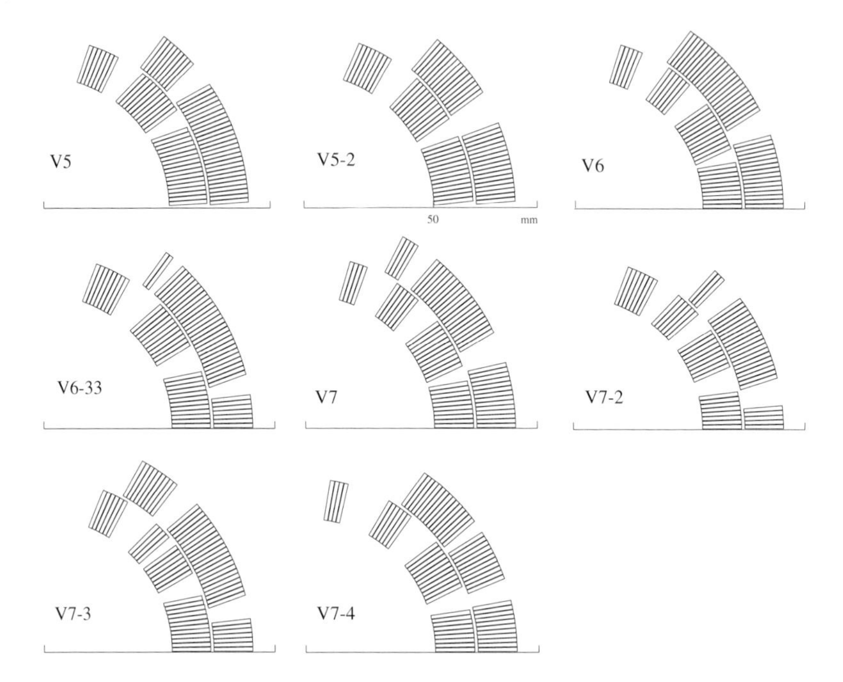

Figure 20.23 5-, 6-, and 7-block coil configuration for a 6 T dipole magnet wound from only one type of cable in both layers. Notice the mid-plane wedge for the 5-block variants.

The least-squares objective function for the V6-1 design shown in Figure 20.23 takes the form

$$\min\left\{t_1(b_3 - 4.0)^2 + t_2(b_5 + 0.8)^2 + t_3(b_7 - 0.3)^2 + t_4(b_9 + 0.15)^2\right\}. \quad (20.118)$$

The deterministic optimization algorithm EXTREM was used to minimize Eq. (20.118). For the eight remaining degrees of freedom (inclination and positioning angles of the outer coil blocks) typically 100 function evaluations must be carried out. The weighting factors t_1–t_4 were chosen as $t_1 = 1$, $t_2 = 5$, $t_3 = 10$, $t_4 = 30$ to account for the different numerical values of the objectives.

Table 20.7 shows the results of the coil-block optimizations for three different coil variants. The 6-block coil cross section selected for the series production of the LHC dipoles, marked as V6-1, is shown in the middle of Figure 20.24. Compared to an earlier design with five coil blocks, the V6-1 variant has a 0.1 T higher maximum field, while the number of turns has been reduced to 40 per coil. The geometric multipoles partly compensate for the persistent current effect at the injection field level. For both six-block alternatives the outward-directed (binormal) electromagnetic force has been reduced considerably.

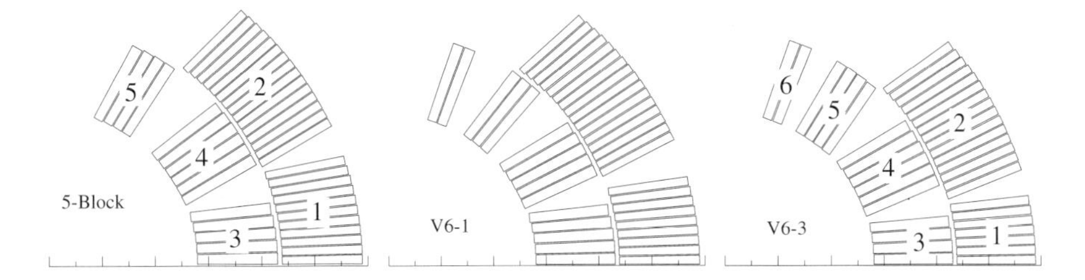

Figure 20.24 Coil block distributions for different variants of the LHC main dipole coil. Designs were found by applying the genetic algorithm with niching, with subsequent fine-tuning using deterministic methods. Left: 5-block coil. Middle: V6-1 variant selected for the series production (one cable less than in the 5-block variant). Right: Alternative with a saving of two cable turns (V6-3). The final choice of the coil cross section for the LHC depended on mechanical considerations, sensitivity to manufacturing tolerances, and coil-winding tests.

Table 20.7 Characteristic data for the optimized coil cross sections and comparison to the 5-block design variant, shown in Figure 20.24 (left) (linear calculations with an ideal iron yoke with inner radius of 98 mm and $\mu_r = 2000$).

	V6-3	V6-1	5-block
Number of turns per coil	38	40	41
Working point (%) on the load line (inner layer)	81.05	84.92	82.5
Working point (%) on the load line (outer layer)	86.15	85.64	86.5
Peak field/main field ratio (outer layer)	0.83	0.89	0.87
Peak field/main field ratio (inner layer)	1.03	1.03	1.052
I nominal (A) at 8.33 T (linear calculation)	11 836.	11 490.	11 183.
B_{ss} (T) (linear calculation)	9.70	9.76	9.65
Self inductance L (mH/m)	6.64	7.17	7.47
Pole angle (degree)	70.5	70.99	57.4
Pole size (mm)	7.1	7.43	8.7
Maximum binormal force (N/m)	16 400.	17 239.	33 877.

The superior magnetic and mechanical performance of the V6-1 design was demonstrated with the production and cold test of nine short-model dipoles; see Figure 20.25. Comparison of the dipole field reached after the first and 15th quench reveals a better training characteristic for the 6-block coil. The electromagnetic forces push a cable into a more stable position and therefore the subsequent quenches are slightly higher. The field obtained after the 15th training quench confirms the prediction (9.76 T) for the V6-1 version, whereas the 5-block versions show a lower average than the predicted 9.65 T. This is due to mechanical problems resulting from the strong outward-directed electromag-

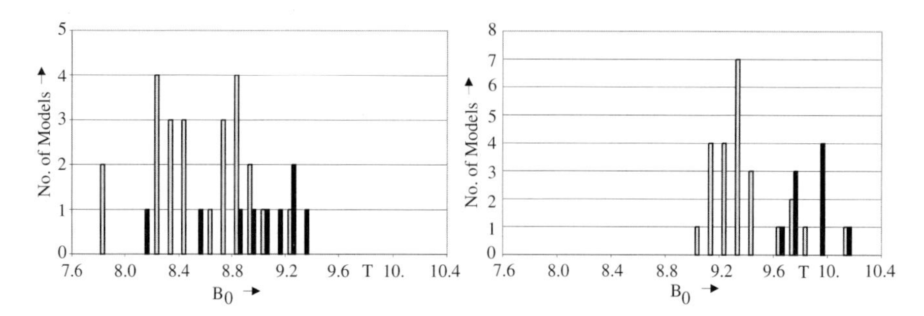

Figure 20.25 Maximum dipole field reached after the first (left) and the 15th training quench (right) for 5-block and V6-1 coil versions. Gray: 5-block version, Black: V6-1 design.

netic forces; see Table 20.7. Because of the insufficient clamping of some of the coil parts, the windings moved slightly under the influence of these forces. This limits the performance of the magnet to below the predicted *short sample field*, which is the aperture field that triggers a natural quench.

20.10.3
Yoke Design as a Material-Distribution Problem

For the iron yoke structure different designs were proposed and optimized using deterministic methods. Solving material-distribution problems addressed in [8] yields a more flexible design tool than shape optimization. This provided the impetus for testing the application of genetic algorithms to the optimization of yoke structures, which was defined as a material-distribution problem. The influence of the iron magnetization is taken into account by means of the FE calculation based on the reduced vector-potential formulation.

The material properties (iron or air) of 170 elementary areas around the superconducting coil are the design variables that allow for separate-collar structures as well as for combined collars.[17] Figure 20.26 shows the area in the iron yoke in which the material properties of predefined quadrilateral facets are free to change. The facets cannot be made too small, since their effect on the objective function may be masked by numerical errors.

If the optimization is performed at a single excitation, for example, at the injection field level of 0.53 T or the nominal field level of 8.33 T, most of the results will show jagged structures and checkerboard-type material distributions. Similar effects have been reported in pole-shape optimization [51]. These structures are not only impossible to construct, but they have only numerical significance (small objective function), since these structures are sen-

17 Up/down and left/right symmetries apply.

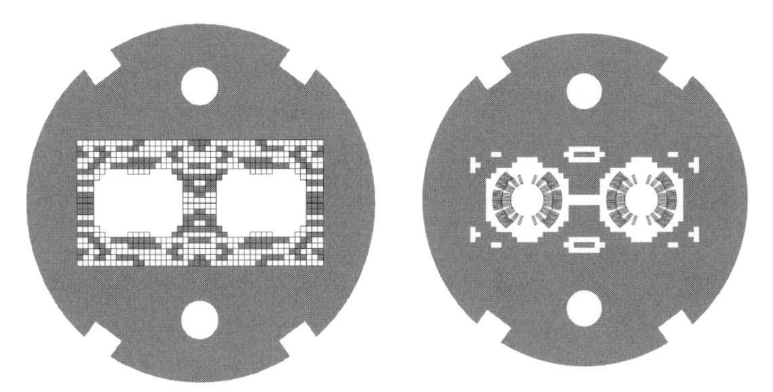

Figure 20.26 Iron yoke with an area where quadrilateral facets are free to change their material properties between iron and air. Result after 12 000 function evaluations.

sitive to local saturation effects. Because synchrotron accelerator magnets are ramped between the injection and nominal field levels, the objective is to minimize the saturation-induced multipole errors. The minimization of these field errors also precludes physically meaningless material distributions.

Figure 20.26 (right) shows the material distribution after about 12 000 function evaluations. The emphasis was put on a maximum main field. The result is not surprising, as can be seen from the comparison with iron cross sections that have been designed using shape-optimization techniques.

The usefulness of the additional small holes in the iron yoke has been shown in [1].

20.10.4 Shape Optimization of the Iron Yoke

As we have explained on different occasions, the design of superconducting magnets can be split into the optimization of the coil and the optimal design of the iron yoke, which influences only the variation in the lower-order multipoles. The yoke optimization is presented here as an example of a real-world vector optimization problem and will show the application of payoff tables, and the Lagrange multiplier estimation. For the shape optimization of the yoke, between five and ten design variables are taken into account. They are shown for three different designs in Figure 20.27 (left). The objectives for the yoke optimization are

- Maximum short-sample field B_{ss}.
- Small b_2 and b_4 component at injection field level.
- Low variation in the b_2, b_3, and b_4 components between injection and nominal field levels. This objective requires a number of FE calculations with different excitations for each objective function evaluation.

- Small outer yoke radius.
- Low sensitivity of the field harmonics to tolerances in the yoke geometry.

The deterministic search-routine EXTREM is used for the optimization. Approximately 150 function evaluations must be performed. The results are given in Figure 20.27. The top figure shows the MBP1 design published in the LHC design report [20] of 1995 with a combined aluminum collar and a ferromagnetic insert that turned out to be difficult to realize with the required precision. The bottom figure shows the MBP2 design selected for the series production [23]. The MBSCA design variant with separate austenitic steel collars is shown in the middle.

The geometry selected for the series production of the magnets features a combined austenitic steel collar designed with two different elliptical "shoulders" in the center and outer part; see Figure 20.27 (bottom). The elliptical shapes provided more than sufficient degrees of freedom for the optimization. The result was that the iron insert and the small hole in the center of the magnet were no longer needed to control the variation in the quadrupole component inherent in the two-in-one design. The geometry of these series magnets is also less sensitive to manufacturing errors than that of the MBP1 design.

Given the mechanical strength of austenitic steel it would have been feasible to use a separated-collar structure as shown in Figure 20.27 (middle). Advantages would have included the decoupling of magnetic flux of the two apertures at injection-field level (thus eliminating crosstalk) and better control of the symmetry with respect to the beam axes, the latter achieved by flipping collar packs during assembly to avoid asymmetries with respect to the beam axes due to punching errors. In order to balance, at nominal excitation, the flux in the center and outer parts of the magnet, and therefore reduce the quadrupole field component, a drop-shaped hole would be punched into the iron insert between the collars. For a variety of scheduling and production reasons this approach could not be implemented.

The results of the optimizations are shown in Table 20.8: For the three design variants the field harmonics at injection and nominal field level are presented together with the short-sample quenching field B_{ss} and the nominal current at 8.33 T.

The multipole errors at injection field level in the case of the separated-collar (MBSCA) geometry are extremely low compared with those in the other geometries, except for the value of b_3, which shows a positive offset of 5 units. This can, however, be easily counterbalanced by the coil design. Other advantages of the separated-collar design are the reduction of the nominal operation current by 370 A and an increase in the short-sample field value B_{ss} by 0.08 T.

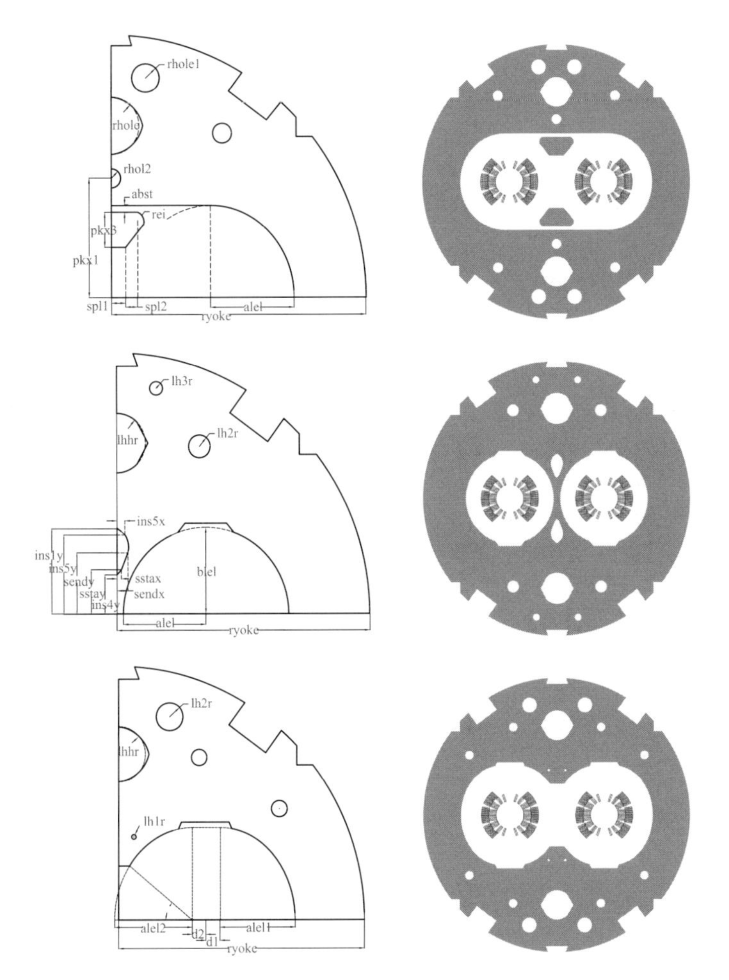

Figure 20.27 Left: Yoke geometries with design variables for the optimization and sensitivity analysis. Top: Design as presented in the LHC design report (Yellow book) [20] with combined aluminum collar and ferromagnetic insert (MBP1). Bottom: Design selected for the series production of the LHC magnets [23]. Middle: Design variant with separate austenitic steel collars (MBSCA). Right: Corresponding optimized yoke cross sections.

20.10.5
Payoff Tables for Dipole Designs

Tables 20.9 and 20.10 are the payoff tables for the MBP2 (series dipole) magnet geometry with combined collars and the separated-collar geometry as shown in Figure 20.27. With constraints applied to the variation in the multipole errors, the maximum attainable field in the case of the combined-collar structure

Table 20.8 Comparison of main parameters of the MBP1 and MBP2 magnet cross sections and the separated-collar geometry MBSCA (relative field components given in units of 10^{-4} at 17 mm reference radius).

	MBP1 (insert)		MBP2		MBSCA (sep. collars)	
I (A) @ 8.33 T	11 729		11 828		11 459	
B_{ss} (T)	9.69		9.68		9.75	
Field comp.	inj	nom	inj	nom	inj	nom
b_2	0.538	2.406	0.646	0.816	0.058	0.034
b_3	1.920	3.452	5.775	6.176	11.008	11.440
b_4	0.532	0.726	0.216	0.062	0.000	0.183
b_5	0.749	0.810	0.913	0.914	1.146	1.024
b_6	0.023	0.006	0.001	0.004	0.000	0.013
b_7	0.611	0.623	0.632	0.635	0.610	0.616
b_8	0.001	0.000	0.000	0.000	0.000	0.000
b_9	0.100	0.101	0.103	0.103	0.098	0.100

is 9.73 T. An increase in the main field can be achieved by bringing the iron yoke closer to the coil, which helps to obtain not only a higher main field but also a better transfer function B/I. Increased main field is attained by smaller collars, however, at the expense of a larger variation in the quadrupole and sextupole field components, and an octupole component at injection exceeding 0.1 units.

Table 20.9 Payoff table for the LHC main dipole magnet geometry.

Objective	B_{ss}	Δb_2	Δb_3	Δb_4	b_4^{inj}	ryoke	Design
Constraint	> 8.2	< 2.0	< 3.0	< 0.2	< 0.1	< 275.	
max. B_{ss}	**-9.73**	0.758	1.464	0.086	0.103	284.8	$\mathbf{x}_1$
min Δb_2	-9.66	**0.036**	0.551	0.077	-0.116	284.2	$\mathbf{x}_2$
min Δb_3	-9.65	3.847	**0.002**	0.267	0.085	266.1	$\mathbf{x}_3$
min Δb_4	-9.66	0.822	0.721	**0.057**	-0.047	284.9	$\mathbf{x}_4$
min b_4^{inj}	-9.65	1.265	0.306	0.114	**0.000**	277.3	$\mathbf{x}_5$
min ryoke	-9.68	1.099	0.483	0.238	0.095	**265.5**	$\mathbf{x}_6$

Design	alel1	alel2	lh2r	leang	houdh	blel1	I_{ss}
	mm	mm	mm	deg	mm	mm	A
$\mathbf{x}_1$	75.84	76.08	7.22	22.14	4.04	95.0	13 647
$\mathbf{x}_2$	92.94	84.44	9.15	33.92	9.21	102.2	13 892
$\mathbf{x}_3$	86.73	88.23	17.24	41.69	4.65	102.6	13 936
$\mathbf{x}_4$	86.39	95.94	7.03	34.26	11.99	105.0	13 961
$\mathbf{x}_5$	88.52	89.02	13.43	39.75	6.91	103.5	13 921
$\mathbf{x}_6$	82.96	79.31	16.14	29.88	10.54	97.4	13 836

Table 20.10 Payoff table for the separated-collar geometry MBSCA.

Objective	B_{ss}	Δb_2	Δb_3	Δb_4	b_4^{inj}	ryoke	Design
Constraint	> 8.2	< 2.0	< 3.0	< 0.2	< 0.1	$< 275.$	
max. B_{ss}	**-9.92**	1.38	2.51	0.16	-0.0003	282.1	x_1
min Δb_2	-9.78	**0.76**	2.43	0.20	-0.0000	275.0	x_2
min Δb_3	-9.74	2.21	**0.05**	0.12	0.0002	280.1	x_3
min Δb_4	-9.77	3.91	1.62	**0.03**	-0.0001	284.8	x_4
min b_4^{inj}	-9.77	1.41	0.27	0.22	**0.0000**	272.3	x_5
min ryoke	-9.75	1.22	2.21	0.22	0.0000	**270.0**	x_6

Design	alel1	blel1	sstax	sendx	I_{ss}
	mm	mm	mm	mm	A
x_1	70.27	73.06	2.63	7.31	13 074
x_2	88.86	84.97	4.40	12.40	13 467
x_3	92.47	94.35	4.22	12.94	13 630
x_4	84.59	99.25	4.55	11.10	13 625
x_5	87.83	89.78	4.36	10.10	13 546
x_6	85.80	97.85	6.78	11.94	13 656

The separate-collar geometry (MBSCA) provides ideal shielding between the apertures at injection field level, and therefore the octupole field component is extremely low. For this design the maximum attainable field is 9.9 T, which compares to 9.73 T reached with the MBP2 geometry, limited by the active constraints on the b_4 component at injection field level.

The above results can be summarized as follows: The hidden resource for the main field is higher for the separated-collar geometry than for the combined-collar magnet.

20.10.6 Lagrange Multiplier Estimation

Similar conclusions to those of the previous section can be drawn from the Lagrange multiplier estimation in the optimal point of the problem as defined by the constraint formulation max $\{B_{SS}(\mathbf{x})\}$ subject to $\Delta b_2 < 2.0$, $\Delta b_3 < 3.0$, $\Delta b_4 < 0.2$, $b_{4_{inj}} < 0.1$. The design variables are shown in Figure 20.27. For the MBSCA geometry (first row in Table 20.10) the Lagrange multiplier for the quadrupole variation is 3.4 and for the sextupole variation 0.57. This indicates that the variation in the quadrupole term is the limiting factor for a further increase in the main field. The constraints on the octupole are not active; increasing the main field component by reducing the size of the nonmagnetic collars reduces the octupole component at the same time.

For the MBP2 geometry (first row in Table 20.9) the multipliers are 0.43 for the quadrupole variation, 0.69 for the sextupole variation, and 0.30 for the octupole component at injection. The octupole component, already relatively high, is thus the limiting factor for a further increase in the main field. The payoff between main field and quadrupole variation is lower for the MBP2 than for the MBSCA design, since less magnetic material between the beam channels is exposed to the field changes during the ramping of the magnets.

20.10.7 Manufacturing Tolerances

Random multipole errors are calculated with a sampling technique using 500 uniformly distributed random errors on the block positioning and inclination angles, as well as radial shifts of the coil blocks. The range of these coil-block displacements is ± 0.05 mm.

Analysis of the multipole content of the 500 samples yields a normal distribution function. The mean value and the standard deviation σ can be calculated by

$$\mu = \frac{1}{500} \sum_{i=1}^{500} x_i, \qquad \sigma^2 = \frac{1}{500} \sum_{i=1}^{500} (x_i - \mu)^2, \tag{20.119}$$

where μ is the mean value and σ^2 the variance. The x_i term presents the calculated relative field components b_n and a_n for the ith sample. Figure 20.28 shows the density function

$$f(x) = \frac{1}{\sigma\sqrt{2\pi}} \exp\left[\frac{-(x-\mu)^2}{2\sigma^2}\right] \tag{20.120}$$

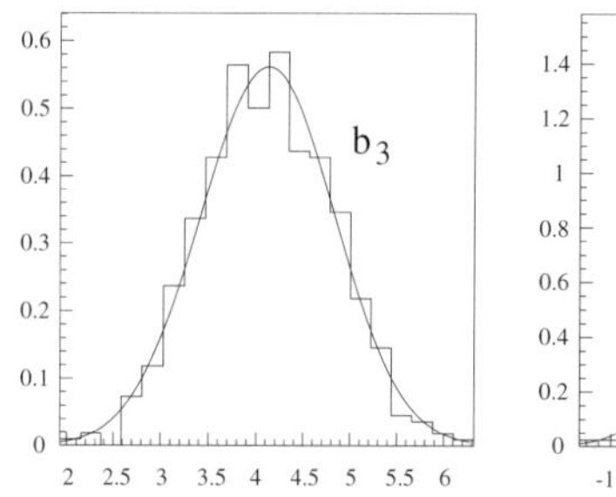

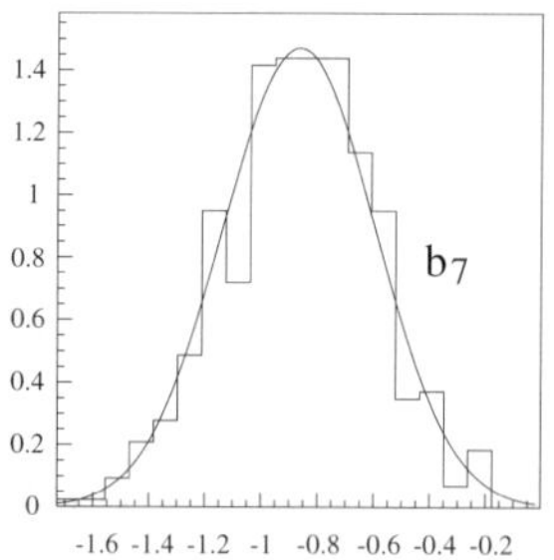

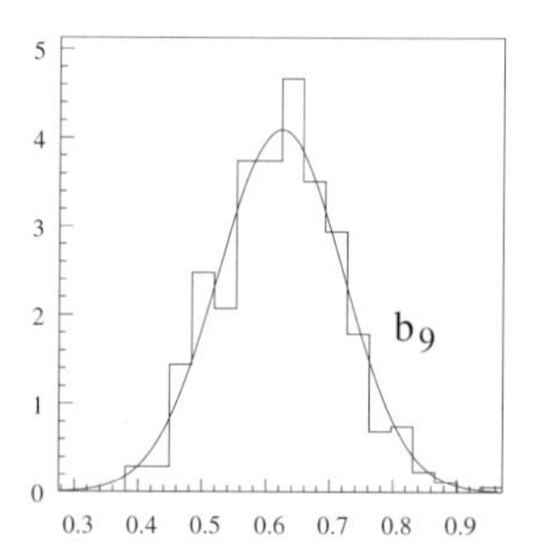

Figure 20.28 Histograms of 500 values, normalized for a total surface of 1, and density functions of the (from left to right) b_3, b_5, and b_7 field component for the V6-1 coil (in units of 10^{-4} at 17 mm radius) for 500 identically distributed random displacements between $\pm$ 0.05 mm from the nominal coil-block positions.

of the b_3, b_5, and b_7 field components for the V6-1 coil. Table 20.11 gives the numerical results. Here we use *Taguchi's loss function* [36], defined by

$$L_{a_i,b_i} := k\left[\sigma^2_{a_i,b_i} + (\mu_{a_i,b_i} - T_i)^2\right], \qquad (20.121)$$

where T_i are the nominal values of the objectives. For our case we set the weighting factor k to 1. If two designs have the same variance, the design with the mean value closer to the design value has a smaller loss function. If two designs have the same mean value, but different variance, then the design with the lower variance has the smaller loss function.

Table 20.11 Mean and standard deviations for 500 identically distributed random displacements between $\pm$ 0.05 mm relative to the nominal coil-block positions ($\sum L$ is the sum of all Taguchi loss functions for the different multipoles of order 1–11).

	V6-3		V6-1		5-block	
μ_{b_2}, σ_{b_2}	0.002	1.009	0.008	0.985	−0.020	1.383
μ_{b_3}, σ_{b_3}	4.331	0.753	4.127	0.710	0.148	0.895
μ_{b_4}, σ_{b_4}	0.012	0.449	0.014	0.436	0.041	0.571
μ_{b_5}, σ_{b_5}	−1.532	0.292	−0.866	0.271	−1.686	0.350
μ_{b_6}, σ_{b_6}	0.011	0.188	0.006	0.165	0.009	0.230
μ_{b_7}, σ_{b_7}	0.329	0.108	0.623	0.097	0.803	0.122
μ_{b_8}, σ_{b_8}	0.001	0.046	0.000	0.042	−0.002	0.066
μ_{b_9}, σ_{b_9}	−0.119	0.028	0.101	0.024	−0.718	0.048
μ_{a_1}, σ_{a_1}	−0.026	1.105	−0.018	1.074	−0.159	1.814
μ_{a_2}, σ_{a_2}	−0.075	1.089	−0.065	1.027	−0.021	1.321
μ_{a_3}, σ_{a_3}	−0.029	0.743	−0.021	0.686	0.023	0.855
μ_{a_4}, σ_{a_4}	−0.014	0.522	−0.017	0.467	−0.035	0.582
μ_{a_5}, σ_{a_5}	0.003	0.275	0.004	0.253	0.028	0.333
μ_{a_6}, σ_{a_6}	0.001	0.151	0.003	0.155	0.001	0.201
μ_{a_7}, σ_{a_7}	0.004	0.087	0.006	0.092	−0.004	0.115
μ_{a_8}, σ_{a_8}	−0.002	0.065	−0.001	0.063	−0.003	0.066
μ_{a_9}, σ_{a_9}	0.001	0.036	0.001	0.036	−0.001	0.036
$\sum_i(L_{a_i} + L_{b_i})$	5.2756		4.7871		9.5749	

Notice that the mean relative values are not the expected intrinsic values. This is due to the fact that the field changes are a nonlinear function of the block perturbation. A shift of a coil block toward the horizontal median plane results in a higher field distortion than a displacement by the same amount away from the horizontal median plane.

The Jacobi matrix for the tolerances in coil-block positioning, coil size, and coil asymmetries resulting from the collaring procedure is too large to be shown here; however, important conclusions can be drawn:

- A 0.1 mm increase in the coil azimuthal size produces a sextupole of 3.5 units and a decapole of 0.25 units.

- A 0.1 mm inner radius difference between lower and upper poles of the same aperture produces a skew quadrupole of 0.75 units.

These two examples show that the sensitivity to coil deformations is an order of magnitude higher than the sensitivity to yoke tolerances, all of which emphasizes the importance of a stringent control during the coil-manufacturing process.

20.10.8
Tuning Range

Table 20.12 shows the convergence of the optimization procedure for the tuning range by solving for the two extreme cases:

- Full compensation of the persistent current effects at injection.
- No compensation of any effects resulting from persistent currents.

Table 20.12 Tuning range toward a full compensation of the persistent-current-induced multipoles at injection (left columns); toward a zero compensation (right columns).

	V6-3	V6-1	5-block	V6-3	V6-1	5-block
$f(\mathbf{x}^*)$	8.82	5.98	13.97	8.57	4.29	9.64
b_3^* (required)	6.	6.	6.	0.	0.	0.
b_5^* (required)	−1.61	−1.19	−1.60	0.	0.	0.
b_7^* (required)	0.60	0.48	0.82	0.	0.	0.
b_9^* (required)	−0.20	−0.23	−0.46	0.	0.	0.
b_3 (achieved)	6.000	6.001	5.547	−0.007	0.000	0.000
b_5 (achieved)	−1.610	−1.189	−1.600	−0.002	0.006	0.026
b_7 (achieved)	−0.271	0.925	0.343	−0.206	0.150	0.002
b_9 (achieved)	−0.206	−0.154	−0.875	−0.324	0.138	−0.475
Active constraints	3	3	4	4	3	4
Rank of J	4	3	3	4	3	3

The resulting objective function $f(\mathbf{x})$ to be minimized reads in both the cases:

$$\min\{f(\mathbf{x})\} = \min\left\{2\,\Delta b_3(\mathbf{x})^2 + 5\,\Delta b_5(\mathbf{x})^2 + 10\,\Delta b_7(\mathbf{x})^2 + 20\,\Delta b_9(\mathbf{x})^2\right\}, \tag{20.122}$$

where $\Delta b_n = b_n - b_n^*$; b_n denoting the calculated and b_n^* the desired geometrical multipole errors for the partial compensation of the persistent-current effects. Some constraints were introduced for the movement of the blocks for mechanical and geometrical reasons. During the optimization process some of these limits are reached and the constraints are active. The number of active constraints for a particular design is given in the table. The rank of the Jacobi matrix gives the number of independent basis vectors building the design space and thus the number of degrees of freedom to tune the magnet performance. The rank is calculated with a singular-value decomposition of the Jacobi matrix. One can see that the V6-1 version has a wider tuning range in both directions (note the achieved values for $f(\mathbf{x})$) than the other two alternatives, whereas the number of linear independent design variables is not higher than for the 5-block version.

20.10.9 Tracing of Manufacturing Errors

Figure 20.29 shows a series of quenches for a long dipole model built in industry. This is a typical training characteristic with successively higher fields at subsequent quenches. The enormous electromagnetic forces push the cables into a more stable position within the mechanical structure. These movements can trigger a training quench. In the prototype magnet presented here, eight quenches occurred well below the nominal LHC operation field of 8.33 T. The subsequent quenches were triggered above the nominal field.

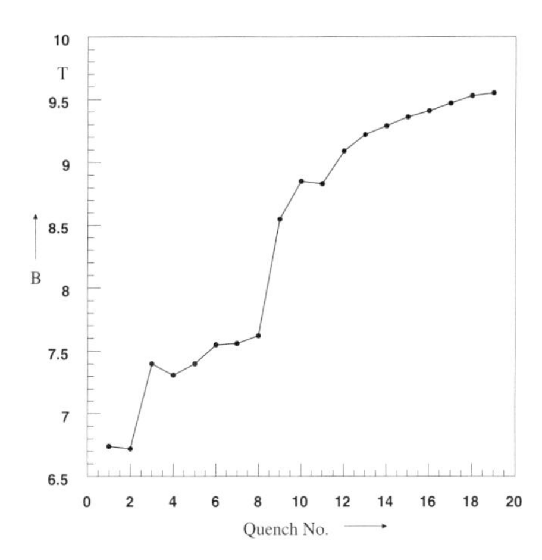

Figure 20.29 Training quenches of a dipole model-magnet built in industry. Notice the improved performance of the magnet after the eighth training quench.

The dimensions of the active parts of the coils after their deformation due to winding, collaring, magnet assembly, cool-down, and excitation are impossible to verify under operational conditions. Measurements of the field quality before and after the cryogenic tests showed that the coil was displaced after the assembly of the magnet and that the quenches pushed the cables toward their nominal positions. This can be seen from Table 20.13 where the field components measured before and after the training quenches are listed together with the calculated nominal errors. It is clear that the field quality improved during the tests. The inverse problem solving consists of optimization routines for finding the distorted coil geometries that exactly reproduce the measured multipole content. This approach allows for the calculation of the displacements of the coil blocks during excitation as the most likely causes of the low-field training-quenches.

Table 20.13 Magnetic field measurements before and after the eighth quench, which resulted in a block movement and a considerably improved training characteristics. The right columns show the intrinsic field errors as computed for nominal dimensions and cable location. All in units of 10^{-4} at a reference radius of 10 mm.

	Before		After		Intrinsic	
n	b	a	b	a	b	a
2	0.378	0.634	0.463	−0.229	0.248	0.000
3	−2.072	0.094	−1.246	0.117	−0.901	0.000
4	−0.055	0.151	−0.028	0.118	0.112	0.000
5	0.247	0.035	0.170	0.013	0.018	0.000
6	0.018	0.006	0.010	0.011	−0.002	0.000
7	0.032	−0.006	0.034	−0.003	0.005	0.000
8	−0.001	0.000	0.000	0.000	0.000	0.000
9	−0.001	0.000	−0.001	−0.001	0.000	0.000

The inverse field computation problem is posed as follows:

$$\min \left\{ \sum_{i=1}^{9} p_i \cdot (b_i^*(\mathbf{x}) - b_i)^2 + q_i \cdot (a_i^*(\mathbf{x}) - a_i)^2 \right\}, \tag{20.123}$$

where $b_i^*(\mathbf{x})$, $a_i^*(\mathbf{x})$ are the measured, b_i, a_i are the calculated multipoles, and p_i, q_i are the weighting factors compensating for the different numerical values of the residuals.

A large number of design variables results for the inverse field problem due to the asymmetric nature of the coil-positioning errors. It is therefore assumed that the positioning errors hold for an entire coil block rather than for individual cables. The design variables are the possible perturbations in radial direction of all 20 coil blocks together with azimuthal displacements in 16 of the 20 blocks. It is assumed that the blocks on both sides of the horizontal median plane are free to move only by the same amount in the azimuthal direction.

Because there are more degrees of freedom than objectives, the problem is ill-posed. A regularization term

$$P_{\text{reg}} := \sum_{i=1}^{36} r_i \cdot x_i^2 \tag{20.124}$$

is added to Eq. (20.123) to assure that the coil block displacements remain as small as possible.

The Levenberg–Marquard algorithm is now applied to the optimization problem [42]. Originally developed for nonlinear regression problems using least-squares objective functions, it can be applied efficiently to the minimization of the distance function. The number of function evaluations is in the range of 800–1000. The relative movement of the coil block during the test can be estimated from the calculated coil-block displacements before and after the test. The displacements are shown in Figure 20.30.

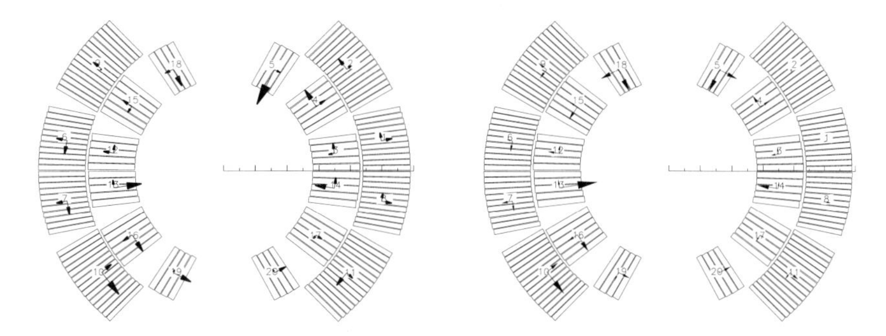

Figure 20.30 Displacement of the coil blocks before (left) and after (right) the cold test with training quenches. The radial displacement of block 5 (left) is 0.183 mm before the eighth quench; all other displacements are to scale. The largest movement occurred in block 5, which was shifted outward by 0.07 mm during the quench.

The calculated displacements show an outward movement of block 5 by 0.07 mm during the quench (toward its nominal position). This proves that the mechanical stability of this block is not guaranteed due to its inclination with respect to the radial direction. The 6-block coil configuration was thus chosen for the series production of the LHC main dipole magnets.

During the production of the coldmasses (1232 for the machine plus spare magnets) the assembly faults were intercepted by magnetic field measurements performed on the collared coils, followed by the inverse field analysis. This turned out to be a very effective tool for controlling the quality of production. Detected manufacturing faults included the mounting of a double coil-protection sheet, a missing outer coil shim, and the mounting of copper wedges in the incorrect order. In addition, twelve cases of an inward movement of block 6 were detected [47,53] .

References

1 Ahlback, J., et. al: Electromagnetic and mechanical design of a 56 mm aperture model dipole for the LHC, IEEE Transactions on Magnetics, Vol. 30, LHC-Note 247, CERN, Geneva, 1994

2 Andreev, N., Artoos, K., Bottura, L., Rodriguez-Mateos, F., Russenschuck, S., Sigel, N., Siemko, A., Sonnemann, F., Tommasini, D., Vanenkov, I.: Proceedings of the 1999 Particle Accelerator Conference, 1999

3 Armstrong, A.G.A.M. , Fan, M.W., Simkin, J., Trowbridge, C.W.: Automated optimization of magnet design using the boundary integral method, IEEE Transactions on Magnetics, 1982

4 Bellesia, B. et al.: Short circuit localization in the LHC main dipole coils by means of room temperature measurements, IEEE Transactions on Applied Superconductivity, 2006

5 Brent, R.P.: Algorithms for Minimization Without Derivatives, Prentice-Hall, Englewood Cliffs, NJ, 1973

6 Charnes, A., Cooper, W.: Management Models and Industrial Applications of Linear Programming, Wiley, New York, 1961

7 Cohon, J. L.: Multiobjective Programming and Planning, Academic Press, New York, 1978

8 Dyck, D.N., Lowther, D.A.: The optimization of electromagnetic devices using sensitivity information, Proceedings of the 6th International IGTE Symposium, Graz Technical University, 1994

9 Fandel, G.: Optimale Entscheidung bei mehrfacher Zielsetzung, Springer, Berlin, 1972

10 Fiacco, A.V., McCormick, G.P.: Sequential Unconstrained Minimization Techniques, Wiley, New York, 1968

11 Fogel, D.B.: An introduction to simulated evolutionary optimization, IEEE Transactions on Neural Networks, 1994

12 Goldberg, D.E.: Genetic Algorithms in Search, Optimization and Machine Learning, Addison-Wesley, Reading, MA, 1989

13 Halbach, K.: A program for inversion of system analysis and its application to the design of magnets, Proceedings of the International Conference on Magnet Technology (MT2), The Rutherford Laboratory, 1967

14 Hestenes, M.R., Stiefel, E.: Methods of conjugate gradients for solving linear systems, Journal of Research of the National Bureau of Standards, 1952

15 Himmelblau, D.M.: Applied Nonlinear Programming, McGraw-Hill, New York, 1972

16 Holland, J.H.: Genetic Algorithms, Scientific American, New York, NY, 1992

17 Holland, J.H.: Adaptation in natural and artificial systems, University of Michigan Press, Ann Arbor, 1975

18 Jacob, H.G.: Rechnergestützte Optimierung Statischer und Dynamischer Systeme, Springer, Berlin, 1982

19 Kuhn, H.W., Tucker, A.W.: Nonlinear programming, Proceedings of the 2nd Berkeley Symposium on Mathematical Statistics and Probability, University of California, Berkeley, 1951

20 LHC study group, The Large Hadron Collider, Conceptual Design, CERN/AC/95-05

21 Karush, W.: Minima of Functions of Several Variables with Inequalities as Side Constraints, M.Sc. Dissertation, University of Chicago, 1939

22 Koski, J.: Multicriterion optimization in structural design, in Artrek et al. (eds.), New Directions in Optimum Structural Design, Wiley, New York, 1984

23 LHC design report, Vol. 1, the LHC main ring, CERN-2004-003, 2004

24 The LHC study group, Large Hadron Collider, The Accelerator Project, CERN/AC/93-03

25 The LHC study group, The Large Hadron Collider, Conceptual Design, CERN/AC/95-05

26 Luenberger, D.G.: Introduction to Linear and Nonlinear Programming, Addison-Wesley, Reading, MA, 1973

27 Mahfoud, S.W.: Niching Methods for Genetic Algorithms, PhD Thesis, University of Illinois, Urbana-Champaign, 1995

28 Mahfoud, S.W.: A comparison of parallel and sequential niching methods, Proceedings of the Sixth International Conference on Genetic Algorithms, Morgan Kaufmann, CA, 1995

29 Marglin, S.A.: Objectives of water-resource development in Maass, A. et al.: Design of Water-Resource Systems, Cambridge, 1966

30 Michalewicz, Z.: Genetic algorithms and data structures = evolution programs, Springer, Berlin, Heidelberg, revised edition 1996

31 Morgan, C: Use of an elliptical aperture to control saturation in closely-coupled, cold iron, superconducting dipole magnets, IEEE Transactions on Nuclear Science, 1985

32 Moritz, G. et al.: Towards fast pulsed superconducting synchrotron magnets, Proceedings of the 2001 Particle Accelerator Conference, Chicago, 2001

33 Moritz, G.: Fast-pulsed SC magnets, Proceedings of the 2004 European Particle Accelerator Conference, Vienna, 2004

34 Pareto, V.: Cours d'Economie Politique, Pouge 1896 or translation by Schwier, A.S.: Manual of Political Economy, Macmillan, London, 1971

35 Park, I.-h., Lee, B.-t., Hahn, S.-Y.: Design sensitivity analysis for nonlinear magnetostatic problems using finite element method, IEEE Transactions on Magnetics, 1992

36 Pignatiello, J.J.: An overview of the strategy and tactics of Taguchi, IIE Transactions, 1988

37 Ramberger, S., Russenschuck, S.: Genetic algorithms with niching for conceptual design studies, IEEE Transactions on Magnetics, 1998

38 Ramberger, S., Russenschuck, S.: Genetic algorithms for the optimal design of superconducting accelerator magnets, EPAC 98, Stockholm, Sweden, 1998

39 Ramberger, S., Russenschuck, S.: Genetic algorithms with niching for the solving of material distribution problems, Conference on the Computation of Electromagnetic Fields, COMPUMAG, Sapporo, 1999

40 Rechenberg, I.: Evolutionsstrategie: Optimierung technischer Systeme nach Prinzipien der biologischen Evolution, Frommann Verlag, 1973

41 Russenschuck, S., Riese, I. , Gesche, R.: Berechnung und Optimierung von Hohlleiterfiltern, Frequenz, 1990

42 Russenschuck, S., Tortschanoff, T., Ijspeert, A., Siegel, N., Perin, R.: Tracing back measured magnetic field imperfections in LHC magnets by means of the inverse problem approach, IEEE Transactions on Magnetics, 1994

43 Rockafellar, R.T.: The multiplier method of Hestenes and Powell applied to convex programming, Journal of Optimization Theory and Applications, 1973

44 Rosenbrock, H.H.: An automatic method for finding the greatest or least value of a function, Computer Journal, 1960

45 Russenschuck, S.: Application of Lagrange-multiplier estimation to the design optimization of permanent magnet synchronous machines, IEEE Transactions on Magnetics, 1992

46 Sareni, B., Krähenbühl, L., Nicolas, A.: Niching genetic algorithms for optimization in electromagnetics, IEEE Transactions on Magnetics, 1998

47 Savary, F. et al.: Description of the main features of the series production of the LHC main dipole magnets, MT20, IEEE Transactions on Applied Superconductivity, 2008

48 Schwefel, H.P.: Numerische Optimierung von Computer-Modellen mittels der Evolutionsstrategie, Birkhäuser Verlag, 1977

49 Schwefel, H.P.: Numerical Optimization for Computer Models, John Wiley & Sons, Chichester, UK, 1981

50 Stadler, W.: Preference optimality and application of Pareto-optimality, in Marzollo, Leitmann: Multicriterion Decision Making, CISM Courses and Lectures, Springer, Berlin, 1975

51 Subramaniam, S., Arkadan, A.A., Hoole S.R.: Optimization of a magnetic pole face using linear constraints to avoid jagged contours, IEEE Transactions on Magnetics, 1994

52 Uler, G.F., Mohammed, O.A., Koh, C.S.: Design optimization of electrical machines using genetic algorithms, IEEE Transactions on Magnetics, 1995

53 Völlinger, C., Todesco, E.: Identification of assembly faults through the detection of magnetic field anomalies in the production of the LHC dipoles, IEEE Transactions on Applied Superconductivity, 2006

54 Wilde, D.G.: Optimum seeking methods, Prentice-Hall, Englewoood Cliffs, NJ, 1964

55 Wolpert, D. H., Macready, W. G.: No free lunch theorems for optimization, IEEE Transactions on Evolutionary Computation, 1997

56 Zoutdendijk, G.: Methods of feasible directions: a study in linear and nonlinear programming, Elsevier, Amsterdam, 1960

57 Zitzler, E., Thiele, L.: An evolutionary algorithm for multiobjective optimization: The strength Pareto approach, TIK-Report, Swiss Federal Institute of Technology, Zürich, 1998

Appendix A
Material Property Data for Quench Simulations

On two occasions I have been asked,-
"Pray, Mr. Babbage, if you put into the machine
wrong figures, will the right answer come out?"
I am not able rightly to apprehend the kind of confusion
of ideas that could provoke such a question.

C. Babbage (1791–1871).

The numerical simulation of quenching superconducting magnets relies on the electrical and thermal properties of materials in a wide range of temperatures and applied magnetic flux densities. The peak temperature in the coil is principally determined by the *thermal conductivity* λ and *volumetric heat capacity* ρc_p of the conductors and the coolant. The electrical resistivity ρ of all magnet components is required for the calculation of ohmic heating and induced eddy currents.

Remark: The empirical laws presented in this chapter shall be parsed as numerical-value expressions of the physical quantities expressed in SI units. We will write, for example, $B \equiv \{B\}_\mathrm{T}$ as shorthand for the numerical value of the flux density expressed in tesla. In treating thermal, mechanical, and electrical properties, we must live with the ambiguity of the symbol ρ for both the mass density and electrical resistivity. □

A.1
Mass Density

Some material properties, such as the heat capacity, are often given as a quantity per unit mass. In simulation programs, it is easier to deal with volumetric properties that require the *mass density* $\rho := m/V$, where m is the mass and V

Field Computation for Accelerator Magnets. Stephan Russenschuck

ISBN: 978-3-527-40769-9

the volumetric extend. The physical unit of the mass density is $[\rho] = 1\,\mathrm{kg\,m^{-3}}$. Table A.1 contains data for the most commonly used materials in SC magnets.

Table A.1 Mass density of solids, $[\rho] = 1\,\mathrm{kg\,m^{-3}}$.

Material	Symbol	$\rho\,(T = 300\,\mathrm{K})$	Material	Symbol (Grade)	$\rho\,(T = 300\,\mathrm{K})$
Aluminum	Al	2700	Niobium–titanium	Nb–Ti	6000
Copper	Cu	8960	Niobium–tin	Nb_3Sn	8400
Iron	Fe	7830	Austenitic steel	316LN	7890
Niobium	Nb	8580	Polyimide		1420
Tin	Sn	7280	Epoxy		1100–1300
Tantalum	Ta	16 690	Concrete		2200–2400

The mass density of solids at cryogenic temperatures can be calculated from the linear contraction factor $\Delta L/L$; see Table A.2. Thus the change of mass density is given by

$$\frac{\Delta\rho}{\rho(T = 300\,\mathrm{K})} = 1 - \left(1 + \frac{\Delta L}{L}\right)^3 . \tag{A.1}$$

The density of the confined helium in the Rutherford cable is a function of the temperature and pressure, which rise simultaneously during a quench.

Table A.2 Thermal contraction data $\Delta L/L$; source [21].

Element/Material	Symbol	20 K	80 K	200 K
Copper	Cu	.00323	.00302	.00149
Brass	65Cu-35Sn	.00380	.00350	.00169
Aluminum	Al	.00412	.00390	.00201
Lead	Pb	.00700	.00577	.00263
Invar	Fe-Ni36	.00046	.00048	.00020
Araldite		.01050	.00880	.00500
Titanium	Ti	.00151	.00142	.00073
Austenitic steel	316LN	.00297	.00278	.00138

For quench simulations, we apply a simplified model: After passing the lambda point, the pressure in the coldmass is assumed to rise linearly with temperature from the operating to the maximum pressure determined by release valves; see Figure A.1 (left). Above $T = 6\,\mathrm{K}$ the pressure stays constant during the quench process. Combining the helium property data from [21] with the pressure values results in the helium mass density versus temperature graph as shown in Figure A.1 (right).

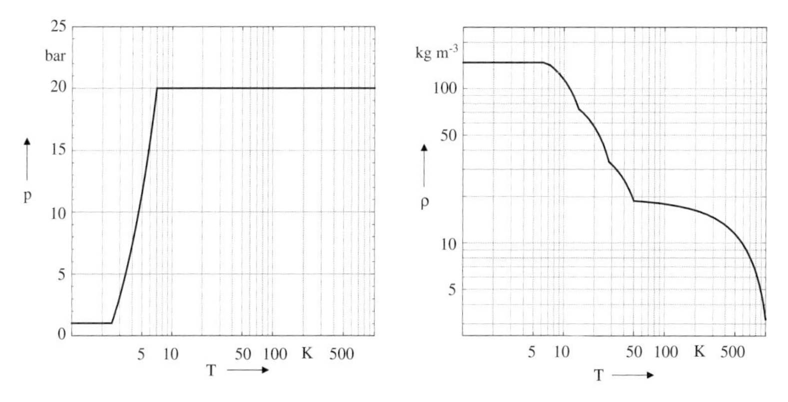

Figure A.1 Left: Assumed pressure rise in a quenching superconducting LHC dipole as a function of temperature. Right: Piecewise linear approximation for the mass density of helium for the assumed pressure rise.

A.2 Electrical Resistivity

The electrical resistivity ρ according to the generalized Ohm's law is defined as $\rho = E/J$, with the physical unit $[\rho] = 1\,\mathrm{V\,m\,A^{-1}} = 1\,\Omega\,\mathrm{m}$. The electrical resistivity of selected materials is given in Table A.3. For metals, the electrical resistivity depends on the length of the *mean free path* of the conduction electrons. This path length is limited by lattice imperfections and vibrations [21]. The rate of collision is further increased by an applied magnetic field, an effect known as *magnetoresistivity* [5].

Table A.3 Electrical resistivity ρ of materials at 273 K, $[\rho] = 1\,\Omega\,\mathrm{m}$

Element/Material	Symbol	ρ (273 K)	RRR
Copper	Cu	1.55×10^{-8}	up to 30 000
Aluminum	Al	2.50×10^{-8}	25–1000
Niobium	Nb	16.10×10^{-8}	30
Lead	Pb	19.20×10^{-8}	1000
Tin	Sn	11.15×10^{-8}	100
Tantalum	Ta	12.40×10^{-8}	10
Titanium	Ti	39.40×10^{-8}	
Iron	Fe	8.66×10^{-8}	100 to 200
Austenitic steel	304	1.02×10^{-6}	2
Austenitic steel	304L	0.70×10^{-6}	1.42

The resistivity of a metal at zero magnetic field consists of two contributions, according to the *Matthiessen rule*,[1]

$$\rho(T) = \rho_0 + \rho_\mathrm{i}(T)\,, \tag{A.2}$$

where the *residual resistivity* is ρ_0 and the *intrinsic resistivity* is ρ_i. This rule can be seen as the motivation for the Kamerlingh Onnes experiment that led to the discovery of superconductivity in 1911. At ambient temperature, the residual resistivity is a negligible fraction of the total resistivity and thus nearly equal to the intrinsic resistivity, which is determined by thermal vibrations of the copper lattice.

The ratio between the resistivity at ambient temperature and the residual resistivity is called the *residual resistivity ratio*, abbreviated as *RRR*, which is a measure of the material's purity. According to [16] the *RRR* value is defined by

$$RRR := \frac{\rho(T{=}273\,\mathrm{K})}{\rho(T{=}4\,\mathrm{K})}\,. \tag{A.3}$$

For high purity, oxygen-free copper $\rho(T = 273\,\mathrm{K}) \approx 1.5 \times 10^{-8}\ \Omega\,\mathrm{m}$, and the *RRR* reaches values of around 2000. For superconductors the lower temperature for the *RRR* measurement is chosen just above the transition temperature [20]. If the *RRR* is derived from the bulk resistivity of a sample, the thermal contraction coefficient must be taken into account.

Copper The temperature-dependence of the electrical resistivity of copper at zero magnetic flux density can be represented within $\pm$ 15% by [16]:

$$\rho(T, RRR) = \rho_0 + \rho_\mathrm{i} + \rho_{\mathrm{i}0}\,, \tag{A.4}$$

where

$$\rho_0 = \frac{1.553 \times 10^{-8}}{RRR}\,, \tag{A.5}$$

$$\rho_\mathrm{i} = \frac{P_1 T^{P_2}}{1 + P_1 P_3 T^{(P_2 - P_4)} \exp\left[-\left(\frac{P_5}{T}\right)^{P_6}\right]}\,, \tag{A.6}$$

$$\rho_{\mathrm{i}0} = P_7 \frac{\rho_\mathrm{i}\rho_0}{\rho_\mathrm{i} + \rho_0}\,, \tag{A.7}$$

and where $\rho \equiv \{\rho\}_{\mathrm{V\,m\,A^{-1}}}$ and $T \equiv \{T\}_\mathrm{K}$. The effect of impurities on the electrical resistivity can therefore be predicted from two resistance measurements at 4 and 273 K. The term $\rho_{\mathrm{i}0}$ represents a deviation from the Matthiesson rule.

1 Augustus Matthiessen (1831–1870).

The fit constants are given by $P_1 = 1.171 \times 10^{-17}$, $P_2 = 4.49$, $P_3 = 3.841 \times 10^{10}$, $P_4 = 1.14$, $P_5 = 50$, $P_6 = 6.428$, and $P_7 = 0.4531$. Figure 18.1 graphically represents the resistivity of copper as a function of temperature and RRR.

The dependence of the electrical resistivity on a transverse magnetic field can be visualized by plotting the magnetoresistivity

$$\frac{\Delta\rho}{\rho} := \frac{\rho(B, T, RRR) - \rho(B{=}0\,\mathrm{T}, T, RRR)}{\rho(B{=}0\,\mathrm{T}, T, RRR)} \tag{A.8}$$

versus $B \cdot S(T, RRR)$ in a log–log plot, known as the Kohler diagram. $S(T, RRR)$ is the ratio between the ice-point resistance and the resistance at the test temperature

$$S(T, RRR) := \frac{\rho(B = 0\,\mathrm{T}, T = 273\,\mathrm{K}, RRR)}{\rho(B = 0\,\mathrm{T}, T, RRR)}. \tag{A.9}$$

For most metals the graph in the log–log diagram is a straight line [18] and can therefore be expressed as a polynomial in $\log(B\ S(T, RRR))$:

$$\log\left(\frac{\Delta\rho}{\rho}\right) = \sum_{n=0}^{N} a_n \left(\log(B\ S(T, RRR))\right)^n. \tag{A.10}$$

The resistance ratio $S(T, RRR)$ makes it possible to compare, in one diagram, the data of different specimens for varying levels of impurities and cold-work, tested at different temperatures. The electrical resistivity, which depends on temperature, magnetic field, and RRR, can be obtained by solving Eq. (A.10) for $\rho(B, T, RRR)$.

We can quantify the field-dependence of the copper resistivity by means of the following polynomial [16]:

$$p(B\ S(T, RRR)) = \sum_{n=0}^{4} a_n \left(\log(B\ S(T, RRR))\right)^n, \tag{A.11}$$

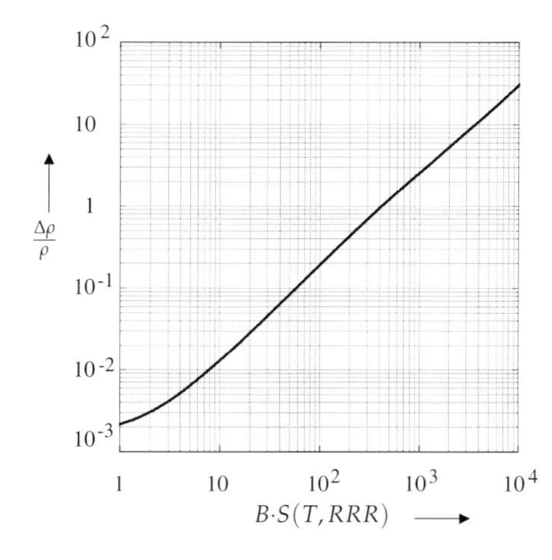

Figure A.2 Kohler diagram for the calculation of the magnetoresistivity of copper [16].

where $B \equiv \{B\}_T$ and $a_0 = -2.662$, $a_1 = 0.3168$, $a_2 = 0.6229$, $a_3 = -0.1839$, $a_4 = 0.01827$, and thus by

$$\rho(B, T, RRR) = \rho(B{=}0\,\mathrm{T}, T, RRR)\,(1 + 10^p) \tag{A.12}$$

Figure A.2 shows the polynomial (A.11) in the Kohler diagram.

Superconductors The electrical resistivity of Nb–Ti can be approximated in the normal-conducting state by a fit [3] based on data given in [5] for a 53.5Nb–46.5Ti alloy,

$$\rho\,(T) = (0.0558\,T + 55.668) \times 10^{-8}, \tag{A.13}$$

where $\rho \equiv \{\rho\}_{\mathrm{V\,m\,A^{-1}}}$ and $T \equiv \{T\}_{\mathrm{K}}$. For the electrical resistivity of Nb_3Sn we can interpolate the data collected in [17], which are based on measurements presented in [5]; see Table A.4.

Table A.4 Electrical resistivity of normal-conducting Nb_3Sn; source [5].

T	ρ	T	ρ	T	ρ	T	ρ
2	2.33×10^{-7}	22	2.46×10^{-7}	52	2.67×10^{-7}	202	3.81×10^{-7}
7	2.36×10^{-7}	32	2.52×10^{-7}	102	3.07×10^{-7}	252	4.06×10^{-7}
12	2.39×10^{-7}	42	2.59×10^{-7}	152	3.47×10^{-7}	297	4.10×10^{-7}

For current densities close to the critical current density, the electrical resistance in a superconductor rises according to the power law,

$$\rho(B, J, T) = \frac{E_\mathrm{c}}{J_\mathrm{c}(B, T)} \left(\frac{J}{J_\mathrm{c}(B, T)} \right)^{n-1}. \tag{A.14}$$

Figure A.3 shows the resistivity of Nb–Ti and Nb_3Sn as a function of the temperature and magnetic flux density. The current density varies between 0.5 and 1.1 times the critical current density, which is assumed to be $J_\mathrm{c}(8.4\,\mathrm{T}, 1.9\,\mathrm{K}) = 2.6 \times 10^9\,\mathrm{A\,m^{-2}}$ for Nb–Ti and $J_\mathrm{c}(15\,\mathrm{T}, 4.2\,\mathrm{K}) = 1.4 \times 10^9\,\mathrm{A\,m^{-2}}$ for Nb_3Sn. The graphs are given for an n-index of 40 for Nb–Ti and $n = 20$ for Nb_3Sn.

A.3 Thermal Conductivity

The thermal conductivity λ is defined as the ratio of the heat flux density ϕ_q/a to the temperature gradient:

$$\lambda := \frac{\phi_\mathrm{q}}{a\,|\,\mathrm{grad}\,T|}, \tag{A.15}$$

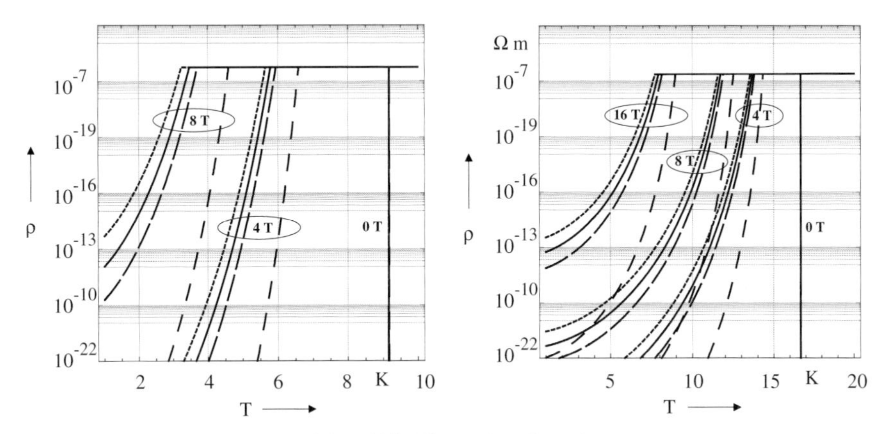

Figure A.3 Left: Electrical resistivity of Nb–Ti as a function of temperature, magnetic flux density, and current density. The current density varies between 0.5 and 1.1 times the critical current density. Right: Electrical resistivity of Nb_3Sn.

where ϕ_q is the heat flux and a is the cross-sectional area of the conductor. The physical unit of λ is $1\,W\,m\,K^{-1}$. The thermal conductivity λ can be expressed as the sum of electronic and lattice conductivities, $\lambda = \lambda_e + \lambda_l$.

According to the Wiedemann–Franz–Lorenz (WFL) law, the thermal conductivity of pure metals and dilute alloys can be calculated form the electrical resistivity by

$$\lambda = \frac{L_0 T}{\rho}, \tag{A.16}$$

where the Lorenz number L_0 is $2.44 \times 10^{-8}\,W\,\Omega\,K^{-2}$. Hence, the determination of the *RRR* value will lead directly to the thermal conductivity via the WFL law. The measurement of the electrical resistivity is easier, however, the Lorenz number is not a universal constant, but varies for intermediate temperatures and high *RRR* values; moreover, it is field-dependent [18].

For insulators with a very high electrical resistivity, the thermal conductivity is dominated by lattice vibrations (phonon transport). Since the transport of phonons is less efficient than the electron transport in metals, the thermal conductivity of insulators is several orders of magnitude smaller [21].

Copper The temperature-dependence of the thermal conductivity of copper at zero magnetic flux density can be represented within ± 10% by the empirical formula [16]:

$$\lambda(T, RRR) = (W_0 + W_i + W_{i0})^{-1}, \tag{A.17}$$

where

$$W_0 = \frac{\beta}{T}, \qquad W_{i0} = P_7 \frac{W_i W_0}{W_i + W_0}, \tag{A.18}$$

and

$$W_i = \frac{P_1 T^{P_2}}{1 + P_1 P_3 T^{P_2+P_4} \exp\left(-\left(\frac{P_5}{T}\right)^{P_6}\right)}, \tag{A.19}$$

again employing the convention $\lambda \equiv \{\lambda\}_{\text{W m K}^{-1}}$ and $T \equiv \{T\}_{\text{K}}$. The fit constants are $\beta = 0.634/RRR$, $P_1 = 1.754 \times 10^{-8}$, $P_2 = 2.763$, $P_3 = 1102$, $P_4 = -0.165$, $P_5 = 70$, $P_6 = 1.756$, and $P_7 = 0.838/\beta_r^{0.1661}$, where $\beta_r = \beta/0.0003$. The thermal conductivity of copper as a function of temperature is shown in Figure A.4 (left).

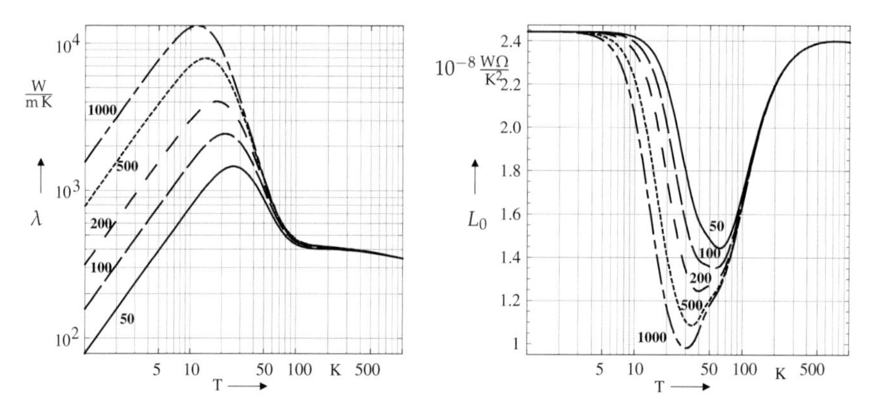

Figure A.4 Left: Thermal conductivity of copper for zero magnetic flux density. Right: Variation in the Lorenz number of copper as a function of *RRR*, calculated from the thermal conductivity and electrical resistivity functions published in [16].

In very pure metals, the thermal conductivity is dominated by the electronic conduction. Lorentz forces due to applied magnetic fields will thus cause a considerable decrease in thermal conductivity. Using the WFL law, it is possible to calculate the thermal conductivity from the electrical conductivity. Figure A.4 (right) shows the field-dependence of the Lorenz number as a function of *RRR*, computed from the thermal conductivity [16].

Niobium–titanium The thermal conductivity of Nb–Ti can be approximated by

$$\lambda(T) = \sum_{n=0}^{6} a_n T^n, \tag{A.20}$$

where $\lambda \equiv \{\lambda\}_{\mathrm{W\,m\,K^{-1}}}$, $T \equiv \{T\}_{\mathrm{K}}$. According to Reference [3], the fit constants are $a_0 = 6.6 \times 10^{-2}$, $a_1 = 4.56 \times 10^{-2}$, $a_2 = 3 \times 10^{-4}$, $a_3 = -3 \times 10^{-6}$, $a_4 = 6 \times 10^{-9}$, $a_5 = 1.5 \times 10^{-11}$, and $a_6 = -5 \times 10^{-14}$.

Niobium-3-tin Measured data presented for Nb_3Sn in [3] are interpolated and summarized in Table A.5.

Table A.5 Thermal conductivity of Nb_3Sn; source [3].

T	λ	T	λ	T	λ	T	λ
2	0.01	22	2.53	102	2.13	297	1.41
7	0.18	32	2.78	152	1.79		
12	0.98	42	2.74	202	1.63		
17	2.02	52	2.64	252	1.53		

Helium The thermal conductivity of superfluid helium He II is very high. In [13] a value of $100\,\mathrm{kW\,m^{-1}\,K^{-1}}$ is given. This is 10–100 times the thermal conductivity of copper. The thermal conductivity of He II is considered perfect, requiring no temperature gradient but limited to a maximum heat flux; see Figure A.6. When traversing the lambda point or exceeding the critical flux [13], helium retains only a small thermal conductivity of $0.02\,\mathrm{W\,m^{-1}K^{-1}}$.

Polyimide The thermal conductivity of polyimide can be modeled by

$$\lambda(T) = 10^{p(T)} \quad \text{with} \quad p(T) = \sum_{n=0}^{7} a_n (\log T)^n \tag{A.21}$$

where the empirical parameters are given by $a_0 = 5.73101$, $a_1 = -39.5199$, $a_2 = 79.9313$, $a_3 = -83.8572$, $a_4 = 50.9157$, $a_5 = -17.9835$, $a_6 = 3.42413$, and $a_7 = -0.27133$.

A.4 Heat Capacity

The *heat capacity* is defined as the amount of energy needed to raise the temperature of a material by 1 K. In terms of thermodynamic quantities, the heat capacity can be written as the derivative of the internal energy U under constant pressure or constant volume:

$$C_\mathrm{V} = \left(\frac{\mathrm{d}U}{\mathrm{d}T}\right)_{V=\text{const.}}, \qquad C_\mathrm{p} = \left(\frac{\mathrm{d}U}{\mathrm{d}T}\right)_{p=\text{const.}}, \tag{A.22}$$

where $[C_V] = [C_p] = 1\,\mathrm{J\,mol^{-1}K^{-1}}$. The internal energy is the sum of the kinetic energy associated with the motion of molecules, the potential energy due to the vibration of atoms in molecules and crystals, the energy of chemical bonds, and the energy of the free conduction electrons in metals.

We employ the *volumetric heat capacity* ρc_p, which can be calculated from the heat capacity by means of the mass density, $[\rho c_p] = 1\,\mathrm{J\,m^{-3}K^{-1}}$. The volumetric heat capacity (VHC) of selected materials is given in Table A.6.

Table A.6 Volumetric heat capacity (VHC) of selected materials; source [14]. $[\rho c_p] = 1\,\mathrm{J\,K^{-1}\,m^{-3}}$.

Material	Symbol	$\rho c_p(2\,\mathrm{K})$	$\rho c_p(10\,\mathrm{K})$	$\rho c_p(100\,\mathrm{K})$	$\rho c_p(300\,\mathrm{K})$
Copper	Cu	0.25×10^3	7.7×10^3	2.28×10^6	3.46×10^6
Aluminum	Al	0.3×10^3	3.78×10^3	1.3×10^6	2.44×10^6
Niobium	Nb	1.54×10^3	18.8×10^3	1.73×10^6	2.3×10^6
Tin	Sn	0.34×10^3	58×10^3	1.375×10^6	1.62×10^6
Tantalum	Ta	1.13×10^3	19×10^3	1.9×10^6	2.33×10^6
Titanium	Ti				2.35×10^6
Austenitic steel[a]	316LN	1.59×10^3	4.22×10^3	2.14×10^6	3.85×10^6
Epoxy					$1.1–2 \times 10^6$

[a] At 4 K instead of 2 K.

Copper The VHC of copper depends only on temperature and is given by an empirical fit of measured data [12]:

$$\rho c_p(T) = \sum_{n=0}^{4} a_n T^n \,, \tag{A.23}$$

where $\rho c_p \equiv \{\rho c_p\}_{\mathrm{JK^{-1}m^{-3}}}$ and $T \equiv \{T\}_{\mathrm{K}}$. The fit parameters can be found in Table A.7. Figure 18.3 (right) shows the VHC of copper and other materials as a function of temperature.

Table A.7 Fit parameters for the volumetric heat capacity of copper; source [12].

Range	a_4	a_3	a_2	a_1	a_0
$T < 10$	-3.08×10^{-2}	7.23×10^0	-2.13×10^0	1.02×10^2	-2.56×10^0
$10 \le T < 40$	-3.05×10^{-1}	2.99×10^1	-4.56×10^2	3.47×10^3	-8.25×10^3
$40 \le T < 125$	4.19×10^{-2}	-1.40×10^1	1.51×10^3	-3.16×10^4	1.78×10^5
$125 \le T < 300$	-8.48×10^{-4}	8.42×10^{-1}	-3.26×10^2	6.06×10^4	-1.29×10^6
$300 \le T < 500$	-4.80×10^{-5}	9.17×10^{-2}	-6.41×10^1	2.04×10^4	1.03×10^6

Superconductors The VHC of Nb–Ti in the normal-conducting state is given by an empirical fit to measurements [22]:

$$\rho c_{\mathrm{p}}(T) = \sum_{n=0}^{4} a_n T^n \,, \tag{A.24}$$

where $\rho c_{\mathrm{p}} \equiv \{\rho c_{\mathrm{p}}\}_{\mathrm{J K^{-1} m^{-3}}}$ and $T \equiv \{T\}_{\mathrm{K}}$. The fit parameters are listed in Table A.8. The function (A.24) is displayed in Figure A.5 (right).

Table A.8 Fit parameters for the volumetric heat capacity of Nb–Ti.

Range	a_4	a_3	a_2	a_1	a_0
$T_c < T < 20$	—	1.624×10^1	—	9.28×10^2	—
$20 \leq T < 50$	-2.18×10^{-1}	1.198×10^{-1}	5.5371×10^2	-7.8461×10^3	4.14×10^4
$50 \leq T < 175$	-4.82×10^{-3}	2.976×10^0	-7.163×10^2	8.3022×10^4	-1.53×10^6
$175 \leq T < 500$	-6.29×10^{-5}	9.296×10^{-2}	-5.166×10^1	1.3706×10^4	1.24×10^6

Measured data for Nb_3Sn are listed in Table A.9 and graphically displayed in Figure A.5 (right).

In the superconducting state, the VHC of Nb–Ti depends on the temperature and applied magnetic flux density:

$$\rho c_{\mathrm{p}}(T, B) = a\,T^3 + b\,T\,B, \tag{A.25}$$

where $a = 49.1$ and $b = 64$. The graph of this function is shown in Figure 18.3 (right). In [15] the VHC of Nb_3Sn is given as

$$\rho c_{\mathrm{p}}(T) = a\,T^3 + b, \tag{A.26}$$

where $a = 22.68$ and $b = 988.2$.

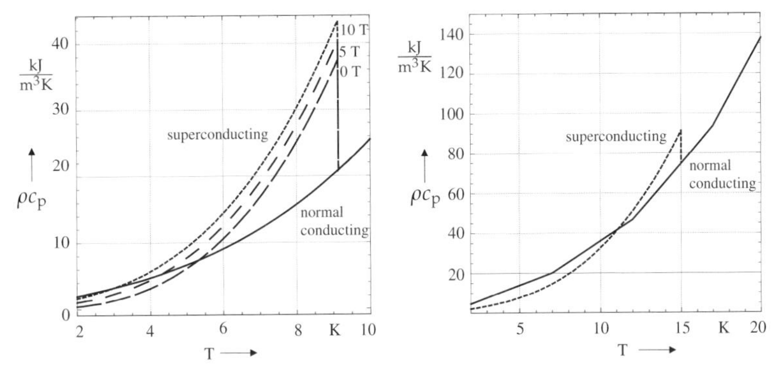

Figure A.5 Volumetric heat capacity. Left: Nb–Ti; source [22]. Right: Nb_3Sn; source [15].

Table A.9 Volumetric heat capacity of Nb_3Sn; source [3]. $[\rho c_\mathrm{p}] = 1\,\mathrm{J\,K^{-1}\,m^{-3}}$.

T	ρc_p	T	ρc_p
2	4.66×10^3	72	1.14×10^6
7	2.01×10^4	102	1.62×10^6
12	4.66×10^4	162	1.90×10^6
17	9.36×10^4	212	2.02×10^6
22	1.67×10^5	262	2.07×10^6
42	6.09×10^5	297	2.10×10^6

Helium Using the assumptions discussed in Section A.1, the data taken from [21] can be scaled to yield the VHC graph displayed in Figure A.6 (right).

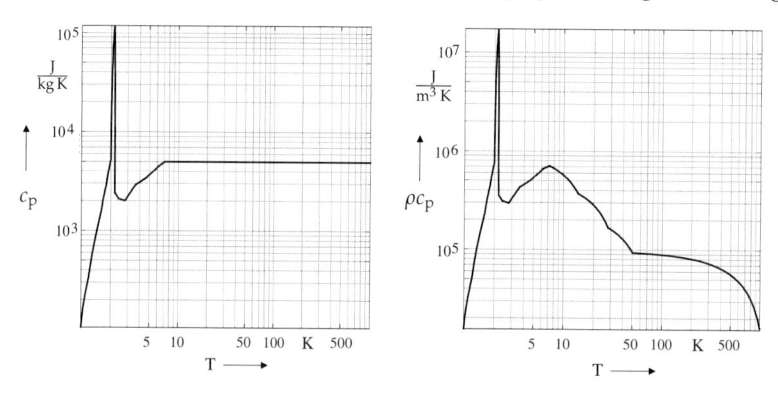

Figure A.6 Approximation of the thermal properties of helium with varying pressure. Left: Heat capacity of helium. Right: Volumetric heat capacity.

Polyimide The VHC of polyimide can be expressed as

$$\rho c_\mathrm{p}(T) = 10^{p(T)} \quad \text{with} \quad p(T) := \sum_{n=0}^{7} a_n (\log T)^n , \tag{A.27}$$

where $a_0 = -1.3684$, $a_1 = 0.65892$, $a_2 = 2.8719$, $a_3 = 0.42651$, $a_4 = -3.0088$, $a_5 = 1.9558$, $a_6 = -0.51998$, and $a_7 = 0.051574$. The function is shown in Figure 18.3 (right).

References

1 Amati, M., Sorbi, M., Volpini, G.: Material properties – a comparison of sources, EDMS doc. no. 555753, CERN, 2004

2 Arenz, R. W., Clark, C. F., Lawless, W. N.: Thermal conductivity and electrical resistivity of copper in intense magnetic fields at low temperatures, Physical Review B, 1982[AQ: Please provide the missing volume and page number for Refs. [2, 4, 6, 7, 8, 9].

3 Bauer, P., Rajainmaki, H., Salpietro, E.: EFDA Material Data Compilation for Superconductor Simulation, EFDA CSU, Garching, 2007

4 Boutboul, T., Le Naour, S., Leroy, D., Oberli, L., Previtali, V.: Critical current density in superconducting Nb–Ti strands in the 100 mT to 11 T applied field range, IEEE Transactions on Applied Superconductivity, 2006.

5 Brechna, H.: Superconducting Magnet Systems, Springer, Berlin, 1973

6 Bruzzone, P.: The index n of the voltage–current curve, in the characterization and specification of technical superconductors, Physica C: Superconductivity, 2004

7 Clark, A. F., Ekin, J. W., Radebaugh, R., Read, D. T.: The Development of Standards for Practical Superconductors, IEEE Transactions on Magnetics, 1979

8 Charifoulline, Z: Residual resistivity ratio (RRR) measurements of LHC superconducting NbTi cable strands, IEEE Transactions on Applied Superconductivity, 2006

9 Chorowski, M., Lebrun, Ph., Serio, L., van Weelderen, R.: Thermohydraulics of quenches and helium recovery in the LHC prototype magnet strings, Cryogenics, 1998

10 Cobine, J.D.: Gaseous Conductors, McGraw-Hill, New York, 1941

11 Dresner, L.: Stability of Superconductors, Plenum, New York, 1995

12 Floch, E.: Specific Heat, Thermal Conductivity and Resistivity of Cu and NbTi – A Bibliography, AT-MTM note, CERN, 2003

13 Iwasa, Y.: Case Studies in Superconducting Magnets, Plenum, New York, 1994

14 Jensen, J. E., Tuttle, W. A., Stewart, R. B., Brechna, H., Prodell, A.G.: Selected Cryogenic Data Book, Brookhaven National Laboratory, 1980

15 Kim, S-W.: Material Properties for Quench Simulation (Cu, NbTi and Nb3Sn), FNAL TD-00-041, Batavia, 2000

16 Properties of Copper and Copper Alloys at Cryogenic Temperature, National Institute of Standards, 1992.

17 Rossi, L., Sorbi, M.: MATPRO: A Computer Library of Material Property at Cryogenic Temperature, INFN/TC-06/02, 2002

18 Seeber, B., editor: Handbook of Applied Superconductivity, Institut of Physics Publishing, Bristol, 1998

19 Superconductivity – part 1: critical current measurements – DC critical current of Nb–Ti composite superconductors. International Standard IEC 61788-1:2006.

20 Superconductivity – part 4: Residual resistance ratio measurement – residual resistance ratio of Nb–Ti composite superconductors. International Standard IEC 61788-4.

21 Van Sciver, S. W.: Helium Cryogenics, Plenum, New York, 1986

22 Verweij, A.: CUDI : A Model for Calculation of Electrodynamic and Thermal Behaviour of Superconducting Rutherford Cables, Departmental Report, CERN, 2006

Appendix B
The LHC Magnet Zoo

B.1
Superconducting Magnets

In the more than twenty years of R&D for the magnet system of the Large Hadron Collider (LHC), a great many magnet variants were designed and tested. An even greater number of studies were carried out in the domain of field computation. One aim of this book is therefore to document some of this design work, whenever it can serve as an example for the described analytical and numerical methods. It may also be useful for future projects to give a brief overview of the electromagnetic design of the magnets installed in the main ring of the LHC.

The LHC magnet system consists of 1232 main dipoles (MB) and 386 main quadrupoles (MQ) together with various types of magnets for insertion and correction. The need for the variety of magnets in the LHC machine and the requirements on field strength and field quality are explained in Chapter 11.

The figures show the result of numerical field calculation using the CERN field calculation program ROXIE and display either the modulus of the magnetic vector potential (field lines) or the modulus of the magnetic flux density.

Main magnets in the arc The main dipoles have been the most challenging components of the LHC from the point of view of machine performance, technology, and cost. The magnets were assembled at three European companies with CERN-supplied components in order to save costs and guarantee a high uniformity in the three production lines. The cross section is shown in Figure B.1. Design parameters of the main dipoles are given in Table B.1. The coils are made of two layers with cable of different widths in order to adjust the current density to the magnetic flux density inside the coil. The aperture diameter is 56 mm, and the beam separation distance is 194 mm at cryogenic temperatures.

The main quadrupole coldmass consists of four coils with two layers each, mounted in separate austenitic steel collars, surrounded by a common yoke punched from a single low-carbon steel sheet, and mounted in an austenitic steel cylinder; see Figure B.1 (right). The cable used for the main quadrupoles

Field Computation for Accelerator Magnets. Stephan Russenschuck

ISBN: 978-3-527-40769-9

Table B.1 Parameters of the LHC main dipoles.

Injection / nominal / ultimate field	0.535 / 8.33 / 9.0 T
Injection / nominal/ ultimate current	763 / 11 850 / 12 840 A
Operating temperature	1.9 K
Coil aperture at 293 K	56 mm
Magnetic length at 1.9 K, 763 A	14.312 m
Coil turns per aperture: Inner / outer shells	30 / 50
Distance between aperture axes at 1.9 K	194 mm
Outer diameter of coldmass at 293 K	570 mm
Overall length of coldmass at 293 K	15.18 m
Overall mass of cryomagnet	27 t
Stored energy in both channels at 8.33 T	6.9 MJ = 1.92 kWh
Self inductance for both channels at nom. field	98 mH
Cold mass sagitta at 293 K	9.14 mm

is identical with the outer-layer cable of the main dipole. The cable is the same for both coil layers and therefore allows a winding technique referred to as *double-pancake winding*, avoiding interlayer soldering joints of the two different cable types.

Dispersion suppressor and matching section quadrupoles The tuning[1] of the LHC insertions is provided by the individually powered quadrupoles, named MQM, in the dispersion suppressor and matching sections. The coils of these magnets are wound from a single length of 8.8-mm-wide insulated cable as

1 The optical functions of the magnets are explained in Chapter 11.

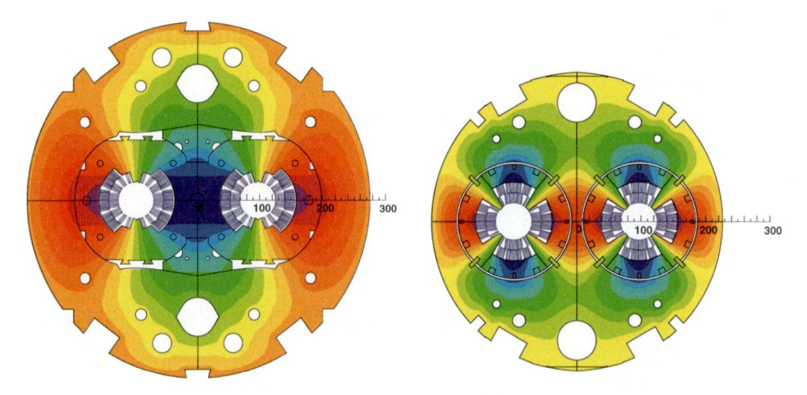

Figure B.1 Display of the magnetic vector potential A_z in the cross sections of the LHC main dipole and quadrupole magnets. Left: Dipole (MB). Right: Quadrupole (MQ).

a double pancake without soldering joints and then cured in a single cycle. The LHC requires three magnetic lengths of MQM tuning quadrupoles. The magnets are individually powered with a nominal current rating of 6 kA and operate in liquid helium at 4.5 or 1.9 K, depending on the location in the tunnel. The cross section of the MQM quadrupoles is shown in Figure B.2 (left).

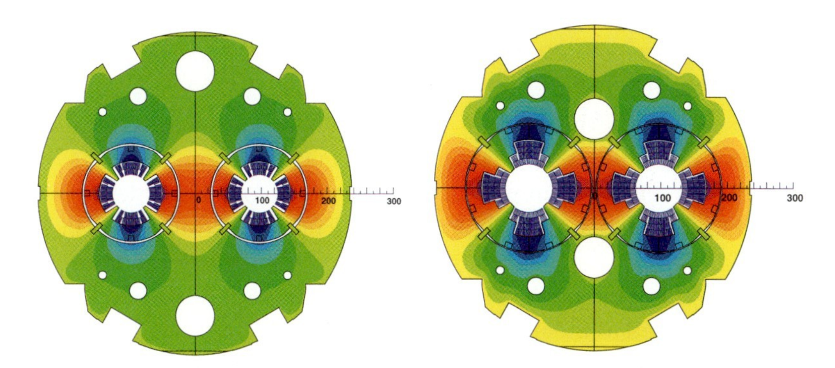

Figure B.2 Left: Insertion quadrupole coldmass (MQM). Right: Wide-aperture quadrupole for the matching section (MQY).

The MQYs, see Figure B.2 (right), are wide-aperture (70 mm diameter) superconducting quadrupoles for the matching sections. The quadrupoles consist of two collared coils assembled in a common magnetic yoke. The yoke has the same outer contour as the MQM quadrupoles. The coils are wound in four layers using 8.3-mm-wide cables with two different width to allow for current grading. The magnets are operated in liquid helium at 4.5 K.

Magnets for the insertion regions The LHC insertion regions are composed of the following sections: a free space of 23 m at each side of the interaction point (with a particle adsorber required to protect the superconducting magnets from the secondary particles), the low-β (or inner) quadrupole triplet, a pair of separation/recombination dipoles, and the matching (or outer) quadrupole triplet. The inner triplet comprises four identical 5.5-m-long, 70-mm-aperture quadrupoles powered in series. The outer-triplet quadrupoles have the same cross section as the main arc quadrupoles but are slightly longer.

The low-β triplet is composed of four single-aperture quadrupoles (MQX) with a coil aperture of 70 mm. The magnets are cooled to 1.9 K using an external heat exchanger system for the extraction of up to 10 W m^{-1} of power deposited in the coils by secondary particles emanating from the proton collisions. Two types of magnets, developed and procured at KEK (Japan) and FNAL (USA), are used in the low-β triplet. They both have a nominal field gradient of 215 T m^{-1} but different cables and consequently different operational currents. The four magnets are powered in series from a 7 kA power supply, with an additional inner loop of 5 kA for the MQXB magnets.

- The cross section of the 5.5-m-long MQXB magnet from FNAL [1] is shown in Figure B.3 (right). The American magnets have two coil layers with 15.4-mm-wide cables (37 strands in the inner layer and 46 strands in the outer layer) excited with a nominal current of 11 950 A. The coils are assembled in freestanding collars which provide the prestress needed to take the electromagnetic forces during excitation.
- The cross section of the 6.370-m-long MQXA magnet from KEK [7] is shown in Figure B.3 (left). The Japanese magnets operate at 7149 A, using smaller cables (11 mm wide) but feature a four-layer design. As opposed to their function in the main quadrupoles and the MQXB, the role of the collars here is simply to act as spacers, and the required prestress is provided by a yoke structure that consists of horizontally split laminations keyed at the median plane.

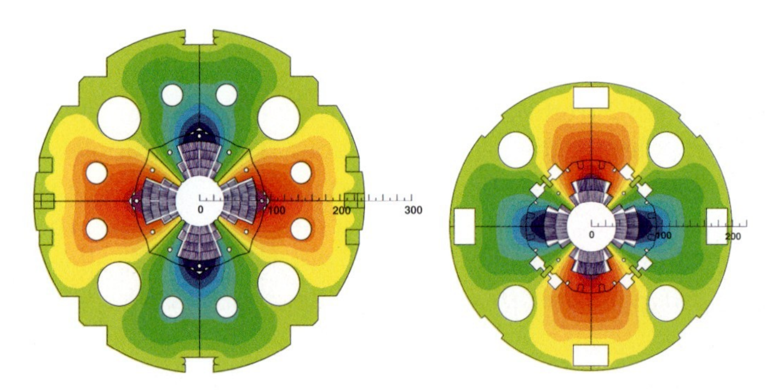

Figure B.3 Single-aperture quadrupoles for the low-β triplets. Left: MQXA. Right: MQXB.

Spool piece and lattice correctors The corrector magnets in the arcs can be grouped in two categories:

- The lattice corrector magnets are housed on both sides of the main quadrupole magnets in the cryostat of the short straight section (SSS). They are powered in 600 A circuits through busbars routed in an external cryogenic distribution line (Line N).
- The spool-piece[2] corrector magnets are mounted on the extremities of the main dipole coldmasses.

2 The term spool-piece was invented by Kuchnir et al. for a modular component containing two correction packages consisting of three superconducting correction elements each [2].

The arc corrector magnets are powered independently for the two beams. Sextupole spool-piece correctors (MCS) are single-aperture magnets installed downstream in the coldmass of each dipole, one for each aperture. The cross section is shown in Figure B.4. The spool-piece sextupole magnets are designed to compensate for the integrated sextupole field of the main dipoles mainly arising from persistent current and saturation effects. The spool-piece sextupoles thus correct for each LHC machine sector the impact of the field errors on the linear chromaticity. The coils of the spool-piece sextupole magnets are clamped by central posts made from copper, which is completely amagnetic and therefore not represented in the numerical model.

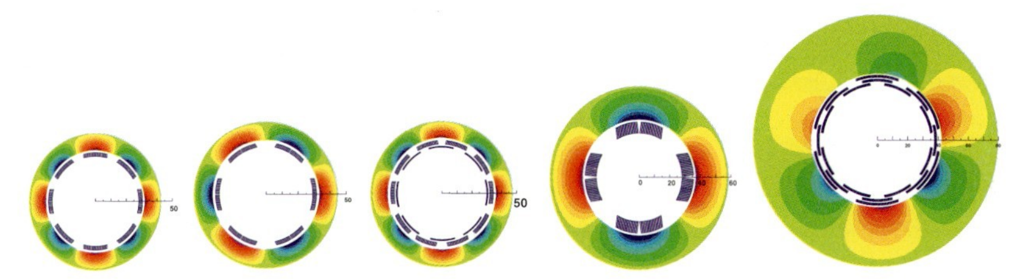

Figure B.4 From left to right: (1) Octupole module. (2) Sextupole spool-piece corrector module (MCS). (3) Nested decapole-octupole corrector MCDO. (4) Tuning quadrupole module MQT. (5) Nested multipole correctors for A_3, A_4, and B_4, MCSOX. Magnets not to scale with the main magnets.

The function of the MCDO magnets is to preserve the dynamic aperture of the LHC at injection. The MCDO are single-aperture assemblies of spool-piece correctors consisting of an octupole magnet and a decapole magnet concentrically mounted (nested) inside an aluminum shrinking cylinder, as shown in Figure B.4. Nested coils help to make compact combined-function correctors; however, the performance and field quality of the coil layers are mutually interdependent; see Chapter 16.

The Landau damping (or lattice) octupoles are made from two octupole modules displayed in Figure B.4, mounted in a common support structure and installed upstream of the main quadrupole in the inertia tube of the arc SSS.

The tuning quadrupoles (MQT) consist of a single-layer quadrupole coil surrounded by a laminated yoke and a shrinking cylinder; see Figure B.4 (right). Two modules (one per aperture) are mounted in a common support structure. The skew quadrupole corrector magnets (MQS) are built from the same magnet module as the MQT but rotated by 45°. They serve for the compensation of the coupling coefficient due to systematic skew quadrupole field errors in the main dipoles. MQT or MQS are installed upstream of the main quadrupoles in the coldmass of the arc SSS. The tuning quadrupoles are powered in series in two families per arc and per beam.

The MCBH and MCBV are horizontal and vertical dipole orbit correctors. They are made from the same single-aperture module, rotated by 90° in the case of the vertical orbit corrector. They are installed at each focusing and defocusing quadrupole; in total 23 or 24 orbit correctors per ring, per arc, and per transverse plane. They achieve a maximum orbit kick of 80.8 μrad at 7 TeV for a nominal current of 55 A. They are assembled in twin-aperture yokes and form the different configurations MCBCA(B,C,D). The MCBCB version is shown in Figure B.6 (left).

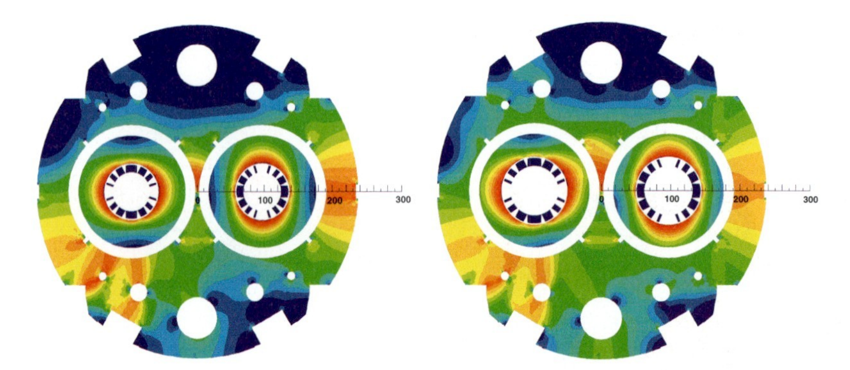

Figure B.5 Display of the magnitude of the magnetic flux density in the cross section of dipole orbit correctors. Left: MCBCB subassembly (orbit corrector associated with MQM, MQML or MQMC+MQM). Vertical orbit corrector in the external aperture and horizontal corrector in the internal aperture. Right: MCBYB, wide-aperture orbit corrector associated with MQY.

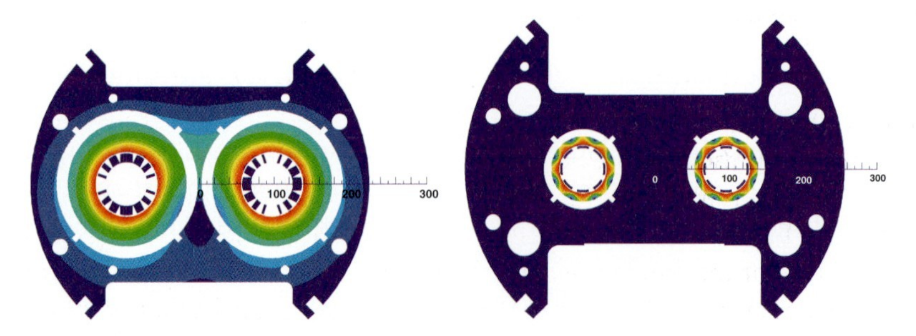

Figure B.6 Left: Cross section of the MSCBB subassembly (consecutively combined sextupole and dipole corrector) installed in the arc SSS. Right: Octupole lattice corrector cross section (MO) installed in the arc SSS.

The chromaticity or lattice sextupoles and lattice skew sextupoles are built from the same magnet module, rotated by 30° in the case of the skew version, and mounted in a twin-aperture structure. The assemblies are installed on the downstream side of the main quadrupole in the inertia tube of the arc SSSs, with the quadrupole coldmasses in half-cell 11 of the dispersion suppressor. The sextupole modules are powered in series, forming one skew sextupole family and four normal sextupole families per arc and per beam. The lattice skew sextupole correctors compensate for the chromatic coupling induced by the skew sextupole field errors in the dipole magnets.

The dipole component of the MSCB assembly is shown in Figure B.6 (left). The MCBC is a superconducting twin-aperture, dipole corrector-magnet assembly in a common support structure; see Figure B.5 (left). The MCBY are wide-aperture dipole corrector magnet assemblies; see Figure B.5 (right).

The MCBX is a combined single-aperture concentric (nested) dipole corrector, with one horizontal dipole inside and one vertical dipole outside. They are associated with the Q1 and with the Q2 quadrupoles in the insertion region; see Figure B.7. The coils have a 90 mm aperture, providing space to nest another corrector magnet package.

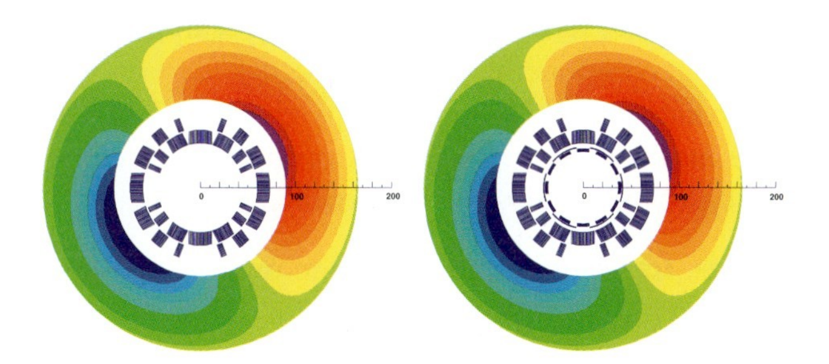

Figure B.7 Left: MCBX, single-aperture orbit corrector associated with MQX close to Q2 and Q3. Right: MCBXA subassembly with sextupole (MCSX) and dodecapole (MCTX) inserts in the MCBX magnet.

Separation dipoles In four insertion regions (IR1, IR2, IR5, IR8), the separation dipoles bring the particle beams into a colliding orbit at the interaction point and then separate them again to the nominal distance of 194 mm in the LHC arcs; see Figure B.8. The cross section of the MBX separation dipoles is identical to that of the single-aperture RHIC dipoles [8]. The cryostat of this magnet is also identical with the RHIC main dipole cryostat, but the coldmass is straight, without the 47 mm sagitta of the RHIC magnets. Unlike the RHIC dipoles, the MBX magnets are operated at 1.9 K.

The MBRB and MBRC separation dipoles are twin-aperture dipoles with 194 and 188 mm beam separation distance, respectively. The magnetic fields in both apertures are of the same polarity. The crosstalk between the two apertures is reduced by the oval shape of the coldmass, which reduces the flux density on the median plane. In the vertical plane the size of the coldmass

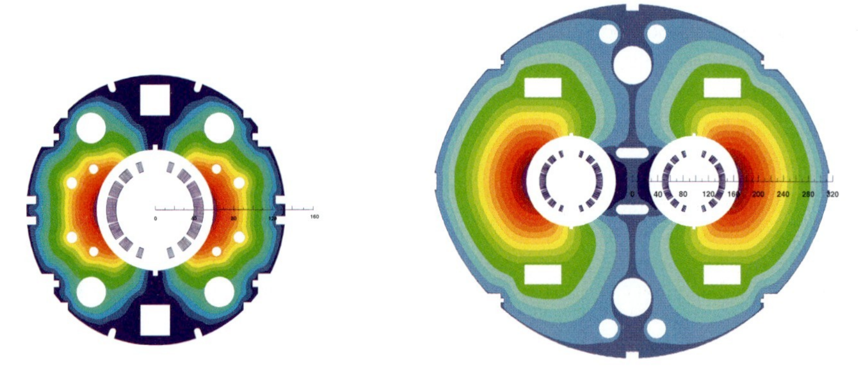

Figure B.8 Separation dipoles. Left: MBX, single-aperture magnet, identical to the dipoles for RHIC. Right: MBRM, twin-aperture magnet with same polarity of the flux density in both apertures. Magnets not to scale.

is identical to the LHC main dipole, which allows the use of standard LHC support posts and other cryostat components.

A summary of the superconducting LHC magnets is given in Table B.2. Columns A and C give the number of apertures and connection terminals, respectively. Four different types of magnets can be identified:

- Single-aperture magnets, for example, the insertion quadrupole MQY with one aperture and one pair of terminals ($A = 1, C = 1$).
- Twin-aperture magnets with one pair of connection terminals, for example, the main dipole MB ($A = 2, C = 1$).
- Twin-aperture magnets, with two pairs of connection terminals for individual powering of the apertures, such as the main quadrupole MQ ($A = 2, C = 2$).
- Individually powered magnet modules assembled in two-in-one subassemblies, such as the octupole corrector MO and the skew quadrupole MQS ($A = 2, C = 2$).

This classification scheme avoids having to distinguish between twin-aperture magnets with a common iron yoke (more or less magnetically coupled) and magnet modules in a common support structure (magnetically decoupled), which in some cases also allow individual cold testing of the modules.

The arrangements of the magnet assemblies and the position of their connection terminals are shown in Figures B.9–B.12. These figures define the installation direction of the magnets. If the magnets, or magnet assemblies, are installed in the LHC tunnel with the connection terminals pointing in opposite directions, the assembly is marked with an asterisk (*) in the electrical layout database and layout drawings; see Section 12.6.

Table B.2 Superconducting magnets for the LHC[a].

Magnet	F	Description	N	T	I_{nom}	l_{m}	L	A	C	D
Units				K	A	m	H			
Main magnets in the arcs										
MB	B_1	Main dipole	1232	1.9	11850	14.3	0.102	2	1	u
MQ	B_2	Lattice quadrupole in the arc	392	1.9	11870	3.1	0.0056	2	2	u
Separation and recombination dipoles										
MBRB	B_1	Twin-aperture (194 mm) D4	2	4.5	5520	9.45	0.052	2	1	u
MBRC	B_1	Twin-aperture (188 mm) D2	8	4.5	6000	9.45	0.052	2	1	u
MBRS	B_1	Single-aperture D3	4	4.5	5520	9.45	0.026	1	1	u
MBX	B_1	Single-aperture D1	4	1.9	5800	9.45	0.026	1	1	u
Lattice correctors										
MCBCH	B_1	Orbit corr. in MCBCA(B,C,D)	84	1.9/4.5	100^+	0.904	2.84	1	1	d
MCBCV	A_1	Orbit corr. in MCBCA(B,C,D)	84	1.9/4.5	100^+	0.904	2.84	1	1	d
MCOSX	A_4	Skew octupole in MCSOX	8	1.9	100	0.138	0.0032	1	1	u
MCOX	B_4	Octupole associated to MCSOX	8	1.9	100	0.137	0.0044	1	1	u
Spool piece corrector magnets										
MCS	B_3	Sextupole corrector	2464	1.9	550	0.11	0.0008	1	1	d
MCD	B_5	Decapole corr. in MCDO	1232	1.9	550	0.066	0.0004	1	1	u
MCO	B_4	Octupole corr. in MCDO	1232	1.9	100	0.066	0.0004	1	1	u
MCSSX	A_3	Skew sextupole in MCSOX	8	1.9	100	0.132	0.0078	1	1	u
MCSX	B_3	Sextupole in MCBXA	8	1.9	100	0.576	0.0047	1	1	u
MCTX	B_6	Dodecapole in MCBXA	8	1.9	80	0.615	0.0292	1	1	u
Insertion quadrupoles and correctors										
MQM	B_2	Insertion region quad.	38	1.9/4.5	5390^*	3.4	0.0151	2	2	u
MQMC	B_2	Insertion region quad.	12	1.9/4.5	5390^*	2.4	0.0107	2	2	u
MQML	B_2	Insertion region quad.	36	1.9/4.5	5390^*	4.8	0.0213	2	2	u
MQY	B_2	Insertion wide-apert. quad.	24	4.5	3610	3.4	0.074	2	2	u
MQTLH	B_2	MQTL (Half Shell Type)	24	4.5	400	1.3	0.120	2	2	d
MQTLI	B_2	MQTL (Inertia Tube Type)	36	1.9	550	1.3	0.120	2	2	d
Inner triplets and associated correctors										
MQXA	B_2	Single-aper t. triplet quad. Q1, Q3	16	1.9	6450	6.37	0.090	1	1	u
MQXB	B_2	Single-aper t. triplet quad. Q2	16	1.9	11950	5.5	0.019	1	1	u
MCBXH	B_1	Horizontal orbit corr. in MCBX(A)	24	1.9	550	0.45	0.287	1	1	u
MCBXV	A_1	Vertical orbit corr. in MCBX(A)	24	1.9	550	0.48	0.175	1	1	u
MQSX	A_2	Skew quadrupole Q3	8	1.9	550	0.223	0.014	1	1	u
Lattice corrector magnets										
MCBH	B_1	Arc orbit corr. in MSCBA(B,C,D), hor.	376	1.9	55	0.647	6.02	1	1	d
MCBV	A_1	Arc orbit corr. in MSCBA(B,C,D), vert.	376	1.9	55	0.647	6.02	1	1	d
MCBYH	B_1	Orbit corr. in MCBYA(B)	38	4.5	72	0.899	5.27	1	1	d
MCBYV	A_1	Orbit corr. in MCBYA(B)	38	4.5	72	0.899	5.27	1	1	d
MQS	A_2	Skew quad. lattice corr. in arc SSS	32	1.9	550	0.32	0.031	2	2	u
MQT	B_2	Tuning quad. in arc SSS	160	1.9	550	0.32	0.031	2	2	u
MS	B_3	Arc sext. corr. next to MCBH, MCBV	688	1.9	550	0.369	0.036	1	1	u
MSS	A_3	Arc skew sextupole corr. in MCBH	64	1.9	550	0.369	0.036	1	1	u
MO	B_4	Octupole lattice corr. in arc SSS	168	1.9	550	0.32	0.00015	2	2	u

[a] Kickers and experimental magnets are excluded. Column F = multipole, N = number of magnets, T = operation temperature, I = nominal current, l_{m} = magnetic length, L = self inductance, A = number of apertures, C = number of pairs of connection terminals, D = normal position of the connection terminals, u: upstream, d: downstream. (*) 4310 A at 4.5 K. (+) 80 A at 4.5 K. Reference: LHC functional layout database.

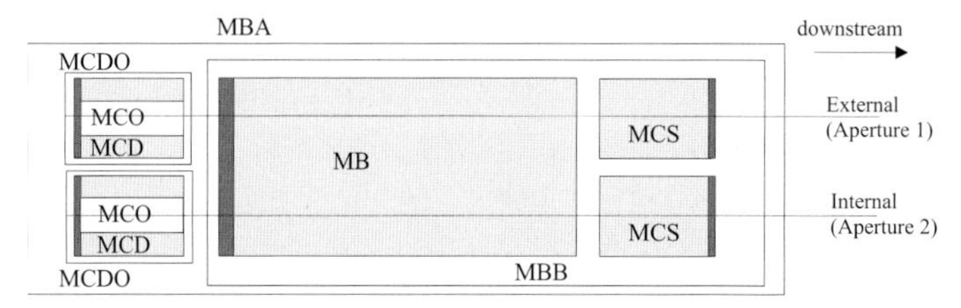

Figure B.9 Arrangement of the magnet assemblies in the MBA and MBB dipole cryomagnets.

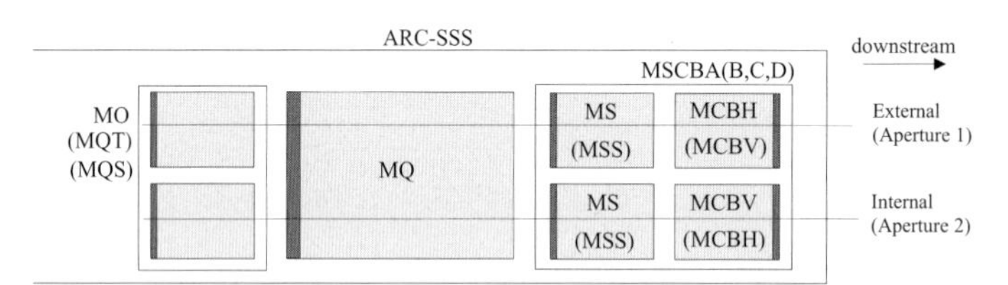

Figure B.10 Arrangement of the magnet assemblies in the arc SSSs. Different combinations of magnets, polarities, and the presence (or not) of jumper connections to the cryogenic transfer line and of pressure plugs result in 40 variants [4].

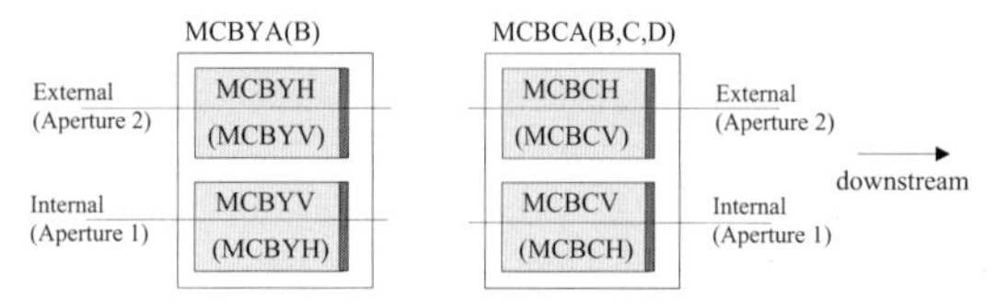

Figure B.11 Arrangement of the magnet assemblies MCBYA(B) and MCBCA(B,C,D). Notice that the normal position of the connection terminals is downstream (in the direction of Beam 1) so that Aperture 2 is on the right, seen from the connection end of the magnet (with the connections at the bottom), i.e, on the external side.

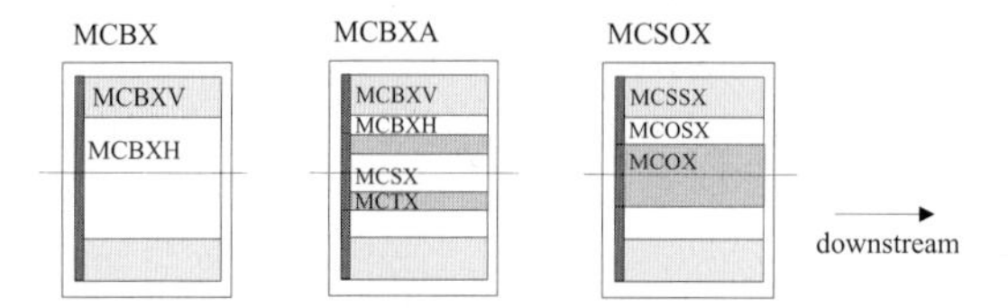

Figure B.12 Arrangement of the magnet assemblies MCBX, MCBXA, and MCSOX.

B.2
Normal-Conducting Magnets

In the beam-cleaning insertions (IR3 and IR7), the high radiation levels caused by scattered particles from the collimation system do not allow the installation of superconducting magnets. The matching quadrupoles installed in these regions consist of a group of six normal-conducting quadrupole magnets shown in Figure B.13. Five of these twin-aperture quadrupoles (MQW) are operated in an excitation mode with opposite optical functions (FD or DF). One is configured with both apertures having the same optical functions (FF or DD) for the correction of asymmetries in the group of five [4]. The construction principle of the MQW magnets is known as the *figure of eight* quadrupole. Both excitation modes are shown in Figure B.13. In order to achieve the required field quality, the separation between the poles is adjusted to within 0.1 mm by tightening-rods along the length of the magnets.

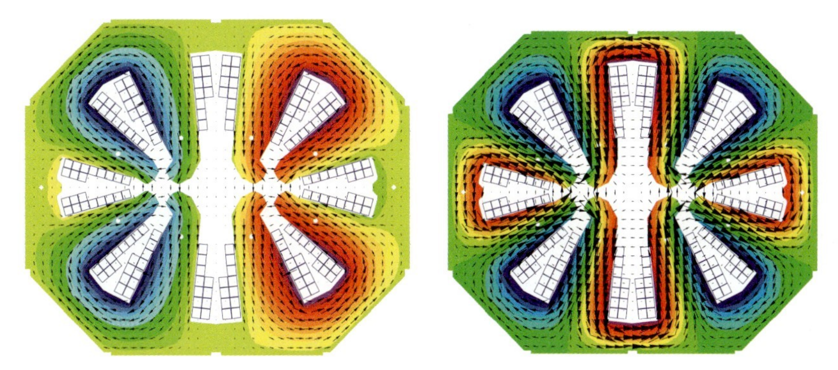

Figure B.13 Left: Powering of the MQWB magnet in FF mode (for Beam 1 in the external aperture) with current entering the A-terminal. Nominal current is 600 A. Right: Powering of the MQWA magnet in FD mode (for Beam 1 in external aperture) with current entering the A-terminal. Nominal current is 710 A.

Radiation-hard, normal-conducting separation dipoles are required in the high-luminosity insertions (IR1 and IR5). The coil design of the separation dipole magnets (MBW), shown in Figure B.14 (left), is described in Section 7.10. For cooling purposes, the coils are constructed of three pancakes wound from hollow rectangular copper conductor. The yoke of the separation dipoles is laminated from magnetic steel sheets of 1.5 mm thickness, manufactured as two half-cores that are clamped together with studs and nuts at the side-cover plates. This allows for the verification and adjustment of the yoke geometry during the production process. The requirements are a twist not exceeding 1 mrad and a sagitta of less than 0.5 mm.

Normal-conducting magnets are also used for the orbit correctors in the insertion regions IR3 and IR7. The orbit correctors act only on one beam. In the case of the horizontal orbit correctors (MCBWH) the second beam passes through a passive aperture; see Figure B.14 (right). In the case of the vertical orbit corrector MCBWV the second beam passes outside the magnet. From a mechanical point of view, the horizontal and vertical magnet types are identical. The main parameters of these magnets are a magnetic length of 1.7 m, a beam separation of 224 mm, a gap height of 52 mm, and a nominal field of 1.1 T at 550 A. The coil is wound from conductors of a cross section of

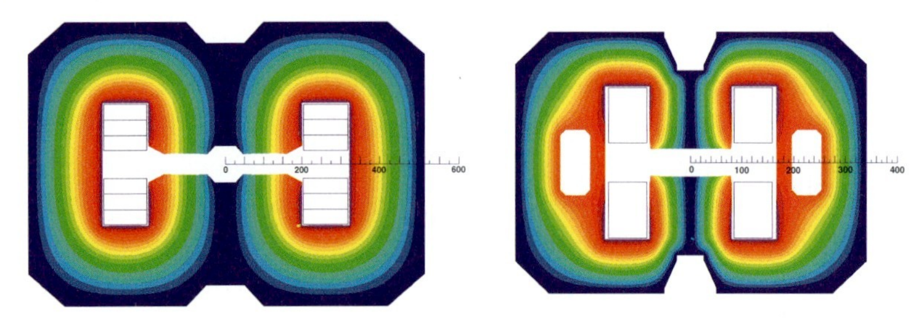

Figure B.14 Left: Separation dipole MBW. Right: Orbit corrector MCBWH. Magnets are not to scale.

16×10 mm^2 and a cooling hole diameter of 5 mm. The dissipated power is 14 kW at nominal excitation [4].

Normal-conducting magnets compensate for the effect of the ALICE and LHCb spectrometer dipoles on the beam. The compensators for the ALICE experiment are refurbished magnets from the SPS accelerator complex, originally built for the beam lines of the intersecting storage rings (ISR).

An overview on the normal-conducting magnets in the LHC main ring is given in Table B.3.

Table B.3 Normal-conducting magnets for the LHC[a].

Magnet		Description	N	I_{nom}	l_m	L	R	A	C	D
Units				A	m	H	Ω			
Separation and correction dipoles										
MBW	B_1	Twin-aperture D3, D4 in IR3,7	20	720	3.4	0.18	0.055	2	1	u
MBXW	B_1	Single-aperture D1 in IR1,5	24	750	3.4	0.145	0.06	1	1	u
MCBWH	B_1	Orbit corrector, horizontal	8	550	1.7	0.05	0.043	1	1	u
MCBWV	A_1	Orbit corrector, vertical	8	550	1.7	0.05	0.043	1	1	u
Compensator dipoles for ALICE and LHCb experiments										
MBWMD	A_1	Compensator for ALICE, IR2	1	550	2.62	0.639	0.172	1	1	u
MBXWH	B_1	Compensator for LHC-b, IR8	1	750	3.4	0.145	0.04	1	1	u
MBXWS	B_1	Compensator for LHC-b, IR8	2	780	0.78	0.04	0.05	1	1	u
MBXWT	A_1	Compensator for ALICE, IR2	2	600	1.53	0.08		1	1	u
Twin-aperture quadrupoles in the cleaning insertions										
MQWA	B_2	Quad. in IR3,7 (FD or DF)	40	710	3.108	0.028	0.037	2	1	u
MQWB	B_2	Quad. in IR3,7 (FF or DD)	8	600	3.108	0.028	0.037	2	1	u
Septa										
MSDA	A_1	Ejection septum, module A	10	880	4.088	0.036	0.027	1	1	u
MSDB	A_1	Ejection septum, module B	10	880	4.088	0.056	0.034	1	1	u
MSDC	A_1	Ejection septum, module C	10	880	4.088	0.079	0.041	1	1	u
MSIA	B_1	Injection septum, module A	4	950	3.73	0.010	0.011	1	1	u
MSIB	B_1	Injection septum, module B	6	950	3.73	0.024	0.0164	1	1	u

[a] N = number of magnets in the LHC, I = Nominal current, l_m = magnetic length, L = self inductance, R = ohmic resistance (if apertures are individually powered, R and L are given for one aperture), A = number of apertures, C = number of pairs of connection terminals, D = normal position of the connection terminals, u: upstream, d: downstream. Reference: LHC functional layout database.

References

1 Andreev, N. et al.: Status of the inner triplet quadrupole program at Fermilab, IEEE-Transactions on Applied Superconductivity, 2001
2 Kuchnir, M., Walker, R. J., Fowler, W. B., Mantsch, P. M.: Spool piece testing facility, IEEE Transactions on Nuclear Science, 1981
3 LHC Design Report, Vol. 1, The LHC main ring, CERN-2004-003, 2004
4 LHC Functional specification: MBXW resistive dipole magnets for the LHC insertions, LHC-MBXW-CA-0001
5 LHC Functional specification: The series production of the twin aperture quadrupoles for the LHC, LHC-MQW-CA-0001
6 The LHC study group, The Large Hadron Collider, Conceptual Design, CERN/AC/95-05
7 Shintomi, T. et al.: Progress of the LHC low-b quadrupole magnets for the LHC insertions, IEEE Transactions of Applied Superconductivity, 2001
8 Willen, E. et al.: Superconducting dipole magnets for the LHC insertion regions, Proceedings of the EPAC Conference, Vienna, 2000

Appendix C
Ramping the LHC Dipoles

The field quality in accelerator magnets is influenced by magnetization effects and time-transient phenomena such as decay and snap-back of persistent currents, interstrand coupling currents, and interfilament coupling currents. The current ramp for the LHC main dipoles has thus been optimized to minimize the time-transient effects while considering the boundary conditions imposed by the power supplies.[1]

At the injection plateau, there is a decay of about 0.8 units in the relative sextupole component generated by superconductor magnetization. The time for this decay is in the range of 1000 s. After an increase of only 30 A the magnetization and thus the field errors bounce back to their original values, an effect known as snap-back. The time constant of this effect is inversely proportional to the ramp rate. Thus a very slow ramp rate is desirable after the injection plateau. In order to reduce the overall ramp time, the current ramp of the LHC main dipoles is described by a C^1 function, piecewise defined by four sections: a **p**arabolic, an **e**xponential, a **l**inear, and again a **p**arabolic function joined to the injection and nominal current plateaus, known by the acronym PELP. The ramp is determined by the parameters given in Table C.1. The highly inductive magnet load on the power converters with their limited output voltage, together with the required accuracy of the output current, requires not only a continuous ramp function but a smooth function with continuous ramp rates.

Table C.1 Parameters for the excitation function of the LHC main dipoles.

t_{inj}	Initial time	D	Deceleration parameter
I_{inj}	Injection current	R	Ramp rate
I_{nom}	Nominal current	t_e	Start time of the ramp
A	Acceleration parameter		

1 Burla, P., King, Q., Pett, J.G.: Optimization of the current ramp for the LHC, Proceedings of the PAC Conference, New York, 1999.

Field Computation for Accelerator Magnets. Stephan Russenschuck

ISBN: 978-3-527-40769-9

The first parabolic function segment for $t_{\mathrm{inj}} < t < t_{\mathrm{e}}$ defines the current ramp from the injection plateau at I_{inj} according to

$$I_{\mathrm{a}}(t) = \frac{A}{2}(t - t_{\mathrm{inj}})^2 + I_{\mathrm{inj}}. \tag{C.1}$$

The acceleration A has been chosen such that the snap back region is traversed in about 67 s, thus allowing for correction schemes to be applied. For the second segment, $t_{\mathrm{e}} < t < t_{\mathrm{l}}$, the exponential function is

$$I_{\mathrm{e}}(t) = a\,e^{bt}, \tag{C.2}$$

where the parameters a and b are determined by the continuity conditions at t_{e},

$$I_{\mathrm{e}}(t_{\mathrm{e}}) = I_{\mathrm{a}}(t_{\mathrm{e}}), \qquad \frac{\mathrm{d}I_{\mathrm{e}}}{\mathrm{d}t}(t_{\mathrm{e}}) = \frac{\mathrm{d}I_{\mathrm{a}}}{\mathrm{d}t}(t_{\mathrm{e}}), \tag{C.3}$$

from which it follows that

$$b = \frac{1}{I_{\mathrm{a}}(t_{\mathrm{e}})}\frac{\mathrm{d}I_{\mathrm{a}}}{\mathrm{d}t}(t_{\mathrm{e}}), \qquad a = e^{-bt_{\mathrm{e}}}\frac{\mathrm{d}I_{\mathrm{a}}}{\mathrm{d}t}(t_{\mathrm{e}}). \tag{C.4}$$

The exponential segment ends at t_{l} when $\mathrm{d}I/\mathrm{d}t$ attains the design ramp rate R, which is about 10 A s^{-1}. The condition $\frac{\mathrm{d}I_{\mathrm{e}}}{\mathrm{d}t}(t_{\mathrm{l}}) = R$ implies

$$t_{\mathrm{l}} = \frac{1}{b}\ln\left(\frac{R}{ab}\right). \tag{C.5}$$

The linear segment is described by

$$I_{\mathrm{l}}(t) = R(t - t_{\mathrm{l}}) + I_{\mathrm{e}}(t_{\mathrm{l}}). \tag{C.6}$$

The fourth segment is a parabolic deceleration defined by

$$I_{\mathrm{d}}(t) = -\frac{D}{2}(t_{\mathrm{nom}} - t)^2 + I_{\mathrm{nom}}, \tag{C.7}$$

where t_{nom} is the time when the flat-top is reached. The parabolic deceleration starts at t_{d} in order to match the ramp rate. This yields

$$t_{\mathrm{nom}} = t_{\mathrm{d}} + \frac{D}{R}. \tag{C.8}$$

The parabolic deceleration time is given by the continuity condition $I_{\mathrm{l}}(t_{\mathrm{p}}) = I_{\mathrm{d}}(t_{\mathrm{d}})$, which implies

$$t_{\mathrm{d}} = \frac{I_{\mathrm{nom}} - I_{\mathrm{l}}}{R} + t_{\mathrm{l}} - \frac{R}{2D}. \tag{C.9}$$

Table C.2 Parameters for the ramping of the LHC main dipoles.

$I_a(t_e) = A/2(t_e - t_{inj})^2 + I_{inj}$	Current after parabolic rise
$\frac{dI_a}{dt}(t_e) = A(t_e - t_{inj})$	Current ramp-rate at t_e
$b = \frac{dI_a}{dt}(t_e)/I_a(t_e)$	Parameter
$a = I_a(t_e)\exp[-bt_e]$	Parameter
$t_l = 1/b \ln(R/ab)$	Start time of linear ramp
$I_l = \exp[bt_l]$	Current at t_l
$t_d = (I_{nom} - I_l)/R + t_l - R/2D$	Deceleration start time
$t_{nom} = t_d + D/R$	Deceleration end time

The current cycle is thus defined by the set of (dependent) parameters given in Table C.2. With this set of parameters, the complete excitation function is given by

$$I(t) = \begin{cases} I_{inj} & \text{for } t < t_{inj}, \\ \frac{A}{2}(t - t_{inj})^2 + I_{inj} & \text{for } t_{inj} \leq t < t_e, \\ a\,e^{bt} & \text{for } t_e \leq t < t_l, \\ R(t - t_l) + I_l & \text{for } t_l \leq t < t_d, \\ -\frac{D}{2}(t_{nom} - t)^2 + I_{nom} & \text{for } t_d \leq t < t_{nom}, \\ I_{nom} & \text{for } t \geq t_{nom}. \end{cases} \tag{C.10}$$

Appendix D
SI (MKSA) Units

Physical quantity	Unit	Symbol	Conversion
Length	meter	m	
Mass	kilogram	kg	
Time	second	s	
Electric current	ampere	A	
Thermodynamic temperature	kelvin	K	
Amount of substance	mole	mol	
Luminous intensity	candela	cd	
Frequency	hertz	Hz	s^{-1}
Force	newton	N	$kg\,m\,s^{-2}$
Pressure, Stress	pascal	Pa	$N\,m^{-2}$
Energy, Work	joule	J	$N\,m = W\,s$
Power	watt	W	$J\,s^{-1} = m^2\,kg\,s^{-3}$
Torque	newton meter	N m	
Electric charge	coulomb	C	A s
Electric potential	volt	V	
Capacitance	farad	F	$C\,V^{-1}$
Permettivity	farad/meter	$F\,m^{-1}$	$A\,s\,V^{-1}\,m^{-1}$
Electric field strength	volt/meter	$V\,m^{-1}$	
Magnetic flux	weber	Wb	V s
Magnetic flux density	tesla	T	$Wb\,m^{-2} = V\,s\,m^{-2}$
Inductance	henry	H	$Wb\,A^{-1} = V\,s\,A^{-1}$
Permeability	henry/meter	$H\,m^{-1}$	$V\,s\,A^{-1}\,m^{-1}$
Magnetic scalar potential	ampere	A	
Magnetic vector potential	weber/meter	$Wb\,m^{-1}$	T m
Magnetic field strength	ampere/meter	$A\,m^{-1}$	
Resistance	ohm	Ω	$V\,A^{-1}$

Field Computation for Accelerator Magnets. Stephan Russenschuck

ISBN: 978-3-527-40769-9

Appendix E
Glossary

Arc: Portion of the ring accelerator occupied by regular half-cells. Each arc contains 46 half-cells. The arc does not contain the dispersion suppressor.

Arc cell: Cell consisting of two arc half-cells presenting the basic period of the optic functions.

Arc half-cell: Periodic part of the LHC arc lattice, consisting of a string of three twin-aperture main dipole magnets and one short straight section (SSS).

ATLAS: Acronym for A Toroidal LHC Apparatus, installed at Point 1 of the LHC.

ALICE: Acronym for A Large Ion Collider Experiment, installed at Point 8 of the LHC.

Beam 1 and Beam 2: The two LHC beams: Beam 1 circulates clockwise in Ring 1 seen from above and Beam 2 circulates counter clockwise in Ring 2.

Beam cleaning: Removal of the large amplitude (larger than 6 sigma) particles from the beam halo. The LHC has two beam cleaning insertions: One dedicated to the removal of particles with large transverse oscillation amplitudes (IR7) and one dedicated to the removal of particles with large longitudinal oscillation amplitudes (IR3). These insertions are also referred to as the *betatron* and *momentum cleaning* insertions.

Beam screen: Perforated tube inserted into the cold bore of the superconducting magnets in order to protect the cold bore from synchrotron radiation and beam losses.

Beam (pilot): Single bunch of 0.5×10^{10} protons, corresponding to the maximum beam current that can be lost without inducing a magnet quench.

Field Computation for Accelerator Magnets. Stephan Russenschuck

ISBN: 978-3-527-40769-9

Beam (nominal): Beam required to reach the design luminosity of $L = 10^{34}\ \text{cm}^{-2}\text{s}^{-1}$ with $\beta^* = 0.55$ m ($\rightarrow$ normalized emittance $\epsilon_n = 3.75\ \mu\text{m}$; $N_b = 1.15 \times 10^{11}$; $n_b = 2808$).

Bean model: See critical state model.

Big Bang: Theory proposed by George Lemaitre (1894–1966) suggesting that the evident expansion in time requires that the universe contracted backwards in time until all the mass of the universe fused into a single point, forming what Lemaitre called a *primeval atom*. The term Big Bang was first used by Fred Hoyle (1915–2001) in a statement seeking to question the credibility of the theory.

BPM: Acronym for beam-position monitor.

Bunch: Collection of particles captured within one RF bucket.

Busbar: Main conductor that carries the current for powering the magnets outside the magnet coil.

CEBAF: Continuous Electron Beam Accelerator Facility at Thomas Jefferson National Accelerator Facility, USA

Class 1 magnets: Large superconducting magnets for applications in fusion, magnetic energy storage, and physics experiments.

Class 2 magnets: Superconducting magnets with a high field and high current density.

CMS: Compact Muon Solenoid installed at Point 5 of the LHC.

Coldmass: Part of a magnet that requires cooling by the cryogenic system, i.e., the assembly of magnet coils, collars, iron yoke, and helium vessel.

Copper-to-superconductor ratio: Area ratio of the copper stabilizer to the superconductor in the strands made of composite material like Nb–Ti.

Critical current: Maximum current a superconductor can carry in the superconducting state.

Critical state model (CSM): Model for the shielding currents in type-II superconductors exposed to an applied field.

Critical surface: Graph of the function $J_c(B, T)$, where J_c is the critical current density, B is the modulus of the magnetic flux density, and T is the operation temperature.

Cryomagnet: Complete magnet system integrated into one cryostat, comprising the main magnet coils, collars and cryostat, correction magnets, and busbars.

CTF: Acronym for coil-test facility, a magnet assembly which was used for validating the manufacturing process of the LHC main dipoles.

Dispersion suppressor: Transition between the LHC arcs and insertions, aiming at a reduction of the machine dispersion inside the insertions. Each LHC dispersion suppressor consists of four individually powered quadrupole magnets, which are separated by two dipole magnets.

Dogleg magnet: Special dipole magnet used for increasing the separation of the two machine channels from standard arc separation, named for pole pieces resembling a dog's leg. Dogleg magnets are installed in the cleaning insertions IR3 and IR7 and the RF insertion IR4.

Dynamic aperture: Maximum initial oscillation amplitude that guarantees stable particle motion over a given number of turns, normally expressed in multiples of the RMS beam size (σ) and the associated number of turns.

EMF: Electromotive force.

Energy swing: Ratio of the injection to nominal energy in an accelerator.

Experimental insertion region: Insertion region that hosts one of the four LHC experiments.

FAIR: Facility for Antiproton and Ion Research, at GSI, Darmstadt, Germany.

Filament: Fine wires of bulk superconducting material with typical thickness in the range of a few microns. The superconducting filaments are embedded in the resistive matrix in a strand.

Fishbone: Glass-fiber reinforced ULTEM spacer positioned between the two coil layers with slots to provide channels for the superfluid helium.

FP420: Forward Physics at 420 m, proposed LHC experiment installed at the available spaces located 420 m on either side of the ATLAS and CMS detectors.

Fringe field: Term for stray field outside the aperture of the magnets.

HEP: Acronym for High Energy Physics

HERA: Hadron-Electron Ring Accelerator, particle accelerator at DESY in Hamburg, Germany, operated 1992–2007, where electrons or positrons were collided with protons at a center of mass energy of 318 GeV.

He II: Superfluid helium phase of He^4.

Insertion region (IR): Machine region between the dispersion suppressors of two neighboring arcs.

Interaction point (IP): Middle of the insertion region (except for IP8), where the two LHC beams cross over the IP, indicating the point where the two LHC beams can intersect.

IFCC: Interfilament coupling currents within the superconducting strands.

ISCC: Interstrand coupling currents induced in the loops formed by the strands in Rutherford-type cables.

ITER: International Thermonuclear Reactor.

Lattice correction magnets: Correction magnets that are installed inside the short straight section (SSS) assembly.

LEP: Large Electron Positron Collider at CERN, decommissioned in the year 2000.

LHC: Large Hadron Collider at CERN in Geneva, Switzerland.

LHCb: Large Hadron Collider "beauty experiment" installed at Point 8 of the LHC.

LHCf: Large Hadron Collider forward experiment for astroparticle (cosmic ray) physics, sharing IP1 with the general-purpose ATLAS experiment.

Long straight section (LSS): The sections between the upstream and downstream dispersion suppressors of an insertion, including the separation/recombination dipole magnets.

Luminosity: The number of events per unit area and time, multiplied by the opacity of the target.

LTS: Low-temperature superconductor.

Machine cycle: One complete operation cycle of a machine, consisting of injection, ramp up, collision flat top, ejection, and ramp down.

Magnetostriction: Phenomenon in which a ferromagnetic specimen changes its dimensions by some parts per million when it is magnetized.

Main lattice magnets: Main magnets of the LHC arcs, comprising the arc dipole and quadrupole magnets.

Matching section: Arrangement of quadrupole magnets located between the dispersion suppressor and the triplet magnets (or the IP for those insertions without triplet magnets) and comprising two matching sections: One upstream and one downstream from the IP.

Mechanical aperture: Minimum aperture of the storage ring, normally expressed in multiples of the RMS beam size (σ).

MIIT: Heat load of the normal zone of a quenching magnet. MIIT $\equiv 10^6\,\mathrm{A}^2\,\mathrm{s}$.

MMF: Magnetomotive force.

MRI: Magnetic resonance imaging.

NIST: National Institute of Standards and Technology, Bolder, CO, USA.

NMR: Nuclear magnetic resonance.

NEG: Acronym for nonevaporable getter.

Nominal bunch: Bunch parameters required to reach the design luminosity of $L = 10^{34}\,\mathrm{cm}^{-2}\,\mathrm{s}^{-1}$, $\beta^* = 0.55$ m, nominal bunch intensity $N_\mathrm{b} = 1.15 \times 10^{11}$ protons.

Nominal powering: Hardware powering required to reach the design beam energy of 7 TeV.

N-value of a superconductor: Exponent obtained in a specific range of electric field or resistivity when the voltage/current curve is approximated by $U = I^n$.

Octant: Portion of the accelerator between two arc centers straddling the insertion and experimental area.

Persistent currents: Eddy currents with infinitely long time constants that flow in the superconducting filaments and shield their interior from the applied magnetic field.

PET: Polyethylene terephthalate.

Quench: Irreversible transition from the superconducting- to the normal-conducting state.

Racetrack coil: Coil resembling the Indianapolis Motor Speedway (USA) but without the banking in the turns. The coils can have two straight sections with two turns or a rectangular oval track shape with four straight sections and four turns.

Resistive matrix: One of the two main constituents of the strand, embedding the filaments and providing a low-resistance current shunt in the event of a quench (transition of the superconducting material to the normal state).

Resistive transition index: See N-value.

RHIC: Relativistic Heavy Ion Collider at Brookhaven National Laboratory, USA.

Ring 1 and Ring 2: Two rings in the LHC, one ring per beam, Ring 1 corresponding to Beam 1, which circulates clockwise, and Ring 2 corresponding to Beam 2, which circulates counter clockwise seen from above.

RRR: Residual resistivity ratio: The ratio of the electrical resistivity of non-superconducting material at 273 K to that at 4.2 K.

SC: Acronym for superconducting.

Sector: Portion of a ring between two successive insertion points (IP) is called a sector. Sector 1–2 is situated between IP1 and IP2.

Separation/recombination magnets: Special dipole magnets left and right of the triplet magnets that generate the beam crossings in the experimental insertions.

Short straight section (SSS): Assembly of the arc-quadrupole and the lattice corrector magnets, consisting of one quadrupole magnet, one beam-position monitor (BPM), one orbit-corrector dipole (horizontal deflection for focusing and vertical deflection for defocusing quadrupoles), one lattice correction element (trim- or skew-quadrupole elements or octupole magnets), and one lattice sextupole or skew sextupole magnet.

SMES: Superconducting magnetic energy storage.

Spool-piece correction magnets: Correction magnets directly attached to the main dipole magnets and included in the dipole cryostat assembly.

Stacking factor: Ratio of the length of the iron lamination to the total length of the yoke pack. It is 0.985 for the LHC dipole yokes.

Strand: Composite wire containing several thousands of superconducting filaments dispersed in a matrix with suitably small electrical resistivity properties. The LHC strands have Nb–Ti as superconducting material and copper as resistive matrix.

Superconducting cable: Cable formed from several superconducting strands in parallel, geometrically arranged in the cabling process to achieve well controlled cable geometry and dimensions. The LHC cables are flat, keystoned cables of the so-called Rutherford type, resembling the Roebel bar known in the domain of electrical machinery.

Tune: Number of particle trajectory oscillations during one revolution in the storage ring (transverse and longitudinal).

Transfer function: The current-to-field correspondence in accelerator magnets, influenced by superconductor magnetization and iron saturation.

Triplet: Assembly of three quadrupole magnets used for a reduction of the optical β-functions at the IPs.

Twin-aperture magnet: Magnet constructed with two coil pairs for two apertures in one common iron yoke.

Two-in-one magnet: Assembly of two magnetically and mechanically separate magnet modules in a common support structure.

Upstream and downstream: Terms that refer to the direction of one of two beams, Beam 1 or Beam 2, with Beam 1 taken as the default when neither beam is specified.

VUV-FEL: Vacuum Ultraviolet Free Electron Laser at DESY, Germany

VHC: Volumetric heat capacity.

Index

Field Computation for Accelerator Magnets. Stephan Russenschuck

ISBN: 978-3-527-40769-9